AF357326

Grassland Ecosystem Services: Research Advances and Future Directions for Sustainability

Editors

Michael Vrahnakis
Yannis (Ioannis) Kazoglou
Manuel Pulido Fernádez

Basel • Beijing • Wuhan • Barcelona • Belgrade • Novi Sad • Cluj • Manchester

Editors

Michael Vrahnakis
University of Thessaly
Karditsa
Greece

Yannis (Ioannis) Kazoglou
University of Thessaly
Karditsa
Greece

Manuel Pulido Fernádez
Universidad de Extremadura
Cáceres
Spain

Editorial Office
MDPI
St. Alban-Anlage 66
4052 Basel, Switzerland

This is a reprint of articles from the Special Issue published online in the open access journal *Land* (ISSN 2073-445X) (available at: https://www.mdpi.com/journal/land/special_issues/Grassland_Ecosystem_Services).

For citation purposes, cite each article independently as indicated on the article page online and as indicated below:

Lastname, A.A.; Lastname, B.B. Article Title. *Journal Name* **Year**, *Volume Number*, Page Range.

ISBN 978-3-7258-0077-3 (Hbk)
ISBN 978-3-7258-0078-0 (PDF)
doi.org/10.3390/books978-3-7258-0078-0

Cover image courtesy of Michael Vrahnakis

Contents

About the Editors

Michael Vrahnakis

Dr. Michael S. Vrahnakis is a Professor of "Rangeland Science" in the Department of Forestry, Wood Sciences and Design at the University of Thessaly, Greece. He has 30 years of experience in higher educational institutes, specializing in rangeland management plans, the organization of agroforestry systems, and the management plans of protected areas. His current research interests are the management, utilization, and ecology of rangelands; the quantification of floristic diversity, agroforestry, rangeland conditions and rangeland health; and the management and monitoring of protected areas. He has published over 200 peer-reviewed papers and articles in various international and national conferences, scientific and sectoral journals, articles, undergraduate books, and collective volumes. He has participated in several international and national scientific committees, and he is also a reviewer for international conferences and journals, as well as a member of academic and scientific committees of journals. He has successfully coordinated more than fifteen research projects, including those in Europe and nationally, both competitive or self-funded, and he has participated in more than twenty other respective projects as a member of a research team. He has been an instructor at many national and international seminars on grassland management, and has participated in scientific trips to Nigeria and China. He has participated as an expert in thematic groups of the Hellenic Ministries, and he has participated in corresponding thematic groups abroad. He is a member of many scientific organizations and associations, and he has participated in the administration of the international scientific organization of the Eurasian Dry Grassland Group (2009–2019).

Ioannis (Yannis) Kazoglou

Dr. Yannis Kazoglou has been an Associate Professor at the Department of Forestry, Wood Sciences and Design at the University of Thessaly, Greece, since April 2019. His research and teaching interests focus on biodiversity conservation and the positive effects that specific primary sector activities, such as extensive grazing, organic agriculture, breeding of autochthonous livestock, and wet meadow management, may have on the biodiversity of sites belonging to the the E.U. Natura 2000 network of protected areas. An important part of his work aims at sustaining local communities and economies in remote areas and halting the loss of biodiversity due to land abandonment. Prior to his academic position, he actively participated in more than 25 research projects and applied restoration and monitoring projects in Greece (one of which was awarded among the five "Best of the Best" LIFE–Nature projects in the E.U., out of a total of 398 projects ending between 2007 and 2008), in the transboundary Prespa Park (Albania, Greece, and North Macedonia) and in Azerbaijan. Between 2015 and 2018, he contributed to important lawmaking acts for the protected areas of Greece and for grazing management as representative of the Hellenic Range and Pasture Society and, later, as a special advisor of the Deputy Minister of the Environment. Yannis currently participates in the European CAP Network Focus Group, "Competitive and resilient mountain areas", and in national research projects aiming at identifying how grazing management can be beneficial for the conservation of specific habitat types and threatened *Lepidoptera* and *Orthoptera* species within protected areas. He has published more than 50 papers in peer-reviewed scientific journals, collective volumes, and national and international conference proceedings, as is a member of five scientific and non-governmental organizations.

Manuel Pulido Fernández

Dr. Manuel Pulido Fernández is an Associate Professor of Physical Geography in the Department of Arts and Geography at the University of Extremadura, Spain. He has worked for more than 15 years in the holistic assessment (i.e., soil quality, water availability, pasture yield, tree health, biodiversity, farm management, etc.) of woody rangelands, named *dehesas* in Spanish and *montados* in Portuguese. He has authored or co-authored more than 100 peer-reviewed publications in the form of articles (mostly in indexed journals) and book chapters. In addition, he has contributed to more than 25 national and international relevant conferences, including the organization of special events and topic sessions. He has been the principal investigator of some regional research projects, as well as being responsible for work packages included in European Union research projects. He has also participated in many other research projects as a regular researcher, including collaborations with entities belonging to the private sector. He has also been a mentor to two outstanding researchers, Ubaldo Marín Comitre and Jesús Barrena González, and the host supervisor of several visiting researchers from Algeria, Brazil, the Netherlands, the Czech Republic, Italy, etc.

Preface

Modern grassland management concerns the sustainable safeguarding of grassland ecosystem services with regard to the use of their resources. This volume contains a number of research and review article reprints published in the Special Issue "Grassland Ecosystem Services: Research Advances and Future Directions for Sustainability" (Land journal, MDPI Publishing). They discuss the recent advances in and future directions for research in grassland ecology and management, which are supported by research carried out in several grassland types around the world. In them, emphasis is placed on biodiversity, the issues and threats related to grasslands, and on the ecosystem services they provide. The aim of this volume is to outline the significance of the structural and functional components of grassland ecosystems and to highlight the imperative need to ensure the protection of grassland resources. We hope that these research articles will offer solutions and ideas to grassland practitioners, stock breeders, and land and protected area managers and policy makers while advancing contemporary research.

We are thankful to all of the authors who kindly contributed to this reprint by presenting their findings, the reviewers who thoroughly reviewed the manuscripts and significantly improved their content, and the Section Managing Director, Ms. Zita Zhang, who supported us during the whole process of editing this volume. Finally, we would like to thank MDPI Publishing for offering us the opportunity to serve as Guest Editors in this Special Issue.

Michael Vrahnakis, Yannis (Ioannis) Kazoglou, and Manuel Pulido Fernádez
Editors

Article

Effects of Ski-Resort Activities and Transhumance Livestock Grazing on Rangeland Ecosystems of Mountain Zireia, Southern Greece

Apostolos P. Kyriazopoulos [1], Maria Karatassiou [2,*], Zoi M. Parissi [3], Eleni M. Abraham [3] and Paraskevi Sklavou [3]

[1] Laboratory of Range Science, Department of Forestry and Management of the Environment and Natural Resources, Democritus University of Thrace, 68200 Orestiada, Greece
[2] Laboratory of Rangeland Ecology (P.O 286), School of Forestry and Natural Environment, Aristotle University of Thessaloniki, 54124 Thessaloniki, Greece
[3] Laboratory of Range Science (P.O 236), School of Forestry and Natural Environment, Aristotle University of Thessaloniki, 54124 Thessaloniki, Greece
* Correspondence: karatass@for.auth.gr; Tel.: +30-2310-992302

Abstract: The objective of the present study was to assess the impacts in time of the ski-resort infrastructure and transhumance livestock grazing on floristic composition, diversity, and rangeland health indices related to ecosystem stability and function. The study was carried out at a site under the pressure of ski resorts and livestock grazing (Ano Trikala) and a site only under the pressure of livestock grazing (Sarantapicho), both located at Mt Zireia, Southern Greece. The plant cover was measured at each site, and the floristic composition was calculated and classified into four functional groups: grasses, legumes, forbs, and woody species. Species richness, ecosystem function and stability landscape indices, diversity indices, and forage value were calculated. According to the results, the development of the ski resort in Ano Trikala had a neglectable negative impact on plant cover (reduced by 5%), while it had a minor impact on species richness and floristic diversity. Livestock grazing had a positive impact on maintaining plant cover in high values. These results suggest that livestock grazing can counterbalance the effects of ski resorts and related activities on plant cover and floristic diversity. Besides the relatively limited effects on the vegetation community, the ski resort significantly negatively impacted landscape composition, function, and stability. Forage value was 25% lower close to the ski resort, mainly due to the significantly lower percentage of legumes. Transhumance livestock grazing should be used as a management tool in ski-resort areas, as it benefits floristic diversity.

Keywords: diversity indices; forage value; species richness; vegetation cover; landscape stability

Citation: Kyriazopoulos, A.P.; Karatassiou, M.; Parissi, Z.M.; Abraham, E.M.; Sklavou, P. Effects of Ski-Resort Activities and Transhumance Livestock Grazing on Rangeland Ecosystems of Mountain Zireia, Southern Greece. *Land* **2022**, *11*, 1462. https://doi.org/10.3390/land11091462

Academic Editor: Manuel Pulido Fernádez

Received: 26 July 2022
Accepted: 30 August 2022
Published: 2 September 2022

Publisher's Note: MDPI stays neutral with regard to jurisdictional claims in published maps and institutional affiliations.

1. Introduction

Mountainous rangelands provide a wide range of valuable ecosystem services [1]. They are a source of high-quality forage for livestock [2], especially for the transhumance livestock farming system [3,4]. Moreover, mountainous rangelands have a crucial role in the conservation of biodiversity and landscape preservation [5] as well as in climate mitigation and water regulation [6]. In addition, rangelands as cultural landscapes [7] and protected natural reserves are attractive for recreational and touristic activities [8]. For all these reasons, rangeland ecosystems have substantial direct or indirect impact on local economies in mountainous areas.

The management status of these ecosystems has changed rapidly in recent decades due to social and economic changes that have led to land use/land cover changes. On the one hand, the transhumant livestock activity has decreased [9,10], while on the other hand, tourist pressure has increased. All these changes have environmental and economic impacts on the ecosystem, its services, and local communities [11].

Transhumance livestock grazing activities in mountainous areas have declined significantly in recent decades due to various socioeconomic reasons [9,10,12,13]. As a result, the extent and structure of mountain rangelands have changed [14,15], as livestock grazing has maintained them for centuries [16]. Woody species encroachment due to transhumance livestock grazing abandonment is among the essential changes that has occurred in these rangelands' lands [17–20] which negatively affects biodiversity [21].

Since the 1970s, human pressure on mountain ecosystems has increased in several developed countries due to the development of ski resorts [22] to cover the demands for recreational activities by the urban population. These activities generally have a positive economic impact on local communities in mountainous regions [23], but ski resorts were significantly correlated with adverse changes in the rangelands. Their development includes using heavy machinery to construct runs and constructing and maintaining access roads and other infrastructure. Harsh conditions in high altitudes and the mechanical damage caused to plants by the construction and maintenance of ski-resort infrastructure retards the recovery of the vegetation cover [24]. As a result, these infrastructures have been reported to cause a reduction in species richness and plant cover in the rangelands [25], increasing the risk of soil erosion as well as changing soil properties [26,27], which in turn may have negative impacts on ecosystem functioning and stability [28]. On the other hand, Allegrezza and coworkers [16] did not find any differences in floristic diversity among undisturbed alpine grassland and grasslands with ski runs covered with natural and artificial snow. It seems that differences in altitude, slope, different management practices applied, and time passed after the ski resort was built are among the factors affecting the floristic diversity in these ecosystems [29].

In some mountainous areas in the Mediterranean region, ski resorts coexist with extensive pasture-based livestock farming. However, there are a limited number of studies investigating both the effects of ski resorts and livestock grazing on plant communities and ecosystem function [30]. In this regard, Goñi and Gúzman [31] proposed that livestock grazing can prevent reductions in plant diversity caused due to ski resorts. A useful tool for assessing the impact of different management regimes on the rangeland ecosystem's function is the indices of rangeland health [32–34].

Recently, the goal of public governance and local communities is the sustainable development of rural areas, which includes three axes: the economic viability of local communities, social cohesion, and environmental sustainability. These three axes collide in some cases but are recognized as having the same weight and importance in ensuring sustainable development [35].

In this respect, the coexistence of tourism with pasture-based livestock is desirable and essential for the sustainability of the less favorable mountainous areas [36]. Moreover, tourist activities related to skiing and transhumance livestock are two non-rivalrous activities in terms of time, since skiing is carried out during the winter months, while livestock farming is carried out from spring to autumn. The question that arises is to what extent the creation and operation of ski centers conflict with transhumance livestock spatially and to what extent the use of rangelands by both activities ultimately leads to their degradation.

In as much as, to the best of our knowledge, there are no similar studies on the combined effect of pastoralism and touristic activities on rangeland ecosystems, we conducted the current study to gain insights on the impact of these activities on both floristic diversity and ecosystem function. The outcome of our analyses, as far as the interaction of pastoralism and touristic activities is regarded and their effects on the ecosystem as a whole, can provide a basis for designing and establishing strategies for the sustainable development of mountainous areas. This research was conducted in a mountainous area of Southern Greece, traditionally used by transhumance, while a ski resort has been present since the 2000s. We assessed the impact of the ski-resort infrastructure and livestock grazing on the rangeland ecosystems; specifically, as far as (1) floristic composition and diversity and (2) rangeland health indices related to ecosystem stability and function are concerned. We tested the hypothesis that ski resorts and transhumance livestock farming coexistence in Mediterranean mountainous areas results in severe degradation of the ecosystem.

2. Materials and Methods

2.1. The Study Area

The study was conducted in Mount (Mt) Zireia (Kyllini), which is located west of Korinthos city in the Peloponnese peninsula, Southern Greece in 2014, 2015, and 2019 (Figure 1). During the study period, the mean annual temperature was 12.67 ± 0.11 °C, and the mean monthly precipitation was 65.66 ± 5.66 mm. The climatic data (precipitation, temperature) were obtained from the nearest meteorological station ($38°00'00''$ N, $22°50'00''$ E, 1077 m a.s.l). The climate is classified as Mediterranean, with warm winters and dry, and very hot summers, according to the bioclimatogram of Emberger and as Csa in the Köppen–Geiger classification (http://www.en.climate-data.org, 23 May 2022). The most important economic activities in the area are agriculture and livestock production. The transhumant sheep and goat system existed in the study area for centuries, but in the last decades significantly decreased. In the 1960s, 245 herders' families with 38,230 sheep and goats followed the transhumance system, while in 2020, only 54 families with 13,717 animals [9] continued to follow it.

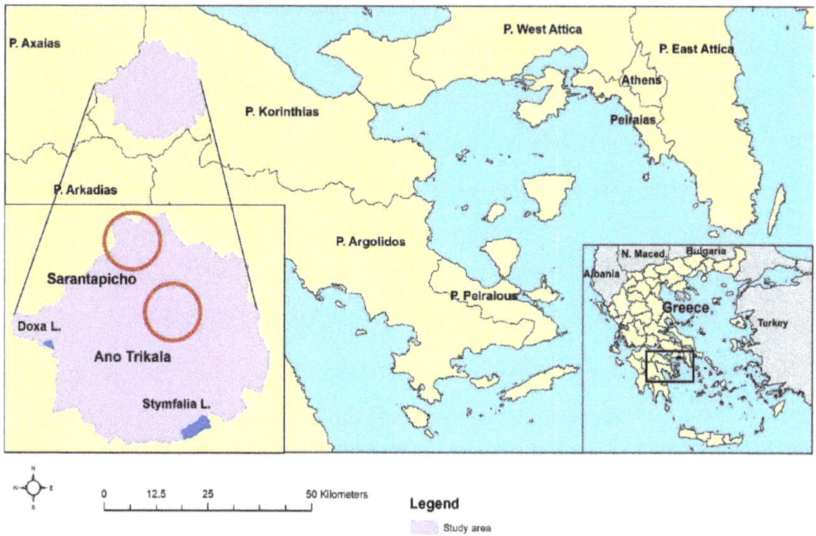

Figure 1. The study area in Mt Zireia in the Peloponnese peninsula. The red circles indicate the selected sites.

Mount Zireia is the second highest mountain of Peloponnese (2374 m). The rangelands of Mt Zireia, both grasslands, and shrublands, are public, and they are communally grazed, from April to October, by transhumant small ruminant flocks in a continuous grazing system. The ski resort was established in the site of Ano Trikala in 2007. After that, the local ski center and the artificial Lake Doxa attract visitors, inducing the development of tourism facilities.

Two sites were selected at the mountainous rangelands of Mt Zireia. The first was close to the village Ano Trikala ($38°58'07''$ N, $22°25'17''$ E) in the area of the ski center and was grazed by sheep and goats from May to September. The second was close to the village Sarantapicho ($38°01'30''$ N, $22°23'05''$ E) and was used by the livestock for the same period (Figure 1). The two sites were at about the same altitude (1350–1450 m), slope (less than 15%), exposure (NW), and similar grazing pressure (Figure 1).

2.2. Vegetation Data Collection and Analysis

Due to the homogeneity of the habitats, six experimental transects of 25 m each were established at each site, in a distance of 80–100 m between them. The plant cover was measured at the end of the growing season of 2014, 2015, and 2019 in each transect according to the line-and-point method, which is widely used in rangeland studies [37]. Transect lines are placed in a way so that every point has similar elevation. Transects were set up in vegetation and 100 recordings (per 25 cm) were conducted per transect. When the pin hit the canopy of a species, it was recorded. If the pin hit rocks, bare soil, or litter, the corresponding measurement was also recorded. The total number of live plant species hits was the plant cover. The floristic composition was calculated from plant cover measurements and classified into four functional plant groups: grasses, legumes, forbs, and woody. Legumes were presented separately from forbs because of their nutritional importance for small ruminants [38].

Floristic diversity, evenness, and dominance were determined for each transect [39] by the following indices [40–42]:

The Shannon–Wiener diversity index (H′) was calculated following the formula in Equation (1) below:

$$H' = -\sum_{i=1}^{S} p_i \ln p_i \tag{1}$$

where S is the maximum recorded number of taxa, and p_i is the population frequency of the i-th taxa.

The Simpson diversity index (D) was calculated following the formula in Equation (2) below:

$$D = \frac{1}{C} \quad \text{where} \quad C = 1 - \sum_{i}^{Sobs} p_i^2 \tag{2}$$

The Pielou evenness index (J) was calculated following the formula in Equation (3) below:

$$J = \frac{H'}{\log(S)} \tag{3}$$

where H′ is the Shannon–Wiener diversity index.

The Buzas and Gibson evenness (E) was calculated following the formula in Equation (4) below:

$$E = \frac{eH'}{S} \tag{4}$$

The Margalef richness index (M) was calculated following the formula in Equation (5) below:

$$M = \frac{S-1}{\ln(N)} \tag{5}$$

where N is the number of individuals of all taxa.

The Berger–Parker dominance index (d) was calculated following the formula in Equation (6) below:

$$d = \frac{N_{max}}{N_T} \tag{6}$$

where N_{max} is the number of records of the dominant taxon and N_T is the total number of records.

2.3. Development of Indices of Landscape Stability, Composition, and Function

Three ecosystem variables, including landscape composition, function, and stability, were utilized to create indices of rangeland health based on empirical data collected annually from each rangeland. Empirical data collected at the same time next to the six experimental transects from each rangeland were used to develop indices of rangeland health in terms of three ecosystem attributes: landscape composition, landscape function, and landscape stability [33,43,44].

Six attributes were used to calculate these indices (Table 1). The possible range of each attribute was divided into a number of ecologically meaningful classes (usually 5 or 6), and each class was then assigned a value according to its perceived effect upon composition, function, or stability. Thus, for example, the percentage of plant cover, which is a crucial component of composition and stability, was divided into five classes, thus: 0–10%—1, 10–25%—2, 25–50%—3, 50–75%—4, and >75%—5. Accordingly, a site with 65% of the soil covered by vegetation would receive a value of 4 for 'plant cover'. For 'function', herbage production was divided into five classes, thus: 0–700 kg ha^{-1}—1, 701–1400 kg ha^{-1}—2, 1401–2100 kg ha^{-1}—3, 2101–2800 kg ha^{-1}—4, and >2801 kg ha^{-1}—5, while soil erosion was also divided into five classes: very severe—1, severe—2, moderate—3, slight—4, and insignificant—5. Data on woody, legumes, and species richness were used as inputs for the composition and function indices such that a higher score indicated a greater cover of woody and legumes and / or a greater diversity of species. 'Species richness' was divided into five classes: 1–5 species—1, 6–10 species—2, 11–15 species—3, 16–20 species—4, and >21 species—5. Total score was calculated by adding the score of each attribute.

Table 1. Attributes, possible scores, and maximum scores used for calculating indices of landscape composition, function, and stability.

Attributes	Landscape Indices		
	Composition	Function	Stability
Plant cover (%)	1–5		1–5
Woody cover (%)	1–5		
Species richness	1–5		
Erosion		1–5	1–5
Herbage production (kg ha^{-1})		1–5	
Legumes (%)		0–5	
Range of scores	3–15	2–15	2–10
Total score		5–30	

2.4. Forage Value Index

The forage value index (FV) was used as an assessment of the plant community's nutritive value. The estimation of FV was based on the Klapp–Stählin index [45,46] after it was weighted with the species percentage in floristic composition. This index indicates the preference of the grazing animals for a plant species in relation to its abundance in the plant community. It was calculated as FV = p_i*FI$_i$, where pi is the percentage of i-th species in the floristic composition and FI is the forage index of i-th species ranging from 0 (unpalatable species) to 8 (preferable species) [47].

2.5. Statistical Analysis

A two-way analysis of variance (ANOVA) was performed to examine the influence of the factor site and the factor treatment (years after the ski resort establishment), and their interaction on the univariate measures: (1) plant cover, (2) functional group composition, (3) diversity indices, (4) rangeland health indices, and (5) forage value index. Data sets consisting of percentage values were arcsine-transformed to degrees prior to analysis [48]. The LSD at the 0.05 probability level was used to detect the differences among means [49]. All statistical analyses were performed using the SPSS statistical package v. 27.0 (IBM Corp. in Armonk, NY, USA).

3. Results

Significant differences ($p \leq 0.05$) between sites were recorded for plant cover, the functional groups legumes and forbs, the Simpson index, and the Berger–Parker dominance index (Table 2). Additionally, significant differences ($p \leq 0.05$) were recorded among the years for the functional groups grasses, legumes, and forbs, the species richness, the Simpson, Shannon, Margalef indices, and the Berger–Parker dominance index. The interaction of site and year was significant ($p \leq 0.05$) for forbs, the species richness, the Simpson, Shannon, and Margalef diversity indices, and the Berger–Parker dominance index (Table 2).

Table 2. Statistical significance of F ratios from the analysis of variance for plant cover, functional group composition, and diversity indices.

	Site	Year	Site * Year
Plant cover	*	NS	NS
Grasses	NS	*	NS
Legumes	*	*	*
Forbs	*	*	NS
Woody	NS	NS	NS
Species richness	NS	*	*
Simpson (D)	*	*	*
Shannon (H')	NS	*	*
Buzas and Gibson (E)	NS	NS	NS
Margalef (M)	NS	*	*
Pielou (J)	NS	NS	NS
Berger–Parker (d)	*	*	*

* Significant (F Test at $p \leq 0.05$); NS $p > 0.05$.

The plant cover (across years) was higher in Sarantapicho rangeland. Functional group composition was differentiated between sites. The percentage of legumes was higher in Sarantapicho, while more forbs were presented in Ano Trikala. There was a slight trend of higher floristic diversity in Sarantapicho, without significant differences. Only the Simpson index was significantly higher, while the Berger–Parker dominance index was significantly lower in Sarantapicho (Table 3).

Table 3. Effects of site (across years) on plant cover, functional group composition, and diversity indices.

Attributes	Sites	
	Ano Trikala	Sarantapicho
Plant cover (%)	89.1 b *	93.8 a
Grasses (%)	29.9 a	36.5 a
Legumes (%)	10.0 b	18.3 a
Forbs (%)	43.9 a	31.0 b
Woody (%)	16.1 a	14.2 a
Species richness	17.1 a	17.6 a
Simpson (D)	8.2 b	9.4 a
Shannon (H')	2.4 a	2.5 a
Buzas and Gibson (E)	0.69 a	0.71 a
Margalef (M)	3.6 a	3.6 a
Pielou (J)	0.87 a	0.88 a
Berger–Parker (d)	0.25 a	0.21 b

* Means within each row followed by the same letter are not significantly different ($p > 0.05$).

The percentage of forbs progressively increased, and it was significantly higher in 2019 compared to those recorded in 2014 and 2015 (Table 4). An opposite trend was recorded for grasses and woody species but without producing significant results. The percentage of legumes was significantly lower in 2014. Floristic diversity indices (species richness, Simpson, Shannon, Margalef) were significantly higher in 2015 compared to 2014, while in 2019, these indices had intermediate values without significantly differentiating from the other years. The Berger–Parker index of dominance followed an opposite trend.

Table 4. Effects of years (across sites) on plant cover, functional group composition, and diversity indices.

Attributes	Year			LSD$_{0.05}$
	2014	2015	2019	
Plant cover (%)	90.5 a *	93.3 a	90.5 a	
Grasses (%)	40.1 a	30.4 b	29.2 b	8.15
Legumes (%)	9.0 b	18.6 a	14.9 a	5.41
Forbs (%)	30.8 b	35.5 b	46.0 a	7.09
Woody (%)	20.1 a	15.5 a	10.0 a	
Species richness	15.5 b	19.2 a	17.3 ab	2.41
Simpson (D)	8.0 b	9.8 a	8.7 ab	1.74
Shannon (H′)	2.4 b	2.6 a	2.5 ab	0.16
Buzas and Gibson (E)	0.69 a	0.73 a	0.70 a	
Margalef (M)	3.2 b	4.0 a	3.6 ab	0.53
Pielou (J)	0.86 a	0.89 a	0.87 a	
Berger–Parker (d)	0.24 a	0.21 b	0.24 a	0.03

* Means within each row followed by the same letter are not significantly different ($p > 0.05$).

The percentage of legumes in Sarantapicho was significantly lower in 2014 compared to 2015 and 2019, while it was higher in 2015 compared to 2019 in Ano Trikala. The percentages of legumes were significantly higher in Sarantapicho than in Ano Trikala in 2015 and 2019, while no significant differences ($p > 0.05$) were detected between sites in 2014 (Figure 2).

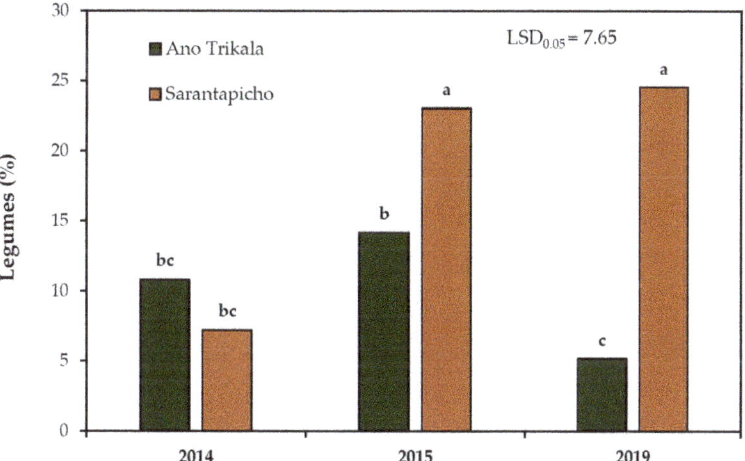

Figure 2. Effects of site and year on legumes percentage. Columns followed by the same letter are not significantly different ($p > 0.05$).

The species richness, Simpson, Shannon, and Margalef indices in Ano Trikala were significantly lower in 2014 compared to 2015 and 2019, while in Sarantapicho, no differences were recorded among years. The values of the Berger–Parker dominance index followed the opposite trend. Those indices were significantly (higher in Sarantapicho than in Ano Trikala only in 2014, while no significant differences ($p > 0.05$) were detected among sites in 2015 and 2019 (Table 5). Berger–Parker dominance index was significantly lower in Sarantapicho than in Ano Trikala in 2014 and 2015.

Table 5. Effects of site and year on the Species richness, Simpson (D), Shannon (H$'$), Margalef (M), and the Berger–Parker (d) dominance index.

Sites	Year	Species Richness	Indices			
			D	H$'$	M	d
Ano Trikala	2014	13.0 c *	6.6 b	2.16 b	2.7 c	0.28 a
	2015	20.3 a	9.8 a	2.66 a	4.3 a	0.23 b
	2019	18.0 ab	8.9 a	2.53 a	3.8 ab	0.24 ab
Sarantapicho	2014	18.0 ab	10.1 a	2.55 a	3.7 ab	0.20 bc
	2015	18.0 ab	9.9 a	2.54 a	3.7 ab	0.18 c
	2019	16.7 b	8.5 a	2.44 a	3.5 b	0.24 ab
LSD$_{0.05}$		3.4	1.8	0.22	0.75	0.04

* Means within each column followed by the same letter are not significantly different ($p > 0.05$).

Significant differences ($p \leq 0.05$) between sites were recorded for all the indices of landscape composition, function, stability, and forage value index (Table 6). Additionally, significant differences ($p \leq 0.05$) for the landscape function index and the total score of indices were recorded over the years. The interaction of site and year was significant ($p \leq 0.05$) for the landscape function index and the forage value index (Table 5).

Table 6. Statistical significance of F ratios from the analysis of variance for indices of landscape composition, function, and stability, and forage value index.

Landscape Indices	Site	Year	Site * Year
Total score	*	*	NS
Composition	*	NS	NS
Function	*	*	*
Stability	*	NS	NS
Forage value	*	NS	*

* Significant (F Test at $p \leq 0.05$); NS $p > 0.05$.

All the indices of landscape composition, function, stability, and forage value index were significantly ($p \leq 0.05$) higher in Sarantapicho compared to Ano Trikala (Table 7).

Table 7. Effects of site (across years) on indices of landscape composition, function, stability, and forage value index.

Landscape Indices	Sites	
	Ano Trikala	Sarantapicho
Total score	20.2 b *	24.48 a
Composition	12.8 b	13.5 a
Function	7.4 b	10.9 a
Stability	9.0 b	9.5 a
Forage Value	2.9 b	3.87 a

* Means within each row followed by the same letter are not significantly different ($p > 0.05$).

The landscape function index and the total score of indices were significantly lower in 2014 compared to those recorded in 2015 and 2019 (Table 8).

The landscape function index in Sarantapicho was significantly lower in 2014 compared to 2015 and 2019, while in Ano Trikala, no significant differences ($p > 0.05$) were detected among years. The landscape function index was significantly higher in Sarantapicho than in Ano Trikala in all the years (Figure 3).

The forage value index in Ano Trikala was significantly lower in 2019 compared to 2014, while in Sarantapicho no significant differences ($p > 0.05$) were detected among years.

The forage value index was significantly higher in Sarantapicho than in Ano Trikala only in 2019, while no significant differences were detected between sites in 2014 and 2019 (Figure 4).

Table 8. Effects of years (across sites) on the indices of landscape composition, function, stability, and forage value.

Landscape Indices	Year			LSD$_{0.05}$
	2014	**2015**	**2019**	
Total score	20.9 b *	23.3 a	22.7 a	1.4
Composition	12.8 a	13.5 a	13.2 a	
Function	8.2 b	9.8 a	9.5 a	1.2
Stability	9.2 a	9.2 a	9.3 a	
Forage Value	3.35 a	3.61 a	3.19 a	

* Means within each row followed by the same letter are not significantly different ($p > 0.05$).

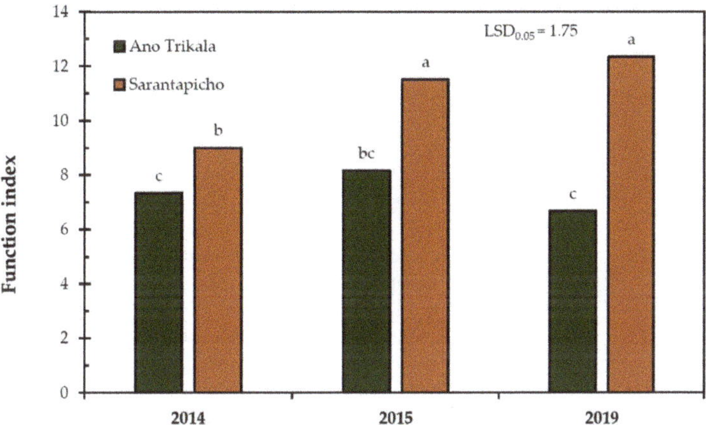

Figure 3. Effects of site and year on landscape function index. Columns followed by the same letter are not significantly different ($p > 0.05$).

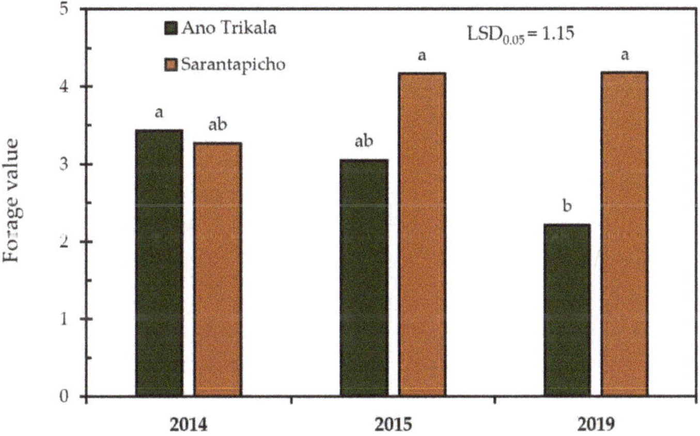

Figure 4. Effects of site and year on forage value index. Columns followed by the same letter are not significantly different ($p > 0.05$).

4. Discussion

Ski resorts and transhumance livestock farming are activities that often coexist in mountainous areas, which may strongly affect the structure and function of rangeland ecosystems. Nevertheless, our data and analyses cannot support the negative impact hypothesis for a representative mountain of South Mediterranean area.

The current study indicates that the development of the ski resort in Ano Trikala had a minimal negative impact on rangelands' plant cover. Ski-resort infrastructure and the increased number of visitors usually cause considerable trampling and other disturbances, leading to a decrease in plant cover [26,30,50]. In such sites, the soils are compacted, and as a result, plant growth is limited, infiltration rates and water-storage capacity are reduced, and soil erosion risk is high [51]. However, this was not the case in the ski resort in Ano Trikala, possibly because of the low-intensity touristic activities in the specific site. Transhumance livestock grazing has been reported to positively impact maintaining plant cover in high values [38]. This result is in accordance with those of Goñi and Gúzman [31], who also reported a minimum increase in bare soil in a grazed ski resort area in Spain. A slight reduction in plant cover was detected in Ano Trikala compared to Sarantapicho; plant cover was generally high in both mountainous rangelands, remaining stable over the years.

Regarding the vegetation composition, the percentage of legumes was lower in the mountainous rangeland close to the ski resort of Ano Trikala, while more forbs were present compared to the rangeland located away from the ski resort. These results are in accordance with previous findings in Spain [30]. Legumes have deeper root systems in general than grasses and forbs. The presence of ice, the reduced soil microporosity that causes poor aeration, and the trampling caused by visitors had increased negative effects on the legume roots compared to those of the other plant functional groups [30,52]. Thus, legumes were significantly less in the rangeland close to the ski resort. The negative impact of the ski resort on legumes is further confirmed by their significant reduction from 2014 to 2019 when the opposite trend was recorded for the rangelands located away from the ski resort. Forbs (across sites) were significantly more in 2019 compared to those recorded in 2014 and 2015, while an opposite trend was recorded for grasses. The differences in the contribution of these plant groups are probably related to grazing and the relative higher preference of sheep for legumes than for forbs when they are available [53], as well as to the differences in climatic over the years [9,38,54,55].

The ski-resort development in Ano Trikala generally had minimal effects on floristic diversity. However, it has been noted that the Simpson index was significantly lower, while the Berger–Parker dominance index was significantly higher compared to those recorded in the mountainous rangeland located away from the ski resort, indicating a slight decrease in floristic diversity at the expense of the increasing abundance of the dominant species near the ski resort. Previous studies [24,56] have reported reduced species richness and diversity close to ski resorts. However, there is evidence that livestock grazing can counterbalance the adverse effects of ski vehicles and visitors through micro-depressions due to trampling and exozoochoria, which can maintain diversity [22]. Thus, Goñi and Gúzman [31] recorded higher plant diversity in grazed ski runs than in non-grazed ones. Barrantes and coworkers [30] found that floristic diversity in grazed skiing areas increased between 1972 and 2005. The present study confirms this result as floristic diversity was also recorded to increase in 2019 compared to 2014 in rangelands close to the Ano Trikala ski resort, while in Sarantapicho biodiversity indices did not change among years. Floristic diversity (across sites) was higher in 2015 compared to 2014 and 2019, probably because of the annual fluctuations in rainfall and temperature [30].

The landscape was negatively affected by the ski resort and related activities. All the indices of landscape composition, function, and stability were significantly higher in the rangeland located in Sarantapicho, away from the ski resort. The increased soil erosion, reduced plant cover and herbage production, and the decreased legume percentage recorded in rangelands close to the ski resort in Ano Trikala constitute the main reasons for this result. Increased risk of soil erosion close to ski resorts has been reported in previous studies [26,57]. Moreover, significant decreases in productivity in rangelands in proximity to ski resorts have been reported by Gartzia and coworkers [58]. It has to be noted that the landscape function index remained low in this site during the years, while it progressively increased in the rangeland located away from the ski resort.

Forage value was significantly lower close to the ski resort. This result is related to the significantly lower percentage of legumes, which have high forage value, and the higher percentage of less palatable forbs. The lower forage value implies this area's decreased grazing capacity and a need for a reduced stocking rate. It is in accordance with the results of previous studies [30,31] that also reported lower pasture quality close to ski resorts. The result of the present study is further confirmed by the fact that forage value remained stable from 2014 to 2019, when it increased in the rangelands located away from the ski resort.

5. Conclusions

Proximity to ski resorts did not reduce plant cover seriously, while it had a minor impact on species richness and floristic diversity. These results suggest that livestock grazing can counterbalance the effects of ski resorts and related activities on plant cover and floristic diversity. On the other hand, the floristic composition was modified in the ski center area. These differences in species composition are reflected in forage value, which is reduced in the ski-resort area. Besides the relatively limited effects on the vegetation community, the ski resort had a significant negative impact on landscape composition, function, and stability.

Transhumance livestock grazing should be used as a management tool in ski-resort areas as it benefits floristic diversity. Low-intensity touristic activities have relatively limited effects on such areas, but some restoration activities may be needed in the future. Furthermore, the two activities could be complementary for the employment of the residents of local communities, as tourism is an activity most preferred by young people, while livestock farming is mainly employed by elderly people. Finally, monitoring these areas at both levels of landscape and vegetation is necessary for managers to make the appropriate decisions at the right time. In this respect, the landscape indices that were used in the present study could be a useful tool for future monitoring and management decisions.

Author Contributions: Conceptualization, A.P.K., M.K., Z.M.P. and E.M.A.; methodology, A.P.K., M.K., Z.M.P. and E.M.A.; software, A.P.K., M.K., Z.M.P. and E.M.A.; validation, A.P.K., M.K., Z.M.P., E.M.A. and P.S.; formal analysis, A.P.K., M.K. and E.M.A.; investigation M.K., Z.M.P., E.M.A. and P.S.; resources, M.K.; data curation, A.P.K., M.K., Z.M.P., E.M.A. and P.S.; writing—original draft preparation, A.P.K., M.K., Z.M.P. and E.M.A.; writing—review and editing. A.P.K., M.K., Z.M.P., E.M.A. and P.S.; visualization, A.P.K., M.K, Z.M.P. and E.M.A.; supervision, M.K.; project administration, M.K.; funding acquisition, M.K. All authors have read and agreed to the published version of the manuscript.

Funding: This research was part of the project «PACTORES, T8EPA2-00022» and was co-funded by the European Union (European Social Fund) and General Secretariat for Research and Innovation (GSRI) through the Action "ERA-Net".

Institutional Review Board Statement: Not applicable.

Informed Consent Statement: Not applicable.

Data Availability Statement: The data presented in this study are available in figures and tables provided in the manuscript.

Conflicts of Interest: The authors declare no conflict of interest.

References

1. Reid, W.V.; Mooney, H.A.; Cropper, A.; Capistrano, D.; Carpenter, S.R.; Chopra, K.; Dasgupta, P.; Dietz, T.; Duraiappah, A.K.; Hassan, R. *Ecosystems and Human Well-Being-Synthesis: A Report of the Millennium Ecosystem Assessment*; Island Press: Washington, DC, USA, 2005.
2. Chan, K.M.A.; Shaw, M.R.; Cameron, D.R.; Underwood, E.C.; Daily, G.C. Conservation Planning for Ecosystem Services. *PLoS Biol.* **2006**, *4*, e379. [CrossRef]
3. Parissi, Z.M.; Karatassiou, M.; Sklavou, P. Chemical composition of a *Trifolium repens* L. population in a grazed mountainous grassland in Central Greece. *Options Méditerr. A* **2016**, *115*, 603–607.
4. Parissi, Z.M.; Karatassiou, M.; Abraham, E.M.; Gioufoglou, C.; Andreopoulos, S.; Drakopoulou, A.T. Forage quality of dominant plant species of mountainous grasslands in northern Greece. *Options Méditerr. A* **2021**, *126*, 79–84.
5. Ferrer, C.; Barrantes, O.; Broca, A. La noción de biodiversidad en los ecosistemas pascícolas españoles. *Pastos* **2011**, *31*, 129–184.
6. Bengtsson, J.; Bullock, J.M.; Egoh, B.; Everson, C.; Everson, T.; O'Connor, T.; O'Farrell, P.J.; Smith, H.G.; Lindborg, R. Grasslands—more important for ecosystem services than you might think. *Ecosphere* **2019**, *10*, e02582. [CrossRef]
7. Urban, E. *The Rural Landscape of Europe: How Man Has Shaped European Nature*; The Swedish Research Council Forms: Stockholm, Sweeden, 2009; 384p.
8. Feng, L.; Gan, M.; Tian, F.P. Effects of Grassland Tourism on Alpine Meadow Community and Soil Properties in the Qinghai-Tibetan Plateau. *Pol. J. Environ. Stud.* **2019**, *28*, 4147–4152. [CrossRef]
9. Karatassiou, M.; Parissi, Z.M.; Stergiou, A.; Chouvardas, D.; Mantzanas, K. Patterns of transhumant livestock system on Mount Zireia, Peloponnese, Greece. *Opt. Méditerr. A* **2021**, *126*, 197–200.
10. Ragkos, A. Transhumance in Greece: Multifunctioality as a Asset for Sustainable Development. In *Grazing Communities: Pastoralism on the Move and Biocultural Heritage Frictions*; Letzia, B., Ed.; Berghahn Books: Oxford, NY, USA, 2022; Volume 29, pp. 23–43.
11. Pătru-Stupariu, I.; Hossu, C.A.; Grădinaru, S.R.; Nita, A.; Stupariu, M.S.; Huzui-Stoiculescu, A.; Gavrilidis, A.A. A Review of Changes in Mountain Land Use and Ecosystem Services: From Theory to Practice. *Land* **2020**, *9*, 336. [CrossRef]
12. Teira, A.G.; Peco, B. Modelling Oldfield Species Richness in a Mountain Area. *Plant Ecol.* **2003**, *166*, 249–261. [CrossRef]
13. Chatzimichali, A. *Sarakatsanoi*, 2nd ed.; Angeliki Chatzimichali Foundation: Athens, Greece, 2007.
14. Barrio, I.C.; Bueno, C.G.; Nagy, L.; Palacio, S.; Grau, O.; Munilla, I.; García, M.B.; Garcia-Cervigón, A.I.; Gartzia, M.; Gazol, A.; et al. Alpine Ecology in the Iberian Peninsula: What Do We Know, and What Do We Need to Learn? *Mt. Res. Dev.* **2013**, *33*, 437–442. [CrossRef]
15. Chrysanthopoulou, G. Synergy of Climate and Grazing in the Evolution of Vegetation on Mount Kyllini. Master's Thesis, Aristotle University of Thessaloniki, Thessaloniki, Greece, 2020.
16. Allegrezza, M.; Cocco, S.; Pesaresi, S.; Courchesne, F.; Corti, G. Effect of snowpack management on grassland biodiversity and soil properties at a ski resort in the Mediterranean basin (central Italy). *Plant Biosyst.* **2017**, *151*, 1101–1110. [CrossRef]
17. Gartzia, M.; Alados, C.L.; Pérez-Cabello, F. Assessment of the effects of biophysical and anthropogenic factors on woody plant encroachment in dense and sparse mountain grasslands based on remote sensing data. *Prog. Phys. Geogr.* **2014**, *38*, 201–217. [CrossRef]
18. Nasiakou, S.; Vrahnakis, M.; Chouvardas, D.; Mamanis, G.; Kleftoyanni, V. Land Use Changes for Investments in Silvoarable Agriculture Projected by the CLUE-S Spatio-Temporal Model. *Land* **2022**, *11*, 598. [CrossRef]
19. Sidiropoulou, A.; Karatassiou, M.; Galidaki, G.; Sklavou, P. Landscape Pattern Changes in Response to Transhumance Abandonment on Mountain Vermio (North Greece). *Sustainability* **2015**, *7*, 15652–15673. [CrossRef]
20. Sklavou, P.; Karatassiou, M.; Parissi, Z.; Galidaki, G.; Ragkos, A.; Sidiropoulou, A. The Role of Transhumance on Land Use/Cover Changes in Mountain Vermio, Northern Greece: A GIS Based Approach. *Not. Bot. Horti Agrobot. Cluj-Napoca* **2017**, *45*, 589–596. [CrossRef]
21. Porqueddu, C.; Ates, S.; Louhaichi, M.; Kyriazopoulos, A.P.; Moreno, G.; del Pozo, A.; Ovalle, C.; Ewing, M.A.; Nichols, P.G.H. Grasslands in 'Old World' and 'New World' Mediterranean-climate zones: Past trends, current status and future research priorities. *Grass Forage Sci.* **2016**, *71*, 1–35. [CrossRef]
22. Isselin-Nondedeu, F.; Rey, F.; Bédécarrats, A. Contributions of vegetation cover and cattle hoof prints towards seed runoff control on ski pistes. *Ecol. Eng.* **2006**, *27*, 193–201. [CrossRef]
23. Genovese, D.; Culasso, F.; Giacosa, E.; Battaglini, L.M. Can Livestock Farming and Tourism Coexist in Mountain Regions? A New Business Model for Sustainability. *Sustainability* **2017**, *9*, 2021. [CrossRef]
24. Wipf, S.; Rixen, C.; Fischer, M.; Schmid, B.; Stoeckli, V. Effects of ski piste preparation on alpine vegetation. *J. Appl. Ecol.* **2005**, *42*, 306–316. [CrossRef]
25. Pignatti, E.; Pignatti, S.; Lucchese, F. Plant communities of the Stirling Range, Western Australia. *J. Veg. Sci.* **1993**, *4*, 477–488. [CrossRef]
26. Barni, E.; Freppaz, M.; Siniscalco, C. Interactions between Vegetation, Roots, and Soil Stability in Restored High-altitude Ski Runs in the Alps. *Arct. Antarct. Alp. Res.* **2007**, *39*, 25–33. [CrossRef]
27. Rixen, C.; Stoeckli, V.; Ammann, W. Does artificial snow production affect soil and vegetation of ski pistes? A review. *Perspect. Plant Ecol.* **2003**, *5*, 219–230. [CrossRef]

28. Mulder, C.P.H.; Uliassi, D.D.; Doak, D.F. Physical stress and diversity-productivity relationships: The role of positive interactions. *Proc. Natl. Acad. Sci. USA* **2001**, *98*, 6704–6708. [CrossRef] [PubMed]

29. Waldhardt, R.; Otte, A. Indicators of plant species and community diversity in grasslands. *Agric. Ecosyst. Environ.* **2003**, *98*, 339–351. [CrossRef]

30. Barrantes, O.; Reiné, R.; Ferrer, C. Changes in Land Use of Pyrenean Mountain Pastures—Ski Runs and Livestock Management—Between 1972 and 2005 and the Effects on Subalpine Grasslands. *Arct. Antarct. Alp. Res.* **2013**, *45*, 318–329. [CrossRef]

31. Goñi, D.; Gúzman, D. Cambios en la vegetación debidos a una estación de esquí alpino en el Pirineo. *Pirineos* **2001**, *156*, 87–118. [CrossRef]

32. Henderson, A.E.; Davis, S.K. Rangeland health assessment: A useful tool for linking range management and grassland bird conservation? *Rangel. Ecol. Manag.* **2014**, *67*, 88–98. [CrossRef]

33. Pellant, M.; Pyke, D.A.; Shaver, P.; Herrick, J.E. *Interpreting Indicators of Rangeland Health: Version 4*; US Department of the Interior, Bureau of Land Management, National Science and Technology Center, Branch of Publishing Services: Denver, CO, USA, 2005.

34. Pyke, D.; Herrick, J.; Shaver, P.; Pellant, M. Rangeland Health Attributes and Indicators for Qualitative Assessment. *J. Range Manag.* **2002**, *55*, 584–597. [CrossRef]

35. Nastis, S.A.; Papanagiotou, E. Dimensions of sustainable rural development in mountainous and less favored areas: Evidence from Greece. *J. Geogr. Inst. Jovan Cvijic SASA* **2009**, *59*, 111–131. [CrossRef]

36. Marín-Yaseli, M.L.; Martínez, T.L. Competing for meadows. *Mt. Res. Dev.* **2003**, *23*, 169–176. [CrossRef]

37. Cook, C.W.; Stubbendieck, J. *Range Research: Basic Problems and Techniques*; Society for Range Management: Denver, CO, USA, 1986; 341p.

38. Karatassiou, M.; Parissi, Z.M.; Panajiotidis, S.; Stergiou, A. Impact of Grazing on Diversity of Semi-Arid Rangelands in Crete Island in the Context of Climatic Change. *Plants* **2022**, *11*, 982. [CrossRef] [PubMed]

39. Abraham, E.M.; Aftzalanidou, A.; Ganopoulos, I.; Osathanunkul, M.; Xanthopoulou, A.; Avramidou, E.; Sarrou, E.; Aravanopoulos, F.; Madesis, P. Genetic diversity of Thymus sibthorpii Bentham in mountainous natural grasslands of Northern Greece as related to local factors and plant community structure. *Ind. Crops Prod.* **2018**, *111*, 651–659. [CrossRef]

40. Henderson, P.A. *Practical Methods in Ecology*; John Wiley & Sons: Hoboken, NJ, USA, 2009.

41. Magurran, A.E. Measuring biological diversity. *Curr. Biol.* **2021**, *31*, R1174–R1177. [CrossRef]

42. Magurran, A.E.; McGill, B.J. *Biological Diversity: Frontiers in Measurement and Assessment*; OUP Oxford: Oxford, UK, 2010.

43. Eldridge, D.J.; Koen, T.B. Detecting environmental change in eastern Australia: Rangeland health in the semi-arid woodlands. *Sci. Total Environ.* **2003**, *310*, 211–219. [CrossRef]

44. Noss, R.F. Indicators for Monitoring Biodiversity: A Hierarchical Approach. *Conserv. Biol.* **1990**, *4*, 355–364. [CrossRef]

45. Gusmeroli, F.; Della Marianna, G.; Puccio, C.; Corti, M.; Maggioni, L. Foraging indices of woody and herbaceous Alpine species for goat livestock. *Quad. SoZooAlp* **2007**, *4*, 73–82.

46. Werner, W.; Paulissen, D. Program VEGBASE database of indicator values of vascular plants after Ellenberg and their evaluation to the personal computer. In *Indicator Values of Plants in Central Europe. Scripta Geobotanica Scripta Geobotanica 18*, 1st ed.; Ellenberg, H., Weber, H.E., Düll, R., Wirth, V., Werner, W., Paulissen, D., Eds.; Goltze, V.W.: Gottingen, Germany, 1992; pp. 238–248.

47. Novak, J. Evaluation of grassland quality. *Ekol. Bratisl.* **2004**, *23*, 127–143.

48. Snedecor, G.W.; Cochran, W.G. *Statistical Methods*, 8th ed.; Iowa State University Press: Ames, IA, USA, 1989; Volume 54, pp. 71–82.

49. Steel, R.G.D.; Torrie, J.H. *Principles and Procedures of Statistics, a Biometrical Approach*, 2nd ed.; McGraw-Hill Kogakusha, Ltd.: New York, NY, USA, 1980.

50. Thompson, J.D.; Hutchinson, I. Cohabitation of Species in an Artificial Grass-Legume Community on Ski-Slopes on Whistler Mountain, British Columbia, Canada. *J. Appl. Ecol.* **1986**, *23*, 239–250. [CrossRef]

51. Roux-Fouillet, P.; Wipf, S.; Rixen, C. Long-term impacts of ski piste management on alpine vegetation and soils. *J. Appl. Ecol.* **2011**, *48*, 906–915. [CrossRef]

52. Delgado, R.; Sánchez-Marañón, M.; Martín-García, J.M.; Aranda, V.; Serrano-Bernardo, F.; Rosúa, J.L. Impact of ski pistes on soil properties: A case study from a mountainous area in the Mediterranean region. *Soil Use Manag.* **2007**, *23*, 269–277. [CrossRef]

53. Manousidis, T.; Iatridou, S.; Lempesi, A.; Kyriazopoulos, A.P. Grazing behavior and dietary preferences of sheep grazing a Mediterranean rangeland. *Options Méditerr. A* **2021**, *125*, 373–376.

54. Karatassiou, M.; Parissi, Z.M.; Sklavou, P. Interaction of climatic conditions and transhumant livestock system on two mountainous rangelands in Greece. *Option Mediter. A* **2016**, *115*, 661–665.

55. Koerner, S.E.; Collins, S.L.; Blair, J.M.; Knapp, A.K.; Smith, M.D. Rainfall variability has minimal effects on grassland recovery from repeated grazing. *J. Veg. Sci.* **2014**, *25*, 36–44. [CrossRef]

56. Forbes, B.C. Tundra Disturbance Studies, I: Long-term Effects of Vehicles on Species Richness and Biomass. *Environ. Conserv.* **1992**, *19*, 48–58. [CrossRef]

57. Gros, R.; Jocteur Monrozier, L.; Bartoli, F.; Chotte, J.L.; Faivre, P. Relationships between soil physico-chemical properties and microbial activity along a restoration chronosequence of alpine grasslands following ski run construction. *Appl. Soil Ecol.* **2004**, *27*, 7–22. [CrossRef]

58. Gartzia, M.; Pérez-Cabello, F.; Bueno, C.G.; Alados, C.L. Physiognomic and physiologic changes in mountain grasslands in response to environmental and anthropogenic factors. *Appl. Geogr.* **2016**, *66*, 1–11. [CrossRef]

 *land*

Article

Simulating Soil-Plant-Climate Interactions and Greenhouse Gas Exchange in Boreal Grasslands Using the DNDC Model

Daniel Forster [1,*], Jia Deng [2], Matthew Tom Harrison [3] and Narasinha Shurpali [1,*]

[1] Natural Resources Institute Finland (Luke), Halolantie 31 A, 71750 Maaninka, Finland
[2] Earth Systems Research Center, Institute for the Study of Earth, Oceans and Space, University of New Hampshire, 39 College Road, Durham, NH 03824, USA
[3] Tasmanian Institute of Agriculture, University of Tasmania, Newnham Drive, Launceston, TAS 7248, Australia
* Correspondence: danforster.grassland@hotmail.com (D.F.); narasinha.shurpali@luke.fi (N.S.)

Abstract: With global warming, arable land in boreal regions is tending to expand into high latitude regions in the northern hemisphere. This entails certain risks; such that inappropriate management could result in previously stable carbon sinks becoming sources. Agroecological models are an important tool for assessing the sustainability of long-term management, yet applications of such models in boreal zones are scarce. We collated eddy-covariance, soil climate and biomass data to evaluate the simulation of GHG emissions from grassland in eastern Finland using the process-based model DNDC. We simulated gross primary production (GPP), net ecosystem exchange (NEE) and ecosystem respiration (Reco) with fair performance. Soil climate, soil temperature and soil moisture at 5 cm were excellent, and soil moisture at 20 cm was good. However, the model overestimated NEE and Reco following crop termination and tillage events. These results indicate that DNDC can satisfactorily simulate GHG fluxes in a boreal grassland setting, but further work is needed, particularly in simulated second biomass cuts, the (>20 cm) soil layers and model response to management transitions between crop types, cultivation, and land use change.

Keywords: ecophysiological modelling; boreal agriculture; greenhouse gases; model evaluation; DNDC; soil organic carbon; net-zero

Citation: Forster, D.; Deng, J.; Harrison, M.T.; Shurpali, N. Simulating Soil-Plant-Climate Interactions and Greenhouse Gas Exchange in Boreal Grasslands Using the DNDC Model. *Land* **2022**, *11*, 1947. https://doi.org/10.3390/land11111947

Academic Editors: Michael Vrahnakis, Yannis (Ioannis) Kazoglou and Manuel Pulido Fernádez

Received: 17 October 2022
Accepted: 29 October 2022
Published: 1 November 2022

Publisher's Note: MDPI stays neutral with regard to jurisdictional claims in published maps and institutional affiliations.

1. Introduction

Global temperature rise is placing increased pressure on boreal lands from agricultural land-use expansion and intensification to meet the needs of a burgeoning population [1]. Boreal regions also present significant opportunities for greenhouse gas (GHG) mitigation and may have the potential to act as further sinks of atmospheric carbon [2]. United nations sustainable development goals [3] have identified sustainable agriculture (goal 2) and protection of ecosystems (goal 15) as key components of the overarching strategy to address climate change. We are thus at a critical juncture in time in which holistic assessments and planning decisions regarding management and land-use trade-offs will be key to ensuring long-term sustainability [4]. Given the risks and opportunities afforded by climate change in boreal areas, decision making needs to balance the manifold factors involved to balance increasing societal requirements for the long-term preservation of the natural capital on which our agri-food systems depend [5–8].

Agroecological models have increasingly been used to simulate the effects of management [9,10] and land use conversion [11] on biomass production [12–14], canopy-level physiology [7,8], GHG emissions [15–18], profitability [19] and soil carbon/nitrogen cycling [20]. When properly calibrated, such models have been able to accurately simulate these agriculturally and environmentally important variables and thus have the potential to assist in management planning while promoting both production and environmental protection, including climate change mitigation by reducing GHG emissions [21–23].

The DeNitrification-DeComposition (DNDC) model [24] has been long in use worldwide, and has demonstrated sufficient accuracy in modelling crop growth and GHG emissions in cooler regions [11,25–33]. The extensive use of this model in colder climates and its successful simulation of relevant outputs would make it seem a sensible choice for application in Scandinavia, although significant uncertainty remains around the performance of DNDC in the challenging boreal environment due to the lack of studies measuring GHG and other variables for use in model calibration in regions with extensive durations of sub-zero temperatures [34].

Of the few agroecological modelling studies that have been conducted in boreal zones, results for DNDC have been promising. For example a study by He et al. [31] modelled the effects of manuring on N_2O emissions in Canadian grasslands, evaluating against measured data and while most metrics were graded "good" to "fair", soil water and N simulations were only "acceptable", and the authors also recommended improvements to the models soil freeze-thaw simulations, as well as soil microbial and water processes in grasslands. Abdalla et al. [27] used DNDC to evaluate soil respiration in the Republic of Ireland from grassland and conventionally managed arable fields under three climate scenarios: a baseline of measured climate data and both high and low temperature sensitivity scenarios. They indicated that DNDC could effectively model soil respiration in both pasture and arable, underestimating annual CO_2 efflux by only 13% and 8% respectively. Another study in the Republic of Ireland [35] examined management effects on N_2O emissions from grasslands using a two year (2008–2009) dataset. The study showed that flux estimates tended to be higher than those estimated using IPCC emissions factors, and the authors suggested that soil parameters needed further calibration for optimum performance. A study in Northern Ireland [36] used DNDC95 to evaluate SOC density and annual changes in temperate long-term grassland soils. They found that the model underestimated SOC by 0.9 t C ha^{-1}, yr^{-1}, a difference which was explained by differences in supplied N and differences in soil C, rainfall, and air temperature as well as soil physiochemical variables. Because most of the studies with DNDC have been conducted in temperate or cool temperate regions with more moderate temperatures, different soils and different land management, such results may not translate to Scandinavian conditions, hence the need for a DNDC study in Finland.

The purpose of this study was to evaluate DNDC performance in simulating soil microclimate, biomass production and GHG fluxes against eddy-covariance measured NEE, here defined as the net exchange of CO_2 between the ecosystem and atmosphere in kg C ha^{-1}, and associated soil and plant data in a legume grassland in eastern Finland with attention to model accuracy in simulating freeze-thaw cycles.

2. Materials and Methods

2.1. Site Description

This study was conducted at the Antilla field site, located in Maaninka, eastern Finland, (63°09′ N, 27°140′ E, 89 m a.s.l.); a location with mean annual temperature (1981–2010) of 3.2 °C and mean annual precipitation of 612 mm year^{-1}. In the study period, mean, maximum and minimum temperatures were 5.1, 25.0 and –26.5 °C, respectively, and annual rainfall was 613 mm, 515 mm, and 532 mm for 2017, 2018 and 2019, respectively (Figure 1). The study site was a 6.3 ha agricultural field where the mineral soil is classified as a haplic cambisol (silt loam: clay 25 ± 7.8%, silt 53 ± 9%, sand 22 ± 7.8%) based on the USDA classification system (Table 1).

The field was cultivated with a mix of timothy (*Phleum pratense* L. cv Nuuti), meadow fescue (*Festuca pratensis*) and red clover (*Trifolium pratense*) at a rate of 15 and 5 kg ha^{-1} for grasses and legumes respectively in 2015, was reseeded in May 2017, and ploughed in the autumn of 2018 when glyphosate was also applied. The grassland was renewed in Spring 2019 with a cover crop of barley (*Hordeum vulgare* L.) in a fresh rotation. Mineral fertilizer was applied (106 kg N, 28 kg P and 50 kg K/ha^{-1}) divided evenly over two applications at the start of the growing season and after first cut in 2017 and 2018, with a single application in the renewal year of 45 kg N, 20 kg P, 38 kg K/ha^{-1}. Cuts were carried out twice annually in late June and mid-August for 2017 and 2018, and a single cut in early August in the renewal year with a disc mower to remove biomass material to 8 cm, which was then swathed, baled, and removed from the field site.

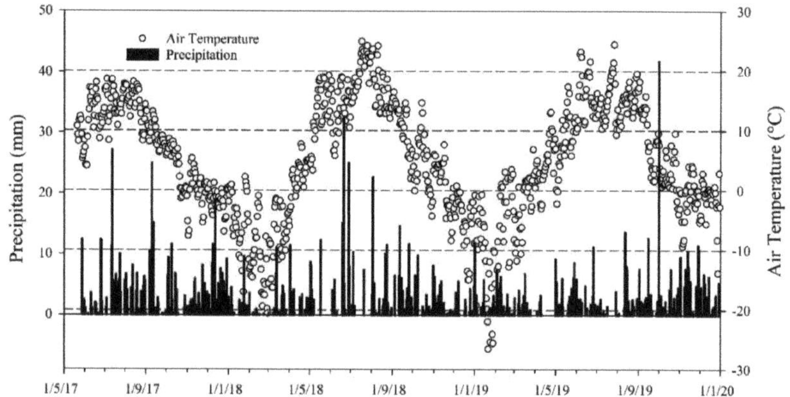

Figure 1. Measured air temperature (°C) and precipitation (mm) per day over the modelled period (Data obtained from Finnish Meteorological Institute (FMI).

Table 1. Topsoil (0–5cm) measurements in the Antilla site.

Unit	Value
Soil pH	5.8 ± 0.19
EC mS m^{-1}	14 ± 2.4
SOM%	5.2 ± 0.9
SOC%	3.0 ± 0.52
C/N ratio	15 ± 0.4
Total N%	0.2 ± 0.03
P mg L^{-1}	5.4 ± 1.28
K mg L^{-1}	104 ± 12.9

2.2. Eddy Covariance Data

An eddy covariance (EC) tower was setup in the centre of the study area in 2017. The CO_2 and H_2O measurements were performed by a closed-path EC system with adjacent weather station providing supporting climate and meteorological data. The EC system was a Li-7000 infrared gas analyser (IRGA, for CO_2 and H_2O mixing ratios, Li-COR inc., Lincoln, NE, USA), a sonic anemometer (wind velocity, sensible heat flux and sonic temperature components R3-50, Gill Instruments Ltd., Lymington, UK) mounted 2.5 m above the soil surface on an instrument tower. Air samples pass through a heated intake tube at a flow rate of 10 L min^{-1}, (a PTFE tube, internal diameter 6 mm, length 8 m) with two 1.0 μm pore size filters (Gelman®). The IRGA was housed in a climate-controlled cabin and calibrated monthly during the growing season. Supporting climate data were net radiation, relative humidity, photosynthetically active radiation, soil temperature, volumetric soil water content (at 5 and 20 cm depths) and air pressure. A CR3000 (Campbell Scientific Inc., Logan, UT, USA) 10 Hz data logger collected raw EC data. Missing supporting meteorological data were gap filled using data from the Maaninka weather station (Finnish Meteorological Institute), located 6 km southeast of the field site. Eddy covariance data processing was carried out using EddyUH [37]. Annual EC data for the period of the experiment (20 May 2017 to 31 May 2020) were employed for calibration and evaluation purposes.

2.3. The DNDC Model

The DeNitrification-DeComposition (DNDC) model is a process-based biogeochemical model developed for quantifying C sequestration as well as emissions of C and N gases from agricultural ecosystems [38–40]. The model is comprised of six sub-models: soil climate, plant growth, decomposition, nitrification, denitrification, and fermentation. The soil climate, plant growth, and decomposition sub-models convert the primary model drivers, such as climate, soil properties, vegetation, and anthropogenic activity, into soil environmental factors (e.g., soil temperature and moisture, pH, redox potential) and concentrations of substrates of relevant biogeochemical processes. The nitrification, denitrification, and fermentation sub-models simulate C and N transformations that are mediated by soil microbes and controlled by soil environmental factors and concentrations of relevant substrates [40,41]. The DNDC model adopted in this study was further improved to simulate surface energy exchange, soil frost and thaw dynamics, and C gas fluxes in cold regions [42–44].

The DNDC model requires daily climate data, including minimum and maximum temperatures ($^\circ$C) and rainfall (mm), as well as humidity (%), windspeed (m/s), and solar radiation (MJ/m^2). The model also requires rainfall N concentrations, atmospheric NH_3 and CO_2 concentrations, and annual increases of atmospheric CO_2, although model default values can be used when such information is not available. In addition, we used measurements of actual snow depth to drive the model [43]. Using measured snow depth in this way improves the surface energy balance and hence the simulated soil temperature and water filled pore space (WFPS cm^3/cm^3). Soil input data include soil properties according to the USDA soil classification system, as well as information on initial soil organic carbon (SOC), pH and other soil physiochemical factors. The input parameters of farming management practices, including crop types, planting and harvest dates, tillage, fertilization, residue return, and irrigation, were taken from Li et al. (*unpublished*). The Antilla site using soil data from Lind et al. [45]. Initial model output evaluations were carried out against EC data for net ecosystem exchange (NEE), gross primary production (GPP, defined as the total CO_2 taken up by the ecosystem in photosynthesis in kg C ha^{-1}) and ecosystem respiration (Reco, defined here as the total ecosystem respiration (sum of aboveground plant and root (autotrophic) and heterotrophic respiration) in kg C ha^{-1}). Output data assessment and analysis was conducted in R version 4.1.1 (R development core team 2021) and RStudio (version 1.4.1106).

2.4. Model Initialisation Calibration and Evaluation

The model was calibrated using data in 2017 and 2018 and then ran continuously from 2017 to 2019, with 2019 used to evaluate the model. The calibration process included optimisation of crop phenological parameters (thermal days to maturity, biomass fraction, root: shoot ratios, stem: leaf: grain fraction, water demand, N fixation index, and optimum temperature, Table 2). For modelling purposes, a 16-year spin-up period was introduced to allow time for simulated soil carbon stocks to stabilise.

The crop setup consisted of two systems, the first consisted of a perennial grass ley (land put down to grass and/or clover for a limited period) calibrated to simulate a grass/legume mixture. The second system simulated the same, with the addition of a barley cover crop, though no cuts were carried out in 2020 (Table 3). Input parameters for soil and management are as described in Section 2.1, though fertiliser applied (106 kg N/ha/year) was divided equally among the number of fertilisation events for that year.

Both calibration and evaluation were conducted using base R and the package 'Hydro-GOF' [46]. Four evaluation metrics were used to evaluate model performance against measured GHG fluxes. These were Spearman's rho (ρ, Equation (1)), where ρ is the Spearman's rank correlation coefficient, d_i is the difference between the two ranks of individual measured and corresponding simulated data pairs, and n is the number of observations. Mean absolute error (MAE, Equation (2)) which assesses the size of prediction errors at the individual level, but does not allow comparison between positive and negative predictors. Root mean square error (RMSE, Equation (3)) measures absolute quadratic prediction error.

S_i is the simulated, and M_i are the measured variables. Percent bias (*pBIAS*, Equation (4)) gives a relative bias estimation to determine over or underestimation in the simulation.

Table 2. Crop calibration parameters used in this study. Figures in grey are automatically produced model outputs in response to calibration, and not directly subject to manipulation.

Perennial Grass	Grain	Leaf	Stem	Root
Max. biomass production (kg C/ha/yr)	400			
Biomass fraction	0.04	0.28	0.28	0.4
Biomass C/N ratio	35			
Annual N demand (kg N/ha/yr)	143			
Thermal degree days to maturity	1500			
Water demand (g water/g dry matter (DM))	150			
N fixation index (crop N/N from soil)	1.5			
Optimum temperature (°C)	18			
Barley				
Max. biomass production (kg C/ha/yr)	2496			
Biomass fraction	0.3	0.23	0.23	0.23
Biomass C/N ratio	45	75	75	85
Annual N demand (kg N/ha/yr)	129			
Thermal degree days to maturity	1500			
Water demand (g water/g DM)	150			
N fixation index (crop N/N from soil)	1			
Optimum temperature for crop growth (°C)	18			

Table 3. Management details used in DNDC simulations in the present paper based on Li et al. (*unpublished*). 'Model setup' indicates the division between calibration and evaluation datasets. The full model was run until May 2020, although that year was not used in model evaluation and is shown in grey to reflect this.

Year	1 (2017)	2 (2018)	3 (2019)	4 (2020)
Model setup	Calibration dataset		Evaluation dataset	NA
Crop: perennial grass	18 May 2017–30 October 2018		4 June 2019–31 May 2020	
Cover crop: barley	NA	NA	4 June 2019–31 May 2020	
Cuts	29 June	26 June	6 August	-
	16 August	7 August		
Overseeding/reseeding	18 May	NA	16 August	-
Fertilisation	22 May & 3 July	22 May & 2 July	2 July	* 22 May & 2 July
Tillage	NA	30 September (crop killing till), 30 October	3 June	-

$$\rho = 1 - \frac{6 \sum d_i^2}{n(n^2 - 1)} \tag{1}$$

$$MAE = \frac{\sum_{i=1}^{n} |S_i - M_i|}{n} \tag{2}$$

$$RMSE = \sqrt{\frac{\sum_{i=1}^{n} (S_i - M_i)^2}{(n)}} \tag{3}$$

$$pBIAS = 100 \frac{\sum_{i=1}^{n} (S_i - M_i)}{\sum_{i=1}^{n} M_i} \tag{4}$$

Following setup and calibration we again ran the model continuously from 2000 to 2020 and used 2019 for model evaluation against gross primary production (GPP kg C ha^{-1}), net ecosystem exchange (NEE kg C ha^{-1}) and ecosystem respiration (Reco kg C ha^{-1}), soil temperature (°C) at 5 cm, and soil moisture (WFPS cm^3/cm^3) at 5 cm and 20 cm according to a pre-determined criteria (Table 4).

Table 4. Model evaluation based on classifications as per [27,31,34]. Overall scores are calculated as the mean performance of variables across all four metrics.

Evaluation Method	Poor	Fair	Good	Excellent
Spearman's rank correlation (ρ)	0.30	0.50	0.70	1.00
MAE	4.0+	3.0–3.9	2.0–2.9	1.0–1.9
RMSE	$\geq$40	20–39	10–19	0–10
pBias%	>20%	15–20%	11–15%	<10%

3. Results

Greenhouse gas exchange (Figure 2), soil climate simulations (Figure 3) and crop biomass simulations (Figure 4) were generally in agreement with measurements through most of the modelled timeframe. For GPP, there was a good correlation ρ between simulated and measured data, though MAE was poor, RMSE was fair, and the model pBias% underestimated and scored poor. For NEE, there was a good ρ correlation between simulated and measured data, though MAE was poor, RMSE was fair, and the model scored good overall, with a small pBias% underestimate. For Reco, there was a good ρ correlation between simulated and measured data, though MAE was poor, RMSE was good, but model pBias% underestimated and scored poor (Table 5).

Table 5. Evaluation results for simulated versus measured GHG exchange and soil microclimate. The 'Mean score' column represents the overall assessment when all four measures are accounted for.

	ρ	MAE	RMSE	pBias%	Mean Score
GPP (kg C ha^{-1})	0.80 ($p < 0.001$)	21.3	35.1	−20.7%	Fair
NEE (kg C ha^{-1})	0.72 ($p < 0.001$)	16.6	26.7	−14.2%	Fair
Reco (kg C ha^{-1})	0.85 ($p < 0.001$)	10.9	14.2	−22.8%	Fair
Soil Temp (°C)	1.00 ($p < 0.001$)	0.1	1.2	18.2%	Excellent
WFPS (cm^3/cm^3) 5 cm	0.73 ($p < 0.001$)	0.1	0.1	−11.2%	Excellent
WFPS (cm^3/cm^3) 20 cm	0.25 ($p < 0.001$)	0.1	0.1	−5.0%	Good

For soil temperature there was a correlation ρ score of excellent, MAE was excellent, RMSE was excellent and pBias% scored fair, and showed that the model overestimated compared to measured data. Soil water (WFPS) at 5 cm scored "good" ρ for Spearman's correlation, MAE was excellent, and RMSE was also excellent, while there was a small, but good underestimation for pBias%. For WFPS at 20 cm there was a poor ρ correlation, whereas MAE was excellent and RMSE was excellent, whilst there was only a small pBias% underestimation, which scored excellent (Table 5).

The DNDC model simulated seasonal patterns of GHG exchange (Figure 2) and soil climate (Figure 3) well. Seasonally, GHG's tended to be close to 0 between October and April, although in the evaluation year eddy covariance did not show the simulated uptick in ecosystem respiration and NEE (Figure 2b,c) until six weeks later in mid-June, and DNDC did not pick up on this and also did not pick up on wintertime ecosystem respiration (Figure 2c).

Soil temperature at 5cm (Figure 3a) was very similar to measured data, although WFPS at 5 cm (Figure 3b) showed less accuracy in January and May, and in the 20 cm layer DNDC overestimated in May 2018 and underestimated between December 2018 and May 2019 (Figure 3c).

Biomass (DM kg/ha^{-1}) simulation were compared with measured cuts using an independent sample t-test. Measured DM (mean = 3117, sd = 915) was compared to simulated DM (mean = 4517, sd = 926) and there was a significant difference between the two t (8) = −2.4, $p < 0.05$.

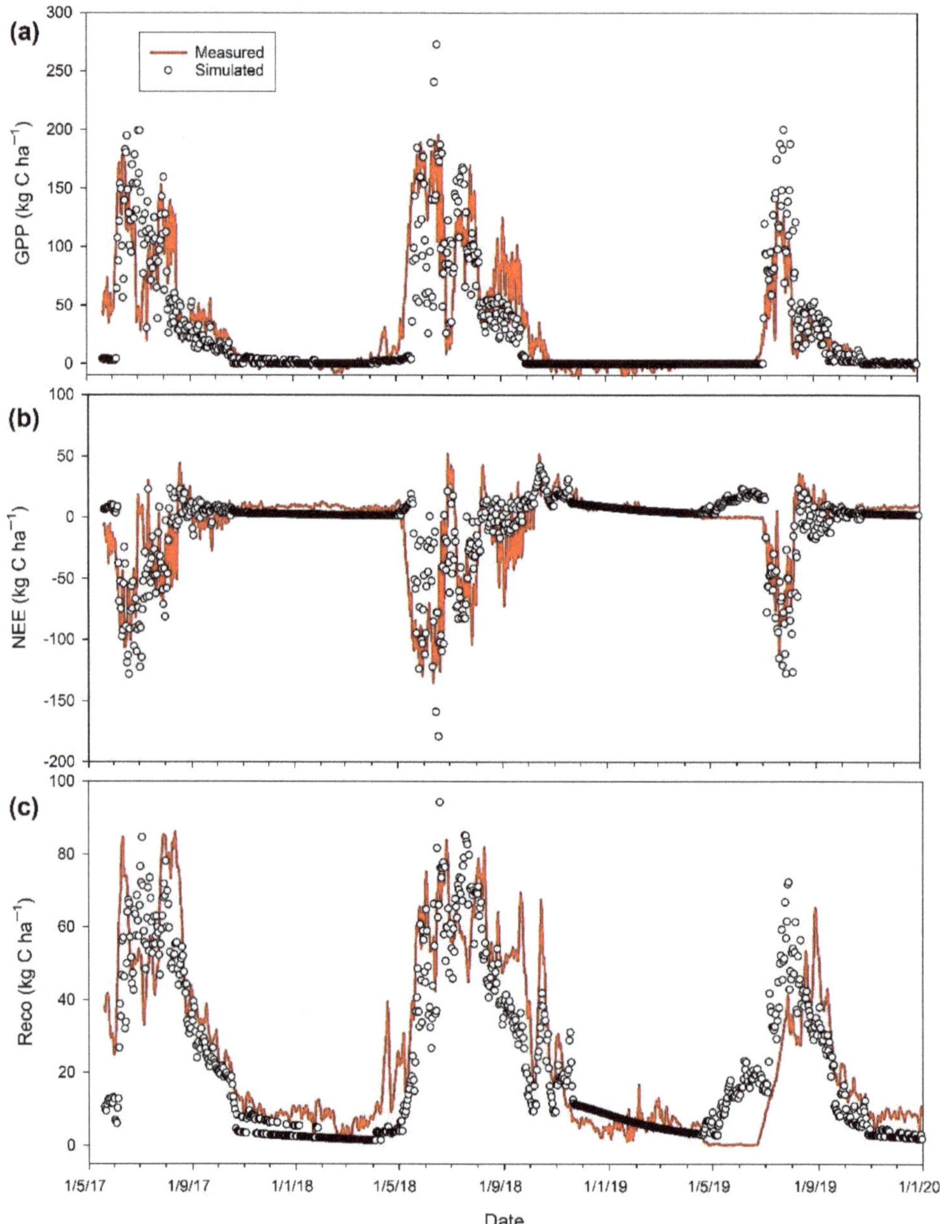

Figure 2. Simulated and measured GPP (**a**), NEE (**b**) and Reco (**c**) over the three modelled growing seasons.

Total annual GHG exchange was measured for the calibration and evaluation years 2018 and 2019 (Table 6) indicating that DNDC simulated GPP, NEE and Reco followed a similar pattern to measured eddy-covariance figures and both field measurement and simulations showed reduced respiration in 2019 compared to 2018.

The DNDC model predicted that the Antilla site would be an overall sink of atmospheric C by an average of -1.17 T C ha^{-1} yr^{-1}, although eddy covariance indicated that in 2019 the field site was a small source of C, but still an overall sink.

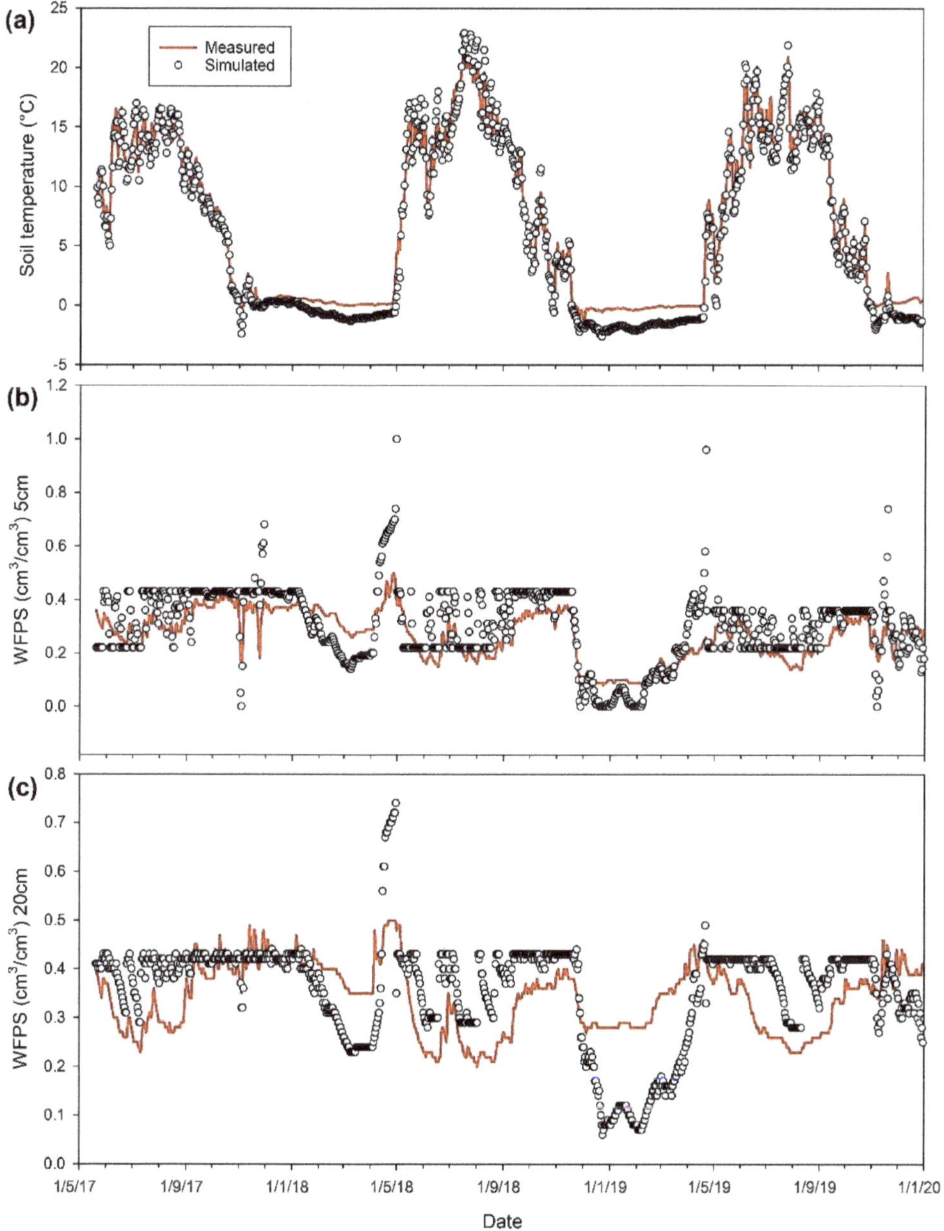

Figure 3. Comparison of soil temperature (°C) at (**a**) 5 cm, WFPS at (**b**) 5 cm, and (**c**) WFPS at 20 cm.

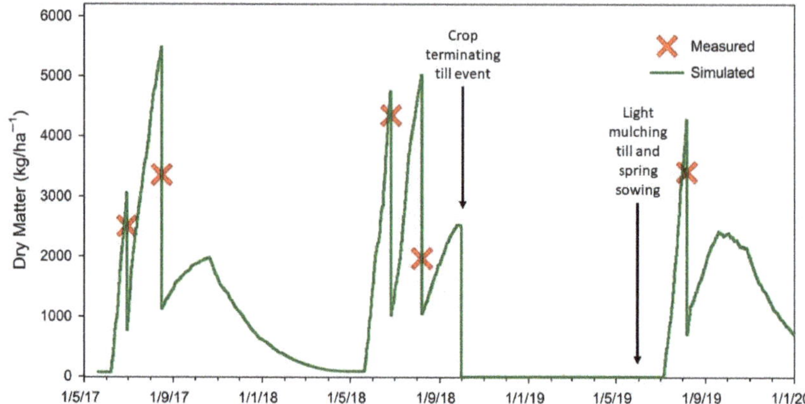

Figure 4. Comparison of simulated (—) and measured ($\times$) biomass production (DM kg/ha^{-1}) for growing seasons 2017 to 2019. For 2017, the difference for cut 1 was 5.6% and cut 2 was 21.3%. For 2018 cut 1 was 4.0% and cut 2 was 30.5%, and for 2019 the simulated cut was 8.7% higher than measured DM and the black arrows ('ploughing event' and 'tillage and spring sowing').

Table 6. Comparison of total annual GHG exchanges for calibration (2018) and evaluation (2019) years. The 2017 year is omitted as measurement did not begin until May.

Calibration/Evaluation Years	Measured	Simulated	Difference
GPP (T C ha^{-1} yr^{-1})			
2018	14.79	10.09	4.70
2019	4.15	5.19	1.04
Mean	9.47	7.64	−1.83
NEE (T C ha^{-1} yr^{-1})			
2018	−3.64	−1.97	1.68
2019	0.22	−0.35	−0.57
Mean	−1.71	−1.17	0.55
Reco (T C ha^{-1} yr^{-1})			
2018	11.15	8.12	−3.03
2019	4.58	4.84	0.26
Mean	7.87	6.48	−1.38

4. Discussion

This study begins to address the dearth of model evaluations for boreal managed grasslands by presenting a comparison of model simulations for DNDC with eddy-covariance GHG flux, soil climate and biomass dry-matter production data. There are relatively few papers containing model evaluations for grasslands in boreal areas, but since model testing during their creation is necessarily limited to regions available to the model creators, it does not follow that they are applicable elsewhere and a careful assessment of regional conditions is necessary to understand how a model might behave in novel environments.

A number of papers have assessed the DNDC model in cool weather regions in Canada [29,31] and Northern Europe [27,43,44], making it an ideal candidate for evaluation with a view to using in the boreal agricultural landscapes of Scandinavia. Furthermore, the present study adds to this by demonstrating that the model can produce fair estimations of the key GHG fluxes and good estimations of soil climate conditions when compared to measured field data, but that there remain a number of uncertainties that would benefit from further elucidation, for example crop parameterisation and root processes, soil moisture simulations and GHG responses to changes in management.

4.1. GHG Exchange

In general, the DNDC model simulated GHG exchange fairly well, there was a tendency to underestimate GPP and Reco compared to measured data. This may be linked to the model simulating little or no respiration during the colder months, and to both increase more slowly and decrease more rapidly at the commencement and ending of the growing season compared to measured data. Using the improved DNDC model [42] improved model performance which was able to simulate GHG satisfactorily.

In the evaluation dataset, spikes of GPP and Reco in August 2018 and 2018 were probably related to increased rainfall noted in those periods (Figure 1), and a simulated increase in NEE and Reco beginning in May 2019 was not matched by observed data (Figure 2b,c). This may reflect a modelled increased soil respiration since the model simulated a 'crop terminating till' in October 2018, followed by a 'light mulching till' and seeding on 3 and 4 June respectively (Table 3). The residue incorporation following crop termination and tillage event increased the modelled soil respiration. However, they simulated an increase in soil respiration before crop germination in contrast to the findings of Oertel et al. [47] who found that bare soils tended to have lower GHG flux than other land-cover types.

The June 2019 sowing also included a cover crop of barley that was absent from previous years, although the use of cover crops has been shown to increase soil microbial activity [48]. However, there we no observed differences related to barley addition (Figure 2), and in the 2019 (evaluation period) growing season there was a month-long difference between observed and simulated uptick of NEE and Reco (Figure 2b,c), which was not observed in the 2018 (calibration) season and requires some explanation. According to Khan, [49], tillage, which is the standout feature of the 2019 (evaluation) period, can stimulate soil microbial activity and thus respiration. Nevertheless, simulated NEE was not significantly different from observed (−14%), in line with the findings of Deng et al. [42] who also reported a good match for NEE, and of Abdalla et al. [27] who reported a corelation of R = 0.6 for NEE simulations compared to measured data on permanent grassland in the Republic of Ireland.

4.2. Soil Climate

Model performance in terms of simulating temperature at the soil surface (5 cm) was exceptional, which was perhaps to be expected given the version of the DNDC model we used was aimed specifically at improving surface exchange of energy fluxes and soil frost/thaw dynamic simulations [41,42], as is evident from the close correlation between simulated and measured outputs (Table 5). Soil moisture (WFPS) at 5cm followed a similar, though less striking trend and tended to underestimate by 11.2%, (Table 5). On the other hand, at 20 cm WFPS simulation quality was much lower, and although the two datasets matched closely the spread of the data was such that it was not possible to make a strong correlation. This discrepancy may be due to the model sensitivity to the soil water/ice status to changes in soil temperature when this was close to zero since a variation of ± 1 °C above or below freezing is small for soil temperature but makes the difference between liquid water and ice in the soil.

4.3. Biomass

The DNDC model was able to simulate biomass production accurately for the first cut in the two-cut system used in Finnish pastures although the second cut tended to underestimate. Model performance in the first cut was closer to measured figures than in the second (2.8 $\pm$ 0.008% and 25.9 $\pm$ 0.05% respectively, Figure 4) and tended to assume higher growth rates after the first cut than were observed in the field. Testing across all years indicated that there was a significant difference between simulated and measured biomass, meaning that the model underperformed as a grass biomass prediction tool when calibrated to greenhouse gas fluxes.

5. Conclusions

This study demonstrated that the DNDC model is able to simulate GHG fluxes, soil climate conditions in a boreal grassland on a mineral soil within reasonable levels of accuracy, albeit at a trade-off in accuracy of crop biomass prediction. Future work using DNDC could be aimed at improving crop phenology (accounting for accurate onset and end of the growing seasons), interactions among plant species and potential benefits of legume crops in legume grassland systems, and improving the characterization of heat and water exchange at the soil surface layer to determine key factors influencing simulated GHG exchanges. Overall, however, our results suggest that the model is suitable for modelling crop, soil and GHG exchange from boreal grasslands.

Author Contributions: D.F.—data analysis, model calibration and evaluation, original draft. J.D.—model calibration and evaluation, editing and advisory. M.T.H.—review, editing and advisory. N.S.—review, editing, advisory and project supervision. All authors have read and agreed to the published version of the manuscript.

Funding: This work was supported by funding from the Ministry of Agriculture and Forestry Finland (Helsinki, FI) (Project: Clover for biogas, Project NC-GRASS: VN/28562/2020-MMM-2).

Data Availability Statement: The data presented in this study are available on request from the corresponding authors.

Conflicts of Interest: We declare no conflict of interest.

References

1. Unc, A.; Altdorff, D.; Abakumov, E.; Adl, S.; Baldursson, S.; Bechtold, M.; Cattani, D.J.; Firbank, L.G.; Grand, S.; Guðjónsdóttir, M.; et al. Expansion of Agriculture in Northern Cold-Climate Regions: A Cross-Sectoral Perspective on Opportunities and Challenges. *Front. Sustain. Food Syst.* **2021**, *5*, 663448. [CrossRef]
2. Lind, S.E.; Virkajärvi, P.; Hyvönen, N.P.; Maljanen, M.; Kivimäenpää, M.; Jokinen, S.; Antikainen, S.; Latva, M.; Räty, M.; Martikainen, P.J.; et al. Carbon dioxide and methane exchange of a perennial grassland on a boreal mineral soil. *Boreal Environ. Res.* **2020**, *25*, 1–17.
3. United Nations Department for Economic and Social Affairs. *Sustainable Development Goals Report 2022*; S.l.: United Nations: New York, NY, USA, 2022.
4. Harrison, M.T.; Cullen, B.R.; Mayberry, D.E.; Cowie, A.L.; Bilotto, F.; Badgery, W.B.; Liu, K.; Davison, T.; Christie, K.M.; Muleke, A.; et al. Carbon myopia: The urgent need for integrated social, economic and environmental action in the livestock sector. *Glob. Chang. Biol.* **2021**, *27*, 5726–5761. [CrossRef] [PubMed]
5. Fleming, A.; O'Grady, A.P.; Stitzlein, C.; Ogilvy, S.; Mendham, D.; Harrison, M.T. Improving acceptance of natural capital accounting in land use decision making: Barriers and opportunities. *Ecol. Econ.* **2022**, *200*, 107510. [CrossRef]
6. Harrison, M.T. Climate change benefits negated by extreme heat. *Nat. Food* **2021**, *2*, 855–856. [CrossRef]
7. Harrison, M.T.; Evans, J.R.; Moore, A.D. Using a mathematical framework to examine physiological changes in winter wheat after livestock grazing: 1. Model derivation and coefficient calibration. *Field Crops Res.* **2012**, *136*, 116–126. [CrossRef]
8. Harrison, M.T.; Evans, J.R.; Moore, A.D. Using a mathematical framework to examine physiological changes in winter wheat after livestock grazing: 2. Model validation and effects of grazing management. *Field Crops Res.* **2012**, *136*, 127–137. [CrossRef]
9. Romera, A.J.; Cichota, R.; Beukes, P.C.; Gregorini, P.; Snow, V.O.; Vogeler, I. Combining Restricted Grazing and Nitrification Inhibitors to Reduce Nitrogen Leaching on New Zealand Dairy Farms. *J. Environ. Qual.* **2017**, *46*, 72–79. [CrossRef]
10. Vogeler, I.; Beukes, P.; Burggraaf, V. Evaluation of mitigation strategies for nitrate leaching on pasture-based dairy systems. *Agric. Syst.* **2013**, *115*, 21–28. [CrossRef]
11. Grant, B.; Smith, W.N.; Desjardins, R.; Lemke, R.; Li, C. Estimated N_2O And CO_2 Emissions as Influenced by Agricultural Practices in Canada. *Clim. Change* **2004**, *65*, 315–332. [CrossRef]
12. Höglind, M.; Cameron, D.; Persson, T.; Huang, X.; van Oijen, M. BASGRA_N: A model for grassland productivity, quality and greenhouse gas balance. *Ecol. Model.* **2020**, *417*, 108925. [CrossRef]
13. Korhonen, P.; Palosuo, T.; Persson, T.; Höglind, M.; Jégo, G.; Van Oijen, M.; Gustavsson, A.-M.; Bélanger, G.; Virkajärvi, P. Modelling grass yields in northern climates–A comparison of three growth models for timothy. *Field Crops Res.* **2017**, *224*, 37–47. [CrossRef]
14. Riedo, M.; Grub, A.; Rosset, M.; Fuhrer, J. A pasture simulation model for dry matter production, and fluxes of carbon, nitrogen, water and energy. *Ecol. Model.* **1998**, *105*, 141–183. [CrossRef]
15. Ehrhardt, F.; Soussana, J.-F.; Bellocchi, G.; Grace, P.; McAuliffe, R.; Recous, S.; Sándor, R.; Smith, P.; Snow, V.; de Antoni Migliorati, M.; et al. Assessing uncertainties in crop and pasture ensemble model simulations of productivity and N2O emissions. *Glob. Chang. Biol.* **2018**, *24*, e603–e616. [CrossRef]

16. Graux, A.-I.; Gaurut, M.; Agabriel, J.; Baumont, R.; Delagarde, R.; Delaby, L.; Soussana, J.-F. Development of the pasture simulation model for assessing livestock production under climate change. *Agric. Ecosyst. Environ.* **2011**, *144*, 69–91. [CrossRef]
17. Harrison, M.T.; Christie, K.M.; Rawnsley, R.P.; Eckard, R.J. Modelling pasture management and livestock genotype interventions to improve whole-farm productivity and reduce greenhouse gas emissions intensities. *Anim. Prod. Sci.* **2014**, *54*, 2018–2028. [CrossRef]
18. Sándor, R.; Ehrhardt, F.; Brilli, L.; Carozzi, M.; Recous, S.; Smith, P.; Snow, V.; Soussana, J.-F.; Dorich, C.D.; Fuchs, K.; et al. The use of biogeochemical models to evaluate mitigation of greenhouse gas emissions from managed grasslands. *Sci. Total Environ.* **2018**, *642*, 292–306. [CrossRef] [PubMed]
19. Ho, C.K.M.; Jackson, T.; Harrison, M.T.; Eckard, R.J. Increasing ewe genetic fecundity improves whole-farm production and reduces greenhouse gas emissions intensities: 2. Economic performance. *Anim. Prod. Sci.* **2014**, *54*, 1248–1253. [CrossRef]
20. Bilotto, F.; Harrison, M.T.; Migliorati, M.D.A.; Christie, K.M.; Rowlings, D.W.; Grace, P.R.; Smith, A.P.; Rawnsley, R.P.; Thorburn, P.J.; Eckard, R.J. Can seasonal soil N mineralisation trends be leveraged to enhance pasture growth? *Sci. Total Environ.* **2021**, *772*, 145031. [CrossRef]
21. Christie, K.M.; Smith, A.P.; Rawnsley, R.P.; Harrison, M.T.; Eckard, R.J. Simulated seasonal responses of grazed dairy pastures to nitrogen fertilizer in SE Australia: N loss and recovery. *Agric. Syst.* **2020**, *182*, 102847. [CrossRef]
22. Sándor, R.; Ehrhardt, F.; Grace, P.; Recous, S.; Smith, P.; Snow, V.; Soussana, J.-F.; Basso, B.; Bhatia, A.; Brilli, L.; et al. Ensemble modelling of carbon fluxes in grasslands and croplands. *Field Crops Res.* **2020**, *252*, 107791. [CrossRef]
23. Taylor, C.A.; Harrison, M.T.; Telfer, M.; Eckard, R. Modelled greenhouse gas emissions from beef cattle grazing irrigated leucaena in northern Australia. *Anim. Prod. Sci.* **2016**, *56*, 594. [CrossRef]
24. Li, C.; Frolking, S.; Crocker, G.J.; Grace, P.R.; Klír, J.; Körchens, M.; Poulton, P.R. Simulating trends in soil organic carbon in long-term experiments using the DNDC model. *Geoderma* **1997**, *81*, 45–60. [CrossRef]
25. Abdalla, M.; Song, X.; Ju, X.; Topp, C.F.E.; Smith, P. Calibration and validation of the DNDC model to estimate nitrous oxide emissions and crop productivity for a summer maize-winter wheat double cropping system in Hebei, China. *Environ. Pollut.* **2020**, *262*, 114199. [CrossRef]
26. Abdalla, M.; Hastings, A.; Chadwick, D.; Jones, D.; Evans, C.; Jones, M.; Rees, R.; Smith, P. Critical review of the impacts of grazing intensity on soil organic carbon storage and other soil quality indicators in extensively managed grasslands. *Agric. Ecosyst. Environ.* **2018**, *253*, 62–81. [CrossRef]
27. Abdalla, M.; Kumar, S.; Jones, M.; Burke, J.; Williams, M. Testing DNDC model for simulating soil respiration and assessing the effects of climate change on the CO_2 gas flux from Irish agriculture. *Glob. Planet. Chang.* **2011**, *78*, 106–115. [CrossRef]
28. Abdalla, M.; Wattenbach, M.; Smith, P.; Ambus, P.; Jones, M.; Williams, M. Application of the DNDC model to predict emissions of N_2O from Irish agriculture. *Geoderma* **2009**, *151*, 327–337. [CrossRef]
29. Congreves, K.; Grant, B.; Dutta, B.; Smith, W.; Chantigny, M.; Rochette, P.; Desjardins, R. Predicting ammonia volatilization after field application of swine slurry: DNDC model development. *Agric. Ecosyst. Environ.* **2016**, *219*, 179–189. [CrossRef]
30. Guest, G.; Kröbel, R.; Grant, B.; Smith, W.; Sansoulet, J.; Pattey, E.; Desjardins, R.; Jégo, G.; Tremblay, N. Model comparison of soil processes in eastern Canada using DayCent, DNDC and STICS. *Nutr. Cycl. Agroecosyst.* **2017**, *109*, 211–232. [CrossRef]
31. He, W.; Dutta, B.; Grant, B.; Chantigny, M.; Hunt, D.; Bittman, S.; Tenuta, M.; Worth, D.; VanderZaag, A.; Desjardins, R.; et al. Assessing the effects of manure application rate and timing on nitrous oxide emissions from managed grasslands under contrasting climate in Canada. *Sci. Total Environ.* **2020**, *716*, 135374. [CrossRef]
32. Levy, P.E.; Mobbs, D.C.; Jones, S.K.; Milne, R.; Campbell, C.; Sutton, M.A. Simulation of fluxes of greenhouse gases from European grasslands using the DNDC model. *Agric. Ecosyst. Environ.* **2007**, *121*, 186–192. [CrossRef]
33. Sansoulet, J.; Pattey, E.; Kröbel, R.; Grant, B.; Smith, W.; Jégo, G.; Desjardins, R.; Tremblay, N. Comparing the performance of the STICS, DNDC, and DayCent models for predicting N uptake and biomass of spring wheat in Eastern Canada. *Field Crops Res.* **2014**, *156*, 135–150. [CrossRef]
34. Forster, D.; Helama, S.; Harrison, M.T.; Rotz, C.A.; Chang, J.; Ciais, P.; Pattey, E.; Virkajärvi, P.; Shurpali, N. Use, calibration and verification of agroecological models for boreal environments: A review. *Grassl. Res.* **2022**, *1*, 14–30. [CrossRef]
35. Rafique, R.; Peichl, M.; Hennessy, D.; Kiely, G. Evaluating management effects on nitrous oxide emissions from grasslands using the process-based DeNitrification-DeComposition (DNDC) model. *Atmos. Environ.* **2011**, *45*, 6029–6039. [CrossRef]
36. Khalil, M.I.; Fornara, D.A.; Osborne, B. Simulation and validation of long-term changes in soil organic carbon under permanent grassland using the DNDC model. *Geoderma* **2020**, *361*, 114014. [CrossRef]
37. Mammarella, I.; Peltola, O.; Nordbo, A.; Järvi, L.; Rannik, Ü. 'EddyUH: An Advanced Software Package for Eddy Covariance Flux Calculation for a Wide Range of Instrumentation and Ecosystems. *Gases/Situ Meas. /Data Process. Inf. Retr.* **2016**. *preprint.* [CrossRef]
38. Li, C.; Frolking, S.; Harriss, R. Modeling carbon biogeochemistry in agricultural soils. *Glob. Biogeochem. Cycles* **1994**, *8*, 237–254. [CrossRef]
39. Li, C.; Frolking, S.; Frolking, T.A. A model of nitrous oxide evolution from soil driven by rainfall events: 2. Model applications. *J. Geophys. Res. Atmos.* **1992**, *97*, 9777–9783. [CrossRef]
40. Li, C.S. Modeling trace gas emissions from agricultural ecosystems. *Nutr. Cycl. Agroecosyst.* **2000**, *58*, 259–276. [CrossRef]
41. Li, C.; Salas, W.; Zhang, R.; Krauter, C.; Rotz, A.; Mitloehner, F. 'Manure-DNDC: A biogeochemical process model for quantifying greenhouse gas and ammonia emissions from livestock manure systems. *Nutr. Cycl. Agroecosyst.* **2012**, *93*, 163–200. [CrossRef]

42. Deng, J.; Xiao, J.; Ouimette, A.; Zhang, Y.; Sanders-DeMott, R.; Frolking, S.; Li, C. Improving a Biogeochemical Model to Simulate Surface Energy, Greenhouse Gas Fluxes, and Radiative Forcing for Different Land Use Types in Northeastern United States. *Glob. Biogeochem. Cycles* **2020**, *34*, e2019GB006520. [CrossRef]

43. Deng, J.; Li, C.; Frolking, S.; Zhang, Y.; Bäckstrand, K.; Crill, P. Assessing effects of permafrost thaw on C fluxes based on multiyear modeling across a permafrost thaw gradient at Stordalen, Sweden. *Biogeosciences* **2014**, *11*, 4753–4770. [CrossRef]

44. Deng, L.; Zhu, G.; Tang, Z.; Shangguan, Z. Global patterns of the effects of land-use changes on soil carbon stocks. *Glob. Ecol. Conserv.* **2016**, *5*, 127–138. [CrossRef]

45. Lind, S.E.; Shurpali, N.J.; Peltola, O.; Mammarella, I.; Hyvönen, N.; Maljanen, M.; Räty, M.; Virkajärvi, P.; Martikainen, P.J. Carbon dioxide exchange of a perennial bioenergy crop cultivation on a mineral soil. *Biogeosciences* **2016**, *13*, 1255–1268. [CrossRef]

46. Zambrano-Bigiarini, M. *hzambran/hydroGOF*, v0.4-0. 2020. Available online: https://github.com/hzambran/hydroGOF/tree/v0.3-10 (accessed on 16 October 2022).

47. Oertel, C.; Matschullat, J.; Zurba, K.; Zimmermann, F.; Erasmi, S. Greenhouse gas emissions from soils—A review. *Geochemistry* **2016**, *76*, 327–352. [CrossRef]

48. Kim, N.; Zabaloy, M.C.; Guan, K.; Villamil, M.B. Do cover crops benefit soil microbiome? A meta-analysis of current research. *Soil Biol. Biochem.* **2020**, *142*, 107701. [CrossRef]

49. Khan, A.R. Influence of Tillage on Soil Aeration. *J. Agron. Crop Sci.* **1996**, *177*, 253–259. [CrossRef]

Article

Ecosystem Services Provided by Pastoral Husbandry: A Bibliometric Analysis

Juan Manuel Mancilla-Leytón [1,2,*], Djamila Gribis [1], Claudio Pozo-Campos [1], Eduardo Morales-Jerrett [3], Yolanda Mena [3], Jesús Cambrollé [1,2] and Ángel Martín Vicente [1]

[1] Departamento de Biología Vegetal y Ecología, Facultad de Biología, Universidad de Sevilla, 41012 Seville, Spain
[2] Instituto Interuniversitario Andaluz en Robótica y Sistemas Inteligentes, "Robotics and Intelligent Systems" (RIS), Universidad de Sevilla, C/Torricelli 18–20, 41092 Seville, Spain
[3] Departamento de Agronomía, Escuela Técnica Superior de Ingeniería Agronómica, Universidad de Sevilla, 41013 Seville, Spain
[*] Correspondence: jmancilla@us.es

Abstract: The ecosystem services provided by the age-old activity of husbandry are presently declining or seriously endangered. The situation is particularly serious for regulation services and for certain cultural services given their growing dependence on external inputs. This work performs a bibliometric analysis for the purpose of identifying the certainties and gaps associated with the different ecosystems generated by pastoral husbandry, and confirms the pressing challenges that the livestock industry is facing in the current context of global change. Two different tools, Scopus and VOSviewer, have been implemented to analyze 2230 documents published between 1961 and 2021 that include the terms "grazing" and "service". The information required for the bibliometric analysis of authorship, country of origin, field of study and number of citations, among other categories, was drawn from the documents to the effect of evidencing their general thematic relationships. Finally, the current state of the ecosystem services currently provided by pastoral husbandry—provisioning, regulation, cultural and support services—was assessed. The results showed a greater abundance of scientific literature on provisioning and regulation services than on cultural and support services. An increase in the number of publications from the beginning of the 21st century was confirmed. The United States stands out as the country with the largest scientific production, and environmental sciences is the most prominent field in the study of ecosystem services. A recent larger academic effort to encourage the promotion of ecosystem services from the institutions has also been observed, as well as to include them as a factor in the development of environmental policies, which is described as the greatest challenge for the future of this discipline. Among other possible solutions, the new European Union agricultural subsidies—the so-called eco-schemes—appear to be essential for that effort to bear fruit as soon as possible.

Keywords: pastoral husbandry; provisioning; regulation; biodiversity; greenhouse gases

Citation: Mancilla-Leytón, J.M.; Gribis, D.; Pozo-Campos, C.; Morales-Jerrett, E.; Mena, Y.; Cambrollé, J.; Vicente, Á.M. Ecosystem Services Provided by Pastoral Husbandry: A Bibliometric Analysis. *Land* **2022**, *11*, 2083. https://doi.org/10.3390/land11112083

Academic Editors: Michael Vrahnakis, Yannis (Ioannis) Kazoglou, Manuel Pulido Fernádez and Shiliang Liu

Received: 12 October 2022
Accepted: 16 November 2022
Published: 18 November 2022

Publisher's Note: MDPI stays neutral with regard to jurisdictional claims in published maps and institutional affiliations.

1. Introduction

The first truly scientific approach to the concept of "ecosystem services" was developed by United States researchers at the end of the 1960s and the beginning of the 1970s, coinciding with the emergence of the environmental movement [1]. In 1970, the *Study of Critical Environmental Problems*, sponsored by the Massachusetts Institute of Technology, provided the first list of "environmental services" [1] (Table 1). From then on, various attempts were made, on the one hand, to properly define the concept of "ecosystem services" and, on the other, to enlist and categorize those services [2]. Daily [1] provided an important definition when she specified that ecosystem services are "the conditions and processes through which natural ecosystems, and the species that make them up, sustain and fulfill human life". Constanza et al. [3] affirmed that "ecosystem goods (such as food) and services (such as waste assimilation) represent the benefits human populations derive, directly or indirectly, from ecosystem functions". These authors also made the first step towards classifying those services and presented a more comprehensive list of 17 ecosystem services (Table 1) that derive from "natural capital", a concept coined a few years earlier by Constanza and Daily [4] to describe "the stock that yields the sustainable flow" of natural benefits. But only in the work by De Groot et al. [5] did the list become a hierarchical and systematized classification of 23 ecosystem functions from which goods and services derive. Those functions are divided into four categories: regulation, habitat, production and information functions (Table 1).

Table 1. Comparative analysis of the different lists of ecosystem services proposed since the term was coined. In the case of the Millennium Ecosystem Assessment, the category to which each service belongs is provided: provisioning (p), regulation (r), cultural (c) and support (s) services.

Study of Critical Environmental Problems [1]	Constanza et al. [3]	De Groot et al. [5]
	Refuge	Refuge function
		Breeding function
Atmosphere composition	Gas regulation	Gas regulation
Climate regulation	Climate regulation	Climate regulation
Flood control	Disturbance regulation	Disturbance prevention
	Water regulation	Water regulation
	Water provision	Water provision
	Waste treatment	Waste treatment
Soil retention	Erosion control and sediment retention	Soil retention
Pest regulation	Biological control	Biological control
Insect pollination	Pollination	Pollination
Soil formation	Soil formation	Soil formation
Matter cyding	Nutrient cycling	Nutrient regulation
Fisheries	Food production	Food
	Raw materials	Raw materials
	Genetic resources	Genetic resources
		Medicinal resources
		Ornamental resources
		Aesthetic information
	Culture	Cultural and artistic information
	Leisure	Leisure
		Spiritual and historical information
		Science and education

Obviously, the force of the idea of ecosystem services lies in its potential as a conceptual tool for the implementation of policies to counteract human impact on the planet—a societal demand that has increased over time and is today more central than ever. The publication of the *Millennium Ecosystem Assessment* report in 2005 provided the final endorsement and the term has become part of the international ecological action and environmental policy development vocabulary. The assessment focuses on the linkages between ecosystems (understood as dynamic complexes of plant, animal, and microorganism communities and the nonliving environment interacting as functional units) and human well-being [6]. The report described ecosystem services as "the benefits people obtain from ecosystems" and established four categories, which were very similar to the ones proposed by De Groot et al. [5], for a total of 30 types of services: (i) provisioning services (products obtained from the ecosystems); (ii) regulation services (benefits obtained from the regulation of ecosystem processes); (iii) cultural services (non-material benefits obtained from the ecosystems); and (iv) support services (required for the production of other ecosystem services) [6] (Table 1).

As evidenced by later works, the definition and categorization of the ecosystem services still need some refinement [7]. The concept is still ambiguous and a clearer distinction between the mechanisms through which the services are obtained and the services themselves is required [8]. The discussion on the consistency of the category of cultural services is a recurring one, because it appears to be subjective, but the fact that it is included in most models proves that, even if open to qualification, those services need to be considered [9]. Support services are also questioned by authors who see them as intertwined with other categories, particularly with regulation services [2]. However, all the contributions of the Millennium Ecosystem Assessment are useful because they are widely acknowledged, recognizable and easily comprehensible [6].

In summary, ecosystem services have come to fill up an argumentative void in environmental protection from a perspective that may be described as utilitarian, but is also transcendent in a world characterized by the market economy [2]. In their work, Constanza et al. [3] drew attention to that need after observing how decision-makers ultimately dismissed natural capital services. This way, the ongoing effort of authors working in the field of sustainability sciences to provide evidence of the link between human wellbeing and natural capital and to propose a theory of value beyond the monetary has allowed not only to fill up the argumentative void in the academic sphere, but to permeate the political and administrative discourse so that it can truly be useful [10].

However, the current valorization of ecosystem services should not make us think that they did not exist in the past. Even if science and academia did not use the concept before, human beings have always perceived the environmental retribution they could profit from through their relationship with the environment. An obvious example is animal domestication. Moreover, through their relationship to domesticated animals, human beings began to transform their environment, and the maintenance of livestock suddenly became a reason to manage and shape that environment to take advantage of everything it could offer. A paradigmatic example is that of the *dehesa*, a Mediterranean ecosystem where human intervention led to the spatial dispersion of trees in order to enable the spreading out of the livestock [11]. According to the Millennium Ecosystem Assessment in Spain (2011), the *dehesa* is an example of agroecosystem, i.e., a "type of ecosystem modified and managed by humans for the purpose of obtaining food, fiber and other materials of biotic origin". Agroecosystem diversity is very high across the world in accordance with the intensity of human intervention, and ranges from extremely simplified agroecosystems with a very high level of inputs—such as intensive agriculture or livestock breeding—to highly complex low-intensity systems—including family vegetable gardens or transhumance [12]. Therefore, we need to keep in mind that, when anthropic actions take into consideration the specificities of the territory, as in agroecosystems, they may have an essential role to play in the provision of ecosystem services [13].

The above-mentioned domestication of wild animals was one of the achievements of the Neolithic revolution, whether it happened earlier or later on during that process. To again use the example of the Mediterranean region and, more specifically, that of the Iberian Peninsula, radiocarbon dating identifies the first human groups with domestic sheep and caprine stocks in the Mediterranean coasts around 5600 AD [14]. If those stocks and the interaction with them have been maintained until today, it is simply because of the benefits they provide to human beings. Through good management practices in pastoral husbandry, a proper management and conservation of the ecosystems that the stocks graze is possible. Therefore, livestock plays an important role as a provider not only of ecosystem services, but of ecosystem regulation and maintenance services, such as the control of accumulated biofuel, the dispersion of Mediterranean species, the decomposition of litter, the enhancement of the nutrient cycle, the balance between native and invasive species, landscape conservation, etc. [15]. However, as advanced by the conclusions of the Millennium Ecosystem Assessment [6], the ecosystem services rendered by age-old husbandry practices are currently in decline or seriously endangered. Regulation services—conservation of habitats of interest to other species, soil fertilization or pasture improvement—and some cultural services—gastronomic traditions, identity and sense of belonging to a territory, or landscape beauty—are especially at risk because of the increasing dependence on external inputs in food production services [15–17].

It is thus possible to affirm that pastoral husbandry is on the wane. The intensification of stockbreeding activities for the purpose of developing more productive systems has had a negative impact on biodiversity and, consequently, on the ecosystem services provided. In particular, this process is affecting the marginal areas that pastoral husbandry has traditionally used, progressively relegating it to a much less relevant role [18]. However, all of these problems could be solved in a context like the current one, where the sustainability of animal production is being questioned [19], and the demand for food produced in ways that respect animal wellbeing and helps to preserve ecosystems and biodiversity is growing.

The aim of this study is to carry out a systematic assessment, through a bibliometric analysis, of the different ecosystem services provided by pastoral husbandry. We have attempted to answer the following research questions, which we believe could be of interest to researchers in this field: (i) what are the global trends of scientific publications on the topic of grazing services?; (ii) which institutions, together with their corresponding collaboration networks, work more intensely on this issue?; and (iii) which discipline publishes the most on this topic? Finally, the implications of this reality, the challenges ahead and the possible future lines of research within this field are also discussed in this review article.

2. Materials and Methods

Bibliometric analysis is a popular and rigorous method for exploring and analyzing large volumes of scientific data [20]. It differs from traditional narrative reviews in that it implements a replicable, scientific and transparent procedure based on exhaustive literature searches of published studies [21]. Unlike systematic literature reviews that tend to rely on qualitative techniques, which may be marred by the interpretation bias of scholars from different academic backgrounds, a bibliometric methodology uses quantitative techniques to analyze bibliometric data and can thus avoid or mitigate that bias. Bibliometric analysis can offer a balance between comprehensively identifying a larger pool of publications and systematically identifying a smaller set of studies that fit criteria for inclusion [22].

First, a search was made in different databases (PubMed, Scopus, Google Scholar, etc.) of articles on the topic using certain keywords in English, such as "ecosystem services", "pastoralism", "grazing" or "husbandry services", as well as combinations of those terms and possible translations of them into Spanish. This initial step, prior to the actual bibliometric analysis, was intended to provide a first overview of the status of ecosystem services in pastoral husbandry in academic publications. For the development of the biometric analysis, data were gathered from the Scopus database of Elsevier. In order to cover full calendar years, 2021 was set as the end date of the search period; consequently, the search yielded results for the period 1961–2021. An exhaustive search was carried out in Scopus using [TITLE-ABS-KEY (grazing AND service) AND (EXCLUDE (PUBYEAR, 2022))] as the search field. The final number of articles found in the search was 2229. Subsequently, a specific function of Scopus was used to collect information from the set of articles for the bibliometric analysis of authorship, country of origin, field of study and number of citations, among other categories. The information obtained from this database was analyzed and processed with VOSviewer software (Leiden University and CWTS) for the purpose of revealing the general thematic relationships among previously obtained manuscripts [20]. A bibliometric map with four clusters was thus obtained by VOSviewer software. Cluster size was determined by the number of keywords within the cluster, the frequency of occurrence of those keywords, and their similarity index. The frequency of co-occurrence was calculated on the basis of keywords repeated more than 40 times.

Finally, after the bibliometric analysis, the articles were subject to a thorough review. This review enabled the assessment of the general situation of ecosystem services currently related to pastoral activities, including provisioning, regulation, cultural and support services [6].

3. Results

The distribution of the documents per year identified is shown in Figure 1. As indicated above, the academic use of the concept of ecosystem services did not take root until the 1970s. The few references to it during the 1960s are interpreted as mere coincidences, explained by their appearing in articles on agriculture; this is confirmed by reading the abstracts of those articles. However, despite the emergence of the term in the environmental debate, during the last three decades of the 20th century its use was far from common, with an average of 5 publications per year (between 1961 and 2000, see Figure 1). Only after the benchmark work by Daily [1] and other contemporary works were published did the topic gain relevance and gather sufficient attention to cause a noticeable change of trend around year 2001. In the 21st century, the production of works referring to ecosystem services in pastoral husbandry greatly increased to reach the amount of 2230 documents available in 2021, rising from five to 96 yearly publications in the period 2001–2021 (Figure 1). These trends were determined by using adjusted coefficients of determination for both periods ($R2 > 0.84$ and $R2 > 0.98$, respectively).

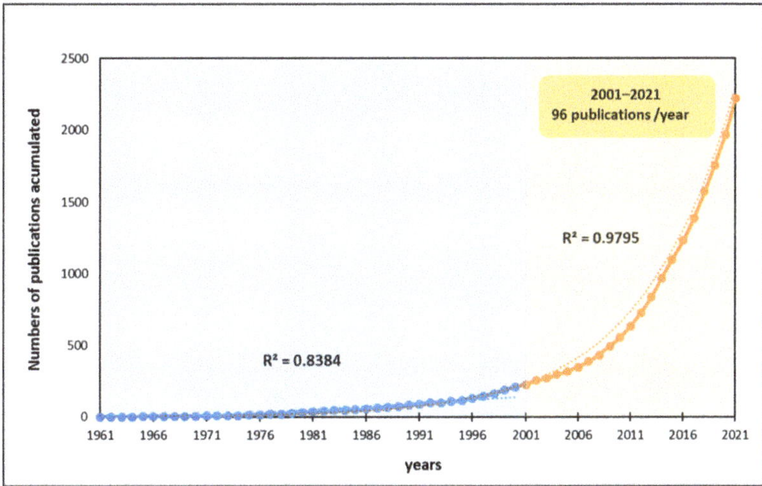

Figure 1. Evolution of the number of publications per year and accumulated in the period 1961–2021, resulting from the search in Scopus of the terms "grazing" and "service". The adjusted coefficient of determination used in each stage is shown.

Most of the documents analyzed (81%) were published as scientific works; the rest were reviews (7.2%), book chapters (6.2%), papers presented at conferences on the topic (3.8%), and other texts classified in rarer categories, each accounting for less than 1%. As usual within the scientific community, the "de facto lingua franca" was English, in which 93.7% of the documents analyzed were written.

With regard to the institutions funding this type of work, it was observed that the main one was the National Natural Science Foundation of China, closely followed by similar bodies attached to the governments of the United States and European Commission. The Chinese Academy of Sciences and the United States Department of Agriculture (USDA) are the two institutions whose researchers have, up to now, published more works on this topic. This explains why most documents resulting from the search (727) originate in the United States (Figure 2), making its territory, where livestock production is so prominent, a relevant object of analysis in relation to ecosystem services. The contribution of the following countries is also noteworthy: China (214 documents), the United Kingdom (197), Australia (183) and Germany (166) are on a second level of importance, followed by Spain (109) and France (102), among others, on the third level. The authors leading these types of studies was the Australian David J. Eldridge, (10 documents), although there are prominent researchers from other countries, for instance, the American Justin D. Derner and Leslie M. Roche, Sandra Lavorel from France, and Wolde Mekuria from Ethiopia, who have each published nine documents.

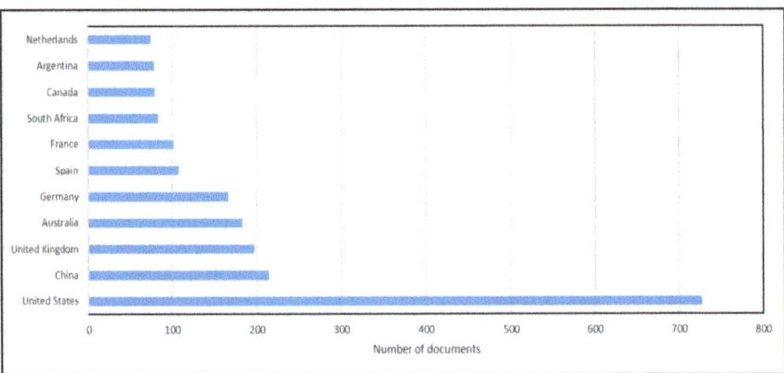

Figure 2. Scientific production addressing the topic of ecosystem services in pastoral husbandry by country, as obtained from the search in Scopus. The countries included are those in which more than 70 documents have been published.

As for the fields of study, agricultural and biological sciences are the ones with a larger production of documents (34%) (Figure 3), followed by environmental sciences (32%) and social sciences, with a much smaller academic output (8% of the total); the prevalence of the two first areas is evident. The results are not surprising given that those sciences belong within the environmental field, to which the concept of "ecosystem services" is attached.

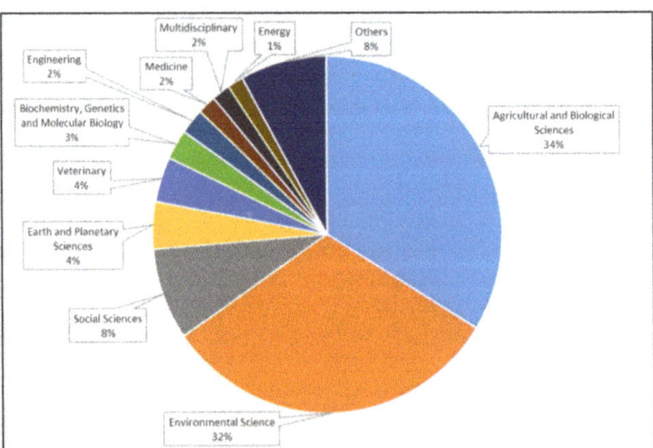

Figure 3. Main fields of study to which the documents subject to the bibliometric analysis are related. Only the fields where more than 60 documents have been produced are shown.

Figure 4 shows the connection between the documents obtained from the search according to the more or less frequent presence of certain keywords in them. The resulting map includes four major fields, which are differentiated by color, where the keywords "ecosystem service", "grazing" and "animals" stand out at the core of the relationships between documents.

Each of these four groups differentiated by the software were drawn in the form of lists in order to compare them (Table 2). Cluster 1, including concepts such as "carbon sequestration", "nutrient cycling" or those referring to the soil, was closely connected to regulation services. Cluster 3, which comprised terms like "crop production" or "milk production", was mostly related to provisioning services. Clusters 2 and 4 included a wide variety of terms that reflected different types of services and, in contrast with clusters 1 and 3, were not closely connected to any of them. The conclusion that can be drawn is that some types of services—regulation and provisioning—receive greater attention and are more thoroughly studied than others, such as cultural and support services, showing an imbalance that may be due to the apparent profitability of each service.

Table 2. Keywords found in the documents analyzed according to the clustering made with VOSviewer. The minimum number of times each term included in those clusters needed to appear was set at 40.

Cluster 1	Cluster 2	Cluster 3	Cluster 4
Agriculture	Abundance	Agricultural production	Adaptation
Carbon	Anthropogenic effect	Animal husbandry	Climate change
Carbon sequestration	Biodiversity	Animals	Decision making
Ecology	Conservation of natural resources	Bovine	Forage
Ecosystems	Ecosystem function	Cattle	Grazing pressure
Environmental protection	Ecosystem service	Controlled study	Land management
Forestry	Fires	Crop production	Livestock grazing
Forests	Grassland	Food supply	Pastoralism
Land use	Grazing	Milk production	Pasture
Nitrogen	Herbivory	Physiology	Rangelands
Nutrient cycling	Invasive species	Procedures	Semiarid region
Remote sensing	Landscape	Reproduction	Sustainability
Restoration ecology	Plant community	Seasons	
Soil conservation	Species diversity	Sheep	
Soil fertility	Species richness		
Soils	Vegetation structure		
Water quality			
Water supply			
Wetlands			

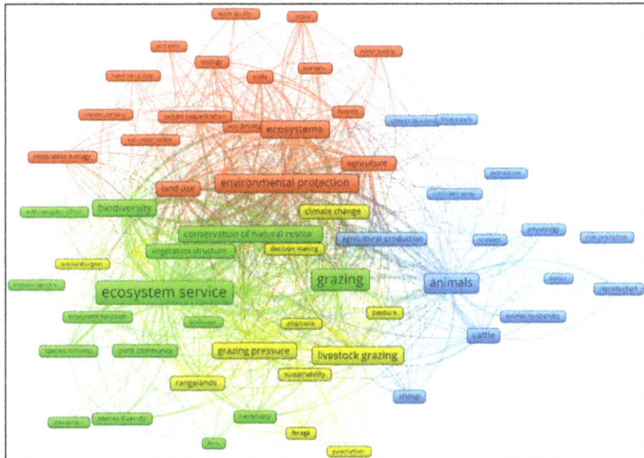

Figure 4. Clusters of keywords appearing in articles about grazing services with a frequency greater than 40 (obtained using VOSviewer software) Four major fields are defined and differentiated by color. Cluster 1 is depicted in red, cluster 2 in green, cluster 3 in blue, and cluster 4 in yellow. The complete list of keywords in each group is shown in Table 2.

4. Discussion

The documents gathered in the search may be divided into: (i) those focused on the valuation of services in a specific context; (ii) those that describe the compensations between services occurring in ecosystems; and (iii) those that examine policy implementation plans and the way they are perceived by society. The growing number of documents on this topic is based on the development of these three categories.

Based on the analysis of ecosystem functions and services made by Gómez-Baggethun and De Groot [10], Table 3 presents a list of functions, goods and services that may be attributed to natural ecosystems where livestock breeding is developed.

Table 3. Ecosystem functions, goods and services associated with pastoral husbandry. From Gómez-Baggethun and De Groot [10].

Functions	Examples of Goods and Services
Refuge and breeding site	Fire prevention Maintenance of the biodiversity of domesticated wild species, seed dispersal
Atmospheric and climate regulation	Carbon sink
Nutrient regulation	Maintenance of soil health and of productive ecosystems, improvement of soil fertility
Raw materials	Energy and natural fertilizers
Recreation	Ecotourism
Medicinal resources	Medicinal plants
Ornamental elements	Materials used in craftwork: leather and fur
Aesthetic information	Landscape enjoyment
Artistic, cultural and historical information	Depictions of nature in books, films, cultural heritage, etc.
Science and education	Environmental education, scientific purposes
Housing	A place to live, maintenance of rural populations
Agriculture	Food and raw materials, functional food
Tourism infrastructures	Reduction of combustible biomass through the development of tourism activities: hiking

Various authors [23,24] have concluded that well-adapted local breeds provide most provisioning, regulation and cultural services. Therefore, the current discourse on provisioning services highlights the importance of reinforcing the husbandry of endangered or less commercially available autochthonous breeds, which not only provide meat, milk and dairy products, fibers and textiles, but also improve the genetic heritage of the species, which is a diversity value in itself as well as a source of protection for the breed in the face of difficulties [25]. As pointed out by Martín-Collado et al. [26], within the field of animal science it is essential to distinguish the exact role that breed, species and breeding practices play in the supply of ecosystem services. Husbandry practices (including the choice of species and breeds) modify the structure and functioning of the ecosystem, which may cause ecosystem disservices (for instance, the reduction of water availability).

The different species and breeds, most of them autochthonous, living in one territory are part of its ecosystem and provide a livelihood for people inhabiting those rural areas, where it is sometimes very difficult to pursue an economic activity other than extensive livestock rearing. In the Mediterranean region and, more specifically, in Andalusia, sheep breeds (such as the Merina, Merina de Grazalema or Segureña), cattle breeds (such as the Retinta, Berrenda or Pajuna), goat breeds (such as the Payoya, Malagueña, Murciano-Granadina, Florida, Blanca Andaluza or Negra Serrana), and pork breeds (mainly the

Ibérica), are good examples of the rich genetic diversity of the *dehesas* and mountain areas, although many of those breeds, as well as the production models developed around them, are currently endangered [27].

Despite the important role played by ecosystem services, the FAO [28] has stressed the need for appropriate legislation on their management in developing countries because of the lack of formal recognition of their value on the part of the ruling classes. Only in European Union countries has this need been adequately fulfilled [29]. It is important to underline, in relation to the protection and promotion of the above-mentioned breeds, that they provide access to a whole market, like the current one, where value-added products are sought because of their gastronomic quality and sustainability [30]. They produce benefits that go beyond those associated with ecosystem services, for they encourage business and overall development in environments where husbandry is supported and promoted [31]. Another important factor in favor of protecting autochthonous breeds is livestock diversification, because diversity enhances multifunctionality [32] and the integration of agents that may provide services to the agroecosystem.

Support services are often mentioned in relation to other services. Primary production, photosynthesis or nutrient cycling, among other support services, are usually taken into consideration for the purpose of subsequently connecting them, for instance, to a provisioning service such as the production of forage or the quality of animal products—which is mainly the result of the way animals are fed [31], thus creating a feedback loop. Regarding primary production, it has been found that in Mediterranean ecosystems pastoral livestock stimulates the production of pastures giving rise to what is called the "pastoral paradox": the most appetizing and nutritious pasture species, which are usually the most grazed ones, increase their abundance thanks to grazing. Mediterranean species have evolved together with pastoral livestock, so they can withstand grazing much more than others due to their greater capacity for regrowth or trampling support. In this sense, they have competitive advantages [33]. Likewise, pastoral livestock can facilitate the improvement of soil resources through: (i) trampling, which activates the recycling of nutrients and the conservation of poor soils by improving their structure, favoring drainage and reducing erosion; (ii) the movement of livestock during grazing, which connects different agroecosystems, thus facilitating horizontal fertility and contributing to seed dispersal; and (iii) access to stubble and crop residues, which favors the recirculation of nutrients and avoids additional work (droppings increase the biological activity of soils and the presence of detritivores, mycorrhizae, fungi, etc.).

It is important to underline that, very often, the contribution of pastoral husbandry to this type of ecosystem services is described in critical terms due to the risk of overgrazing. Only through lighter, less intensive forms of pastoralism, based on the application of a mixed-method approach to agroecosystems, can the balance between the services provided by pastoral husbandry and the damaging effects of this activity on the environment [34,35].

Other services examined in the documents analyzed are the so-called regulation services, which include, for instance, climate regulation, fire prevention, plant species control or pollination. With regard to climate regulation, Teague et al. [34] point out that ruminants are usually accused of being the source of greenhouse gas emissions—mostly via methane production, but through proper pasture management these animals may actually help increase soil carbon sequestration, thus offsetting the emissions. According to McDermot and Elavarthi [36], agroecosystems have the capacity to sink 1.2 to 3.1 billion tons of carbon per year over a period of 50 years. These authors consider that this degree of carbon sequestration is capable of offsetting a third of the annual increase in atmospheric CO_2, which is estimated at 3.3 PgC/year. One ton of carbon sequestered in the soil allows the elimination of 3.67 tons of atmospheric CO_2, in addition to improving agronomic productivity and enhancing soil resilience. According to Bork et al. [37], the key lies in the activity of the roots of herbaceous species: their response to grazing is to produce more roots and more exudates through them. The higher root mass produced in grazed grasslands partially explains why grazing tends to concentrate more carbon in the soil. In spite of this, the relationship between pastoralism and organic soil carbon is not linear, and a thorough study of each agroecosystem seems necessary if any conclusions are to be drawn [38,39]. The success of grazing management will depend on how well the increase in livestock efficiency is balanced with the need to maintain the chemical, physical, hydrological and biological properties of each type of soil (a key element of the ecosystem). It is important to note that, although the capacity of vegetation to act as a carbon sink is well documented in the literature [38,40,41], grazing lands associated with livestock production, which could significantly balance the net greenhouse gas emissions emission values, has been rarely considered to date [40]. This is due, to a large extent, to the difficulties in measuring it. It is therefore necessary to develop methodologies that facilitate the quantification of the carbon sequestered by grasslands and stored in their soils, and of the methane they oxidize, and to propose an emission model that is closer to the complex reality of these production systems.

The use of livestock to prevent fires is known around the world and has been demonstrated throughout human history [42]. Fire prevention is one of the most valued and frequently remunerated ecosystem services provided by pastoral husbandry. Pastoral husbandry helps reduce combustible biomass, consequently decreasing the risk of fire [43]. Certainly, as with many other services provided by agroecosystems, it requires that pastoral activities are properly organized in order to be fully effective. The work developed by the Red de Áreas Pasto-Cortafuegos de Andalucía (RAPCA, Andalusian Network of Pasture-Firebreak Areas) exemplifies such an attempt. Through a controlled management of the livestock by the shepherds' activities, it helps clear the grass and brush from strategically designed fire lanes [44], and contributes to diversifying the income of farmers, who receive economic compensation, albeit minor, for their work [45].

Plant species control through herbivory cannot be highly targeted, but is important enough to prevent the spread of invasive species or to reduce the density of herbal or woody species competing among themselves and hindering landscape heterogeneity [46]. Grazing changes the abundance and diversity of flowers, thus affecting the structure and dynamics of the entire community of interactions. [47]. The effect of grazing on pollinators and their pollination services can vary from positive to negative depending on the way in which livestock modify the vegetation and on whether the observed foraging intensity increases or decreases the floral resources used by pollinators [47,48].

Cultural services are more connected to human subjectivity than to objective matter. In general terms, society recognizes the importance of the ecosystem services provided by livestock farming [29], valuing each category—even the relatively invisible support services, but it particularly appreciates cultural services, which are perceived as elements of the common identity [29,49]. However, despite this positive reception, the scientific community has not truly delved into the study of cultural services, which have been barely characterized. It seems necessary to take these services out of the purely subjective level in order to examine them more thoroughly, because being the most "human" of all ecosystem services, they can help society understand the general concept more accurately.

5. Conclusions and Future Prospects

The bibliometric analysis and the assessment of the documents revealed the great scientific and academic interest in the role that pastoral husbandry plays in current ecosystems. Most documents analyzed insist on the need to encourage the promotion of ecosystem services from the institutions and to take them into consideration as a factor in the development of environmental policies. In this sense, the different approaches on ecosystem services can be an obstacle in themselves, given the general lack of agreement on: (i) their definition and classification; (ii) the way to integrate them in land management [7,50]; and (iii) the actual steps to be taken in decision-making processes. Ultimately, those approaches remain mere theoretical interpretations [51]. The Ecosystem Service Database and other platforms created for the purpose of making the most important studies on the topic available to researchers, or the application ARIES, developed to evaluate ecosystem services using artificial intelligence, are tools—sponsored, in these two cases, by the University of

Vermont—which are expected to improve scientific communication and to help implement measures to solve the problems associated with those services [7]. On the other hand, some studies have placed emphasis on the reduction of services associated with bad land management practices and on how this is causing massive economic losses across the world [52]. Providing a more appropriate framework for action should support rather than undermine the efforts to propose practical measures.

Although it is true that in European Union countries ecosystem services are acknowledged at a national and even regional level, this recognition is still limited and does not ease the supply and flow of ecosystem services in general and those related to pastoral husbandry in particular [29]. One example of European Union policies aimed not only at protecting populations dependent on agriculture, but also, in principle, at managing ecosystem services are the subsidies provided through the Common Agrarian Policy (CAP) mechanism. However, they are constantly being reformed because their allocation methods are often controversial. To solve this situation, in 2021 a debate was initiated on a new type of subsidy known as an eco-scheme. Eco-schemes are released following the fulfillment of stricter requirements related to the maintenance of specific ecosystem services [53].

There is no doubt that, when well-managed, pastoral husbandry and the use of autochthonous breeds are an opportunity, even a necessity, for livestock farming to continue providing innumerable ecosystem services. There can be many reasons for the absence of concrete actions: (i) lack of recognition to and remuneration of the ecosystem services associated with this model of husbandry; (ii) because of the low prices at source of meat and milk (the main source of income for animal farms), which most often do not compensate the costs of production; (iii) due to changes in food habits, including the decrease in the demand of ruminant meat resulting from the "bad press" that livestock breeding has gained in recent times; or (iv) to the lack of real and effective institutional support for this production model. The reality is that no measures are taken and, within a few years, extensive and pastoral husbandry will most probably disappear or become so residual that most ecosystem services provided by them will be lost. In addition to putting food sovereignty at risk, this will have a negative effect on the regulation of ecosystems associated with the Mediterranean forest and will certainly contribute to the depopulation and abandonment of many rural areas. This section is not mandatory but can be added to the manuscript if the discussion is unusually long or complex.

Author Contributions: Conceptualization and Methodology, J.M.M.-L., D.G. and C.P.-C.; Formal analysis and Data curation, J.M.M.-L., C.P.-C., E.M.-J. and Y.M.; Investigation and writing edition, J.M.M.-L., D.G., C.P.-C., J.C. and Á.M.V. Supervision J.M.M.-L. and J.C. All authors have read and agreed to the published version of the manuscript.

Funding: This research received no external funding.

Informed Consent Statement: Not applicable.

Data Availability Statement: Not applicable.

Acknowledgments: Special thanks to M. M. Ballesteros Martín for her help in using the VOSviewer software. The authors thank the anonymous reviewers for helpful suggestions concerning presentation of this manuscript. T. Muñoz Sebastián (Lengua fértil) revised the English version of the manuscript.

Conflicts of Interest: The authors declare that they have no conflict of interest.

References

1. Daily, G.C. Nature's services: Societal dependence on natural ecosystems. In *The Future of Nature*; Yale University Press: New Haven, CT, USA, 1997; pp. 454–464.
2. Camacho-Valdez, V.; Ruiz-Luna, A. Marco conceptual y clasificación de los servicios ecosistémicos. *Rev. Bio Cienc.* **2012**, 1. [CrossRef]
3. Costanza, R.; d'Arge, R.; De Groot, R.; Farber, S.; Grasso, M.; Hannon, B.; Limburg, K.; Naeem, S.; O'Neill, R.V.; Paruelo, J.; et al. The value of the world's ecosystem services and natural capital. *Nature* **1997**, *387*, 253–260. [CrossRef]

4.   Costanza, R.; Daly, H.E. Natural capital and sustainable development. *Conserv. Biol.* **1992**, *6*, 37–46. [CrossRef]
5.   De Groot, R.S.; Wilson, M.A.; Boumans, R.M. A typology for the classification, description and valuation of ecosystem functions, goods and services. *Ecol. Econ.* **2002**, *41*, 393–408. [CrossRef]
6.   Millennium Ecosystem Assessment. *Ecosystems and Human Well-Being: Synthesis*; Island Press: Washington, DC, USA, 2005.
7.   De Groot, R.S.; Alkemade, R.; Braat, L.; Heina, L.; Willemenad, L. Challenges in integrating the concept of ecosystem services and values in landscape planning, management and decision making. *Ecol. Complex.* **2010**, *7*, 260–272. [CrossRef]
8.   Wallace, K.J. Classification of ecosystem services: Problems and solutions. *Biol. Conserv.* **2007**, *139*, 235–246. [CrossRef]
9.   Daniel, T.C.; Muhar, A.; Arnberger, A.; Aznar, O.; Boyd, J.W.; Chan, K.M.A.; Costanza, R.; Elmqvist, T.; Flint, C.G.; Gobster, P.H.; et al. Contributions of cultural services to the ecosystem services agenda. *Proc. Natl. Acad. Sci. USA* **2012**, *109*, 8812–8819. [CrossRef]
10.   Gómez-Baggethun, E.; De Groot, R. Capital natural y funciones de los ecosistemas: Explorando las bases ecológicas de la economía. *Rev. Ecosistemas* **2007**, *16*.
11.   Martín-Vicente, Á.; Fernández-Alés, R. Long term persistence of dehesas. Evidences from history. *Agrofor. Syst.* **2006**, *67*, 19–28. [CrossRef]
12.   Altieri, M.A.; Koohafkan, P. *Enduring Farms: Climate Change, Smallholders and Traditional Farming Communities*; Third World Network: George Town, Malaysia, 2008.
13.   Nieto-Romero, M.; Oteros-Rozas, E.; González, J.A.; Martín-López, B. Exploring the knowledge landscape of ecosystem services assessments in Mediterranean agroecosystems: Insights for future research. *Environ. Sci. Policy* **2014**, *37*, 121–133. [CrossRef]
14.   Tejedor-Rodríguez, C.; Moreno-García, M.; Tornero, C.; Hoffmann, A.; de Lagrán, G.-M.; Arcusa-Magallón, H.; Garrido-Pena, R.; Royo-Guillén, J.I.; Díaz-Navarro, S.; Peña-Chocarro, L.; et al. Investigating Neolithic caprine husbandry in the Central Pyrenees: Insights from a multi-proxy study at Els Trocs cave (Bisaurri, Spain). *PLoS ONE* **2021**, *16*, e0244139. [CrossRef] [PubMed]
15.   Morales-Jerrett, E.; Mancilla-Leytón, J.M.; Delgado-Pertíñez, M.; Mena, Y. The contribution of traditional meat goat farming systems to human wellbeing and its importance for the sustainability of this livestock subsector. *Sustainability* **2020**, *12*, 1181. [CrossRef]
16.   Castro, A.J.; Verburg, P.H.; Martín-López, B.; Garcia-Llorente, M.; Cabello, J.; Vaughnb, C.C.; López, E. Ecosystem service trade-offs from supply to social demand: A landscape-scale spatial analysis. *Landsc. Urban Plan.* **2014**, *132*, 102–110. [CrossRef]
17.   Turkelboom, F.; Leone, M.; Jacobs, S.; Kelemen, E.; García-Llorente, M.; Baró, F.; Termansen, M.; Barton, D.N.; Berry, P.; Stange, E.; et al. When we cannot have it all: Ecosystem services trade-offs in real-life planning contexts. *Ecosyst. Serv.* **2017**, *29*, 566–578. [CrossRef]
18.   Dong, S.; Wen, L.; Liu, S.; Zhang, X.; Lassoie, J.P.; Yi, S.; Li, X.; Li, J.; Li, Y. Vulnerability of worldwide pastoralism to global changes and interdisciplinary strategies for sustainable pastoralism. *Ecol. Soc.* **2011**, *16*, 10. [CrossRef]
19.   Steinfeld, H.; Gerber, P.; Wassenaar, T.D.; Castel, V.; Rosales, M.; Rosales, M.; de Haan, C. *Livestock's Long Shadow: Environmental Issues and Options*; Food and Agriculture Organization: Rome, Italy, 2006.
20.   Donthu, N.; Kumar, S.; Mukherjee, D.; Pandey, N.; Lim, W.M. How to conduct a bibliometric analysis: An overview and guidelines. *J. Bus. Res.* **2021**, *133*, 285–296. [CrossRef]
21.   Linnenluecke, M.K.; Marrone, M.; Singh, A.K. Conducting systematic literature reviews and bibliometric analyses. *Aust. J. Manag.* **2020**, *45*, 175–194. [CrossRef]
22.   Tranfield, D.; Denyer, D.; Smart, P. Towards a methodology for developing evidence-informed management knowledge by means of systematic review. *Br. J. Manag.* **2003**, *14*, 207–222. [CrossRef]
23.   Marsoner, T.; Vigl, L.E.; Manck, F.; Jaritz, G.; Tappeiner, U.; Tasser, E. Indigenous livestock breeds as indicators for cultural ecosystem services: A spatial analysis within the Alpine Space. *Ecol. Indic.* **2018**, *94*, 55–63. [CrossRef]
24.   Leroy, G.; Hoffmann, I.; From, T.; Hiemstra, S.J.; Gandini, G. Perception of livestock ecosystem services in grazing areas. *Animal* **2018**, *12*, 2627–2638. [CrossRef]
25.   Buller, H.; Morris, C.; Jones, O.; Hopkins, A.; Wood, J.D.; Whittington, F.M.; Kirwan, J. Eating biodiversity: An investigation of the links between quality food production and biodiversity protection. In *The Science of Beef Quality*; British Society of Animal Science: Scotland, UK, 2005; p. 57.
26.   Martín-Collado, D.; Boettcher, P.; Bernués, A. Opinion paper: Livestock agroecosystems provide ecosystem services but not their components–the case of species and breeds. *Animal* **2019**, *13*, 2111–2113. [CrossRef] [PubMed]
27.   ARCA. Sistema Nacional de Información de Razas. Available online: https://www.mapa.gob.es/es/ganaderia/temas/zootecnia/razas-ganaderas/razas/catalogo/default.aspx (accessed on 15 July 2022).
28.   FAO. The Second Report on the State of the World's Animal Genetic Resources for Food and Agriculture. Available online: https://agris.fao.org/agris-search/search.do?recordID=XF2016015790 (accessed on 1 July 2022).
29.   Leroy, G.; Baumung, R.; Boettcher, P.; Besbes, B.; From, T.; Hoffmann, I. Animal genetic resources diversity and ecosystem services. *Glob Food Secur.* **2018**, *17*, 84–91. [CrossRef]
30.   Wood, J.D.; Richardson, R.I.; Scollan, N.D.; Hopkins, A.; Dunn, R.; Buller, H.; Whittington, F.M. Quality of meat from biodiverse grassland. In *High Value Grassland*; Hopkins, J.J., Duncan, A.J., McCracken, D.I., Peel, S., Tallowin, J.R.B., Eds.; British Grassland Society: Cirencester, UK, 2007; pp. 107–116.

31. Bullock, J.M.; Jefferson, R.G.; Blackstock, T.H.; Pakeman, R.J.; Emmett, B.A.; Pywell, R.J.; Grime, J.P.; Silvertown, J. Semi-natural grasslands. In *The UK National Ecosystem Assessment Technical Report*; UK National Ecosystem Assessment, UNEP-WCMC: Cambridge, UK, 2011.
32. Wang, L.; Delgado-Baquerizo, M.; Wang, D.; Isbell, F.; Liu, J.; Feng, C.; Liu, J.; Zhong, Z.; Zhu, H.; Yuan, X.; et al. Diversifying livestock promotes multidiversity and multifunctionality in managed grasslands. *Proc. Natl. Acad. Sci. USA* **2019**, *116*, 6187–6192. [CrossRef]
33. Hernández Díaz-Ambrona, C.G. Ecología productiva de la dehesa. *Agricultura* **2003**, *72*, 38–42.
34. Teague, W.R.; Apfelbaum, S.; Lal, R.; Kreuter, U.P.; Rowntree, J.; Davies, C.A.; Conser, R.; Rasmussen, M.; Hatfield, J.; Wang, T.; et al. The role of ruminants in reducing agriculture's carbon footprint in North America. *J. Soils Water Conserv.* **2016**, *71*, 156–164. [CrossRef]
35. Fan, F.; Liang, C.; Tang, Y.; Harker-Schuchd, I.; Porteref, J.R. Effects and relationships of grazing intensity on multiple ecosystem services in the Inner Mongolian steppe. *Sci. Total Environ.* **2019**, *675*, 642–650. [CrossRef] [PubMed]
36. McDermot, C.; Elavarthi, S. Rangelands as Carbon Sinks to Mitigate Climate Change: A Review. *J. Earth Sci. Clim. Chang.* **2014**, *5*, 221.
37. Bork, E.W.; Raatz, L.L.; Carlyle, C.N.; Hewins, D.B.; Thompson, K.A. Soil carbon increases with long-term cattle stocking in northern temperate grasslands. *Soil Use Manag.* **2020**, *36*, 387–399. [CrossRef]
38. Soussana, J.F.; Loiseau, P.; Vuichard, N.; Ceschia, E.; Balesdent, J.; Chevallier, T.; Arrouays, D. Carbon cycling and sequestration opportunities in temperate grasslands. *Soil Use Manag.* **2004**, *20*, 219–230. [CrossRef]
39. Eldridge, D.J.; Delgado-Baquerizo, M. Continental-scale impacts of livestock grazing on ecosystem supporting and regulating services. *Land Degrad. Dev.* **2017**, *28*, 1473–1481. [CrossRef]
40. Muñoz Vallés, S.; Mancilla-Leytón, J.M.; Morales-Jerrett, E.; Mena, Y. Natural Carbon Sinks Linked to Pastoral Activity in S Spain: A Territorial Evaluation Methodology for Mediterranean Goat Grazing Systems. *Sustainability* **2021**, *13*, 6085. [CrossRef]
41. Terrer, C.; Phillips, R.P.; Hungate, B.A.; Rosende, J.; Pett-Ridge, J.; Craig, M.E.; Jackson, R.B. A trade-off between plant and soil carbon storage under elevated CO_2. *Nature* **2021**, *591*, 599–603. [CrossRef] [PubMed]
42. Bowman, D.M.; Balch, J.; Artaxo, P.; Bond, W.J.; Cochrane, M.A.; D'antonio, C.M.; Swetnam, T.W. The human dimension of fire regimes on Earth. *J. Biogeog.* **2011**, *38*, 2223–2236. [CrossRef] [PubMed]
43. Mancilla-Leytón, J.M.; Pino Mejías, R.; Martín Vicente, A. Do goats preserve the forest? Evaluating the effects of grazing goats on combustible Mediterranean scrub. *Appl. Veg. Sci.* **2013**, *16*, 63–73. [CrossRef]
44. Olivera-García, R.; de Miguel García, Y.; Varela-Redondo, E.; Ruiz-Mirazo, J.; Robles Cruz, A.B.; González Rebollar, J.L.; Caballero Sánchez, J. Red de Áreas Pasto-Cortafuegos de Andalucía (RAPCA): El Pastoreo Controlado Como Herramienta de Prevención de Incendios Forestales. Available online: https://digital.csic.es/handle/10261/42949 (accessed on 10 July 2022).
45. Mena, Y.; Ruiz-Mirazo, J.; Ruiz, F.A.; Castel, J.M. Characterization and typification of small ruminant farms providing fuelbreak grazing services for wildfire prevention in Andalusia (Spain). *Sci. Total Environ.* **2016**, *544*, 211–219. [CrossRef]
46. Clark, E.A. Benefits of re-integrating livestock and forages in crop production systems. *J. Crop Improv.* **2004**, *12*, 405–436. [CrossRef]
47. Lázaro, A.; Tur, C. Los cambios de uso del suelo como responsables del declive de polinizadores. *Ecosistemas* **2018**, *27*, 23–33. [CrossRef]
48. Sjödin, N.E.; Bengtsson, J.; Ekbom, B. The influence of grazing intensity and landscape composition on the diversity and abundance of flower-visiting insects. *J. App. Ecol.* **2008**, *45*, 763–772. [CrossRef]
49. Bernués, A.; Rodríguez-Ortega, T.; Ripoll-Bosch, R.; Alfnes, F. Socio-cultural and economic valuation of ecosystem services provided by Mediterranean mountain agroecosystems. *PLoS ONE* **2014**, *9*, e102479. [CrossRef]
50. Nahlik, A.M.; Kentula, M.E.; Fennessy, M.S.; Landers, D.H. Where is the consensus? A proposed foundation for moving ecosystem service concepts into practice. *Ecol. Econ.* **2012**, *77*, 27–35. [CrossRef]
51. Martínez-Harms, M.J.; Bryan, B.A.; Balvanera, P.; Lawa, E.A.; Rhodesd, J.R.; Possinghamae, H.P.; Wilsona, K.A. Making decisions for managing ecosystem services. *Biol. Conserv.* **2015**, *184*, 229–238. [CrossRef]
52. Costanza, R.; De Groot, R.; Sutton, P.; van der Ploeg, S.; Anderson, S.J.; Kubiszewski, I.; van der Ploeg, S.; Anderson, S.J.; Kubiszewski, I.; Farber, S.; et al. Changes in the global value of ecosystem services. *Glob. Environ. Chang.* **2014**, *26*, 152–158. [CrossRef]
53. MAPA (Ministerio de Agricultura, Pesca y Alimentación) Ecoesquema 1: Mejora de la Sostenibilidad de los Pastos, Aumento de la Capacidad de Sumidero de Carbono y Prevención de Incendios Mediante el Impulso del Pastoreo Extensivo. Available online: https://www.agrodigital.com/wp-content/uploads/2020/12/eco1c.pdf (accessed on 15 July 2022).

 *land*

Article

Identifying the Spatiotemporal Transitions and Future Development of a Grazed Mediterranean Landscape of South Greece

Dimitrios Chouvardas, Maria Karatassiou *, Afroditi Stergiou and Garyfallia Chrysanthopoulou

Laboratory of Rangeland Ecology (P.O. 286), School of Forestry and Natural Environment, Aristotle University of Thessaloniki, 54124 Thessaloniki, Greece
* Correspondence: karatass@for.auth.gr (M.K.); Tel.: +30-2310-992302

Abstract: Spatiotemporal changes over previous decades in grazed Mediterranean landscapes have taken the form of woody plant encroachment in open areas (e.g., grasslands, open shrublands, silvopastoral areas), altering its structure and diversity. Demographic and socioeconomic changes have played a significant role in landscape transformations, mainly by causing the abandonment of traditional management practices such as pastoral activities, wood harvesting, and agricultural practices in marginal lands. This study aimed to quantify and evaluate the spatiotemporal changes in a typical grazed Mediterranean landscape of Mount Zireia during 1945–2020, and to investigate the effect of these changes on the future development (2020–2040) of land use/land cover (LULC) types. Cartographic materials such as aerial orthophotos from 1945, land use maps of 1960, Corine Land Cover of 2018, and recent satellite images were processed with ArcGIS software. To estimate the future projection trends of LULC types, logistic regression analyses were considered in the framework of CLUE modeling. The results indicated that the strongest trend of spatiotemporal changes were forest expansion in open areas, and grasslands reduction, suggesting that the LULC types that were mainly affected were forest, grasslands, and silvopastoral areas. Future development prediction showed that forests will most probably continue to expand over grassland and silvopastoral areas, holding a high dynamic of expansion into abandoned areas. The reduction in grasslands and silvopastoral areas, independent of environment and biodiversity implications, represents a major threat to sustainable livestock husbandry based on natural grazing resources.

Keywords: land abandonment; pastoral activities; forest expansion; grassland reduction; CLUE modeling framework; logistic regression

Citation: Chouvardas, D.; Karatassiou, M.; Stergiou, A.; Chrysanthopoulou, G. Identifying the Spatiotemporal Transitions and Future Development of a Grazed Mediterranean Landscape of South Greece. *Land* **2022**, *11*, 2141. https://doi.org/10.3390/land11122141

Academic Editor: Xiangzheng Deng

Received: 17 October 2022
Accepted: 24 November 2022
Published: 28 November 2022

Publisher's Note: MDPI stays neutral with regard to jurisdictional claims in published maps and institutional affiliations.

1. Introduction

Mediterranean landscapes are considered highly diverse areas in terms of history, geography and land uses. Several civilizations from ancient times have left a rich cultural heritage promoting this variety [1]. The Mediterranean landscapes, as a result of their long history of human activities, with a unique combination of topographic and climatic variability, have generated a rare combination of unique, but fragile, diverse species-rich ecosystems [2,3]. The Mediterranean basin is the second largest biodiversity hotspot in the world, holding more than 25,000 plant species [4]. The long history of human intervention in this area has formed plant communities that are considered as "man-made" and composed of natural components, a fact that has a significant value in setting goals and methodology for sound conservation interventions [5]. The last 75 years of technological advances, such as the introduction of heavy machinery in farming activities [6], trade globalization, the creation of the European Economic Community [2], and the Common Agricultural Policy (CAP) [7,8], have driven dramatic changes in these ecosystems unlike those experienced in the past [9]. In recent years, climate changes, along with unbalanced land use activities (e.g., coastalization, undergrazing, and land abandonment), have facilitated Mediterranean ecosystems change [10]. Two opposite trends of landscape evolution have occurred in the Mediterranean region in recent decades. Forest cover increased around the northern edge of the Mediterranean region (south European countries) and decreased around its southern edge (mainly in the Maghreb countries). This increase in forest cover in northern Mediterranean landscapes is mainly attributed to the abandonment of marginal agricultural lands [11,12], while the decrease in forest in the south is attributed to the expansion of cropland in marginal areas initially dominated by woodlands [13]. The above changes have followed the socioeconomic trends of land abandonment in rural areas in the north versus the increased population pressure in rural areas in the south [9,14].

One of the main land use activities in Mediterranean landscapes is pastoral activities [2,9,13,15]. Approximately one-fifth of European agricultural lands are dedicated to extensive livestock grazing, with the majority being situated in southern Mediterranean Europe, including the Balkans. Furthermore, 80% of Europe's sheep and goat flocks are located in Spain, Italy, Greece, and southern France [16]. Grazing is considered a major landscape-changing factor directly related to human activities, especially in Mediterranean areas [17,18]. Greek landscapes have historically been grazed by livestock in quite a similar way as modern practices, and are highly influenced by the changes in traditional pastoral activities [19,20]. Recently, significant changes emerged in the traditional extensive livestock production systems of Greece, mainly related to the reduction in the number of local and transhuman flocks of free-grazing animals (sheep, goats, and cattle) [8,20–23]. These changes follow the land abandonment trend already mentioned for the European part of the Mediterranean region, and they highly contribute to the spatiotemporal transitions occurring in grazed areas. These transitions are taking the form of woody plant expansion in open areas, transforming grasslands, open shrublands, silvopastoral areas and abandoned agricultural areas, into forest or dense shrublands [21,24–28].

The study of land use/land cover (LULC) change provides an important aspect in understanding the history of spatiotemporal transition patterns, derived from landscape changes. Spatiotemporal transition patterns produce useful data for studying the effect of physical and socioeconomic interactions, land use conflicts, and influences on landscape changes [29–31]. Analysis of spatiotemporal changes and transitions is typically conducted within the geographic information systems (GIS) environment [30,32], with visual photointerpretation of a time series set of aerial photographs [33,34] through digital processing of multispectral satellite images [32], or more recently through object-based recognition technics [35]. The development of transition matrices has become an important part of landscape history analysis [36]. New tools and indicators of LULC changes derived from the matrices have emerged, addressing issues related to the annual rate of changes [30], persistence and net changes as quantity difference and swap as allocation difference [37], and identifying systematic or main transitions [29,36].

Predicting the future development of LULC types and transitions is an effective and reliable technique for evaluating both the causes and the significance of past and present conditions, usually under future scenarios [27,38,39]. Several spatiotemporal models for LULC future projections have been proposed over the years [31], including the adoption of empirical models for LULC prediction such as logistic regression approaches (e.g., CLUE modeling framework, LCM, MaxEnt) [39–41]. The use of regression analysis in landscape prediction studies contributes to understanding and describing the change mechanisms and processes of LULC types, provides an advanced statistical environment for analyzing multivariate components, and finally, predicts the LULC changes [14,42]. The above prediction models can also produce accurate results to support policy makers, land managers, and scientists in reaching sustainable landscape management decisions [41].

Spatiotemporal changes have a significant effect on altering landscape structure in terms of landscape composition and configuration [43]. These changes can be easily evaluated with the use of landscape metrics [44,45], applied in spatiotemporal studies of landscape changes [22,24,46,47], or in the future projections of land use changes [48].

Overall, there is a limited amount of published information regarding the spatiotemporal changes in grazed landscapes, especially for the eastern part of the Mediterranean region, and particularly about the influences of land abandonment in the future development of land uses that are related to pastoral activities. Therefore, the present research aims to: (a) quantify and evaluate the spatiotemporal changes of a typical, grazed Mediterranean landscape of south Greece (Mt Zireia landscape), (b) investigate the effect of these changes on the future development of the most significant LULC types, and (c) identify their correlation to a set of landscape driving factors. Finally, the overall effect and interactions of socioeconomic changes are explored, focusing on pastoral activities in LULC transitions and future development.

2. Study Area

Mount (Mt) Zireia (or Kyllini), located in the Peloponnese peninsula (South Greece), was selected for the study. Mount Zireia is the second highest mountain in the Peloponnese, located in the Korinthos prefecture 115 km west of Athens (Figure 1).

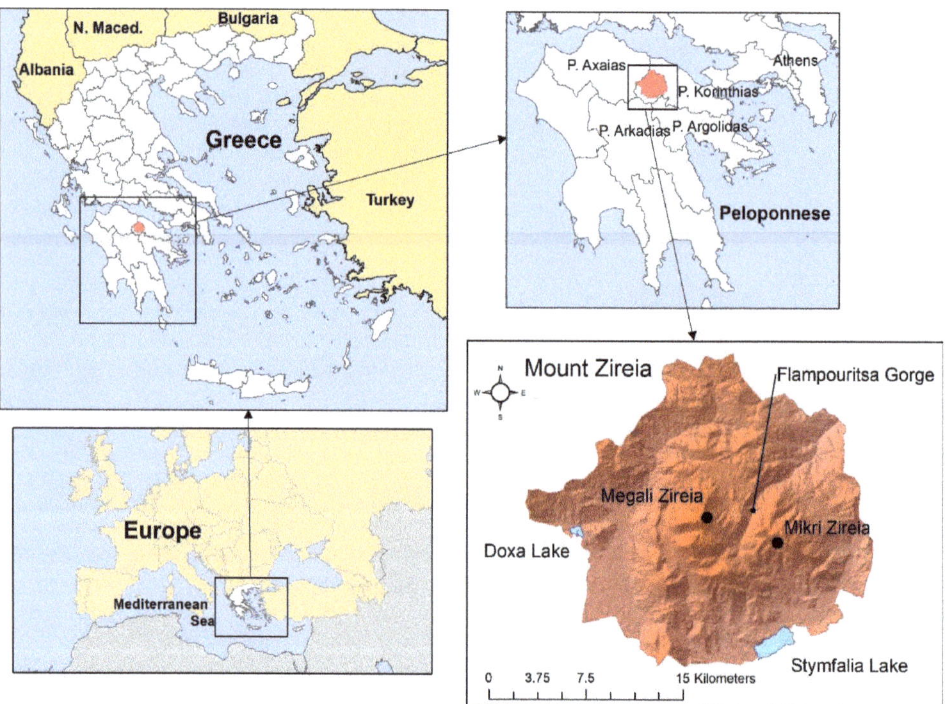

Figure 1. Location of the study area in Mt Zireia, south Greece.

The study area covers 39,762 ha of land inhabiting 3777 people living in 19 village communities–municipalities subdistricts. Elevations in the study range from 310 m to 2374 m a.s.l. A large gorge, called Flampouritsa, divides the mountain into two areas, "Mikri" (small) Zireia and "Megali" (big) Zireia. Mt Zireia, apart from the highest point of 2374 m, has other seven peaks above 2000 m (four in Megali and three in Mikri Zireia). The multiple ridges created by the mountain tops, in combination with valleys and plateaus, create a particularly diverse relief of hills, plains, cliffs, and canyons. More than two-thirds of the study area is part of the network of Natura 2000 protected areas (pSCI, SCI or SAC, SPA) [49]. Two main hydrological basins are found in the area, creating the natural lake Stymfalia (area 15,285 ha) to the south, and the artificial lake Doxa (area 48 ha) in the west

(Figure 1), which greatly affect the microclimatic conditions and facilitate the touristic development of the area. Lake Stymfalia is closely connected to Greek mythology, and especially with Heracles' legendary labors. According to mythology, the lake was full of aggressive man-eating Stymphalian birds, and Heracles' sixth labor was to exterminate them [50].

The climate, according to Köppen–Geiger climate classification, is a hot summer Mediterranean climate (coded as "Csa") [51]. The mean annual precipitation has varied over the last 60 years, from 418.62 mm (in 1993) to 1056 mm (in 2005), while the mean annual temperature varied from 12.59 °C (coldest year in 1976) to 15.55 °C (warmest year in 2010) [52].

The main land uses of the area are forests, rangelands, and agricultural areas. Rangelands include grasslands, shrublands, and silvopastoral areas with less than 40% tree cover and grazed by sheep and goats. Agricultural areas are cultivated mainly with annual crops such as beans, corn, barley, and wheat [52].

According to the official census report derived from the Hellenic Statistical Authority [53], the temporal evolution of socioeconomic data from 1961 (oldest available data) until the most recent available data of 2011, showed that in the last 50 years, the total population, active workforce, and employees in the primary economic sector has rapidly been reducing (Table 1), following the general trend of land abandonment that many researchers have reported for the Mediterranean region [8,44,54,55]. Age structure analysis indicates that the human population is becoming older. Indeed, 62% of the population was under 44 in 1961, versus 47% in 2011 (Table 2).

Table 1. Temporal evolution (1961–2011) of the total population, active workforce, and primary sector employees of the nineteen village communities in the study area.

	Total Population	Total Working Force	Employees in the Primary Sector	% of Primary Sector Employees per Total Working Force
1961	7420	3632	3169	87.25
2011	3777	1354	682	50.37

(Source: Hellenic Statistical Authority.)

Table 2. Temporal evolution (1961–2011) of age structure (as percentages) of the local population of the nineteen village communities in the study area.

	1961		2011	
	% (0–44)	% (45–)	% (0–44)	% (45–)
Total	62.34	37.66	46.93	53.07

(Source: Hellenic Statistical Authority.)

In contrast, the local population over 45 years old increased from 38% to 53%, for the same period. The above data are in line with demonstrated demographic change in the Mediterranean region and the movement of the mainly younger population from rural areas to urban centers [8,11,44].

Census data from the Hellenic Statistical Authority and the Payment and Control Agency for Guidance and Guarantee Community Aid [56], regarding the historical data of transhumans [57], revealed that the number of grazing animals (mainly sheep and goats) and their farms have significantly reduced in the last 50 years (Table 3) [20].

Table 3. Temporal evolution (1961–2020) of grazing animals (sheep and goats) in the study area.

	Number of Animals		Percentage of Change
	1961	**2020**	
Sedentary	28,750	27,595	−4.02
Transhumant	38,230	13,717	−64.12
Total	66,980	41,312	−38.32

The total number of grazing animals decreased from 1961 to 2011 by 38% (Table 3). This reduction was more intensive for transhuman animals (more than 64%) and less for sedentary animals (almost 4%). According to the available inventory data, the number of sedentary animal farms significantly reduced by 80% during a similar period (Hellenic Statistical Authority, 1961 to 2000). This reduction follows the similar trend of change as the number of people that are employed in the primary sector of the economy (Table 1).

3. Materials and Methods

3.1. Land Use/Land Cover Changes. Spatiotemporal Transitions

The following cartographic materials (Figure 2) were considered: (a) digital aerial orthophotographs of 1945 with a spatial resolution of 1 m obtained from the National Cadastre of Greece (georeferenced to the Hellenic Geodetic Reference System 1987-HGRS87); (b) satellite images obtained from the Google Earth Pro program for the years 2017, 2019 and 2020 (georeferenced to HGRS87); (c) maps of forest vegetation and land cover for 1960 (scale 1:20,000), obtained from the Ministry of Agriculture in digital format (shapefile in HGRS87); and, (d) digital maps of Corine Land Cover 2018 (shapefile reprojected in HGRS87).

Aerial orthophotographs from 1945, as well as the recent Google Earth satellite pictures, were digitally processed using the software ArcGIS v.10.8.1, to produce LULC maps for 1945 and 2020. To identify the distinct LULC types, on-screen visual photointerpretation and manual delineation of LULC polygons in shapefile format were performed within the ArcGIS environment (Figure 2). The chosen analysis used a classification scheme consisting of eight categories of LULC types and was based on the Greek Forest Service's LULC classification system (Table 4). According to the chosen classification system, numerous elements on aerial orthophotos and Google Earth images were recognized by using common photographic keys (tone, texture, pattern, shade, form, and size) and feature association [15,21,29,33,34]. Special attention was placed on identifying tree and shrub cover density patterns with the use of crown density scales [58]. The 1960 forest vegetation and land cover maps in shapefile format were a valuable resource for the 1945 LULC mapping, since they served as a reference map and guided the photointerpretation. The minimum mapping unit of the reference map was one hectare, and the same unit was chosen for the 1945 and 2020 mappings. For the 2020 LULC mapping, additional supporting materials were considered from the 2018 Corine Land Cover digital map, and from several elements of the Google Earth application software, such as 3D views and street view images available from many narrow-paved roads between villages of the study area. The visual interpretation was also supported by field sampling verifications from well-experienced human image interpreters with good knowledge of the area. The above cartographic materials were further processed using ArcGIS and Excel to create tables and digital maps of the temporal evolution of LULC types. This approach produced two digital maps of LULC types for 1945 and 2020, as well as a temporal evolution table (Figure 2).

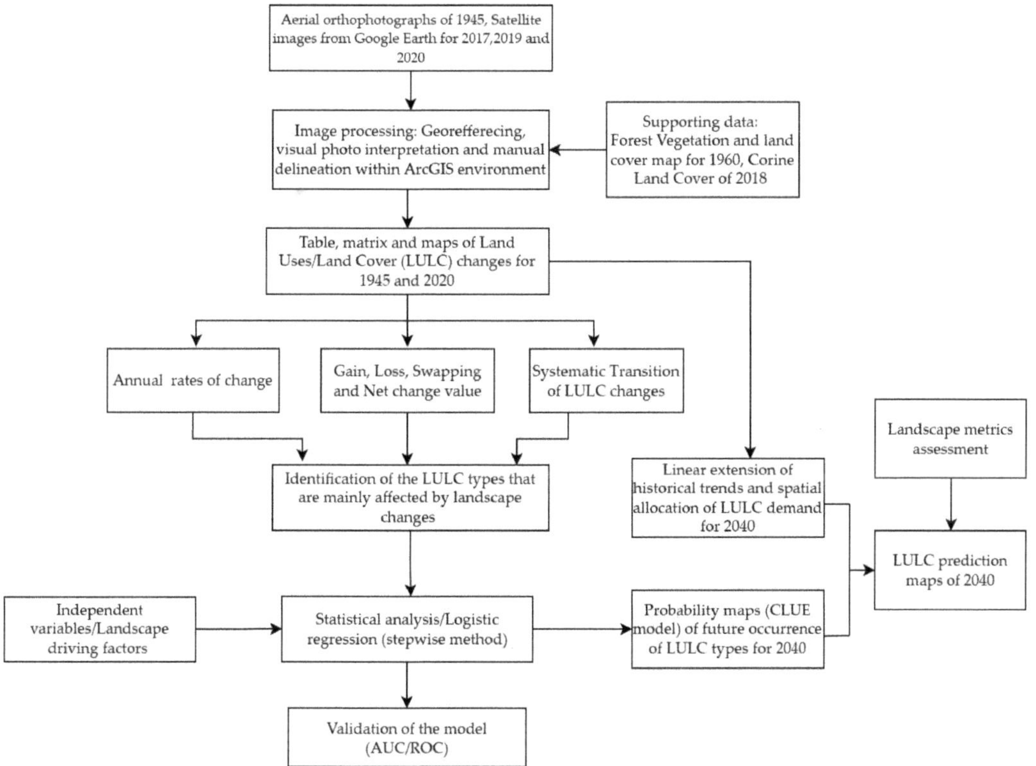

Figure 2. Procedural and methodological workflow chart.

Table 4. Classification scheme of land use/land cover types in the study area.

Land Use/Land Cover Types	Description	Codes
Agricultural areas	Areas covered with annual or permanent crops	AG
Grasslands	Areas dominated by herbaceous plants, with woody vegetation cover of less than 10%	GR
Open shrublands	Areas dominated by sparse woody shrubs with less than 40% cover	OS
Dense shrublands	Areas dominated by dense woody shrubs higher than 40% cover	DS
Silvopastoral areas	Open grazed forest with tree cover between 10 and 40%	SI
Forest	Forest areas with tree cover higher than 40%	FO
Barren areas	Mainly bare lands with little or no vegetation	BA
Urban areas	Areas with man-made features, mainly villages	UR
Lakes	Large area of water surrounded by land	LA

According to Puyravad [59], the annual rate of change of LULC types was calculated for the overall study period (1945–2020). The annual rate calculation Equation (1) was based on the formula developed from compound interest law, and offers a better assessment and biological meaning to the LULC change comparisons because it is insensitive to the different time periods between observation dates [29]:

$$r = \left(\frac{1}{t_2 - t_1} \right) \times \ln\left(\frac{A_2}{A_1} \right) \tag{1}$$

where r is the annual rate of change, and A_1 and A_2 are the LULC class areas at time t_1 and t_2, respectively.

The next phase in the process was to estimate the spatiotemporal transformations of the study area for 2020 as a result of the diachronic transitions of all LULC types from their original surfaces in 1945. This was accomplished by employing a common post-classification comparison (PCC) change detection method across the study's periods of various dates [30]. The PCC method produced a LULC change transition matrix, which was calculated using ArcGIS overlay functions for all time periods. In addition, a map showing the spatiotemporal transition of LULC types was constructed. Additional components of land changes, such as gains and losses, net changes, total changes, and swap [60], were included in the LULC changes study due to transition matrices. The proportion of the landscape that underwent gross gain or loss of LULC type j between times 1 and 2 was represented by the letters P_{+j} and P_{j+}, respectively. The proportion of the landscape that demonstrated the persistence of category j was indicated by the diagonal elements (denoted as P_{jj}) of LULC types [60]. The difference between gain and loss is called net change and was denoted as D_j. Swap is the simultaneous gain and loss of LULC type j, and was calculated as two times the minimum gain and loss (S_j). The sum of the net change and the swap, or the sum of gains and losses for each LULC type j, abbreviated as C_j, is the total change [29,60]. In order to calculate net changes, swaps, and total changes, Equations (2)–(4) were applied:

$$D_j = P_{+j} - P_{j+} \tag{2}$$

$$S_j = 2 \times MIN \left(P_{j+} - P_{jj}, \ P_{+j} - P_{jj} \right) \tag{3}$$

$$C_j = D_j + S_j \tag{4}$$

Recent scientific views have defined net change as quantity difference (or quantity disagreement), and swap as allocation difference (or allocation disagreement) [37].

Identification of the most systematic transitions or dominant signals of change is another critical component in evaluating LULC alterations [29,61]. The most important form of transition can be determined using the transition matrix data by adding the total area of change for each LULC type over the time periods. This technique cannot consider the random process of LULC changes caused by the dominant LULC types and, therefore, interpreting LULC transitions based on their sizes is the correct way to evaluate them [29]. The predicted gains (denoted as G_{ij}) and expected losses (denoted as L_{ij}) that will occur if random changes among the LULC types occur, were computed using a process that was first proposed by Pontius [61] (Equations (5) and (6)):

$$G_{ij} = \left(P_{+j} - P_{jj} \right) \left(\frac{P_{i+}}{100 - P_{j+}} \right) \tag{5}$$

$$L_{ij} = \left(P_{i+} - P_{ii} \right) \left(\frac{P_{+j}}{100 - P_{+i}} \right) \tag{6}$$

The difference between the observed (P_{ij}) and expected (G_{ij} or L_{ij}) transitions in a random process of gain ($P_{ij} - G_{ij}$) or loss ($P_{ij} - L_{ij}$) is indicated as D_{ij}, and the ratios meaning ($P_{ij} - G_{ij})/G_{ij}$ or ($P_{ij} - L_{ij})/L_{ij}$ are denoted as R_{ij}. D_{ij} and R_{ij} values show the tendency of a LULC type j to gain from type i (focus on gains) and the tendency of LULC type i to lose from type j (focus on losses) [60]. Systematic transitions or dominant signals of change are indicated by values having a considerable positive or negative deviation from zero [29]. R_{ij} ratios are equivalent to the (observed value $-$ expected value)/expected value ratios that are used in chi-square tests [61].

The results from the annual rate of change, absolute values of net change, and the main systematic transitions of all LULC types were used to identify the main LULC types that had undergone significant spatiotemporal changes.

3.2. Logistic Regression, Probability Maps, and Future Projection

The future projection trends of the main LULC types that were identified as experiencing the most significant spatiotemporal changes, were determined by logistic regression analyses, under the methodological approach of the CLUE modeling framework [21,62–65]. According to CLUE modeling, a set of landscape driving factors (LDF were used as independent variables in the regression analysis. In this research, 20 LDF variables were identified and collected based on the physiographic, accessibility, and socioeconomic conditions of the study area (Table 5).

Table 5. Type, units, and data sources of the independent variables used in the logistic regression analyses for future projection modeling of land use changes in the study area.

a/a	Independent Variables *	Type/Unit	Data Source
1	Elevation	Continuous/m	DEM Aster 2
2	Slope	Continuous/%	DEM Aster 2
3	Alluvial deposits/very deep soils	Binary/0–1	Soil map of Greece (Nakos 1991)
4	Hard limestone/ahallow to bare soils	>>	>>
5	Limestone colluvium/seep to moderately deep soils	>>	>>
6	Doline-deposition cones/seep soils	>>	>>
7	Tertiary deposits/seep to moderately deep soils	>>	>>
8	Tertiary deposits/shallow soils	>>	>>
9	Schist/shallow soils	>>	>>
10	Schist/deep soils	>>	>>
11	Erosion potential	Continuous/t $\times$ ha^{-1} $\times$ year^{-1}	Soil erosion by water (RUSLE 2015)/ESDAC **
12	Distance from unpaved roads	Continuous/m	Digital files from state Cadastre, Google Earth
13	Distance from paved roads	>>	>>
14	Distance from water courses	>>	Hydrological model from DEM, topographic maps, Google Earth
15	Distance to settlements	>>	Land use map 2020
16	Population density	Continuous/number $\times$ ha^{-1} of the total area	Hellenic Statistical Authority [53]
17	Sheep/cow density	>>	Hellenic Statistical Authority [53], PCAGGCA *** [56],
18	Goat density	>>	Hellenic Statistical Authority [53], PCAGGCA *** [56],
19	Annual mean temperature	Continuous/°C	https://worldclim.org/ (accessed 20 April 2022)
20	Annual precipitation	Continuous/mm	>>

* Landscape driving factors (LDF), ** European Soil Data Cent, *** Payment and Control Agency for Guidance and Guarantee Community Aid.

In addition to the above independent variables, the identified main LULC types were selected as dependent variables. According to the spatial module of the CLUE model, both the LULC types and the independent variables were transformed into digital raster files (ArcGrid format) with a pixel size of 100 m. The raster files of all dependent variables and 8 out of 20 independent variables (Table 5) were binarily rendered. As a result, each pixel of a given main LULC type and the eight independent variables received a value of 1, and those without received a value of 0. All the other independent variables received a continuous value according to their definition. The raster data sets were then transformed into ArcGIS ASCII grids, and with the use of the "File Convert v2" application of Dyna-CLUE modeling version [64], were further transformed into tabular format necessary for entry in the SPSS statistical package v. 27.0 (IBM Corp., Armonk, NY, U.S.A.).

In SPSS, the data were analyzed by the method of binary logistic regression of absence/presence, using forward conditional analysis as a step-by-step regression method. The input and output probabilities of the independent variables in the equation were set to not surpass the significance levels of input = 0.01 and output = 0.02, respectively, during the process [64]. Variables that did not meet the above criteria were rejected as exhibiting a low degree of correlation to the model. This procedure resulted with fewer independent variables from the original selection, and resolved problems due to multicollinearity [66–68]. The regression coefficients (b_i) of the remaining independent variables in the logistic equation were tabulated. Furthermore, the area under the ROC (relative operating characteristic) curve (AUC) was estimated, as a measure of controlling the goodness of fit of the data to the logistic regression model [64,69], and was used to validate the model (Figure 2) [65,70].

The AUC number indicates how well the model can differentiate across classes [71,72]. The greater the value, the better the model's ability to distinguish between classes. AUC values range from 0 to 1, with 0.5 indicating that the model is unable to distinguish between the classes, and 1 indicating that the model is perfectly fitted [73–75]. AUC levels of 0.7 to 0.8 are considered acceptable, 0.8 to 0.9 are considered excellent, and values beyond 0.9 are considered exceptional [76,77]. All the available data concerning the dependent and independent variables and the logistic regression results from SPSS were introduced into the Dyna-CLUE version of the model, to produce a set of land use probability maps. Probability maps represent the distribution of the results of the logistic regression equations in the landscape [64].

The land use demands for 2040 of the identified main LULC types that have undergone significant spatiotemporal changes were computed using linear interpolation of their historical trend (Figure 2). This technique is often used to construct "Business as Usual" model scenarios (BAU scenario) [27,62]. The BAU scenario for a 20-year prediction period (2020–2040) was calculated by adding one-third of the total positive or negative trend of change from the most recent available historic trend, which were the years 1960 and 2020 (60-year trend). The areas of identified main LULC types for 1960 were obtained from the available forest vegetation and land cover map. As Mamanis and coworkers [27] suggested, the one-third ratio was used because the 20-year prediction period is equal to a third of the historical trend (20 years/60 years = 1:3). The projected land use demands under the BAU scenario were spatially allocated into the probability maps based on the higher probability of occurrences, which resulted in the creation of the predicted potential map of the future distribution of the main LULC types. The 2040 prediction maps did not consider LULC interactions.

Finally, the projected results were examined using the ArcGIS Patch Analyst program [21,45,78–80]. The number of patches (NumP) and mean patch size (MPS, ha) as overall measures of landscape fragmentation, edge density (ED, m/ha) as a measure of the number of ecotones [44], and mean nearest neighbor (MNN, m) as a measure of patch isolation, were calculated as indicators of spatial heterogeneity in landscape and class levels. The mathematical formulas for the specified indices can be found in the user manuals for Patch Analyst and Arc Fragstats [45,78,79].

4. Results

4.1. Land Use/Land Cover Changes. Spatiotemporal Transitions

The results of photo interpretation and LULC changes over the 75-year periods in terms of area, percentage, and annual rate of changes are shown in Table 6. The LULC types that increased in the study area were forest (68%), dense shrublands (10%), and urban areas (41%) (Table 6).

Table 6. Temporal evolution and the annual rate of changes (1945–2020) of land use/land cover types in the study area (ha).

Land Use/Land Cover Types	1945	2020	Area Change from 1945 to 2020 (ha)	Percentage Change from 1945 to 2020 (%)	Annual Rate of Change (% per Year)
Agricultural areas	9581.59	8092.84	−1488.75	−15.54	−0.23
Grasslands	8052.24	5893.51	−2158.73	−26.81	−0.42
Open Shrublands	3756.35	3571.97	−184.38	−4.91	−0.07
Dense Shrublands	4390.13	4849.15	459.02	10.46	0.13
Silvopastoral areas	4671.17	2850.80	−1820.37	−38.97	−0.66
Forest	8000.30	13,430.98	5430.68	67.88	0.69
Barren areas	788.44	432.18	−356.26	−45.19	−0.80
Urban areas	317.46	447.05	129.59	40.82	0.46
Lakes	203.87	193.18	−10.69	−5.24	−0.07
Total	39,761.55	39,761.66			

All the other LULC types decreased, with the more important changes being the reduction in silvopastoral areas (39%), grasslands (27%), and agricultural areas (16%) (Table 6). Open shrublands were reduced in area to a limited extent (5%). Barren areas and lakes were also reduced by 45% and 5%, respectively, but they covered only a limited part of the study area. Forest and dense shrubland expansion, at the expense of silvopastoral areas, grasslands, and agricultural areas, demonstrated that in the last 75 years, woody vegetation in the study area had significantly increased. Analyzing in more detail the annual rate of changes of LULC types (Table 6) during the study period 1945–2020, suggested that the most significant changes were the declining trend of barren areas, silvopastoral areas, and grasslands, and the increasing trend of forest and urban areas. Agricultural areas presented a considerable declining trend, but were less severe compared with grasslands and silvopastoral areas.

Gradual conversion of silvopastoral areas and grasslands into forest was observed in the northern areas between the villages of Feneos, Tarsos, Karya, and Trikala (Figure 3). Forest also seemed to have expanded in the southern area near the village of Drosopigi at the expense of shrubland areas. Grasslands decreased over time in the study areas, except for the central area east of the village Goura, where a higher elevation of landscape occurs (>1200 m). Changes in agricultural areas did not have a strong spatiotemporal orientation, suggesting that they covered a broader range of landscape territories.

The LULC transition matrix of the study area showed that between 1945 and 2020, 65.71% of the total landscape remained unchanged, while 34.29% was transformed into a different LULC type (Table 7). According to the matrix, the most important LULC transitions (>2%) were those of silvopastoral areas (SI) into forest (FO); of grasslands (GR) into open shrublands (OS), silvopastoral areas (SI) and forest (FO); and finally, of dense shrublands (DS) into forest (FO). Additional significant changes (1–2%) were presented in agricultural area transition into grasslands (GR) and shrublands (open, OS; or dense, DS); and open shrublands (OS) transition into dense shrublands (DS) and forest (FO). The results of the matrix suggested that the most important changes in the Mt Zireia landscape were woody plant encroachment into open areas such as grasslands, open shrublands, and silvopastoral areas, and to a lesser extent, into agricultural areas (Table 7).

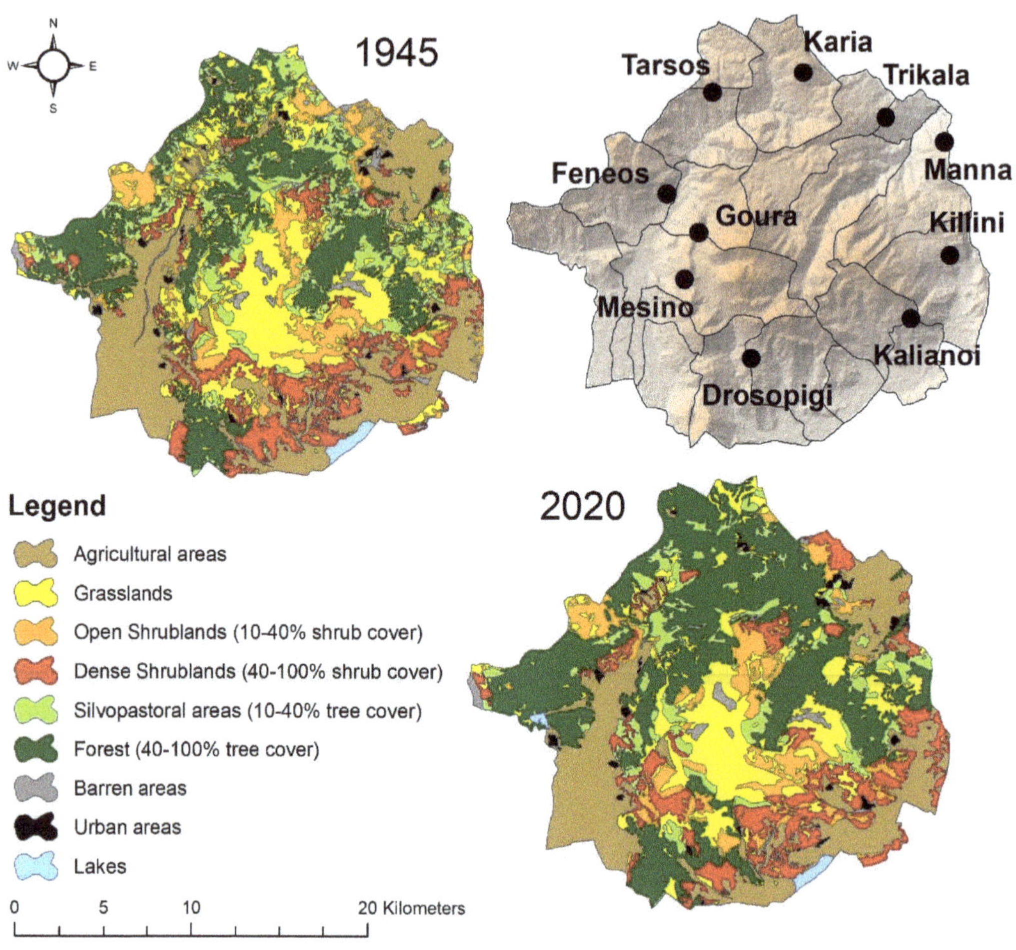

Figure 3. Spatiotemporal distribution of land use/land cover types in the study area for the entire period (1945 to 2020).

Figure 4 presents the map of LULC transitions in the study area. It is notable that silvopastoral and forest expansions were mainly located in the northern parts of the study area. On the other hand, shrubland expansions were mainly observed in the southern and central parts of the landscape. Overall, LULC transitions were observed in all parts of the Zireia landscape, but appeared to be more extensive in the northern parts.

The most significant changes in net values (absolute values) were observed in forest, grasslands, and silvopastoral areas, and to a lesser extent, in agricultural areas (Table 8). These data also indicated that the highest losses in the area were observed in grasslands and silvopastoral areas, and the highest gains in forest areas. Net change values in forest were much higher than in comparison with their swap values, suggesting that forest expansion in new areas (quantity difference) was more significant than their simultaneous exchange of forest areas to other uses (allocation difference). On the contrary, net change and swap value changes appeared to be more balanced in grasslands and silvopastoral areas. Forest, grasslands, and silvopastoral areas were the main LULC types that underwent significant changes, and the recorded woody plant expansion in the landscape focused particularly on forest development (Table 8).

Table 7. Land use/land cover (**LULC**) change transition matrix between 1945 and 2020 (%) in the study area.

1945 LULC	2020 LULC									Total 1945	Loss
	AG	**GR**	**OS**	**DS**	**SI**	**FO**	**BA**	**UR**	**LA**		
AG	19.16	1.12	1.05	1.10	0.35	0.97	0.02	0.26	0.08	24.11	4.95
GR	0.41	10.94	2.21	0.67	2.57	3.27	0.15	0.03	0.00	20.25	9.31
OS	0.06	0.68	4.65	1.81	0.41	1.78	0.01	0.04	0.00	9.44	4.79
DS	0.20	0.17	0.51	7.80	0.20	2.12	0.02	0.02	0.00	11.04	3.24
SI	0.09	0.90	0.38	0.35	2.85	7.16	0.00	0.01	0.00	11.74	8.89
FO	0.04	0.78	0.04	0.24	0.68	18.30	0.00	0.00	0.04	20.12	1.82
BA	0.24	0.24	0.14	0.20	0.10	0.17	0.89	0.01	0.00	1.99	1.10
UR	0.03	0.00	0.00	0.00	0.00	0.01	0.00	0.75	0.00	0.79	0.04
LA	0.13	0.00	0.00	0.02	0.00	0.00	0.00	0.00	0.37	0.52	0.15
Total 2020	20.36	14.83	8.98	12.19	7.16	33.78	1.09	1.12	0.49	100.00	34.29
Gain	1.20	3.89	4.33	4.39	4.31	15.48	0.20	0.37	0.12	34.29	

Note: The values in the shaded box (diagonally) indicate the unchanged LULC types from 1945 to 2020. The underlined values indicate the most important land use/land cover transitions (>1%). AG: agricultural areas, GR: grasslands, OS: open shrublands, DS: dense shrublands, SI: silvopastoral areas, FO: forest, BA: barren areas, UR: urban areas, LA: lakes.

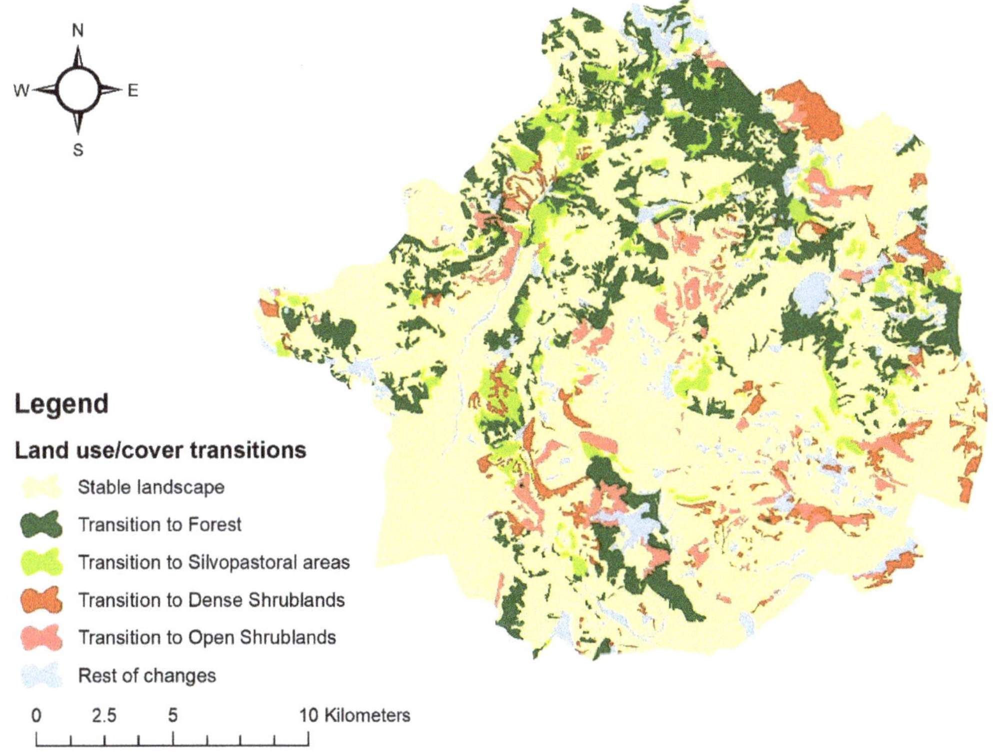

Figure 4. Land use/land cover transitions map (1945 to 2020) of the study area.

Table 8. Temporal evolution of gain, losses, net change, and swap of land use/land cover, in terms of percent, in the landscape of Mt Zireia for the period 1945 to 2020.

Land Use/Land Cover	Percentage of Change				
	Gain	Loss	Total Change	Swap	Absolute Value of Net Change
Agricultural areas	1.20	4.95	6.15	2.40	3.75
Grasslands	3.89	9.31	13.20	7.78	5.42
Open shrublands	4.33	4.79	9.12	8.66	0.46
Dense shrublands	4.39	3.24	7.63	6.48	1.15
Silvopastoral areas	4.31	8.89	13.20	8.62	4.58
Forest	15.48	1.82	17.30	3.64	13.66
Barren areas	0.20	1.10	1.30	0.40	0.90
Urban areas	0.37	0.04	0.41	0.08	0.33
Lakes	0.12	0.15	0.27	0.24	0.03
Landscape	34.29	34.29	34.29	19.15	15.14

Table 9 presents the percentage of the area of the main systematic transitions of LULC changes in terms of gains and losses. The largest positive or negative variation from zero (systematic transitions) appeared to be in the transitions of silvopastoral to forest (SI to FO), grasslands to silvopastoral (GR to SI), grasslands to open shrublands (GR to OS), and open to dense shrublands (OS to DS). These results suggested that forest was systematically gaining area from silvopastoral areas, while dense shrublands gained from open ones (focus on gains). On the other hand, the same results showed that grasslands were systematically losing areas to open shrublands and silvopastoral areas (focus on losses).

Table 9. Area percentage of the main systematic transitions of land use/land cover changes in terms of gains and losses in the landscape of Mt Zireia for the period 1945 to 2020.

Transitions	Gains (%)		Losses (%)	
	D_{ij}	R_{ij}	D_{ij}	R
AG to GR	−0.06	−0.05	0.20	0.22
AG to OS	−0.11	−0.09	0.49	0.88
AG to DS	−0.09	−0.08	0.34	0.45
GR to OS	1.24	1.28	1.23	1.25
GR to SI	1.58	1.59	1.79	2.28
GR to FO	−0.65	−0.17	−0.42	−0.11
OS to DS	1.34	2.87	1.17	1.81
OS to FO	−0.05	−0.03	0.00	0.00
DS to FO	−0.02	−0.01	0.87	0.70
SI to FO	4.88	2.14	3.92	1.21

AG: agricultural areas, GR: grasslands; OS: open shrublands; DS: dense shrublands; SI: silvopastoral areas; FO: forest; BA: barren areas; UR: urban areas; LA: lakes; D_{ij}: the difference between the observed and expected transitions; R_{ij}: the difference between the observed and expected transitions, relative to the expected transitions.

Overall, spatiotemporal changes in the landscape of Mt. Zireia indicated that the most important element of landscape change was the woody plant expansion into open areas. Furthermore, among the different LULC types, the ones that were mainly affected by landscape changes were forest, grasslands, and silvopastoral areas in terms of area, percent, and the annual rate of changes (Table 6), the absolute value of net change (Table 8) and the main systematic transitions (Table 9).

4.2. Logistic Regression, Probability Maps and Future Projection of Forest, Grasslands and Silvopastoral Areas

The logistic regression analyses (forward conditional–stepwise) revealed the influence of each of the 20 included independent variables on the LULC types of forest, grasslands, and silvopastoral areas (dependent variables). The area under the ROC curve (AUC) is presented in Figure 5.

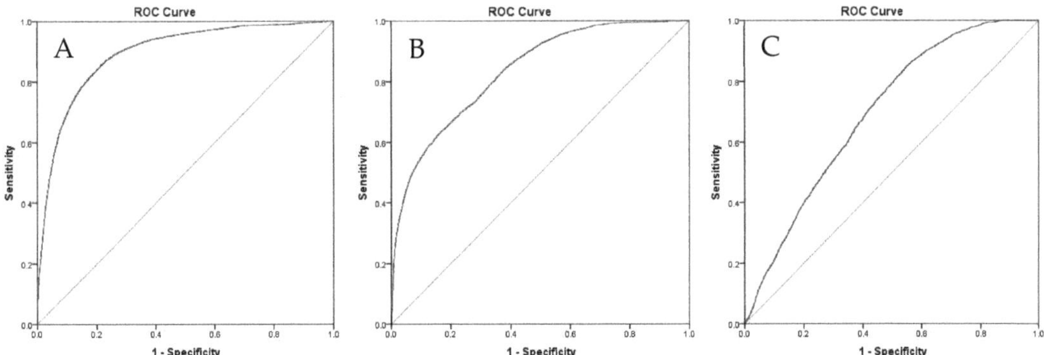

Figure 5. Graphs of the relative operating characteristic (ROC) curves and areas under the curve (AUC) values (spss v27) of stepwise logistic regression analyses for: (**A**) forest (AUC = 0.894), (**B**) grassland (AUC = 0.834), and (**C**) silvopastoral areas (AUC = 0.697), in the landscape of Mt. Zireia.

AUC values for the three main LULC types were above 0.8 for forest and grasslands, which is accepted as an excellent discrimination, and 0.697 (equal to almost 0.7) for silvopastoral areas, which is marginally acceptable discrimination [77], indicating that the logistic regression models possessed significant goodness of fit.

The b-values of the independent variables are presented in Table 10. The cells without data indicate the independent variables which did not show a statistically significant correlation with the LULC types. The regression coefficients of Table 10 (b-values and constant) were entered into the CLUE software environment to build a set of three probability maps and finally complete the landscape change prediction procedure. The produced maps of the probability (%) of future occurrence for forest, grasslands, and silvopastoral areas in the study area are presented in Figure 6.

According to the probability maps (Figure 6), forest possessed a higher possibility of occurrence mainly in the north-northwest parts of the study area, and to a lesser extent, in a restricted area in the south. Grasslands, on the other hand, were found to be highly possible to occupy areas of high altitude in the center part of the study region, covering the grounds of the sub-alpine zones near Mt Zireia's summit. Silvopastoral areas received a lower chance of occurrence compared with forest and grasslands, and these chances of occurrence were scattered around the landscape.

Land use demands of the three LULC types under the BAU scenario for a 20-year prediction period (2020–2040), are presented in Table 11. According to Table 11, if the main factors of change (landscape changing factors–independent variables) continue to be the same between 2020 and 2040, then forest is expected to increase in some areas by 15%, and grasslands and silvopastoral areas to decrease by 11% and 31%, respectively.

Table 10. The logistic regression coefficients (b-values and constant).

Independent Variables	Forest	Grasslands	Silvopastoral
Elevation (m)	0.0010	0.0024	−0.0007
Slopes (%)	0.0306	−0.0289	0.0122
Alluvial deposits/very deep soils	−2.8321	−2.6520	−4.1481
Hard limestone/shallow to bare soils	−0.6249		−0.4752
Limestone colluvium/deep to moderately deep soils	0.6755		
Doline-deposition cones/deep soils	−19.1054	−1.3975	−18.2201
Tertiary deposits/deep to moderately deep soils		−0.3152	0.6172
Tertiary deposits/shallow soils	−0.6550	−0.6041	
Schist/shallow soils	1.6790		−0.4972
Schist/deep soils	2.7273	−1.7426	−2.2187
Erosion (t/ha/year)	−0.2248	0.0324	−0.0059
Distance from unpaved roads (m)	0.0001	−0.0003	
Distance from paved roads (m)	0.0005		
Distance from water courses (m)	−0.0002	−0.0003	
Distance to settlements (m)	−0.0001	−0.0002	−0.0003
Population density (number of people per area (ha)	−7.4041	1.5776	
Sheep/cow density (number of heads per area (ha)	−0.3747	0.1150	0.2747
Goat density (number of heads per area (ha))	−3.0946		
Annual mean temperature (Co)	−0.3339		−0.2338
Annual precipitation (mm)	−0.0113	0.0097	0.0052
Constant	12.8058	−11.7581	−3.7588

Table 11. Area distribution (ha) and rate of change (%) of forest, grasslands, and silvopastoral areas in the study area for the projection period 2020–2040.

Land Use	Area (ha)		Rate of Change %
	2020	2040	
Forest	13,430.98	15,474.87	15.22
Grasslands	5893.51	5228.86	−11.28
Silvopastoral areas	2850.80	1979.33	−30.57
Total	22,175.29	22,683.06	2.29

Spatial allocation of the projected land use demands into the probability maps (Figure 6) produced the predicted potential map (Figure 7) of the spatiotemporal distribution for forest, grasslands and silvopastoral areas for the 2020–2040 period. Forest is expected to continue expanding in the north-northwest parts, and probably will occupy scattered new areas in the central parts of the landscape. The projected grasslands reduction, on the other hand, will most probably force the remaining grassland patches to be limited to the central parts of the landscape. Silvopastoral areas will probably continue to occupy small, scattered areas around the landscape, but with a spatial distribution uneven in size.

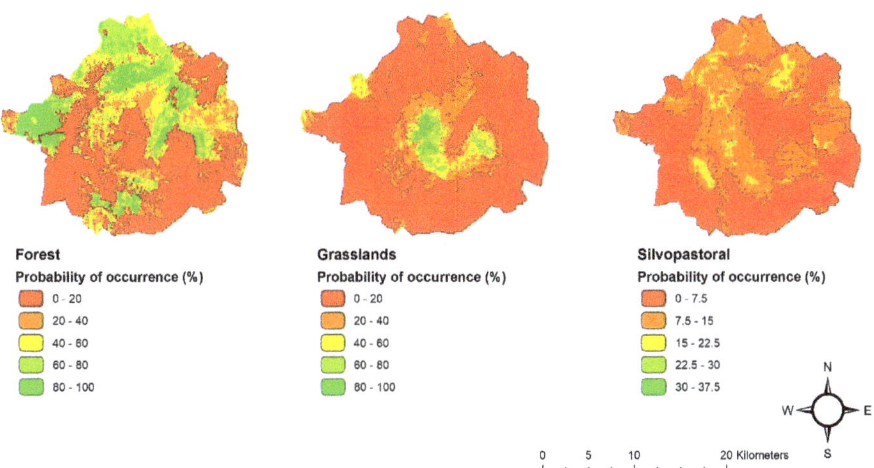

Figure 6. Probability maps of future occurrence of forest, grasslands and silvopastoral areas in the study area.

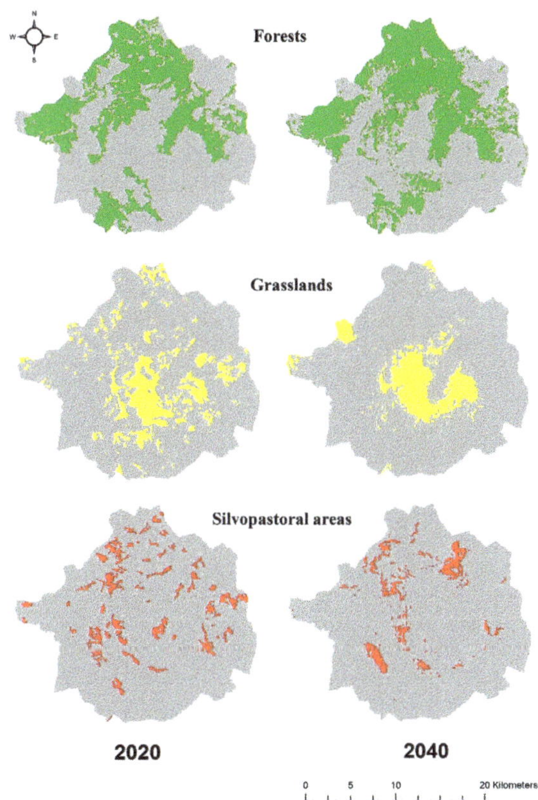

Figure 7. Predicted potential map of the spatiotemporal distribution of forest (green color), grasslands (yellow color), and silvopastoral areas (red color) for the 2020–2040 period in the study area.

The above results can be confirmed by evaluating the landscape structure of the projected maps with the help of landscape metrics (Table 12). According to the metrics during the projected period (2020–2040), the expansion of forests will probably increase their overall fragmentation in the sense that numerous new and smaller sized patches (indicated by the NumP and MPS values) will be created in new areas across the landscape (Table 12). That increase will improve the ED value, creating new forest edges and promoting forest connectivity, as was indicated by the decrease in their MNN value.

Table 12. Landscape metric evaluation for forest (FO), grasslands (GR), and silvopastoral areas (SI) in the landscape of Mt Zireia for the projection period 2020–2040.

	NumP [1]			MPS [2]			ED [3]			MNN [4]		
	FO	GR	SI	FO	GR	SI	FO	GR	SI	FO	GR	SI
2020	30	155	70	448.70	38.11	40.21	14.80	13.48	8.22	348.96	229.83	419.34
2040	200	155	123	77.39	33.90	15.95	20.49	7.74	5.78	192.68	289.07	268.55

[1] Number of patches. [2] Mean patch size (ha). [3] Edge density (m/ha). [4] Mean nearest neighbor (m).

Grassland patches, on the other hand, will probably become smaller in size (decrease in MPS value) and distant from each other, as was indicated by the increase in their MNN value. The latter observation will probably promote patch isolation of the smaller grassland units which occupy the marginal areas around their main distribution in the center of the landscape (Figure 7). Furthermore, the decrease in ED value indicates that a significant reduction in grasslands ecotone is to be expected. Silvopastoral patches are expected to become more fragmented in the future, meaning greater in numbers but smaller in size (indicated by the NumP and MPS differences). Moreover, even though silvopastoral areas would increase their overall connectivity (MNN value), they are expected to greatly reduce their edges (ED value), similar to grasslands.

5. Discussion

Spatiotemporal transition analysis of the landscape of Mt Zireia suggested that the strongest trend of landscape evolution was woody plant expansion in open areas, and grasslands reduction. Among the different LULC transitions, the most systematic ones (Table 9) were forest expansion over silvopastoral areas, of open shrublands over dense shrublands, and of grassland reduction in favor of open shrublands and silvopastoral areas. These results, combined with the data of the total LULC changes in area, percent, and the annual rate and net changes (Tables 6 and 8), suggested that the LULC types that are mainly affected by landscape changes are forests, grasslands and silvopastoral areas. This finding, especially as far as the forest expansion/grasslands reduction trend is concerned, is in line with similar studies conducted in Greece [15,24,25,28] and other Mediterranean countries [2,47,55,81], indicating that special focus should be provided to these specific LULC interactions, especially in the rapidly-changing Mediterranean landscapes [9].

The above trend of LULC interactions can mainly be attributed to land abandonment issues related to socioeconomic conditions. Relevant inventory data from the village communities in the study area (Tables 1 and 2) suggested that socioeconomic changes over the previous decades in the study area had the form of a decrease in local population, population aging and a significant temporal reduction in the percentage of employees in the primary economic sector. These specific types of socioeconomic changes are reported to especially occur in Mediterranean landscapes, as directly related to the abandonment of traditional management practices, such as extensive or semi-extensive pastoral activities (including transhumance pastoralism), wood product collection (e.g., coal and fuel woods) and agricultural practices in less favorable areas (e.g., crop fields in terraces) [8,47,54,55,81–83].

Additional inventory data related to the numbers and farms of grazing animals from the study area supports the notion that land abandonment has affected pastoral activities (Tables 3 and 4). More specifically, the number of sedentary and transhumant grazing animals and farms for the study area were significantly reduced over the previous decades, following a similar trend of change as the number of people that were employed in the primary sector of the economy. The reduction in grazing animals was also reported to follow a similar, more general, trend of reduction for the whole country and for other south European Mediterranean countries [8,11,81]. The collection of forest products seems to be affected by the abandonment of traditional practices. Unpublished data from the PACTORES Project (www.pactores.eu (accessed on 10 December 2021)) indicated that fuel wood collection by local people within the study area has significantly reduced over the years, and in some areas has practically stopped. On the other hand, some of the forest expansion over open areas can be attributed to the afforestation policies of the local forest service to increase the area covered by high forests. Finally, agricultural activities were also affected by land abandonment, but this effect was less severe on the extent of agricultural lands. According to the spatiotemporal analysis of this research, agricultural areas scored as the fourth most important LULC change in terms of total area, percent of change, annual rate, and net changes (Tables 6 and 8) and these changes did not appear to have a strong geographic orientation (Figure 3). Agricultural activities were mainly oriented in plains in favorable and more accessible parts of the Mt Zireia landscape, which, as similar studies have pointed out, are probably less affected by land abandonment [11,84]. All these aspects of socioeconomic effects in the current management of Mediterranean landscapes have been noted throughout the Mediterranean region of Europe [2,13,15,22,24,25,54,55,81,85] and have been identified as the main reasons for landscape change.

Future development for forest, grasslands, and silvopastoral areas based on the BAU scenario of linear extension of land use demands for 2040 and probability maps, suggested that forest will most probably continue to expand in the north-northwest parts, adding new areas scattered mainly in the central parts of the landscape. At the same time, grassland and silvopastoral areas will probably continue to reduce in area, occupying territories mainly at the central part for grasslands, or small scattered territories around the landscape for silvopastoral areas (Figure 7). Similar results of future development of forests and grasslands were very recently reported from a rural landscape study of central Greece under a similar trend of land abandonment [27,65]. Evaluation of the structural developments (Table 12) of the Mt Zireia landscape from the projected maps revealed that forest expansion into new areas, and in many cases, as small patches, will increase their overall dispersal and will create new forest edges (higher ED value). Furthermore, the decrease in MNN value will promote forest connectivity. Grasslands, on the other hand, apart from occupying one large and three smaller core areas in the center and the north-northwest part of the study area (Figure 7), will probably keep only smaller, fragmented, and isolated patches around the landscape, as indicated by the reduction in their MPS and the increase in their MNN values. Moreover, the decrease in ED value will cause a significant reduction in grassland edges. Silvopastoral patches, similar to grasslands, will became more fragmented with reduced edges. These findings correlate with the response of many other landscapes around the world, showing that forest expansion usually leads to increased forest patch connectivity promoting forest edges, while open habitat reduction usually creates the opposite trend of a reduction in connectivity and edges [86,87]. These results could be alarming for sustaining the environmental integrity of the Mt Zireia landscape, as many researchers have linked grasslands fragmentation and the loss of connectivity and boundary lengths of open habitats, to the decline of species richness and mountainous biodiversity [47,86,88]. The results of landscape metrics evaluation on the future development of LULC types can serve as evidence of the great dynamic of expansion that forest patches possess over grassland and silvopastoral patches, independent of the environment and biodiversity implications.

The findings of this study are consistent with the common pattern of woody cover expansion over open regions in many Mediterranean landscapes that suffer from land abandonment [89]. Environmental integrity, biodiversity, and cultural heritage may be positively or negatively impacted by land abandonment, which can additionally benefit forest ecosystems by fostering it at minimal cost and on a larger scale [65]. Forest recovery promotes carbon sequestration, erosion reduction, and several other ecosystem services such as climate and water regulation, wood production, and recreation [12]. On the other hand, land abandonment, especially in the Mediterranean region, results in declining biodiversity and loss of traditional cultural landscapes [7,81,84], and is often linked to an increased risk of wildfires and decreased river flows [83,90,91]. The land abandonment effect can also be associated with the loss of important cultural elements and services related to traditional pastoral activities, such as cultural heterogenic pastoral landscapes, gastronomical heritage, and folklore elements [92,93].

Developmental planning must take into consideration the spatiotemporal trends and the future projection of LULC types recorded in this study. Forest expansion over grassland and silvopastoral areas, apart from the environmental and cultural implications, would have a strong negative effect on the future of sustainable development of livestock husbandry in the study area. Grassland and silvopastoral areas are considered important natural resources necessary for applying extensive pastoral practices, especially the transhuman livestock system, and the reported threat status could have a damaging effect on keeping the ecological integrity and the social benefits that people expect from pastoral landscapes [62,94,95].

Author Contributions: Conceptualization, D.C. and M.K.; methodology, D.C., M.K., A.S. and G.C.; software, D.C. and A.S.; validation, D.C., M.K., A.S. and G.C.; formal analysis, D.C., M.K., A.S. and G.C.; investigation D.C., M.K., A.S. and G.C.; resources, D.C. and M.K.; data curation, D.C., M.K., A.S. and G.C.; writing—original draft preparation, D.C., M.K. and A.S.; writing—review and editing. D.C., M.K. and A.S.; visualization, D.C., M.K., A.S. and G.C.; supervision, M.K.; project administration, M.K.; funding acquisition, M.K. All authors have read and agreed to the published version of the manuscript.

Funding: This research was part of the project «PACTORES» and was co-funded by the European Union (European Social Fund, ERANETMED2-72-303) and General Secretariat for Research and Innovation (GSRI, T8EPA2-00022) through the Action "ERA-Net".

Data Availability Statement: The data presented in this study are available in figures and tables provided in the manuscript.

Conflicts of Interest: The authors declare no conflict of interest.

References

1. Sluiter, R.; de Jong, S.M. Spatial patterns of Mediterranean land abandonment and related land cover transitions. *Landsc. Ecol.* **2007**, *22*, 559–576. [CrossRef]
2. Sirami, C.; Nespoulous, A.; Cheylan, J.-P.; Marty, P.; Hvenegaard, G.T.; Geniez, P.; Schatz, B.; Martin, J.-L. Long-term anthropogenic and ecological dynamics of a Mediterranean landscape: Impacts on multiple taxa. *Landsc. Urban Plan.* **2010**, *96*, 214–223. [CrossRef]
3. Marignani, M.; Chiarucci, A.; Sadori, L.; Mercuri, A.M. Natural and human impact in Mediterranean landscapes: An intriguing puzzle or only a question of time? *Plant Biosyst.* **2016**, *151*, 900–905. [CrossRef]
4. Ecosystem, C. Ecosystem Profile: Mediterranean Basin Biodiversity Hotspot. Critical Ecosystem Partnership Fund. *BirdLife Int.* **2017**, 309. Available online: https://www.cepf.net/sites/default/files/mediterranean-basin-2017-ecosystem-profile-english_0.pdf (accessed on 23 November 2022).
5. Perevolotsky, A. Integrating landscape ecology in the conservation of Mediterranean ecosystems: The Israeli experience. *Isr. J. Plant Sci.* **2005**, *53*, 203–213. [CrossRef]
6. Papanastasis, V. Traditional vs contemporary management of Mediterranean vegetation: The case of the island of Crete. *J. Biolog. Res.* **2004**, *1*, 39–46.
7. Delattre, L.; Debolini, M.; Paoli, J.C.; Napoleone, C.; Moulery, M.; Leonelli, L.; Santucci, P. Understanding the Relationships between Extensive Livestock Systems, Land-Cover Changes, and CAP Support in Less-Favored Mediterranean Areas. *Land* **2020**, *9*, 518. [CrossRef]
8. Nori, M.; Farinella, D. *Migration, Agriculture and Rural Development: IMISCOE Short Reader*; Springer Nature: Berlin/Heidelberg, Germany, 2020.
9. Papanastasis, V.P. Land Use Changes. In *Mediterranean Mountain Environments*; John Wiley & Sons: Hoboken, NJ, USA, 2012; pp. 159–184. [CrossRef]
10. Guarino, R.; Vrahnakis, M.; Rojo, M.P.R.; Giuga, L.; Pasta, S. Grasslands and shrublands of the Mediterranean region. In *Encyclopedia of the World's Biomes*; Goldstein, M., DellaSala, D., Eds.; Elsevier: Amsterdam, The Netherlands, 2020; Volume 3, pp. 635–638.

11. Levers, C.; Müller, D.; Erb, K.; Haberl, H.; Jepsen, M.R.; Metzger, M.J.; Meyfroidt, P.; Plieninger, T.; Plutzar, C.; Stürck, J.; et al. Archetypical patterns and trajectories of land systems in Europe. *Reg. Environ. Chang.* **2015**, *18*, 715–732. [CrossRef]

12. Nocentini, S.; Travaglini, D.; Muys, B. Managing Mediterranean Forests for Multiple Ecosystem Services: Research Progress and Knowledge Gaps. *Curr. For. Rep.* **2022**, *8*, 229–256. [CrossRef]

13. Peñuelas, J.; Sardans, J. Global Change and Forest Disturbances in the Mediterranean Basin: Breakthroughs, Knowledge Gaps, and Recommendations. *Forests* **2021**, *12*, 603. [CrossRef]

14. Millington, J.D.A.; Perry, G.; Romero-Calcerrada, R. Regression Techniques for Examining Land Use/Cover Change: A Case Study of a Mediterranean Landscape. *Ecosystems* **2007**, *10*, 562–578. [CrossRef]

15. Papanastasis, V.P.; Chouvardas, D. Application of the state-and-transition approach to conservation management of a grazed Mediterranean landscape in Greece. *Isr. J. Plant Sci.* **2005**, *53*, 191–202. [CrossRef]

16. Nori, M. Redressing Policy Making in Pastoral Areas of the Mediterranean Region. *J. Policy Gov.* **2022**, *2*, 21–32. [CrossRef]

17. Glasser, T.; Hadar, L. Grazing management aimed at producing landscape mosaics to restore and enhance biodiversity in Mediterranean ecosystems. *Options Méditerr.* **2014**, *109*, 437.

18. Psyllos, G.; Hadjigeorgiou, I.; Dimitrakopoulos, P.G.; Kizos, T. Grazing Land Productivity, Floral Diversity, and Management in a Semi-Arid Mediterranean Landscape. *Sustainability* **2022**, *14*, 4623. [CrossRef]

19. Hadjigeorgiou, I. Past, present and future of pastoralism in Greece. *Pastor. Res. Policy Pract.* **2011**, *1*, 24. [CrossRef]

20. Karatassiou, M.; Parissi, Z.M.; Stergiou, A.; Chouvardas, D.; Mantzanas, K. Patterns of transhumant livestock system on Mount Zireia, Peloponnese, Greece. *Opt. Mediterr. Ser. A* **2021**, *126*, 197–200.

21. Chouvardas, D. Estimation of Diachronic Effects of Pastoral Systems and Land Uses in Landscapes with the Use of Geographic Information Systems (GIS). Ph.D. Thesis, School of Forestry and Natural Environment, Aristotle University of Thessaloniki, Thessaloniki, Greece, 2007.

22. Sidiropoulou, A.; Karatassiou, M.; Galidaki, G.; Sklavou, P. Landscape Pattern Changes in Response to Transhumance Abandonment on Mountain Vermio (North Greece). *Sustainability* **2015**, *7*, 15652–15673. [CrossRef]

23. Ragkos, A.; Abraham, E.M.; Papadopoulou, A.; Kyriazopoulos, A.P.; Parissi, Z.M.; Hadjigeorgiou, I. Effects of European Union agricultural policies on the sustainability of grazingland use in a typical Greek rural area. *Land Use Policy* **2017**, *66*, 196–204. [CrossRef]

24. Zomeni, M.; Tzanopoulos, J.; Pantis, J.D. Historical analysis of landscape change using remote sensing techniques: An explanatory tool for agricultural transformation in Greek rural areas. *Landsc. Urban Plan.* **2008**, *86*, 38–46. [CrossRef]

25. Sklavou, P.; Karatassiou, M.; Parissi, Z.; Galidaki, G.; Ragkos, A.; Sidiropoulou, A. The Role of Transhumance on Land Use /Cover Changes in Mountain Vermio, Northern Greece: A GIS Based Approach. *Not. Bot. Horti Agrobot. Cluj-Napoca* **2017**, *45*, 589–596. [CrossRef]

26. Rapti, D.; Chouvardas, D.; Parissi, Z. Study of Temporal Evolution of Olympus Landscape. In Proceedings of the 9th Panhellenic Rangeland Congress, Larisa, Greece, 9–12 October 2018; pp. 82–89.

27. Mamanis, G.; Vrahnakis, M.; Chouvardas, D.; Nasiakou, S.; Kleftoyanni, V. Land Use Demands for the CLUE-S Spatiotemporal Model in an Agroforestry Perspective. *Land* **2021**, *10*, 1097. [CrossRef]

28. Nasiakou, S.; Chouvardas, D.; Vrahnakis, M.; Kleftoyanni, V. Temporal Changes Analysis of Mouzaki's Landscape, Western Thessaly, Greece (1960–2020). In Proceedings of the 10th Panhellenic Rangeland Congress, Florina, Greece, 4 March 2022; pp. 147–152.

29. Teferi, E.; Bewket, W.; Uhlenbrook, S.; Wenninger, J. Understanding recent land use and land cover dynamics in the source region of the Upper Blue Nile, Ethiopia: Spatially explicit statistical modeling of systematic transitions. *Agric. Ecosyst. Environ.* **2013**, *165*, 98–117. [CrossRef]

30. Munthali, M.; Botai, O.; Davis, N.; Abiodun, A. Multi-temporal Analysis of Land Use and Land Cover Change Detection for Dedza District of Malawi using Geospatial Techniques. *Int. J. Appl. Eng.* **2019**, *14*, 1151–1162.

31. Abbas, Z.; Yang, G.; Zhong, Y.; Zhao, Y. Spatiotemporal Change Analysis and Future Scenario of LULC Using the CA-ANN Approach: A Case Study of the Greater Bay Area, China. *Land* **2021**, *10*, 584. [CrossRef]

32. Attri, P.; Chaudhry, S.; Sharma, S. Remote Sensing & GIS based Approaches for LULC Change Detection—A Review. *Int. J. Curr. Eng. Techn.* **2015**, *5*, 3126–3137.

33. Carta, A.; Taboada, T.; Müller, J.V. Diachronic analysis using aerial photographs across fifty years reveals significant land use and vegetation changes on a Mediterranean island. *Appl. Geogr.* **2018**, *98*, 78–86. [CrossRef]

34. Kiziridis, D.A.; Mastrogianni, A.; Pleniou, M.; Karadimou, E.; Tsiftsis, S.; Xystrakis, F.; Tsiripidis, I. Acceleration and Relocation of Abandonment in a Mediterranean Mountainous Landscape: Drivers, Consequences, and Management Implications. *Land* **2022**, *11*, 406. [CrossRef]

35. Feizizadeh, B.; Alajujeh, K.M.; Lakes, T.; Blaschke, T.; Omarzadeh, D. A comparison of the integrated fuzzy object-based deep learning approach and three machine learning techniques for land use/cover change monitoring and environmental impacts assessment. *GIScience Remote Sens.* **2021**, *58*, 1543–1570. [CrossRef]

36. Zhang, B.; Zhang, Q.; Feng, C.; Feng, Q.; Zhang, S. Understanding Land Use and Land Cover Dynamics from 1976 to 2014 in Yellow River Delta. *Land* **2017**, *6*, 20. [CrossRef]

37. Cribari, V.; Strager, M.P.; Maxwell, A.E.; Yuill, C. Landscape Changes in the Southern Coalfields of West Virginia: Multi-Level Intensity Analysis and Surface Mining Transitions in the Headwaters of the Coal River from 1976 to 2016. *Land* **2021**, *10*, 748. [CrossRef]

38. Verburg, P.H.; Overmars, K.P.; Huigen, M.G.A.; de Groot, W.T.; Veldkamp, A. Analysis of the effects of land use change on protected areas in the Philippines. *Appl. Geogr.* **2006**, *26*, 153–173. [CrossRef]
39. Mas, J.-F.; Kolb, M.; Paegelow, M.; Olmedo, M.T.C.; Houet, T. Inductive pattern-based land use/cover change models: A comparison of four software packages. *Environ. Model. Softw.* **2014**, *51*, 94–111. [CrossRef]
40. Amici, V.; Marcantonio, M.; La Porta, N.; Rocchini, D. A multi-temporal approach in MaxEnt modelling: A new frontier for land use/land cover change detection. *Ecol. Inform.* **2017**, *40*, 40–49. [CrossRef]
41. Nwaogu, C.; Benc, A.; Pechanec, V. Prediction Models for Landscape Development in GIS. In *Dynamics in GIscience*; Ivan, I., Horak, J., Inspektor, T., Eds.; Springer International Publishing AG 2018: Cham, Switzerland, 2018; pp. 289–304.
42. Minetos, D.; Polyzos, S. Multivariate statistical methodologies for testing hypothesis of land-use change at the regional level. A review and evaluation. *J. Environ. Prot.* **2009**, *10*, 834–866.
43. Feng, Y.; Liu, Y.; Tong, X. Spatiotemporal variation of landscape patterns and their spatial determinants in Shanghai, China. *Ecol. Indic.* **2018**, *87*, 22–32. [CrossRef]
44. Farina, A. The Cultural Landscape as a Model for the Integration of Ecology and Economics. *Bioscience* **2000**, *50*, 313–320. [CrossRef]
45. McGarigal, K. *FRAGSTATS Help*; University of Massachusetts: Amherst, MA, USA, 2015; Volume 182.
46. Kumar, M.; Denis, D.M.; Singh, S.K.; Szabó, S.; Suryavanshi, S. Landscape metrics for assessment of land cover change and fragmentation of a heterogeneous watershed. *Remote Sens. Appl. Soc. Environ.* **2018**, *10*, 224–233. [CrossRef]
47. Ameztegui, A.; Morán-Ordóñez, A.; Márquez, A.; Blázquez, C.; Pla, M.; Villero, D.; García, M.B.; Errea, M.P.; Coll, L. Forest expansion in mountain protected areas: Trends and consequences for the landscape. *Landsc. Urban Plan.* **2021**, *216*, 104240. [CrossRef]
48. Motlagh, Z.K.; Lotfi, A.; Pourmanafi, S.; Ahmadizadeh, S.; Soffianian, A. Spatial modeling of land-use change in a rapidly urbanizing landscape in central Iran: Integration of remote sensing, CA-Markov, and landscape metrics. *Environ. Monit. Assess.* **2020**, *192*, 695. [CrossRef]
49. Natura 2000. Network Viewer. Available online: https://natura2000.eea.europa.eu (accessed on 10 May 2022).
50. Wikipedia. Heracles. Available online: https://en.wikipedia.org/wiki/Heracles#Labours_of_Heracles (accessed on 10 May 2022).
51. Climate Data for Cities Worldwide. Available online: https://en.climate-data.org (accessed on 15 January 2018).
52. Chrysanthopoulou, G. Synergy of Climate and Grazing in the Evolution of Vegetation on Mount Kyllini. Aristotle University of Thessaloniki: Thessaloniki, Greece, 2021.
53. Hellenic Statistical Authority. Digital Library (ELSTAT). General and Special Publications. Available online: http://dlib.statistics.gr/portal/page/portal/ESYE (accessed on 15 April 2022).
54. Ispikoudis, I.; Chouvardas, D. Livestock, land use and landscape. In *Animal Production and Natural Resources Utilisation in the Mediterraneanmountain Areas*; Georgoudis, A., Rosati, A., Moscani, C., Eds.; Wageningen Academic Publishers: Wageningen, The Netherlands, 2005; Volume 115, pp. 151–157.
55. Lasanta-Martínez, T.; Vicente-Serrano, S.M.; Cuadrat-Prats, J.M. Mountain Mediterranean landscape evolution caused by the abandonment of traditional primary activities: A study of the Spanish Central Pyrenees. *Appl. Geogr.* **2005**, *25*, 47–65. [CrossRef]
56. PCAGGCA. *Payment and Control Agency for Guidance and Guarantee Community Aid-Registry of Farms and Farmers*; Ministry of Rural Development and Food: Athens, Greece, 2020.
57. Chatzimichali, A. *Sarakatsanoi*, 2nd ed.; Angeliki Chatzimichali Foundation: Athina, Greece, 2007.
58. Paine, D.P.; Kiser, J.D. *Aerial Photography and Image Interpretation*; John Wiley & Sons: Hoboken, NY, USA, 2012.
59. Puyravaud, J.P. Standardizing the calculation of the annual rate of deforestation. *For. Ecol. Manag.* **2003**, *177*, 593–596. [CrossRef]
60. Viana, C.M.; Rocha, J. Evaluating Dominant Land Use/Land Cover Changes and Predicting Future Scenario in a Rural Region Using a Memoryless Stochastic Method. *Sustainability* **2020**, *12*, 4332. [CrossRef]
61. Pontius, R.G.; Huffaker, D.; Denman, K. Useful techniques of validation for spatially explicit land-change models. *Ecol. Modell.* **2004**, *179*, 445–461. [CrossRef]
62. Chouvardas, D.; Vrahnakis, M. A semi-empirical model for the near future evolution of the lake Koronia landscape. *J. Environ. Prot. Ecol.* **2009**, *10*, 867–876.
63. Verburg, P.H.; Overmars, K.P. Combining top-down and bottom-up dynamics in land use modeling: Exploring the future of abandoned farmlands in Europe with the Dyna-CLUE model. *Landsc. Ecol.* **2009**, *24*, 1167–1181. [CrossRef]
64. Verburg, P. The CLUE modelling framework: The conversion of land use and its effects. In *Manual for the CLUE-S Model; The CLUE Group; Department of Environmental Sciences*; Wageningen University: Wageningen, The Netherlands, 2010; p. 53.
65. Nasiakou, S.; Vrahnakis, M.; Chouvardas, D.; Mamanis, G.; Kleftoyanni, V. Land Use Changes for Investments in Silvoarable Agriculture Projected by the CLUE-S Spatio-Temporal Model. *Land* **2022**, *11*, 598. [CrossRef]
66. Temme, A.; Verburg, P.H. Modelling of Intensive and Extensive Farming in CLUE. *Wettelijke Onderz. Nat. Milieu* **2010**. Available online: https://edepot.wur.nl/139499 (accessed on 23 November 2022).
67. Sohl, T.L.; Sleeter, B.M.; Sayler, K.L.; Bouchard, M.A.; Reker, R.R.; Bennett, S.L.; Sleeter, R.R.; Kanengieter, R.L.; Zhu, Z. Spatially explicit land-use and land-cover scenarios for the Great Plains of the United States. *Agric. Ecosyst. Environ.* **2012**, *153*, 1–15. [CrossRef]

68. SPSS Stepwise Regression—Simple Tutorial. Resolving Multicollinearity with Stepwise Regression. Available online: https://www.spss-tutorials.com/stepwise-regression-in-spss-example (accessed on 15 September 2022).

69. Verburg, P.H.; Veldkamp, A. Projecting land use transitions at forest fringes in the Philippines at two spatial scales. *Landsc. Ecol.* **2004**, *19*, 77–98. [CrossRef]

70. Majnik, M.; Bosnić, Z. ROC analysis of classifiers in machine learning: A survey. *Intell. Data Anal.* **2013**, *17*, 531–558. [CrossRef]

71. Pontius Jr, R.G.; Si, K. The total operating characteristic to measure diagnostic ability for multiple thresholds. *Int. J. Geogr. Inf. Sci.* **2014**, *28*, 570–583. [CrossRef]

72. Talukdar, S.; Singha, P.; Mahato, S.; Shahfahad; Pal, S.; Liou, Y.-A.; Rahman, A. Land-Use Land-Cover Classification by Machine Learning Classifiers for Satellite Observations—A Review. *Remote. Sens.* **2020**, *12*, 1135. [CrossRef]

73. DeFries, R.S.; Rudel, T.; Uriarte, M.; Hansen, M. Deforestation driven by urban population growth and agricultural trade in the twenty-first century. *Nat. Geosci.* **2010**, *3*, 178–181. [CrossRef]

74. Yoshikawa, S.; Sanga-Ngoie, K. Deforestation dynamics in Mato Grosso in the southern Brazilian Amazon using GIS and NOAA/AVHRR data. *Int. J. Remote Sens.* **2011**, *32*, 523–544. [CrossRef]

75. Yirsaw, E.; Wu, W.; Shi, X.; Temesgen, H.; Bekele, B. Land Use/Land Cover Change Modeling and the Prediction of Subsequent Changes in Ecosystem Service Values in a Coastal Area of China, the Su-Xi-Chang Region. *Sustainability* **2017**, *9*, 1204. [CrossRef]

76. Mandrekar, J.N. Receiver Operating Characteristic Curve in Diagnostic Test Assessment. *J. Thorac. Oncol.* **2010**, *5*, 1315–1316. [CrossRef]

77. Hosmer, D.W.; Lemeshow, S.; Sturdivant, R.X. *Applied Logistic Regression*; John Wiley & Sons: Hoboken, NJ, USA, 2013; Volume 398.

78. McGarigal, K.; Marks, B.J. Spatial pattern analysis program for quantifying landscape structure. In *General Technical Report. PNW-GTR-351*; US Department of Agriculture, Forest Service, Pacific Northwest Research Station: Portland, OR, USA, 1995; pp. 1–122.

79. Elkie, P.C.; Rempel, R.S.; Carr, A. *Patch Analyst User's Manual: A Tool for Quantifying Landscape Structure*; Ontario Ministry of Natural Resources, Boreal Science, Northwest Science & Technology: Thunder Bay, Canada, 1999.

80. Leitão, A.; Miller, J.; Ahern, J.; McGarigal, K. *Measuring Landscapes: A Planner's Handbook*; Island Press: Washington, DC, USA, 2006.

81. Pelorosso, R.; Leone, A.; Boccia, L. Land cover and land use change in the Italian central Apennines: A comparison of assessment methods. *Appl. Geogr.* **2009**, *29*, 35–48. [CrossRef]

82. Levers, C.; Schneider, M.; Prishchepov, A.V.; Estel, S.; Kuemmerle, T. Spatial variation in determinants of agricultural land abandonment in Europe. *Sci. Total. Environ.* **2018**, *644*, 95–111. [CrossRef]

83. Quintas-Soriano, C.; Buerkert, A.; Plieninger, T. Effects of land abandonment on nature contributions to people and good quality of life components in the Mediterranean region: A review. *Land Use Policy* **2022**, *116*, 106053. [CrossRef]

84. Van der Sluis, T.; Kizos, T.; Pedroli, B. Landscape Change in Mediterranean Farmlands: Impacts of Land Abandonment on Cultivation Terraces in Portofino (Italy) and Lesvos (Greece). *J. Landsc. Ecol.* **2014**, *7*, 23–44. [CrossRef]

85. Henkin, Z. The role of brush encroachment in Mediterranean ecosystems: A review. *Isr. J. Plant Sci.* **2021**, *69*, 1–12. [CrossRef]

86. Sitzia, T.; Semenzato, P.; Trentanovi, G. Natural reforestation is changing spatial patterns of rural mountain and hill landscapes: A global overview. *For. Ecol. Manag.* **2010**, *259*, 1354–1362. [CrossRef]

87. Campagnaro, T.; Frate, L.; Carranza, M.L.; Sitzia, T. Multi-scale analysis of alpine landscapes with different intensities of abandonment reveals similar spatial pattern changes: Implications for habitat conservation. *Ecol. Indic.* **2017**, *74*, 147–159. [CrossRef]

88. Howell, P.E.; Terhune, T.M.; Martin, J.A. Edge density affects demography of an exploited grassland bird. *Ecosphere* **2021**, *12*, e03499. [CrossRef]

89. Ustaoglu, E.; Collier, M. Farmland abandonment in Europe: An overview of drivers, consequences, and assessment of the sustainability implications. *Environ. Rev.* **2018**, *26*, 396–416. [CrossRef]

90. Bentley, L.; Coomes, D.A. Partial river flow recovery with forest age is rare in the decades following establishment. *Glob. Chang. Biol.* **2020**, *26*, 1458–1473. [CrossRef]

91. Mantero, G.; Morresi, D.; Marzano, R.; Motta, R.; Mladenoff, D.J.; Garbarino, M. The influence of land abandonment on forest disturbance regimes: A global review. *Landsc. Ecol.* **2020**, *35*, 2723–2744. [CrossRef]

92. Muñoz-Ulecia, E.; Bernués, A.; Casasús, I.; Olaizola, A.M.; Lobón, S.; Martín-Collado, D. Drivers of change in mountain agriculture: A thirty-year analysis of trajectories of evolution of cattle farming systems in the Spanish Pyrenees. *Agric. Syst.* **2021**, *186*, 102983. [CrossRef]

93. National Inventory of the Intangible Cultural Heritage of Greece. Transhumant Livestock Farming. Available online: https://ayla.culture.gr/wp-content/uploads/2017/07/TRANSHUMANCE_GREECE_TRANSL.pdf (accessed on 30 May 2022).

94. Porqueddu, C.; Franca, A.; Lombardi, G.; Molle, G.; Peratoner, G.; Hopkins, A. *Grassland Resources for Extensive Farming Systems in Marginal Lands*; Major Drivers and Future Scenarios: Alghero, Italy, 2017; p. 681.

95. Surová, D.; Ravera, F.; Guiomar, N.; Sastre, R.M.; Pinto-Correia, T. Contributions of Iberian Silvo-Pastoral Landscapes to the Well-Being of Contemporary Society. *Rangel. Ecol. Manag.* **2018**, *71*, 560–570. [CrossRef]

land

Article

Valuing Ecosystem Services Provided by Pasture-Based Beef Farms in Alentejo, Portugal

Manuel P. dos Santos *, Tiago G. Morais, Tiago Domingos and Ricardo F. M. Teixeira

MARETEC—Marine, Environment and Technology Centre, LARSyS, Instituto Superior Técnico, Universidade de Lisboa, Av. Rovisco Pais, 1, 1049-001 Lisbon, Portugal
* Correspondence: manueldossantos@tecnico.ulisboa.pt

Abstract: This work aims to measure and value the ecosystem services of grasslands and croplands covered by pasture-based beef farms in Alentejo. It combines pixel-level data from the Portuguese Mapping and Assessment of Ecosystem Services study and farm-level data from 40 farms. Five ecosystem services were considered: soil protection, carbon sequestration, support to extensive animal production, plant food production and fiber production. Two different approaches for service quantification were used: an "average class" method and a "buffer" approach. Double counting issues were avoided by applying a specific methodology developed for this study. The results obtained were similar for both approaches in the case of grasslands, with an average value between 146 and 176 €/ha/year. For croplands, the average service value oscillated between 40 and 166 €/ha/year. Soil protection was the most valuable service, with over 90% of the total value. Extrapolating these results for the entire region, the five ecosystem services were estimated to be worth between 173 M€ (class method) and 223 M€ (buffer approach). These results suggest that pasture-based beef farms in Alentejo help to provide a significant number of ecosystem services with positive environmental effects that are currently not remunerated by the market.

Keywords: economic valuation; ecosystem services; grasslands; beef farms; pasture; agriculture; environment; Alentejo; *montado; dehesa*

Citation: dos Santos, M.P.; Morais, T.G.; Domingos, T.; Teixeira, R.F.M. Valuing Ecosystem Services Provided by Pasture-Based Beef Farms in Alentejo, Portugal. *Land* **2022**, *11*, 2238. https://doi.org/10.3390/land11122238

Academic Editors: Michael Vrahnakis, Yannis (Ioannis) Kazoglou and Manuel Pulido Fernádez

Received: 11 November 2022
Accepted: 5 December 2022
Published: 8 December 2022

Publisher's Note: MDPI stays neutral with regard to jurisdictional claims in published maps and institutional affiliations.

1. Introduction

Livestock provides a relevant fraction of the protein that humans consume [1]. Within the livestock sector, beef farming has gained a reputation as one of the most polluting food production systems [2]. However, beef production includes a vast range of production systems and subsystems; a proper quantitative impact evaluation should take into account, as far as possible, local conditions and mitigating factors [3–5]. The absence of such an analysis, at local and global scales, gives rise to the risk of acting on general observations and heuristics, which may cause unwanted results. Environmentally friendly local/regional systems may get categorized as unsustainable simply due to the product they produce. It is especially important to avoid these kinds of perverse effects in areas where livestock is key for managing and enhancing ecosystems and their services [6].

Ecosystem services were first defined as all of the benefits that human populations derive directly or indirectly from ecosystem functions [7]. This concept has developed to include certain hidden contributions of ecosystems and gained momentum after the Millennium Ecosystem Assessment [8]. However, along with services, ecosystems also provide dis-services that affect well-being by reducing productivity and increasing production costs [9]. The net ecosystem services, i.e., positive services minus negative services, affect human well-being. The concept is especially relevant in agroecosystems that are important providers of ES, in the sense that they provide a more diversified set of economic, environmental, cultural and social goods and services [10]. The ecosystem services (ES) approach has a great potential to link ecosystem conservation and the sustainable use of natural resources; however, due to limited funding and resources, the concept has not been widely implemented [11].

One attempt to measure ES and link them with sustainability issues in Portugal was the Portuguese Mapping and Assessment of ES (PT-MAES) [12]. This work was a pilot study within the framework of the European Union Biodiversity Strategy of 2020, a pan European initiative launched by the European Commission. The PT-MAES report and results focus on the Alentejo region. Alentejo, located in the south of Portugal, is the largest NUTS II region in the country, accounting for 54% of national utilized agricultural area [13]. The Mediterranean climate, soils and topographic characteristics of the region favor extensive beef cattle production, which is one of the main agricultural outputs from the region [14]. A significant fraction of the territory used for cattle production is part of an agroforestry system called *dehesa* in Spain and *montado* in Portugal. This is a mixed agricultural/pasture ecosystem coexisting with medium/low densities of trees (cork/holm oak and oak-based). It has been shown to be very well adapted to livestock production and ensures diversification of farmer income through the supply of additional forestry outputs like cork while enhancing the provision of ecosystem services [15]. Alentejo is therefore a prime example of a region where animal management may be key to the health and management of ecosystems.

In this work, we perform economic and environmental assessments of the effects of pasture-based beef farms (PBBF) in Alentejo as providers of ES. The values from PT-MAES were combined with real farm data to account for ES supply level. The ES considered were soil protection (SP), in terms of avoided erosion, carbon sequestration (CS), support to extensive animal production (SEAP), plant food production (PFP) and fiber production (FP). Two different methods were used to determine the ES level of the PBBF of the dataset. In the first approach, for each ES, we applied the average regional service class level. In the second approach, we estimated the average value of each service level, considering a buffer around the geographic coordinates of each farm. A regional extrapolation was then performed, departing from a farm-level assessment to estimate the total ES value promoted by PBBF in the Alentejo region.

2. Materials and Methods

2.1. Characterization of the Study Region

2.1.1. Study Region

The Alentejo region makes up a significant fraction of the south of Portugal. Of the approximately 2 million ha of useful agricultural area in Alentejo, 67% comprises permanent pastures. There are approximately 1.4 million bovines in mainland Portugal, 56% of which are in Alentejo [16]. The typical pasture-based beef production system consists of raising the calves on the farm. On average, pasture-based beef farms in Alentejo have 182 ha and 98 livestock units (LU) and occupy 0.39 full-time equivalent annual work units [17]. There are additional specialized fattening farms in the region, based on grazing plus roughage or concentrate feed regimes until animals reach the required weight for slaughter [18].

2.1.2. Farm Data

Farm data were collected in the context of the Animal Future project (SusAn/0001/2016). The purpose of this project was to study ways to increase the sustainability of animal production systems in Europe. This was mainly achieved by three actions: measuring the different aspects of sustainability in animal farming to make an inventory of innovative practices applied in different European regions, evaluating the impacts of innovations on sustainability, and devising strategies to promote the adoption of innovation practices. Data from farms were collected by the authors between May and October of 2019 through personal interviews with farmers. In total, 40 farms were sampled in this work, whose location can be observed in Figure 1, where the names of the farms were omitted due to privacy issues. Contact with farmers was established mainly through former collaborations with scientific projects (47%) and by producer association referencing (39%). About 35% of the surveyed farms have land within the Natura 2000 Network and 32% produce according to organic production standards. Each farmer was interviewed individually using the survey in Appendix A, which included 119 questions that address general farm information (17 questions) and environmental (46), economic (28) and social (28) dimensions. Examples of general questions were "area of the farm", "legal form" and "years of experience". Specific environmental questions were related to, among other things, indicators of herd characteristics, fertilizers used and energy consumption. The economic component mainly regarded outputs and costs. The social dimension comprised questions such as hours worked per week, work-life balance or succession in the farm. From the total number of sampled farms, 31 answered the social component completely while the other 13 farmers preferred not to do so. In the environmental and economic compartments, in the absence of information or refusal from the farmer to answer, regional and/or representative defaults were applied.

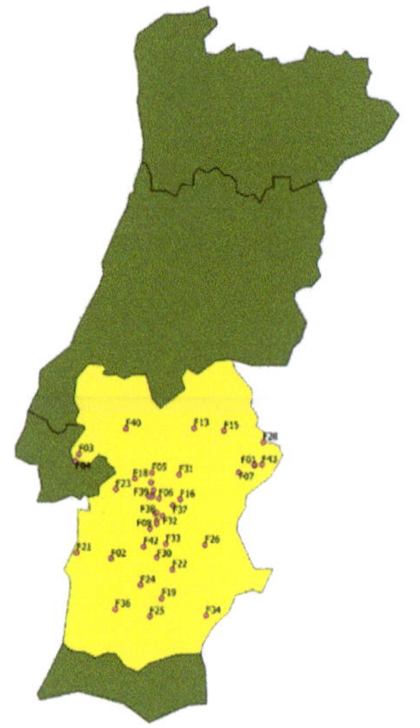

Figure 1. Location of surveyed farms within the Alentejo region (NUTS II), Portugal.

2.2. Portuguese Mapping and Assessment of Ecosystem Services (PT-MAES)

In this study, the ecosystem service (ES) levels were obtained from a PT-MAES assessment considering five ES services. Soil protection (SP) level was estimated through the contribution to reducing soil erosion by comparing it with the worst-case scenario (i.e., land cover that would generate the highest erosion rate at a given point). Soil erosion rates were estimated using the Universal Soil Loss Equation.

Carbon sequestration (CS) was estimated as the balance of gains and losses of carbon in biomass (above and belowground), considering the land use transitions that occurred between 1990 and 2007 (assuming these transitions occurred at a constant rate). Emission/sequestration coefficients were obtained from the Portuguese National Inventory Report (NIR) for greenhouse gases emissions for each land use transition [19].

Support to Extensive Animal Production (SEAP) was quantified and mapped by determining average livestock densities (for calves, dairy cattle and sheep) in pasture areas within the study region, using official national statistics at the Municipality level. As it was impossible to geographically identify pastures where each type of animal production occurs, average livestock density was estimated considering the two main species together (cows and sheep) for the given pasture area in the national statistics for each municipality.

Plant food production (PFP) was assessed based on the establishment of a correspondence between the main crops in the study area with the European Nature Information

System (EUNIS) ecosystems. Once the correspondence with land use classes had been made, the average productivity was estimated for each ecosystem.

Fiber production (FP) was estimated through the mean annual increments of forest trees, as presented in the Portuguese National Inventory Report, deducting biomass losses due to natural mortality.

2.3. Modelling Approaches

To evaluate the ES levels at a farm level and at the entire dataset level, two different approaches were applied. The "class approach" assumed that the surveyed farms were representative of the PBBF of Alentejo and used the average ES level of the entire region for each relevant ecosystem present in the sampled farms, namely grasslands, *dehesa* and croplands. The "buffer approach" used the coordinates of each farm and generated a buffer with a radius of 1 km. From that buffered area, an average ES level for each relevant ecosystem in the farm was computed. A more detailed explanation of each approach is available in the following sections. The two approaches were then subjected to economic valuation and a comparison of results.

2.3.1. Modelling Approach by Average Class (AC)

In PT-MAES, the studied ecosystems are named according to the EUNIS classification system [20]. Only the relevant areas in the context of PBBF, pastures and croplands, were analyzed under this approach. To match these areas from sampled farms with the EUNIS classification, a classification key was produced according to the characteristics of each ecosystem. According to this classification, all pasture areas (natural and sown) were assumed to correspond to the EUNIS class "dry grasslands or *dehesa*" and all croplands (for animal consumption or not) to "arable land and market gardens".

For each ES, PT-MAES presents a scale of service level by ecosystem. This scale was divided by classes, with a correspondent fraction of area that presents each level of service. We computed an average for the overall service level by multiplying each fraction of area with the mid-point value of its class or, in its absence, with the upper value of the service level. The proportion between grassland and *dehesa* presented in PT-MAES was taken into account when computing the average values for the pasture areas of the surveyed farms.

2.3.2. Modelling Approach by Buffer Generation (BG)

In this approach, we combined the location of each farm with PT-MAES data to estimate the level of each ES. The geographic coordinates of each farm were obtained directly during the visit to the farm and/or the interviews. Since real farm limits were unavailable, we generated a buffer with a radius of 1 km around the geographic coordinates of each farm headquarters. We computed the average ES level from this area for each relevant ecosystem.

Data for ES were based on a map with a scale of 1:100,000. To calculate the average ES level in the 1 km buffer area of each farm, a geographic information system software was used (QGIS software version 3.16.2, Open Source Geospatial Foundation Project, Chicago, IL, USA. http://qgis.osgeo.org (accessed on 10 November 2022)).

2.4. Ecosystem Services Economic Valuation

For each ES, different data sources were used to convert environmental benefit/damage into an economic value. For SP, the economic value was 5.03 € per ton of avoided erosion [21]. This value includes sediment and nutrient losses and is a previous replacement cost estimation. Since the reference value was for the year 2007 (SP2007), it was updated to 2020 price levels (SP2020) using the yearly inflation rate between 2007 and 2019 (IFi):

$$SP_{2020}\left(\frac{€}{t}\text{of avoided erosion}\right) = SP_{2007} \times \prod_{i=2007}^{2020} (1 + IF_i). \tag{1}$$

For CS valuation, the most common practice is to apply the social cost of carbon. This term represents the economic cost caused by an additional ton of carbon dioxide emissions or its equivalent and varies according to the applied discount rate, productivity growth and temperature sensitivity [22]. There is no academic consensus about this value [23], as it depends on many variables, assumptions and on the context. Other options are to use the carbon market price or the shadow price. In Table 1, we present the list of values applied for the CS valuation [24].

Table 1. Carbon price estimates based on different explicit carbon pricing mechanisms [24].

Pricing Mechanism	Source	Details	Value €/tCO$_2$eq
Market Price	Daily Carbon Prices (ember-climate.org)	European market quotation for June 2021 (average)	53
Social Cost	[25]	Estimates for Social Cost of Carbon	37.5
Shadow Price	[26]	Upper bound value for the 2030 range estimate	92

For the estimation of the value of support to extensive animal production (SEAP), we used data from regional agricultural accountancy statistics for beef farms [13]. We started by calculating the average rent value per hectare without subsidies or taxes, i.e., 14 €/ha/year. The average added value per ha includes three sources of added value: animal production, vegetal production and other production. Here, we considered the proportion related to animal production (67%) and applied it to the calculated rent value per ha, resulting in 8.72 €/ha of rent due to animal production. As the SEAP service level is measured in terms of LU, we applied the regional value of 0.5 LU/ha to stipulate an overall price of 17.43 €/LU for this ES.

To estimate the value of plant food production, a similar method to that explained for SEAP was applied. Departing from the same rent value, 14 €/ha, here, we used the proportion of added value generated by vegetal production (14%). We estimated a value of 1.89 €/ha of rent due to vegetal production. Considering the average productivity of the relevant crop basket for the region, i.e., 4.91 t/ha [16], we estimated a price of 0.39 € per ton for the valuation of this ES.

For fiber production, we considered the volume of the main three species used for fiber production in the region: eucalyptus, maritime pine and stone pine [27]. We used the same source to infer the regional proportion of each species in terms of biomass volume. We then multiplied the proportion of each species by the respective price (based on the terrain prices). This calculation delivered a regional price of 21.5 € per m3 for the valuation of fiber production.

To perform the valuation, the estimated ES levels (ESlevel) were multiplied by the values explained in the previous section (ESprice) to calculate the ESvalue of each ES, as follows:

$$\mathrm{ES}_{\mathrm{value}}\left(\frac{€}{\mathrm{ha.year}}\right) = \mathrm{ES}_{\mathrm{level}}\left(\frac{i}{\mathrm{ha.year}}\right) \times \mathrm{ES}_{\mathrm{price}}\left(\frac{€}{i}\right), \tag{2}$$

where i represents the relevant unit for each ES.

As mentioned in Section 2.3.1, the farm-level dataset included information about agricultural occupations and respective areas, which were organized into four main groups: natural pastures, sown pastures, crops and other cultures. The first two were assumed to correspond to ecosystem grasslands or *dehesa* (Ag&d), while the other two corresponded to croplands (Acrop). Summing the ESvalue applicable for each ES and ecosystem, we obtained the Total Ecosystem Value (TEV) per ecosystem. It is also possible to estimate the total value of the ES provided by each farm (ESfarm), as follows:

$$\mathrm{ES}_{\mathrm{farm}}\left(\frac{€}{\mathrm{year}}\right) = \mathrm{TEV}_{\mathrm{g\&d}}\left(\frac{€}{\mathrm{ha.year}}\right) \times \mathrm{A}_{\mathrm{g\&d}}(\mathrm{ha}) + \mathrm{TEV}_{\mathrm{crop}}\left(\frac{€}{\mathrm{ha.year}}\right) \times \mathrm{A}_{\mathrm{crop}}(\mathrm{ha}). \tag{3}$$

For the modelling approach of a buffer around farm location, the ES values provided yearly by each farm (ES_{farm}) were obtained directly by the multiplication of the valuation per unit (i) by the average service level of the buffered area:

$$ES_{farm}\left(\frac{\text{\euro}}{\text{year}}\right) = ES_{price}\left(\frac{\text{\euro}}{i}\right) \times ES_{level}\left(\frac{i}{\text{ha.year}}\right) \times A_{farm}(ha). \qquad (4)$$

With the previously estimated values, we performed an extrapolation of the ES value provided by PBBF for the entire Alentejo region. Multiplying the area of permanent pastures in the region reported by national statistics by the calculated TEV (€/ha/year), we estimated the total value of ES provided by PBBF in Alentejo (assuming that all permanent pastures are used by PBBF). Moreover, assuming the same proportion of grasslands and *dehesa* vs. cropland area verified in the farm-level dataset applied to the region, it was possible to estimate the value of ES provided by the area of croplands associated with PBBF by summing the two components:

$$\text{Value of ES on Alentejo}\left(\frac{\text{\euro}}{\text{year}}\right) = TEV_{g\&d}\left(\frac{\text{\euro}}{\text{ha.year}}\right) \times A_{g\&d\_Alentejo}(ha) + TEV_{crop}\left(\frac{\text{\euro}}{\text{ha.year}}\right) \times A_{crop\_PBBF\_Alentejo}(ha). \quad (5)$$

3. Results

This section is divided by subheadings. It is intended to provide a concise and precise description of the experimental results, their interpretation, as well as the experimental conclusions that can be drawn.

3.1. Characterization of the Sampled Farms

Figure 2 presents the sampled area in terms of land use class, ecosystem type and total area per farm. The sampled farms are heterogenous in terms of total area but also in terms of the land use classes. The median area of grasslands is about 602 ha per farm, ranging between about 23 ha (Farm 16) and 3450 ha (Farm 03). Among the sampled farms, 35% (14 farms) only have natural pastures and only 2 have sown pastures, while all others have both pasture systems (60%—24 farms). Farm 03 has the highest area of grasslands, i.e., 3450 ha of sown pasture (this farm does not have natural pasture). Farm 16 only uses natural pastures, and it has also the lowest grassland area among the sampled farms. There are 9 farms with grassland only. The median area of croplands is 33 ha, ranging between zero and 300 ha (Farm 22). Among the farms with cropland, about 87% (16 farms) have crops for animal consumption, and only 40% (13 farms) have other plant production.

The land use class with the largest variation (interquartile distance) in area between farms is the natural pastures class, i.e., about 270 ha, while the class with the lowest variation is the other plant production class, with only 10 ha. The interquartile distances of sown pastures and crops for animal consumption are 165 ha and 61 ha, respectively. Due to the higher variation of both pasture systems in comparison with the variation of crops production classes, the aggregated land use classes of grassland have a significantly higher variation than of croplands, with interquartile distances of 288 ha and 67 ha, respectively.

Regarding the complete set of sampled farms, grasslands and *dehesa* classes account for 91% of the total area of the dataset (about 22,370 ha). Cropland accounts for 9% (about 2232 ha). Within grasslands, natural pastures have a higher proportion than sown pastures, i.e., 66% (14,757 ha) and 34% (7613 ha), respectively. In croplands, crops for animal consumption represent 86% of the area (1998 ha), while other productions represent only 14% (234 ha).

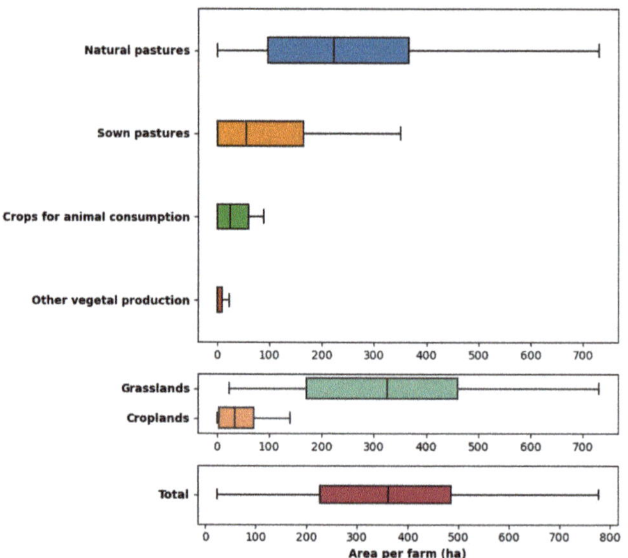

Figure 2. Areas in the sample for each land use class, ecosystem type (grassland and cropland), and total area per farm. "Grassland" area is the sum of "Natural pastures" and "Sown pastures", and "Croplands" is the sum of "Crops for animal consumption" and "Other vegetal production".

3.2. Ecosystem Services Valuation

Table 2 presents the average ES level per unit of area for the two modelling approaches. In general, the buffer approach leads to higher values than the class approach. The only exception is for SEAP in the grassland and *dehesa* land use (−0.1 LU/ha). The most significant difference occurs in the SP service. For the grassland and *dehesa* land use, the difference is 20 t/ha/year (buffer approach: 34.5 t/ha/year; class approach: 24.5 t/ha/year). For croplands, the difference is 6.4 t/ha/year (buffer approach: 34.5 t/ha/year; class approach: 28.1 t/ha/year). The difference between modelling approach for other ES is not significant. For example, excluding SP, the highest difference is only 1.8 t/ha in PFP in croplands. The two modelling approaches also lead to equal ES levels for 3 out 5 ES for grasslands (CS, PFP and FP) and 1 out of 5 for croplands, i.e., SEAP.

Table 2. Average ecosystem service level per unit of area for the two modelling approaches.

Ecosystem Service	Units	Approach by Average Class		Approach by Buffer Generation	
		Grasslands and *Dehesa*	Croplands	Grasslands and *Dehesa*	Croplands
Soil protection (SP)	t/ha/year	28.1	14.5	34.5	34.5
Carbon sequestration (CS)	t C/ha/year	−0.2	−0.6	−0.2	−0.3
Support to extensive animal production (SEAP)	LU/ha	0.4	0.0	0.3	0
Plant food production (PFP)	t/ha	0.0	1.5	0	3.3
Fiber production (FP)	m³/ha	0.4	0.0	0.4	0.2

Analysing the results for PBBF per ha (Table 3), the average of the total ES valuation (sum of all ES) is different for the two modelling approaches. The valuation is nearly 34% higher with the buffer generation approach (131 €/ha vs. 175 €/ha). Observing each ES separately, SP dominates the total ES valuation for both approaches: 131.7 €/ha using the class approach and 173.4 €/ha using buffers. CS is the only ES with a negative average valuation in both modelling approaches (class: −13.7 €/ha; buffer: −12.2 €/ha, applying the market price). The average value for SEAP in the two modelling approaches is similar between approaches, but the minimum estimated value differs significatively (class: 2.9 €/ha; buffer: 0 €/ha). PFP present very low values under both approaches, i.e., between 0 €/ha an 0.03 €/ha in the approach by class and between 0 €/ha and 1.3 €/ha in the second method. For FP, both approaches present similar average values (class: 7.4 €/ha; buffer: 8.3.6 €/ha), but the buffer approach presents a much higher variation, ranging between 0 €/ha and 53.1 €/ha, while under the class approach, it ranges between 4.1 €/ha and 8.5 €/ha).

Table 3. Ecosystem services valuation in €/ha for the PBBF of the dataset.

Ecosystem Service	Average Value Per ha		Minimum Value Per ha		Maximum Value Per ha	
	AC	BG	AC	BG	AC	BG
SP	131.7 €	173.4 €	104.1 €	18.4 €	141.3 €	1153.9 €
CS	−13.7 €	−12.2 €	−22.6 €	−54.2 €	−10.5 €	86.8 €
SEAP	5.5 €	5.1 €	2.9 €	0.0 €	6.4 €	8.8 €
PFP	0.1 €	0.2 €	0.0 €	0.0 €	0.3 €	1.3 €
FP	7.4 €	8.3 €	4.1 €	0.0 €	8.5 €	53.1 €
Total	131.0 €	174.9 €	88.8 €	16.4 €	145.8 €	1166.1 €

SP—Soil protection; CS—Carbon sequestration; SEAP—Support to extensive animal production; VFP—Vegetal food production; FP—Fiber production; AC—Approach by average class; BG—Buffer generation.

Figure 3 presents the average, minimum and maximum valuations per ES at the farm level, as well as the totals resulting from the sum of the individual valuations of all ES within each farm. SP is clearly the ES that contributes the most to ES valuation at the farm level, i.e., 99% for the class approach and 102% for the buffer approach (over 100% due to the negative contribution of CS). CS presents a negative value for all farms under the approach by class, but under the buffer approach, 10 out of the 40 farms present a positive value up to a maximum of 8794 € for a single farm. SEAP has a similar proportion in both approaches (4% for both). On average, PFP only contributes 0.04% and 0.08% to the total valuation at the farm level for the class and buffer approaches, respectively. The minimum value for PFP is 0 €/farm for both approaches, but the maximum per farm is more than double that of the buffer approach (170 €/ha vs. 460 €/ha). The inverse situation occurs with FP, representing 6% of the total ES value in the class approach compared to only 4% in the buffer approach. Despite these differences, the average total value of the ES provided per farm is similar for both methods, i.e., 81,719 € for the class approach and 86,663 € with the buffer approach. The total value of ES provided by a single PBBF ranges between 3352 € and 502,923 € in the first method and between 928 € and 851,271€ in the second.

Table 4 presents the sum of ES valuations for the entire dataset considered in this work. SP is by far the most valuable ES provided by PBBF, accounting for almost the total value in both methods. SEAP represents 4% of the ES value in both approaches, being the second highest value with the buffer approach but the third using the class approach, in which FP presents the second highest value. PFP represents a very low value, nearly 0%, for both approaches. CS has a negative effect in both methods, from 6% to 18% of the total dataset ES value, depending on the considered carbon valuation. The total ES value attributable to the PBBF of the dataset adds up to 3,205,429 € applying the class approach and 3,466,401 € using buffers (applying carbon market prices).

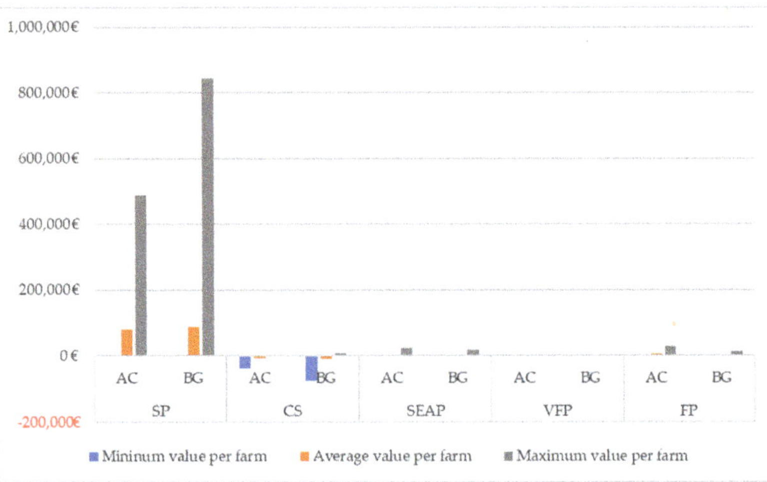

Figure 3. Graphical comparison of the economic values of each ecosystem service per farm. SP—Soil protection; CS—Carbon sequestration; SEAP—Support to extensive animal production; VFP—Vegetal food production; FP—Fiber production; AC—Approach by average class; BG—Buffer generation.

Table 4. Total ecosystem services value for all farms in the dataset on aggregate, using the class (AC) and buffer (BG) approaches.

Ecosystem Service	Approach by Average Class (AC)		Approach by Buffer Generation (BG)	
	AC	% of Total by AC	BG	% of Total by BG
Soil protection (SP)	3,245,142 €	99%	3,548,257 €	102%
Carbon sequestration (CS), market price	−304,189 €	(9%)	−342,031 €	(10%)
Support to extensive animal production (SEAP)	139,423 €	4%	132,882 €	4%
Plant food production (PFP)	1301 €	0%	2707 €	0%
Fiber production (FP)	123,752 €	6%	124,585 €	4%
Total	3,205,429 €	100%	3,466,401 €	100%
Carbon sequestration (CS), social cost	−215,228 €	(6%)	−242,003 €	(7%)
Carbon sequestration (CS), shadow price	−528,025 €	(17%)	−593,714 €	(18%)

3.3. Extrapolation for the Regional Level

The TEV for each ecosystem is presented in Table 5. The TEV calculated for grasslands and *dehesa* lies between 146 €/ha/year and 176 €/ha. Here, the estimated values are mainly due to SP, followed by FP and then by SEAP. The TEV calculated for croplands lies between 41 € and 166 €. This wide range is due to the fact that the estimated ES values differ a lot across approaches, with the most significant difference occurring in SP. While in the class approach SP is valued at 73 €/ha, this value is 138% higher according to the buffer approach, which yielded 173 €/ha. For croplands in the region, PFP is almost irrelevant with both approaches (between 0.6 €/ha and 1.3 €/ha). At the TEV level, CS continues to present a negative contribution in both modelling approaches and ecosystems.

From national statistics, the area of permanent pastures in the region is 1,151,238 ha [13]. If this area corresponds to grasslands and *dehesa*, and the proportion of land uses verified on the dataset applies, there is an estimated area of 112,305 ha of croplands associated with PBBF in Alentejo. Taking these areas into consideration and applying them together Equation (5) with the TEV presented in Table 5, we arrive at the regional results presented in Figure 4, for the class approach. The results for BG are obtained directly from each ES value multiplied by the total area considered for Alentejo.

Table 5. Total Ecosystem Value (in euros—€/ha/year) per considered ecosystem.

Ecosystem Service	Approach by Average Class		Approach by Buffer Generation	
	Grasslands and *Dehesa*	Croplands	Grasslands and *Dehesa*	Croplands
Soil protection (SP)	141.3 €	72.8 €	173.4 €	173.4 €
Carbon sequestration (CS), market price	−10.5 €	−32.8 €	−12.0 €	−13.3 €
Support to extensive animal production (SEAP)	6.4 €	-	6.0 €	-
Plant food production (PFP)	-	0.6 €	-	1.3 €
Fiber production (FP)	8.6 €	0.3 €	8.7 €	4.8 €
Total ecosystem value	**145.8 €**	**40.8 €**	**176.0 €**	**166.2 €**
Carbon sequestration (CS), social cost	−7.5 €	−23.2 €	−8.5 €	−9.4 €
Carbon sequestration (CS), shadow price	−18.3 €	−57.0 €	−20.8 €	−23.0 €

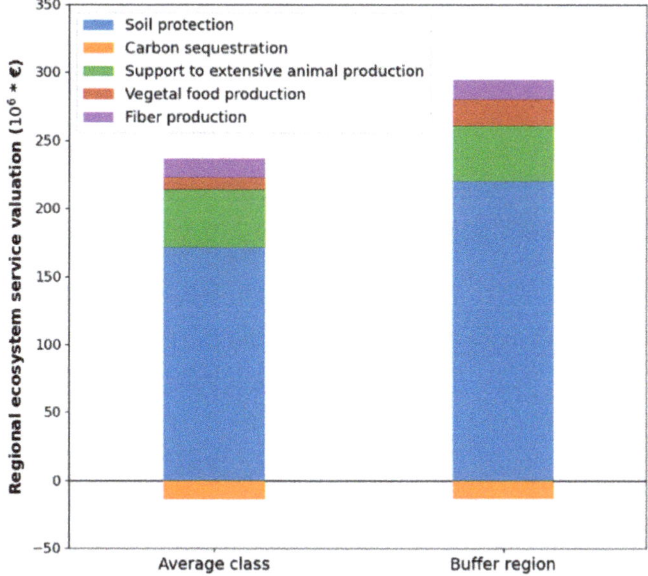

Figure 4. Estimated value of ecosystem services (€) from pasture-based beef farming in Alentejo after regional extrapolation for both modelling approaches.

Figure 4 depicts the total ES estimated value promoted by PBBF in Alentejo according to both modelling approaches. SP is the most valuable service, accounting for around 171 M€ and 220 M€ with class and buffer approaches, respectively. FP is the second most valuable ES, accounting for nearly 9.9 M€ in the first approach and 10.5 M€ in the second. It is followed by SEAP, that accounts for nearly 7.4 M€ in the class approach but just 6.9 M€ with the buffer approach. PFP presents a residual value of 68 K€ in the class approach and 154 K€ in the buffer approach. CS present similar negative values with the two methods, with values between −16 M€ and −15.5 M€. In total regional terms, ES provided by PBBF are estimated to be worth around 173 M€ using the class approach and around 223 M€ using the buffer approach (nearly 29% more than class approach).

4. Discussion

In both modelling approaches, SP is the ES that contributes most to the total value. This means that the main positive effect of animal production in Alentejo is soil protection, which is especially relevant, given that this is a region prone to desertification. SP is followed by FP, SEAP and then PFP. Concerning SP, land cover characteristics are, in general, very satisfactory in the territories where PBBF develop their activity, so the high service levels obtained are aligned with expectations based on knowledge and experience from the field. In the approach by class, SP presented significant differences between grasslands/*dehesa* ecosystems and cropland ecosystems (28.1 vs. 14.5 t/ha). The buffer approach considers the exact location of each farm and thus the local characteristics of the terrain in terms of soil and slope; accordingly, it returned a higher level of service of 34.5 t/ha of avoided erosion for both grasslands and croplands. These results suggest that in the studied sample of PBBF, both grasslands/*dehesa* and croplands provide overall higher levels of SP than grasslands/*dehesa* and croplands in general in the region.

The SEAP service is influenced by the average livestock density of cattle and sheep calculated at the municipality level, so it does not represent the real service level at the farm's location. A limitation of this indicator is that it does not take into account the proportion of livestock that is effectively in extensive production (or not). Nevertheless, in regional aggregated terms, it can be considered as a useful indicator, since the predominant animal production system in Alentejo is extensive. The values estimated for grasslands and *dehesas* are very similar across approaches, i.e., 0.4 and 0.3 LU per ha, both of which are below the regional average of 0.5 LU per ha, which suggests a lower livestock density in PBBF than in the remaining cattle and sheep production systems. The valuation applied for this ES is one of the main innovations of this work, since it intends to avoid double counting, a common mistake made in many similar studies. By using the rent values per ha (without subsidies) and associating it with the corresponding fraction of the animal production output, it is possible to ensure that the valuation refers exclusively to the added value provided by the land and its intrinsic characteristics. Other types of valuations usually fail to separate the added value coming from the different production factors as capital and labor, which can lead to double counting. There would also be another possible approach for this ES valuation, based on willingness to pay (WTP). In this case, there is a reference to an additional WTP for beef from pasture of 1.5 € per kg of meat [5]. However, that estimation was obtained in a study that measured WTP for a bundle of management practices. It is impossible to distinguish the fraction of that WTP that is specifically attributable to the support of extensive animal production, among the other benefits that beef from pasture may provide, which could lead to an overcounting problem.

Concerning PFP, the service levels obtained come from an aggregated regional basket of different crops. In terms of grasslands and *dehesa*, the PFP service level is zero, regardless of the approach. For croplands, the approach by class presented an overall service level of 1.5 t/ha, and the result through buffer generation was 3.3 t/ha. This evidences that croplands near the studied PBFF present a higher productivity than croplands in general in Alentejo. As many of the considered crops from the reference basket are not usually representative of PBBF, one improvement for future work would be to refine the crop basket to make it more representative of the most common crops associated with PBBF. The valuation technique applied here was the same as in SEAP, but in this case, it considered the fraction of vegetal production output.

For FP, the estimated service levels are 0.4 m^3/ha in grasslands and *dehesa* for both approaches. Using the class approach for croplands, there is a null service level (0 m^3/ha), while for the buffer approach, the same ecosystem has a service level of 0.2 m^3/ha. This suggests that croplands associated with PBBF territories present a higher service level. The FP service level was estimated directly through the mean annual increments of forests trees, deducting biomass losses due to natural mortality. The monetary value only considers the price of the main trees explored for timber, in regional terms. This means that all the other forest products were not considered in this indicator. In the specific case of the

Alentejo region, this could be a relevant limitation, since nearly 71% of the forest area is devoted to alternative productions, such as cork (very economically important), feed for autochthonous pigs, etc. Increasing the scope of the FP service in the future could be an interesting way for better accountancy of the ES provided by the forest fraction on PBBF. For now, this was impossible due to a lack of regionalized data on the other products from forests.

CS is the only ES that presents a negative contribution for the ES total value under both approaches. The CS for croplands was estimated at -0.6 tC/ha using the class approach and at -0.3 tC/with the buffer approach. For grasslands and *dehesa*, the obtained service levels are 0.2 tC/ha for both approaches, suggesting that the areas covered by these ecosystems emit small quantities of carbon instead of sequestering it, as could be expected. These results can be explained by the fact that the method for carbon sequestration calculations only considers land use change transitions as reported in the Portuguese NIR [19]. This means that any other aspects concerning the territory characteristics and/or specific practices that are applied in PBBF to promote carbon sequestration were not considered for this service level calculation. This limitation derives mainly from the high heterogeneity that characterizes PBBF and the territories where they develop their activity.

In this study, the average value of ES of grasslands was estimated as 146 €/ha/year (class approach) or 176 €/ha/year (buffer approach). Those are relatively low estimates compared with those in the relevant literature. For example, [28] estimated a global ES value for grasslands of 232 €/ha/year. Some other studies performed at a macro level estimated values of ES from grasslands of between 249 €/ha/year and 2352 €/ha/year [29], but in this case accounting for additional ES such as water treatment, supply and purification, gene pool protection (conservation) and cultural services. Finally, a study also based on PT-MAES data presented a value of 331 €/ha/year for permanent pastures in a natural park in Portugal [30]. Here, the different values are mostly explained by the different valuation methods applied.

Extrapolating those results to the entire region, the total estimated value for the considered ES provided by PBBF in Alentejo would amount to 173 or 223 M€, depending on the approach. The relevant difference between approaches suggests that the areas around the studied PBBF generate a higher overall ES value, when comparing with the same ecosystem areas in general. Here, it is necessary to point out that the total area considered for regional extrapolation does take in account pastures under tree cover, a relevant fraction of the area of the region [31]. Consideration of these areas would probably increase the overall ES levels and value.

According to collected data (and farmers' perspectives), there is no relevant PFP or FP in most of the surveyed farms, so the importance of these ES, as well as the relevance of SP, which is usually not perceived, can be considered another of the most surprising outcomes of this work. The absence of biodiversity indicators in this study is also notable, since many of the concerned areas present high nature value [32]. These limitations probably make the presented values fall short of the real value of ES provided by PBBF in Alentejo. The inclusion of biodiversity also likely made the valuation of ES in this study increase to levels like those cited in the literature. Nevertheless, comparing the scale of values obtained with the regional extrapolation—comprising nearly half of the national output of the beef cattle sector [13]—indicates the relevance of the value generated by the studied ES.

A potential future improvement in the continuation of this work would be to match the buffers generated to the correspondent real area of each farm. In fact, any in situ measurements that could be taken would improve the reliability of the results. From the farmers' perspective, it could be interesting to assess the main ES provided in each farm, especially if with that, the farmer could valorize his/her production and/or apply better management practices. When designing policies and incentives, the farm-size related particularities should also be considered to ensure equity and parity across the sector. It could also be argued that despite the generated ES, PBBF have significant environmental impacts that are also not valued by the market. This is the case particularly for beef systems with methane and nitrous oxide emissions from enteric fermentation and manure management. A valuation of those emissions was beyond the scope of this work but should be carried out in the future and compared with the ES value obtained here.

5. Conclusions

The goal of this paper was to estimate the value of five ecosystem services provided by the areas covered by pasture-based beef farms in Alentejo. We obtained values for those ecosystem services of 146–176 €/ha/year for grasslands and *dehesa* (*montado*) and 41–166 €/ha/year for croplands. These results were robust to methodological choices, as we used two options for joining pixel-level ecosystem services valuations with farm-level production data. The two approaches provided similar results for the five studied services. Soil protection was the most valuable service, with the other studied services making significantly lower contributions. This led us to conclude that the main benefit of animal production for ecosystems in the region is the avoidance of soil loss. This result is particularly significant in Alentejo, which a highly desertified part of the country and of Europe. Regional extrapolation allowed us to estimate the overall value of ecosystem services associated with pasture-based beef farms in Alentejo: between 173 M€ and 223 M€, i.e., almost half of the national production value of the beef cattle sector. Thus, our results suggest that the maintenance of grasslands and croplands in pasture-based beef farms in Alentejo generates positive externalities for society. We therefore conclude that this production system is very important for the region in terms of the value generated by the studied ecosystem services. This work explored innovative ways of valuating ecosystem services, presenting two methodologies based on data available at the European level and a valuation method that avoids double counting. The present research is intended to address to the growing need for ecosystem services accounting in environmental and sustainability studies. Future studies should compare the positive contributions of the systems quantified here with the overall environmental impacts of the animal production systems in Alentejo. Starting with the conclusions of this study about the importance of pasture-based beef farms systems in the region, further work should also explore and engineer innovative and creative ways of maximizing the positive effects of those systems, as well as matching remuneration and incentives to these farmlands with the true value they generate for society.

Author Contributions: Conceptualization, M.P.d.S. and R.F.M.T.; methodology, M.P.d.S., T.G.M. and T.D.; formal analysis, data curation and writing—original draft preparation, M.P.d.S.; writing—review and editing, R.F.M.T., T.G.M. and T.D.; supervision, R.F.M.T. and T.D.; project administration, R.F.M.T.; funding acquisition, R.F.M.T. and T.D. All authors have read and agreed to the published version of the manuscript.

Funding: This work was supported by Fundação para a Ciência e Tecnologia through project "LEAnMeat—Lifecycle-based Environmental Assessment and impact reduction of Meat production with a novel multi-level tool" (PTDC/EAM-AMB/30809/2017), and by grants SFRH/BD/147158/2019 (dos Santos, M.P.) and CEECIND/00365/2018 (Teixeira, R.). The work was also supported by FCT/MCTES (PIDDAC) through project LARSyS—FCT Pluriannual funding 2020–2023 (UIDB/EEA/50009/2020).

Data Availability Statement: Not applicable.

Conflicts of Interest: The authors declare no conflict of interest. The funders had no role in the design of the study; in the collection, analyses, or interpretation of data; in the writing of the manuscript; or in the decision to publish the results.

Appendix A

Table A1. Animal future Questionnaire items.

General Information	Social Compartment	Economic Compartment	Animal Compartment
Own Land (ha)	Number Of Working Hours Rate	Total Output Vegetal	LivestockType (Cattle/Sheep)
Rented Land (ha)	Weekends Off	Total Output Animal	Categor (per age)
Legal Form (Individual/Company/Others)	Sundays Off	Total SpecificCosts	Average Number (No.)
Farmer Since (years of experience)	Days Off Holidays	Total Farming Overheads	Number Of Produced Animals (No.)
Family Labour (hours/year)	Workload Rate	Taxes	Age At End of Fattening (months)
Hired Labour (hours/year)	Hazardous Chemicals	Total Subsides On Crops	Number Of Sold Animals (No.)
Conventional or Organic	Physical Work	Total Subsides On Livestock	Live Weight At Sale (kg)
Area In Natura 2000 (ha)	Overwork Stress Rate	Total Support For RD	**Diet Compartment**
Area In Conservation Land No Natura 2000 (ha)	Activities Besides Farm	Decoupled Payments	Diet name (for all herd/fattenning/others)
Area Under Agro Environmental Measures (ha)	Work Life Balance Rate	Depreciation	Diet component name (ex: hay)
Area Of Specific Natural Habitat (ha)	Working Atmosphere Rate	Wages Paid	Diet Component Quantity (kgs/animal/year)
Overall Satisfaction being a Farmer (1–5)	Farm Economically Viable In 10 ys	Rent Paid	Protein Content Fraction (0–1)
Other Information	Expectation Farm Succession	Interests Paid	Considered number of animals (No.)
Outdoor Access Animals (y/n)	If Over 45 Succession Expected	Total Assets	Diet Component Quantity (kgs/farm/year)
Days Outdoor (0–365)	Training Days—Family Workers	Total Assets ExclLand	**Crop Areas Compartment**
Additional Enrichment (y/n)	Training Days—Staff	Liabilities	Surface (ha)
Animal Social Contact (y/n)	Highest Educational Degree	Own Labour Force Persons	Yield (t/ha)
Non Curative Treatments (y/n.)	Highest Agricultural Educational Degree	Own labour Force Hours	Fraction Of Legumes (0–1)
Resources Utilization	Public Access To Farm	Hired Labour Force Persons	Dry Matter Content (0–1)
Diesel Consumption (l)	Visits To Farm	Hired Labour Force Hours	**Crop Protection Agents**
Electricity Consumption (kw)	Professional Organisations Besides Farm	Imputed Labour Costs	Type (natural/artificial)
Renewable Energy Fraction (0–1)	What Professional Organisations Besides Farm	Rented Farm Land	Name
Irrigated Area (ha)	Non Professional Organisations Besides Farm	Rental For Farm Land	Quantity (m^3 or t/ha)
Water Use Animals (m^3)	Direct Selling or Tasting	Own Farm Land	**Fertilizers**
Water Use Irrigation (m^3)	Labelling Schemes	Interest Rate	Type (natural/artificial)
Water Use Total (m^3)	What Labelling Schemes	Profit	Total Quantity (m^3 or t/ha)
	Other Activities On Farm	Total Subsidies	N content (0–1)
	What Other Activities On Farm	Total hours	

References

1. Wu, G.; Fanzo, J.; Miller, D.D.; Pingali, P.; Post, M.; Steiner, J.L.; Thalacker-Mercer, A.E. Production and Supply of High-Quality Food Protein for Human Consumption: Sustainability, Challenges, and Innovations. *Ann. N. Y. Acad. Sci.* **2014**, *1321*, 1–19. [CrossRef] [PubMed]
2. Hocquette, J.F.; Ellies-Oury, M.P.; Lherm, M.; Pineau, C.; Deblitz, C.; Farmer, L. Current Situation and Future Prospects for Beef Production in Europe—A Review. *Asian-Australas. J. Anim. Sci.* **2018**, *31*, 1017–1035. [CrossRef] [PubMed]
3. Accatino, F.; Tonda, A.; Dross, C.; Léger, F.; Tichit, M. Trade-Offs and Synergies between Livestock Production and Other Ecosystem Services. *Agric. Syst.* **2019**, *168*, 58–72. [CrossRef]

4. Herrero, M.; Havlík, P.; Valin, H.; Notenbaert, A.; Rufino, M.C.; Thornton, P.K.; Blümmel, M.; Weiss, F.; Grace, D.; Obersteiner, M. Biomass Use, Production, Feed Efficiencies, and Greenhouse Gas Emissions from Global Livestock Systems. *Proc. Natl. Acad. Sci. USA* **2013**, *110*, 20888–20893. [CrossRef] [PubMed]

5. Teixeira, R.F.M.; Proença, V.; Crespo, D.; Valada, T.; Domingos, T. A Conceptual Framework for the Analysis of Engineered Biodiverse Pastures. *Ecol. Eng.* **2015**, *77*, 85–97. [CrossRef]

6. Bugalho, M.N.; Caldeira, M.C.; Pereira, J.S.; Aronson, J.; Pausas, J.G. Mediterranean Cork Oak Savannas Require Human Use to Sustain Biodiversity and Ecosystem Services. *Front. Ecol. Environ.* **2011**, *9*, 278–286. [CrossRef] [PubMed]

7. Costanza, R.; D'Arge, R.; de Groot, R.; Farber, S.; Grasso, M.; Hannon, B.; Limburg, K.; Naeem, S.; O'Neill, R.V.; Paruelo, J.; et al. The Value of the World's Ecosystem Services and Natural Capital. *Nature* **1997**, *387*, 253–260. [CrossRef]

8. World Resources Institute. Millennium Ecosystem Assessment Ecosystems and Human Well-Being: Opportunities and Challenges for Business and Industry. In *Millennium Ecosystem Assessment (MEA)*; World Resources Institute: Washington, DC, USA, 2005.

9. Escobedo, F.J.; Kroeger, T.; Wagner, J.E. Urban Forests and Pollution Mitigation: Analyzing Ecosystem Services and Disservices. *Environ. Pollut.* **2011**, *159*, 2078–2087. [CrossRef] [PubMed]

10. Lynch, D.H.; Sumner, J.; Martin, R.C. Framing the Social, Ecological and Economic Goods and Services Derived from Organic Agriculture in the Canadian Context. In *Organic Farming, Prototype for Sustainable Agricultures: Prototype for Sustainable Agricultures*; Springer: Dordrecht, The Netherlands, 2014; ISBN 9789400779273.

11. Li, C.; Zheng, H.; Li, S.; Chen, X.; Li, J.; Zeng, W.; Liang, Y.; Polasky, S.; Feldman, M.W.; Ruckelshaus, M.; et al. Impacts of Conservation and Human Development Policy across Stakeholders and Scales. *Proc. Natl. Acad. Sci. USA* **2015**, *112*, 7396–7401. [CrossRef] [PubMed]

12. Marta-Pedroso, C.; Domingos, T.; Mesquita, S.; Capelo, J.; Gama, I.; Laporta, L.; Alves, M.; Proença, V.; Canaveira, P.; Reis, M. (Eds.) *Mapeamento e Avaliação Dos Serviços de Ecossistema Em Portugal—Relatório Final*; Instituto Superior Técnico: Lisboa, Portugal, 2014.

13. GPP Estrutura Das Explorações Agrícolas Diagnóstico. 2019. Available online: https://www.gpp.pt/images/DiagnosticoEstruturalExploraesAgricolas.pdf (accessed on 10 November 2022).

14. Marques, C.; Carvalho, M. A Agricultura E Os Sistemas De Produção Da Região Alentejo De Portugal: Evolução, Situação Atual E Perspectivas. *Rev. De Econ. E Agronegócio* **2017**, *15*, 439. [CrossRef]

15. Batista, T.; de Mascarenhas, J.M.; Mendes, P. Montado's Ecosystem Functions and Services: The Case Study of Alentejo Central-Portugal. *Probl. Landsc. Ecol.* **2017**, *44*, 15–27.

16. INE. *Estatísticas Agrícolas 2018*; Instituto Nacional de Estatística, I.P.: Lisboa, Portugal, 2019; ISBN 9789892504957.

17. GPP Rede de Informação de Contabilidades Agrícolas. 2017. Available online: https://www.gpp.pt/images/Estatisticas_e_analises/SistemasInfornacao/RICA/PubRICA_2017.pdf (accessed on 1 November 2022).

18. Dos Santos Carreira, E.R. Eficácia de Modalidades de Recria/Engorda Em Bovinos de Carne. Master's Thesis, Universidade de Évora, Évora, Portugal, 2016.

19. Costa Pereira, T.; Amaro, A.; Borges, M.; Silva, R.; Seabra, T.; Caveira, P. *Portuguese National Inventory Report on Greenhouse Gases, 1990–2020*; Portuguese Environment Agency: Amadora, Portugal, 2022.

20. Davies, C.E.; Moss, D.; Hill, O.M. EUNIS Habitat Classification Revised 2004. Report to: European Environment Agency-European Topic Centre on Nature Protection and Biodiversity. 2004, pp. 127–143. Available online: https://inpn.mnhn.fr/docs/ref_habitats/Davies_&_Moss_2004_EUNIS_habitat_classification.pdf (accessed on 2 November 2022).

21. Marta-Pedroso, C.; Domingos, T.; Freitas, H.; de Groot, R.S. Cost-Benefit Analysis of the Zonal Program of Castro Verde (Portugal): Highlighting the Trade-off between Biodiversity and Soil Conservation. *Soil Tillage Res.* **2007**, *97*, 79–90. [CrossRef]

22. Nordhaus, W.D. Revisiting the Social Cost of Carbon. *Proc. Natl. Acad. Sci. USA* **2017**, *114*, 1518–1523. [CrossRef] [PubMed]

23. Moore, F.C.; Baldos, U.; Hertel, T.; Diaz, D. New Science of Climate Change Impacts on Agriculture Implies Higher Social Cost of Carbon. *Nat. Commun.* **2017**, *8*, 1–8. [CrossRef] [PubMed]

24. Laporta, L.; Domingos, T.; Marta-Pedroso, C. It's a Keeper: Valuing the Carbon Storage Service of Agroforestry Ecosystems in the Context of CAP Eco-Schemes. *Land Use Policy* **2021**, *109*, 105712. [CrossRef]

25. Rose, S.K.; Diaz, D.B.; Blanford, G.J. Understanding the Social Cost of Carbon: A Model Diagnostic and Inter-Comparison Study. *Clim. Chang. Econ.* **2017**, *8*, 1750009. [CrossRef]

26. World Bank Report of the High-Level Commission on Carbon Prices. 2017. Available online: https://static1.squarespace.com/static/54ff9c5ce4b0a53decccfb4c/t/59b7f2409f8dce5316811916/1505227332748/CarbonPricing_FullReport.pdf (accessed on 2 November 2022).

27. Instituto da Conservação da Natureza e das Florestas (ICNF) 6; Inventário Florestal Nacional (IFN6). *2015 Relatório Final*; Instituto da Conservação da Natureza e das Florestas: Lisboa, Portugal, 2019.

28. Costanza, R.; de Groot, R.; Sutton, P.; van der Ploeg, S.; Anderson, S.J.; Kubiszewski, I.; Farber, S.; Turner, R.K. Changes in the Global Value of Ecosystem Services. *Glob. Environ. Change* **2014**, *26*, 152–158. [CrossRef]

29. De Groot, R.; Kumar, P. Estimates of monetary values of ecosystem services. In *Ecological and Economic Foundation*; EarthScan: London, UK, 2010; pp. 367–402.

30. Marta-Pedroso, C.; Laporta, L.; Gama, I.; Domingos, T. Economic Valuation and Mapping of Ecosystem Services in the Context of Protected Area Management (Natural Park of Serra de São Mamede, Portugal). *One Ecosyst.* **2018**, *3*, e26722. [CrossRef]

31. Surová, D.; Pinto-Correia, T.; Marušák, R. Visual Complexity and the Montado Do Matter: Landscape Pattern Preferences of User Groups in Alentejo, Portugal. *Ann. For. Sci.* **2014**, *71*, 15–24. [CrossRef]

32. Ferraz-de-Oliveira, M.I.; Azeda, C.; Pinto-Correia, T. Management of Montados and Dehesas for High Nature Value: An Interdisciplinary Pathway. *Agrofor. Syst.* **2016**, *90*, 1–6. [CrossRef]

 *land*

Article

Ecological Niche Modelling and Potential Distribution of *Artemisia sieberi* in the Iranian Steppe Vegetation

Hamidreza Mirdavoudi [1,*], Darush Ghorbanian [2], Sedigheh Zarekia [3], Javad Miri Soleiman [4], Mashaalaah Ghonchepur [5], Eileen Mac Sweeney [6] and Andrea Mastinu [6,*]

[1]   Forests and Rangelands Research Department, Markazi Agricultural and Natural Resources Research and Education Center, AREEO, Arak 38135, Iran

[2]   Forests and Rangelands Research Department, Semnan Agricultural and Natural Resources Research and Education Center, AREEO, Semnan 35131, Iran

[3]   Forests and Rangelands Research Department, Yazd Agricultural and Natural Resources Research and Education Center, AREEO, Yazd 89158, Iran

[4]   Forests and Rangelands Research Department, North Khorasan Agricultural and Natural Resources Research and Education Center, AREEO, Bojnord 94714, Iran

[5]   Forests and Rangelands Research Department, Kerman Agricultural and Natural Resources Research and Education Center, AREEO, Kerman 78514, Iran

[6]   Department of Molecular and Translational Medicine, Division of Pharmacology, University of Brescia, Viale Europa 11, 25123 Brescia, Italy

*   Correspondence: h.mirdavoudi@areeo.ac.ir (H.M.); andrea.mastinu@unibs.it (A.M.)

Citation: Mirdavoudi, H.; Ghorbanian, D.; Zarekia, S.; Soleiman, J.M.; Ghonchepur, M.; Sweeney, E.M.; Mastinu, A. Ecological Niche Modelling and Potential Distribution of *Artemisia sieberi* in the Iranian Steppe Vegetation. *Land* **2022**, *11*, 2315. https://doi.org/10.3390/land11122315

Academic Editors: Michael Vrahnakis, Yannis (Ioannis) Kazoglou and Manuel Pulido Fernádez

Received: 26 November 2022
Accepted: 13 December 2022
Published: 16 December 2022

Publisher's Note: MDPI stays neutral with regard to jurisdictional claims in published maps and institutional affiliations.

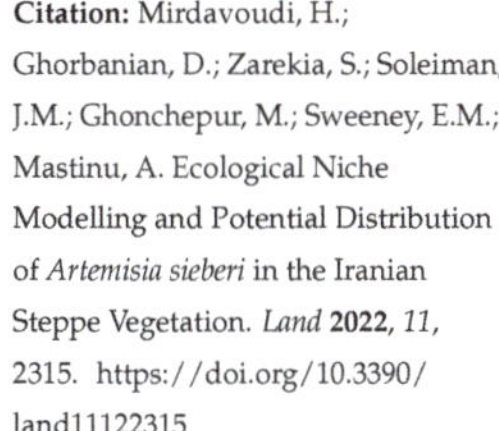

Abstract: *Artemisia sieberi* Besser occurs in many parts of the Irano-Turanian floristic region, which is mostly distributed throughout the Iranian plateau, especially in Iran. This study aimed to identify the effect of the soil and topography variables on *A. sieberi* distribution. We used canopy cover data to fit models using generalized additive models (GAMs). The results showed that the response pattern of *A. sieberi* along with the gradient of soil clay, soil saturation moisture, soil nitrogen and soil acidity followed the monotonic increase model, and its canopy cover percentage augmented by increasing the values of the factors. Conversely, the *A. sieberi* canopy cover percentage decreased by increasing the amount of soil sand, bare soil and the geographic aspect. The *A. sieberi* responses are in contrast to the niche theory. The relationship between the species response pattern and the gradients of soil silt, soil salinity, lime percentage, organic carbon, altitude, land slope, litter, gravel percentage, stone percentage, mean annual precipitation and mean annual temperature followed a unimodal model (consistent with the niche theory). The optimal growth limits for these factors were 32%, 1.75 ds/m, 35%, 1.3%, 2000 m, 43%, 10%, 32%, 250 mm and 15 °C, respectively. Our results highlight that environmental factors, such as soil texture, amount of soil lime, mean annual precipitation, altitude and land slope, had quantifiable effects on the performance of *A. sieberi*. Our findings could provide useful information for improvement, restoration and conservation programs. However, a further comprehension of the species–environment relationship is needed to predict the effects of climate change on the species habitat.

Keywords: canopy cover; ecological factors; Generalized Additive Models; Irano-Turanian region; optimum growth; species distribution

1. Introduction

Artemisia sieberi Besser (Asteraceae) is a shrubby aromatic plant distributed in Palestine, Syria, Iraq, Afghanistan, Pakistan, Central Asia and Iran (in the Irano-Turanian floristic region) [1,2]. It occurs in many arid and semiarid rangelands of Iran (from the Alborz mountains southern slopes in the north to the Saharo-Sindian floristic region boundary in the south and from the Zagros Mountains eastern slopes in the west to the east of Iran). More than 25,423,578 hectares of vegetation types of the Iranian natural resources (about 33.9%) are characterized by dominant and subdominant *A. sieberi*, which are mostly distributed throughout central Iran [3]. This species grows on different soil types, and it has a vast ecological distribution. *A. sieberi* is used for animal feeding, and because of its healing properties, it is also used as a medicinal plant [4,5]. Unfortunately, in recent years, disturbances such as grazing, land use changes and climate changes have caused the destruction of *A. sieberi* habitats. These threats have led to the destruction of vegetation, loss of biological diversity and erosion of the soil in many rangelands of Iran [6]. Destroyed rangelands are an important topic; in particular, conservation and rehabilitation of rangelands, monitoring of the vegetation dynamics and determination of suitable plant species that can be planted in areas with different environmental conditions should be considered [5]. Abiotic factors are one of the main components of the environmental niche [7,8]. Indeed, they have a major impact on the distribution and performance of plant species. In ecology, the geographical distribution of a species is predicted by quantifying the relationship between the species and the environment. On this basis, multiple hypotheses can be formulated in relation to the control of environmental factors that influence the distribution of the species [9]. It is crucial to understand the effects of the soil and topographic factors on the distribution of plants and vegetation restoration to improve the fragile ecosystems of damaged land. Knowledge of Artemisia response to environmental variables, and modeling of Artemisia occurrence, is essential for using this species in the reclamation of arid and semi-arid lands. For the species niche modeling, many analytical approaches can be used to answer the fundamental question 'what are the environmental factors controlling the species performance and distribution?' [10]. Canonical correspondence analysis and generalized additive models are among the methods most used for analyzing the reaction of plant species to environmental factors [11,12]. Generalized additive models [13,14], a powerful extension of GLM, are increasingly used for species modelling [15–19] because they do not assume any general shape of the response prior to the estimation [20].

In Iran, several studies have been conducted to investigate the factors that affect the distribution of *A. sieberi* as a result of environmental changes [5,21–27]. Jalili and colleagues studied the *A. sieberi* habitats in Iran and recognized important ecological differences in the functional characteristics of diploids and polyploids [28]. Mousaei Sanjerehei and Rundel determined that the most significant factors influencing the distribution of *A. sieberi* in Iran were annual precipitation and annual mean temperature [29]. In this research, we studied the role of the soil and topography in the function of *A. sieberi* in the Irano-Turanian floristic region of Iran. The following question was discussed: What are the responses and ecological requirements of *Artemisia sieberi* regarding some of the environmental variables?

2. Materials and Methods

Sixteen research sites, which presented different vegetation types of *A. sieberi* [30], were selected in Iran to have the maximum range of changes in altitude and geographical dispersion (29°23′–37°16′ N, 49°35′–60°05′ E). The range of altitude was between 560 m a.s.l. in Zirkooh Ghayenat (south Khorasan province) and 3000 m a.s.l in the mountainous areas of Hossein Abad Rain (Kerman province) (Figure 1).

A systematic-random method was used for the vegetation sampling [31,32]. Thirty sampling plots were positioned along five transects (6 sampling plots, with a random starting point, were placed in each transect) so that samples were collected in the range of the *A. sieberi* distribution in each site (480 sampling plots in total) during 2018 to 2020. The geographical coordinates of the plot locations were recorded using the Global Positioning System (GPS). The number of individuals per plot was evaluated and a visual estimation of each plot was performed to determine the vascular plant density and the canopy cover percentage. Additionally, stone and gravel percentage, bare soil percentage and litter percentage were estimated in each plot. The minimal area of the sampling plots was calculated for each sampling site and was selected to be 25 m² [33]. Vascular plants were identified using the Flora Iranica [34] and Flora of Iran [35].

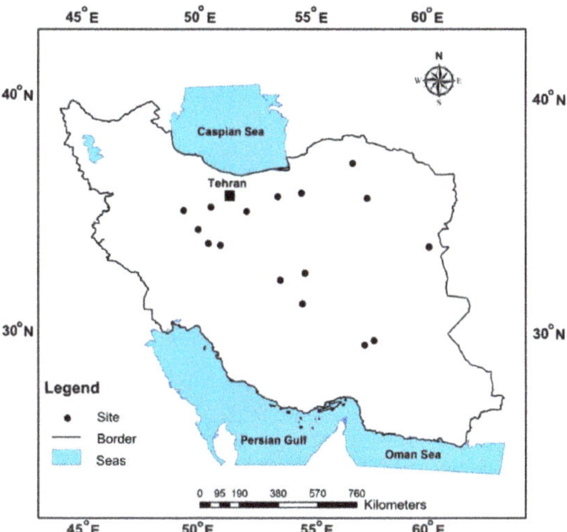

Figure 1. Location of studied sites in Iran.

The topographic characteristics (altitude, land slope percentage and geographical direction) of each plot were investigated after the analysis of the sampling network in each site. Four directions (90, 180, 270, 360 degrees) were used to register the geographical direction [36]. Then, the equation of Beers et al. [37] was used to enter the geographical direction in the analysis (Equation (1)).

$$A' = \text{Cos}\,(45 - A) + 1 \tag{1}$$

where A' is the converted value of direction and A is the azimuth of aspect in degrees, measured clockwise from north. The influence of the soil on the *A. sieberi* distribution was evaluated by sampling the soil to the root depth of the plant (0–30 cm above the soil surface) in each plot (a total of 480 samples). The soil sample of each plot was taken in three replications (composite sample).

The soil's physical and chemical properties were measured. In particular, soil texture was determined through the hydrometer method [38], EC was measured using the saturated extract and EC meter [39], pH was established using the saturated mud and pH meter [40], TNV% was calculated through the titration method, organic carbon% was measured through the Walkley-Black method [41], total nitrogen was evaluated using the Kjeldahl method, and soil saturation moisture (%) was determined by drying saturated mud inside the oven [42].

Prior to data analysis, spatial autocorrelation of plots was performed using the Mantel test [43] with PC-ORD 4.17 software [44].

The *A. sieberi* response to environmental factors changes was predicted using the generalized additive model (GAM) (Equation (2)) [36]. The canopy cover percentage measurement is easier when the species habitat is less degraded, because of the high correlation between the species yield and canopy cover percentage (Pearson correlation = 0.856, $p = 0.000$). This factor was measured as a response variable [45,46].

$$g(\mu) = \alpha + \sum_{j=1}^{p} f_j(X_j) \tag{2}$$

where f_j is unknown and smooth functions and X_j are predictive variables. Using the advanced scatter plot smoothing techniques, f_j is evaluated from the data. Log link function analysis and Poisson error distribution were used to fit the generalized additive model (they entered the model individually to avoid an over-fitting of the predictive variables).

Smooth terms were fitted using a cubic spline smoother with three degrees of freedom, and a quasi-Poisson distribution was used for the Poisson [8,47].

The variables that influenced the *A. sieberi* performance were classified through the Akaike information criterion (AIC) [48]. The models showing higher parsimony (lower Akaike information criterion, AIC), when compared to the null model, were selected [8]; the smaller the AIC value was, the more appropriate the proposed model for fitting the species response curve (Equations (3) and (4)) [49].

$$RMSE = \sqrt{\frac{\sum_{i=1}^{n}(Qi - \hat{Q}i)^2}{n}} \tag{3}$$

$$AIC = n\ln(RMSE) + 2p \tag{4}$$

where *RMSE* is the root-mean-square deviation, n is the number of observations, Qi is the observed value, $\hat{Q}i$ is the fitted value, and p is the number of model variables. Canoco software version 4.5 [50] was used for data analysis.

3. Results

Table 1 illustrates the average values of the environmental factors studied in the habitats.

Table 1. The average values of the environmental factors in the *A. sieberi* habitats.

Environmental Factors	Mean ± Stdev.	Environmental Factors	Mean ± Stdev.
Sand (%)	63.6 ± 12.8	Altitude (m)	1796.6 ± 525.7
Silt (%)	20.6 ± 8.7	Aspect (converted value of azimuth)	1.07 ± 0.75
Clay (%)	17.1 ± 7.5	Slope (%)	11.4 ± 9.3
EC (dsm-1)	2.4 ± 3.9	Bare soil (%)	45.4 ± 29.1
pH	7.9 ± 0.3	Stone and gravel (%)	33.1 ± 22.2
Lime (%)	21.8 ± 8.6	Litter (%)	5.4 ± 5.3
Organic carbon (%)	0.32 ± 0.3	Mean annual temperature (°C)	15.3 ± 1.7
Total nitrogen (%)	0.03 ± 0.3	Mean annual precipitation (mm)	185.7 ± 90.3
Soil saturation moisture (%)	27.7 ± 6.7	Latitude (Decimal degrees)	33.69 ± 2.68

Table 2 illustrates that most of the studied variables had a significant effect on the *A. sieberi* canopy cover percentage ($p < 0.01$), as shown by the generalized additive model with Poisson error distribution.

Table 2. The results of the generalized additive model for each of the significant explanatory variables.

Environmental Variable	F *	P *	AIC Value	Environmental Variable	F *	P *	AIC Value
Sand (%)	18.9	0.0000 **	1840.3	Altitude (m)	44.5	0.0000 **	1590.4
Silt (%)	87.2	0.0000 **	1262.3	Aspect	4.2	0.0001 **	2066.8
Clay (%)	23.2	0.0000 **	1812.4	Slope (%)	24.1	0.0000 **	1754.8
pH	6.6	0.0003 **	2035.7	Bare soil (%)	19.2	0.0000 **	1847.2
EC (dsm-1)	11.5	0.0000 **	1966.2	Stone and gravel (%)	3.1	0.027 *	2095
Lime percentage (%)	80.8	0.0000 **	1296.4	Litter (%)	6.4	0.0003 **	2034.8
Organic carbon (%)	24.3	0.0000 **	1771.4	Mean annual temperature (°C)	52.4	0.0000 **	2178.9
Total nitrogen (%)	24.5	0.0000 **	1761.6	Mean annual precipitation (mm)	125.7	0.0000 **	1572.6
Soil saturation moisture (%)	16.1	0.0000 **	1769.3	Latitude (Decimal degrees)	17	0.0000 **	2594.3

* Significance at the 5% level. ** Significance at the 1% level.

The A. sieberi response curve to each of the effective environmental variables was evaluated (Figure 2).

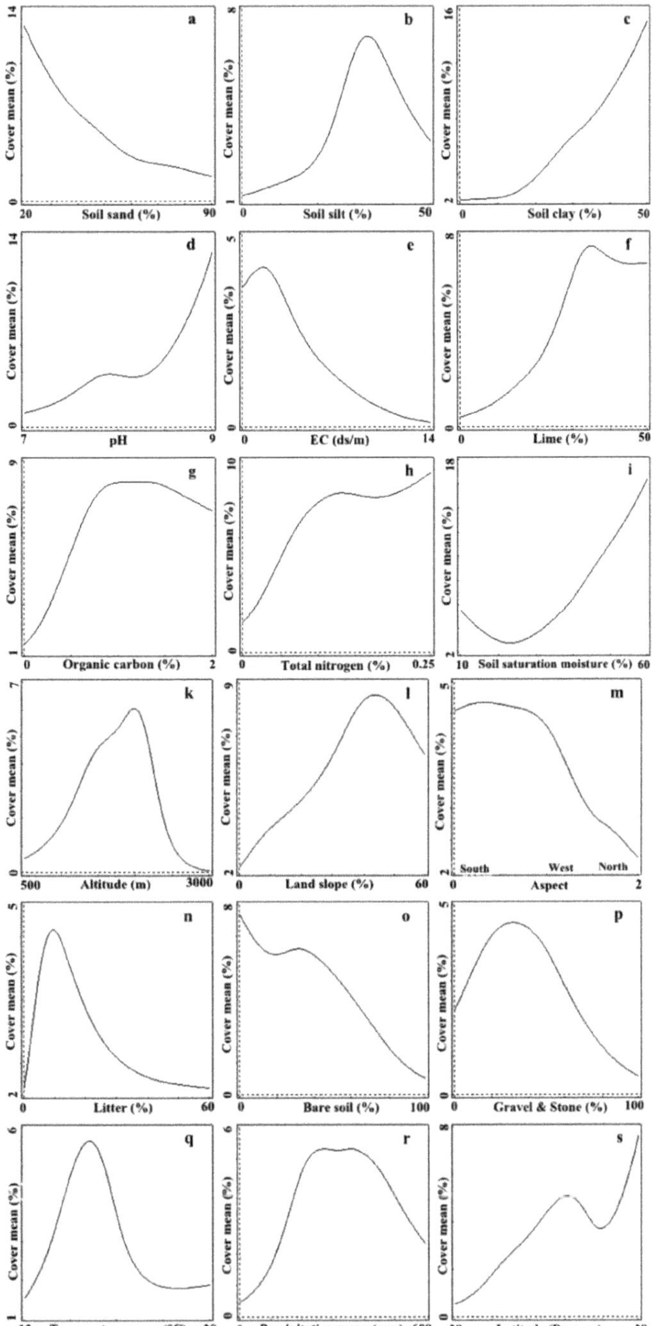

Figure 2. *A. sieberi* responses along the gradient of environmental factors.

The *A. sieberi* unimodal response to the soil silt (Figure 2b), altitude (Figure 2k), land slope (Figure 2l), gravel and stone (Figure 2p), litter (Figure 2n), temperature (Figure 2q) and precipitation gradient (Figure 2r) spanned its central borders, respectively from 1.6 to 43% (optimum 32%), 560 to 2788 m (optimum 2000 m), 0 to 60% (optimum 43% (on south-facing slopes)), 0 to 97% (optimum 32%), 0 to 54% (optimum 10%), 12 to 19.7 °C (optimum 15 °C) and 78.9 to 600 mm (optimum 250 mm). Increases in the amount of these factors, more than the optimum, resulted in a reduction of the *A. sieberi* canopy cover percentage. The species' performance augmented up to a soil salinity concentration of 1.75 ds/m; the canopy cover percentage decreased with further increases in the soil salinity (ranging between 0.09 and 12.8 ds/m). Therefore, *A. sieberi* is distributed in soils with relatively low salinity. The *A. sieberi*'s response to the soil lime percentage almost followed the asymmetric unimodal (Figure 2f): the plant performance initially increased slowly and then rapidly, approaching an exponential growth rate. However, when the amount of soil lime increases (more than 35%, ranging between 1.7 and 48.7%), the *A. sieberi* canopy cover percentage decreases until it stabilizes at a certain value.

Considering the soil organic carbon gradient, *A. sieberi* had its maximum response at 1%; increases in the organic carbon percentage from 0.02 to about 1.3% increased the species canopy cover percentage; an organic carbon percentage higher than 1.3% reduced the species distribution and presence (Figure 2g). According to the data of the *A. sieberi* response to the pH, this species is distributed in alkaline soils (following the positive exponential model).

The soil sand percentage showed a negative exponential trend (Figure 2a): as the soil sand value becomes larger, the *A. sieberi* performance approaches zero. Conversely, the soil clay and soil saturation moisture gradients showed a positive exponential trend (Figure 2c,i): increases in these factors increased the canopy cover percentage of the studied species. Considering the bare soil gradient, the *A. sieberi* response curve showed an almost negative linear trend: *A. sieberi* had the lowest performance with the highest percentage of bare soil (Figure 2o).

The *A. sieberi*'s response to the total nitrogen percentage almost followed the monotonic increase model, and the canopy cover percentage increased with increasing values of nitrogen (Figure 2h). In mountainous areas, changes in the geographical direction of the slope also affected the performance of the studied species; in particular, the *A. sieberi* performance reached its highest value in the southern facing slopes (Figure 2m), while the lowest performance was observed in the north facing slopes.

The species' response to changes in latitude was bimodal. Increases in latitude up to 35° resulted in an increased canopy cover percentage, while increases in latitude from 35° to 36° led to a reduction of the canopy cover percentage; however, further increases in latitude increased the canopy cover percentage. The lower limit of the species' presence was observed at latitude 28° (Figure 2s).

4. Discussion

The determination of the species response to the studied environmental factors represents one of the basic issues in ecology. Response curves allow the estimation of the species optimum and the niche width (tolerance) [51–55]. Statistical models play an important role in this regard [56]. *Aremisia sieberi* belongs to the steppe vegetation of Iran and is important in terms of forage production and medicinal properties [5,26,57]. However, the *A. sieberi* habitats are at risk due to the extra exploitation of these ecosystems during the past decades [58]. In this research, the GAM model was used to investigate the *A. sieberi* response to gradients of different environmental factors. The studied species' ecological needs were also studied.

In the studied habitats, *A. sieberi* has a significant yield; this aspect is highlighted by the average dry forage production, which is 66 kg per hectare. The forage production differences are due to both the habitat potential and the presence of disturbances, such as livestock grazing; this issue has also been mentioned by Khosravi Mashizi and Sharafatmandrad [58].

As reported also by other researchers [5,26,59], the data of this study showed that the *A. sieberi* performance was significantly affected by 18 variables, which were soil texture, lime, mean annual precipitation, altitude and land slope.

The *A. sieberi* response to the main environmental factors that affected the plant performance was exponential or unimodal; specifically, the response curves were mostly asymmetric unimodal.

According to the soil texture response curves, *A. sieberi* was found in areas that were characterized by a high soil clay percentage. This soil textural class causes the production of meso- and micro-porosities; indeed, these areas present the greatest amount of plant available water. As reported by other researchers [60,61], *A. sieberi* needs relatively more water compared to other species. Additionally, the lowest performance of *A. sieberi* was recorded in areas with the highest soil sand percentage and lower water retention [5].

Soil texture has an important function in the plants distribution and is involved in the soil humidity percentage, water holding capacity, plant elements availability and plant ventilation and rooting [62–64].

A. sieberi showed a positive exponential response to the soil acidity gradient, therefore, it has a strongly alkaline ecological niche. Piri Sahragard and Zare Chahouki [26] reported that *A. sieberi* is more compatible with soils characterized by a higher pH; moreover, soil acidity has a major role in the control of nutrients solubility [65]. Consequently, the pH level could be considered an important factor for the species distribution in the area. Our results illustrated that *A. sieberi* followed an asymmetric unimodal model in response to soil salinity. A low salinity rate (1.75 ds/m) caused significant abundance of *A. sieberi*. However, a higher soil salinity level caused an increased soil osmotic pressure (osmotic stress). This led to a reduction in the plant yield due to a lower water absorption and reduced plant metabolism; ultimately, the enzyme activity was also inhibited (ion toxicity). The studied areas presented a strongly different soil lime content (1.7–48.7%), and the *A. sieberi* response to the level of lime followed an asymmetric hump-shaped curve (the optimum was at 35%). The data of our study showed that *A. sieberi* tolerates various lime concentrations, although has a higher distribution in soils with a relatively high lime percentage, as reported also by Hosseini et al. [5]. Hence, *A. sieberi* is positively correlated with the soil lime percentage. It can be deduced that *A. sieberi* is a calcicole plant, because lime increases soil alkalinity. The soil lime content also decreases the growth of plants that need acidic soil [66]. The soil lime content is also a source of calcium and magnesium and is related to higher water retention [67,68]. All these aspects influence the distribution of *A. sieberi* in the studied region. *A. sieberi*'s response to the soil nitrogen percentage almost followed the monotonic increase model: the plant performance initially increased rapidly and then augmented slowly. A high soil nitrogen percentage (the optimum was at 0.25%) is related to increased soil microorganism activity, litter decomposition and plant growth and is considered a fertility indicator [69]. An increased soil organic carbon content (the optimum was at 1.3%) causes the formation of larger aggregates in the soil; the higher soil structure stability and increased number of large pores derived lead to an increased permeability coefficient and reduced erosion, as illustrated also by Lado et al. [70].

Considering *A. sieberi*'s response to the land slope percentage, the data of our study showed that plant performance was higher in the southern moderate slopes in mountainous areas. Davis et al. [71] illustrated that slopes have an average correlation with the species composition. The reason for the decreased presence of *A. sieberi* on slopes higher than 43% seems to be the reduction of the establishment of *A. sieberi* seeds due to an increased percentage of bare soil, stone and water erosion; this was also confirmed by other researchers such as Alavi et al. [72]. The soil's porosity in the moderate slopes is increased and more seeds are exposed to moisture; consequently, a higher number of seeds germinate, leading to increased reproduction [73]. According to the data about the performance of *A. sieberi* in response to altitude changes, this species was more present in the middle altitudes (about 2000 m). Hosseini et al. [5] illustrated that the optimum altitude was about 2300 m in the Poshtkouh area (local study), in central Iran. The increased presence of *A. sieberi* in the middle altitudes could be due to a combination of factors, such as adequate ecological factors and high levels of species turnover. As discussed also by other authors [74–76], this highlights the importance of the balance between communities and environment.

The highest canopy cover percentage was observed in the south aspects (south, southwest and southeast). Moreover, the studied plant distribution was also affected by the geographical direction due to the amount of plant available water and light [77]. As stated by other authors [19,78–81], the *A. sieberi* distribution was influenced by terrain and climate-related factors. It seems that *A. sieberi* is a xerophyte and helliophyta species.

In addition, the *A. sieberi* maximum response was recorded with the highest soil saturation moisture percentage ($\geq$60%). The texture and structure of the soil influence this factor and affect the pore size distribution of the soil, soil water storage and plant available water [82,83]. The response of the studied species to the soil saturation moisture percentage is similar to the response given to the soil clay percentage.

A decreased permeability (decreased soil moisture storage) and soil erosion could affect the low *A. sieberi* presence and canopy cover percentage recorded in the areas with the highest percentage of bare soil. As reported by other researchers such as Carcey Hincz and Diaz Aguilar [29], Wassie et al. [84] and Laris and Wardell [85], this influences the germination deficiency and vegetation establishment.

The *A. sieberi* growth decreased with a high stone and gravel percentage (>30%), which characterizes the Zagros Mountains [67].

With an increasing amount of litter up to 10%, the abundance of *A. sieberi* increased; a further increase in the litter amount led to a decreased presence of this species. It seems that the early life stages of this species may benefit from a low amount of litter; indeed, with medium and high litter amounts the frequency of *A. sieberi* decreased, which may be due to reduced germination. This was in accordance with the results of a previous study on other plant species [86].

The relationship between *A. sieberi* performance and mean annual temperature followed a hump shape: the minimum temperature was 12 °C and the maximum was 19.7 °C (the optimum was at 15 °C), which may suggest an adaptation to a narrow climate niche (stenothermal); this was also reported by other researchers [87,88].

The probability of *A. sieberi* presence in the Iranian steppe vegetation, with an annual precipitation of about 200–400 mm, was the highest. Canopy cover and presence of the species dramatically decreased with an annual precipitation higher than 400 mm or lower than 200 mm. Sanjerehei and Rundel (2017) [87] and Amiri et al. [27] also reported a decrease in the *A. sieberi* yield with increasing precipitation.

The increase of latitude up to 34° resulted in an increased canopy cover percentage, while increases in latitude from 34° to 36° led to a canopy cover reduction; however, further increases in latitude increased the *A. sieberi* canopy cover percentage. The lower limit of the species' presence was observed at 28° of latitude. In general, *A. sieberi* tended to grow at the middle latitude (34–35°) and middle-lower altitude (1700–2200 m) with middle lower precipitation (200–400 mm) and middle temperatures (14–16°). However, this species also had a higher yield at a higher latitude (38°) with lower altitudes in the Iranian steppe vegetation. These data are in accordance with the results of a previous study [3]. The species' response to changes in latitude was bimodal. This model shows that a superior competitor has displaced this species from its optimum in a certain range of environmental conditions [89]. However, the nature of the environmental process, which may link indirect variables to direct variables, can affect the ecological relationship [12]. Therefore, it seems that the use of indirect predictors with a known relationship to direct variables that affect plant performance is often not suitable for statistical modeling. In conclusion, environmental variations (precipitation, temperature, sunlight, etc.) resulting from latitudinal gradients had a great influence on the spatial distribution of *A. sieberi* in the studied areas.

5. Conclusions

The potential plant species response to environmental variables changes can be predicted based on the species distribution, using niche models. Consequently, adequate species habitats can be defined; this has been increasingly used by natural resource managers to plan biodiversity conservation management, to assess climate impact and for land use activities. Because of the importance of the presence of *A. sieberi* in the steppe vegetation of Iran [26,88], up to date knowledge of its distribution is necessary for the ecosystem management planning. According to our results, the distribution of *A. sieberi* is described by the GAM and a reduction in abundance and performance of the studied species is explained by the constraints of environmental factors. The response of *A. sieberi* to environmental factors can be used to explain abiotic limitations, range management, conservation, improvement and restoration of degraded habitats of *A. sieberi*. We concluded that environmental factors, specifically soil texture, soil lime amount, mean annual precipitation, altitude and land slope, had quantifiable effects on the performance of *A. sieberi*. An improved understanding of these species-environment relationships in the context of natural climatic fluctuations will also aid in better prediction of the effects of climate change on the species habitats. In addition, due to the contribution of biotic interactions, interaction between variables and extreme environmental stress to the species response shape, it seems that further studies are needed to develop modeling approaches that consider these factors in relation to the species distribution. These findings will be used to complete the quantification of its ecology and the risk of anthropogenic activities.

Author Contributions: Conceptualization, H.M.; methodology, D.G., S.Z. and J.M.S.; formal analysis, D.G., S.Z., J.M.S. and M.G.; investigation, H.M., D.G., S.Z., J.M.S. and M.G.; data curation, E.M.S.; writing—original draft preparation, H.M.; writing—review and editing, A.M.; supervision, A.M.; funding acquisition, A.M. All authors have read and agreed to the published version of the manuscript.

Funding: This research received no external funding.

Data Availability Statement: The data will be made available upon request to the corresponding authors.

Conflicts of Interest: The authors declare no conflict of interest.

References

1. Podlech, D. Artemisia. In *Flora Iranica, Compositae, VI-Anthemideae, Akademische Druck-u*; Rechinger, K.H., Ed.; Verlagsanstalt: Graz, Austria, 1986; Volume 158, pp. 159–223.
2. Mahboubi, M. Artemisia sieberi Besser essential oil and treatment of fungal infections. *Biomed. Pharmacother.* **2017**, *89*, 1422–1430. [CrossRef] [PubMed]
3. Jalili, A. *Ecology, Evolution and Biogeography of Artemisia L.*; Research Institute of Forests and Rangelands: Tehran, Iran, 2016; p. 493.
4. Nigam, M.; Atanassova, M.; Mishra, A.P.; Pezzani, R.; Devkota, H.P.; Plygun, S.; Salehi, B.; Setzer, W.N.; Sharifi-Rad, J. Bioactive Compounds and Health Benefits of Artemisia Species. *Nat. Prod. Commun.* **2019**, *14*, 1–17. [CrossRef]
5. Hosseini, S.; Kappas, M.; Chahouki, M.A.Z.; Gerold, G.; Erasmi, S.; Emam, A.R. Modelling potential habitats for Artemisia sieberi and Artemisia aucheri in Poshtkouh area, central Iran using the maximum entropy model and geostatistics. *Ecol. Inform.* **2013**, *18*, 61–68. [CrossRef]
6. Akhani, H.; Mahdavi, P.; Noroozi, J.; Zarrinpour, V. Vegetation patterns of the Irano-Turanian steppe along a 3000 m al-titudinal gradient in the Alborz Mountains of Northern Iran. *Folia. Geobot.* **2013**, *48*, 229–255. [CrossRef]
7. Whittaker, R.H. *Direct Ggradient Analysis. Handbook of Vegetation Science 5: Ordination and Classification of Communities*; Whittaker, R.H., Ed.; Junk Publishers: Hague, The Netherlands, 1973; pp. 19–50.
8. Traoré, S.; Zerbo, L.; Schmidt, M.; Thiombiano, L. Acacia communities and species responses to soil and climate gradients in the Sudano-Sahelian zone of West Africa. *J. Arid. Environ.* **2012**, *87*, 144–152. [CrossRef]
9. Gogina, M. Investigation of Interrelations between Sediment and Near-Bottom Environmental Parameters and Macrozoobenthic Distribution Patterns for the Baltic Sea. Ph.D. Thesis, Ernst Moritz Arndt University of Greifswald, Greifswald, Germany, 2010.
10. Moisen, G.G.; Frescino, T.S. Comparing five modelling techniques for predicting forest characteristics. *Ecol. Model.* **2002**, *157*, 209–225. [CrossRef]
11. Cajo, J.F.; Ter Braak, C. Canonical Correspondence Analysis: A New Eigenvector Technique for Multivariate Direct Gradient Analysis. *Ecology* **1986**, *67*, 1167–1179. [CrossRef]

12. Austin, M.; Belbin, L.; Meyers, J.; Doherty, M.; Luoto, M. Evaluation of statistical models used for predicting plant species distributions: Role of artificial data and theory. *Ecol. Model.* **2006**, *199*, 197–216. [CrossRef]

13. Hastie, T.; Tibshirani, R. *Generalised Additive Models*; Chapman and Hall: London, UK, 1990.

14. Kosicki, J.Z. Generalised Additive Models and Random Forest Approach as effective methods for predictive species density and functional species richness. *Environ. Ecol. Stat.* **2020**, *27*, 273–292. [CrossRef]

15. Yee, T.W.; Mitchell, N.D. Generalized additive models in plant ecology. *J. Veg. Sci.* **1991**, *2*, 587–602. [CrossRef]

16. Austin, M.P. The potential contribution of vegetation ecology to biodiversity research. *Ecography* **1999**, *22*, 465–484. [CrossRef]

17. Guisan, A.; Zimmermann, N.E. Predictive habitat distribution models in ecology. *Ecol. Model.* **2000**, *135*, 147–186. [CrossRef]

18. Leathwick, J.R.; Whitehead, D. Soil and atmospheric water deficits and the distributions of New Zealand's indigenous tree species. *Funct. Ecol.* **2001**, *15*, 233–242.

19. Vetaas, O.R.; Grytnes, J.A. Distribution of vascular plant species richness and endemic richness along the Himalayan elevation gradient in Nepal. *Glob. Ecol. Biogeogr.* **2002**, *11*, 291–301. [CrossRef]

20. Austin, M.P.; Meyers, J.A. Current approaches to modelling the environmental niche of eucalypts: Implication for management of forest biodiversity. *For. Ecol. Manag.* **1996**, *85*, 95–106. [CrossRef]

21. Azarnivand, H.; Jafari, M.; Moghaddam, M.R.; Jalili, A.; Zare Chahouki, M.A. Investigation on environmental variables affecting habitat distribution of A. sieberi and A. aucheri in rangelands of Vardavard, Garmsar and Semnan. *Iran. J. Nat. Resour.* **2006**, *56*, 93–99.

22. Yaghmaei, L.; Soltani Koupaei, S.; Khodagholami, M. Effect of climatic factors on distribution of Artemisia sieberi and Artemisia aucheri in Isfahan province using multivariate statistical methods. *Water Soil Sci.* **2008**, *12*, 359–371.

23. Abdollahi, J.; Naderi, H. Soil and topographical variation influencing the growing factors of Artemisia sieberi in steppic rangeland, Nodoushan-Yazd. *Watershed Manag. Res. Pajouhesh Sazandegi* **2011**, *97*, 52–62.

24. Sahragard, H.P.; Ajorlo, M.; Karami, P. Modeling habitat suitability of range plant species using random forest method in arid mountainous rangelands. *J. Mt. Sci.* **2018**, *15*, 2159–2171. [CrossRef]

25. Chahouki, M.A.Z.; Azarnivand, H.; Jafari, M.; Tavili, A. Multivariate statistical methods as a tool for model-based prediction of vegetation types. *Russ. J. Ecol.* **2010**, *41*, 84–94. [CrossRef]

26. Piri Sahragard, H.; Zare Chahouki, M.A. Modeling of Artemisia sieberi Besser habitat distribution using maximum entropy method in desert rangelands. *J. Rangel. Sci.* **2016**, *6*, 93–101.

27. Amiri, M.; Tarkesh, M.; Jafari, R.; Jetschke, G. Bioclimatic variables from precipitation and temperature records vs. remote sensing-based bioclimatic variables: Which side can perform better in species distribution modeling? *Ecol. Inform.* **2020**, *57*, 101060. [CrossRef]

28. Jalili, A.; Rabie, M.; Azarnivand, H.; Hodgson, J.G.; Arzani, H.; Jamzad, Z.; Asri, Y.; Hamzehee, B.; Ghasemi, F.; Hejazi, S.H.; et al. Distribution and ecological consequences of ploidy variation in Artemisia sieberi in Iran. *Acta. Oecologica* **2013**, *53*, 95–101. [CrossRef]

29. Mousaei Sanjerehei, M.; Rundel, P.W. The Impact of Climate Change on Habitat Suitability for Artemisia sieberi and Artemisia aucheri (Asteraceae), a Modeling Approach. *Pol. J. Ecol.* **2017**, *65*, 97–109.

30. Khodagholi, M.; Saboohi, R.; Esfahani, E.Z. A comparison of vegetative climate of Artemisia sieberi Besser and Artemisia aucheri Boiss in Iran. *Arch. Meteorol. Geophys. Bioclimatol. Ser. B* **2021**, *146*, 921–939. [CrossRef]

31. Krebs, C.J. *Ecological Methodology*, 2nd ed.; Addison Wesley Longman: Menlo Park, CA, USA, 1999; p. 620.

32. Arzani, H.; Abedi, M. *Rangeland Assessment: Vegetation Measurement*, 2nd ed.; University of Tehran: Tehran, Iran, 2015.

33. Küchler, A.W.; Mueller-Dombois, D.; Ellenberg, H. Aims and Methods of Vegetation Ecology. *Geogr. Rev.* **1976**, *66*, 114. [CrossRef]

34. Iranica, F.; Druck-u, A.; Rechinger, K.H. (Eds.) *Flora Iranica: Facts and Figures and a List of Publications by k.h. Rechinger on Iran and Adjacent Areas*; Verlagsanstalt: Graz, Austria, 2015; Volume 181.

35. Research Institute of Forests and Rangelands. *Flora of Iran*. Assadi, M., Ed.; Research Institute of Forests and Rangelands: Tehran, Iran, 2019; Volume 149. Available online: https://irannature.areeo.ac.ir/article_119036.html?lang=en (accessed on 25 November 2022).

36. Palmer, M.W. Putting Things in Even Better Order: The Advantages of Canonical Correspondence Analysis. *Ecology* **1993**, *74*, 2215–2230. [CrossRef]

37. Beers, T.W.; Dress, P.E.; Wensel, L.C. Aspect trans-formation in site productivity research. *J. For. Res.* **1966**, *64*, 691–692.

38. Bouyoucos, G.J. Hydrometer Method Improved for Making Particle Size Analyses of Soils. *Agron. J.* **1962**, *54*, 464–465. [CrossRef]

39. Black, C.A.; Evans, D.D.; Dinauer, R.C. *Methods of Soil Analysis*; American Society of Agronomy: Madison, NC, USA, 1965.

40. McLean, E.; Ph, S.; Requirement, L. *Methods of Soil Analysis: Part 2*; Chemical and Microbiological Properties: Agronomy Monographs, 1982; pp. 199–224. Available online: https://acsess.onlinelibrary.wiley.com/doi/book/10.2134/agronmonogr9.2.2 ed (accessed on 25 November 2022).

41. Walkley, A.; Black, I.A. Estimation of soil organic carbon by the chromic acid titration method. *Soil Sci.* **1934**, *37*, 29–38. [CrossRef]

42. Ryan, J.G.; Estefan, G.; Rashid, A. *Soil and Plant Analysis Laboratory Manual*; ICARDA: Aleppo, Syria, 2002.

43. Mantel, N. The detection of disease clustering and a generalized regression approach. *Cancer Res.* **1967**, *27*, 209–220. [PubMed]

44. McCune, B.; Mefford, M.J. *Multivariate Analysis of Ecological Data*, version 4.17; MjM Software; Gleneden, OR, USA, 1999. Available online: https://www.scirp.org/(S(351jmbntvnsjt1aadkposzje))/reference/ReferencesPapers.aspx?ReferenceID=551732 (accessed on 25 November 2022).

45. Smilauer, P.; Leps, J. *Multivariate Analysis of Ecological Data Using CANOCO 5*, 2nd ed.; Cambridge University: Cambridge, UK, 2014.
46. Akaike, H. A new look at the statistical model identification. *IEEE Trans. Autom. Control* **1974**, *19*, 716–723. [CrossRef]
47. Dawson, C.W.; Abrahart, R.J.; See, L.M. HydroTest: A web-based toolbox of evaluation metrics for the standardized assessment of hydrological forecasts. *Environ. Model. Softw.* **2007**, *22*, 1034–1052. [CrossRef]
48. ter Braak, C.J.F.; Smilauer, P. *Canoco, Reference Manual and CanoDraw for Windows User's Guide: Software for Canonical Community Ordination, version 4.5*; Microcomputer Power: Ithaca, Greece, 2002.
49. Gaston, K.J.; Blackburn, T.M. A Critique for Macroecology. *Oikos* **1999**, *84*, 353. [CrossRef]
50. Toner, M.; Keddy, P. River hydrology and riparian wetlands: A predictive model for ecological assembly. *Ecol. Appl.* **1997**, *7*, 236–246. [CrossRef]
51. Wiser, S.K.; Peet, R.K.; White, P.S. Prediction of rare-plant occurrence: A southern Appalachian example. *Ecol. Appl.* **1998**, *8*, 909–920. [CrossRef]
52. Witkowski, E.T.F.; Lamont, B. Does the rare Banksia goodii have inferior vegetative, reproductive or ecological attributes compared with its widespread co-occurring relative B. gardneri? *J. Biogeogr.* **1997**, *24*, 469–482. [CrossRef]
53. Olden, J.D.; Jacksona, D.A. Comparison of statistical approaches for modelling fish species distributions. *Freshw. Biol.* **2002**, *47*, 1–20. [CrossRef]
54. Bagheri, R.; Chaichi, M.R.; Mohseni-Saravi, M.; Amin, G.R.; Zahedi, G. Grazing affects essential of compositions of Artemisis sieberi Besser. *Pak. J. Biol. Sci.* **2007**, *5*, 810–813. [CrossRef]
55. Khosravi Mashizi, A.; Sharafatmandrad, M. Assessing Impact of Anthropogenic Disturbances on Forage Production in Arid and Semiarid Rangelands. *J. Rangel. Sci.* **2019**, *9*, 234–245.
56. Vereecken, H.; Diels, J.; Van Orshoven, J.; Feyen, J.; Bouma, J. Functional evaluation of pedotransfer functions for the estimtion of soil hydraulic properties. *Soil Sci. Soc. Am. J.* **1992**, *56*, 1371–1379. [CrossRef]
57. Aryanpour, H.; Shorafa, M. Cultivation impact on soil available water in different soil textures using pore size distribution. *J. Soil Manag. Sustain. Prod.* **2013**, *3*, 131–148.
58. Wahba, S.A.; Abdel Rahman, S.I.; Cairo, M.Y.T.; Matyn, M.A. Soil moisture, salinity, water use efficiency and sunflower growth as influenced by irrigation bitumen mulch and plant density. *Soil Technol.* **1990**, *3*, 33–44. [CrossRef]
59. Rangel, T.F.L.V.B.; Diniz-Filho, J.A.F.; Bini, L.M. Towards an integrated computational tool water use efficiency and sun-flower growth as influenced by irrigation bitumen mulch and plant density. *Soil Technol.* **2006**, *3*, 33–44.
60. Moradi, P.; Aghajanloo, F.; Moosavi, A.; Monfared, H.; Khalafi, J.; Taghiloo, M.; Khoshzaman, T.; Shojaee, M.; Mastinu, A. Anthropic Effects on the Biodiversity of the Habitats of Ferula gummosa. *Sustainability* **2021**, *13*, 7874. [CrossRef]
61. Janisova, M. Vegetation-environment relationship in dry calcareous grassland. *Ekológia-Bratisl* **2005**, *24*, 25–44.
62. Mahmodi, S.; Hakymian, M. *Fundamental of Soil Science*; Tehran University: Tehran, Iran, 2007.
63. Mohtashamnia, S.; Zahedi, G.H.; Arzani, H. An investigation on synecology of semi-steppe vegetation in relation to edaphic and physiographical factors (case study: Eghlid rangelands of Fars). *J. Agric. Nat. Resour.* **2008**, *14*, 111–123.
64. Mseddi, K.; Al-Shammari, A.; Sharif, H.; Chaieb, M. Plant diversity and relationships with environmental factors after rangeland exclosure in arid Tunisia. *Turk. J. Bot.* **2016**, *40*, 287–297. [CrossRef]
65. Brahim, N.; Blavet, D.; Gallali, T.; Bernoux, M. Application of structural equation modeling for assessing relationships between organic carbon and soil properties in semiarid Mediterranean region. *Int. J. Environ. Sci. Technol.* **2011**, *8*, 305–320. [CrossRef]
66. Lado, M.; Paz, A.; Ben-Hur, M. Organic Matter and Aggregate-Size Interactions in Saturated Hydraulic Conductivity. *Soil Sci. Soc. Am. J.* **2004**, *68*, 234–242. [CrossRef]
67. Davies, K.W.; Bates, J.D.; Miller, R.F. Vegetation Characteristics Across Part of the Wyoming Big Sagebrush Alliance. *Rangel. Ecol. Manag.* **2006**, *59*, 567–575. [CrossRef]
68. Alavi, S.J.; Nouri, Z.; Zahedi Amiri, G.H. The response curve of beech tree (Fagus orientalis lipsky.) in relation to enviromental variables using generalized additive model in Khayroud forest, Nowshahr. *J. Wood For. Sci. Technol.* **2017**, *24*, 29–42.
69. Rahimi, A. Autecology of Onobrychis Chorassanica Bunge. In Khorasan Razavi Province. Master's Thesis, Ferdowsi University of Mashhad, Mashad, Iran, 1999.
70. Balent, G.; Stafford Smith, D.M. Conceptual model for evaluating the consequences of management practices on the use of pastoral resources. In Proceedings of the fourth International Rangeland Congress, Montpellier, France, 22–26 April 1991; pp. 1158–1164.
71. Wang, G.; Zhou, G.; Yang, L.; Li, Z. Distribution, species diversity and life-form spectra of plant communities along an altitudinal gradient in the northern slopes of Qilianshan Mountains, Gansu, China. *Plant Ecol.* **2002**, *165*, 169–181. [CrossRef]
72. Grytnes, J.A. Ecological interpretations of the mid-domain effect. *Ecol. Lett.* **2003**, *6*, 883–888. [CrossRef]
73. Moghaddam, M.R. *Ecology of Terrestrial Plants*; Tehran University: Tehran, Iran, 2005.
74. Chaplygin, V.A.; Rajput, V.D.; Mandzhieva, S.S.; Minkina, T.M.; Nevidomskaya, D.G.; Nazarenko, O.G.; Kalinitchenko, V.P.; Singh, R.; Maksimov, A.Y.; Popova, V.A. Comparison of Heavy Metal Content in Artemisia austriaca in Various Impact Zones. *ACS Omega* **2020**, *5*, 23393–23400. [CrossRef] [PubMed]
75. Mark, A.F.; Dickinson, K.J.M.; Hofstede, R.G.M. Alpine vegetation, plant distribution, life forms, and environments in a humid New Zealand region: Oceanic and tropical high mountain affinities. *Arct. Antarct. Alp. Res.* **2018**, *32*, 240–254. [CrossRef]
76. Ranjbar, A.; Vali, A.; Mokarram, M.; Taripanah, F. Investigating variations of vegetation: Climatic, geological substrate, and topographic factors-a case study of Kharestan area, Fars Province, Iran. *Arab. J. Geosci.* **2020**, *13*, 597–614. [CrossRef]

77. Zare, M.; Mohammady, M.; Pradhan, B. Modeling the effect of land use and climate change scenarios on future soil loss rate in Kasilian watershed of northern Iran. *Environ. Earth Sci.* **2017**, *76*, 305. [CrossRef]

78. Childs, E.C. The Use of Soil Moisture Characteristics in Soil Studies. *Soil Sci.* **1940**, *50*, 239–252. [CrossRef]

79. Vogel, H.J. A numerical experiment on pore size, pore connectivity, water retention, permeability, and solute transport using network models. *Eur. J. Soil Sci.* **2000**, *51*, 99–105. [CrossRef]

80. Carcey Hincz, P.A.; Diaz Aguilar, I. Impact of Grazing on Soil Mesofauna Diversity and Community Composition in Deciduous Forested Rangelands of Northwest Alberta. Available online: https://open.alberta.ca/dataset/303a7691-a457-4b7e-a32b-f39b034 060ec/resource/1883d8df-abe6-4848-a1bb-ba6d46cc6f66 (accessed on 7 April 2016).

81. Wassie, A.; Sterck, F.J.; Teketay, D.; Bongers, F. Effects of livestock exclusion on tree regeneration in church forests of Ethiopia. *For. Ecol. Manag.* **2009**, *257*, 765–772. [CrossRef]

82. Laris, P.; Wardell, D.A. Good, bad or 'necessary evil'? Reinterpreting the colonial durning experiment in savanna landscapes of West Africa. *Geogr. J.* **2006**, *172*, 271–290. [CrossRef]

83. Olliff-yang, R.L.; Case, E.J. The effects of leaf litter on germination in the serpentin endemic Boechera constancei. *Madroño* **2018**, *65*, 159–167.

84. Bakåsmoen, H. Endemic Alpine Plant Facing Local Extinction in Context of Climate and Landuse Change in Hardanger, Norway, Nich Modeling of Artemisis Norvegica Fr. Master's Thesis, University of Bergen, Bergen, Norway, 2021; p. 102.

85. Austin, M. Spatial prediction of species distribution: An interface between ecological theory and statistical modelling. *Ecol. Model.* **2002**, *157*, 101–118. [CrossRef]

86. Osborne, P.; Alonso, J.; Bryant, R. Modelling landscape-scale habitat use using GIS and remote sensing: A case study with great bustards. *J. Appl. Ecol.* **2001**, *38*, 458–471. [CrossRef]

87. Gustafson, E.J.; Lietz, S.M.; Wright, J.L. Predicting the spatial distribution of aspen growth potential in the upper Great Lakes region. *For. Sci.* **2003**, *49*, 499–508.

88. Holloway, G.J.; Griffiths, G.H.; Richardson, P. Conservation strategy maps: A tool to facilitate biodiversity action planning illustrated using the heath fritillary butterfly. *J. Appl. Ecol.* **2003**, *40*, 413–421. [CrossRef]

89. Cabeza, M.; Araújo, M.B.; Wilson, R.J.; Thomas, C.D.; Cowley, M.J.R.; Moilanen, A. Combining probabilities of occurrence with spatial reserve design. *J. Appl. Ecol.* **2004**, *41*, 252–262. [CrossRef]

 *land*

Article

Seasonal Trap Abundance of Two Species of *Psilochalcis* Kieffer (Hymenoptera: Chalcididae) in Rangelands of the Eastern Great Basin of Utah, USA

Mark J. Petersen [1,*], Val J. Anderson [1], Robert L. Johnson [2] and Dennis L. Eggett [3]

[1] Department of Plant and Wildlife Sciences, Brigham Young University, 4105 LSB, Provo, UT 84602, USA
[2] Department of Biology, Brigham Young University, 4102 LSB, Provo, UT 84602, USA
[3] Department of Statistics, Brigham Young University, 2152F WVB, Provo, UT 84602, USA
* Correspondence: markjp89@byu.edu

Abstract: Two species of *Psilochalcis* (Hymenoptera: Chalcididae) wasps occurring in the Great Basin region of the western United States were sampled from three locations in central Utah (USA) over a two-year period using Malaise traps. Each location is composed of four contiguous habitat types: pinyon/juniper (*Pinus edulis* or *P. monophylla* and *Juniperus osteosperma*), sagebrush (*Artemisia tridentata*), cheatgrass (*Bromus tectorum*), and crested wheatgrass (*Agropyron cristatum*). Seasonal trap abundance for each *Psilochalcis* species was determined. *Psilochalcis minuta* Petersen and *Psilochalcis quadratis* Petersen occur in highest abundance from mid-May to early August. *Psilochalcis minuta* demonstrates a significant association with pinyon/juniper habitat, specifically at the Utah; Juab County, Yuba Valley sample site, whereas *P. quadratis* demonstrates a significant association with cheatgrass (*Bromus tectorum*) habitat at the same location.

Keywords: chalcidid wasps; malaise trap; pinyon/juniper; cheatgrass; crested wheatgrass; habitat; ecological relationships

Citation: Petersen, M.J.; Anderson, V.J.; Johnson, R.L.; Eggett, D.L. Seasonal Trap Abundance of Two Species of *Psilochalcis* Kieffer (Hymenoptera: Chalcididae) in Rangelands of the Eastern Great Basin of Utah, USA. *Land* **2023**, *12*, 54. https://doi.org/10.3390/land12010054

Academic Editors: Michael Vrahnakis, Yannis (Ioannis) Kazoglou and Manuel Pulido Fernádez

Received: 29 November 2022
Revised: 20 December 2022
Accepted: 24 December 2022
Published: 25 December 2022

1. Introduction

The Great Basin is a region of north and south running mountain ranges and valleys that extend from the Wasatch Mountains of Utah in the east to the Sierra Nevada Mountains in the west. The northern border is the Snake River Plain extending south to the Mohave Desert. It is characterized as a cold desert with hot summers and freezing winters. Precipitation ranges from 125 to 500 mm annually [1]. Two common habitat types found throughout the region are pinyon/juniper (*Pinus edulis* Engelm. or *P. monophylla* Torr. and Frem. and *Juniperus osteosperma* (Torr.) Little) woodland and sagebrush (*Artemisia tridentata* Nutt.) steppe. In Utah, these two habitat types compose nearly 40 percent of the semi-arid region of the state [2]. These habitats have become fragmented due to frequent wildfires, leaving a habitat mosaic across the region. Subsequent to wildfire, they are often replaced by introduced exotic species, especially annual cheatgrass (*Bromus tectorum* L.), which is native to Eurasia. A regional model suggests that nearly one-third of the Great Basin (210,000 km^2) has cheatgrass cover of at least 15 percent [3]. Crested wheatgrass (*Agropyron cristatum* (L.) Gaertn.), a perennial grass native to Russian and Siberia, has been used extensively in post wildfire reseeding efforts across the region. While it is an introduced species, it is considered preferable to cheatgrass. Like cheatgrass, it forms monotypic stands. The impact of these habitat alterations from native to nonnative plant communities on insect communities is poorly understood. We are interested in the effect this might have on chalcidid wasps that occur in these habitats.

The family Chalcididae is a taxon of parasitic Hymenoptera. These wasps most often parasitize pupae of Lepidoptera [4]. As such, they are typically studied for their potential use as biological control agents in areas where their lepidopteran hosts are important economic insect pests. The taxonomy of Chalcididae has changed very little in the past

150 years, including divisions into subfamilies and genera [4], a 1992 revision of the new world Chalcididae being the most recent [5]. Recently, three species of chalcidid wasps were described from rangelands of the eastern Great Basin in Utah, two of which were collected in enough abundance to warrant further investigation [6]. These two species belong to the subfamily Haltichellinae.

Haltichellinae is comprised of six chalcidid genera that occur in the United States and Canada [4]. The most recent report of Haltichellinae species that occur in the Great Basin describes the distribution of species in the genus *Psilochalcis* Kieffer [6].

The Universal Chalcidoidea Database [7] currently reports 60 species of *Psilochalcis* worldwide. Old world and new world distributions are not known to overlap [8]. The literature suggests that less than 20 species of *Psilochalcis* are known from the Western Hemisphere [5]. In general, *Psilochalcis* species occur in arid to semi-arid areas across the southwest and western United States [6]. Biological observations are rare, with only a few host species reported [8,9].

Of the eight species of *Psilochalcis* wasps known to occur in the United States [6], four are associated with agricultural environments. *Psilochalcis brevialata* Grissell and Johnson is known from a culled fig warehouse in California [8]. *Psilochalcis deceptor* (Grissell and Schauff), *P. threa* (Grissell and Schauff), and *P. usta* (Grissell and Schauff) occur in cultivated peanut crop in Oklahoma and Texas [9]. *Psilochalcis hesphenheidei* (Boŭcek) occurs in natural areas across the western United States [10], but no habitat associations are reported. *Psilochalcis adenticulata* Petersen is known from multiple natural habitat types based on label information from collection sites in New Mexico, Nevada, and Utah [6].

The paucity of ecological data associated with *Psilochalcis* is understandable, due to their being rarely collected and the lack of interest in adding ecological data to collection labels. This paper is unique, relative to all other publications on North American *Psilochalcis*, because it links two *Psilochalcis* species with specific ecological data. Using Malaise trap capture data, we evaluate the changes in seasonal abundance over time for *P. minuta* Petersen and *P. quadratis* Petersen and their associations within distinct habitat types of the eastern Great Basin.

2. Materials and Methods

In 2019, we were able to examine and extract chalcid wasps from historic Malaise trap samples collected in 2006 and 2007. These samples were originally used to study different insect groups with the remaining material stored in 500 ml Nalgene bottles with 70% ethanol at 2 °C. The sample sites were originally selected in areas where native and non-native plant communities formed contiguous boundaries representing; native shrubland, native woodland, introduced annual grassland, and introduced perennial grassland.

2.1. Site Descriptions

In 2006, a study was established to examine insect diversity in native pinyon/juniper and sagebrush habitats relative to those in the non-native conversion habitats of cheatgrass or crested wheatgrass. From these samples, we are able to test the effect of habitat on the presence of *Psilochalcis* wasps. Malaise traps were set up in four contiguous habitat types common to the Great Basin at three different locations. Locations are (1) Utah, Juab County, Tintic Valley; Utah, Juab County, Yuba Valley; and Utah, Sanpete County, Antelope Valley. The habitat types at these locations are defined by the dominant plant species, namely (1) pinyon/juniper (*Pinus edulis* or *P. monophylla* and *Juniperus osteosperma*), (2) sagebrush (*Artemisia tridentata*), (3) cheatgrass (*Bromus tectorum*) and (4) crested wheatgrass (*Agropyron cristatum*) (Figure 1). Both the cheatgrass and crested wheatgrass sites were either native shrubland or woodland prior to wildfires. Crested wheatgrass sites were reseeded after fire disturbance whereas cheatgrass sites were untreated or failed reseedings. These two altered plant communities formed the dominant habitat type at each location with sagebrush comprising much smaller remnant patches. The overall topography was relatively flat except where dry washes bisected the study site or where slopes gradually gain elevation.

The Antelope Valley site has abundant pinyon/juniper on the adjacent slopes to the west and north. At the Tintic Valley site, pinyon/juniper is abundant to the north, and at the Yuba Valley site, pinyon/juniper is abundant to the east (Figure 1). Within habitat types, exact trap locations were based partly on road accessibility and the logistics of regular retrieval of trap samples. Though some trap sites appear close to habitat boundaries, they were still placed a minimum distance of 100 meters from the habitat edge, to reduce the effect of edge bias.

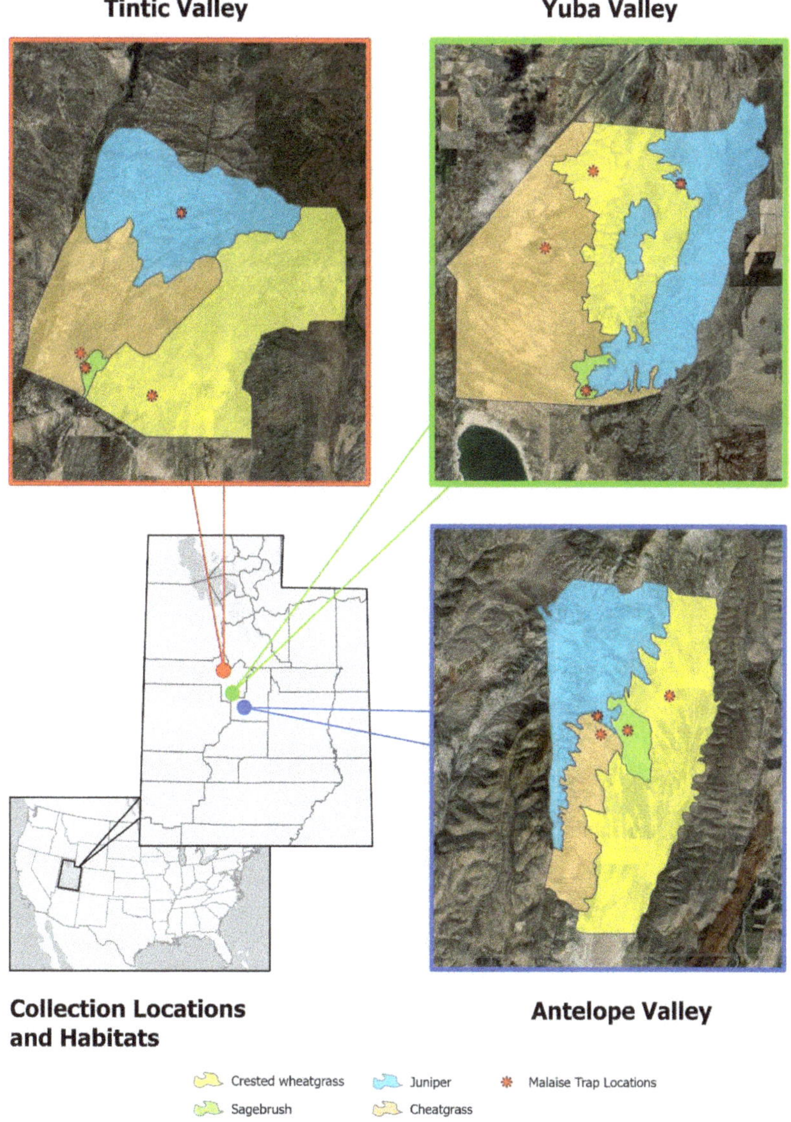

Figure 1. Placement of Malaise traps within four contiguous habitat types at three locations in central Utah. Tintic Valley and Yuba Valley are in Juab County. Antelope Valley is in San Pete County.

2.2. Sampling Method

Townes-style malaise traps [11,12] were installed at three different sites. At each site, three traps were set up in each of the four habitats. Within each habitat, the three traps were installed 120° opposing each other thus effecting a full 360° sample orientation. A total of 36 traps were installed in 2006 and repeated in 2007. Samples from each trap were retrieved biweekly from 1 April to 1 October (Spring through Fall). Trap setup required two eight-foot T-posts pounded into the soil on either side of trap, securing trap to posts, and staking the trap to the ground (Figure 2).

Figure 2. Malaise trap setup. Trap shown in pinyon/juniper habitat.

Malaise trapping is the prevailing method for collecting chalcidid wasps [13–15]. Malaise traps have been shown to be more effective in trapping chalcidoids than other style traps and are preferred when time and/cost are major constraints [16]. Due to the remoteness of our trap locations, it was not feasible to check traps more frequent than biweekly. Malaise traps passively capture through flight interception, which allows to them to be left unattended between sample retrieval for longer periods relative to other style traps without the concern of sample degradation.

Preliminary examination of chalcidid wasp captures indicated two *Psilochalcis* species almost exclusively occurring in pinyon/juniper and cheatgrass habitats. This discovery prompted us to set up a few additional traps during the summers of 2020 through 2022 to see if we could predictably recapture the two wasp species in the same habitats but from different locations. Eleven traps were set up in pinyon/juniper and four traps in cheatgrass.

2.3. Plant Composition

In 2007, aerial plant cover was estimated at each trap location using a half-square meter quadrat placed every 5 m (excluding point 0) along a 45 m transect in each cardinal direction from the Malaise trap center. This yielded 32 sample quadrats for each trap with the sample area being a 45-m radius around each trap.

2.4. Seasonal Abundance

Malaise trap data were used to calculate species trap abundance by location, habitat type and collection year. For each *Psilochalcis* species, we produced annual species abundance graphs using Microsoft Excel 2016 to visualize seasonal trends.

2.5. Statistical Analysis

The trap abundance data for both *P. minuta* and *P. quadratis* were highly skewed having a large number of zeros from several traps across habitat type, site, and collection year. The data were transformed to a log scale before analysis to account for this. A separate two-way analysis of variance (ANOVA) and logistic regression were conducted for *P. minuta* and *P. quadratis* using the Statistical Analysis System (SAS) version 9.4 to determine the likelihood of each species occurring at each location and habitat type.

2.6. Climate Data Analysis

To assess any impact of climate on seasonal abundance, we analyzed both county [17] and local [18] temperature and precipitation data to visualize climate trends at collection sites. The weather stations in closest proximity to each location were chosen. Stations are Little Sahara for Tintic Valley, Scipio for Yuba Valley, and Manti for Antelope Valley.

3. Results

3.1. Seasonal Trap Abundance

In 2006, 99.24 percent of *P. minuta* specimens were collected in pinyon/juniper habitat, while 0.76 percent of specimens were collected in crested wheatgrass habitat. Total number of specimens collected was 131, and only at the Yuba Valley site. In 2006, 96.75 percent of *P. quadratis* specimens were collected in cheatgrass habitat, 3.0 percent in crested wheatgrass habitat and 0.25 percent in sagebrush habitat. Total number of specimens collected was 401, with 400 specimens collected at Yuba Valley, and 1 collected at Tintic Valley. The 2006 seasonal trap abundance for both *P. minuta* (Figure 3) and *P. quadratis* (Figure 4) peaked from mid-July to early August.

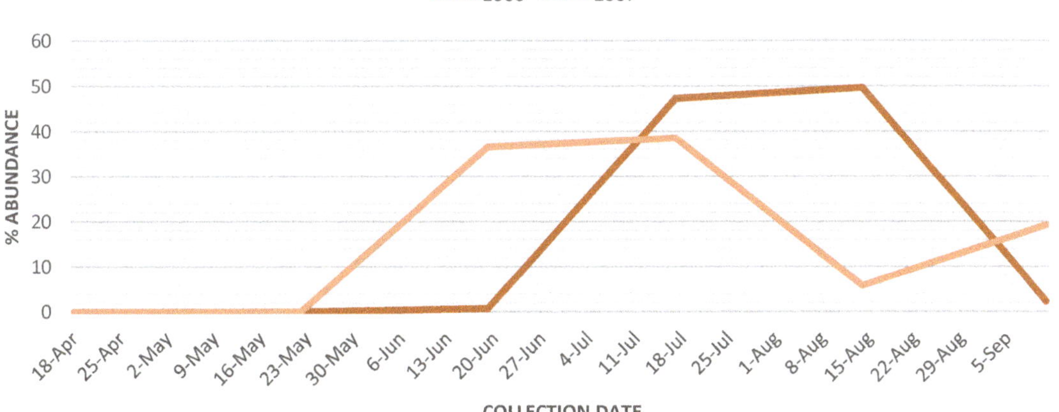

Figure 3. Relative seasonal trap abundance for *Psilochalcis minuta* over a two-year period at Utah; Juab County, Yuba Valley. Note that peak abundance shifted one month earlier between years.

Collection data for 2007 showed a decrease in seasonal abundance and a time shift in the peak abundance for both *Psilochalcis* species. The total number of *P. minuta* specimens collected in 2007 decreased to 52, with all specimens collected in pinyon/juniper habitat at Yuba Valley. The seasonal abundance for *P. minuta* peaked one month earlier in mid-June (Figure 3). The total number of *P. quadratis* specimens collected decreased to 33, with 75.8 percent of specimens collected in cheatgrass habitat, 12.1 percent in crested wheatgrass habitat and 12.1 percent in sagebrush habitat. Thirty *P. quadratis* specimens were collected at Yuba Valley, with 2 specimens collected at Antelope Valley and 1 specimen collected at Tintic Valley. The seasonal abundance for *P. quadratis* peaked two months earlier in mid-May (Figure 4).

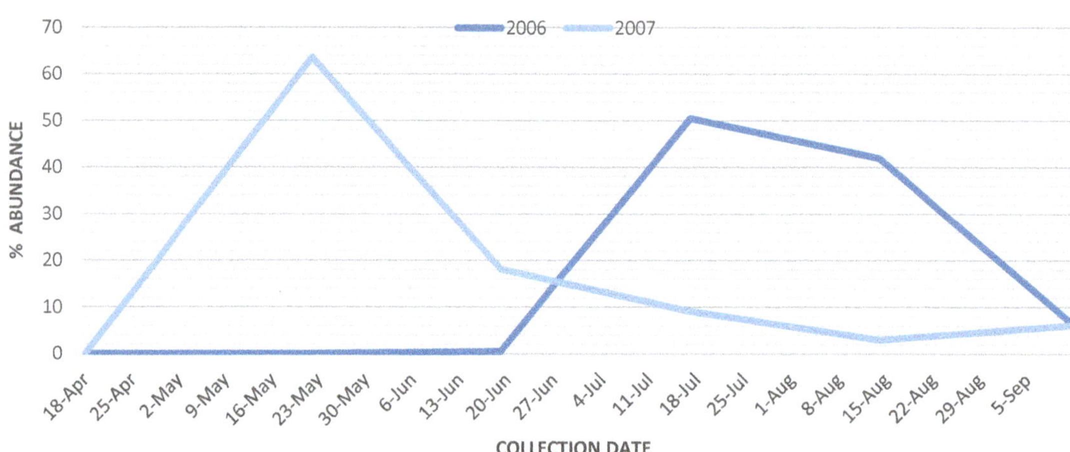

Figure 4. Relative seasonal trap abundance for *Psilochalcis quadratis* over a two-year period at Utah; Juab County, Yuba Valley. Note that peak abundance shifted two months earlier between years.

Subsequent Malaise trapping from other locations in 2020 through 2022 revealed similar *Psilochalcis* affinities to habitat. *Psilochalcis minuta* was predominantly found in traps placed in pinyon/juniper habitat. All eleven traps placed in pinyon/juniper yielded specimens of *P. minuta* totaling 348. Five *P. minuta* specimens were retrieved from two traps placed in cheatgrass habitat. Three of four traps placed in cheatgrass habitat yielded specimens of *P. quadratis* totaling 10. Additionally, 18 *P. quadratis* specimens were retrieved from five traps placed in pinyon/juniper.

3.2. Plant Composition

Pinyon/juniper habitat sites are characterized by the dominant species Utah juniper (*Juniperus osteosperma*). The percent cover of Utah Juniper was 10.9% at Tintic Valley, 19.4% at Yuba, and 17.3% at Antelope Valley. Only the Antelope Valley site had cover of pinyon pine (Pinus edulis) at 6.0%. Cheatgrass is present in the understory at all three locations but with extremely low cover. Tintic Valley had cheatgrass cover of 0.1%, Yuba Valley had 1.5% and Antelope Valley had 0.1%. Other plants varied amongst the three locations (Table 1).

Cheatgrass habitat sites are characterized by the dominant species cheatgrass (*Bromus tectorum*). The percent cover of cheatgrass at Tintic Valley was 16.1%, 1.3% at Yuba, and 42% at Antelope Valley. *Eremopyrum triticeum* (Gaertn.) Nevski had a slightly greater percent cover than cheatgrass at the Yuba site. It should be noted that the Yuba cheatgrass site experienced cheatgrass die-off in 2007, a phenomenon that occurs sporadically but regularly throughout the Great Basin [19]. No other species were sampled at the Antelope Valley site. Other plants varied amongst the other two locations (Table 2).

Table 1. Species Composition in Pinyon/Juniper Habitat Type.

Antelope Valley	Percent Cover
Bromus tectorum L.	0.05%
Ceratocephala testiculata (Crantz) Besser	0.27%
Chaetopappa ericoides (Torr.) G.L. Nesom	0.02%
Chrysothamnus viscidiflorus (Hook.) Nutt.	0.03%
Eriocoma hymenoides (Roem. & Schult.) Rydb.	0.01%
Gutierrezia sarothraea (Pursh) Britton & Rusby	0.04%
Hesperostipa comata (Trin. & Rupr.) Barkworth	0.01%
Juniperus osteosperma (Torr.) Little	17.26%
Pinus edulis Engelm.	6.77%
Total	24.46%
Tintic Valley	
Alyssum desertorum Stapf	0.15%
Artemisia tridentata Nutt.	0.31%
Astagalus eurekensis M.E. Jones	0.08%
Astragalus calycosus Torr. ex S. Wats.	0.25%
Astragalus lentiginosus Douglas	0.04%
Bromus tectorum L.	0.09%
Chaenactis douglasii (Hook.) Hook. & Arn.	0.01%
Elymus elymoides (Raf.) Swezey	0.21%
Eriocoma hymenoides (Roem. & Schult.) Rydb.	0.61%
Juniperus osteosperma (Torr.) Little	10.86%
Pseudoroegneria spicata (Pursh) Á. Löve	0.85%
Total	13.46%
Yuba Valley	
Artemisia tridentata Nutt.	0.31%
Bromus tectorum L.	1.48%
Ceratocephala testiculata (Crantz) Besser	0.03%
Chaetopappa ericoides (Torr.) G.L. Nesom	0.01%
Chrysothamnus viscidiflorus (Hook.) Nutt.	0.07%
Eriocoma hymenoides (Roem. & Schult.) Rydb.	0.02%
Hesperostipa comata (Trin. & Rupr.) Barkworth	0.01%
Juniperus osteosperma (Torr.) Little	19.40%
Linanthus pungens (Torr.) J.J. Porter & L. A. Johnson	0.05%
Oreocarya humilis (A. Gray) Greene	0.02%
Total	21.40%

Table 2. Species Composition in Cheatgrass Habitat Type.

Antelope Valley	Percent Cover
Alyssum desertorum Stapf	0.03%
Bromus tectorum L.	42.14%
Ceratocephala testiculata (Crantz) Besser	0.01%
Total	42.18%
Tintic Valley	
Agropyron cristatum (L.) Gaertn.	0.31%
Bromus tectorum L.	16.14%
Eriocoma hymenoides (Roem. & Schult.) Rydb.	0.01%
Pascopyrum smithii (Rydb.) Barkworth& D.R. Dewey	0.26%
Pseudoroegneria spicata (Pursh) Á. Löve	0.31%
Salsola kali L.	0.01%
Sisymbrium altissimum L.	0.01%
Total	17.05%
Yuba Valley	
Bromus tectorum L.	1.31%
Ceratocephala testiculata (Crantz) Besser	0.84%
Ceratoides lanata Krascheninnikovia lanata (Pursh) A.D.J Meeuse & Smit	0.08%
Chorispora tenella (Pall.) DC	0.14%

Table 2. *Cont.*

Antelope Valley	Percent Cover
Elymus elymoides (Raf.) Swezey	0.20%
Eremopyrum triticeum (Gaertn.) Nevski	1.36%
Eriocoma hymenoides (Roem. & Schult.) Rydb.	0.05%
Total	3.98%

In general, sagebrush habitat sites are characterized by the dominant species Big Sagebrush (*Artemisia tridentata*) comprising slightly more than 12% of the total plant cover. Cheatgrass (*Bromus tectorum*) is present in the understory comprising approximately 7.5% total cover. Other forbes found were *Elymus elymoides* and *Chrysothamnus viscidiflorus*, both approximately 0.3% total cover.

Crested wheatgrass habitat sites are characterized by the dominant species Crested wheatgrass (*Agropyron cristatum*) comprising slightly more than 4% of the total plant cover. Cheatgrass (*Bromus tectorum*) is present in the understory comprising approximately 0.7% total cover. Other forbes found were *Stipa hymenoides*, slightly less than 1%, and *Elymus elymoides*, less than 0.5 % total cover.

3.3. Statistical Results

Psilochalcis minuta demonstrated a significant interaction in trap abundance for habitat type by location ($F_{6,\,60}$ = 120.97 (p < 0.0001). The least square means analysis shows a significant interaction with pinyon/juniper habitat at Yuba Valley (p < 0.0001, t value 38.53). All other habitat type and location combinations were insignificant. An odds ratio estimate was not obtained for *P. minuta* due to it being collected only at the Yuba Valley location.

Psilochalcis quadratis also demonstrated a significant interaction in trap abundance for habitat type by location ($F_{6,\,60}$ = 20.06 (p < 0.0001). The least square means combination of cheatgrass at Yuba Valley shows a significant interaction of these two variables (p < 0.0001, t value 16.83). Crested wheatgrass at Yuba Valley also shows significance (p < 0.0001) but with a much lower t value (t value 5.07). With zeros so prevalent in our data set, this significant interaction between Crested wheatgrass habitat at Yuba Valley is to be expected. The number of *P. quadratis* specimens collected from Crested wheatgrass habitat at Yuba Valley were 16 out of 401. *Psilochalcis quadratis* was collected at all three locations. From the logistic regression we obtained an odds ratio estimate that *P. quadratis* is 63.46 times more likely to be present at Yuba Valley than Antelope Valley (confidence interval from 6.13 to 656.88), and 29.73 times more likely to be present at Yuba Valley than Tintic Valley (confidence interval from 4.53 to 195.08).

3.4. Climate Data Results

For Juab County Utah, monthly mean temperatures for 2006 and 2007 were compared to the 20th century average. In 2006, February and March temperatures were at or slightly below the 20th century average. April through July temperatures were consistently warmer. In 2007, the February temperature was slightly above average and rose more above average in March. April through July temperatures were consistently above average. The temperatures for these months were slightly higher when compared to the same months in 2006. In 2006, monthly precipitation was greater from February through September when compared to 2007.

The three weather stations closest to each of our collection sites showed the same trends with minor differences becoming apparent. In 2006, all three stations were lower than the 30-year normal, the greatest deviation occurred at Little Sahara, which was 5.5 degrees below the 30-year normal. Scipio recorded a difference of 3.8 degrees and Manti was 4.1 degrees lower. In 2007, all three stations were higher than the 30-year normal. Temperatures ranged from 4.6 degrees higher at Manti up to 5.0 higher at Little Sahara. April through July in both years were consistently above the 30-year normal. The 30-year normal, 2006, and 2007 mean maximum temperatures from July through December were very similar with only minor fluctuations. The three weather stations reported lower monthly precipitation than the 30-year normal. All three stations also reported slightly higher precipitation in March and April of 2006 compared to the same months in 2007. Scipio showed the greatest difference between years, followed by Manti. Overall, the precipitation records for Little Sahara were spotty for both years, so a comparison was not easily made.

4. Discussion

We collected *P. minuta* predominantly from areas of pinyon/juniper habitat, with rare collections in cheatgrass. *Psilochalcis quadratis* was collected primarily from cheatgrass habitat, with occasional collections in pinyon/juniper. The occasional occurrence of *Psilochalcis* species in adjacent habitat types is either due to their close proximities or the host plant of their host moth co-occurring in both habitats. Both *Psilochalcis* species are highly mobile. There is no reason not to expect flight across plant community boundaries. Our results suggest biological and/or ecological factors are driving the associations of *P. minuta* within pinyon/juniper habitat and *P. quadratis* within cheatgrass habitat and not trap placement within the habitat, but we are uncertain what the key host plant is within either habitat.

Plant composition in the same habitat across sites was not always uniform. The variability seen in trap capture between sites is likely due to differences in plant community compositions. The presence or absence of a particular plant species in a given habitat could be the primary factor influencing the presence of *P. minuta* and *P. quadratis* in those areas. This might account for the stark absence of *P. minuta* and *P. quadratis* in Antelope Valley, it being the most different in plant composition.

Psilochalcis are known parasitoids of lepidopterans in the families Pyralidae and Gelechiidae [4]. One particular species, *P. brevialata*, has been documented parasitizing the pupal stage of pyralid moths infesting stored figs [20]. Reports of host associations between dominant plant species occurring in our sampled habitats and pyralid moths are lacking, however; it is interesting to note that the pyralid *Dioryctria albovitella* (Hulst) was reported as attacking pinyon pine [21]. The pupal stage of *D. albovitella* has been documented as occurring from mid-July through September [22]. This timing of pupal abundance coincides with the peak abundance of *P. minuta* and *P. quadratis* adult wasps observed in our study. We observed both pyralid and gelichiid moth species in the same Malaise trap samples in which *P. minuta* and *P. quadratis* specimens were retrieved though they are yet to be quantified or identified to species. They were similarly observed in our subsequent Malaise trap samples. It is therefore likely that the abundance of *P. minuta* in pinyon/juniper habitat and *P. quadratis* in cheatgrass habitat is directly related to the abundance of their lepidopteran hosts occurring in each of these habitat types.

There are many variables affecting the phenology of living organisms, in particular environmental factors. Temperature and precipitation have been shown to play a role in the phenology of plants [23]. This pattern extends to plant/insect associations. The countywide differences in temperature and precipitation between 2006 and 2007 could account for the shift in observed phenology in *P. minuta* and *P. quadratis*. In 2007, warmer than average temperatures were recorded in February and March as well as from July through September when compared to 2006. Less than average precipitation was recorded from January through October when compared to 2006. We hypothesize that the warmer and drier spring of 2007 accelerated the phenology of the plant communities at the Yuba location. This could have altered the timing of the peak abundance of the host moth pupae within those habitats, and in turn resulted in the shift in seasonal abundance observed for both *P. minuta* and *P. quadratis* that parasitize those moths.

5. Conclusions

Psilochalcis minuta and *P. quadratis* are the first North American *Psilochalcis* species for which seasonal abundance has been associated with a specific habitat type. While *P. minuta* is strongly associated with pinyon/juniper habitat and *P. quadratis* is with cheatgrass habitat, the exact plant species accounting for these relationships remains unknown. The associations with habitat type were however, further confirmed by the additional Malaise traps placed in multiple areas beyond the original trap sites in multiple subsequent years.

We recognize there are still many unknown ecological relationships of *Psilochalcis* species. While Malaise traps can be a useful tool for determining species/habitat associations, trap costs and trap maintenance are often unfeasible. The simple addition of habitat information to collection labels would significantly contribute to our increased understanding of *Psilochalcis* and our ability to predict future occurrences through the development of species distribution models. We encourage collectors to include ecological data along with location data to increase our understanding of the ecological relationships of *Psilochalcis* species in their natural habitats. This study uniquely ties two *Psilochalcis* species' seasonal abundance to particular habitats of the Great Basin.

Author Contributions: Conceptualization, M.J.P., V.J.A. and R.L.J.; methodology, M.J.P., V.J.A. and R.L.J.; formal analysis, M.J.P. and D.L.E.; resources, V.J.A.; writing—original draft preparation, M.J.P.; writing—review and editing, M.J.P., V.J.A., R.L.J. and D.L.E.; supervision, V.J.A.; project administration, V.J.A.; funding acquisition, M.J.P. and V.J.A. All authors have read and agreed to the published version of the manuscript.

Funding: This research was funded by Western North American Naturalist, grant number 2019-MLBM-2 and Brigham Young University. The APC was funded by Land.

Data Availability Statement: Not applicable.

Acknowledgments: The authors wish to thank Madelyn Beehler and Alora Gubler for their assistance on the trap fieldwork team and Teresa Gomez for her assistance formatting the location/habitat maps.

Conflicts of Interest: The authors declare no conflict of interest.

References

1. Belesky, D.P.; Malinowski, D.P. Grassland communities in the USA and expected trends associated with climate change. *Acta Agrobot.* **2016**, *69*, 1–25. [CrossRef]
2. Banner, R.E.; Baldwin, B.D.; Leydsman-McGinty, E.I. *Rangeland Resources of Utah*; USU Cooperative Extension: Logan, UT, USA, 2009; pp. 1–188.
3. Bradley, B.A.; Curtis, C.A.; Fusco, E.J.; Abatzoglou, J.T.; Balch, J.K.; Dadashi, S.; Tuanmu, M.N. Cheatgrass (*Bromus tectorum*) distribution in the intermountain Western United States and its relationship to fire frequency, seasonality, and ignitions. *Biolog. Invas.* **2018**, *20*, 1493–1506. [CrossRef]
4. Bouček, Z.; Halstead, J.A. Chalcididae. In *Annotated Keys to the Genera of Nearctic Chalcidoidea (Hymenoptera)*; Gibson, G.A., Huber, J.T., Woolley, J.B., Eds.; NRC Research Press: Ottawa, ON, Canada, 1997; pp. 151–164.
5. Bouček, Z. The new world genera of Chalcididae. In On the New World Chalcididae (Hymenoptera). *Mem. Amer. Entomo. Inst.* **1992**, *53*, 49–117.
6. Petersen, M.J.; Johnson, R.L.; Anderson, V.J. A review of known *Psilochalcis* Kieffer (Hymenoptera: Chalcidoidea: Chalcididae) from the western United States with descriptions of three new species from Utah and surrounding states. *West N. Am. Nat.* **2022**, *82*, 704–718.
7. Universal Chalcidoidea Database. Available online: http://www.nhm.ac.uk/chalcidoids (accessed on 15 December 2022).
8. Johnson, J.A.; Grissell, E.E.; Gokhman, V.E.; Valero, K.A. Description, biology, and karyotype of a new *Psilochalcis* Kieffer (Hymenoptera: Chalcididae) from Indainmeal Moth pupae (Lepidoptera: Pyralidae) associated with culled figs. *Proc. Entomol. Soc. Wash.* **2001**, *103*, 777–787.
9. Grissell, E.E.; Schauff, M.E. New Nearctic Invreia (Hymenoptera: Chalcididae) from lepidopterous pests of peanut. *Proc. Entomol. Soc. Wash.* **1981**, *83*, 1–12.
10. Bouček, Z. On Schwarzella, Invreia and some other Hybothoracini (Hym. Chalcididae). *Boll. Lab. Entomol. Agrar. Silvestri' Portici* **1984**, *41*, 53–64.
11. Townes, H.E. Design for a Malaise trap. *Proc. Entomol. Soc. Wash.* **1962**, *64*, 253–262.
12. Malaise, R. A new insect-trap. *Entomol. Tidskr.* **1937**, *58*, 148–160.
13. Preethi, N.; Menon, P.L.D. Diversity and distribution of chalcid wasps in Kerala: Key biological control agents in cultivated ecosystems. In *Arthropod Diversity and Conservation in the Tropics and Sub-Tropics*; Chakravarthy, A.K., Sridhara, S., Eds.; Springer: Singapore, 2016; pp. 213–226.
14. Tavakoli Roodi, T.; Fallahzadeh, M.; Lotfalizadeh, H. Fauna of chalcid wasps (Hymenoptera: Chalcidoidea, Chalcididae) in Hormozgan province, southern Iran. *J. Insect Bio. Syst.* **2016**, *2*, 155–166.
15. Noyes, J.S. Collecting and preserving chalcid wasps (Hymenoptera: Chalcidoidea). *J. Nat. Hist.* **1982**, *16*, 315–334. [CrossRef]

16. Schoeninger, K.; Souza, J.L.; Krug, C.; Oliveira, M.L. Diversity of parasitoid wasps in conventional and organic guarana (*Paullinia cupana* var. sorbilis) cultivation areas in the Brazilian Amazon. *Acta Amaz.* **2019**, *49*, 283–293. [CrossRef]

17. USAFACTS. Available online: https://usafacts.org/issues/climate/state/utah/county/juab-county (accessed on 16 December 2022).

18. National Weather Service. Available online: https://weather.gov/wrh/climate (accessed on 18 December 2022).

19. Boyte, S.P.; Wylie, B.K.; Major, D.J. Mapping and monitoring cheatgrass dieoff in rangelands of the Northern Great Basin, USA. *Range. Ecol. Manag.* **2015**, *68*, 18–28. [CrossRef]

20. Johnson, J.A.; Valero, K.A.; Hannel, M.M.; Gill, R.F. Seasonal occurrence of postharvest dried fruit insects and their parasitoids in a culled fig warehouse. *J. Econ. Ento.* **2000**, *93*, 1380–1390. [CrossRef] [PubMed]

21. Negron, J.F. Cone and Seed Insects Associated with Piñon Pine. In Proceedings of the Desired Future Conditions for Piñon-Juniper Ecosystems, Flagstaff, AZ, USA, 8–12 August 1994; Volume 258, pp. 97–104.

22. Jenkins, M.J. Seed and cone insects associated with Pinus monophylla in the Raft River Mountains, Utah. *Great Basin Nat.* **1984**, *44*, 349–356.

23. Zachmann, L.J.; Wiens, J.F.; Franklin, K.; Crausbay, S.D.; Landau, V.A.; Munson, S.M. Dominant Sonoran Desert plant species have divergent phenological responses to climate change. *Madroño* **2021**, *68*, 473–486. [CrossRef]

 *land*

Article

Potential of Forage Grasses in Phytoremediation of Lead through Production of Phytoliths in Contaminated Soils

Múcio Magno de Melo Farnezi [1], Enilson de Barros Silva [2,*], Lauana Lopes dos Santos [2], Alexandre Christofaro Silva [1], Paulo Henrique Grazziotti [1], Luís Reynaldo Ferracciú Alleoni [3], Wesley Costa Silva [2], Angela Aparecida Santos [2], Flávio Antônio Fernandes Alves [2], Iracemá Raquel Santos Bezerra [2] and Li Chaves Miranda [4]

[1] Department of Forestry, Federal University of the Jequitinhonha and Mucuri Valley (UFVJM), Campus JK, Diamantina 39100-000, Minas Gerais State, Brazil
[2] Department of Agronomy, Federal University of the Jequitinhonha and Mucuri Valley (UFVJM), Campus JK, Diamantina 39100-000, Minas Gerais State, Brazil
[3] Department of Soil Science, ESALQ, University of São Paulo, Piracicaba 13418-900, São Paulo State, Brazil
[4] Engineering, Federal Institute of Minas Gerais, Campus Bambuí, Bambuí 38900-000, Minas Gerais State, Brazil
* Correspondence: ebsilva@ufvjm.edu.br

Abstract: Phytoremediation has become a promising technique for cleaning Pb-contaminated soils. Grasses have a phytoextractor potential for extracting metal from soil by transporting it and accumulating it in high concentrations in their shoots, and they have the ability to immobilize and inactivate it via phytoliths. The objective of this work was to evaluate the phytoremediation potential of forage grasses through the production of phytoliths and the occlusion of Pb in the phytoliths cultivated in Pb-contaminated soils. Three greenhouse experiments were conducted in a completely randomized design, separated by soil type (Typical Hapludox, Xanthic Hapludox and Rhodic Hapludox), in a 3×4 factorial scheme consisting of three forage grasses (*Megathyrsus maximus*, *Urochloa brizantha* and *Urochloa decumbens*) and four Pb rates (0, 45, 90 and 270 mg kg^{-1}) with four repetitions. The forage grasses were influenced by increases in the Pb concentrations in the soils. The higher Pb availability in Typic Quartzipsamment promoted Pb toxicity, as indicated by the reduced dry weights of the shoots, increased phytolith production in the shoots, increased Pb in the shoots and Pb occlusion in the phytoliths of the forage grasses. The production and Pb capture in the phytoliths in the grasses in the Pb-contaminated soils were related to the genetic and physiological differences in the forage grasses and the Pb availability in the soils. *Urochloa brizantha* was the most tolerant forage to the excess Pb, with a higher production of phytoliths and higher Pb occlusion in the phytoliths, making it a forage grass that can be used in the future for the phytoremediation of Pb-contaminated soils.

Keywords: *Urochloa*; *Megathyrsus maximus*; lead; biomineralization; Entisol; Oxisol

Citation: Farnezi, M.M.d.M.; Silva, E.d.B.; Santos, L.L.d.; Silva, A.C.; Grazziotti, P.H.; Alleoni, L.R.F.; Silva, W.C.; Santos, A.A.; Alves, F.A.F.; Bezerra, I.R.S.; et al. Potential of Forage Grasses in Phytoremediation of Lead through Production of Phytoliths in Contaminated Soils. *Land* **2023**, *12*, 62. https://doi.org/10.3390/land12010062

Academic Editors: Michael Vrahnakis, Yannis (Ioannis) Kazoglou and Manuel Pulido Fernádez

Received: 29 November 2022
Revised: 14 December 2022
Accepted: 17 December 2022
Published: 26 December 2022

1. Introduction

Plants are targets for a wide range of pollutants that vary in concentration, speciation and toxicity. Lead (Pb) is among the most toxic pollutants that affect plants [1–6], being found in all environmental compartments, such as soil, water, the atmosphere and living organisms [5,7]. The capacity for environmental contamination by Pb results from its persistence in soil, which it can achieve via its low mobility and non-degradation in soil [6,8–12].

Lead tends to accumulate on the soil surface, and its concentration is reduced deeper in the soil [5,6,8–12]. Lead, which plants easily uptake and accumulate [1,13,14], is considered a subtle and slow-acting general protoplasmic poison, and it is highly toxic, even in small concentrations [6,8–12]. Lead toxicity causes changes in enzymatic, nutritional, hormonal and water balances and changes in photosynthesis, respiration and membrane structure and permeability, resulting in a reduction in growth, chlorosis and the darkening of the root system of plants [1,6,15]. High concentrations of Pb in a plant may eventually promote cell death, resulting in reduced crop yields becoming a serious problem for agriculture [1,6].

The remediation of Pb-contaminated soils is a major challenge for many industries and government agencies. Lead-contaminated sites have been remedied through a relatively narrow range of engineering-based technologies [1]. Phytoremediation uses plants to remediate areas contaminated with heavy metals, and it is a promising technique with great potential for cleaning Pb-contaminated soils [1,5]. Phytoremediation may involve several processes, including phytoextraction, which is based on easy cultivation and the use of fast-growing plants capable of extracting heavy metal from soil by transporting it and accumulating it in high concentrations in their shoots [1]. Grasses have a potential phytoextractor of heavy metals due to tolerance and accumulation of metals in tissues, besides presenting high growth rate, biomass production and abundant root system [5,16] combined with their ability to produce phytoliths [17–19]. Plant tolerance to Pb depends on genotype and physiological characteristics [1,5,20], with the existence of different defense strategies that provide protection against its harmful effects [1,6,21]. Phytolith production is one of these defense strategies that enable plant survival under such extreme conditions [17–19,22,23].

Phytoliths are amorphous silica particles between 1 and 250 µm in size, resulting from the uptake of silicic acid from the soil solution by plant roots [24,25]. Structures form via the polymerization process of silicic acid, which causes amorphous silica to precipitate along with metals in the cells of some plants [17–19,26]. Silica bodies can occlude harmful metal ions in some parts of plants, reducing the stress caused by these metals in many terrestrial plants, especially in species of the Cyperaceae and Poaceae families [17–19,22,24]. They also reduce soil-soluble metals, mainly in contaminated areas and, most importantly, without the risk of contamination of the food chain due to its stability [12,22,23,25,27–29].

Studies have reported that phytoliths produced in a plant can contribute to the immobilization and subsequent inactivation of the plant's tolerance to toxic metals [22,25,27]. The production and concentration of plant phytoliths depend on phylogenetic characteristics, such as species and genus, and they vary with plant phenology, as well as soil type [17–19,22,23]. Some grass species exhibit characteristics of a high growth rate and biomass yield capacity, and they can tolerate and accumulate toxic metal [28] with a high phytolith yield capacity [17,19,22]. The grasses evaluated in this study are *Megathyrsus maximus* and *Urochloa* genera, with high biomass production and rapid growth [5], classified according to their management and soil fertility requirements [5,16]. *Megathyrsus maximus* is known worldwide for having high productivity and for being able to adapt to different soil and climatic conditions [16], and *Urochloa* grasses stand out for being more rustic and having a good ability to adapt to diverse environments [16], but the phytolith yield capacities of these grasses are unknown. Thus, the present study aimed to evaluate the phytoremediation potential of forage grasses through the production of phytoliths and the occlusion of Pb in the phytoliths cultivated in Pb-contaminated soils.

2. Material and Methods

2.1. Soil Characterization and Experimental Design

Three greenhouse experiments were conducted in Diamantina (18°15′ S, 43°36′ W, 1250 m a.s.l.), Minas Gerais, Brazil. The soils were a Typic Quartzipsamment (TQ), a Xantic Hapludox (XH) and a Rhodic Hapludox (RH), classified according to Soil Taxonomy [30], with different chemical and textural characteristics. The soils were collected in a condition of native "Cerrado" to ensure the absence of metal contamination of the surface soil layer (0–0.2 m depth). A subsample was air-dried and sieved (2.0 mm) for chemical analyses and soil texture determination [31] (Table 1). The available Pb concentrations in the soils were determined by using the USEPA 3052 method [32] (Table 1).

Table 1. Soil attributes before application of basic fertilizer and Pb rates.

Attribute	Unit	Soil [6]		
		TQ	**XH**	**RH**
pH [1] $_{water}$	-	5.1	5.4	5.5
P [2]	mg kg^{-1}	0.2	0.1	0.2
K [2]	mmol$_c$ kg^{-1}	0.4	0.1	0.2
Ca [3]	mmol$_c$ kg^{-1}	6.7	4.50	8.1
Mg [3]	mmol$_c$ kg^{-1}	3.5	1.8	3.9
Al [3]	mmol$_c$ kg^{-1}	7.8	4.2	1.6
Cation-exchange capacity	mmol$_c$ kg^{-1}	40.6	71.4	49.2
Organic carbon	g kg^{-1}	3.5	5.8	5.2
Pb [4]	mg kg^{-1}	0.0	0.0	0.0
Maximum P adsorption	mg kg^{-1}	100	200	250
Sand [5]	g kg^{-1}	830.0	580.0	310.0
Loam [5]	g kg^{-1}	110.0	70.0	180.0
Clay [5]	g kg^{-1}	60.0	350.0	510.0

[1] Soil:water 1:2.5. [2] Mehlich-1 extractor. [3] KCl 1 mol L^{-1} extractor. [4] USEPA 3052. [5] Pipette method.
[6] TQ: Typic Quartzipsamment. XH: Xantic Hapludox. RH: Rhodic Hapludox.

The liming of soils was carried out with dolomitic limestone to increase the base saturation to 45%. The lime requirement (LR) was calculated as LR (Mg ha^{-1}) = ((V$_2$ − V$_1$) × CEC)/100, where V$_2$ is the base saturation recommended for the grasses (45%), V$_1$ is the base saturation in the soil analysis, and CEC is the cation-exchange capacity (Table 1). The soils were incubated for 30 days under field capacity conditions, controlled by daily weighing and maintained throughout the experiment.

The basal fertilization rates were 180 mg N (Urea, NH$_4$H$_2$PO$_4$, (NH$_4$)$_2$SO$_4$, Pb(NO$_3$)$_2$), 150 mg K (KCl), 50 mg S ((NH$_4$)$_2$SO$_4$), 1 mg B (H$_3$BO$_3$), 1.5 mg Cu (CuCl$_2$), 5.0 mg Fe (FeSO$_4$.7H$_2$O-EDTA), 4.0 mg Mn (MnCl$_2$.H$_2$O) and 4 mg Zn (ZnCl$_2$) per kg of soil. The phosphate fertilization was based on the maximum P adsorption capacity of each soil (Table 1), estimated from the data of the Langmuir isotherm second region [33]. The applied phosphorus rate was 200 mg for the TQ, 350 mg for the XH and 450 mg for the RH per kg of soil with source NaH$_2$PO$_4$. Nutrients were applied as pure reagents for analyses, and they were mixed completely and incubated for 15 days in each soil. The Pb rates were applied after liming, and a basic fertilizer was applied as a pure lead nitrate reagent for planting and soil incubation for 15 days.

Three experiments were conducted in a completely randomized design, with a 3 × 4 factorial scheme and three replications. The factors were three forage grasses (*Urochloa decumbens* (Stapf) R.D. Webster cv. Basilisk; *Urochloa brizantha* (Hochst. ex A. Rich.) R.D. Webster cv. Marandu e *Megathyrsus maximus* (Jacq.) B.K. Simon & S.W.L. Jacobs cv. Mombaça) and four Pb rates (0, 45, 90 and 270 mg per kg of soil) examined in three soils. The choice of forage grasses was due to their rusticity, ability to adapt to different environments, easy handling and good market acceptance [16]. The Pb rates were based on the investigation values of the soils [34].

The forage grasses were seeded in pots with 3 kg of soil. Then, 7 days after seedling emergence, thinning was performed, leaving only one plant per pot. The four top-dressing fertilizations of 30 mg N (urea) per kg soil were carried out every 15 days, after the thinning of the grasses. The pots were irrigated daily with distilled water to maintain soil moisture at field capacity, which was checked daily by weighing the plots.

2.2. Measurements and Analytical Determinations

The plant shoots were harvested 120 days after the thinning of the forage grasses. The shoot samples were dried in a forced-air oven at 65 °C until constant weight, and they were ground and weighed to determine their dry weights. The Pb was extracted in a microwave oven (CEM Mars^{TM} 6) with nitric acid (65% *v/v* - Merck), and its concentration was determined using atomic absorption spectrometry in a graphite oven (AAnalyst 800, Perkin-Elmer, Waltham, MA, USA). The quality of the Pb analysis of the plant tissue was assured by using certified reference material (NIST SRM 1573a tomato leaf) and reagent blanks.

The phytoliths in the shoots of the forage grasses were prepared and separated using the process detailed in [35]. The separated phytoliths were opened by using the USEPA 3052 method [32]. The Pb concentrations in the filtered solutions were determined using atomic absorption spectrometry and a graphite oven (AAnalyst 800, Perkin-Elmer, Waltham, MA, USA).

2.3. Statistics

The data were subjected to an analysis of joccccint variance, which consisted of a study of the Pb rates and the forage grasses within each soil type. The means of the forage grasses and soil types were compared by using Tukey's test at a 5% significance level. The regression equations were adjusted for the variables in the function of Pb rates.

3. Results

The shoot dry weights of the forage grasses decreased with an increase in the Pb rates applied to the soils ($p < 0.01$), and the forages that were grown in the TQ presented lower reductions in their shoot dry weights than those that were grown in the other soils (Figure 1). Based on the regression coefficients (Figure 1), *Urochloa brizantha* had the highest tolerance to Pb than the other forages evaluated and the lowest reduction in dry matter yield when grown in all evaluated soils. The higher tolerance of the forages in the TQ and the lower sensitivity to Pb presented by the forage *Urochloa brizantha* may be related to the characteristics of the species.

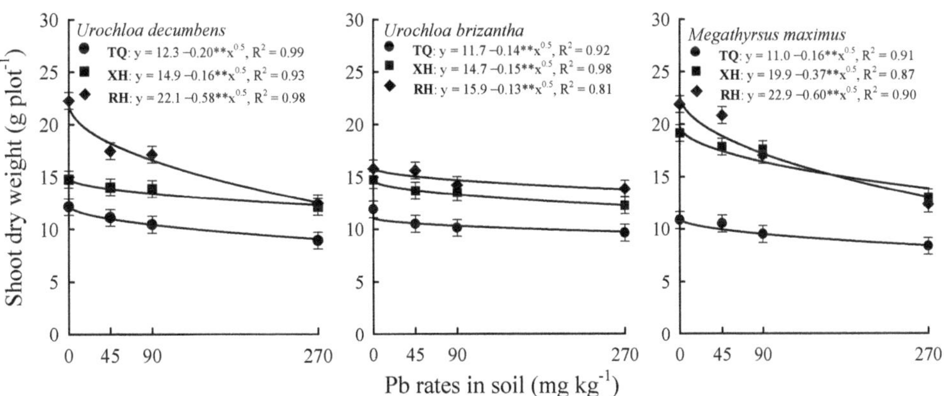

Figure 1. Shoot dry weights of forage grasses with increases in Pb rates within 120 days of thinning in three soils (TQ: Typic Quartzipsamment. XH: Xantic Hapludox. RH: Rhodic Hapludox) (significant at ** $p = 0.01$).

Phytolith production in the forage shoots increased with an increase in the Pb rates applied to the soils ($p < 0.01$). The three forages presented phytolith production, independent of soil and species, with *Urochloa brizantha* presenting the highest phytolith production (Figure 2). This may also explain the higher phytolith production observed in the forages that were grown in the TQ than in the forages that were grown in the other evaluated soils (Figure 2). Pb toxicity caused a greater increase in the phytoliths in *Urochloa brizantha* than in the other forages evaluated, with the higher yield reflecting the genetic and physiological differences between the forage grasses in producing phytoliths.

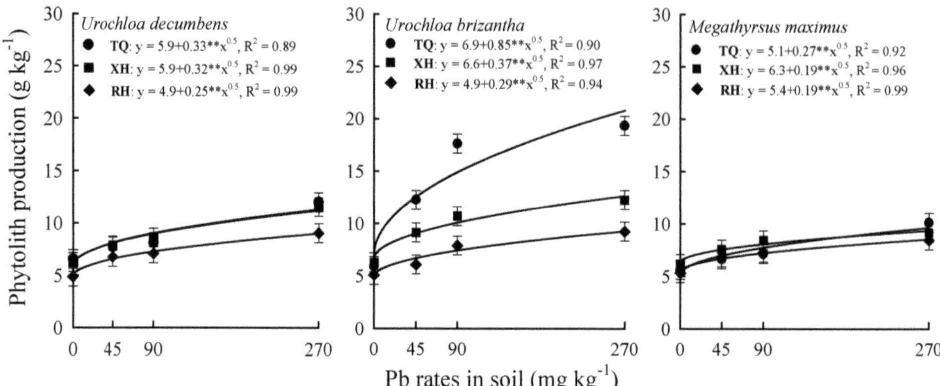

Figure 2. Phytolith production in forage grasses with increases in Pb rates within 120 days of thinning in three soils (TQ: Typic Quartzipsamment. XH: Xantic Hapludox. RH: Rhodic Hapludox) (significant at ** $p = 0.01$).

The lead concentrations in the shoots and phytoliths were evaluated to verify the Pb uptake and Pb occlusion by the grasses, respectively. The Pb addition to the soils linearly increased the Pb concentrations in the shoots (Figure 3) and phytoliths (Figure 4). The forages grasses only differed in Pb uptake (Figure 3) and Pb occlusion in the phytoliths (Figure 4) when cultivated in the TQ.

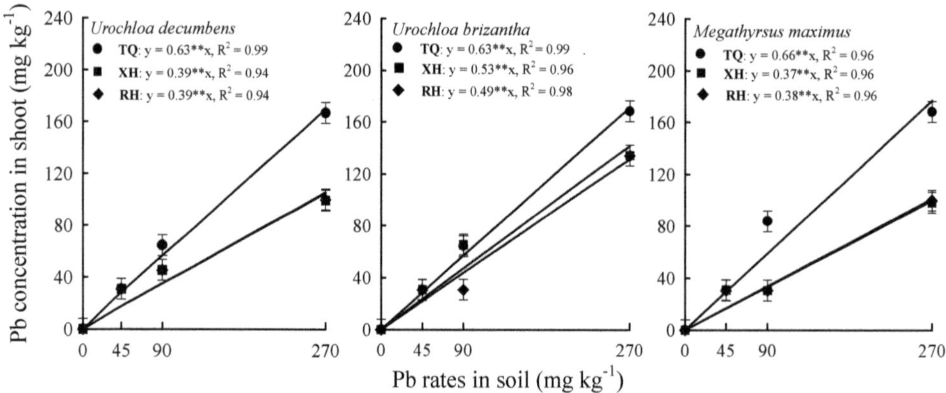

Figure 3. Pb concentrations in shoots of forage grasses with increases in Pb rates within 120 days of thinning in three soils (TQ: Typic Quartzipsamment. XH: Xantic Hapludox. RH: Rhodic Hapludox) (significant at ** $p = 0.01$).

Higher Pb concentrations in the shoots were found in the grasses that were cultivated in the TQ than in the grasses that were cultivated in the other soils (Figure 3); this was due to the higher Pb availability in the TQ, being a soil with a sandy texture (Table 1), reflecting the toxic effect of Pb on the dry matter yield of grasses (Figure 1). On average, the Pb concentration in the shoot of *Urochloa brizantha* was 17% higher than that in the other two forage grasses when grown in the soils at the highest applied rate (270 mg kg^{-1}).

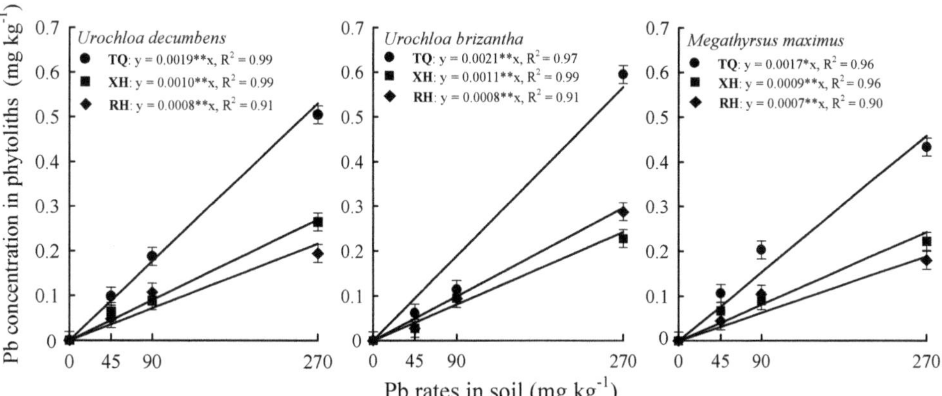

Figure 4. Pb concentrations in phytoliths of forage grasses with increases in Pb rates within 120 days of thinning in three soils (TQ: Typic Quartzipsamment. XH: Xantic Hapludox. RH: Rhodic Hapludox) (significant at ** $p = 0.01$).

The Pb occlusion in the phytoliths was higher in the grasses that were grown in the TQ than in the grasses that were grown in the other soils (Figure 4); this occurred in response to the greater toxic effect of Pb due to the higher availability of Ni in the TQ, which is related to this soil having a sandier texture than the other evaluated soils (Table 1). *Urochloa brizantha* had a higher Pb occlusion in the phytoliths than the other evaluated forage grasses, which may be related to the tolerance mechanism of this grass in response to Pb toxicity.

The lead concentrations in the shoots increased as a function of the Pb rates applied to the three soils ($p < 0.01$) for all the forages evaluated (Figure 3). The lead concentrations in the phytoliths increased with an increase in the Pb concentrations in the three soils ($p < 0.01$). The phytoliths produced in the forages in the different soils (Figure 2) were able to capture Pb metal in the three soils (Figure 4). A higher Pb concentration was observed in the phytoliths in the forages that were grown in the TQ, with *Urochloa brizantha*, in general, being the forage that presented the highest Pb occlusion in the phytoliths, based on the regression coefficients (Figure 4), with a 17% greater P occlusion compared to the other two forage grasses at the highest Pb rate in the soil (270 mg kg^{-1}).

The increase in the Pb concentrations in the soils, in general, provided the greatest reduction in the production of shoot dry matter (Figure 1), the greatest production of phytoliths (Figure 2) and the highest Pb concentrations in the shoots of the forage grasses (Figure 3) and in the phytoliths (Figure 4). *Urochloa brizantha*, in general, was the forage that showed the lowest reduction in biomass production in the shoot (Figure 1), the highest production of phytoliths (Figure 2) and the highest concentration of Pb in the phytoliths (Figure 4). The forages cultivated in the TQ showed the lowest dry matter production in the shoots (Figure 1), the highest production of phytoliths (Figure 2), the highest Pb concentrations in the shoots (Figure 3) and the greatest Pb occlusion in the phytoliths (Figure 4) at the maximum dose of Pb applied (270 mg kg^{-1}).

4. Discussion

The low Pb supply rate reduced the shoot dry weights of the forage grasses in the evaluated soils (Figure 1). The results prove the toxic effect of Pb, since the growth and development of plants grown in Pb environments are affected by several negative effects that occur after Pb absorption by plants [1–6]. The nutritional imbalance caused by Pb may have occurred in the forage grasses evaluated in the present study, contributing to the reductions in the dry weights of the shoots (Figure 1). Nutritional imbalance in different plant species [1], such as reductions in the absorption, distribution and accumulation of macro- and micro-nutrients in various bean crop organs [3], affects their biomass production and development [2,3,5,6].

Lead has a higher availability in sandy soils [5], but the forages cultivated in the TQ showed the lowest reductions in the dry weights of the shoots (Figure 1). The results confirm that the phytotoxicity of Pb [2,3,5,6] depends not only on its availability and concentration [6,8–12] but also on the period of exposure to the metal, the species and its physiological characteristics, the affected organ or tissue and the tolerance mechanisms [1,6,7,21]. Studies have shown that Pb affects plants differently, with the characteristics of each species resulting in different sensitivities to the adverse effects of Pb [1,13,14].

The phytolith production in the forages proves that they are potential producers of silica bodies (Figure 2) in soils with different textures (Table 1), compositions and Pb concentrations. Studies have claimed that Poaceas are major phytolith producers [17–19,22,24]. The forages exhibited increased phytolith production when higher Pb rates were applied to the soils (Figure 2), confirming that a higher concentration and availability of Pb in soil may influence phytolith production in plant organs [22,26]. That is, the higher phytolith production in the forages was related to the higher Pb concentrations and availability [6,8–12] in the soils in the present study. Soils that are more sandy with a low clay concentration, a cation-exchange capacity and organic matter are known to have a higher bioavailability of Pb [6,8–12], which indicates the importance of these factors in the adsorption and desorption of Pb to soil and, consequently, in the production of phytoliths in plants [22].

Urochloa brizantha produced more phytoliths with increases in the Pb rates applied to the three soils (Figure 2). Plant phytolith production depends on phylogenetic characteristics, such as genus and species, but species phenology, soil characteristics and environmental conditions are also factors that influence the production of silica bodies [12,22,26]. Under experimental conditions, the results indicate that the higher tolerance of *Urochloa brizantha*, demonstrated by the lower reduction in the dry weights of the shoots, may reflect higher phytolith production (Figure 2). Silica bodies called phytoliths provide mechanical strength and help protect against physical, chemical and biological stresses on the plant [17–19,22,23].

The forages evaluated could extract Pb from the soil, and their absorption was not limited by increases in the Pb rates (Figure 3), noting their phytoextracting capacity; it is possible that they were tolerant because they have effective physiological and biochemical mechanisms in place to reduce Pb toxicity in tissues [1,13,14].

At first, the forages evaluated may not be considered plants of interest for Pb phytoremediation due to their reduced growth when the Pb concentrations increased in the soils (Figure 1). In addition, Pb-hyperaccumulating plants can extract, tolerate and accumulate high Pb concentrations in tissues when grown in soils contaminated with Pb. Lead tolerance may be at concentrations exceeding 1.000 mg kg^{-1} of Pb in dry matter [5], and these plants may be able to extract and accumulate it in tissues, reaching Pb concentrations of up to 1% of the dry matter produced [11], although this was not observed in the present study.

In general, almost all hyperaccumulating plants reported in the literature have high concentrations of heavy metals in their dry mass and produce a low biomass, which results in a low metal uptake per area [20]. In this sense, the high dry mass production of the forages [5,16] and their Pb accumulation capacity demonstrate their potential use in Pb phytoremediation programs. In addition, the phenotype of metal overaccumulation in shoots is an extreme plant response to soils with high metallic concentrations and is acquired throughout a plant's evolution [15]. Therefore, the evaluated forages are bioindicator plants, since this classification is given to plants that absorb toxic metals and have internal metal concentrations that reflect external concentrations [20], and, therefore, they may be plants with the potential for Pb phytoextraction in contaminated soils.

The most well-known toxic metal tolerance mechanisms are summarized as mechanisms that act to expel absorbed metal or prevent root entry and detoxification by sequestering the metal into plant-specific organelles, particularly vacuoles [1]. However, other potential mechanisms, such as the intra- and extra-cellular binding of toxic metals and their isolation in a non-vital compartment, are the subject of speculation, discussion and study [12,27]. The potential mechanisms include phytolith production and metal occlusion [17–19,22,23]. In the present study, the Pb applied to soils provided increases in the Pb concentrations in shoots and Pb occlusion in phytoliths (Figure 4), with a higher phytolith production (Figure 2) in the forages. It is possible that the capture and accumulation of Pb by phytoliths are due to a forage defense mechanism, which may have helped the evaluated forages to reduce Pb toxicity.

Among the functions attributed to plant phytolith production [17–19] is the relief of physiological stresses on plant growth due to heavy metal toxicity from the capture and immobilization of these metals by the silico phytoliths [12,22,23,25,27–30]. A higher production of phytoliths (Figure 2) and Pb occlusion in the phytoliths were observed when the forages were grown in the TQ (Figure 4), possibly due to the higher availability and/or concentrations of Pb and Si in this soil, because its low clay concentration, cation-exchange capacity and organic matter (Table 1) are characteristics that increase soil Pb availability [5]. In contrast, in the clay soils with higher Fe oxide concentrations (XH and RH) (Table 1), the forages presented lower Pb toxicity with lower reductions in the dry weights of the shoots (Figure 1) when compared to the sandy soil (TQ). The lower effect of Pb phytotoxicity can be attributed to the lower Pb availability in clay soils, whose components have a strong adsorptive capacity, and to the higher Pb binding energy in soils with higher clay mineral concentrations, which reduces the availability of Pb to plants and, consequently, ensures a toxic potential that is lower than that of sandy soils [5,6,8–12].

The available Pb concentration is adsorbed in an exchangeable form in soil, indicating high Pb mobility and immediate bioavailability [6,8–12], while concentrations resulting from intense chemical bonds are present in fractions of organic matter and oxides of amorphous and crystalline Mn and Fe, and they indicate that the metal is immobilized, poorly mobile in the environment and has a low availability to plants, presenting a lower risk of environmental contamination [6,8–12]. In addition to the availability of Pb in the soil, the production and chemical composition of phytoliths can be influenced by metal uptake, climatic conditions, the silicon concentration in the soil, plant species, location, disease resistance and fertilizer requirements [22,25,26].

5. Conclusions

The production of phytoliths by forage grasses in Pb-contaminated soils can increase the tolerance of forages to Pb through the detoxification, immobilization and inactivation of Pb due to the stable nature of these siliceous bodies. *Urochloa brizantha* can be a future forage grass used for the phytoremediation of Pb-contaminated soils. However, there is a need for further studies to evaluate the role of phytolith formation in Pb sequestration in forage grasses.

Author Contributions: Conceptualization, E.d.B.S.; Methodology, M.M.d.M.F., E.d.B.S., L.L.d.S., A.C.S., W.C.S., A.A.S., F.A.F.A., I.R.S.B. and L.C.M.; Formal analysis, M.M.d.M.F., E.d.B.S., W.C.S., A.A.S., F.A.F.A., I.R.S.B. and L.C.M.; Investigation, M.M.d.M.F., E.d.B.S. and L.L.d.S.; Resources, M.M.d.M.F., E.d.B.S., A.C.S., P.H.G., L.R.F.A., W.C.S., A.A.S., F.A.F.A., I.R.S.B. and L.C.M.; Writing—original draft, M.M.d.M.F., E.d.B.S., L.L.d.S., A.C.S., P.H.G., L.R.F.A., W.C.S., A.A.S., F.A.F.A., I.R.S.B. and L.C.M.; Writing—review & editing, M.M.d.M.F., E.d.B.S., L.L.d.S., A.C.S., P.H.G., L.R.F.A., W.C.S., A.A.S., F.A.F.A., I.R.S.B. and L.C.M.; Visualization, M.M.d.M.F. and E.d.B.S.; Supervision, E.d.B.S.; Project administration, E.d.B.S.; Funding acquisition, E.d.B.S., and L.R.F.A. All authors have read and agreed to the published version of the manuscript.

Funding: This research was funded by National Council for Scientific and Technological Development (CNPq) grant number 303599/2015-4 and Research Supporting Foundation for the State of Minas Gerais (Fapemig) grant number CAG-PPM-00480-16.

Institutional Review Board Statement: Not applicable.

Informed Consent Statement: Not applicable.

Data Availability Statement: Not applicable.

Acknowledgments: We acknowledge The Coordination for the Improvement of Higher Level Personnel (Capes) provided a graduate student stipend. The Department of Soil Science, ESALQ, University of São Paulo, provided the laboratory analysis of soil samples. The Federal University of the Jequitinhonha and Mucuri Valley (UFVJM) provided the infrastructure needed to conduct this study.

Conflicts of Interest: The authors declare no conflict of interest.

References

1. Sperdouli, I. Heavy metal toxicity effects on plants. *Toxics* **2022**, *10*, 715. [CrossRef] [PubMed]
2. Van Der Merwe, M.J.; Osorio, S.; Moritz, T.; Nunes-Nesi, A.; Fernie, A.R. Decreased mitochondrial activities of malate dehydrogenase and fumarase in tomato lead to altered root growth and architecture via diverse mechanisms. *Plant Physiol.* **2009**, *149*, 653–669. [CrossRef] [PubMed]
3. Hossain, M.A.; Piyatida, P.; Silva, J.A.T.; Fujita, M. Molecular mechanism of heavy metal toxicity and tolerance in plants: Central role of glutathione in detoxification of reactive oxygen species and methylglyoxal and in heavy metal chelation. *J. Bot.* **2012**, *2012*, 1–37. [CrossRef]
4. Cannata, M.G.; Carvalho, R.; Bertoli, A.C.; Bastos, A.R.R.; Carvalho, J.G.; Freitas, M.P.; Augusto, A.S. Effects of lead on the content, accumulation, and translocation of nutrients in bean plant cultivated in nutritive solution. *Commun. Soil Sci. Plant Anal.* **2013**, *44*, 939–951. [CrossRef]
5. Nascimento, S.S.; Silva, E.B.; Alleoni, L.R.F.; Grazziotti, P.H.; Fonseca, F.G.; Nardis, B.O. Availability and accumulation of lead for forage grasses in contaminated soil. *J. Soil Sci. Plant Nutr.* **2014**, *14*, 783–802. [CrossRef]
6. Fontenele, N.M.B.; Otoch, M.D.L.O.; Gomes-Rochette, N.F.; Menezes Sobreira, A.C.; Barreto, A.A.G.C.; Oliveira, F.D.B.; Melo, D.F. Effect of lead on physiological and antioxidant responses in two Vigna unguiculata cultivars differing in Pb-ac.cumulation. *Chemosphere* **2017**, *176*, 397–404. [CrossRef]
7. Charkiewic, A.E.; Backstrand, J.R. Lead Toxicity and Pollution in Poland. *Int. J. Environ. Res. Public Health* **2020**, *17*, 4385. [CrossRef]
8. Miranda, L.S.; Anjos, J.A.S.A. Occupational impacts and adaptation to standards in accordance with Brazilian legislation: The case of Santo Amaro, Brazil. *Saf. Sci.* **2018**, *104*, 10–15. [CrossRef]
9. Ng, C.C.; Nasrulhaq, A.; Rahman, M.M.; Abas, M.R.B. Tolerance Threshold and Phyto-assessment of Cadmium and Lead in Vetiver Grass, *Vetiveria zizanioides* (Linn.) Nash. *Chiang Mai J. Sci.* **2017**, *44*, 1367–1378.
10. Yang, Y.; Jiang, M.; Liao, J.; Luo, Z.; Gao, Y.; Yu, W.; He, R.; Feng, S. Effects of simultaneous application of double chelating agents to pb-contaminated soil on the phytoremediation efficiency of *Indocalamus decorus* Q. H. Dai and the soil environment. *Toxics* **2022**, *10*, 713. [CrossRef]
11. Ali, H.; Khan, E.; Sajad, M.A. Phytoremediation of heavy metals-concepts and applications. *Chemosphere* **2013**, *91*, 869–881. [CrossRef] [PubMed]
12. Adrees, M.; Ali, S.; Rizwan, M.; Zia-ur-Rehman, M.; Ibrahim, M.; Abbas, F.; Farid, M.; Qayyum, M.F.; Kashiflrshad, M. Mechanisms of silicon-mediated alleviation of heavy metal toxicity in plants: A review. *Ecotox. Environ. Saf.* **2015**, *119*, 186–197. [CrossRef] [PubMed]
13. Inam, F.; Deo, S.; Narkhede, N. Analysis of minerals and heavy metals in some spices collected from local market. *J. Pharm. Biol. Sci.* **2013**, *8*, 40–43. [CrossRef]
14. Usman, K.; Abu-Dieyeh, M.H.; Zouari, N.; Al-Ghouti, M.A. Lead (Pb) bioaccumulation and antioxidative responses in *Tetraena qataranse*. *Sci. Rep.* **2020**, *10*, 1–10. [CrossRef] [PubMed]
15. Singh, S.; Parihar, P.; Singh, R.; Singh, V.P.; Prasad, S.M. Heavy metal tolerance in plants: Role of transcriptomics, proteomics, metabolomics, and ionomics. *Front. Plant Sci.* **2016**, *6*, 1143. [CrossRef] [PubMed]
16. Pezzopane, J.R.M.; Santos, P.M.; Mendonça, F.M.; Araujo, L.C.; Cruz, P.G. Dry matter production of Tanzania grass as a function of agrometeorological variables. *Pesq. Agropec. Bras.* **2012**, *47*, 471–477. [CrossRef]
17. Pan, W.; Song, Z.; Liu, H.; Müeller, K.; Yang, X.; Zhang, X.; Li, Z.; Liu, X.; Qiu, S.; Hao, Q.; et al. Impact of grassland degradation on soil phytolith carbon sequestration in Inner Mongolian steppe of China. *Geoderma* **2017**, *308*, 86–92. [CrossRef]
18. Ru, N.; Yang, X.; Song, Z.; Liu, H.; Hao, Q.; Liu, X.; Wu, X. Phytoliths and phytolith carbon occlusion in aboveground vegetation of sandy grasslands in eastern Inner Mongolia, China. *Sci. Total Environ.* **2018**, *625*, 1283–1289. [CrossRef]
19. Yang, X.; Song, Z.; Liu, H.; Van Zwieten, L.; Song, A.; Li, Z.; Hao, Q.; Zhang, X.; Wang, H. Phytolith accumulation in broadleaf and conifer forests of northern China: Implications for phytolith carbon sequestration. *Geoderma* **2018**, *312*, 36–44. [CrossRef]
20. Yan, A.; Wang, Y.; Tan, S.N.; Yusof, M.L.M.; Ghosh, S.; Chen, Z. Phytoremediation: A promising approach for revegetation of heavy metal-polluted land. *Front. Plant Sci.* **2020**, *11*, 359. [CrossRef]
21. Gupta, D.K.; Huang, H.G.; Corpas, F.J. Lead tolerance in plants: Strategies for phytoremediation. *Environ. Sci. Pollut. Res. Int.* **2013**, *20*, 2150–2161. [CrossRef] [PubMed]
22. Buján, E. Elemental composition of phytoliths in modern plants (Ericaceae). *Quatern. Int.* **2013**, *287*, 114–120. [CrossRef]

23. Anala, R.; Nambisan, P. Study of morphology and chemical composition of phytoliths on the surface of paddy straw. *Paddy Water Environ.* **2015**, *13*, 521–527. [CrossRef]
24. Olonova, M.V.; Gudkova, P.D.; Shiposha, V.D.; Kriuchkova, E.A.; Mezina, N.S.; Blinnikov, M. Phytoliths from some grasses (Poaceae) in arid lands of Xinjiang, China. *Acta Biol. Sibirica.* **2021**, *7*, 345–361. [CrossRef]
25. Song, Z.; McGrouther, K.; Wang, H. Occurrence, turnover and carbon sequestration potential of phytoliths in terrestrial ecosystems. *Earth-Sci. Rev.* **2016**, *158*, 19–30. [CrossRef]
26. Oliva, S.R.; Mingorance, M.D.; Leidi, E.O. Effects of silicon on copper toxicity in Erica andevalensis Cabezudo and Rivera: A potential species to remediate contaminated soils. *J. Environ. Monit.* **2011**, *13*, 591–596. [CrossRef]
27. Fernandes-Horn, H.M.; Sampaio, R.A.; Horn, A.H.; Oliveira, E.S.A.; Lepsch, I.F.; Bilal, E. Use of Si-Phytoliths in depollution of mining areas in the Cerrado-Caatinga region, MG, Brazil. *Int. J. Geomate* **2016**, *11*, 2216–2221.
28. Su, R.; Ou, Q.; Wang, H.; Luo, Y.; Dai, X.; Wang, Y.; Chen, Y.; Shi, L. Comparison of phytoremediation potential of *Nerium indicum* with inorganic modifier calcium carbonate and organic modifier mushroom residue to lead-zinc tailings. *Int. J. Environ. Res. Public Health* **2022**, *19*, 10353. [CrossRef]
29. Bhat, S.A.; Bashir, O.; Ul Haq, S.A.; Amin, T.; Rafiq, A.; Ali, M.; Américo-Pinheiro, J.H.P.; Sher, F. Phytoremediation of heavy metals in soil and water: An eco-friendly, sustainable and multidisciplinary approach. *Chemosphere* **2022**, *303*, 134788. [CrossRef]
30. Soil Survey Staff. *Soil Taxonomy: Keys to Soil Taxonomy*, 12th ed.; Department of Agriculture, Natural Resources and Conservation Service: Washington, DC, USA, 2014; p. 360.
31. Teixeira, P.C.; Donagemma, G.K.; Fontana, A.; Teixeira, W.G. *Manual for Methods of Soil Analysis*, 3rd ed.; Embrapa: Brasília, Brazil, 2017; p. 573.
32. USEPA. United States Environmental Protection Agency. *Microwave Assisted Acid Digestion of Sediments, Sludges, Soils and Oils—Method 3052—SW—846*; EPA: Washington, DC, USA, 1994. Available online: http:www.epa.gov/epaosver/hazwaste/test/3052.pdf (accessed on 8 October 2018).
33. Santos, S.R.; Silva, E.B.; Alleoni, L.R.F.; Grazziotti, P.H. Citric acid influence on soil phosphorus availability. *J. Plant Nut.* **2017**, *40*, 2138–2145. [CrossRef]
34. Conama; National Environmental Council of Brazil. Resolution no 420/2009. It Provides Criteria and Guiding Values of Soil Quality for the Presence of Chemical Substances and Establishes Guidelines for the Environmental Management of Areas Contaminated by These Substances as a Result of An-thropic Activities. Conama, Brasília, Brazil. 2009. Available online: http://www2.mma.gov.br/port/conama/legiabre.cfm?codlegi=620 (accessed on 8 February 2022).
35. Parr, J.F.; Lentfer, C.J.; Boyd, W.E. A comparative analysis of wet and dry ashing techniques for the extraction of phytoliths from plant material. *J. Archaeol. Sci.* **2001**, *28*, 875–886. [CrossRef]

Article

A Topographic Perspective on the Propensity for Degradation of Plateau Swampy Meadows in Maduo County, West China

Xilai Li [1,*], Jing Zhang [1] and Jay Gao [2,*]

[1] Key Laboratory of the Alpine Grassland Ecology in the Three Rivers Region at Ministry of Education, College of Agriculture and Animal Husbandry, Qinghai University, Xining 810016, China

[2] School of Environment, University of Auckland, New Zealand Private Bag 92019, Auckland 1142, New Zealand

* Correspondence: xilai-li@163.com or 1985990024@qhu.edu.cn (X.L.); jg.gao@auckland.ac.nz (J.G.)

Abstract: The swampy meadows atop the vast Qinghai–Tibet Plateau in West China fall into alpine, pediment, valley, floodplain, terrace, lacustrine, and riverine types according to their hydro-geomorphic properties. They have suffered degradation to various levels of severity due to climate change and external disturbance. In this paper, we studied the propensity of these types of swampy meadows to degrade from the topographic perspective. Evaluated against four degradation indicators of vegetation, hydrology, soil erosion, and pika (*Ochotona curzoniae*) damage, degradation severity at 106 swampy meadows representing all types of wetlands was graded to one of four levels, from which the field-based propensity to degrade (PtD) index value was derived. Judged against this index, terrace and alpine swampy meadows are the most prone to degradation while valley, lacustrine, and riverine swampy meadows are the least. The index value of a given swampy meadow type bears a close relationship ($R^2 = 0.916$) with its rate of change during 1990–2013, which confirms the validity of the proposed index in predicting the propensity of swampy meadows to change. The observed differential PtD of different types of swampy meadows is attributed primarily to elevation ($R^2 = 0.746$; $p = 0.027$) and, secondarily, to surface morphology ($R^2 = 0.696$; $p = 0.039$). Thus, the elevation at which a swampy meadow is situated is a more important factor to its PtD than its surface morphology. In particular, swampy meadows located at a higher elevation with a convex surface are much more prone to degradation than those at a lower elevation of a concave slope. Such findings can guide the proper management of different types of swampy meadows to achieve sustainable animal husbandry.

Keywords: swampy meadow type; degradation propensity; severity assessment; topographic influence; Qinghai–Tibet Plateau

Citation: Li, X.; Zhang, J.; Gao, J. A Topographic Perspective on the Propensity for Degradation of Plateau Swampy Meadows in Maduo County, West China. *Land* **2023**, *12*, 80. https://doi.org/10.3390/land12010080

Academic Editors: Michael Vrahnakis, Yannis (Ioannis) Kazoglou and Manuel Pulido Fernádez

Received: 2 December 2022
Revised: 19 December 2022
Accepted: 24 December 2022
Published: 27 December 2022

1. Introduction

Swampy meadows around the world provide several important eco-services, such as balancing regional ecology, conserving biodiversity, trapping pollutants, and being important habitats for the wildlife. As a consequence of global climate change, nutrient enrichment, salinization, and pollution with pesticides and heavy metals, swampy meadows around the world are facing a mounting risk of degradation, with millions of hectares lost over the last few decades [1,2]. Swampy meadow degradation is a highly complex phenomenon that has been defined in terms of hydrology, e.g., shrunk water areas and declined water regulation capacity [3], decreased vegetative cover and its interannual variability [4], changed plant community structure and species diversity [5], and soil properties [6]. In this paper, the degradation of swampy meadows on the Qinghai–Tibet Plateau is defined as the reduction in water reserves to such a level that their ecological functions are adversely impacted, including reduced water regulating capacity, reduced protection of the underlying soil, and reduced grazing value. Of these changes, the change in the hydrological conditions of swampy meadows is considered the most fundamental, as other changes (e.g., change in grass species composition and even the advent of soil erosion) are secondary in that they are triggered by it. Therefore, the propensity of a swampy meadow to degrade is best studied through its hydrological state, especially its water/moisture level.

It is very important to study swampy meadow degradation and understand its causes because it can lead to grave consequences, such as dissolved carbon dynamics [7], reduced carbon uptake and increased global warming potential [8], reduced spawning grounds for fish, extinction of wild flora and fauna, and reduced capability of erosion control and sediment trapping [9]. Due to their environmental sensitivity and vulnerability, the swampy meadows atop the Qinghai–Tibet Plateau have been studied by a number of scientists. Wang et al. evaluated the changes in swampy meadow components, spatial pattern, and hydro-ecologic functions [10]. Niu et al. validated the gross primary production of the alpine swampy meadow on the Tibet Plateau from MODIS satellite data [11], while Yang et al. monitored grassland degradation with the assistance of a remote sensing-derived index in Shangri-La of China [12]. Wu et al. studied the associations between environmental factors in alpine marshy meadows and shifts in plant and soil C, N, and P concentrations and C:N:P stoichiometry [13]. Wu et al. examined the change in the microtopography of swampy meadows in Sanjiangyuan via inferring vegetation and soil properties [14]. Li et al. studied how the degradation of alpine marshy meadows affected ecosystem respiration and its components [15], while Lin et al. explored how the degradation succession of alpine marshy meadows impacted soil organic carbon and total nitrogen in the Yellow River source zone [16]. The potential risk of swampy meadow degradation in the Mt. Qomolangma National Nature Reserve was evaluated from annual mean temperature, settlements, and proximity to roads [17]. However, it still remains unknown why certain types of swampy meadows on the Qinghai–Plateau are more prone to degradation and degrade more seriously than others.

Dependent upon its type and geographic location, a swampy meadow can be degraded by different factors. The common causes of peatland and coastal wetland degradation and disappearance are attributed to land drainage and reclamation for agriculture [18]. The accelerated degradation of lacustrine swampy meadow was caused mainly by constructions in the concerned area and warmer temperature, while annual precipitation and evapotranspiration exerted little influence [19]. However, these causes are not applicable to the plateau setting where grassy wetlands occur mainly as swampy meadows. The causes of their degradation are identified as overgrazing, climate change, and external disturbances [20,21]. Regionally, they have caused widespread degradation and shrinkage of the swampy meadows on the Qinghai–Tibet Plateau to various levels [22].

At a finer local scale, both climate and external disturbance can be assumed to be uniform. Why one type of swampy meadow is more prone to degradation than another is dependent largely on its topography in the landscape. Topographic settings govern the distribution of solar energy and moisture on a slope and, hence, the propensity of swampy meadow to degrade. So far, topography has been considered in predicting sites of future coastal marsh loss [23] and in detecting swampy meadows using a topographic wetland index from multitemporal optical satellite data [24]. Chignell et al. recognized the importance of elevation to the nature of Afroalpine wetland of the Bale Mountains in Ethiopia [25]. Namely, wetlands located at over approximately 3800 m a.s.l. are likely to be ephemeral, and those at lower elevations tend to be perennial. Nungesser analyzed the temporal and spatial changes of a patterned peatland in relation to topography [26]. Nevertheless, nobody has examined the influence of topography on swampy meadow change and its propensity to degrade.

Of particular note, in the plateau setting topography plays an especially decisive role in affecting surface water distribution (e.g., melting of permafrost, evaporation of moisture, and water flow) and, hence, potential degradation of swampy meadows on the Qinghai–Tibet Plateau. How topography affects a swampy meadow's propensity to degrade (PtD) has not been explored yet. This study aimed to bridge this knowledge gap by ascertaining why different types of swampy meadows atop the Qinghai–Tibet Plateau have been degraded to various levels of severity, even though they have undergone the same environmental change over the last few decades. The specific objectives were: (1) to devise an index for realistically assessing the PtD by swampy meadow type based on field data; (2) to determine how the PtD varies among different types of swampy meadow; and (3) to assess the relative influence of elevation and surface morphology on the PtD of swampy meadows in Maduo County on the Qinghai–Tibet Plateau. The knowledge about the topographic influence on the PtD of different types of swampy meadows can guide their proper grazing to achieve sustainable animal husbandry.

2. Study Area

Situated in southern Qinghai Province, Maduo County ($33°50'$ N–$96°50'$ E to $35°40'$ N–$99°20'$ E) has a dimension of 228 km by 207 km, covering an area of 25,253 km^2 (Figure 1). It has a frigid alpine continental climate, with the annual temperature averaging only 1.2 °C. This perennially low temperature regime causes the growing season to be limited to June–September. Most of the county lies between 4500 and 5000 m a.s.l., at which there is no distinct seasonality. Distributed atop the tall mountains are snow and glaciers. Natural vegetation at lower elevations comprises mostly alpine meadows, with grasslands making up 87.5% of the entire county, including marshy and swampy meadows [22].

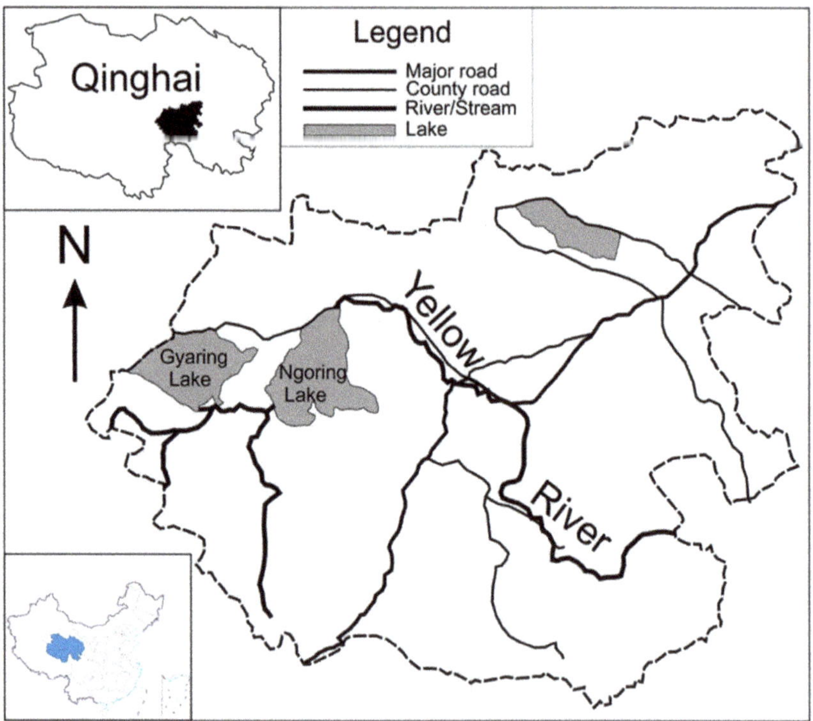

Figure 1. Location of the study area in Qinghai Province, West China.

Maduo receives an annual rainfall of only 303.9 mm per annum, a fraction of the annual evaporation of 1260 mm. Despite this huge deficit, it is bountiful in water resources owing to the injection of water via numerous rivers. In addition, thousands of freshwater lakes are distributed throughout the County at a combined area of 1674 km^2. Associated with the rivers, lakes, and glaciers are swampy meadows of various sizes and types. These swampy meadows are inherently fragile and vulnerable to degradation due to the harsh environment (e.g., strong solar radiation and winds, low precipitation). Swampy meadows have declined in the past, even though they have showed signs of recovery over recent years [10]. This county was selected for study because it encompasses a variety of swampy meadows. The high elevation of the county makes them extremely sensitive to topography and external disturbance. More importantly, the swampy meadows of this area have been widely degraded to various levels as a consequence of overgrazing and climate change [27]. If not properly managed, meadow degradation will worsen with more swampy meadows eventually lost to become ordinary meadows.

3. Grading of Swampy Meadow Degradation Severity

3.1. Swampy Meadow Types

Inland swampy meadows have been classified as alpine, lacustrine, riverine, and swampy based on wetland hydrology, plants, and soil [28]. Since the swampy meadows on the Qinghai–Tibet Plateau have drastically differing internal structures, such a broad classification is not conducive to revealing how they can be properly restored in case of degradation. In particular, no consideration is given to their geomorphic uniqueness. This deficiency has been overcome with the hydro-geomorphic classification in which these swampy meadows are categorized into valley, terrace, floodplain, piedmont, alpine, lacustrine, and riverine [29]. Alpine swampy meadows are small, irregularly shaped marshy meadows distributed in the middle and lower slopes on a tall mountainside. Confined to the bottom of a valley, valley swampy meadows are flanked by mountains or mountainous ranges on both sides, or partially encircled by them if they join. Piedmont swampy meadows are located at the foot of a mountain (range) that has a gentler slope than the mountain slope. Very extensive in area, they usually lie parallel to the mountain (range) in an elongated shape. Floodplain swampy meadows are distributed on a floodplain of a river between the terrace and the channel. Terrace swampy meadows are situated on the higher river terrace due to tectonic activities or channel incision. Spatially, they are further away from the channel than floodplain swampy meadows. Both floodplain and terrace swampy meadows are hydrologically replenished by the river water during flooding. Lacustrine swampy meadows refer to the narrow band of the land–water interface of lakes, within which grassy plants are distributed. Thus, the deeper water devoid of grasses is excluded from consideration. Riverine swampy meadows are those small grassy wetlands located amid inactive or stagnant channels or in the riverbank.

3.2. Selection of Degradation Indicators

Selection of the most appropriate degradation indicators is a prerequisite to constructing a reliable and reasonable grading scheme of degradation severity. Yu and Zhou developed a wetland degradation geoindicatior system involving cause indicators, state indicators, and result indicators [30]. The state and result indicators are identified as land degradation, reduced water reserve, and vegetation degradation [31]. Although vegetative cover and aboveground biomass are significantly lower in degraded swampy meadows than at intact sites [6], biomass is not a reliable indicator due to the varying proportion of surface water area in a swampy meadow. In contrast, the composition of the plant community and the emergence of indicator species are useful clues for assessing swampy meadow degradation [6]. For example, *Kobresia tibetica* is dominant in intact swampy meadows, but is replaced by *Pedicularis* at the advanced stage of degradation. The advent of a completely new plant community comprising mostly pioneer species and alien species is a sure sign of peatland degradation [32].

Denudated ground area and vegetation cover can be used to predict meadow condition and associated ecological thresholds [33]. The percentage of vegetative cover and soil moisture are more reliable indicators than pika burrow density, even though neither is perfect [34]. Based on these findings, four indicators (vegetation, hydrology, soil erosion, and pika damage) were selected for grading the degradation severity of swampy meadows (Table 1). As the most sensitive indicator, vegetation encompasses two subvariables of cover (%) and species composition. A low cover indicates a high severity of degradation. The presence of *Kobresia tibetica* signifies a sound state. The advent of drought-tolerant species that have replaced it suggests severe degradation. Similarly, hydrology also encompasses two subvariables of water reserve and soil moisture. An abundant water reserve is indicative of a healthy state while a dry surface with a moisture content of 25–40% signifies that the swampy meadow is under stress (Table 1). In case of reduced water reserve, the soil moisture at 10 cm below the surface is also used. Soil conditions are indicative of the bioproductivity of degraded swampy meadows and their potential for recovery. The more damage is done to the sod layer, the more likely the underlying soil will be eroded, and the more vulnerable the remaining vegetation will be to erosion, all diminishing the chance of vegetation regeneration and growth, a sign of severe degradation. As a kind of external disturbance to the swampy meadows, pikas (*Ochotona curzoniae*) are an active agent in exacerbating swampy meadow degradation [27]. Pika damage accelerates degradation from the slight to the advanced stage quickly [35]. Since it is difficult to accurately census pika population, the density of active pika burrows was used as a proxy for this indicator.

Table 1. Indicators of plateau swampy meadow degradation and criteria for grading degradation severity of swampy meadows in the study area.

Severity Level	Vegetation		Hydrology		Soil Erosion	Pest Damage (Pika Burrows/9 m^2)
	Cover (%)	Indicator Species *	Water Reserve	Moisture Content at 10 cm		
Reference	>90	K. tibetica	Ponds & pools	>50%	Absent	<1
Slight	>80	K. pygmaea, K humilis	Small pools	>40%	Sod layer damaged	2–3
Moderate	≥50	Poaannua, Stipacapillata	Wet surface	≥25%	Piles of loosened soil	4–5
Severe	<50	Pedicularis	Dry surface	<25%	>50 sod layer gone	≥5

*¹ The exact indicator species vary with wetland type. These are based mostly on swampy meadows.

3.3. Grading of Degradation Severity

After the indicators of swampy meadow degradation have been selected, criteria must be established to grade degradation severity that is enumerated at four levels of intact, slight, moderate, and severe (Table 1). Intact refers to the original, ideal, pristine state of swampy meadows with few signs of external disturbance (Figure 2A). It can serve as the reference state, against which the degradation severity of the same type of swampy meadow is judged. Intact swampy meadows are healthy with abundant forage (mostly *Kobresia tibetica*) for productive grazing. Occasionally, there may be one pika burrow present, but it is mostly innocuous as the soil surrounding it is still not affected. Slight degradation means an 80–90% cover of mostly *Kobresia pygmaea* and *K. humilis* vegetation. Surface water is also reduced to small pools with a corresponding drop in soil moisture (Figure 2B). Some of the original soil has been exposed by pika whose burrows are numbered 2–3 per 9 m^2. By the moderate stage of degradation, surface vegetation cover is reduced further to about 50% (Figure 2C). Although the swampy meadow surface is still wet, the moisture content at 10 cm below the surface drops to just above 25%. By this stage, pika burrow density has risen to 4–5 per 9 m^2. Pika have caused noticeable damage to the soil and partially destroyed the top crust. At the severe stage, <50% of the original vegetation remains, with the remaining vegetation either disappeared or replaced by exotic, unpalatable species of grass, such as *Pedicularis* (Figure 2D). The meadow surface is rather dry with a moisture content <25%. The original turf has been mostly eroded, resulting in a low soil fertility. In extreme cases, only pebbles and sands are left behind, within which pika burrows total more than 5 per 9 m^2.

(A)   (B)
(C)   (D)

Figure 2. Typical severity levels of swampy meadow degradation in the study area. The original state (**A**) can be used as the reference state against which the severity level of degradation is judged (Table 1). (**A**) Intact; (**B**) Slight degradation; (**C**) Moderate degradation; (**D**) Severe degradation.

4. Data and Analysis

4.1. Data Collection

Field work was carried out in late August of 2011. Swampy meadows distributed in a diverse range of elevations were sampled, subject to site accessibility. In total, samples were collected at 106 randomly selected sites encompassing all seven types of swampy meadows. Sample size is proportional to swampy meadow prevalence. Namely, the more predominant types of swampy meadows are better represented in the samples (e.g., having a larger sample size) than the rare ones. At each site, the swampy meadow type was identified first. Afterwards, a sample plot of 3 m by 3 m in size was randomly laid out on the ground. Together with surface water area, vegetative cover within it was estimated visually to an accuracy of 5% by three experts independently, and the average of the three estimates was used as the final result. The grass species and their richness were recorded. After the number of pika burrows was counted, the soil condition (e.g., portion of denudated patches and the remaining sod layer) was assessed, and the slope gradient measured. The surface morphology was identified as one of three forms (linear, concave, and convex), with the general morphologic setting (e.g., curvature) noted. At each site, soil moisture was measured at 10 cm below the surface using the Delta-T ML2x ThetaProbe sensor to an accuracy of $\pm 1\%$. The measurement was replicated thrice at three spots within each plot, and the mean was used as the final reading. Finally, the location of each site was logged with a Garmin GPS_{map} 60 CS_{x} receiver in the stationary mode. Owing to the absence of any obstruction (e.g., no trees and no buildings nearby), horizontal positions were logged at the best accuracy of <10 m, and the vertical height had a much lower accuracy (GPS readings were not differentially corrected, only averaged).

4.2. Data Analysis

The collected data were analyzed to grade the degradation severity at each site into one of the four levels (Table 1). In order to compare the PtD of different types of swampy meadows objectively, the observed number of swampy meadows at each severity level was converted to a numerical weight (e.g., s_4 = severe, s_3 = moderate, s_2 = slight, s_1 = intact). PtD_j of swampy meadow type j (j = 1, 2,..., 7) in a given year was calculated from the summed product of weighted severity of degraded sites (s_i) and their quantity, divided by the total number of sampling sites N_j, namely:

$$PtD_j = \Sigma(\, s_i \times n_i)/N_j \tag{1}$$

where n_i refers to the number of sites at a given severity i (i = 1, 2, 3, 4); s_i is the weight assigned to the severity (e.g., s_4 = 4, s_3 = 3, s_2 = 2, s_1 = 1) (Table 2).

Table 2. The observed number of sampled swampy meadows (n_i) that have been degraded to various severity levels (s_i), the calculated PtD score by swampy meadow type, its mean elevation, and the numerical value assigned to surface morphology of the seven types of swampy meadow for the purpose of regression analysis.

Swampy Meadow Type	Severity of Degradation (s_i)				Sum (N)	PtD Score	Elevation (m)	Morphology	Tendency to Degrade
	Intact (1)	Slight (2)	Moderate (3)	Severe (4)					
Terrace *			2	3	5	3.60	4248	0.2	
Alpine		2	3	2	7	3.00	4310	0.05	Vulnerable
Piedmont	10	8	2	7	27	2.22	4269	0.3	
Floodplain	6	3	4	2	15	2.13	4243	0	Stable
Valley	7	1			8	1.13	4252	−0.5	
Lacustrine	23	5	1	1	30	1.33	4230	−0.2	Resilient
Riverine	12	2			14	1.14	4221	−0.3	

*: Since terrace swampy meadows have been degraded to the moderate level and beyond, they are virtually ordinary meadows and, hence, excluded from further analysis.

The proposed PtD_j index was validated against the rate of swampy meadow change during 1990–2013 via regression analysis. It was determined from overlay analysis of swampy meadow distribution maps visually interpreted from multitemporal Landsat satellite images in a geographic information system (for more information, refer to [22]). After the elevation of the same type of swampy meadow samples was averaged, the influence of elevation and surface morphology on PtD was statistically analyzed through regression analysis individually. Prior to the analysis, each type of linear, concave, and convex surfaces was assigned a weight proportional to its ability to retain water within the swampy meadow. Namely, a positive value was assigned to a convex surface (e.g., piedmont swampy meadow) as it causes water/moisture to diverge from the swampy meadow, reducing its water reserve and increasing its propensity to degrade. Conversely, a negative value was assigned to a concave surface because it facilitates convergence of water/moisture to the swampy meadow. The exact value was proportional to surface curvature (Table 2). A more concave morphology (e.g., valley) receives a higher weight. A weight of 0 was assigned to linear or flat surfaces that neither encourage nor discourage the accumulation of water within the swampy meadow, such as floodplain swampy meadows (Table 2).

5. Results

5.1. Propensity for Degradation by Swampy Meadow Type

Of all the samples, lacustrine swampy meadows are the most represented (30), followed by piedmont (27), while terrace (5), alpine (7), and valley (8) are less represented due to their subordinance in the landscape (Table 2). A swampy meadow type is construed to be more prone to degradation if it has a higher proportion of more severely degraded sample sites and vice versa. The calculated PtD score (Table 2) ranges from 1.13 for valley swampy meadows to 3.60 for terrace swampy meadows. Valley, lacustrine, and riverine swampy meadows are the least prone to degradation with a PtD value < 1.5 (Table 2). They are considered resilient. Except for lacustrine swampy meadows, they have not been degraded beyond the moderate level. Lacustrine swampy meadows have been degraded to all three levels, even though those moderately and severely degraded ones are truly rare, accounting for only 6.7% of the total. The degradation was caused and exacerbated by the frequent trampling of livestock along lakeshores, as deducted from their hoof prints on the ground. Such differential PtD is attributed to water reserve. Both riverine and lacustrine swampy meadows have a large water reserve that enables them to withstand short-term environmental fluctuations without showing obvious signs of degradation. Moreover, the lakeshores and riverbanks are not prone to pika attacks because pika burrows can be easily inundated during rains or flooding. Valley swampy meadows are not so prone to degradation because of their relative abundance of water. The high moisture content of the ground makes them immune to pika attacks.

Piedmont and floodplain swampy meadows are considered stable as they have a PtD value between 2 and 3. Both have experienced degradation at all severity levels (Table 2), with moderately and severely degraded swampy meadows comprising roughly one third of the total sites. Their moderate vulnerability is attributed to their low water reserve and limited chances of hydrologic replenishment. Although floodplain swampy meadows have a higher water reserve, they are not rehydrated frequently. Apart from the direct recharge by rainwater, their primary source of replenishment is river water during infrequent flooding. In contrast, piedmont swampy meadows are constantly replenished via surface and subsurface inflows from upland. However, there is also a high rate of outflow.

Terrace and alpine swampy meadows are the most vulnerable and prone to degradation with a PtD value ≥ 3 (Table 2). Terrace swampy meadows are the most degraded due to their remoteness from the water flow from upslopes. Their high ground from the river channel means that they have a limited chance of being replenished by river water even during flooding. Although saturated with moisture, alpine swampy meadows are still prone to degradation for three reasons despite the fact they are the least subject to grazing due to their high elevation. First, their small extent and a highly limited water reserve make them sensitive to climate fluctuation. A minor drought can trigger degradation. Once their moisture level drops below a certain threshold, alpine swampy meadows become the ideal candidate for pika attacks. Second, they are located at the steepest terrain among all types of swampy meadows. Any effects caused by external disturbances are magnified disproportionately here and can trigger severe degradation easily. Third, located at the highest elevation among all the types of swampy meadows (Table 2), they have the smallest moisture/water contributing area.

5.2. Validation of the Propensity to Degrade Index

The observed PtD of the six types of swampy meadows (terrace swampy meadows were excluded from further study because they did not experience any change, e.g., no change from swampy meadow to ordinary meadow) is correlated closely with their annual rate of change during 1990–2013 (Figure 3) that had been detected from satellite images [22]. The regression relationship between the two can be represented as:

$$\text{Annual rate of change} = 12.31 - 9.009\text{PtD} \ (R^2 = 0.916) \tag{2}$$

The negative coefficient of 9.009 means that those swampy meadows having a larger PtD will be lost at a higher rate than those with a lower PtD. This close relationship indicates that the derived PtD is credible as it can show the propensity of a swampy meadow to degrade. Namely, those swampy meadows more prone to degradation disappeared at a faster pace than those having a lower PtD value during 1990–2013. Conversely, those more resilient ones actually gained more. For instance, alpine swampy meadows having the (second) highest PtD score of 3 suffered the highest rate of loss at 16 km^2 per annum (Figure 3). Having the lowest PtD score of 1.13, valley swampy meadows expanded by 3.3 km^2 annually. They are only one of the two types of swampy meadows whose area increased during 1990–2013. Given their high mean elevation of 4252 m, their area should have shrunk instead of expanded. The explanation is the climate-enhanced melting of glaciers and possibly permafrost that causes more water to converge on the valley floor. The warmed climate in this region over the last two decades [22] accelerated snow melting and permafrost thawing, both of which facilitated the expansion of valley swampy meadows.

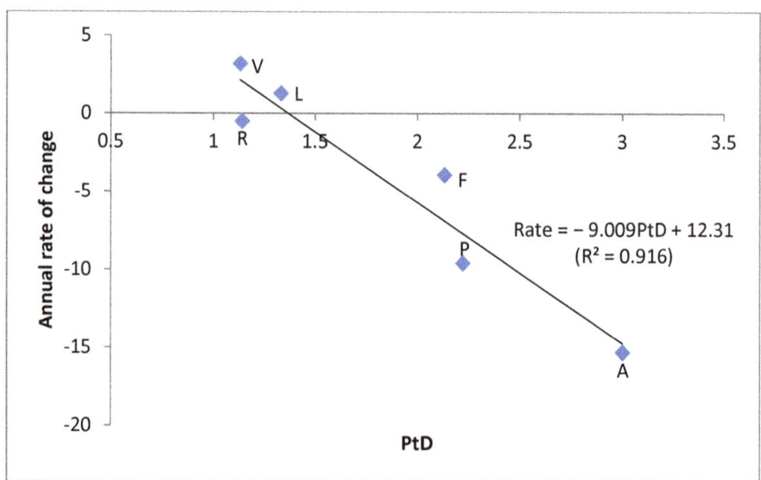

Figure 3. Regression relationship between the annual rate [(Area$_{2013}$−Area$_{1990}$)/(2013–1990), unit: km^2 per annum] of swampy meadow change during 1990–2013 derived from satellite images [22] and the derived propensity for degradation. A-alpine; F-floodplain; L-lacustrine; P-piedmont, R-riverine; V-valley (the same in Figures 4 and 5).

The high R^2 value (0.916) of Equation (2) validates that the derived PtD is able to reveal the propensity of swampy meadows to change reliably. It can be used to predict future changes based on the past environmental settings. In addition, the degradation indicators and the severity grading criteria in Table 1 used to derive the PtD score are appropriate and reasonable. They should be applicable to other areas with a similar setting.

5.3. Influence of Topographic Variables on Degradation Propensity

The influence of topographic variables on the propensity of a swampy meadow type to degrade verbally described above was quantitatively analyzed, and the results are presented in this section. As shown in Figure 4a, the propensity of a given type of swampy meadow to degrade is related positively to its mean height (R^2 = 0.746, p = 0.027):

$$PtD_H = 0.0202Height - 84.273 \ (R^2 = 0.746; p = 0.027) \tag{3}$$

The relationship between height and PtD is perfect for alpine, piedmont, lacustrine, and riverine swampy meadows (Figure 4a). As elevation rises, their PtD also rises linearly. The lowest propensity occurs in lacustrine and riverine swampy meadows whose elevation is lower than 4230 m a.s.l., while alpine swampy meadows are the most prone to change because of their highest elevation (≥4269 m a.s.l.) among all the swampy meadow types. This importance of elevation to PtD is consistent with the finding that changes in elevation will reduce habitat quality within the salt marshes in the San Francisco Estuary [36]. Nevertheless, the regression model is marred by two anomalies, the unusually low PtD of valley swampy meadows, and the slightly above the propensity trend of floodplain swampy meadows. Given their rather high elevation (4252 m a.s.l.), valley swampy meadows should receive a higher PtD score than their elevation suggests of 1.13 while floodplain swampy meadows' PtD should be lower than the current PtD value of 2.13 due to their lower elevation (4243 m a.s.l.). These anomalies can be explained by the surface morphology to be discussed below.

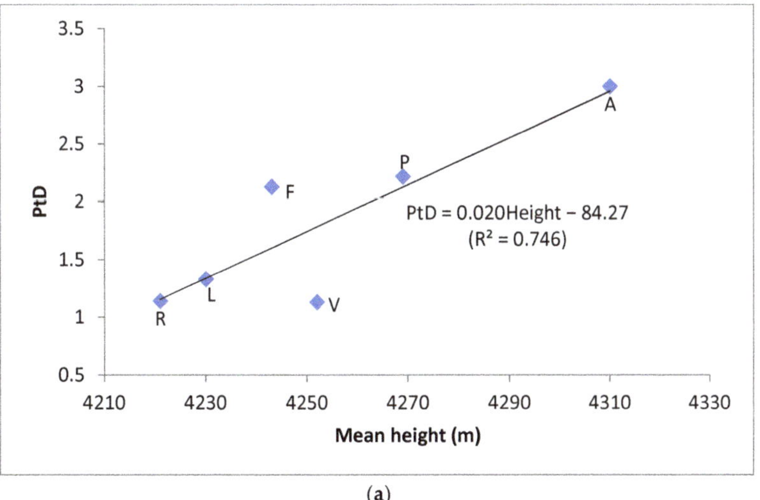

(**a**)

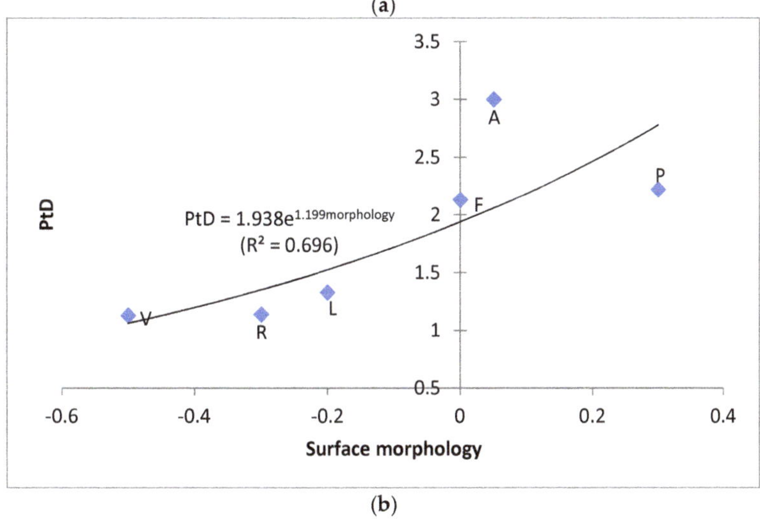

(**b**)

Figure 4. Regression relationship between propensity to degrade (PtD) and topographic features for six types of swampy meadows. (**a**) Mean height (m) above sea level; (**b**) surface morphology.

If PtD is nonlinearly estimated from the weighted surface morphology, 69.6% of its variations can be accounted for by surface morphology (Equation (4)). This proportion is 5.0% lower than that of height. Hence, elevation is a more reliable predictor of a swampy meadow's PtD than surface morphology. The same conclusion can be drawn from the larger p value (0.039 versus 0.027). As shown in Figure 4b, riverine, lacustrine, and valley swampy meadows located in a concave topography all possess a low degree of PtD, which is explained by the accumulation of melted snow and glacier water converging inside them. In contrast, alpine and piedmont swampy meadows deviate from the general trend widely owing to their indistinct morphology or unusually high elevation.

$$\mathrm{PtD}_\mathrm{M} = 1.938e^{1.199\mathrm{Morphology}} + 2.0484 \; (R^2 = 0.696, p = 0.039) \tag{4}$$

6. Discussion

6.1. PtD and Topography

The established relationship between PtD and topography can be traced to water/moisture movement and water balance on a slope and in the catchment. Both are inherently affected by elevation in that water/moisture always flows from a higher ground to a lower one. A higher elevation is synonymous with a smaller catchment size and, hence, lower chances of rehydration. Admittedly, a swampy meadow at a higher elevation is grazed less intensively than its counterpart at a lower elevation, the reduced biomass exerts only a secondary impact on water reserve through evaporation in comparison with temperature. A lower elevation corresponds to a warmer temperature regime that enhances evaporation. Thus, elevation exerts the most direct influence on moisture availability and distribution at the local (e.g., watershed) scale and is, thus, the primary influential controller of PtD of plateau swampy meadows. Dissimilar to elevation that affects all types of swampy meadows indiscriminately, morphology dictates the local movement of water and moisture on a slope for only certain types of swampy meadow selectively.

6.2. Reference State of Degradation

Since intact swampy meadows can serve as the reference state of degradation, naturally, the PtD of a given type of swampy meadow can also be judged from the ratio of the number of degraded swampy meadows to the total number of observed swampy meadows. The portion of degraded swampy meadows out of the total samples (%) is treated as the dependent variable in another regression analysis (Figure 5). This variable achieved a high R^2 value of 0.735 (p = 0.029). This value is rather similar to, but slightly lower than, the 0.746 achieved by PtD. In the scatterplot, the position of the six types of swampy meadows in relation to the general trend is identical to that in Figure 1a. Therefore, the percentage of degraded sites is also a reliable indicator of the tendency of a swampy meadow to degrade, even though it is not as accurate as the PtD. The exact way of expressing degradation severity (e.g., enumerated in two versus four levels) does not alter the influence of elevation on a swampy meadow's PtD.

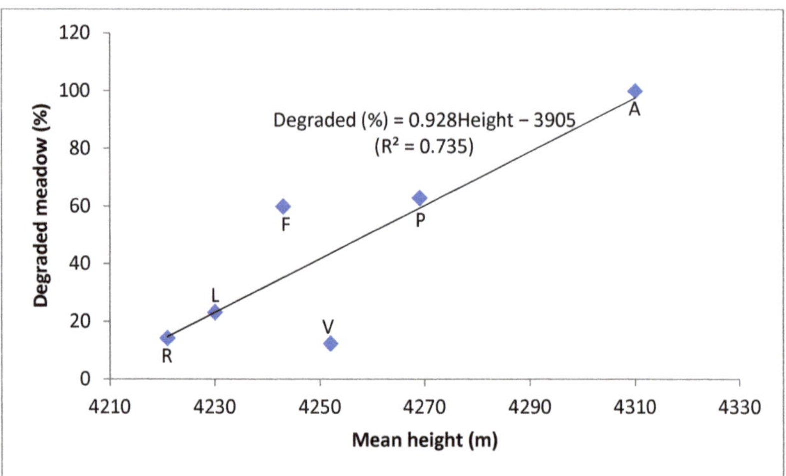

Figure 5. Regression relationship between the ratio of degraded swampy meadow sites to the total sampled sites (%) with their mean height among the six types of swampy meadow.

Due to the lack of the reference state, the 106 samples cannot be analyzed individually in a way similar to Equations (2) and (3). In fact, such regression relationship between topography and PtD may not exist at the individual swampy meadow level because the properties of one type of swampy meadows may overlap with those of another owing to the spatial variation in their topographic features. The relationship becomes more apparent and definite after swampy meadows are grouped by their hydro-geomorphic properties. This grouping is conducive to revealing the topographic influence on the PtD of swampy meadows by type.

7. Conclusions

Derived from swampy meadow degradation severity based on the consideration of vegetation, hydrology, soil erosion, and pika damage, the proposed PtD index of plateau swampy meadows can predict their tendency of change by swampy meadow type in the study area reliably. This conclusion is backed by the close correlation of the calculated PtD score with the 1990–2013 annual rate of swampy meadow change detected from satellite images ($R^2 = 0.916$). The swampy meadows with a higher PtD index value shrank more while those with a lower PtD actually expanded. Of the seven types of swampy meadows, terrace and alpine swampy meadows are vulnerable to degradation judging by their highest PtD value. Both piedmont and floodplain swampy meadows are stable with a moderate PtD value. By comparison, valley, riverine, and lacustrine swampy meadows are resilient in having the lowest value. Such differential PtD is explained mostly by topography. The PtD of a given type of swampy meadow is related inversely to its mean elevation ($R^2 = 0.746$, $p = 0.027$). Elevation is a more effective predictor of the PtD of a swampy meadow type than surface morphology that explains only 69.6% of the variation in PtD ($p = 0.039$), 5% lower than elevation ($p = 0.027$). The level of degradation severity enumeration exerts little influence on the relationship between the mean elevation of a swampy meadow and its PtD. The findings of this study have practical value for proper meadow resource management in that those swampy meadows with a higher PtD value should be grazed less intensively to prevent them from degrading and to achieve sustainable animal husbandry.

Author Contributions: Conceptualization, X.L., J.G. and J.Z.; methodology, J.G. and X.L.; software, J.G.; validation, J.G.; formal analysis, J.G.; investigation, J.G., J.Z. and X.L.; resources, X.L. and J.Z.; data curation, J.G.; writing—original draft preparation, J.G.; writing—review and editing, J.G., J.Z. and X.L.; supervision, J.G.; project administration, X.L.; funding acquisition, X.L. All authors have read and agreed to the published version of the manuscript.

Funding: This research was funded by the Joint Research Project of Sanjiangyuan National Park funded by the Chinese Academy of Sciences and Qinghai Provincial People's Government, grant number LHZX-2020-08, the National Natural Sciences Foundation of China, grant number U21A20191, the 111 Project of China, grant number D18013, and the Qinghai Science and Technology Innovation and Entrepreneurship Team Project titled 'Sanjiangyuan Ecological Evolution and Management Innovation Team', without grant number.

Data Availability Statement: Not applicable.

Acknowledgments: We are appreciated of the constructive comments made by two anonymous reviewers that helped to improve the manuscript.

Conflicts of Interest: The authors declare no conflict of interest.

References

1. Hu, S.; Niu, Z.; Chen, Y.; Li, L.; Zhang, H. Global wetlands: Potential distribution, wetland loss, and status. *Sci. Total Environ.* **2017**, *586*, 319–327. [CrossRef]
2. Mao, D.; Wang, Z.; Wu, J.; Wu, B.; Zeng, Y.; Song, K.; Yi, K.; Luo, L. China's wetlands loss to urban expansion. *Land Degrad. Dev.* **2018**, *29*, 2644–2657. [CrossRef]
3. Wang, G.; Li, Y.; Wang, Y.; Chen, L. Typical alpine wetland system changes on the Qinghai-Tibet Plateau in recent 40 years. *Acta Geogr. Sin.* **2007**, *62*, 481–491. (In Chinese)
4. Li, G.Q.; Kan, A.K.; Wang, X.B.; Li, G.M.; Gao, Z.Y.; Wang, H.; Yong, Z. Distribution of degraded wetlands and their influence factors in Qomolangma National Nature Reserve. *Wetl. Sci.* **2010**, *8*, 110–114. (In Chinese)
5. Hou, Y.; Guo, Z.G.; Long, R.J. Changes of plant community structure and species diversity in degradation process of Shouqu wetland of Yellow River. *Chin. J. Appl. Ecol.* **2009**, *20*, 27–32. (In Chinese)

6.   Gao, Y.H.; Schumann, M.; Zeng, X.Y.; Chen, H. Changes of plant communities and soil properties due to degradation of alpine wetlands on the Qinghai-Tibetan plateau. *J. Environ. Prot. Ecol.* **2011**, *12*, 788–798.
7.   Song, C.C.; Wang, L.L.; Guo, Y.D.; Song, Y.; Yang, G.S.; Li, Y.C. Impacts of natural wetland degradation on dissolved carbon dynamics in the Sanjiang Plain, Northeastern China. *J. Hydrol.* **2011**, *398*, 26–32. [CrossRef]
8.   Ma, W.; Alhassan, A.-R.M.; Wang, Y.; Li, G.; Wang, H.; Zhao, J. Greenhouse gas emissions as influenced by wetland vegetation degradation along a moisture gradient on the eastern Qinghai-Tibet Plateau of North-West China. *Nutr. Cycl. Agroecosyst.* **2018**, *112*, 335–354. [CrossRef]
9.   Bezabih, B. Review on distribution, importance, threats and consequences of wetland degradation in Ethiopia. *Int. J. Water Resour. Environ. Eng.* **2017**, *9*, 64–71. [CrossRef]
10. Wang, C.T.; Long, R.J.; Wang, Q.L.; Jing, Z.C.; Shi, J.J. Changes in plant diversity, biomass and soil C, in alpine meadows at different degradation stages in the headwater region of three rivers, China. *Land Degrad. Dev.* **2009**, *20*, 187–198. [CrossRef]
11. Niu, B.; He, Y.; Zhan, X.; Fu, G.; Shi, P.; Du, M.; Zhang, Y.; Zong, N. Tower-based validation and improvement of MODIS gross primary production in an alpine swamp meadow on the Tibetan Plateau. *Remote Sens.* **2016**, *8*, 592. [CrossRef]
12. Yang, Y.; Wang, J.; Chen, Y.; Cheng, F.; Liu, G.; He, Z. Remote-sensing monitoring of grassland degradation based on the GDI in Shangri-La, China. *Remote Sens.* **2019**, *11*, 3030. [CrossRef]
13. Wu, G.L.; Gao, J.; Li, H.L.; Ren, F.; Liang, D.F.; Li, X.L. Shifts in plant and soil C, N, and P concentrations and C:N:P stoichiometry associated with environmental factors in alpine marshy wetlands in West China. *Catena* **2023**, *221 Pt B*, 106801. [CrossRef]
14. Wu, G.L.; Li, X.L.; Gao, J. The evolution of hummock-depression micro-topography in an alpine marshy wetland in Sanjiangyuan as inferred from vegetation and soil characteristics. *Ecol. Evol.* **2021**, *11*, 3901–3916. [CrossRef] [PubMed]
15. Li, C.Y.; Li, X.L.; Yang, Y.W.; Shi, Y.; Li, H.L.; Yang, P.N.; Duan, C.W. The influence of degradation of alpine marshy wetland on ecosystem respiration and its components. *Wetlands* **2022**, *42*, 62. [CrossRef]
16. Lin, C.Y.; Li, X.L.; Zhang, J.; Sun, H.F.; Zhang, J.; Han, H.B.; Wang, Q.H.; Ma, C.B.; Li, C.Y.; Zhang, Y.X.; et al. Effects of degradation succession of alpine wetland on soil organic carbon and total nitrogen in the Yellow River source zone, west China. *J. Mt. Sci.* **2021**, *18*, 694–705. [CrossRef]
17. Ma, F.; Kan, A.; Li, J.; Lei, G.; Chen, X. Distribution and potential degradation risk evaluation of marsh wetland in the Mt. Qomolangma National Nature Reserve. *J. Geo-Inf. Sci.* **2011**, *13*, 594–600. (In Chinese) [CrossRef]
18. Wójcicki, K.J.; Woskowicz-Ślęzak, B. Anthropogenic causes of wetland loss and degradation in the lower kłodnica valley (southern Poland). *Environ. Socio-Econ. Stud.* **2015**, *3*, 20–29. [CrossRef]
19. Zhou, B.; Jian, Y.; Jin, B.; Lei, Z. Degradation of Wuchang Lake wetland and its causes during 1980–2010. *Acta Geogr.* **2014**, *69*, 1697–1706. [CrossRef]
20. Gao, J. Wetland and its degradation in the Yellow River Source Zone. In *Landscape and Ecosystem Diversity, Dynamics and Management in the Yellow River Source Zone*; Brierley, G.J., Li, X., Cullum, C., Gao, J., Eds.; Springer: Berlin/Heidelberg, Germany, 2016; pp. 209–232.
21. Gao, J.; Li, X.L. Degradation of frigid swampy meadows on the Qinghai–Tibet Plateau: Current status and future directions of research. *Prog. Phys. Geogr.* **2016**, *40*, 794–810. [CrossRef]
22. Li, X.; Xue, Z.P.; Gao, J. Dynamic changes of plateau wetlands in Madou County, the Yellow River Source Zone of China: 1990–2013. *Wetlands* **2016**, *36*, 299–310. [CrossRef]
23. Kearney, M.S.; Roger, A.S. Forecasting sites of future coastal marsh loss using topographical relationships and logistic regression. *Wetl. Ecol. Manag.* **2010**, *18*, 449–461. [CrossRef]
24. Ludwig, C.; Walli, A.; Schleicher, C.; Weichselbaum, J.; Riffler, M. A highly automated algorithm for wetland detection using multi-temporal optical satellite data. *Remote Sens. Environ.* **2019**, *224*, 333–351. [CrossRef]
25. Chignell, S.M.; Laituri, M.J.; Young, N.E.; Evangelista, P.H. Afroalpine wetlands of the Bale Mountains, Ethiopia: Distribution, dynamics, and conceptual Flow Model. *Ann. Am. Assoc. Geogr.* **2019**, *109*, 791–811. [CrossRef]
26. Nungesser, M.K. Reading the landscape: Temporal and spatial changes in a patterned peatland. *Wetl. Ecol. Manag.* **2011**, *19*, 475–493. [CrossRef]
27. Miehe, G.; Miehe, S.; Bach, K.; Nölling, J.; Hanspach, J.; Reudenbach, C.; Kaiser, K.; Wesche, K.; Mosbrugger, V.; Yang, Y.P.; et al. Plant communities of central Tibetan pastures in the Alpine Steppe/Kobresiapygmaea ecotone. *J. Arid. Environ.* **2011**, *75*, 711–723. [CrossRef]
28. Liu, D.; Wang, T.; Shen, W.; Lin, N.; Zou, C. Dynamic of the alpine wetlands and its response to climate change in the YarlungZangbo River Valley in recent 30 years. *J. Ecol. Rural. Environ.* **2016**, *32*, 243–251. (In Chinese) [CrossRef]
29. Gao, J.; Li, X.L.; Brierley, G.; Cheung, A.; Yang, Y.W. Geomorphic-centered classification of wetlands on the Qinghai-Tibet Plateau, Western China. *J. Mt. Sci.* **2013**, *10*, 632–642. [CrossRef]
30. Yu, S.W.; Zhou, A.G. Geoindicator system of wetland degradation. *Geol. Bull. China* **2011**, *30*, 1757–1762. (In Chinese)
31. Hou, Y.; Shang, Z.H.; Ouyang, F.; Wang, L.; Wu, G.L.; Liu, Z.H.; Long, R.J. Analytic hierarchy process on problems and threatening factors of wetland environment in Maqu County, Gansu Province. *Wetl. Sci.* **2009**, *7*, 11–15. (In Chinese)
32. Rebelo, A.J.; Emsens, W.-J.; Meire, P.; Esler, K.J. The impact of anthropogenically induced degradation on the vegetation and biochemistry of South African palmiet wetlands. *Wetl. Ecol. Manag.* **2018**, *26*, 1157–1171. [CrossRef]
33. Li, X.; Perry, G.L.W.; Brierley, G.; Sun, H.; Li, C.; Lu, G. Quantitative assessment of degradation classifications for degraded alpine meadows (heitutan), Sanjiangyuan, Western China. *Land Degrad. Dev.* **2014**, *25*, 417–427. [CrossRef]

34. Gao, J.; Li, X.L.; Cheung, A.; Yang, Y.W. Degradation of wetlands on the Qinghai-Tibet Plateau: A comparison of the effectiveness of three indicators. *J. Mt. Sci.* **2013**, *10*, 658–667. [CrossRef]
35. Hu, G.; Dong, Z.; Lu, J.; Yan, C. Desertification and change of landscape pattern in the source region of Yellow River. *Acta Ecol. Sin.* **2011**, *31*, 3872–3881. (In Chinese)
36. Swanson, K.M.; Drexler, J.Z.; Schoellhamer, D.H.; Thorne, K.M.; Casazza, M.L.; Overton, C.T.; Callaway, J.C.; Takekawa, J.Y. Wetland accretion rate model of ecosystem resilience (WARMER) and its application to habitat sustainability for endangered species in the San Francisco estuary. *Estuaries Coasts* **2014**, *37*, 476–492. [CrossRef]

Article

Multi-Temporal Assessment of Remotely Sensed Autumn Grass Senescence across Climatic and Topographic Gradients

Lwando Royimani [1], Onisimo Mutanga [1,*], John Odindi [1] and Rob Slotow [2]

[1] Discipline of Geography, School of Agricultural, Earth and Environmental Sciences, University of KwaZulu-Natal, P/Bag X01, Scottsville, Pietermaritzburg 3209, South Africa
[2] School of Life Sciences, University of KwaZulu-Natal, Pietermaritzburg 3209, South Africa
* Correspondence: mutangao@ukzn.ac.za

Abstract: Climate and topography are influential variables in the autumn senescence of grassland ecosystems. For instance, extreme weather can lead to earlier or later senescence than normal, while higher altitudes often favor early grass senescence. However, to date, there is no comprehensive understanding of key remote-sensing-derived environmental variables that influence the occurrence of autumn grassland senescence, particularly in tropical and subtropical regions. Meanwhile, knowledge of the relationship between autumn grass senescence and environmental variables is required to aid the formulation of optimal rangeland management practices. Therefore, this study aimed to examine the spatial autocorrelations between remotely sensed autumn grass senescence vis-a-vis climatic and topographic variables in the subtropical grasslands. Sentinel 2's Normalized Difference NIR/Rededge Normalized Difference Red-Edge (NDRE) and the Chlorophyll Red-Edge (Chlred-edge) indices were used as best proxies to explain the occurrence of autumn grassland senescence, while monthly (i.e., March to June) estimates of the remotely sensed autumn grass senescence were examined against their corresponding climatic and topographic factors using the Partial Least Square Regression (PLSR), the Multiple Linear Regression (MLR), the Classification and Regression Trees (CART), and the Random Forest Regression (RFR) models. The RFR model displayed a superior performance on both proxies (i.e., RMSEs of 0.017, 0.012, 0.056, and 0.013, as well as R^2s of 0.69, 0.71, 0.56, and 0.71 for the NDRE, with RMSEs and R^2s 0.023, 0.018, 0.014 and 0.056, as well as 0.59, 0.60, 0.69, and 0.72 for the Chlred-edge in March, April, May, and June, respectively). Next, the mean monthly values of the remotely sensed autumn grass senescence were separately tested for significance against the average monthly climatic (i.e., minimum (T_{min}) and maximum (T_{max}) air temperatures, rainfall, soil moisture, and solar radiation) and topographic (i.e., slope, aspect, and elevation) factors to define the environmental drivers of autumn grassland senescence. Overall, the results indicated that T_{max} ($p = 0.000$ and 0.005 for the NDRE and the Chlred-edge, respectively), T_{min} ($p = 0.021$ and 0.041 for the NDRE and the Chlred-edge, respectively), and the soil moisture ($p = 0.031$ and 0.040 for the NDRE and the Chlred-edge, respectively) were the most influential autumn grass senescence drivers. Overall, these results have shown the role of remote sensing techniques in assessing autumn grassland senescence along climatic and topographic gradients as well as in determining key environmental drivers of this senescence in the study area

Keywords: autumn senescence; grass; climate; remotely sensed; topographic factors

Citation: Royimani, L.; Mutanga, O.; Odindi, J.; Slotow, R. Multi-Temporal Assessment of Remotely Sensed Autumn Grass Senescence across Climatic and Topographic Gradients. *Land* **2023**, *12*, 183. https://doi.org/10.3390/land12010183

Academic Editors: Michael Vrahnakis, Yannis (Ioannis) Kazoglou and Manuel Pulido Fernádez

Received: 28 November 2022
Revised: 29 December 2022
Accepted: 4 January 2023
Published: 6 January 2023

1. Introduction

Climate and topography are key drivers of plant phenology in terrestrial environments [1–7]. Their variability often influences the occurrence, rate, and duration of key phenological stages such as the autumn grassland ecosystem senescence. For instance, [6] noted a variation in the start of grass senescence in the low-lying Inner Mongolian grasslands than the higher Qinghai-Tibetian Plateau. However, the extent and significance of the overlaps between autumn grass senescence and environmental factors such as climate and topography have not been established, especially from a remote sensing point of view. Meanwhile, understanding the relationship between autumn grass senescence and environmental variables is vital, given that senescence markedly decreases photosynthetic activities and plant productivity [8], which, in turn, affects forage quality, production, and availability. Lwando Royimani et al. [9] also noted that senescence can either extend or reduce the floral species growing season with serious implications on forage productivity. In addition, studies [9–11] have noted the socioeconomic and ecological impact of grassland senescence including their regulatory role in the climate–biosphere interactions and potential contribution to land degradation [6]. Given the importance of rangelands and livestock farming for subsistence and commercial purposes, particularly in the developing world [2], knowledge on the implications of senescence on forage productivity in response to climatic and topographic gradients is increasingly becoming a need. This information is required to monitor the impact of autumn senescence on forage productivity [12], hence guiding planning and decision-making on, among others, grazing patterns and stock densities.

Useful assessment of the links between the occurrence of autumn grassland senescence and environmental variables at a landscape-scale requires repeated observations acquired at extensive spatial extents. However, the commonly used methods for assessing plant senescence, such as visual scoring, which monitors changes in leaf color and fall [11], do not effectively satisfy these requirements. Furthermore, these methods are generally not objective and suffer from the time lag effect [13]. Contrarily, remote sensing techniques offer repeated synoptic viewing of the Earth's surface [14–16], which may benefit the assessment of the spatial autocorrelations between grass senescence and environmental factors during the autumn season. Although many studies have examined plant senescence dynamics based on remote sensing techniques [13,17,18], few have focused on the interactions between autumn senescence and environmental parameters. For instance, [6] assessed the impact of temperature, insolation, and precipitation during the dormancy stage on China's temperate biomes using the Normalized Difference Vegetation Index (NDVI) derived over a 30-year period (1981–2011) from the Global Inventory Modeling and Mapping Studies (GIMMS). Their findings showed that temperature is a decisive factor to the end of the growing season. However, the study was generalized across biomes; hence, it did not offer an opportunity for a greater understanding of the autumn-senescence-environmental factors relationship in grassland environments, particularly in the subtropical regions.

In addressing this knowledge gap, the current study examines the spatial autocorrelations between remotely sensed autumn grass senescence and environmental parameters (i.e., climatic and topographic factors) in the subtropical sour-veld grasslands of the Midlands region, KwaZulu-Natal, South Africa, where autumn senescence is a key factor of forage productivity [9]. Such information is critical to ascertain the understanding of the dynamics around the occurrence of autumn grass senescence and to accurately determine grass wilting for improved planning and decision-making on grazing patterns and overall rangeland management. Specifically, a better understanding of the influence of environmental factors on autumn grass senescence will help improve the projection of the onset and duration of autumn grassland senescence, hence reliably determining the period of low- and poor-quality forage for grazing while minimizing the subsequent impact on livestock and wildlife. To achieve this aim, this study adopted two Sentinel-2-derived vegetation indices (i.e., the Normalized Difference NIR/Rededge Normalized Difference Red-Edge (NDRE) and the Chlorophyll Red-Edge (Chlred-edge)) that have been identified as the best proxies for explaining the occurrence of autumn grassland senescence within the study area [10]. Remotely sensed monthly (i.e., March to June) estimates of the autumn grass senescence were assessed for sensitivity against their corresponding climatic (i.e., minimum (T_{min}) and maximum (T_{max}) air temperatures, soil moisture, solar radiation, and rainfall) and topographic (i.e., slope, aspect, and elevation) factors using the Partial Least Squares Regression (PLSR), the Multiple Linear Regression (MLR), the Classification and Regression Trees (CART), and the Random Forest Regression (RFR) models. Next, monthly averages of the remotely sensed autumn grass senescence were tested against monthly mean values of the climatic and topographic variables using Pearson's product-moment correlation approach to understand possible environmental drivers of the autumn grass senescence. We hypothesized that the occurrence of autumn grass senescence in this area can be explained by the dynamics in the micro-climatic and topographic gradients.

2. Materials and Methods

2.1. The Study Site

The study area is situated in Vulindlela, KwaZulu-Natal, South Africa (Figure 1). The total size of the area is 112 km^2 and is characterized by rigid terrain with an elevation ranging between 1273 and 1412 m above sea level (m.a.s.l). The soils are generally loam with random rocky surfaces. Average annual rainfall is around 900 mm [19,20] with mean annual minimum and maximum air temperatures of 6 °C and 22 °C in winter and summer, respectively. Vegetation is mesic subtropical grass, dominated by the Ngongoni (*Aristida junciformis*) of the sour-veld, a mixture of non-native grass species and a random distribution of wattle and pine [10]. Sour-veld grasses are reported to lose their quality through senescence, thus significantly affecting their grazing importance [9]. In addition, grasses in the study area are subjected to regular and uncontrolled livestock grazing patterns, which may have serious implications on the forage. Moreover, irregular fire occurrences are common, especially during the winter season when the grasses are dry due to senescence, in turn affecting forage availability.

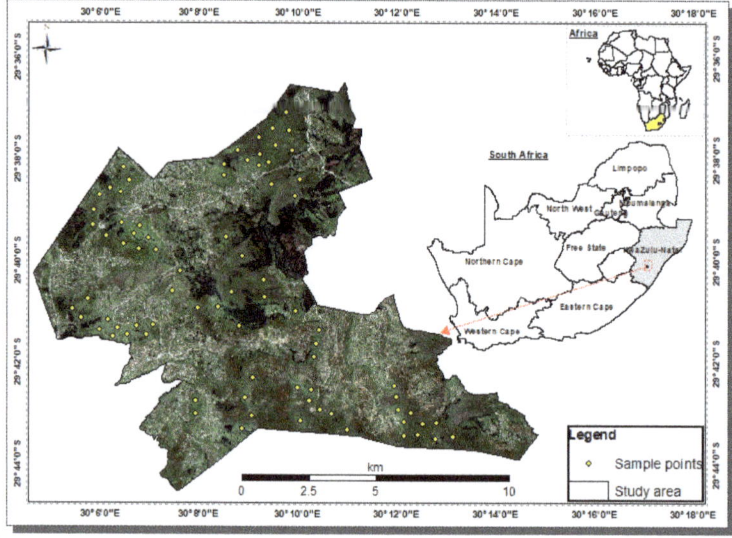

Figure 1. Location of the study area in Vulindlela, KwaZulu-Natal, South Africa and sampling sites.

2.2. Field Data Collection

A purposive sampling approach was used to establish 110 plots measuring about 10 m by 10 m and their center coordinates recorded. The plots were designed to provide a representation of the topography of the study site, particularly with regard to the elevation, aspect, and slope. For instance, some plots were created in low, middle, and high altitudinal areas while considering the effect of south-, east-, west-, and north-facing slopes. Equally, we considered the effect of the slope gradient whereby some plots were designed on steeper while others on gentle slopes. Soil moisture content readings were collected monthly within the plots using the ML3 ThetaProbe Soil Moisture Sensor between the 20 March and 30 June 2021. The ML3 ThetaProbe Soil Moisture Sensor measures soil moisture from the Earth's surface to the depth of 7 cm and the measurements are often expressed in percentage per volumetric water content (%/VWC) [21]. In this study, five measurements were randomly taken within each plot and averaged to obtain a value for the plot, and the points ultimately added up to 110 monthly values. Subsequently, we created four monthly point maps of the soil moisture with the corresponding coordinate points for the months of March, April, May, and June.

2.3. Remotely Sensed Autumn Grass Senescence

Two vegetation indices (i.e., the NDRE and the Chlred-edge), identified as the best proxies in explaining the occurrence of autumn grassland senescence in this area, were adopted [10]. These indices were derived from monthly Sentinel 2 images acquired using the Copernicus Open Access Hub data repository between the 29 March and 25 June 2021. Formulas for these indices are given in Equations (1) and (2). For detailed explanation on the establishment and validation of the named indices, readers are directed to [10]. The considered indices were derived on a monthly basis representing March, April, May, and June 2021. In total, eight vegetation index maps were generated, with four monthly indices generated using the NDRE and the Chlred-edge.

$$NRE = NIR - rededge/NIR + rededge \tag{1}$$

where NIR is the Near-Infrared band and rededge is the red-edge (band).

$$Chlred\text{-}edge = (R_{0.705} - R_{0.740})/(R_{0.783} - R_{0.740}) \tag{2}$$

where $R_{0.705}$ and $R_{0.783}$ correspond to the boundary wavebands while $R_{0.740}$ denotes the center waveband of the red-edge section.

2.4. Climatic and Topographic Variables

Daily rainfall and minimum (T_{min}) and maximum (T_{max}) air temperature data for the study area were acquired from the South African Weather Service (SAWS). The daily rainfall and temperature values were aggregated to obtain monthly records. However, these data were provided as point data for the city of Pietermaritzburg, hence being inadequate for analysis. Therefore, additional monthly T_{min} and T_{max} and rainfall data were downloaded from the KwaZulu-Natal Sugarcane Research Institute (KZN-SRI) website. Whereas the KZN-SRI has many weather stations distributed throughout the province of KwaZulu-Natal, we only used data from 22 stations that are surrounding the study site. The 22 weather stations are in a radius of 10 to 70 km from the central point of the study area across the eastern, northern, southern, and western directions. Next, we interpolated the combined KZN-SRI and SAWS data using the Inverse Difference Weighted (IDW) technique in ArcGIS 10.7 to generate a comprehensive T_{min} and T_{max} as well as rainfall data for the study site. Detailed descriptions of the topographic and climatic factors used are given in Table 1.

Table 1. Topographic plus climatic variables used in this study.

Variable	Units of Measurement	Source
Topographic factors		
Aspect	Degrees North (°N)	ASTER DEM
Elevation	Miters (m)	ASTER DEM
Slope	Degrees (°)	ASTER DEM
Climatic factor		
T_{min}	Degrees Celsius (°C)	SAWS, KZN-SRI
T_{max}	Degrees Celsius (°C)	SAWS, KZN-SRI
Rainfall	Millimeters (mm)	SAWS, KZN-SRI
Radiation	Watts Hours per square meter (Wh/m^2)	ASTER DEM

Note: ASTER = Advanced Spaceborne Thermal Emission and Reflection Radiometer, DEM = Digital Elevation Model.

Aspect, slope, elevation, and radiation were derived from a 30 m Digital Elevation Model (DEM) in ArcGIS. Specifically, aspect and slope were, respectively, calculated using the aspect and slope functions under the surface tools in Spatial Analysis Tools, ArcGIS 10.7 (Environmental Systems Research Institute (ESRI), Johannesburg, South Africa). Similarly, radiation was derived using the Area Solar Radiation extension found under surface tools of the Spatial Analysis Tools, ArcGIS 10.7 (Environmental Systems Research Institute (ESRI), Johannesburg, South Africa). Studies show that the application of modeled solar radiation from the DEM is a widely accepted practice in ecological remote sensing [2,22–24].

2.5. Data Processing and Statistical Analysis

To ensure compatibility and consistency in all the monthly maps generated (i.e., Sections 2.3 and 2.4), we applied the nearest-neighbor resampling approach in ArcGIS 10.7 based on the same resolution. We then overlaid all the monthly vegetation indices plus topographic and climatic maps with their respective monthly point maps to extract the corresponding monthly climatic, topographic, and remotely sensed autumn grass senescence information. Although the total number of the corresponding sampling points was 110, during data preparation, we discovered that 10 of those were outliers and were, hence, discarded in the analysis. Ultimately, we generated four spreadsheets with the monthly climatic and topographic information jointly with corresponding monthly soil moisture contents and remotely sensed autumn grass senescence values. The four monthly spreadsheets were further split into eight spreadsheets based on the vegetation index (i.e., the NDRE or the Chlred-edge) as the predictor variable. The data were separately split into 80 and 20 for calibration and validation, respectively, and imported into R version 4.1.3 ([25] R Core Team) for further analysis (R Core Team, Vienna, Austria). Four popular regression algorithms (i.e., the PLSR, MLR, RFR, and CART) were employed in each monthly NDRE and Chlred-edge spreadsheet to test the association between the remotely sensed autumn grass senescence and the climatic factors and topography. A 10-fold-cross validation approach was used at each stage of analysis to evaluate the model performances based on the obtainable Root-Mean-Square Error (RMSE), the coefficient of determination (R^2), and the Mean Absolute Error (MAE).

2.6. Model Optimization and Identification of Key Environmental Determinants of Autumn Grassland Senescence

According to the performance of the four popular algorithms employed in Section 2.5, one superior model was identified using the RMSE, R^2, and MAE. The model was identified by averaging all the RMSEs, MAEs, and R^2s obtained throughout the four months of investigation. The model that yielded the lowest MAE and RMSE jointly with the highest R^2 was determined to be the best and was, hence, selected for the final prediction of remotely sensed autumn grass senescence with climatic factors and topography. As the superior algorithm, the RFR was adopted and eight final models were built to individually relate the monthly remotely sensed autumn grass senescence values (i.e., NDRE and the Chlred-edge) with their respective monthly climatic and topographic factors. These final models were optimized by tuning their *ntree*, *mtry*, and *nodesize* values. *ntrees* ranged between 300 and 1200, *mtrys* was between 2 and 16, while *nodesizes* was set to 1 throughout the analysis. The final prediction results were judged based on the RMSEs and their R^2s. Next, we averaged all the monthly predictor (i.e., NDRE and the Chlred-edge) and response (i.e., climatic and topographic) variables. The outcome was a set of two spreadsheets, first with the NDRE and second with the Chlred-edge as predictors, along with their monthly averages of topographic and climatic factors. Pearson's product-moment correlation tests were conducted in each set of the spreadsheet to determine the sensitivity of each climatic and topographic factor to the remotely sensed autumn grass senescence. The significance of each topographic or climatic variable in influencing the occurrence of autumn grassland senescence was judged by the *p*-value ($p \leq 0.05$).

3. Results

3.1. Descriptive Statistics

Table 2 provides the descriptive statistics of the remotely sensed autumn grass senescence plus climatic and topographic factors used in this study. Overall, the estimates of autumn grassland senescence based on the NDRE increased with a decrease in the Chlred-edge across the four-month period. In addition, there were no significant variations between the NDRE and the Chlred-edge values of autumn grass senescence from March to June. However, in March, the values of the $NDVI_{705}$-based autumn grass senescence were higher than those of the CHL-RED-EDGE-derived autumn grassland senescence. In addition, monthly means of all the topographic factors (i.e., aspect, elevation, and slope) did not show differences across the four-month period, while monthly means of the climatic variables (i.e., T_{min} and T_{max}, soil moisture, rainfall, and solar radiation) showed notable variations. Specifically, the means of the solar radiation, T_{min} and T_{max}, demonstrated consistent declines throughout the four months, whereas the observable decreases in rainfall and soil moisture from March to May were followed by an increase in June (Table 2).

Table 2. Descriptive statistics of the data gathered and retrieved for analysis.

Month	Variable	Min	Max	Mean	Stdv
March	NDRE	0.248	0.532	0.396	0.057
	Chlred-edge	0.239	0.519	0.357	0.058
	Aspect	7.723	340.649	144.777	87.127
	Elevation	1273	1412	1340	30.359
	Slope	0.512	19.411	5.702	3.860
	T_{max}	25.5	25.85	25.65	0.131
	T_{min}	13.68	14.66	14.13	0.398
	Radiation	22,878	232,161	150,843	65,496.12
	Rainfall	69.44	87.65	79.39	7.095
	Soil moisture	12.5	34.9	22.43	3.764
April	NDRE	0.182	0.477	0.346	0.051
	Chlred-edge	0.266	0.562	0.390	0.056
	Aspect	7.723	340.649	144.777	87.127
	Elevation	1273	1412	1340	30.359
	Slope	0.512	19.411	5.702	3.860
	T_{max}	24.51	25.08	24.78	0.217
	T_{min}	11.25	12.21	11.71	0.387
	Radiation	20,736	256,029	138,918	75,657.96
	Rainfall	58.5	64.74	62.04	2.137
	Soil moisture	10.1	30.1	16.36	4.505
May	NDRE	0.108	0.291	0.223	0.034
	Chlred-edge	0.266	0.562	0.390	0.049
	Aspect	7.723	340.649	144.777	87.127
	Elevation	1273	1412	1340	30.359
	Slope	0.512	19.411	5.702	3.860
	T_{max}	22.2	22.85	22.51	0.262
	T_{min}	8.481	9.672	9.057	0.488
	Radiation	19,653	304,608	137,763	87,583.85
	Rainfall	13.86	15.25	14.64	0.401
	Soil moisture	0.685	21.030	11.269	4.289
June	NDRE	−0.004	0.203	0.113	0.050
	Chlred-edge	0.522	1.076	0.666	0.111
	Aspect	7.723	340.649	144.777	87.127
	Elevation	1273	1412	1340	30.359
	Slope	0.512	19.411	5.702	3.860
	T_{max}	20.43	21.14	20.77	0.283
	T_{min}	6.876	7.919	7.379	0.418
	Radiation	22,430	303,014	131,301	89,098.69
	Rainfall	30.46	37.7	34.34	2.862
	Soil moisture	10.8	26.7	18.97	3.898

3.2. Remotely Sensed Autumn Grass Senescence with Climatic and Topographic Variables

Based on the results from the preliminary analysis (Table 3), the prediction outputs of the four popular regression models (i.e., the PLSR, MLR, CART, and the RFR) adopted in the study were generally significant. Specifically, the RFR outperformed all the other algorithms when using both the NDRE and the Chlred-edge as predictors throughout the four months considered in this investigation. This was demonstrated by the low RMSE and MAE with high R^2. These results (Table 3) further indicated that the CART was the second most important algorithm in the four months of analysis. On the other hand, the performance of the PSLR was generally inferior throughout the various stages of the analysis.

Table 3. Performance of the adopted algorithms based on the R^2, MEA, and the RMSE.

Month	Predictor Variable	Algorithm	RMSE	R^2	MAE
March	NDRE	PLS	0.046	0.39	0.037
		CART	0.042	0.47	0.033
		MLR	0.041	0.46	0.032
		RFR	0.039	0.50	0.031
	Chlred-edge	PLS	0.053	0.38	0.042
		CART	0.045	0.45	0.037
		MLR	0.046	0.46	0.036
		RFR	0.044	0.50	0.035
April	NDRE	PLS	0.038	0.35	0.031
		CART	0.034	0.63	0.028
		MLR	0.038	0.50	0.030
		RFR	0.035	0.62	0.026
	Chlred-edge	PLS	0.042	0.34	0.034
		CART	0.041	0.42	0.031
		MLR	0.043	0.42	0.034
		RFR	0.041	0.55	0.032
May	NDRE	PLS	0.024	0.52	0.020
		CART	0.024	0.50	0.018
		MLR	0.026	0.49	0.021
		RFR	0.022	0.53	0.017
	Chlred-edge	PLS	0.043	0.30	0.033
		CART	0.036	0.46	0.029
		MLR	0.043	0.36	0.036
		RFR	0.036	0.56	0.028
June	NDRE	PLS	0.041	0.36	0.033
		CART	0.046	0.42	0.035
		MLR	0.041	0.47	0.034
		RFR	0.033	0.68	0.026
	Chlred-edge	PLS	0.091	0.35	0.077
		CART	0.082	0.53	0.060
		MLR	0.101	0.33	0.078
		RFR	0.081	0.60	0.058

Moreover, the averaged prediction outputs of the adopted algorithms across the four-month period of the investigation maintained the findings presented in Table 3 that the RFR was the most useful model in associating the remotely sensed autumn grass senescence with climatic and topographic factors (Figure 2). A closer look at Figure 2a–c indicates that the RFR is the only algorithm that had a low RMSE and MAE with a high R^2 followed by CART. On the contrary, the PLSR displayed inferior performance based on two of the three model evaluation matrices (i.e., the R^2 and the MAE).

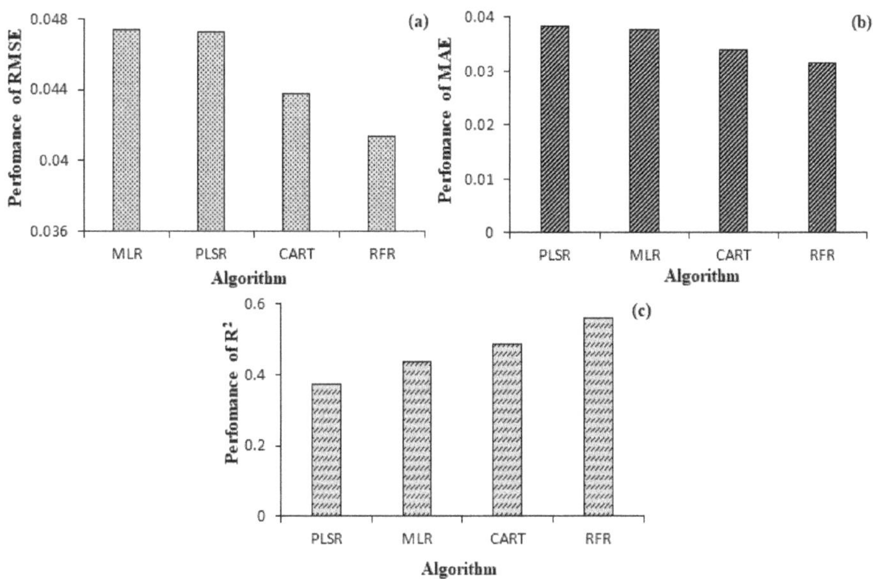

Figure 2. Algorithms' performances based on the (**a**) RMSE, (**b**) MAE, and the (**c**) R^2.

The final RFR models showed an improved explanation of the association between the remotely sensed autumn grass senescence and topographic and climatic factors when using both predictors across the four months considered (Table 4). For instance, when using the NDRE and the climatic and topographic factors in March, the model yielded an RMSE of 0.017 and an R^2 of 0.69 while obtaining an RMSE and an R^2 of 0.023 and 0.59, respectively, when using the Chlred-edge. Likewise, the NDRE recorded an RMSE of 0.012 and an R^2 of 0.71 in April, whereas the Chlred-edge produced an RMSE of 0.018 and R^2 of 0.60. Similarly, both the NDRE and the Chlred-edge reported RMSEs and R^2s of 0.056 and 0.014, as well as 0.56 and 0.69 in May, respectively. Moreover, the NDRE showed an RMSE and R^2 of 0.013 and 0.71, while the Chlred-edge obtained an RMSE of 0.056 and R^2 of 0.72 in June, respectively. Important variables for the final prediction models are presented in Figure 3. The predictive performance of each variable was assessed based on the obtainable Out of Bag error rate, which increases with significance.

Table 4. Optimal RFR results for the relationships between remotely sensed grass senescence and climatic factors and topography.

	NDRE		Chlred-Edge	
Month	RMSE	R^2	RMSE	R^2
March	0.017	0.69	0.023	0.59
April	0.012	0.71	0.018	0.60
May	0.056	0.56	0.014	0.69
June	0.013	0.71	0.056	0.72

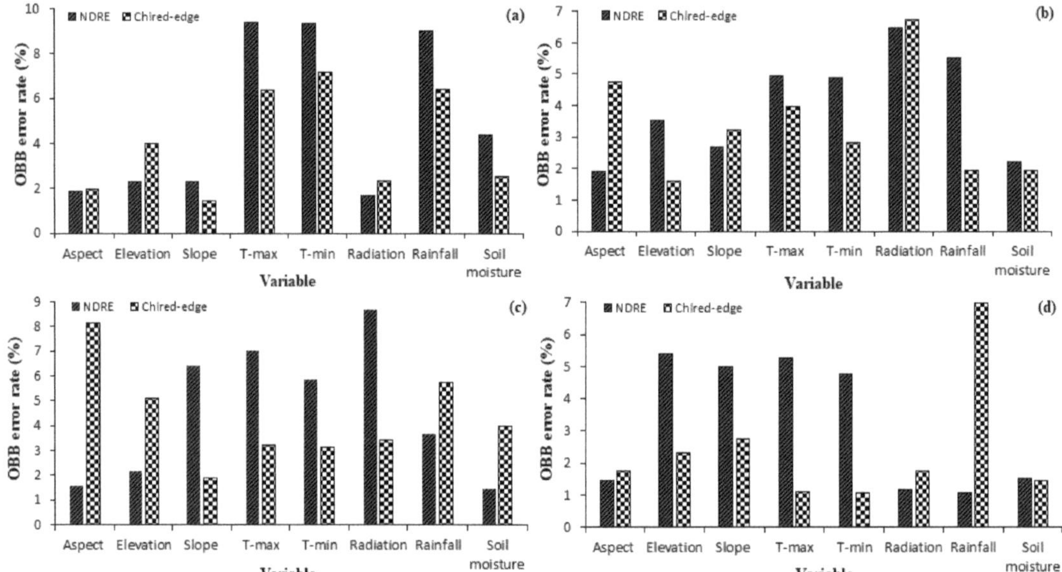

Figure 3. RFR model's variable importance for assessing the response of remotely sensed autumn grass senescence against the climatic and topographic factors in (**a**) March, (**b**) April, (**c**) May, and (**d**) June.

3.3. Climatic and Topographic Drivers of the Autumn Grassland Senescence

Using the monthly averages of the predictors (i.e., the NDRE and the Chlred-edge) against the response variables (i.e., topographic and climatic variables), we identified the key drivers influencing the occurrence of autumn grassland senescence (Table 5). In general, our findings showed that only the climatic factors were sensitive to the occurrence of autumn grassland senescence. Specifically, the T_{min} and T_{max}, jointly with soil moisture, were identified as the most influential factors in the occurrence of autumn grass senescence, as shown by their significance levels ($p \leq 0.05$). Obtainable R^2 values for the three climatic factors that significantly influence the occurrence of autumn grass senescence were 1.00, 0.98, and 0.81 based on the NDRE and -1.00, -0.96, and -0.78 when using the Chlred-edge, respectively. Conversely, even though they displayed good R^2 values (i.e., between 0.76 and 0.93), the insignificant p-values ($p \geq 0.05$) highlighted the poor contribution of these other climatic variables in explaining the occurrence of autumn grass senescence in the study area. With regard to the topographic factors, only the slope showed good R^2 values (i.e., -0.80 and 0.75 when using the NDRE and the Chlred-edge, respectively); otherwise, they were all insignificant when considering the p-values ($p \geq 0.05$). Table 5 shows the contribution of environmental factors on autumn grassland senescence, with significant variables in bold.

The sensitivity of the topographic and climatic factors in influencing the occurrence of autumn grass senescence in the study area was further emphasized by the value of the t-statistics, with higher values signifying the importance and vice versa.

Table 5. Correlations between remotely sensed grass senescence and climatic factors and topography. Influential variables are shown in bold.

Variable	NDRE			Chlred-Edge		
	t-Statistics	*p*-Value	R^2	*t*-Statistics	*p*-Value	R^2
Topographic factors						
Aspect	−0.597	0.611	−0.39	0.492	0.672	0.33
Elevation	0.163	0.886	0.11	−0.276	0.809	−0.19
Slope	−1.865	0.203	−0.80	1.588	0.253	0.75
Climatic factors						
T_{max}	55.095	**0.000**	1.00	−14.388	**0.005**	−1.00
T_{min}	6.832	**0.021**	0.98	−4.806	**0.041**	−0.96
Radiation	3.502	0.073	0.93	−2.852	0.104	−0.90
Rainfall	1.881	0.201	0.80	−1.661	0.239	−0.76
Soil moisture	6.579	**0.031**	0.81	−4.461	**0.040**	−0.78

Figure 4 shows the response of the remotely sensed autumn grass senescence (i.e., NDRE and Chlred-edge) to the most influential variables (i.e., T_{min}, T_{max}, and the soil moisture). Figure 4a–c illustrate the remotely sensed autumn grass senescence based on the NDRE, while Figure 4d–f display the remotely sensed autumn grass senescence based on the Chlred-edge. Overall, the effect of time lag was evident between the occurrence of autumn grassland senescence and the change in sensitive variables. The NDRE-based autumn grass senescence indicated a continuous decline with a decrease in both the T_{min} and T_{max} during the autumn season. On the other hand, a synonymous decline in the NDRE-based autumn grass senescence with soil moisture was followed by a sudden increase in soil moisture in June. Figure 4d–f indicate an inverse relationship between the Chlred-edge-based autumn grass senescence and the influential variables. Generally, the consistent drop in T_{min}, T_{max}, and the soil moisture values was concurrent with the increasing Chlred-edge-based autumn grass senescence estimates.

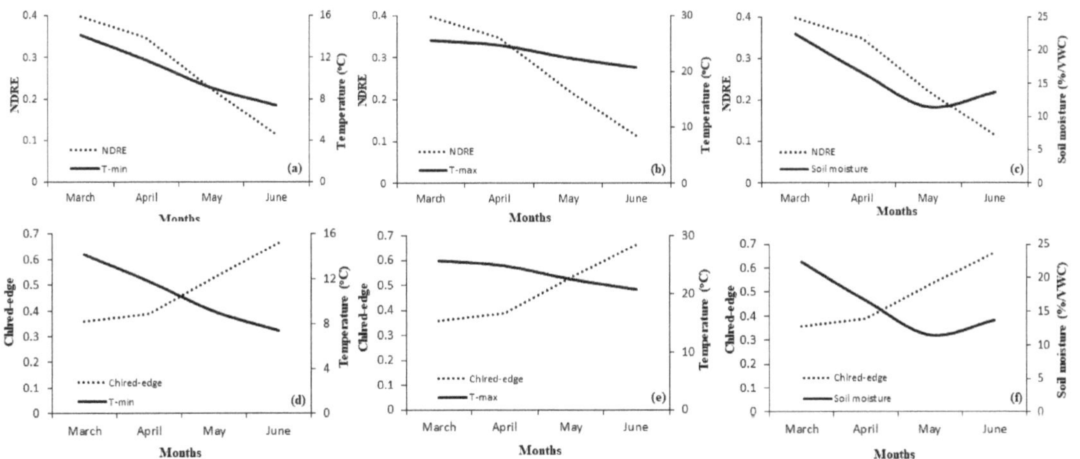

Figure 4. The responses of the (**a–c**) NDRE-based autumn grass senescence to (**a**) T_{min}, (**b**) T_{max}, and (**c**) soil moisture together with those of the (**d–f**) Chlred-edge-based autumn grass senescence to (**d**) T_{min}, (**e**) T_{max}, and (**f**) soil moisture through time.

4. Discussion

The present study has shown the value of the multi-temporal remotely acquired Sentinel 2 satellite data in elucidating the occurrence of autumn senescence along climatic and topographic gradients in the subtropical sour-veld grassland ecosystems. This has been a limitation in understanding the dynamics around the occurrence of autumn senescence as well as the subsequent impact on foraging resource productivity and feed availability in these regions. Our findings indicated that the occurrence of autumn grassland senescence in the present study site is controlled by climatic drivers, particularly the soil moisture, T_{min}, and T_{max} rather than topographic factors ($p \leq 0.05$ in Table 5). Although not pronounced in the current findings, the sensitivity of air temperature variables (i.e., T_{max} and T_{min}) in influencing the occurrence of autumn grassland senescence in the area could be attributed to the reported extremities of these variables [26]. For instance, the observed consistent decline in air temperatures (Table 2) is believed to have promoted irregular frost events, as they are known to be a common phenomenon in the area during this period [27] and, hence, grass senescence. These results concur with studies indicating that extreme-temperature conditions affect the natural processes of photosynthetic enzymes and thereby accelerate or delay chlorophyll deterioration [4,6,28], whereas water shortages are known to influence plant carboxylation reaction, hence fast-tracking chlorophyll degradation and plant senescence [1,5,12,29].

Conversely, although solar radiation and rainfall are known to be key climatic factors influencing plant phenology [2], their impact was not significant (Table 5). However, these results should be discussed with caution, as the observed poor relationship between the remotely sensed autumn grass senescence and rainfall and solar radiation may not be universally constant, i.e., could be site-specific as a result of topographic and micro-climatic conditions. Specifically, the recorded poor correlation between autumn grassland senescence and rainfall in this study may possibly be a consequence of the high variability in rainfall during the same period [26], which could destruct the uniformity in the phenology of the grass. Similarly, the poor relationship notable between the autumn grass senescence and solar radiation could be justified by the relatively uniform topography of the study area, which was observed during field data collection. Meanwhile, our assumption is that meaningful characterization of the links between the remotely sensed autumn grass senescence and the incoming solar radiation and topographic factors such as slope, aspect, and elevation requires heterogeneity in the landscape, which is possible in pronounced mountainous and valley areas [2], also indicated that heterogeneity in topography promotes spatial distinction in vegetation phenology regardless of the similarity in the age of the floral species. Our results further showed the effect of the time lag between the occurrence of autumn grass senescence and the change in sensitive climatic factors (Figure 4), thereby suggesting that the chlorophyll breakdown is not concurrent with, but follows the triggering effect of, the environmental cue. Evidently, the significance of understanding the response of autumn grassland senescence to changes in climatic and topographic factors cannot be over-emphasized, particularly in countries such as South Africa, considering the projected shifts in seasonal patterns [30], which may further alter the current dynamics in phenological stages such as the autumn grassland senescence, leading to potential forage deficiencies, especially during dry seasons. With its ability to either shorten or extend the growing season of the floral species, and hence productivity [8], the understanding of the links between autumn grass senescence and environmental factors may help to strengthen our projections on the possible timing and duration of the autumn grassland senescence, which will, in turn, improve our assessment of fodder bank capacities for quality forage provision. Whereas this highlights the essence of future research on this subject matter, the emphasis of such work should be on multi-year studies conducted on heterogeneous terrains, while fully embracing the potential impact of frost activities in the analysis.

With regard to the performance of the RFR model, our results reinforce the evidence presented in previous studies that this model is robust when explaining ecological problems based on remotely acquired datasets [15,31]. Again, although the findings in Figure 3 may give an impression that the topographic factors were among the important variables in April, May, and June, a correct view is that these variables were only important in displaying the monthly relationship with the tested variables, which does not necessarily reflect the autumn grassland senescence in our case. According to our approach in this study, the autumn grassland senescence was explained based on the averaged performance of the month-to-month contributions of each variable, and the variables that were consistently significant were identified as the environmental drivers of autumn grassland senescence.

5. Conclusions

The present study examined the relationship between remotely sensed autumn grass senescence and the climatic factors and topography in the subtropical sour-veld grasslands of the Midlands region, KwaZulu-Natal, South Africa. The study employed the Sentinel 2 derivatives using the PLSR, MLR, CART, and RFR models, and the RFR model emerged as the superior model. Among the best of the model outputs, RMSEs of 0.017, 0.012, 0.056, and 0.013 as well as R^2s of 0.69, 0.71, 0.56, and 0.71 for the NDRE, with RMSEs and R^2s of 0.023, 0.018, 0.014, and 0.056 as well as 0.59, 0.60, 0.69, and 0.72 for the Chlred-edge in March, April, May, and June, respectively, were obtained. The results further showed that T_{min}, T_{max}, and soil moisture were the most influential factors in the occurrence of autumn grassland senescence at the study site. However, the observable poor relationship between autumn grass senescence and the other climatic factors and topography is believed to be indicative of the micro-climatic conditions and the relative homogeneity in the topography. However, given that the study was carried over a season, it does not reflect the possible year-to-year climatic changes and, hence, cannot be used to draw finality on the relationship between the tested variables. Therefore, for a conclusive understanding of the overlaps between autumn grass senescence and climatic and topographic factors, we suggest further investigation, particularly focusing on multi-year studies conducted in heterogeneous landscapes and taking into account the effect of frost occurrences in the analysis.

Author Contributions: Conceptualization, L.R., O.M. and R.S.; methodology, L.R. and O.M.; software, L.R.; validation, O.M. and J.O.; formal analysis, L.R.; investigation, L.R.; resources, L.R. and O.M.; data curation, L.R.; writing—original draft preparation, L.R.; writing—review and editing, J.O. and O.M.; visualization, L.R.; supervision, O.M. and J.O.; project administration, L.R. and O.M.; funding acquisition, O.M. and R.S. All authors have read and agreed to the published version of the manuscript.

Funding: This study was funded by the National Research Foundation (NRF) of South Africa, Research Chair initiative in Land Use Planning and Management (Grant Numbers: 84157).

Data Availability Statement: The data could not be made available at the present moment because the study is part of an ongoing bigger project and publishing its data at this stage may compromise the scope of the subsequent chapters.

Acknowledgments: The authors would like the thank the South African Weather Services (SAWS) for providing rainfall and temperature data. In addition, recognition is given to Mthembeni Mngadi, Omosalewa Odebiri, Asekho Mantintsilili, Nkululeko Dladla, and Lesogo Hlwayi for their field data collection assistance. In addition, the authors express their words of thanks to Andile Etah Mshengu for her kind support with the processing of finances and booking of vehicles for site visits.

Conflicts of Interest: The authors declare no conflict of interest.

References

1. Tao, Z.; Ge, Q.; Wang, H.; Dai, J. The important role of soil moisture in controlling autumn phenology of herbaceous plants in the Inner Mongolian steppe. *Land Degrad. Dev.* **2021**, *32*, 3698–3710. [CrossRef]
2. Shoko, C.; Mutanga, O.; Dube, T. Remotely sensed C3 and C4 grass species aboveground biomass variability in response to seasonal climate and topography. *Afr. J. Ecol.* **2019**, *57*, 477–489. [CrossRef]
3. Wu, C.; Wang, X.; Wang, H.; Ciais, P.; Peñuelas, J.; Myneni, R.B.; Desai, A.R.; Gough, C.M.; Gonsamo, A.; Black, A.T. Contrasting responses of autumn-leaf senescence to daytime and night-time warming. *Nat. Clim. Change* **2018**, *8*, 1092–1096. [CrossRef]

4. Liu, G.; Chen, X.; Zhang, Q.; Lang, W.; Delpierre, N. Antagonistic effects of growing season and autumn temperatures on the timing of leaf coloration in winter deciduous trees. *Glob. Change Biol.* **2018**, *24*, 3537–3545. [CrossRef] [PubMed]

5. Tao, Z.; Wang, H.; Dai, J.; Alatalo, J.; Ge, Q. Modeling spatiotemporal variations in leaf coloring date of three tree species across China. *Agric. For. Meteorol.* **2018**, *249*, 310–318. [CrossRef]

6. Liu, Q.; Fu, Y.H.; Zeng, Z.; Huang, M.; Li, X.; Piao, S. Temperature, precipitation, and insolation effects on autumn vegetation phenology in temperate China. *Glob. Change Biol.* **2016**, *22*, 644–655. [CrossRef] [PubMed]

7. McKean, J.; Buechel, S.; Gaydos, L. Remote sensing and landslide hazard assessment. *Photogramm. Eng. Remote Sens.* **1991**, *57*, 1185–1193.

8. Gepstein, S.; Sabehi, G.; Carp, M.J.; Hajouj, T.; Nesher, M.F.O.; Yariv, I.; Dor, C.; Bassani, M. Large-scale identification of leaf senescence-associated genes. *Plant J.* **2003**, *36*, 629–642. [CrossRef]

9. Royimani, L.; Mutanga, O.; Dube, T. Progress in remote sensing of plant senescence: A review on the challenges and opportunities. *IEEE J. Sel. Top. Appl. Earth Obs. Remote Sens.* **2021**, *14*, 7714–7723. [CrossRef]

10. Royimani, L.; Mutanga, O.; Odindi, J.; Sibanda, M.; Chamane, S. Determining the onset of autumn grass senescence in subtropical sour-veld grasslands using remote sensing proxies and the breakpoint approach. *Ecol. Inform.* **2022**, *69*, 101651. [CrossRef]

11. Anderegg, J.; Yu, K.; Aasen, H.; Walter, A.; Liebisch, F.; Hund, A. Spectral vegetation indices to track senescence dynamics in diverse wheat germplasm. *Front. Plant Sci.* **2020**, *10*, 1749. [CrossRef] [PubMed]

12. Munné-Bosch, S.; Alegre, L. Die and let live: Leaf senescence contributes to plant survival under drought stress. *Funct. Plant Biol.* **2004**, *31*, 203–216. [CrossRef] [PubMed]

13. Mariën, B.; Dox, I.; De Boeck, H.J.; Willems, P.; Leys, S.; Papadimitriou, D.; Campioli, M. Does drought advance the onset of autumn leaf senescence in temperate deciduous forest trees? *Biogeosciences* **2021**, *18*, 3309–3330. [CrossRef]

14. Royimani, L.; Mutanga, O.; Odindi, J.; Dube, T.; Matongera, T.N. Advancements in satellite remote sensing for mapping and monitoring of alien invasive plant species (AIPs). *Phys. Chem. Earth Parts A/B/C* **2019**, *112*, 237–245. [CrossRef]

15. Royimani, L.; Mutanga, O.; Odindi, J.; Zolo, K.S.; Sibanda, M.; Dube, T. Distribution of *Parthenium hysterophoru* L. with variation in rainfall using multi-year SPOT data and random forest classification. *Remote Sens. Appl. Soc. Environ.* **2019**, *13*, 215–223. [CrossRef]

16. Sibanda, M.; Mutanga, O.; Rouget, M. Comparing the spectral settings of the new generation broad and narrow band sensors in estimating biomass of native grasses grown under different management practices. *GISci. Remote Sens.* **2016**, *53*, 614–633. [CrossRef]

17. Makanza, R.; Zaman-Allah, M.; Cairns, J.E.; Magorokosho, C.; Tarekegne, A.; Olsen, M.; Prasanna, B.M. High-throughput phenotyping of canopy cover and senescence in maize field trials using aerial digital canopy imaging. *Remote Sens.* **2018**, *10*, 330. [CrossRef]

18. Renier, C.; Waldner, F.; Jacques, D.C.; Ebbe, M.A.B.; Cressman, K.; Defourny, P. A dynamic vegetation senescence indicator for near-real-time desert locust habitat monitoring with MODIS. *Remote Sens.* **2015**, *7*, 7545–7570. [CrossRef]

19. Sibanda, M.; Onisimo, M.; Dube, T.; Mabhaudhi, T. Quantitative assessment of grassland foliar moisture parameters as an inference on rangeland condition in the mesic rangelands of southern Africa. *Int. J. Remote Sens.* **2021**, *42*, 1474–1491. [CrossRef]

20. Singh, L.; Mutanga, O.; Mafongoya, P.; Peerbhay, K.Y. Multispectral mapping of key grassland nutrients in KwaZulu-Natal, South Africa. *J. Spat. Sci.* **2018**, *63*, 155–172. [CrossRef]

21. Delta-T Devices. Thetakit User Guide: ML3-Kit. Available online: https://delta-t.co.uk/wp-content/uploads/2016/09/ML3 _Kit_User_Guide_ver_1.0.pdf (accessed on 22 February 2021).

22. Dube, T.; Mutanga, O. The impact of integrating WorldView-2 sensor and environmental variables in estimating plantation forest species aboveground biomass and carbon stocks in uMgeni Catchment, South Africa. *ISPRS J. Photogramm. Remote Sens.* **2016**, *119*, 415–425. [CrossRef]

23. Ruiz-Arias, J.; Tovar-Pescador, J.; Pozo-Vázquez, D.; Alsamamra, H. A comparative analysis of DEM-based models to estimate the solar radiation in mountainous terrain. *Int. J. Geogr. Inf. Sci.* **2009**, *23*, 1049–1076. [CrossRef]

24. Kumar, L.; Skidmore, A.K.; Knowles, E. Modelling topographic variation in solar radiation in a GIS environment. *Int. J. Geogr. Inf. Sci.* **1997**, *11*, 475–497. [CrossRef]

25. R Core Team. *R: A Language and Environment for Statistical Computing*; R Foundation for Statistical Computing: Vienna, Austria, 2021; Available online: https://www.R-project.org/ (accessed on 5 April 2022).

26. Ndlovu, M.; Clulow, A.D.; Savage, M.J.; Nhamo, L.; Magidi, J.; Mabhaudhi, T. An assessment of the impacts of climate variability and change in KwaZulu-Natal Province, South Africa. *Atmosphere* **2021**, *12*, 427. [CrossRef]

27. Ismail, R.; Crous, J.; Sale, G.; Morris, A.R.; Peerbhay, K.Y. Developing a satellite-based frost risk model for the Southern African commercial forestry landscape. *South. For. A J. For. Sci.* **2021**, *83*, 10–18. [CrossRef]

28. Fracheboud, Y.; Luquez, V.; Bjorken, L.; Sjodin, A.; Tuominen, H.; Jansson, S. The control of autumn senescence in European aspen. *Plant Physiol.* **2009**, *149*, 1982–1991. [CrossRef]

29. Sade, N.; del Mar Rubio-Wilhelmi, M.; Umnajkitikorn, K.; Blumwald, E. Stress-induced senescence and plant tolerance to abiotic stress. *J. Exp. Bot.* **2018**, *69*, 845–853. [CrossRef]
30. Van der Walt, A.J.; Fitchett, J.M. Statistical classification of South African seasonal divisions on the basis of daily temperature data. *S. Afr. J. Sci.* **2020**, *116*, 1–15. [CrossRef]
31. Mutanga, O.; Adam, E.; Cho, M.A. High density biomass estimation for wetland vegetation using WorldView-2 imagery and random forest regression algorithm. *Int. J. Appl. Earth Obs. Geoinf.* **2012**, *18*, 399–406. [CrossRef]

Article

Improving Biodiversity Offset Schemes through the Identification of Ecosystem Services at a Landscape Level

Annaêl Barnes [1,2,3,*], Alexandre Ickowicz [1,2], Jean-Daniel Cesaro [1,2], Paulo Salgado [1,2], Véronique Rayot [4], Sholpan Koldasbekova [5] and Simon Taugourdeau [1,2]

[1] CIRAD, UMR SELMET, F-34090 Montpellier, France
[2] UMR SELMET, University of Montpellier, CIRAD, INRAE, Institut Agro, F-34090 Montpellier, France
[3] UMR AMAP, University of Montpellier, CIRAD, CNRS, INRAE, IRD, F-34090 Montpellier, France
[4] Orano Mining, 125 Av. de Paris, 92320 Châtillon, France
[5] KATCO JV LLP, Sauran Street 48, Astana 020000, Kazakhstan
*   Correspondence: annael.barnes@cirad.fr

Abstract: Biodiversity offsets aim to compensate the negative residual impacts of development projects on biodiversity, including ecosystem functions, uses by people and cultural values. Conceptually, ecosystem services (ES) should be considered, but in practice this integration rarely occurs. Their consideration would improve the societal impact of biodiversity offsets. However, the prioritisation of ES in a given area is still limited. We developed a framework for this purpose, applied in rangelands landscapes in Kazakhstan, in the context of uranium mining. We assumed that different landscapes provide different ES, and that stakeholders perceive ES according to their category (e.g., elders and herders) and gender. We performed qualitative, semi-structured interviews with a range of stakeholders. Using the Common International Classification of Ecosystem Services, we identified 300 ES in 31 classes across 8 landscape units. We produced a systemic representation of the provision of ES across the landscapes. We showed a significant link between ES and landscape units, but not between ES and stakeholder categories or gender. Stakeholders mostly identified ES according to the location of their villages. Therefore, we suggest that the biodiversity offsets should target ES provided by the landscape unit where mining activities occur and would be most interesting in the landscapes common to all villages. By performing a systemic representation, potential impacts of some offset strategies can be predicted. The framework was therefore effective in determining a bundle of ES at a landscape scale, and in prioritising them for future biodiversity offset plans.

Keywords: ecosystem services; biodiversity offset; CICES; landscape units; stakeholders; rangelands

Citation: Barnes, A.; Ickowicz, A.; Cesaro, J.-D.; Salgado, P.; Rayot, V.; Koldasbekova, S.; Taugourdeau, S. Improving Biodiversity Offset Schemes through the Identification of Ecosystem Services at a Landscape Level. *Land* **2023**, *12*, 202. https://doi.org/10.3390/land12010202

Academic Editors: Michael Vrahnakis, Yannis (Ioannis) Kazoglou and Manuel Pulido Fernádez

Received: 6 December 2022
Revised: 23 December 2022
Accepted: 1 January 2023
Published: 8 January 2023

1. Introduction

Biodiversity loss and ecosystem degradation are massive environmental problems worldwide [1,2]. To mitigate biodiversity losses due to development projects (such as mining activities), policy makers, governments and the private sector are increasingly adhering to biodiversity offset mechanisms [2–6]. Biodiversity offsets are defined as 'measurable conservation outcomes resulting from actions designed to compensate significant residual adverse biodiversity impacts arising from project development after appropriate prevention and mitigation measures has been taken' [7]. These mechanisms target the residual impacts that cannot be avoided, reduced, and will not be restored, according to the steps of the mitigation hierarchy [2,5,6]. Biodiversity offsets have a goal of no net loss (NNL) of biodiversity or when possible, a net gain (NG) of biodiversity [2,7,8], including species composition, habitat structure, ecosystem function, its use by people and associated cultural value [7]. The methods for implementing offsets are diverse: restoration of degraded ecosystems, creation of new habitats, protection of existing high quality ecosystems at risk of degradation or loss, change of practice in favour of biodiversity on already managed areas, or mitigation bank [3,7,9–11].

Ecosystem services (ES) are defined as 'the aspects of ecosystems utilized (actively or passively) to produce human well-being' [12]. Benefits derive from ES and are 'the contributions to aspects of well-being' (e.g., health) [13]. The conceptualization of the benefits provided by nature has been driven by a loss of ES [14], mostly resulting from fragmentation and loss of habitat after economic development projects [11]. Several classification systems for describing ES exist, for example, the Millennium Ecosystem Assessment [1], the Intergovernmental Science-Policy Platform on Biodiversity and Ecosystem Services (IPBES, [15]) and also the Common International Classification of Ecosystem Services (CICES) [16,17], which has become a common international reference [18,19]. Its hierarchical structure (from least to most detailed: section, division, group and class of ES) allows users to go to the level of detail wanted or needed [16,17]. The CICES can also include abiotic flows, which may be socially important [19], while abiotic structure and processes are less or not explicitly integrated in some other classification frameworks [19,20].

Although it can be argued that biodiversity provides ES and needs them for its persistence [7,21], the relationship between biodiversity and ES provision is not so obvious. Some ES can be provided by a diverse set of species and habitats, while others are strongly linked to specific species or sets of species (e.g., pollination) [22]. Moreover, area considered as important for biodiversity conservation will not necessarily be crucial for ES supply and vice versa [11]. In a study in Iran, Karimi et al. [23] showed a strong relationship between cultural services and biodiversity hotspots, but found a weaker link between provisioning services and biodiversity. Therefore, the relationship between biodiversity and ES provision depends on the ES considered.

From a conceptual perspective, biodiversity offset schemes could consider how changes in biodiversity might influence the provision of ES to different types of stakeholder [7,10,21]. Some methods for integrating ES into biodiversity offsets have already been developed [7], but current offsetting is mostly focused on critical habitats and threatened species [5], rather than on ordinary biodiversity and the services it provides [5,22]. The failure to take proper account of ES is due to several reasons. There is no standardized and systematic methodology for the integration of ES into the mitigation hierarchy [22], or biodiversity offset schemes [5]. Coupled with a lack of legislation for the consideration of ES in offsets programs [24], and Environmental and Social Impact Assessment (ESIA) of economic development projects [11], ES are rarely mentioned in existing offset practices [5], or are a consequence of the chosen offset strategy and not a driver [24]. Nevertheless, integrating ES into offset mechanisms is receiving increasing attention in the international community in recent years [5,22], and within companies responsible for biodiversity offsets [24].

The consideration of ES could improve greatly the societal impact of biodiversity offsets. In some countries, residual direct and indirect impacts of developments projects threaten the survival of local populations due to loss of biodiversity and ES [5]. Some biodiversity offset strategies sometimes negatively affect populations [24]. For example, offsets implemented far from the impacted site (*off-site* offsets [2]) will increase inequalities, as the impacted population will not be compensated in terms of biodiversity or ES provision [4,10]. Moreover, some biodiversity conservation schemes decrease ES access, when implementing a protected area, for example [10,11]. The integration of ES in biodiversity offsets planning could lead to fairer offsets, that consider those kinds of impacts on important ES and livelihoods.

People living in drylands are especially highly dependent on the provision of ES. Indeed, over one-third of the world's population live in drylands [25,26], that are usually vital for the provision of forage for livestock [26]. Other important ES provided are, for example, food production, medicinal plants and fuelwood, water supply (whose availability is limited and variable) for drinking, irrigation and supporting fauna and flora, as well as cultural services related to tourism, spirituality, creation of indigenous knowledge and aesthetics [25–28].

Ecosystem services are therefore essential to people, and their consideration in biodiversity offsets would give a more social, economic and health scope to offsets mechanisms. How could we integrate ES into biodiversity offsets for economic development projects? We are still limited on how diverse ES should be in a specific landscape [22] and on how to prioritise ES [5].Therefore, a systematic identification and prioritisation procedure is needed to ensure that the targeted ES are representative of the area [22].

As ES are attached to beneficiaries, the involvement of stakeholders is vital for identifying bundles of ES to prioritise. The lack of integration of different stakeholders is currently one of the limiting factors for the consideration of ES in biodiversity offsets [11,24], even though it is a recommendation for offsets planning [7]. Stakeholder needs can be assessed through, e.g., questionnaires and individual interviews (e.g., [29–31]), or participatory approaches such as focus groups (e.g., [23,29–32]). However, the perceived ES may vary according to stakeholder categories (e.g., [29]) or gender (e.g., [30]), but these characteristics can be considered through the ES identification process. Identification of ES should also take into account the landscape scale via an ecosystem approach for the design and implementation of biodiversity offsets [7]. Some studies showed already the links between landscape and ES provision (e.g., [30–32]). Identifying the ES provided by the different landscape units should be possible when working in conjunction with local stakeholders.

Our study investigates the prioritisation of ES through a systemic approach across various landscapes. We assumed that different landscape units provide different ES, and that ES are perceived differently depending on stakeholder categories and gender. Our case study takes place in rangelands landscapes in the drylands of southern Kazakhstan, in a context of uranium mining activities. We mobilised the CICES framework for a standardised method, through interviews with a range of local stakeholders. Eventually, our study aims to propose a framework for the identification of ES at a landscape scale, that can be used to integrate ES into the biodiversity offset scheme.

2. Materials and Methods

2.1. Study Site

The study area was located in the Sozak district of the Turkestan province in Kazakhstan ($46°1'18.00''$–$43°23'44.1''$ N and $67°6'10.19''$–$69°20'31.4''$ E). This district comprises the sandy desert of Muyunkum, where the Muyunkum and Tortkuduk uranium mines are located (Figure 1).

The Sozak district is a dry, mid-latitude steppe and mid-latitude desert climate [33]. Mean annual precipitation (MAP) is less than 200 mm/year [34]. According to the Muyunkum Central site project's environmental impact assessment (EIA) carried out in 2011 (hereafter called *project's EIA*), on a more local scale, the climate is continental with temperatures ranging from $-30\ °C$ in winter to $+40\ °C$ in summer.

Our study site is not limited to only desert, but also comprises a mosaic of landscapes, including:

- The Muyunkum sandy desert, where MAP is 155 mm and during summer and the soil surface temperature can reach $60\ °C$ according to the project's EIA. It is composed of dunes and shrublands, especially of *Haloxylon ammodendron.* (C.A. Mey.) Bunge. [34,35]. Muyunkum is considered as good winter pasture because of the shrubby vegetation that can be found, even after heavy snowfalls [34].
- The steppe to the south and west of the sandy desert. Steppes are grazing or grasslands areas are found, that allow livestock farming and wheat cultivation [33].
- On the northern of the sandy desert, the presence of the Shu River, which forms in Kyrgyzstan, has created riparian and flooded zones where reed (*Phragmites* spp.) is the dominant species. Reeds are consumed by cattle but also cut and stored as winter fodder by villagers living along the river [35,36]. The riparian zone includes *Tamarix* spp., and the herbaceous species *Agrophyron* spp., *Festucca* spp. and *Artemisia* spp. are present in the seasonally flooded zones [36].

- The Betpak-Dala steppe that lies north of the Shu river. This steppe is described as a clay desert, comprising sparse vegetation that incudes *Artemisia* spp. and *Salsola* spp. and several annual species, representing an important and rich source of protein for herbivores in the early spring [34–36].
- An area of salty, clay desert between the Shu river and the Muyunkum sandy desert, where several halophyte plant species are present [35].
- The Karatau mountains in the south-west, where the majority of precipitation falls, according to the project's EIA.
- A salty lake area in the south-east, comprising the Kyzylkol lake and sacred pond of the religious site of Baba Tukti Shashty Aziz mausoleum, on both sides of the border of the province of Djambul.

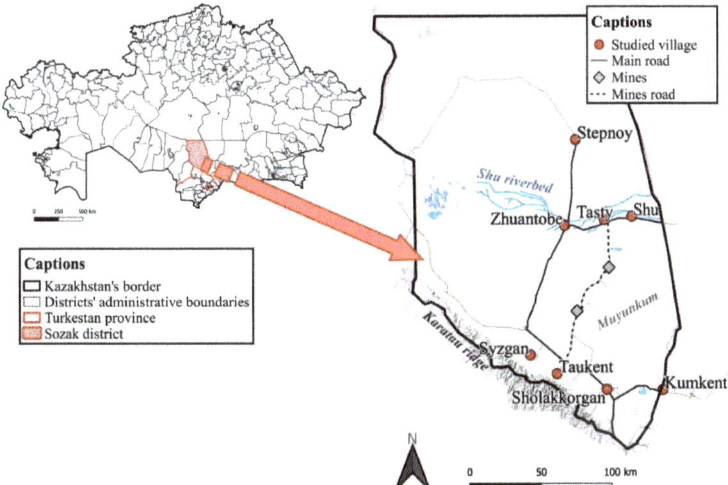

Figure 1. Location of Sozak district in Kazakhstan. Ecosystem services were identified in and around eight villages across the landscape.

2.2. Main Economic Activities: Livestock Farming and Uranium Mining

The population of the Turkestan province is 2.7 million people, resulting in a population density of 23 people/km^2, and the population of the Sozak district is 62,000 people, resulting in a population density of 1.5 people/km^2 [37]. According to the project's EIA, livestock farming is the primary activity in the study area, followed by crop production.

The environment of Kazakhstan is favourable to mobile pastoralism [38]: the rangelands represent 60% of the country, i.e., 189 million hectares [36]. Agriculture, including livestock farming, contributed 4.4% of the country's GDP and accounted for 18% of employment in Kazakhstan in 2017 [39]. Turkestan province is one of the most important regions for livestock farming, with the larger concentration of small-scale farms in 2016 and the highest number of registered agricultural cooperatives and cooperative members in 2018 [39]. Nevertheless, over the past 150 years, nomadic pastoralism has declined and now accounts for very few herders [34]. The use of rangelands for livestock production has undergone major changes, from state-owned farms in the Soviet era, to the development of private livestock systems since the mid-1990s [34,35,40]. Nowadays, the richest herders with large herds can exclusively rent grazing areas with access to wells, via a semi-privatization mechanism [35]. Herders with less livestock tend to keep animals in the proximity of villages [35,40]. Herders can grow and store hay for use in the winter months, mostly as fodder for dairy cows and calves [40,41]. Small livestock (sheep and goats) and large livestock (cows, horses and camels) are raised for different animal products (dairy, meat and wool) [38,41]. In 2009, Turkestan province had 3,415,000 sheep and goats, 716,000 cows and 144,000 horses (no data on camels) [41].

Kazakhstan's subsoil is rich in various mineral resources including oil, gas, uranium, coal, copper, zinc, gold, chromium, manganese, iron and lead [42,43]. In 2016, the extractive sector contributed almost 30% of the country's GDP [44]. The Shu-Syrdarya mining region, in which our study area is located, was discovered and explored between 1971 and 1991, and is now the largest uranium mining region in Kazakhstan [45]. The Kazakh–French joint venture KATCO (Kazatomprom–Orano) owns the mines at the Muyunkum and Tortkuduk deposits in the Muyunkum sandy desert of the Sozak district. Since the 1990s, the in-situ recovery (ISR) technique became widespread for uranium extraction, used for ore located between impermeable soil layers [37]. Compared to older mining methods, this technique has less surface damage, no waste rock or tailings storage and lower remediation costs [45,46]. Nevertheless, there are risks of local underground contamination due to leaching reagents and metals in solutions [46]. In addition, the implementation of mining projects can also have impacts on ES, such as livestock-related ES, by decreasing access to grazing areas, for example.

2.3. Identification of Ecosystem Services to Include in Biodiversity Offsets

Figure 2 summarises the methodology applied. Each methodological point is described in more detail in the following sub-section, with application in our study area.

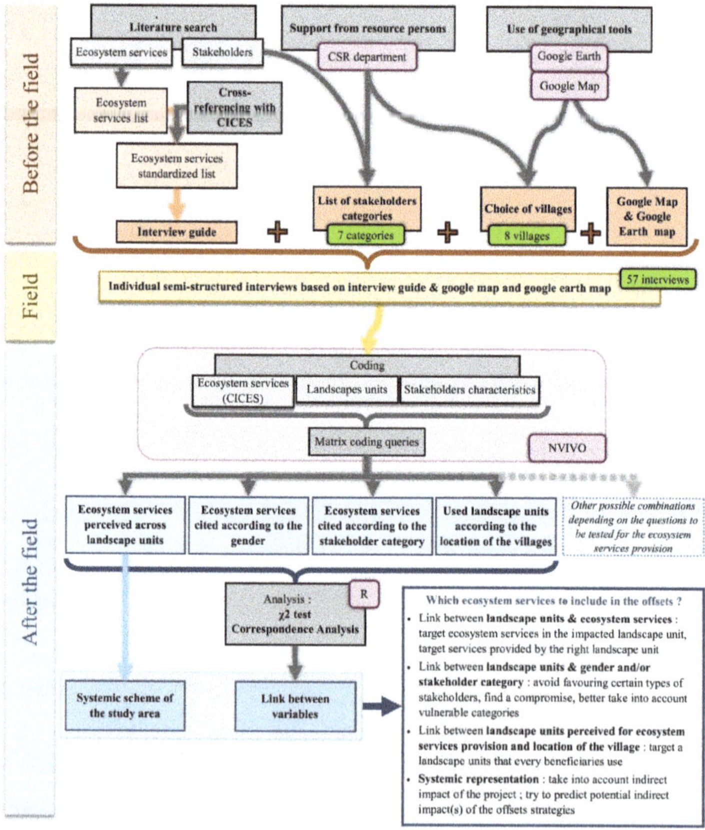

Figure 2. Methodology applied for the identification of ecosystem services to be included in biodiversity offset strategies (green: case study of the Sozak district in Kazakhstan; purple: software and resources used; CSR = corporate social responsibility).

2.3.1. Construction of the Interview Guide

To identify the ES throughout the landscape in the Sozak region, we carried out a literature search and developed a questionnaire before then interviewing local stakeholders. We performed a literature review on the ES provided by pastures in Kazakhstan, Central Asia and drylands. The keywords used for the bibliographic research were: 'ecosystem + services' and: 'Kazakhstan' or 'Central Asia' or 'drylands' or 'arid + lands' or 'pastures' or 'rangelands' or 'grasslands', leading to 11 studies, reports, papers and book chapters [25–28,47–53]. We then compiled the ES and cross-referenced them with the CICES [16] to produce a standardized list of ES. This list was the basis for our interviews with stakeholders. We kept the CICES class level for some ES and remained at the CICES group level for other when the CICES division level was too vague to guide our questions, and the CICES class level too specific (e.g., the Biomass division includes the Wild Plants for food, materials or energy group, which includes the classes Wild plants for food, Wild plants for energy, etc. [16]). The interview guide, which is based on this list, is available in Table S1.

2.3.2. Selection of Local Stakeholders and Villages in Sozak District

Although uranium mines are located in the sandy desert of Muyunkum, this area is not inhabited by a permanent population and is only used as a winter pasture. The majority of the local population live along the Shu river and the foothills of the Karatau mountains. Therefore, we selected stakeholders from eight villages located around Muyunkum: Taukent, Syzgan, Sholakkorgan and Kumkent in the south and Zhuantobe, Tasty, Shu and Stepnoy in the north (Figure 1). Among these eight villages, four (Taukent, Sholakkorgan, Tasty and Shu) work closely with the French–Kazakh Joint-Venture KATCO. KATCO finances infrastructure such as schools, or gifts coal to the inhabitants. We chose stakeholders based on: categories of stakeholders (Table 1) and a balanced gender dimension. Stakeholder categories were determined through a stakeholder mapping report (performed by Central Asia agency for KATCO in 2019), discussions with KATCO Corporate Social Responsibility (CSR) department, and studies about Kazakh livestock systems and management [36,40,54] (Table 1). They represent the different types of local stakeholders in the study area.

Table 1. Stakeholder categories description, number and proportion of participants per category.

Category	Description	Number of Participants	Percentage of Participant (%)
Local authority or his deputy	Akim or his deputy. It is at the level of the Akimat that decisions on land planning are made. It is also at this level that the grievances and various demands of the inhabitants are received	10	18
Elder	These are older men and women. They are respected and may have knowledge of current land and natural resource use, but also of the past. Among them, there are 'veterans', elderly men with a special status: organized as councils, they are asked by inhabitants for their opinion on certain issues and they can act in some decision-making processes related to the life of the village	8	14

Table 1. *Cont.*

Category	Description	Number of Participants	Percentage of Participant (%)
Herder farmer	Herders and farmers in cooperatives: in farms, often a family business, with a big herd. They move their livestock every season. They are key actors in the ecosystem services provided by the different kinds of pastures; but also, stakeholders not necessarily organized in business but with a big herd and willing for production, and whose herd move pasture every season.	7	12
Social and health worker	Health workers are nurses or doctors, for example. People working in the social field deal with isolated people, large families, disabled people, elderly and/or sick people, either through providing legal and financial support or social support. This category can account for villagers' health problems that may be linked to environmental problems, as well as ecosystem services that are important to vulnerable individuals or families.	7	12
Mother with many children	Women with at least six children. They can receive a medal, of different levels depending on the number of children. They are among the categories eligible for social and financial assistance, often in the lists of vulnerable persons. They may have different perceptions and/or needs for ecosystem services for their family.	7	12
Teacher	Teachers of different levels, from school to high school and in different fields. They are educated people who may have other types of knowledge.	8	14
Inhabitant	A random person, regardless of status and occupation: unemployed, driver, veterinarian, shop owner, media, etc.	10	18
Total		57	100

2.3.3. Interviews in Sozak District

Between 28 June 2021 and 17 July 2021, we conducted a total of 57 individual interviews (Table 1). In each of the villages, seven people were interviewed (except for Sholakkorgan where we interviewed eight people). The first person we met in each village was usually the *Akim* (mayor), whose office had identified the interviewees according to the characteristics we were looking for. Interviews were conducted at the *Akimat* (town hall) of each village and were translated simultaneously by an interpreter. We interviewed 34 men and only 23 women, because *Akims* and herder farmers were always men. In total, 86% of the 57 participants were breeders of livestock, some based in agricultural cooperatives (herder farmer in Table 1) and others who were small-scale livestock farmers.

The questionnaire was comprised of boxes, with each box corresponding to a CICES group or class of ES (Table S1). Our questions were not fixed and were semi-structured. For example: 'Are crops grown in the area?' and depending on the answer we then asked: 'For what purpose?' 'Where are they grown?' and so on. The sub-questions, which sought to identify specific details about each ES, depended on the individual we met, their answers and perceptions. For example, we were able to go into detail about the soil quality regulation services with an individual X, but went to another level of detail on the groundwater services with individual Y.

Several studies have shown that the landscapes perceived by local beneficiaries may be different from those identified by academics, e.g., a unit identified by academics may actually be two units for local stakeholders (e.g., [32]). In addition, the use of maps makes it easier to introduce the subject of ES provision [30]. Thus, we tried as far as possible to locate the spot(s) and/or landscape(s) where the ES occurred using printed maps from Google Earth [55] and Google maps [56] (up to a radius of 20 km around the village of the interviewee), or orally, if the interviewee could not locate it on a map.

During the interview, all CICES groups of ES were discussed, even though some ES may not be directly relevant for biodiversity offset mechanisms. In addition, we were not limited to our list of CICES groups: some ES were added when required, as stated by the participant as the discussion progressed. Thereby, we ensured that the main benefits provided by services were not overlooked [22].

2.3.4. Data Analysis

Interview data were processed using NVivo [57], a qualitative data analysis software. Interviews were described according to interviewees' attributes: gender, stakeholder category and village. Then, interviews were coded by ES and landscape units. For ES codes, we used the hierarchical structure of the CICES. From the CICES class level, details on the ES and the benefits from the ES were added. Where necessary, ES categories were added, following CICES recommendations [16]. The landscape unit codes were added according to the landscape described for the provision of ES (maps from Google Earth and Google maps or oral location). However, some ES had no landscape boundaries, such as services related to existence value [22] ('Characteristics or features of living systems that have an existence value' in CICES [16]). In these cases, we coded certain ES in a special category that we called *no landscape frontier*.

From the organisation and coding of data, we could create matrix coding queries with the NVivo software. We used matrix coding queries to analyse in detail the relationships between the following sets of data, and to produce tables of data that included the following information for each cell (x,y):

(i) Ecosystem services (each level of CICES) and landscape units: number of interviewees who cited the ES (x) in the landscape unit (y). Subsequent analysis based on this matrix will verify the hypothesis that different landscapes provide different ES.

(ii) Stakeholder categories and ecosystem services (each level of CICES): number of interviewees from a category (y) who cited the ES (x).

(iii) Gender and ecosystem services (each level of CICES): number of male and female interviewees (y) who cited the ES (x).

Further analysis based on matrix (ii) and (iii) will test the hypothesis that stakeholders perceive ES according to their category and their gender.

(iv) Location of the villages and landscape units: number of interviewees from a village (y) who cited the landscape unit (x). In the case where ES are linked to the landscape units (analysis based on matrix (i)) but the stakeholders' categories and gender (analysis based on matrix (ii) and (iii)) do not guide the perception of ES, we tried to find out what did. Here, we test the location of villages as a criterion.

Before the following analysis, we removed the CICES ES class if it was cited by less than 10% of the participants (i.e., by five or fewer participants).

A χ^2 test was then performed on these data using the software R [58] to check the independence of variables. If the H0 hypothesis (independence between the two variables) was rejected (meaning that the obtained p-value < 0.05), and our variables were significantly related, then we performed a correspondence analysis (CA) to visualise the nature of the relationships. Instead of showing all CA, for the purpose of readability, only the positive relationships according to the CA are given and illustrated by ES section (provisioning, regulation and maintenance or cultural) for all level of CICES according to its hierarchical structure. The positive relationships are based on the respective contribution of variables to CA dimensions (axes).

As an overall result, we show in a systemic way the perception of ES provided by the different landscape units, depending on whether they are provisioning, regulation and maintenance or cultural services. An ES is shown with a landscape unit only if it was cited by at least 10% of the participants as being present at that unit, even if it was at first selected for analysis because it had been cited by 10% of participants in total (i.e., with all landscape units combined).

3. Results

From our literature study, we found a total of 116 mentions of ES. When classed into the CICES system, we organised ES into 10 divisions, 22 groups and 46 classes of ES among the 15 divisions, 36 groups and 90 classes initially present in the CICES. We also added two services; *cultivated plants for fodder* and *wild plants grazed by reared animals*, following the structure of the CICES. These services are not final ES, according to the cascade model [13,17], but are important to take into account since livestock farming and production are major activities in our study area.

From our interviews with stakeholders, 300 ES in total were identified by interviewees, which we grouped into the 61 classes of CICES. By keeping the classes of ES in which 10% participants (i.e., at least six participants), mentioned ES, we obtained 37 CICES classes of ES of interest (Table 2).

Table 2. Number of perceived ecosystem services and CICES class before and after the 10% selection.

Ecosystem Service Section	Number of Perceived Ecosystem Services	Number of Ecosystem Service Classes (CICES)	Selected Ecosystem Services Classes (CICES) (>10%)
Provisioning (biotic)	141	14	11
Provisioning (abiotic)	20	14	8
Regulation and maintenance (biotic)	71	14	6
Regulation and maintenance (abiotic)	12	6	3
Cultural (biotic)	41	10	5
Cultural (abiotic)	15	3	3
Total	300	61	36

By gathering 300 perceived ES into 37 classes of ES, i.e., almost 10 times less, further analysis could be simplified. Provisioning services are the most perceived (total of 161 ES in 19 classes of CICES) than regulation and maintenance (total of 83 ES and 20 classes) and then cultural services (total of 56 ES and 13 classes).

Moreover, participants were generally able to describe the provision of ES across various landscape units. Figure 3 illustrates the 7 landscapes units described by interviewees. In addition, the participants also located ES in a landscape unit termed *village*. The salty clay steppe area between the Shu river and the Muyunkum sandy desert was the only landscape unit not mentioned by stakeholders (see Section 2.1 in Materials and Methods).

In the further results, we keep the terms used by stakeholders to describe the landscape units. The corresponding landscapes (see Section 2.1 in Materials and Methods) are:

- The Muyunkum sandy desert: called *Sand* by participants
- The steppes: called *Steppe*
- The Shu river and riverbanks: called *River*
- The steppes of Betpak-Dala: called *Betpak-Dala*
- The Karatau mountains and its foothills: called *Mountains* and *Foothills*
- The salty lakes area: called *Lakes*

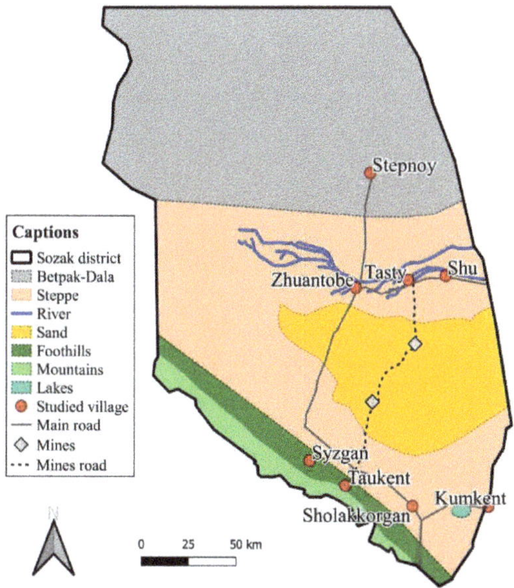

Figure 3. Landscape units as described by interviewees in the context of ecosystem services provision. Main road, Mines and Mines road are not landscape units but supplementary information.

3.1. Provision of Ecosystem Services through Landscape Units

All χ^2 tests rejected the hypothesis H0 of independence between the landscape unit and ecosystem services' variable ($p < 0.05$) both (i) at each level of the CICES (from least accurate to most accurate) and (ii) within each section (provisioning, regulation and maintenance, cultural) (Table S2). Therefore, the distribution of perceived ES provision across landscape units and the differences attributed was not random. Our first hypothesis is verified, meaning that some ES are linked to specific landscape units.

Several correspondence analysis (CA) were performed to visualize this link. The CA between ES section (provisioning, regulation and maintenance, cultural) and landscape unit is provided as an example (Figure 4). Positive links between ES section and landscape units in this CA are explained in Table 3. From this example, we observed that provisioning services are mostly provided by the steppe, the sand and the village for the biotic section (the provisioning biotic section includes, for example, cultivated plant, wild plants grazed by reared animals, and reared animals) and by the foothills and Betpak-Dala for the abiotic part (e.g., surface water or mineral substance). Cultural services are linked to the no landscape frontier category (e.g., existence or bequest value), to the lakes (e.g., spiritual, symbolic and other interactions) and to the river units (e.g., activities promoting health, recuperation, enjoyment). Biotic regulation and maintenance services (e.g., maintaining nursery population and habitats) are mostly linked to no landscape frontier category, whereas abiotic services do not seem to be link to specific landscapes units. The landscapes unit mountains does not provide specific ES at this level of CICES.

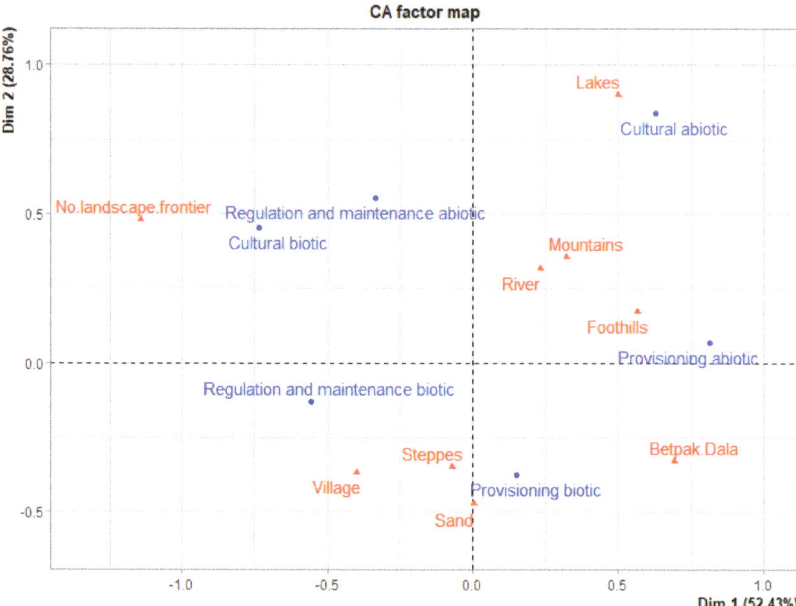

Figure 4. Example of correspondence analysis (CA) factor map. Here is the CA performed on ecosystem services CICES section and landscape units. In blue: ecosystem services CICES section, in red: landscape units.

Table 3. Links between ecosystem services CICES section and landscape units, according to the respective contributions of each variable to dimensions (axes) 1 to 3 (3 axes remained) and their relative position on the contributed axes in the correspondence analysis.

Dimension	Ecosystem Service Section	Landscape Unit
2	Provisioning biotic	Steppe, Sandy area, Village
1 and 3	Provisioning abiotic	Foothills, Betpak-Dala
1	Regulation and maintenance biotic	No landscape frontier
/	Regulation and maintenance abiotic	?
1	Cultural biotic	No landscape frontier
2 and 3	Cultural abiotic	Lakes, River, No landscape frontier

The CA were performed on each of the other CICES levels: division, group, class by service section: provisioning, regulation and maintenance, cultural, resulting in a total of nine additional CA. Positive links between ES and landscape units are illustrated in Figures 5–7 and described in Table S3. We have followed the CICES hierarchical structure for clarity in the illustration. The legend, i.e., to which landscape unit the letters correspond, is available in the figure titles. When a landscape unit box is on an ES box, then this landscape units is more linked than the others to this ES.

The χ^2 test, followed by CA on each level of CICES, and schematized in Figures 5–7, verified and illustrated our hypothesis that different landscapes units provide different ES. Few of the CICES classes of ES were not linked to a specific landscape unit, as wild plants for energy, reared animals for materials (health), wild animals for nutrition, and surface water and groundwater for watering livestock (Figure 5), as well as wind protection, pollination, regulation of temperature and humidity and seed dispersal (Figure 6). Therefore, 75% of the class of ES were provided by specific landscape unit(s) according to the interviewees.

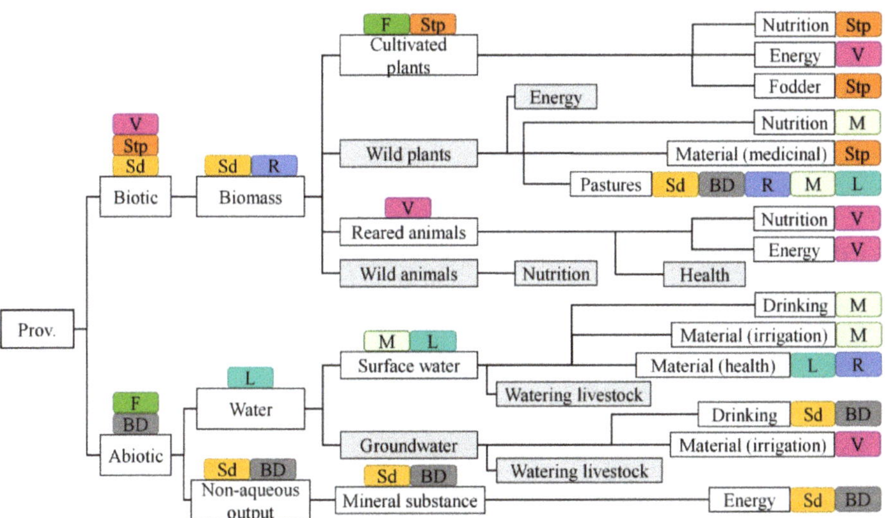

Figure 5. Hierarchical diagram following CICES representing the links between provisioning (Prov.) ecosystem services (ES) and landscape units according to their respective contribution to CA dimensions. When a landscape unit box (in colour) is on an ES box, then this landscape unit is more linked than the others to this ES (Stp = Steppe; Sd = Sand; V = village; M = Mountains; F = Foothills; L = Lakes; R = River; BD = Betpak-Dala). When an ES box is grey, it is not linked to any landscape unit according to its contribution to CA dimensions.

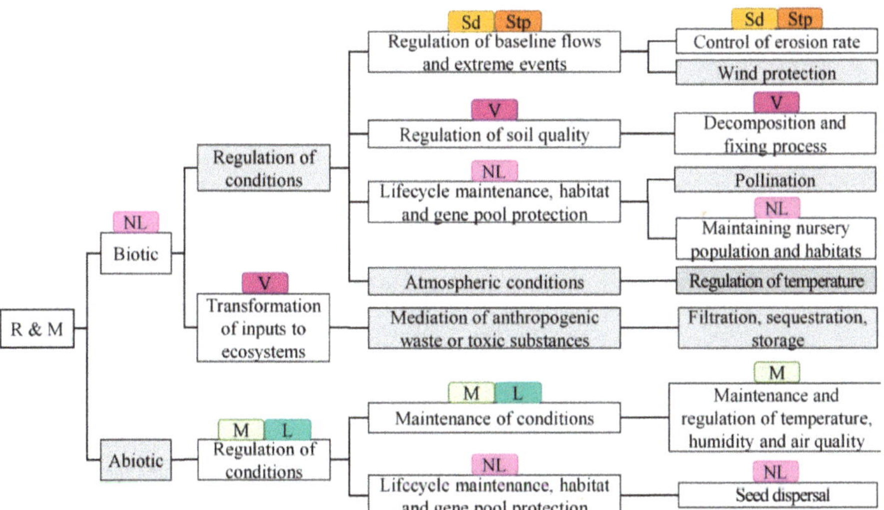

Figure 6. Hierarchical diagram following CICES representing the links between regulation and maintenance (R&M) ecosystem services (ES) and landscape units according to their respective contribution to CA dimensions. When a landscape unit box (in colour) is on an ES box, then this landscape unit is more linked than the others to this ES (Stp = Steppe; Sd = Sand; V = village; M = Mountains; F = Foothills; L = Lakes; R = River; BD = Betpak-Dala; NL = No Landscape Frontier). When an ES box is grey, it is not linked to any landscape unit according to its contribution to CA dimensions.

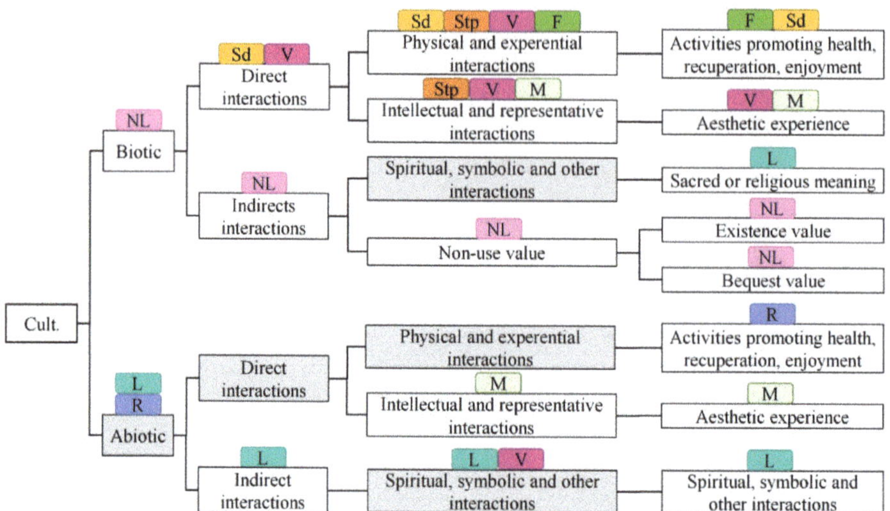

Figure 7. Hierarchical diagram following CICES structure representing the links between cultural (Cult.) ecosystem services (ES) and landscape units according to their respective contribution to CA dimensions. When a landscape unit box (in color) is on an ES box, then this landscape unit is more linked than the others to this ES (Stp = Steppe; Sd = Sand; V = village; M = Mountains; F = Foothills; L = Lakes; R = River; BD = Betpak-Dala; NL = No landscape frontier). When an ES box is grey, it is not linked to any landscape unit according to its contribution to CA dimensions.

3.2. Preference of Ecosystem Services according to Stakeholder Category and Gender

No χ^2 test rejected the hypothesis H0 of independence ($p > 0.05$), no matter the level of CICES. Thus, there was no significant link between stakeholder categories and ES and between gender and ES. Our second hypothesis is rejected, meaning that stakeholders do not perceive ES according to their category or gender. In the case of Sozak district, other criterion(s) guides the perception of ES.

3.3. Links between Stakeholder Village Location, Landscape Unit and Ecosystem Services

The χ^2 test rejected the hypothesis of independence between the village location and landscape unit variables ($p = 2.005 \times 10^{-11} < 0.05$). Therefore, the described landscape units in terms of ES provision are not random and is linked to the location of the interviewees' village.

CA were performed to visualize this link (Figure 8). There was also a contrast between villages in the north and south of the study area. Participants tended to identify more frequently the landscape units they lived close to in terms of ES provision. Interviewees in the south (from Kumkent, Taukent, Sholakkorgan and Syzgan villages) tended to perceive the mountains, foothills and lakes as providing most of the services, while those in the north (Zhuantobe, Tasty, Shu and Stepnoy villages) tended to perceive the river and Betpak-Dala as providing most of the ES. The landscape units no landscape frontier, village, steppe and sand were common to all interviewees, in the context of ES provision. This result showed that stakeholders described the landscape units for ES provision according to the location of their village within the landscapes, rather than the stakeholder category or gender.

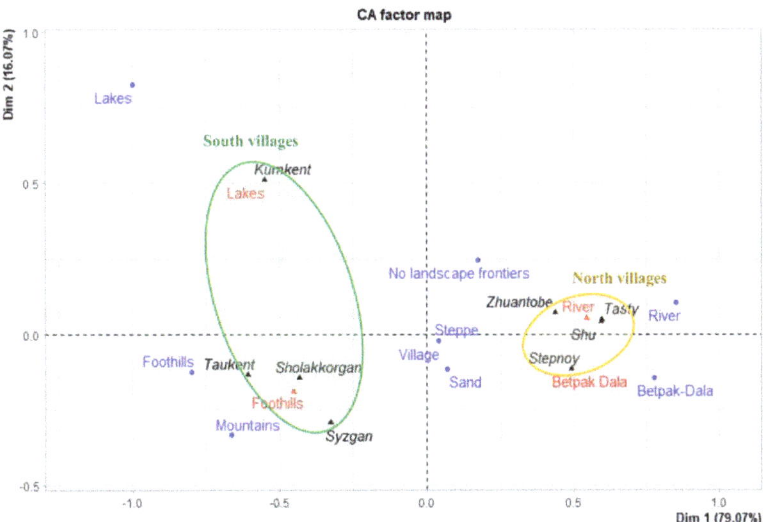

Figure 8. CA performed on participant's village location (in red) and landscape units described for ES provision (in blue). In black: name of the villages as supplementary variable. For blue variables, the unit Foothills comprises Taukent, Sholakkorgan and Syzgan; the unit Lakes comprises the village of Kumkent; the unit River comprises Zhuantobe, Tasty and Shu; and Betpak-Dala comprises the village of Stepnoy. The CA factor map shows a contrast between villages in the south (green ellipse), and those in the north (yellow ellipse).

3.4. Systemic Representation of the Perception of ES Provided by the Different Landscape Units

From the results above, we considered the inter-relation between ES throughout the landscape (Figures 9–11). Not all ES are shown, as some ES were cited by 10% of the total number of participants, but not in all landscape units (e.g., mineral substances for energy were cited by 15% of participants in total, but only by 2% in the steppe, 7% in sand and 7% in Betpak-Dala). In addition, services related to reared animals were linked to species: if less than 10% of participants cited a species for a given service, it was not represented (e.g., sheep manure is used by 22% of participants and so was represented, but horse or cow manure was only used by 2% and so was not shown). A detailed description of stakeholders' perceptions of ES is provided below.

3.4.1. Provisioning Services

All landscape units except the lakes were identified by at least 10% of participants as pasture (Figure 9). The steppe is an important grazing area (85% of participants), in all seasons. The sandy area is an important winter grazing area (42% of participants). Betpak-Dala and the mountains are more of a summer grazing area. The steppe and foothills allow for the cultivation of winter fodder (Figure 9) (42% and 21% of participants, respectively), mainly clover and corn. Other strategies are implemented for winter fodder storage (Figure 9) via cutting and storing wild plants: around the Shu River, people store reeds (17% of participants). Steppe grasses can also be stored as hay (26% of participants). The reared animals are important contributors to the nutrition of the district's population (Figure 9). Cows and sheep are the main contributors: 77% of the participants consume dairy products from cows, 28% beef, and 66% mutton.

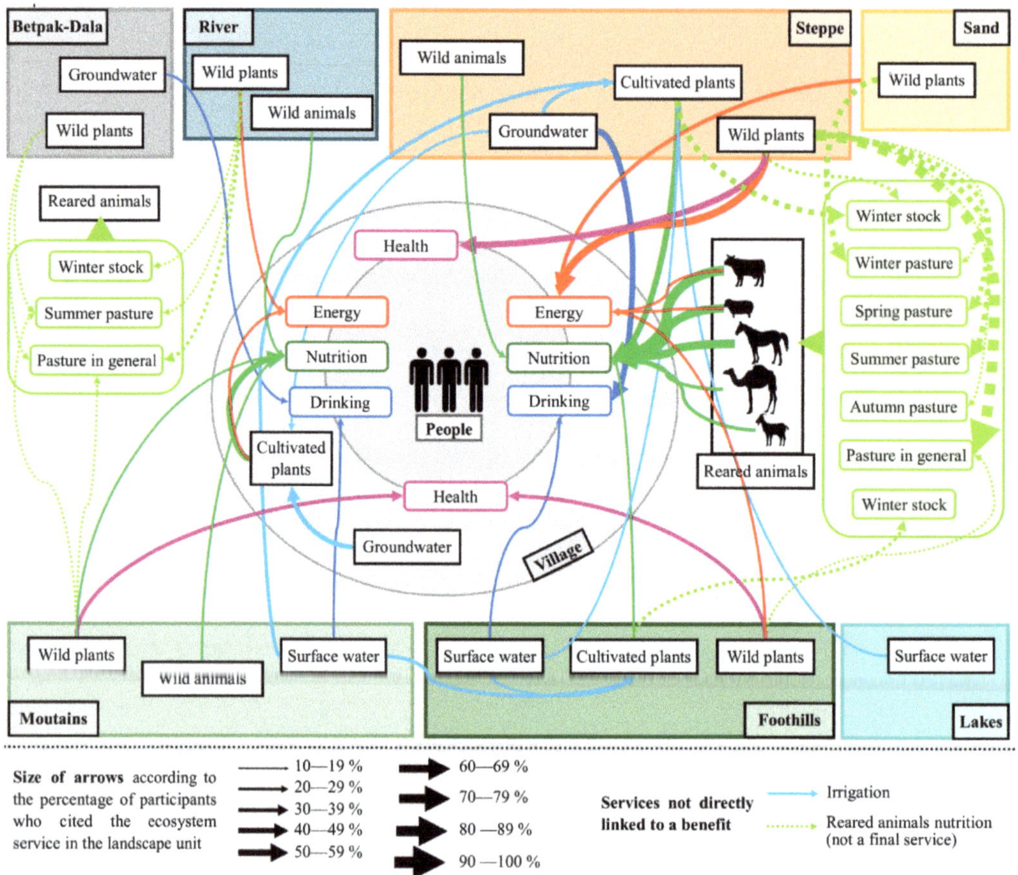

Figure 9. Systemic scheme of provisioning services provision across the landscape units.

The other sources of food are from cultivated plants (Figure 9) in the steppe (mentioned by 49% of the participants), and in the foothills (17%): mainly cultivated melons, watermelons, vegetables, as well as some fruit trees. Additionally, 45% of the participants had a vegetable garden and/or a yard with fruit trees, which highly contribute to the food of the families (Figure 9). Finally, wild plants are used for nutrition; they are mostly collected for this purpose in the mountains (19%), including herbs and wild fruits (Figure 9).

The sources of energy from biomass are wild plants collected mainly in the steppe (57%) and the sandy zone (22%) and include mainly woody plants, e.g., *Bayalich* (*Salsola arbuscula* Pall.), *Djingil* (*Tamarix* spp.) in the steppe and Saxaul (*Haloxylon* spp.) in the *sand* that are used, purchased with permission from the authorities or sometimes cut illegally (Figure 9). Sheep and cow manure are also used as energy sources (by 17% and 12% of participants, respectively) (Figure 9). Moreover, some participants use the fallen and dried branches of the village's ornamental trees (15%) (Figure 9).

Wild plants are also used as medicinal or purifying plants and come mainly from the steppe (49%), mountains and foothills (24% and 26%); the species mostly collected is *Adrespan* (*Peganum harmala* L.), used to purify houses (40% of the participants collect it in the steppe, for example, Figure 9).

Drinking water is either obtained from surface water in the mountains and foothills, or from groundwater in the steppe and Betpak-Dala. These water resources are also used for the irrigation of cultivated plants (Figure 9). For surface water irrigation: mountains contribute 28%, followed by foothills and lakes, according to the participants. For groundwater for irrigation, only the steppe was cited by at least 10% of participants. In the village, families with gardens and yards generally use water from their own well (36%) (Figure 9).

3.4.2. Regulation and Maintenance Services

The air quality service is provided by the village trees (31% of participants), which purify the air and filter and stop the dust brought by strong winds (Figure 10). Among participants, 36% found the air quality to be good, compared to 9% who did not. The structure of the Karatau mountain also contributes to the air quality (12%), as a physical barrier (Figure 10). The trees in the village also contribute to the local regulation of temperature and humidity (28%): they make the local climate more pleasant by providing shade and cool air. The mountains also contribute to this, again due to their structure and altitude (12%) (Figure 10). Among the participants, 19% did not perceive this service, especially in the north towards the Shu river and Betpak-Dala.

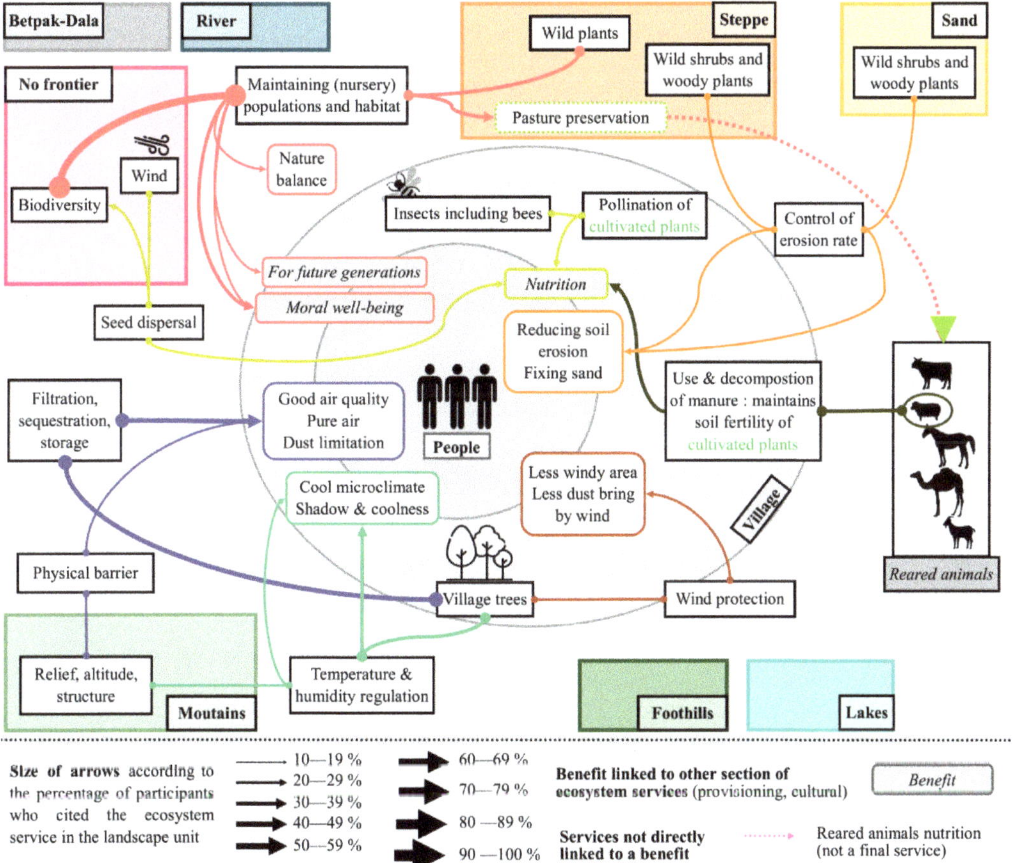

Figure 10. Systemic scheme of regulation and maintenance services provision across the landscape units.

With regard to the soil quality service, sheep manure is used by 22% of the participants (Figure 10) for individual vegetable gardens and thus contributes to the nutrition benefit (Figure 9). Wild woody plants Saxaul (*Haloxylon* spp.) contribute to the control of erosion by fixing the soil, mainly in the steppe (17%) and in the sand (19%) (Figure 10).

Regarding pollination and seed dispersal, they were, respectively, enabled by insect pollinators (perceived by 10% of participants for pollination in the village) and wind (perceived as a disseminator by 18% of participants) (Figure 10). Wind was not attached to any landscape unit, as 'the wind is everywhere'.

Finally, biodiversity, which includes wild flora and fauna, is considered important to maintain and/or protect by 71% of participants, and 66% want biodiversity protected to maintain other services and benefits that cannot be included in landscape boundaries (Figure 10). The majority of the participants wish to maintain it and give it importance for cultural reasons: for future generations (17%), but also because fauna and flora are components of nature, which have as much right as humans to exist (21%). Thus, this ES contributes to the maintenance of cultural services (Figure 11). Additionally, the maintenance of local flora, especially in the steppe, is considered important because it is a source of food for reared animals (17% of participants) (Figure 10).

3.4.3. Cultural Services

The lakes (10%), mountain (19%), foothills (21%) and the Shu river (24%) participate in the rest and enjoyment of the inhabitants, through direct interactions with their abiotic components (Figure 11). Those interactions result in visits to these landscape units, with swimming, resting and picnics, often with the family. The foothills also participate through biotic components (12%) (Figure 11): for example, through picnics and resting in private gardens, and recreational fishing.

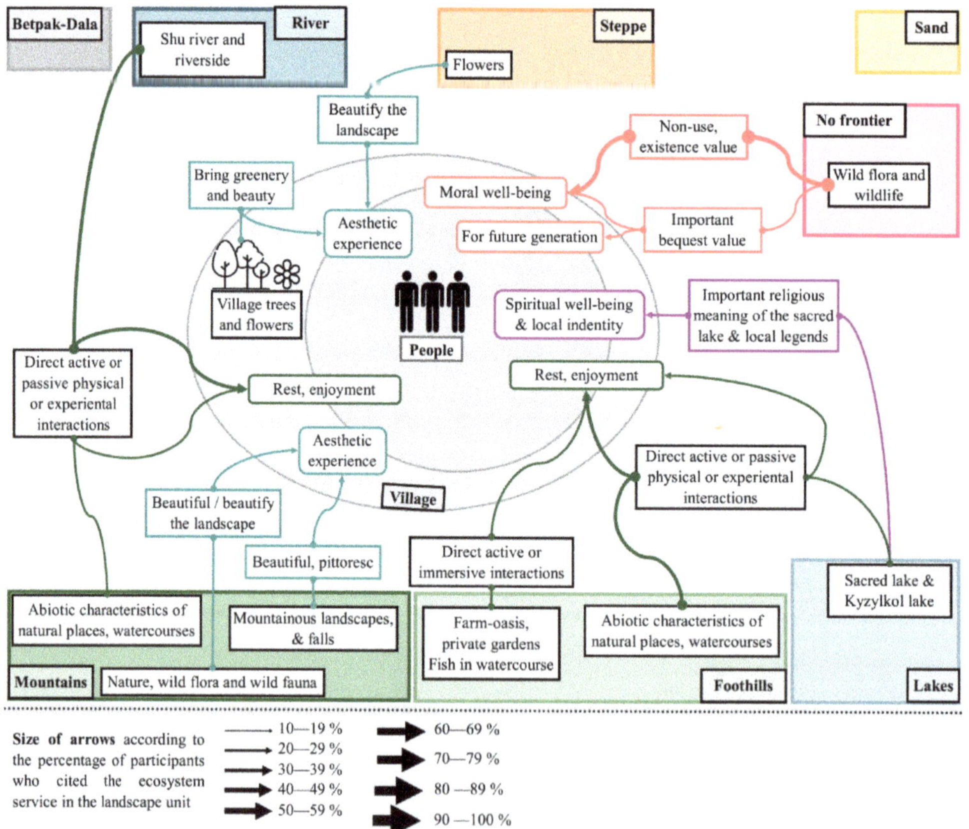

Figure 11. Systemic scheme of cultural services provision across the landscape units.

The mountainous landscape with its waterfalls (14%), but also its fauna and flora (10%), is appreciated for its beauty (Figure 11). The flowers of the steppe that bloom in spring (10%) are also appreciated for their colour (described as a 'multicolor carpet') (Figure 11). The flowers and ornamental trees in the villages are also important for the inhabitants, because they make the village more beautiful and provide it with greenery (19%) (Figure 11).

The lake of Baba Tukti Shashty Aziz mausoleum is sacred and therefore important from a religious and spiritual point of view (Figure 11). Local legends are told about its creation, as well as about the creation of the nearby large lake of Kyzylkol (meaning 'red angel') (importance for 10% of the participants) (Figure 11).

Finally, biodiversity and its maintenance ('maintaining nursery populations and habitats' [16], Figure 10) are distinguished by non-use values: biodiversity has an existence value ('characteristics or features of living systems that have an existence value' [16]), and represents a debt we owe to nature (33%), and a bequest value (14%) ('characteristics or features of living systems that have an option or bequest value' [16]) (Figure 11). It is biodiversity in general, fauna and flora everywhere: there is, therefore, no landscape frontier for these services.

4. Discussion

4.1. Identification of Ecosystem Services and Landscape Units

A large number of services and benefits were perceived by the stakeholders during interviews (Table 2). Among them, several CICES classes had not been identified in our initial literature review, e.g., 'characteristics or features of living systems that have an existence value'. Conversely, some ES found in the literature review were not perceived by the participants, e.g., 'hydrological cycle and water flow regulation'. It was thus necessary to rely on beneficiaries when identifying the priority ES, and to adapt the interviews according to the interviewees' answers. Therefore, the involvement of local stakeholders [7,11,24] and the consideration of all ES during the identification process is of utmost importance so that all priority ES are identified [22]. An example of important ES of drylands found in our literature review, but not perceived by the stakeholders was 'regulation of chemical composition of atmosphere and oceans' [25,26,47,49,51,53], that contributes to 'regulating our global climate' [16]. This ES can be hard to explain during interviews, and interviewees can have difficulties to link global ES and benefits to the local landscapes [30].

By using maps, the participants were able to refer to the majority of landscape units described in the literature for the provision of ES (Figure 3). The vision of local stakeholders is important for the identification of landscape units, even if in our case the units used were not different from those previously identified.

Stakeholder interviews confirmed that livestock farming is one of the main activities of the Sozak region, illustrating that people in drylands depend a lot on natural resources for their livelihood [25–28]. Indeed, reared animals, cultivated and wild plants, surface- and ground-water were major ES, identified by over 60% of interviewees, with almost all stakeholders citing pasture and grazed species when talking about wild plants related ES. Concerning regulation and maintenance services, they were less perceived by stakeholders, in agreement with Costanza et al. [14] (Table 2). Surprisingly, stakeholders were strongly and culturally attached to the biodiversity. The 'other biotic characteristics that have a non-use value' were an important group of ES, through 'characteristics or features of living systems that have an existence value' and 'characteristics or features of living systems that have an option or bequest value' services, (as they perceived that fauna and flora have as much right to exist as humans and should be preserved for future generations). In addition, in a study in Iran, Karimi et al. found a strong association between cultural services and biodiversity hotspots [23]. Such results contradict certain criticisms about the integration of ES into offsetting schemes. The concept of ES is controversial in biodiversity conservation [5], because it is considered difficult to link threatened resources or species with the concept of ES that have a strong utilitarian value and are provided in landscapes dominated by humans [59]. However, we show that, in this very case study, the synergy between biodiversity preservation and cultural services could be exploited through biodiversity offset mechanisms, as also proposed by Sonter et al. [24]. Whether this relationship is generic or highly context-specific should be explored further.

4.2. Exploiting the Link between Landscape, Ecosystem Services and Beneficiaries

4.2.1. Suggestions of Several ES Selection Options: Impacted Ecosystem Services, Greatest Number of Beneficiaries or Common Landscapes

Biodiversity offset mechanisms should provide same services as those impacted during an economic development project (*like-for-like* or *in-kind* biodiversity offsets) to avoid creating or worsening social inequality [22]. Moreover, biodiversity offsets should focus on impacted beneficiaries, based on the number and characteristics of stakeholders, for example, targeting ES that impact the most beneficiaries, or those in a vulnerable category [22]. In addition, care must be taken to avoid bias from stakeholders wishing to maximize the benefits from specific services [5]. In view of the literature, we suggest several options for selecting ES to be integrated into biodiversity offset schemes.

Option 1: like-for-like biodiversity offset. We have shown that it is necessary to work with landscape units to identify which ES to incorporate into offset schemes (Figures 4–7). If like-for-like biodiversity offset was implemented in the Sozak district, services related to the landscape unit sand should be offset, as this is where direct impacts have taken place. Therefore, the following services would be prioritised (excluding mineral substance for energy, that was uranium): (i) 'wild plants grazed by reared animals', (ii) 'groundwater for drinking', (iii) 'control of erosion rate', (iv) 'characteristics of living systems that enable activities promoting health, recuperation or enjoyment through active or immersive interactions'.

As a development project can indirectly impact the provision of ES [5], indirect effects should also be considered in offsetting schemes (Figures 9–11). For example, by impacting winter pastures in the sand, mining can increase the use of cultivated plants for fodder, and so results in a greater use of mountain surface water for irrigation, leading to the negative effect of less surface water for drinking (Figure 9). Reducing winter pasture could also lead to a decrease in animal products. The negative effects of mining activities on wild plants also impacts the service maintaining nursery population and habitat and the associated cultural services (existence value and bequest value, Figure 11).

Option 2: Biodiversity offset based on stakeholder categories and gender. A selection based on beneficiaries could bring another point of view. Contrary to previous studies (e.g., [29,30]), we found no relationship between stakeholder category and gender with ES. This result is because almost all interviewees raised livestock (86%), even on a small scale for their own consumption, and those who did not consumed livestock products. Services such as 'animal reared for nutrition purposes' or 'wild plants grazed by reared animals' were therefore identified by almost all participants. Similarly, services related to biodiversity and its maintenance were identified by more than three quarters of interviewees. Therefore, in this type of rangeland, targeting ES through biodiversity offset will not lead to a bias in stakeholder category or gender inequality, nor will the selection of ES based on vulnerable categories of stakeholders be necessary (such as women or elders [22]).

Option 3: Biodiversity offset based on common landscape units. Our results suggest that from the point of view of equality between beneficiaries across the landscape, offsetting could target ES produced by the four most common landscapes identified by interviewees (Figure 8), i.e., sand, steppe, village and no landscape frontier. If we add a criterion of targeting ES that impact the most beneficiaries [22], we are able to prioritise ES among these four landscape units. We suggest that by cross-referencing the ES related to the four relevant landscape units (Figures 5–7) and targeting ES identified by most interviewees (Figures 9–11), we should prioritise those ES in the following types of landscape:

(i)   Sand: 'Wild plants grazed by reared animals' (Figures 5 and 9) and 'Control of erosion rate' (Figures 6 and 10)

(ii) Steppe: 'Cultivated plants for nutrition', 'Cultivated plant for fodder', 'Material from wild plants' (medicinal use) (Figures 5 and 9) and 'Control of erosion rate' (Figures 6 and 10)

(iii) Village: 'Reared animals for nutrition', 'Reared animals for energy' (manure), 'Groundwater for irrigation' (Figures 5 and 9), 'Decomposition and fixing processes and their effect on soil quality' (Figures 6 and 10), and 'Characteristics of living systems that enable aesthetic experiences' (Figures 7 and 11).

(iv) No landscape frontier: 'Maintaining nursery population and habitat', 'Seed dispersal' (wind-induced) (Figures 6 and 10), 'Characteristics or features of living systems that have an existence value' and 'Characteristics or features of living systems that have an option or bequest value' (Figures 7 and 11).

4.2.2. Choice of Sites for Biodiversity Offset Schemes

Since ES are associated with beneficiaries, off-site biodiversity offset is not an option. In addition, to provide benefits to all stakeholders, common landscape units should be used for offset (Figure 8). For example, if offsetting was implemented around the Shu river or Betpak-Dala steppe, northern villages would be favoured by ES, whereas southern villages would benefit more from offsetting in the mountains, foothills and lakes area. Therefore, the units sand, steppe, village and no landscape frontier should be used for biodiversity offset schemes.

4.3. From Suggestions to Biodiversity Offset Strategies: What to Offset and Where?

By cross-referencing options 1–3 above (Section 4.2.1) with site selection (Section 4.2.2), we found different strategies. (i) The first would prioritises the ES 'wild plants grazed by reared animals' and 'control of erosion rate' in the sand landscape unit. This strategy would be a like-for-like offset solution, as the ES targeted are those that are potentially impacted in the *sand* unit (Figure 5) where the uranium mines are located. Furthermore, biodiversity offsets would be implemented in the landscape unit that initially provided the ES (Figure 5) and is a unit common to all villages (Figure 8). If such an offsetting strategy was not possible to implement, other possibilities would be: (ii) like-for-like offsets but in another landscape unit, such as steppe, that provides similar ES (Figure 5). (iii) Use of the no landscape frontier unit, where suitable solutions to prioritise could be protection measures for the preservation of fauna and flora (Figure 6), contributing to the existence value and bequest value of biodiversity-related ES (Figure 10), that are indirectly impacted by mining activities. As ES provided by no landscape frontier units can be perceived anywhere, protection measures could be implemented in common landscape units such as sand or steppe (Figure 8). However, the site should be chosen in terms of its ability to compensate for biodiversity and it would have to be verified that sufficient ecological gains can be achieved in the steppe or sand landscape units [22]. (iv) *Out-of-kind* offsets (meaning that the ES targeted are not those impacted), such as the planting of trees in villages, would contribute to the service 'characteristics of living systems that enable aesthetic experiences' and reach most of the beneficiaries (Figure 11).

4.4. Service-Based Scenarios and Their Potential Indirect Impacts on Other Provision of ES

Since offsetting schemes themselves can have indirect impacts on the provision of ES [10,11], we can use Figures 9–11 to consider those impacts. (i) and (ii) Ecosystem services such as 'control of erosion rate' could be targeted through the protection and restoration of vegetation [60]. However, such scenarios would initially reduce winter grazing and associated services (wild plant grazed by reared animals and then reared animals for nutrition, material and energy) until the vegetation is re-established (Figure 9). (iii) The implementation of biodiversity protection measures would improve the service 'maintaining nursery population and habitat', and the associated cultural services 'characteristics or features of living systems that have an existence' value and 'characteristics or features of living systems that have an option or bequest value' (Figure 11). Nevertheless, such measures would lead to a decrease of other ES [10,11], either through access to winter pastures if implemented in the sandy desert, or access to medicinal plants in the steppe (Figure 9). (iv) Planting in villages to green and beautify would improve the service 'characteristics of living systems that enable aesthetic experiences' (Figure 11) but would also involve higher consumption of groundwater for irrigation, which would then be less available for vegetable gardens and orchards (Figure 9). Another example of an indirect impact through biodiversity offset, exists already in the Sozak district: saxaul trees are planted in compensation schemes, but are protected and cannot be used as firewood (Figure 9) without the permission of the forest authorities. Therefore, the service 'wild plants for energy' is reduced.

4.5. The Benefits of Considering Ecosystem Services in Biodiversity Offsets

Populations in rangelands rely heavily on ES for their livelihoods [5,24]; therefore, biodiversity offset mechanisms should also compensate for the ES impacted. Our study does not call this into question. We provide a framework for identifying the ES to prioritise for beneficiaries in the right landscape units according to the vision of the local stakeholders. This approach highlights the importance of implementing offset schemes in the landscapes closed to the area impacted, thus providing a fairer offset for the beneficiaries [5]. As offset mechanisms can affect ES or access to ES [10,11,24], we suggest that a systemic approach to ES identification (i.e., at the landscape scale) should be implemented during the offset scenarios' design stage. This approach would avoid the potential indirect impacts sometimes caused by biodiversity offsets.

4.6. Towards a Framework Applicable Worldwide

The developed framework incorporates recommendations from the literature on the integration of ES in biodiversity offset schemes. We propose a systemic approach [22], which considers ES at the landscape scale in order to identify priority ES, access to these different ES by beneficiaries [11] and also to improve understanding of impacts on social needs and preferences [22]. In their study, Souza et al. state that a review of ES is necessary for ES-oriented offsets [11]. Our framework addresses the first two steps of this review: the identification and prioritisation of ES.

The developed framework could be used worldwide for future offset plans. Figure 2 shows the methodology to be applied. The framework is intended to be generic but there are specific parts for each area worldwide and new context: (i) the literature search on ecosystem services to adapt the interview guide to other study areas; (ii) the determination of stakeholder categories, as they should represent the diversity of local stakeholders; (iii) the choice of villages (or settlements, e.g., in the case of mobile populations); and (iv) the Google Earth and Google Map background maps used during the interviews. Through this framework, we test some recommendations for the prioritisation step: targeting ES lost in development and targeting the most beneficiaries [22]. Nevertheless, in future research and other area of the world, other prioritisation criterions can be used. As shown in Figure 2, depending on the local context, other questions and hypothesis can be tested, in addition to the relationship between landscape and ES and stakeholders and ES ('Other possible combinations depending on the questions to be tested for the ecosystem services provision').

4.7. Ecosystem Services-Based Offsets as a Complementary Measure in Biodiversity Offset

In order to effectively integrate ES into biodiversity offset schemes, other elements have to be assessed. Planning offsets for both biodiversity and ES can be challenging. Areas providing ES and ecologically important areas are not always compatible [5,11]. Offset strategies targeting impacted ES may, therefore, not meet the criteria for biodiversity offsets [11]. Furthermore, they may not fulfil the legal requirement. For example, in Kazakhstan in the case of ISR mining 'Subsoil users, when using plots of the state forest fund for uranium mining by the method of underground borehole leaching, shall be obliged, during the first three years of subsoil development, to make compensatory plantings of forest plantations in double the size of the area' (Article 54 of the Forest Code of the Republic of Kazakhstan [61]). However, decisions on the modalities of the plantation are taken at the level of regional authorities [61]. Thus, in order to achieve ecological and biodiversity objectives and meet legal requirements, biodiversity offsets for ES should take the form of additional or complementary measures [5,11].

The implementation of additional measures for ES would avoid some drifts. Indeed, it would be risky to supplant the current approach of biodiversity offsets, based on species and habitats: an approach based only on ES may lead to the substitution of the original species or habitats by other species and habitats providing the same ES as those impacted [5]. Therefore, Jacob et al. [5] propose that the integration of ES should be conducted in a second step, after ensuring that the ecological equivalence required by biodiversity offsets is achieved.

When offsetting ES, the potential biodiversity offset schemes should be differentiated according to the categories of ES considered. Provisioning and cultural ES should be accessible to communities, whereas this is not necessarily the case for regulating and maintenance services [11]. Attention should be paid to increasing access to provisioning services, that can sometimes negatively impact cultural and regulation and maintenance services [5]. When a development project is implemented, it should also be taken into account that some ES are simply not compensable, for example, a unique area that provides important spiritual or aesthetic ES for populations will most likely never be compensated for if lost [5].

5. Conclusions

We developed a systemic approach to integrate ecosystem services (ES) into biodiversity offset schemes to compensate for the negative impacts of economic development projects. We outlined a framework that allowed ES to be identified and prioritised across different landscapes. Interviews with local stakeholders allowed us to determine bundles of ES that impacted as many beneficiaries as possible. Interviewees also efficiently described the landscape units providing those ES. The category (local authority, elder, herder farmer, social and health worker, mother with many children, teacher or inhabitant) and gender of participants did not influence the identification of ES in this specific context of Kazakh rangelands. Stakeholders preferred services provided by the landscape units close to their village. Since ES and landscape units are significantly linked, we suggest that biodiversity offset should target ES provided by the landscape where mining activities occur. We show that to avoid conflict and bias when offset schemes are implemented, ES provided by landscape units accessible to all villagers should be targeted as a priority. In addition, a systemic understanding of the provision of ES across different landscape units would make it possible to consider both the potential direct and indirect impacts of development project and biodiversity offset scenarios on ES.

Supplementary Materials: The following supporting information can be downloaded at: https://www.mdpi.com/article/10.3390/land12010202/s1, Table S1. Interview guide for identification of ecosystem services. Table S2: *p*-value of the 13 χ^2 test performed on the ecosystem services and landscape units matrix tables. A *p*-value < 0.05 rejects the H0 hypothesis (independence between ecosystem services at each level of CICES and landscape unit) and shows that the two variables are related. Table S3: Links between ecosystem services class (C) (or group (G) when class contribution to the axes is not sufficient) and landscape units, according to the respective contributions of each variable to dimensions (axes) and their relative position on the contributed axes of the performed correspondence analysis. Table S3 summarizes the results of 6 correspondence analysis (Group of provisioning ecosystem services and landscape units; group of regulation and maintenance ecosystem services and landscape units; group of cultural ecosystem services and landscape units; class of provisioning ecosystem services and landscape units; class of regulation and maintenance ecosystem services and landscape units; class of cultural ecosystem services and landscape units).

Author Contributions: Conceptualization, A.B., A.I. and J.-D.C.; methodology, A.B., A.I. and J.-D.C.; formal analysis, A.B.; investigation, A.B., A.I. and S.K.; data curation, A.B.; writing—original draft preparation, A.B.; writing—review and editing, A.B., A.I., J.-D.C., P.S., V.R. and S.T.; visualization, A.B.; supervision S.T., V.R. and P.S.; project administration, S.T. and V.R.; funding acquisition, V.R. and S.T. All authors have read and agreed to the published version of the manuscript.

Funding: This research was funded by Orano Mining, France.

Data Availability Statement: The data presented in this study are available on request from the corresponding author.

Acknowledgments: We would like to thank all the participants who agreed to take part in the interview, the employees of the akimats in the eight villages for organising the interviews and welcoming us to their offices, KATCO's CSR department for organising the interviews prior to the fieldwork, our Kazakh-French interpreter Akbope Saparaliyeva, KATCO for the logistics, support and assistance provided in Kazakhstan, Alexia Stokes (native English speaker) who proofread the article, and Orano for granting permission to publish our results. We also thank the anonymous reviewers and editors for their valuable comments and suggestions.

Conflicts of Interest: Véronique Rayot (Orano Mining) reviewed the article as a co-author (contribution in the Writing, review and editing section). Sholpan Koldasbekova (KATCO JV LLP) worked with Annaêl Barnes in the selection of villages and categories of stakeholders (i.e., local authority, elder, herder-farmer, social and health worker, mother of a large family, teachers and villagers). She was in charge of contacting the mayors or mayors' assistants of each village to schedule the interviews. (Contribution in the investigation section).

References

1. MEA. *Ecosystems and Human Well-Being: Synthesis*; Island Press: Washington, DC, USA, 2005; ISBN 978-1-59726-040-4.
2. OECD. *Biodiversity Offsets: Effective Design and Implementation*; OECD: Paris, France, 2016; ISBN 978-92-64-22251-9.
3. Bennett, G.; Gallant, M. *State of Biodiversity Mitigation 2017. Markets and Compensation for Global Infrastructure Development*; Forest Trends' Ecosystem Marketplace: Washington, DC, USA, 2017.
4. Griffiths, V.F.; Bull, J.W.; Baker, J.; Milner-Gulland, E.J. No Net Loss for People and Biodiversity. *Conserv. Biol.* **2019**, *33*, 76–87. [CrossRef] [PubMed]
5. Jacob, C.; Vaissiere, A.-C.; Bas, A.; Calvet, C. Investigating the Inclusion of Ecosystem Services in Biodiversity Offsetting. *Ecosyst. Serv.* **2016**, *21*, 92–102. [CrossRef]
6. Kiesecker, J.M.; Copeland, H.; Pocewicz, A.; McKenney, B. Development by Design: Blending Landscape-Level Planning with the Mitigation Hierarchy. *Front. Ecol. Environ.* **2009**, *8*, 261–266. [CrossRef] [PubMed]
7. BBOP. *Biodiversity Offset Design Handbook-Updated*; BBOP: Washington, DC, USA, 2012; ISBN 978-1-932928-50-1.
8. Bull, J.W.; Suttle, K.B.; Gordon, A.; Singh, N.J.; Milner-Gulland, E.J. Biodiversity Offsets in Theory and Practice. *Oryx* **2013**, *47*, 369–380. [CrossRef]
9. Bigard, C. Eviter—Réduire—Compenser: D'un Idéal Conceptuel Aux Défis de Mise En Oeuvre. In *Une Analyse Pluridisciplinaire et Multi-Échelle*; Ecologie des communautés, Université de Montpellier: Montpellier, France, 2018.
10. Moilanen, A.; Kotiaho, J.S. Fifteen Operationally Important Decisions in the Planning of Biodiversity Offsets. *Biol. Conserv.* **2018**, *227*, 112–120. [CrossRef]
11. Souza, B.A.; Rosa, J.C.S.; Siqueira-Gay, J.; Sánchez, L.E. Mitigating Impacts on Ecosystem Services Requires More than Biodiversity Offsets. *Land Use Policy* **2021**, *105*, 105393. [CrossRef]
12. Fisher, B.; Turner, R.K.; Morling, P. Defining and Classifying Ecosystem Services for Decision Making. *Ecol. Econ.* **2009**, *68*, 643–653. [CrossRef]
13. Potschin, M.; Haines-Young, R. Defining and Measuring Ecosystem Services. In *Routledge Handbook of Ecosystem Services*; Potschin, M., Haines-Young, R., Fish, R., Turner, R.K., Eds.; Routledge: New York, NY, USA, 2016.
14. Costanza, R.; de Groot, R.; Braat, L.; Kubiszewski, I.; Fioramonti, L.; Sutton, P.; Farber, S.; Grasso, M. Twenty Years of Ecosystem Services: How Far Have We Come and How Far Do We Still Need to Go? *Ecosyst. Serv.* **2017**, *28*, 1–16. [CrossRef]
15. Pascual, U.; Balvanera, P.; Díaz, S.; Pataki, G.; Roth, E.; Stenseke, M.; Watson, R.T.; Başak Dessane, E.; Islar, M.; Kelemen, E.; et al. Valuing Nature's Contributions to People: The IPBES Approach. *Curr. Opin. Environ. Sustain.* **2017**, *26*, 7–16. [CrossRef]
16. CICES Common International Classification of Ecosystem Services Version 5.1 2018. Available online: https://cices.eu/ (accessed on 9 March 2021).
17. Haines-Young, R.; Potschin, M. *Common International Classification of Ecosystem Services (CICES) V5.1. Guidance on the Application of the Revised Structure*; Fabis Consulting: Nottingham, UK, 2018; p. 53.
18. Richter, F.; Jan, P.; El Benni, N.; Lüscher, A.; Buchmann, N.; Klaus, V.H. A Guide to Assess and Value Ecosystem Services of Grasslands. *Ecosyst. Serv.* **2021**, *52*, 101376. [CrossRef]

19. van der Meulen, E.S.; Braat, L.C.; Brils, J.M. Abiotic Flows Should Be Inherent Part of Ecosystem Services Classification. *Ecosyst. Serv.* **2016**, *19*, 1–5. [CrossRef]
20. Haines-Young, R.; Potschin, M. Categorisation Systems: The Classification Challenge. In *Mapping Ecosystem Services*; Burkhard, B., Maes, J., Eds.; Pensoft Publishers: Sofia, Bulgaria, 2017; pp. 42–45.
21. BBOP. *Standard on Biodiversity Offsets*; BBOP: Washington, DC, USA, 2012; ISBN 978-1-932928-44-0.
22. Tallis, H.; Kennedy, C.M.; Ruckelshaus, M.; Goldstein, J.; Kiesecker, J.M. Mitigation for One & All: An Integrated Framework for Mitigation of Development Impacts on Biodiversity and Ecosystem Services. *Environ. Impact Assess. Rev.* **2015**, *55*, 21–34. [CrossRef]
23. Karimi, A.; Yazdandad, H.; Fagerholm, N. Evaluating Social Perceptions of Ecosystem Services, Biodiversity, and Land Management: Trade-Offs, Synergies and Implications for Landscape Planning and Management. *Ecosyst. Serv.* **2020**, *45*, 11. [CrossRef]
24. Sonter, L.J.; Gourevitch, J.; Koh, I.; Nicholson, C.C.; Richardson, L.L.; Schwartz, A.J.; Singh, N.K.; Watson, K.B.; Maron, M.; Ricketts, T.H. Biodiversity Offsets May Miss Opportunities to Mitigate Impacts on Ecosystem Services. *Front. Ecol. Environ.* **2018**, *16*, 143–148. [CrossRef]
25. MEA. *Ecosystems and Human Well-Being: Desertification Synthesis*; World Resources Institute: Washington, DC, USA, 2005; ISBN 978-1-56973-590-9.
26. White, R.P.; Nackoney, J. *Drylands, People, and Ecosystem Services: A Web-Based Geospatial Analysis*; World Resources Institute: Washington, DC, USA, 2003; p. 58.
27. Boone, R.B.; Conant, R.T.; Sircely, J.; Thornton, P.K.; Herrero, M. Climate Change Impacts on Selected Global Rangeland Ecosystem Services. *Glob. Chang. Biol.* **2018**, *24*, 1382–1393. [CrossRef] [PubMed]
28. Stringer, L.C.; Dougill, A.J. Drylands. In *Routledge Handbook of Ecosystem Services*; Potschin, M., Haines-Young, R., Fish, R., Turner, R.K., Eds.; Routledge: New York, NY, USA, 2016.
29. Cáceres, D.M.; Tapella, E.; Quétier, F.; Díaz, S. The Social Value of Biodiversity and Ecosystem Services from the Perspectives of Different Social Actors. *Ecol. Soc.* **2015**, *20*, art62. [CrossRef]
30. Cifuentes-Espinosa, J.A.; Feintrenie, L.; Gutiérrez-Montes, I.; Sibelet, N. Ecosystem Services and Gender in Rural Areas of Nicaragua: Different Perceptions about the Landscape. *Ecosyst. Serv.* **2021**, *50*, 11. [CrossRef]
31. Vialatte, A.; Barnaud, C.; Blanco, J.; Ouin, A.; Choisis, J.-P.; Andrieu, E.; Sheeren, D.; Ladet, S.; Deconchat, M.; Clément, F.; et al. A Conceptual Framework for the Governance of Multiple Ecosystem Services in Agricultural Landscapes. *Landsc. Ecol.* **2019**, *34*, 1653–1673. [CrossRef]
32. Sinare, H.; Gordon, L.J.; Enfors Kautsky, E. Assessment of Ecosystem Services and Benefits in Village Landscapes—A Case Study from Burkina Faso. *Ecosyst. Serv.* **2016**, *21*, 141–152. [CrossRef]
33. Brunn, S.D.; Toops, S.W.; Gilbreath, R. *The Routledge Atlas of Central Asian Affairs*; Routledge: London, UK, 2012; ISBN 978-0-415-49752-7.
34. Robinson, S.; Milner-Gulland, E.J. Contraction in Livestock Mobility Resulting from State Farm Reorganisation. In *Prospects for Pastoralism in Kazakhstan and Turkmenistan. From State Farms to Private Flocks*; Kerven, C., Ed.; Routledge: London, UK, 2003; pp. 128–145.
35. Kerven, C.; Robinson, S.; Behnke, R.; Kushenov, K.; Milner-Gulland, E.J. A Pastoral Frontier: From Chaos to Capitalism and the Re-Colonisation of the Kazakh Rangelands. *J. Arid. Environ.* **2016**, *127*, 106–119. [CrossRef]
36. Kerven, C.; Shanbaev, K.; Alimaev, I.; Smailov, A.; Smailov, K. Livestock Mobility and Degradation in Kazakhstan's Semi-Arid Rangelands. In *The Socio-Economic Causes and Consequences of Desertification in Central Asia*; Behnke, R., Ed.; NATO Science for Peace and Security Series; Springer: Dordrecht, The Netherlands, 2008; pp. 113–140. ISBN 978-1-4020-8543-7.
37. KATCO. Corporate Social Responsability Report. Delivering on Our Sustainability Agenda. 2020. Available online: https://www.orano.group/docs/default-source/orano-doc/expertises/producteur-uranium/katco_csr_2020_english.pdf?sfvrsn=13d4e966_2 (accessed on 9 March 2021).
38. Ferret, C. Mobile Pastoralism a Century Apart: Continuity and Change in South-Eastern Kazakhstan, 1910 and 2012. *Cent. Asian Surv.* **2018**, *37*, 503–525. [CrossRef]
39. OECD. *Monitoring the Development of Agricultural Co-Operatives in Kazakhstan*; OECD Publishing: Paris, France, 2019; p. 59.
40. Wright, I.A.; Malmakov, N.I.; Vidon, H. New Patterns of Livestock Management. Constraints to Productivity. In *Prospects for Pastoralism in Kazakhstan and Turkmenistan. From State Farms to Private Flocks*; Kerven, C., Ed.; Routledge: London, UK, 2003; pp. 108–127.
41. Fileccia, T.; Jumabayeva, A.; Nazhmidenov, K. *Highlights on Four Livestock Sub-Sectors in Kazakhstan—Sub-Sectoral Cross-Cutting Features and Issues*; FAO Investment Centre Division: Rome, Italy, 2010; p. 138.
42. Baltic Cleantech Alliance. The Report on Mining for UNCSD 18 (Republic of Kazakhstan). 2016, p. 5. Available online: https://balticcleantech.com/wp-content/uploads/2016/04/mining-Kazakhstan_eng.pdf (accessed on 9 March 2021).
43. Global Business Reports. *Kazakhstan's Mining Industry. Steppe by Steppe*; Engineering and Mining Journal: Jacksonville, FL, USA, 2015; p. 20.
44. Lengellé, J.-F.; Park, Y.; Bloch, F.; Kupina, L.; Olson, O.; Trimouillas, P.-E. *Reforming Kazakhstan: Progress, Challenges and Opportunities*; Published under the responsibility of the Secretary-General of the OECD, OECD: Paris, France, 2018; p. 187. Available online: https://www.oecd.org/eurasia/countries/OECD-Eurasia-Reforming-Kazakhstan-EN.pdf (accessed on 9 March 2021).

45. Fyodorov, G.V. *Uranium Production and the Environment in Kazakhstan*; International Atomic Energy Agency: Vienna, Austria, 2002; Volume 33.
46. Seredkin, M.; Zabolotsky, A.; Jeffress, G. In Situ Recovery, an Alternative to Conventional Methods of Mining: Exploration, Resource Estimation, Environmental Issues, Project Evaluation and Economics. *Ore Geol. Rev.* **2016**, *79*, 500–514. [CrossRef]
47. Archer, S.R.; Predick, K.I. An Ecosystem Services Perspective on Brush Management: Research Priorities for Competing Land-Use Objectives. *J. Ecol.* **2014**, *102*, 1394–1407. [CrossRef]
48. Du, B.; Zhen, Y.; Yan, H.; de Groot, R.; Leemans, R. Comparison of Ecosystem Services Provided by Grasslands with Different Utilization Patterns in China's Inner Mongolia Autonomous Region. *J. Geogr. Sci.* **2018**, *28*, 1399–1414. [CrossRef]
49. Dutilly-Diane, C.; McCarthy, N.; Turkelboom, F.; Bruggeman, A.; Tiedemann, J.; Street, K.; Gianluca, S. *Could Payments for Environmental Services Improve Rangeland Management in Central Asia, West Asia and North Africa?* CGIAR Systemwide Program on Collective Action and Property Rights: Washington, DC, USA, 2007.
50. Fu, Q.; Li, B.; Hou, Y.; Bi, X.; Zhang, X. Effects of Land Use and Climate Change on Ecosystem Services in Central Asia's Arid Regions: A Case Study in Altay Prefecture, China. *Sci. Total Environ.* **2017**, *607*, 633–646. [CrossRef] [PubMed]
51. Li, J.; Chen, H.; Zhang, C.; Pan, T. Variations in Ecosystem Service Value in Response to Land Use/Land Cover Changes in Central Asia from 1995-2035. *PeerJ* **2019**, *7*, e7665. [CrossRef] [PubMed]
52. Murali, R.; Ikhagvajav, P.; Amankul, V.; Jumabay, K.; Sharma, K.; Bhatnagar, Y.V.; Suryawanshi, K.; Mishra, C. Ecosystem Service Dependence in Livestock and Crop-Based Production Systems in Asia's High Mountains. *J. Arid. Environ.* **2020**, *180*, 104204. [CrossRef]
53. Schild, J.E.M.; Vermaat, J.E.; de Groot, R.S.; Quatrini, S.; van Bodegom, P.M. A Global Meta-Analysis on the Monetary Valuation of Dryland Ecosystem Services: The Role of Socio-Economic, Environmental and Methodological Indicators. *Ecosyst. Serv.* **2018**, *32*, 78–89. [CrossRef]
54. Behnke, R.H., Jr. Reconfiguring Property Rights and Land Use. In *Prospects for Pastoralism in Kazakhstan and Turkmenistan. From State Farms to Private Flocks*; Kerven, C., Ed.; Routledge: London, UK, 2003; pp. 75–107.
55. *Google Earth Pro*, version V. 7.3.3.7721; Google: Mountain View, CA, USA, 2021.
56. *Google Maps*, Données cartographiques ©2021; Google: Mountain View, CA, USA, 2021.
57. *NVivo*, version NVivo 12; QSR International Pty Ltd.: Burlington, MA, USA, 2018.
58. *R: A Language and Environment for Statistical Computing*, version 4.2.2.; R Core Team: Vienna, Austria, 2022.
59. Grantham, H.S.; Portela, R.; Alam, M.; Juhn, D.; Connell, L. Maximizing Biodiversity and Ecosystem Service Benefits in Conservation Decision-Making. In *Routledge Handbook of Ecosystem Services*; Potschin, M., Haines-Young, R., Fish, R., Turner, R.K., Eds.; Routledge: New York, NY, USA, 2016; pp. 554–563.
60. Liu, J.; Li, S.; Ouyang, Z.; Tam, C.; Chen, X. Ecological and Socioeconomic Effects of China's Policies for Ecosystem Services. *Proc. Natl. Acad. Sci. USA* **2008**, *105*, 9477–9482. [CrossRef]
61. Republic of Kazakhstan. *Article 54 of Forest Code of Republic of Kazakhstan: Conduction of Works in the State Forest Resources That Are Not Related to Forest Management and Forest Use (2003, Amended in 2021)*; «Institute of legislation and legal information of the Republic of Kazakhstan» of the Ministry of Justice of the Republic of Kazakhstan: Astana, Kazakhstan, 2021.

Review

System Dynamics Tools to Study Mediterranean Rangeland's Sustainability

Jaime Martínez-Valderrama [1,2,*] and Javier Ibáñez Puerta [3]

[1] Estación Experimental de Zonas Áridas, Consejo Superior de Investigaciones Científicas, La Cañada de San Urbano, 04120 Almería, Spain
[2] Instituto Multidisciplinar del Estudio del Medio "Ramón Margalef", Universidad de Alicante, 03690 Alicante, Spain
[3] Departamento de Economía Agraria, Estadística y Gestión de Empresas, Escuela Técnica Superior de Ingeniería Agronómica, Alimentaria y de Biosistemas, Universidad Politécnica de Madrid, 28040 Madrid, Spain
[*] Correspondence: jaimonides@eeza.csic.es

Abstract: Rangelands are a key resource present all over the world and cover half of all emerged lands. They are even more important in drylands, where they cover 48% of the total area. Their intensification and the additional pressure added by climate change push these socio-ecological systems towards desertification. Over the last two decades, we have developed and applied System Dynamics (SD) models for the study of Mediterranean grasslands. In addition, we have designed procedures and analysis tools, such as global sensitivity analysis, stability analysis condition, or risk analysis, to detect the main drivers of these socio-ecological systems and provide indicators about their long-term sustainability. This paper reviews these works, their scientific background, and the most relevant conclusions, including purely technical and rangeland-related ones, as well as our experience as systemic modelers in a world driven by field specialists.

Keywords: drylands; Mediterranean; early warning systems; sensitivity analysis; desertification

Citation: Martínez-Valderrama, J.; Ibáñez Puerta, J. System Dynamics Tools to Study Mediterranean Rangeland's Sustainability. *Land* **2023**, *12*, 206. https://doi.org/10.3390/land12010206

Academic Editors: Michael Vrahnakis, Yannis (Ioannis) Kazoglou and Manuel Pulido Fernádez

Received: 17 December 2022
Revised: 5 January 2023
Accepted: 5 January 2023
Published: 9 January 2023

1. Introduction

Rangelands can be defined as those ecosystems where humans have managed their vegetation cover through the presence of livestock in order to obtain economic benefits [1]. This is the predominant land use in the world, occupying half of all emerged lands. Their extension is about 29 million km^2, of which 63% is located in drylands [2]. They cover about 70% of the needs of domestic ruminants [3] and are a key resource for developing countries, where they are the main support for the 1.2 billion people who survive on less than USD 1 a day [4].

The degradation of grazing systems can therefore affect large areas of the planet and the most vulnerable population. Many of these ecosystems are located on marginal, less fertile land, where increased livestock densities alter ecosystem structure and functions [5], leading to deterioration of their economic and biological productivity. The impact of grazing increases with aridity [6]. Substantial degradation is occurring across the world's arid and semi-arid rangelands [7–9], and the expected increasing frequency and duration of droughts [7,10,11] and the foreseeable aridification of mid-latitudes [12,13], as a consequence of global warming, pose major threats to rangelands.

This work focuses on the Mediterranean region, where rangelands occupy 48% of the territory [14] and the threat of global warming is particularly acute [15,16]. In these ecosystems, the varied floristic diversity is noteworthy, including grasslands and meadows, which occupy 20% of the total [17], and more or less dense shrubs and forests where the main use is for livestock and the dominant species are goats and sheep. The abrupt land-use changes that were triggered in the Mediterranean in the middle of the last century are part of the Great Acceleration [18] and have had a major impact on rangelands. On the one hand, in a process that has advanced from north to south, agricultural systems have intensified due to technological possibilities that respond to the logic of the market. This has led to (i) higher stocking rates (a situation more noticeable on the southern shore of the Mediterranean Basin than in the north, where irrigation is a more characteristic feature of this intensification [19]; (ii) to the collapse of pastoral systems based on the movement of livestock (nomadism and transhumance) [20]; and (iii) to the massive use of animal feed [21,22]. On the other hand, large areas have been abandoned due to a rural exodus that concentrates the population in large cities [2].

The degradation of the Mediterranean rangelands belongs to the scope of desertification, since it occurs in drylands (specifically dry sub-humid, semi-arid, and arid areas) and is a consequence of climatic variations and economic activities, as states the United Nations definition of desertification [23]. Although some authors have questioned the possibility of desertification in Europe [24], the reality is that the European Union considers it a growing threat [25]. In countries such as Spain, agropastoral systems and abandoned rangelands are recognized as already degraded scenarios or at risk of desertification [26,27]. However, in southern Europe, the current threat of desertification is linked to macrofarms [28,29] and extensive livestock farming is more a solution than a problem [30].

This indicates that both desertification and rangeland use are complex issues. One of the main causes of quantitative and/or qualitative degradation is overgrazing as a result of increased livestock loading. Numerous works report the erosion processes triggered after the loss of plant cover, or the loss of fodder species [5,31–34]. Simultaneously, opposite forces operate in the territory, which allows us to glimpse that desertification is a complex phenomenon that requires a very fine adjustment in the intensity of land use. In fact, rural abandonment and, therefore, undergrazing, is another typical source of degradation in the northern Mediterranean. Shrub encroachment and the invasion of woody vegetation give rise to the so-called 'green deserts' [35], as unproductive, from a socio-economic point of view, as the territories where primary productivity has been reduced.

In addition, the rural outflow and lack of grazing that prevented the accumulation of plant biomass, has created enormous extensions of homogeneous forest masses with hardly any discontinuities with increased fuel loads [36], resulting in fire-prone landscapes [37], which combined with global warming, lead to increased higher fire risk, longer fire seasons, and more frequent large, severe fires [38–40]. Although low-intensity and low-frequency fires have always occurred naturally and play a regulatory role in Mediterranean ecosystems (against phytotoxic agents, promoting seed germination, etc.), when their virulence and recurrence increase (median fire return has been reduced from ~30 to ~10 years in some instances [41], they cause serious damage by exposing the soil to heavy rainfall, preventing seeders from replenishing seed banks [42], depleting re-sprouters bud banks [43], and/or favoring invasive species [44].

Modeling studies, while being a simplified representation of actual systems, can provide additional insights by allowing impact analyses of a wide range of farming practices and short- and long-term climate scenarios [45]. In particular, System Dynamics (SD) [46] has been used in the modeling of rangelands because of its ability to bring together the different aspects that concur in these socio-ecosystems. For example, several models have been developed on the relationships between environmental conditions and animal stocks [47–54], specifically taking into account the importance of livestock mobility in some cases [55]. Some studies incorporated economic components [56–61] and analyzed the economic effects of a public land policy [62,63] in their rangeland models of a livestock farm in developed countries. More recently, climate change has driven study on the impact of droughts in rangelands [45,64–69].

However, few studies [70,71] fill the gap between the biophysical relationships and socioeconomic behaviors. Hence, the economic and environmental importance of rangelands and the challenges they face (climate change, intensification, land-use change, or abandonment) is a relevant field of research [30,72,73], with growing concerns about the socioeconomic behaviors and decision making of farmers, such as profit-seeking behaviors [9,74–79]. Under the umbrella of this boy of research, here, (i) we point out the utilities of SD for tackling such a complex problem; (ii) we present the SD models developed for different cases of Mediterranean grasslands (Figure 1), paying special attention to dehesa rangelands, i.e., an agro-silvo-pastoral system resulting from the progressive clearing of the original forest of oaks and/or cork oaks and covering some 90,000 km^2 of the southwest of the Iberian Peninsula [78]; (iii) we describe the analysis procedures developed to study the stability of these socio-ecosystems and the factors that most influence it; and finally (iv), we present some of the main findings in the light of these analyses and modeling carried out.

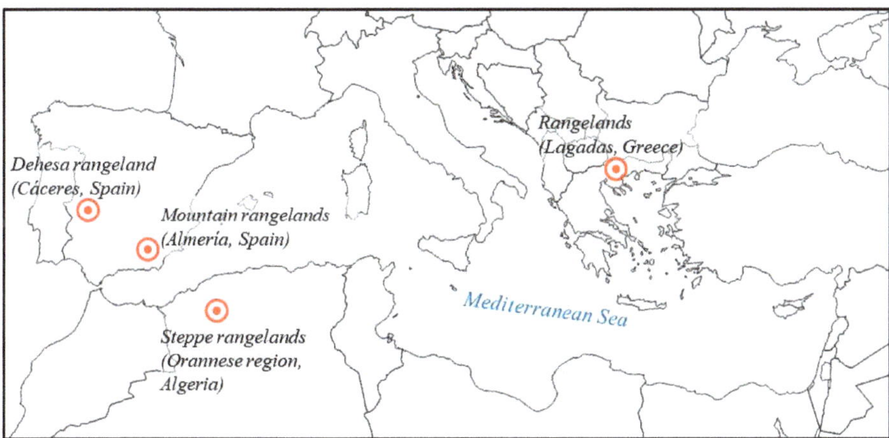

Figure 1. Location of the case studies. Several SD models have been implemented in dehesa rangelands (SW Spain). We have also studied grasslands in the SE of the Iberian Peninsula (Sierra de Filabres), characterized by their aridity and low livestock density. The grazing lands of Lagadas (Greece) have allowed us to apply the models in a more eastern European area. Finally, we have analyzed the degradation processes of the North African steppes, dominated by alfa grass (*Macrochloa tenacissima* L.) steppes.

2. An Appropriate Research Field for System Dynamics

The study of the sustainability of rangelands (or desertification, which would be its opposite) requires the use of comprehensive tools and a multidisciplinary approach, since various disciplines such as ecology, economics, or agronomy are involved in its understanding and management. The need for a holistic approach in complex socio-ecosystems is recurrent [79–86], and SD is a suitable tool for this challenge.

SD is a modeling methodology grounded on the theories of nonlinear dynamical systems and feedback control developed in mathematics, physics, and engineering. SD states that the main, but easily overlooked, cause of the behavior of a complex system lies in its underlying structure of relationships, which includes feedback loops, non-linear relations, delays, and decision rules. Formally, an SD model is a set of first-order ordinary differential equations that makes a stock-and-flow representation of the studied system; stock variables show the state of the system over time, and flow variables represent the processes that change the stocks [46,87]. The main advantages of SD [88,89]: (i) it improves system understanding, and develops system thinking skills, even from the first stage of its development as causal or sketch diagrams; (ii) SD models can incorporate empirical and process-based approaches, and help integrate interdisciplinary knowledge; (iii) the SD literature provides abundant information about related methodologies; and (iv) user-friendly software platforms allow easy access for non-modeler users.

The use of qualitative information is particularly useful in drylands where available data are limited [90]. SD is particularly useful when the system may face situations that have not previously occurred, i.e., its desertification. For such a task it is required to know the full range of behavior of the variables involved in the system. An example that can help to illustrate this critical aspect is the influence of soil quantity on biomass primary productivity (Figure 2). The loss of soil through erosion reduces the moisture content and availability of nutrients, and consequently the production of biomass falls. Usual information available to characterize the soil-productivity relationship covers the central part of the function, i.e., where the system is productive (red line in Figure 2). However, obtaining information on this function at the extremes is not so simple. It is in these uncomfortable parts of the function where the contribution of the SD is paramount, since it allows the implementation of hypotheses about how systems can work in critical situations that were initially inconceivable. On the one hand, we know that primary productivity will not grow indefinitely, however much soil there is, i.e., the function becomes saturated at some point. On the other hand, and this is where the problem of desertification lies, there will be a soil threshold below which the system becomes unproductive. Hence, at some point, the function becomes zero (Soil_{min} in Figure 2). As Sterman (2000) [87] puts it, the relationships between variables expressed by means of multiplicative factors are more realistic than their linear alternative when the equations are subjected to extreme conditions. This is precisely what happens when rangelands are degraded.

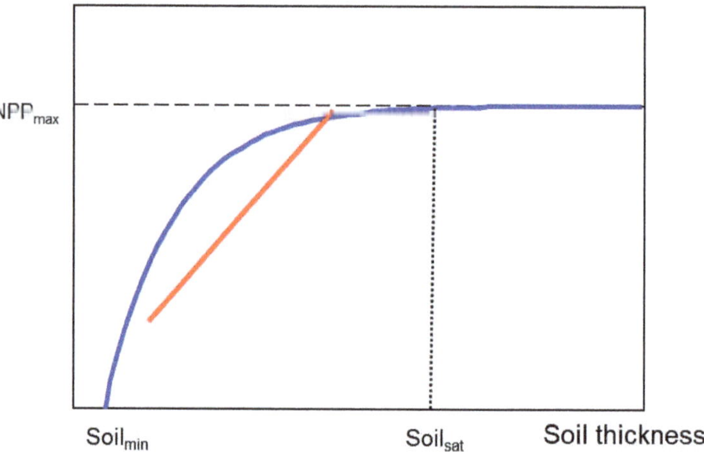

Figure 2. Net Primary Production-Soil thickness non-linear function (blue line) that takes into account minimum soil thickness for grass growth (Soil_{min}) and soil thickness from which grass growth stabilizes (Soil_{sat}), compared to the conventional linear function (red line), which overlooks the behavior of the function for its extreme values.

SD aims to build dynamic, complex, and comprehensive models capable of exploring the long-term impacts of alternative decisions, taking into account the laxity of the laws regulating the behavior of socio-ecological systems and the scarcity of data [91]. In addition, SD is a flexible enough tool to support different data sources and to accommodate multiple analyses. Thus, it is possible to use statistical or stochastic models within its structure and, as we will present later, program routines to implement advanced sensitivity analyses, optimizations, and probability calculations.

3. A Suite of Models for Assessing Rangeland Desertification

3.1. A Generic Desertification Model (GDM)

The conceptual paradigm of the SD models implemented to study rangeland stability is the classic models of predator–prey ecology of Lotka [92] and Volterra [93]. We have followed the work of Noy-Meir, who considered extensive livestock systems as a specific case of predator–prey systems [49,94], and those of Thornes [95,96], who addressed the study of erosion as an ecological relationship of competition for water between eroded soil and plant cover.

These ideas inspired the formulation of a GDM (see complete description in Ibáñez et al. (2008) [97]) that consists of an eight-equation dynamic model of a generic human–resource system. Briefly, the resource (N) plays the role of prey and the consumption units (U) that exploit it are the predators. The N renewal depends on the climate, the N stock, and a limiting factor (S), so called because its level is decisive for the survival of the exploited resource. Reciprocally, the level of N affects the regeneration of S, which also depends on U. In this way, S and N have a common destiny: if one does well, the other also does well, but if the degradation of one of them is triggered, then the other is also dragged along. The exploitation of N generates profits through a production function that also requires capital (K). As profit increases, so does U and N consumption; this mechanism also works in the opposite direction. The evolution of U, K, and N demand per unit of consumption follows a hill-climbing heuristic, i.e., is driven by the pursuit of a dynamic target (e.g., desired U) [87] that depends on profitability and, as in the case of U, on the opportunity cost (O), i.e., the average alternative rent outside the current economic activity.

The GDM supports the cyclical behavior of predators and prey. In nature, the increase in prey makes the predators grow at their expense reducing their number. When they run out of feed, the predator population falls and the prey population recovers, returning to the beginning of the cycle. However, the GDM can reproduce other types of dynamics. In the case of socio-ecological systems, the signs of scarcity are bypassed. The profit generated, which depends on prices, costs, and subsidies, allows using inputs that replace the lost resource (e.g., feed can replace the shortage of grass) and create a sense of prosperity even though the environment is degrading. Guided by a misleading abundance, the resource can be overexploited, hastening its degradation and causing irreparable damage to the system by crossing critical S thresholds (e.g., loss of fertile soil). This alternative behavior manifests itself in the form of unsustainable exponential growths that can lead to the collapse of the resource, that is, to the desertification of the system.

3.2. DESPAS Model

The adaptation of the GDM for the understanding of rangeland grazing, which seeks to study desertification processes due to overexploitation of pasturelands, gave rise to the DESPAS model (the Spanish acronym for desertification by overgrazing) [98,99]. The structure of DESPAS is shown in Figure 3. The model contemplates a single-species livestock herd composed of breeding females, (which serves as the capital, K, of the GDM) with mean and constant physiological states and nutritional requirements. The grass (the predatory resource) consumed by the animals is modulated by its availability. This function, called the functional response of livestock, can adopt various formulations [100] and determines, taking into account the animal's energy requirements, the level of supplementary feeding required (see Section 3.3.3). It is assumed that these are commercial farms (GDM consumption units), and that all the animals meet their caloric needs in order to maximize their yield. Feed consumption determines the profit and loss account of the farm, since the expenditure on supplementary feeding is the most important. There is a feedback loop in which good economic results encourage the arrival of new farmers in the area or the intensification of the stocking density, leading to greater inputs of supplementary feed, reducing profits, and therefore discouraging the growth of livestock farming in the area.

The grass is composed of a single perennial species. Under this condition, along with the uniformity of climatic conditions assumed earlier, the primary production of grass can be satisfactorily represented by means of the logistic function. The outflows from this stock are grazing and grass decomposition, which are linearly proportional to the quantity of biomass present.

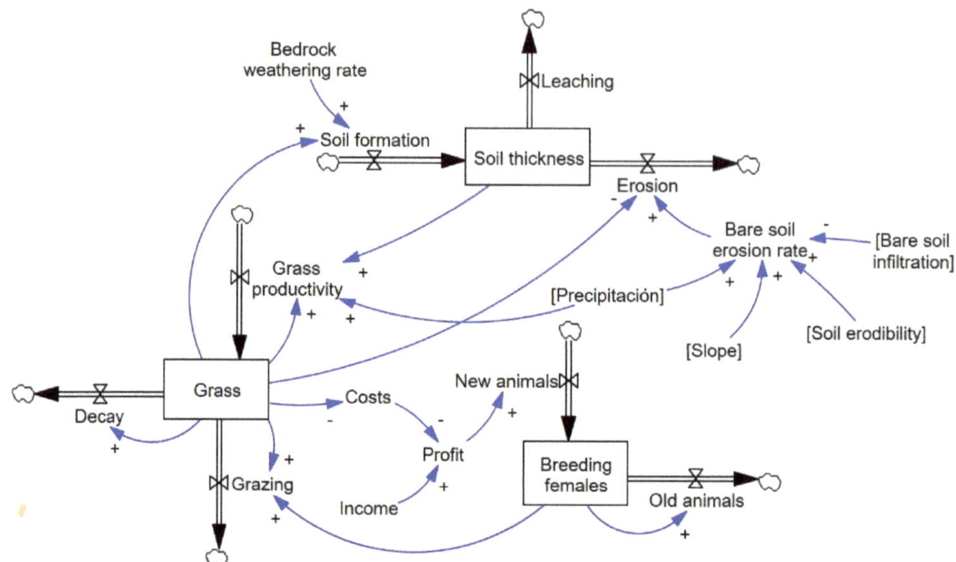

Figure 3. Sketch diagram of DESPAS model. The causal relationships between variables can be positive (i.e., direct, since the changes occur in the same direction: an increase/decrease in the explanatory variable produces an increase/decrease in the explained variable), or negative (i.e., indirect: an increase in the explanatory variable produces a decrease in the explained variable or vice versa). In the case of the flow variables that fill or empty the level variables (box variables), the relationship is, respectively, positive or negative. This network of causal relationships creates feedback loops in the system. Depending on their interaction, one or the other dynamics of the system results.

The reduction in plant cover due to grazing exposes the soil to the erosive effect of rain. Runoff, which is the erosive agent considered, depends on soil infiltration, the slope of the land, and soil erodibility [101–103]. The resulting relationship between plant cover and soil loss is compatible with those given by Elwell and Stocking (1976) [104], a robust empirical relationship in which erosion is maximum with bare soil and declines exponentially as plant cover increases.

Soil thickness depends on two other processes. On the one hand, soil formation from bedrock (weathering rate) and the decomposition of vegetation and, on the other, the leaching rate, i.e., the loss of water-soluble plant nutrients from the soil due to rain. The stock of this limiting factor determines grass productivity, forming a positive feedback mechanism between soil and vegetation. If the soil is kept above certain thicknesses, the system's biomass productivity is reinforced: more soil > more fertility > more plant cover > more protection against erosion > more soil. However, if the soil begins to be lost, the direction of the loop is reversed (less soil > less vegetation cover > less soil), leading to the degradation of the vegetation–soil subsystem.

Changes in livestock stock are based on the economic rationality of the farmer: when the incomes exceed the costs per breeding female (which depend on the amount of supplementary feeding, i.e., the amount of grass consumed by the livestock) then the number of animals is increased and vice versa. Finally, destocking depends on the useful breeding life of females.

3.3. Extensions of the DESPAS Model

The design of a model depends on its purpose. This, however, may change over time as new situations arise. That is why DESPAS has been refined, extended, and sometimes even simplified in order to study different cases. The following sub-models have been implemented: (i) soil moisture, runoff production, and its erosive power; (ii) shrub–grass competition; (iii) supplementary feeding; (iv) farmers' behavior; and (v) the price forming mechanism. In addition, we refer in this section to the temporal and special scales of the models. While the latter has been maintained in all the models, the different methodological developments and processes included have led us to modify the former.

3.3.1. Soil Moisture, Runoff, and Erosion

In DESPAS, soil thickness is used as a limiting factor. However, in drylands, it is more accurate to use water as a limiting factor. For this purpose, the soil moisture level variable was included in the model [105,106]. This makes it possible to implement a water erosion mechanism (Figure 4) inspired by the analogy used by Thornes (1985) [95] to consider that runoff and soil compete for water. In fact, the better the soil absorption conditions and the more spaced the water falls, the less runoff is left to act as an erosive factor.

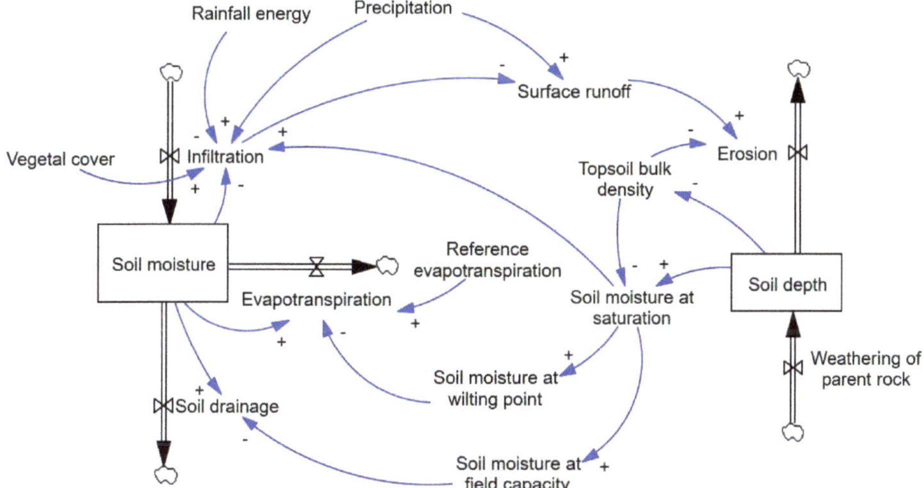

Figure 4. Soil moisture sketch and its influence on the erosion rate. The empirical formulation used in DESPAS, which relates vegetation cover to erosion rate, was replaced by a much more mechanistic approach in which soil water fluxes are detailed in order to determine surface runoff.

Soil moisture results from the balance between infiltration, evapotranspiration, and soil drainage; these three flows are naturally conditioned by the availability of water in the soil. The purpose of this water balance is to determine runoff; that is, water that cannot be trapped by soil pores and circulates freely on the surface. Runoff flow determines the rate of erosion.

The three initial flows are determined by two factors. First, they depend on the free space that the soil has to store the water. If all the pores of the soil are filled with water (soil moisture saturation) then there is no infiltration and all the water that falls becomes runoff and triggers soil erosion. Soil field capacity is the amount of soil moisture or water content held in soil after excess water has drained away; a sandy soil drains more water than a clay soil. Finally, the water used by the plants and reflected by the rate of evapotranspiration is the available water between the field capacity and the wilting point. The second factor is the rainfall torrentiality; i.e., how it is distributed over time. Even if the soil has a large storage capacity, if too much water falls in a short period of time, it cannot absorb it. On the other side, if the rain falls in a more distributed pattern (a lower torrential flow), then a greater fraction of the precipitation can be absorbed.

Finally, vegetation cover continues to play an essential role in erosion control. This is reflected in the model by considering that the infiltration rate is linked to the percentage of vegetation cover. Its protecting capacity follows the exponential behavior described in the previous section. The higher this is, the more precipitation is intercepted and retained, which translates into a greater infiltration rate.

The sub-model adds a further nuance to soil erosion, since it considers that the erosion rate decreases as soil is lost. In other words, the deeper layers of soil exposed by erosion are more compact because they contain fewer pores. Although this implies greater runoff, it also means that the erodibility of the soil is lower and therefore the erosion rate is reduced.

3.3.2. Shrub–Grass Competition

As mentioned above, the degradation of grazing areas in the Mediterranean has two opposing causes. DESPAS considers the most common, i.e., that overgrazing removes vegetation cover and triggers erosion rates. However, the excess of woody vegetation at the expense of pasture resulting from undergrazing is not sustainable either, since it does not allow livestock activity (in this sense, degradation is considered a loss of economic productivity). In both cases, the resulting degradation is difficult to reverse. On the one hand, the global average rate of soil formation is 0.036 mm per year [107], so that recovering 1 cm of soil takes 278 years. On the other hand, once perennial plants are able to establish themselves, they have an inherent advantage over annual plants at the beginning of the growing season. Since the latter have to restart their growth cycle from seed, they lose the competition for nutrients and light to established perennials, which emerge quickly from dormancy at the end of winter or a dry season [108].

We included the interaction between annual and perennial species (Figure 5) to enrich the behavior possibilities of the model and simulate shrub encroachment, which takes place in abandoned European rangelands [109]. For this purpose, herb productivity depends on shrub biomass through a multiplier. It considers that, in the absence of woody species, herb productivity is maximum (depending on rainfall and soil thickness), while as the proportion of woody species increases, annual herb productivity decreases until it is canceled out when woody plants have colonized all the available space. It was assumed that both annual and perennial herbs dry out at the end of the growing season (end of spring) and start growing again the next season (autumn) from seeds, roots, or underground stems. Since only aboveground biomass is considered and the time scale of the model is annual, no stock variable is needed for herbs, as they are annual plants.

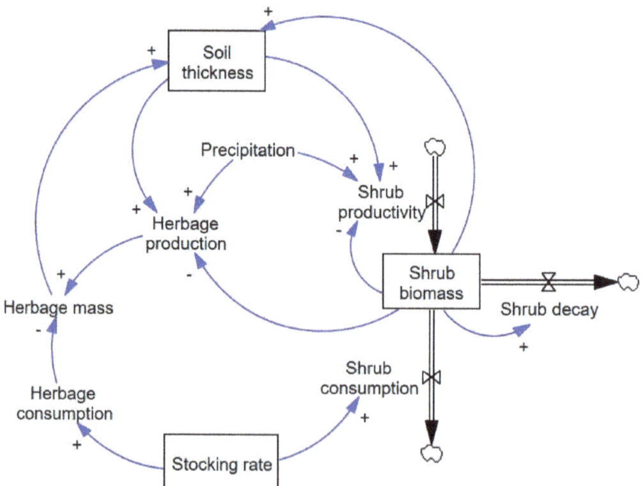

Figure 5. Sketch diagram for the interaction between woody and herbaceous vegetation.

3.3.3. Supplementary Feeding

In commercial rangelands, one of the common strategies for coping with resource scarcity during dry seasons or droughts is the use of supplementary feeding. Sometimes, in addition, another drought-enduring strategy comes into play, such as allowing animals to lose weight during these shortages. Our model, however, assumes that all the energy requirements of the animals are always met and, for this purpose, there is a sub-model dedicated to calculating the amount of supplementary feed required and its cost.

Although the use of animal feed began as a temporary practice, it has been consolidated as a common practice that allows increasing the stocking rate. The basic structure of this sub-model is as follows (Figure 6): the energy gap resulting from the lack of pasture due to (i) the excessive presence of animals; (ii) drought periods; or (iii) reduced soil fertility due to erosion, increases the need for supplementary feeding. This has a negative impact on the benefit of the farmer, which should lead to a reduction in the stocking density. Relieving pressure on pasture leads to its recovery, which brings the situation back to the starting point; i.e., the animals would return to grazing exclusively on pasture and feed costs would disappear. However, fluctuations in feed prices can play an important role and allow for high stocking rates under scenarios of soil and pasture degradation. This sub-model presents one of the ways in which the scarcity signals of the territory are bypassed by the use of external inputs.

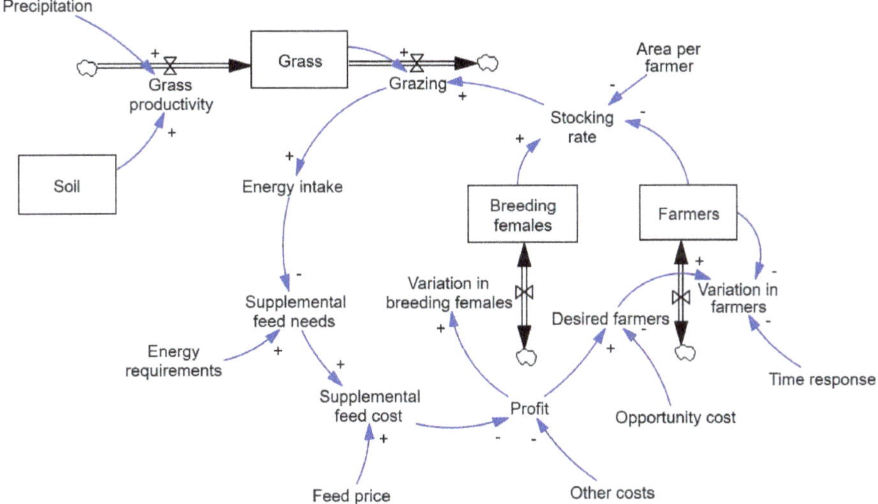

Figure 6. Sketch diagram for supplementary feed dynamics and goal-seeking behavior for the variation in farmers of the modeled area.

3.3.4. Farmers' Behavior

One of the main assumptions of the GDM model is that consumption units (U)—the number of livestock farmers present in the area—depend on the profitability in relation to the opportunity cost; i.e., the alternative profit that would be obtained in another economic sector. To implement this hypothesis, the classical model of "goal-seeking" behavior [87] is used. The discrepancy between the current number of farmers (a stock variable) and the desired ones (the target in the goal-seeking model) is eliminated after a time delay by a positive or negative flow, depending on the sign of the discrepancy (note that this discrepancy is also dynamic, as the target changes) (Figure 6). Desired farmers depend on the profitability–opportunity cost ratio. The former variable is a function, in turn, of income and costs, which are built up from prices, subsidies, sales, and purchase volumes, which include the supplementary food item. The opportunity cost, on the other hand, can have a constant value or be a stochastic variable that follows an exponential probability distribution, i.e., the greater the opportunity cost, the less likely it is. This reflects well the fact that there are more economic actors with low opportunity costs, i.e., with alternative economic activities that offer a lower economic return than their current activity. The adjustment time of the function makes it possible to reflect the behavior of the economic agents involved. It can be more opportunistic (shorter delays) or conservative (longer delays).

3.3.5. Price Forming Mechanism

In the initial versions of the DESPAS model, prices of inputs and outputs of the modeled goods are considered exogenous variables. However, this approach is not very realistic, since the price of raw materials is subject to changes derived from various circumstances, such as the current energy crisis. In addition, the internal dynamics of the system itself is responsible for changes in the prices of the products generated. Depending on the size of the livestock sector in the modeled area, the input and output market may be more or less influenced.

The price formation mechanism is similar for all commodities considered and it is represented in Figure 7. The sub-model assumes that farmers and traders lack complete knowledge of the system, so they use the hill-climbing heuristic to adjust their expectations to reality; i.e., the prices prevailing in the model at each instant are determined by the level variable "Price". A reference price (which may be global or regional) and a price derived from the interplay of supply and demand are involved in the "Indicative price" setting. The product in question (feed, meat, etc.) evolves towards this price with a certain delay ("Adjustment time 1").

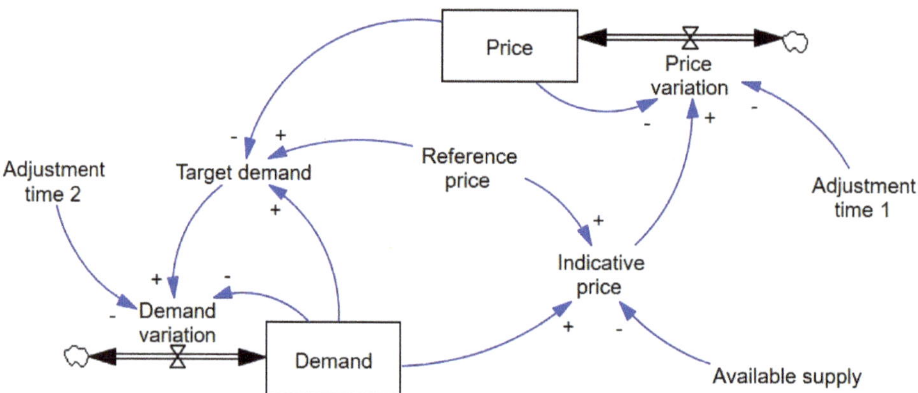

Figure 7. Sketch diagram for the price formation mechanism. The sub-model is based on the "goal-seeking" behavior algorithm. The "target" variable for price dynamics is the Indicative price, which depends on three variables: a reference price, product demand, and available supply.

The price, in turn, determines the "Target demand", towards which the demand converges with another lag, the "Adjustment Time 2". As demand changes, the indicative prices change, and so does the price. This simple structure is capable of generating a great complexity of behavior and, above all, eliminates the simplicity of considering that prices are fixed for the entire simulation period.

3.3.6. Temporal and Spatial Scales

Classical models of ecology do not refer to any specific spatial scale [110]. Our approach is, in this sense, more stringent, since the developed models refer to a spatial unit [89], such as a hectare, but do not distinguish between different parts within that space. The models presented in this section use superimposed two-time scales—short and medium term—to detail processes operating at different resolutions. First, the day is used to model the evolution of soil water, considering variables such as infiltration, saturation, runoff, and evapotranspiration. For purely operational reasons, the time unit is not exactly 1 day. The implementation of the models in the Vensim© software v.5.8 [111] makes it necessary to use time units (when the time unit is the year) that are a multiple of 2.85 days ($\approx$0.0078125 years), as this is the minimum time step allowed by the program.

The year is used to represent processes occurring in the medium term, such as the evolution of the livestock population or the number of economic agents operating in the territory or their profits. Finally, the simulation periods cover several years, tens, or even centuries, since their purpose is to prospect the sustainability of the system, i.e., its long-term stability. For this purpose, it is necessary to study the behavior of variables whose dynamics are much slower (e.g., soil thickness, pasture productivity) and whose effect is felt over several decades.

4. Design and Implementation of Analysis Tools to Explore Rangeland Behavior

In this section, we review the procedures we have designed and applied to analyze the modeled social-ecological systems. The exploitation of a model ranges from running a simple simulation scenario, which is the default use of an SD model, to the implementation of thousands of scenarios to rank the factors involved in a model (Figure 8). As these analyses become more sophisticated, programming routines are needed to automate the process of scenario creation and import, model simulation, and data export [112].

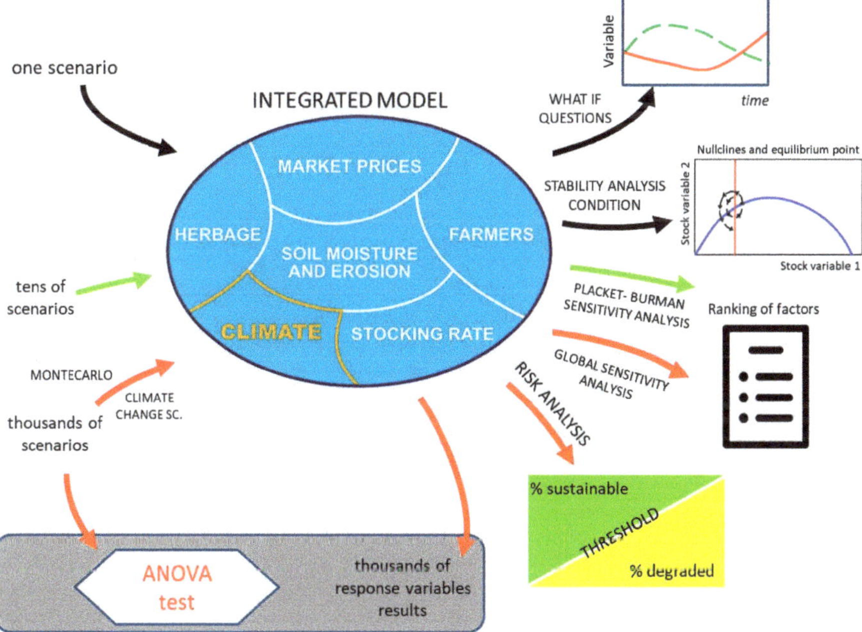

Figure 8. Different options for using SD models, ranging from the simulation of one scenario to the implementation of thousands of scenarios required for a Global Sensitivity Analysis.

4.1. Temporal Trends and "What If" Questions

The standard output of SD models is the time trends of their variables (Figure 9). They respond to the scenario of simulation, i.e., the values of parameters and exogenous variables. Strictly speaking, these trajectories should not be considered predictions, since the background of the equations used is of a socio-ecological nature; i.e., they do not respond to laws of a physical and universal nature. Models that are founded on economic, social, and even biological formulations try to explore the future, but cannot forecast what will happen [113].

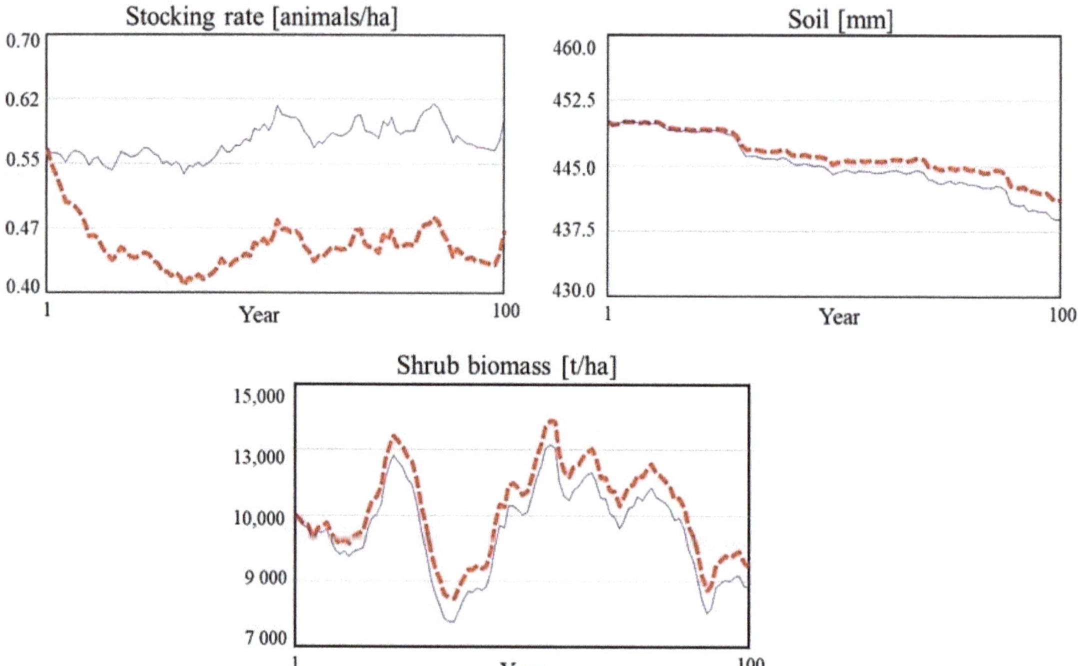

Figure 9. Time trends for three variables in Lagadas under two scenarios (default, blue solid line; and half subsidies, dotted red line).

In this context, it is extremely useful to compare different scenarios, i.e., to answering 'what if' questions to analyze deviations from the baseline scenario. The following example (Figure 9) shows what would happen if subsidies were halved in Lagadas rangelands (Greece) [109]. As can be seen, the stocking rate is falling as the financial support is reduced (although not to the same extent), easing the pressure on the environment and slowing down erosion rates. In the absence of grazing, grasslands are invaded by woody species, which helps to protect the soil but at the same time reduces the productive capacity of grasslands.

Following the line of this exercise, it was interesting to look for the level of subsidy with which the erosion would be canceled. This would require a reduction in subsidies of up to 60%, which would mean a 26% drop in livestock (values not found in the historical record of the area) and a 30% decrease in the gross margin. According to these simulation results, it seems that erosion is inherent in grazing and that limiting soil erosion may in practice mean that farmers will have to close down their businesses.

4.2. Stability Analysis Condition

In order to gain a more precise idea of the long-term sustainability of grazing systems, it is possible to develop procedures that give us a more global vision than the more or less random simulation of scenarios. For this aim, the study of the stability of dynamic systems through the qualitative analysis of their equations [91,114] is the appropriate path to follow. Due to the uncertainty that is usually associated with the parametric values of many systems, and in particular, those referring to the natural environment [115–117], the qualitative analysis of a model can often be of greater interest than its quantitative results [118].

This methodological approach has been applied to dynamic predator–prey systems through the analysis of nullclines, both in linear [119] and non-linear [120] models and, more specifically, to ecological [121] and grazing systems [94,122]. A nullcline is defined as the equilibrium of a level equation (N); i.e., the equation resulting from performing dN/dt = 0. With them, it is possible to anticipate the behavior of a system in the long term by knowing the parametric values of the scenario and the initial values of the stock variables. This gives an idea of where the system is heading under current conditions, which serves as an early warning indicator.

Figure 10 shows the phase plane Pasture–Stocking rate, its nullclines, and the equilibrium point associated with its intersection. The stability of the equilibrium depends on the slope of the Pasture nullcline at the point of cut [122]. In this case, since the slope is positive, we are facing a stable equilibrium. To illustrate the use of these indicators, we used parametric scenarios to recreate three standard extensive livestock farming systems in Spain: cattle and sheep farmed on the dehesas and goats farmed on the south-eastern pastures [98,123].

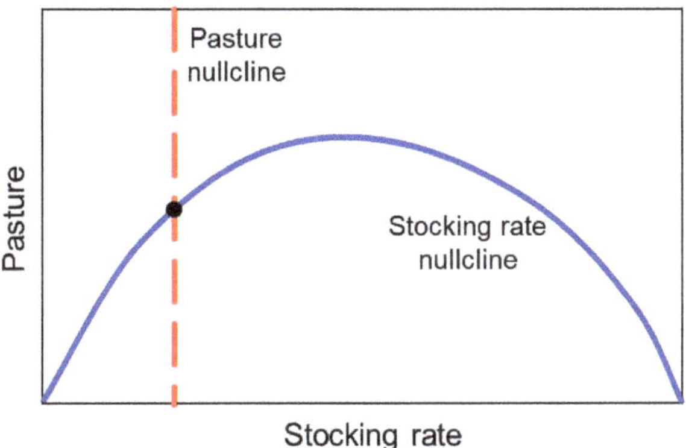

Figure 10. Nullclines and long-term equilibrium point for the subsystem Pasture–Stocking rate. The equilibrium point represents the values of pasture and stocking rate at which the system will stabilize.

4.3. Risk Analysis

The use of nullclines and graphical qualitative analysis is limited by the complexity of the model. Although it is possible to visualize three-dimensional isoclines [98,123], when the SD model has more than three level variables or the formulation of some nonlinear equation is intricate, it is not possible to obtain the nullclines equations. In this case, long-term equilibria are obtained by simulating the model with time horizons long enough to ensure the stabilization of the values. In addition, calculating the nullclines of a system means 'freezing' a scenario and assuming that everything will remain the same in the future. However, conditions fluctuate permanently.

To obtain a more precise idea of where the system is going, further equilibrium points can be calculated by varying the baseline scenario. These scenarios can be randomly generated using the Monte Carlo method by converting some model parameters into stochastic variables. For example, instead of using the mean precipitation, random values can be extracted from a stochastic variable that considers the mean and variance of precipitation. The procedure results in clouds of equilibrium points represented in a scatter plot (Figure 11). The dispersion of the cloud is critical to have a diagnosis. When it is high (Figure 11A), the system's time-trajectory will wander rather erratically; if dispersion is low (Figure 11B), the time-trajectory will be more predictable.

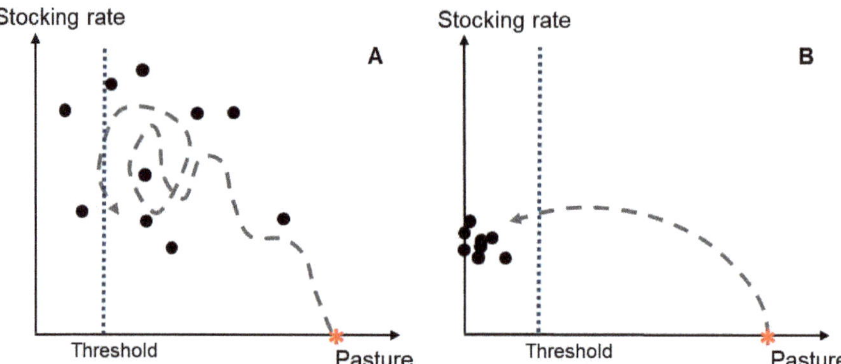

Figure 11. Cloud of long-term equilibriums for the Stocking Rate–Pasture subsystem. In some instances, the clustering of points clearly points towards a region of the scatterplot (**B**), while in others the dispersion of the point cloud will not provide a clear forecast (**A**). The most likely path followed by the system from its original situation (red asterisk) is the one indicated by the dotted line. In the first case, a clearer trajectory (dashed line) can be expected, while in the second case, the dispersion of points predicts an erratic trajectory. Note the threshold (dotted line) separating the degradation region from the sustainable.

To determine the risk of degradation, it is necessary to add degradation thresholds. Our role as modelers has often been to put these tools in the hands of specialists so that they can establish the thresholds they consider appropriate as well as other parametric values of the models. Additionally, to enrich the estimation of risks, the time needed to reach the defined thresholds is evaluated. Bear in mind that a model could show a desertification risk of 100%, but if it occurs after thousands of years (remember that the model simulates the time needed to reach a stable equilibrium), the risk may be negligible. To implement this idea, the model includes equations for computing the time the variables take to exceed their degradation thresholds. In this way, a probability of desertification will be obtained together with a "time to desertification" whose average will provide an estimate in each case.

The application of this methodology allowed the risk of desertification for the five "desertification landscapes" to be estimated [27,112,124] and included in the Spanish Action Plan against Desertification (SAPD; [26]). The results tell us that dehesas are one of the most sustainable land uses. Neither the soil nor the vegetation had an appreciable risk of deterioration over a 100-year time horizon, while for other desertification landscapes, such as groundwater-dependent irrigation systems, the results show that the risk of desertification is 88.2%, and that it will take, on average, 47 years.

4.4. Ranking of Factors

One of the objectives of the models presented is to have a precise idea of the most important factors in the future of the system. Specifically, and within the framework of desertification, it is crucial to distinguish between anthropic and climatic causes [125]. The Plackett–Burman Sensitivity Analysis (PBSA) [126,127] is an excellent option for ranking the factors of a socio-ecological system. This is a sound statistical procedure that measures the effects of each parameter on the target variables in an efficient way in terms of the number of necessary scenarios. An important feature is that the effects of every parameter are not measured with the all-other-things-being-equal assumption but are averaged over variations made in all other parameters. PBSA also enables measuring two-way interactions of pairs of parameters; although, this option was not used in this case.

Fortunately, the analysis capacity of computers is no longer an excuse to simulate a large number of scenarios. This paves the way to implement much more robust and conclusive sensitive analysis, such as Global Sensitivity Analysis (GSA) [128,129]. The most common GSAs are variance-based methods, which decompose the variance of a target variable into terms corresponding to the different parameters and their interactions [130]. Through GSA, we evaluated the sensitivities of key endogenous factors to the same percentage variation in 70 factors, including economic and climate drivers. The analysis considered the behaviors of 288,000 variants of the modeled system, each under a different 300-year driver scenario [131].

Among the main conclusions reached through the establishment of rankings, we have been able to verify the supremacy of climatic factors over the rest. For example (see Table 1 for details), we found [105] that, when the "Mean annual precipitation" is increased by 10%, the time for the soil to be depleted was brought forward by 36.9%, while the effects for economic and behavioral variables were located in the lowest positions in the ranking. On the contrary, a 10% increase in "Mean meat price" delays time for the soil to be depleted only by −1.2%. The explanation for this result is strongly influenced by supplementary feeding, a common practice in commercial rangelands. Although this is one of the major costs of livestock farms, the farmer has enough financial margin to invest in feed and thus maintain production and, therefore, profit. Obviously, this situation may change if or when the prices of raw materials used to manufacture compound feed change.

Table 1. Ranking example for a PBSA conducted in a dehesa [105]. In this case, the objective variable was "Time to loss 20 cm of soil". The higher the negative percentage (red cells) means that the time is shortened, i.e., that the process of soil loss is faster. Positive percentages (green cells) mean a delay in soil loss. As can be seen, the ranking is led by climatic parameters.

Parameter	Impact
Mean annual precipitation	−36.9%
Fraction of annual precipitation that fell in the humid season	−17.5%
Mean annual reference evapotranspiration	12.5%
Fraction of annual evapotranspiration in the wet season	12.2%
Initial mean runoff coefficient soil at wilting point	−9.5%
Coefficient of variation annual precipitation	−8.1%
Coefficient of variation runoff coefficient soil at wilting point	−4.5%
Months when precipitation > ET_o (length of the humid season)	2.4%
Total subsidies per hectare	−1.4%
Costs per female other than the cost of supplemental feed	1.3%
Mean meat price	−1.2%
Weathering rate of the parent rock	1.1%
Average number of years to form gross margin expectations	−0.5%
Mean price of supplemental feed	0.4%
Coefficient of variation supplemental feed	−0.2%
Coefficient of variation meat price	−0.2%
% Increase in breading females if gross margin increased by 10%	0.1%
Secondary income per breeding female	0.0%

4.5. Implementation of ANOVA Test

SD models can be used as virtual laboratories in which to conduct experiments [132]. In this context, a multi-way ANOVA test was coupled to an SD model to evaluate the sensitivity of a valuable type of commercial rangelands to increases in the frequency and intensity of droughts considering climate change scenarios [133]. In particular, the question is whether the current strategy of using feed to mitigate the effects of droughts will continue to be effective in the context of water scarcity that is expected to be particularly relevant in the Mediterranean [16].

For this purpose, 5400 simulation scenarios were generated from two blocking factors and two treatment factors. We have considered three Representative Concentration Pathway (RCP), i.e., scenario of future greenhouse gas emissions, and two downscaling methods, i.e., process by which coarse-resolution Global Climate Models outputs are translated into local climate information. Additionally, three levels were defined for the frequency and intensity of droughts. A hundred simulations (replicates) were run for each of the 3·2·3·3 = 54 cells in the analysis. These were obtained by varying the value of the random seed from 1 to 100.

The scenarios feed the model to generate results and after those inputs and outputs are used to implement the multi-way ANOVA test (see Figure 8, bottom). This has shown that most of the main effects and interactions turned out to be highly significant; although, the sensitivity of response variables to increases in the frequency and severity of droughts under climate change would be low or very low.

5. Findings through SD Modeling

5.1. Learnings from Mediterranean Rangelands Modeling

Agro-silvo-pastoral systems are one of the five desertification landscapes identified in the SAPD [26]. Our main conclusion in this context is that this land use presents a low risk of desertification [27,124]. This is due to the use of feed, which allows for mitigating the scarcity of pasture in dry periods [133]. Even in the context of climate change, with clear decreases in precipitation, it is estimated that the system will cope well with the shortage of pasture with the use of feed. Consistent with this conclusion, sensitivity analyses have revealed that climatic factors are more decisive than socio-economic factors [106]. This reinforces the validity of the use of feed as a drought-enduring strategy that safeguards the system.

However, we cannot think that the use of animal feed is a panacea. In northern Algeria, we have an example of how the progressive replacement of grass with cereal grain has led to the system's collapse [134–136]. In these steppe rangelands, feed initially entered, as is often the case, as a punctual solution for extreme drought situations. Gradually, it became a regular supplement until national policies decided to turn the so-called Alfa seas (the name indicates the density of bushes in the region; Alfa is the name for the esparto grass, *Macrochloa tenacissima*) into an open-air farm. To this end, barley gradually replaced the country's wheat fields. Through a state policy of subsidies, this barley is used to feed the increasingly numerous herds of the steppes. The main mistake was to ignore the sheep's fiber needs, as barley only met their energy requirements. The consequences were devastating: thousands of hungry animals devoured the esparto grass, which was the shrub that helped alleviate periods of grass shortage. In the south of Oran, 700,000 ha out of 1.2 Mha of Alfa grass has completely disappeared and the remaining half million is much sparser (biomass was reduced from 1750 to 100 kg DM ha^{-1}) [137]. The loss of plant cover combined with the strong winds in the area has led to the appearance of dunes. This is a good example to illustrate that desertification is not the advance of a desert, but the creation of a desert-like landscape due to poor land management [138].

The livestock industry, which revolves around the use of compound feed, is a good example of global telecoupling [139], i.e., global supply chains involving large geographical distances and creating environmental pressures (including deforestation and other types of land conversions) remote from the places where the consumption of goods and services take place. Although industrial farming is the main consumer of feed, a more comprehensive assessment of the environmental impact of extensive farming may include the area of soybeans and cereal fields needed to supplement the animals' diet. In the European case, the deforestation of primary forests in South America due to soy imports for feed compounds is especially relevant [140]. For the period 2000–2010, we have estimated that soybean consumption associated with the Spanish feed industry is equivalent to the deforestation of 1220 kha of primary ecosystems in South America, the main exporter of soybeans [141]. The models we have presented can be completed by incorporating the impact of feed consumption on the livestock farms studied in terms of area deforested.

Although socioeconomic drivers have less influence than climate drivers on the sustainability of the rangelands, there are many situations where their role is key. In Lagadas (Greece), we observed how the reduction in subsidies triggered the deterioration of the system [109]. Something similar to the Algerian case described above occurred. The rangeland scientists who helped us with that model expected that the cut in subsidies would lead to a reduction in livestock and thus slow down erosion. However, the model showed the opposite behavior. The conclusion seemed obvious to our colleagues: the model had to be wrong. Analyzing in depth the reasons for such unexpected behavior, we noticed that what was happening was that the livestock, although it was decreasing, did not do so in the proportion in which the subsidies did. Analyzing the causal tree, we could see that the reduction in subsidies meant a reduction in supplementary feeding but not to the same extent of the stocking rate. Consequently, the actual stocking rate was higher than in the baseline scenario, and therefore the animals were forced to consume more grass than was adequate, since the feed given was not sufficient to cover their needs.

Another relevant dynamic of the degradation of the Mediterranean grazing systems has to do with the economic behavior of farmers [142]. We have seen that a few opportunistic farmers, who only seek to maximize their profit by playing with the size of the herds, are enough to trigger degradation rates in the environment. The more cautious behavior of traditional farmers is only effective, in terms of rangeland sustainability, when it is highly dominant.

5.2. Multidisciplinarity: Under the Crossfire of Specialists

The scientific literature is full of recommendations about the need for multidisciplinary studies as the only way to address a multi-faceted and increasingly interconnected reality. Specifically, economics, combined with earth system sciences, is crucial for understanding both positive and negative impacts of alternatives and the trade-offs involved in a sustainable development path [143]. This is especially relevant to the serious environmental problems facing the planet, such as global warming, desertification, or loss of biodiversity. A more harmonious relationship between food systems and the ecological framework on which they are based is called for in order to achieve the Sustainable Development Goals [144]. As a result of this demand, numerous journals specializing in multidisciplinary approaches have emerged, and initiatives such as the EAT-Lancet [145], which bet on the systemic approach, have been launched. New paradigms have also emerged such as the socio-ecological systems [146], ecological economics [80], and the water–food–energy nexus [147], which try to give an integrated vision of nature and human beings.

Our experience during all these years has shown us that the integration of knowledge from different disciplines is difficult, to say the least. Inevitably, multidisciplinary work is evaluated by specialists in each of the subjects that are included in the integrated models. The problem is that, for a specialist, nothing is superfluous in his field and she/he declares she/himself incapable of judging and appreciating the added value of the contributions of other disciplines, which she/he does not know. Thus, for example, we find that an edaphologist misses, in the erosion sub-model, much more detailed equations, pointing out the impossibility of using point models, instead of spatially explicit ones, or considers unacceptable the simplification that involves ignoring the lithological characteristics of the terrain. However, it will be difficult to appreciate that this same model contains equations on the evolution of prices according to changes between supply and demand. Likewise, an economist will miss a more in-depth treatment of the profit and loss account, and a botanist may criticize the fact that the dynamics of each of the species that make up the pasture have not been treated separately. For both the economist and the botanist, it is likely to be superfluous to model runoff in order to calculate erosion rates.

Another practice that we have observed and that seriously penalizes the construction of integrated models is the growing refusal to review this type of work. Again, at least part of the explanation lies in the fact that the review work is carried out by specialists in the different disciplines that the model brings together, but who are not usually familiar with equations, much less with systems of differential equations. This task requires a great deal of time for understanding, as well as a minimum of mathematical knowledge. We are faced with judgments that again do not go beyond the boundaries of the reviewer's discipline. At best, the reviewer assumes that a model with so many equations and references must be right (with all the vagueness that this judgment implies); at worst, the paper runs the risk of being rejected outright if the reviewer in question reads some detail that clashes with his or her perception of the subject.

In our case, we have had work rejected on the basis of arguments that demonstrate a lack of knowledge of the model. It has been said that the model is speculative (indeed it is, as is the case with any model based on a series of hypotheses or speculations), that the time horizons are excessive (in some cases, it is necessary to simulate the model for several hundred years in order to calculate equilibrium points of the system), that it is too simple (in models with more than eighty equations), or that the model is wrong because it does not reflect reality. In this last aspect, we agree since, in the end, "All models are wrong", since they are deliberate simplifications of reality [148]. From our point of view, this type of judgment fits in perfectly with one of the obstacles Sterman points to in properly understanding complex dynamic systems [87]: unscientific reasoning, even among the scientific community itself.

There are notable exceptions to this discourse. For example, there are those who appreciate the connection of aspects as far apart, in principle, as subsidies and erosion rates. This is the culmination of the top-down approach of the systemic approach models. Indeed, one of the major achievements is to complete the model, in the sense of connecting all the elements of the system. It is obvious that these connections can be made more precise and improved over years. The important thing is to make the assumptions very clear and to return to these questions whenever possible and pertinent. On the contrary, the bottom-up approach, which is the immediate consequence of the reductionist approach of the scientific method, tries to aggregate particles of knowledge in such a way that a complex system is generated from the coupling of subsystems. As a consequence, its predictive capacity is quite high. However, this approach has a number of disadvantages [149] that are critical for our objectives: (i) they are models that need a large amount of data for their operation, which, in arid areas, is often not possible; (ii) there is a great risk of error propagation; and (iii) the strategy of trying to capture and replicate all kinds of processes makes bottom-up models hardly reach 'the top'.

Over the years, we have found a number of specialists in fields such as hydrology, ecology, geography, or biology who have joined our working group and become enthusiastic advocates of the systemic approach. We must also acknowledge and thank the valuable contributions of reviewers from outside SD, which have allowed us to improve both models and manuscripts. One of the tools that are useful for involving participants from different disciplines and institutions is Decision Support Systems (DSS). These are very simple computer applications in which the user only has to press a series of buttons to execute tasks such as moving from one screen to another or performing more or less complex calculations. In our case, the DSS allow us to use Vensim© software v.5.8 [111] through Visual Basic. This opens the door to use a widely spread program such as Excel and to simulate SD models remotely, so that both scenarios and results are accessible from a spreadsheet.

DSS can play a key role in expanding scientific production to society, since it allows exploring, in a simple way, sophisticated simulation models and their results, involving the decision-making processes [150] and reducing the resistance that often produces environmental problems such as desertification [151]. Many of the methods developed during these fifteen years have been channeled into a DSS called SAT (the Spanish acronym for Early Warning System) [27,112,124]. SAT implements three SD models to cope with the five desertification landscapes described in the SPAD, two of which are related to the "rangelands affected by erosion" syndrome (Figure 12B). Despite their usefulness, we agree with Oxley (2004) [152] on the limited role of DSS and simulation models and that "decision-support for socio-natural systems is more fruitfully concerned with providing the political actors involved a means of exploration than a set of 'definite' solutions" [152].

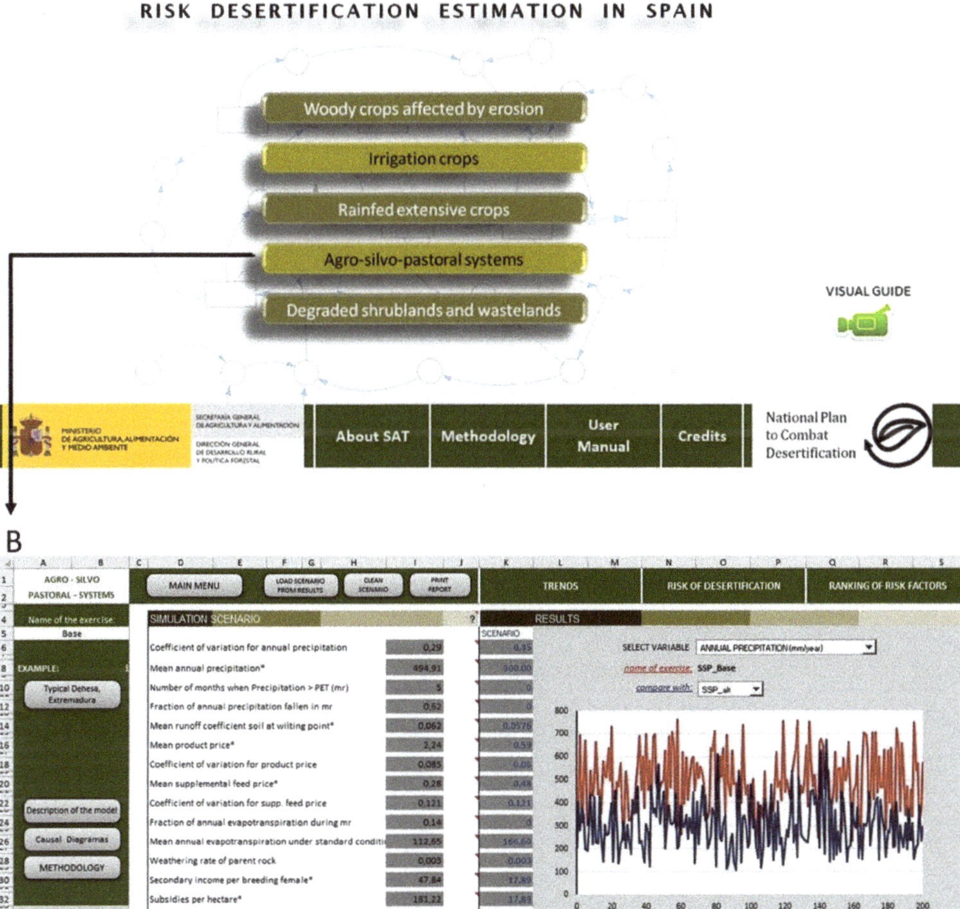

Figure 12. SAT screens: (**A**) Main menu; (**B**) Implementation of SAT for dehesa rangelands, one of the Spanish NAPD landscapes.

6. Conclusions

Simulation models are a vital tool for understanding the multiple dynamics that converge on rangelands. This is the main land use in drylands and is key to the survival of the poorest countries. Over the course of two decades, we have developed integrated SD models to study Mediterranean rangelands and designed analytical tools coupled to them. Our goal was to understand the interactions between the different components of the system, to provide sustainability indicators, and to detect the main drivers of degradation of these socio-ecological systems.

Since the beginning of our research activity, we have addressed the study of rangelands from a holistic approach. Although multidisciplinarity in the study of socio-ecological systems is repeatedly advised, in many cases, the specialist's point of view and reluctance to integrate knowledge from other disciplines still prevails. This is one of the burdens that the modeler must learn to bear, distinguishing constructive criticisms from those that only arise from those who refuse to leave their discipline of comfort and do not admit other points of view.

As we work on models, we encounter new challenges that call for new developments, which has led us to versions that incorporate new elements. Currently, two situations are of particular concern to us. On the one hand, the transformation of rangelands and silvo-pastoral dryland systems to croplands increases the risk of desertification due to increased pressure on the remaining rangelands or to the use of unsustainable cultivation practices. To address this problem, it is necessary to include other land-use dynamics, and one option is to link them with other models that we have implemented for other land uses, such as groundwater-dependent irrigation systems. On the other hand, we have to take into account the effects that move beyond the physical boundaries of rangelands due to feed consumption. These are some of the possible paths that models will follow. At the same time, new models usually require different analytical tools, which we have also been developing over the years.

In an increasingly complex world, it is mandatory to use tools that can deal with it. Simply pointing out the contradictions that arise in land use management and bringing them to the attention of stakeholders and politicians is, in our opinion, a valuable contribution.

Author Contributions: Conceptualization, J.M.-V. and J.I.P.; methodology, J.M.-V. and J.I.P.; software, J.M.-V. and J.I.P.; validation, J.I.P. and J.M.-V.; formal analysis, J.M.-V. and J.I.P.; investigation, J.M.-V. and J.I.P.; writing—original draft preparation, J.M.-V.; writing—review and editing, J.M.-V. and J.I.P.; visualization, J.M.-V.; supervision, J.M.-V.; funding acquisition, J.M.-V. and J.I.P. All authors have read and agreed to the published version of the manuscript.

Funding: This research was funded by DeSurvey (CE-Integrated Project Contract No. 003950), PADEG (CGL2008/01215/BTE), and BIODESERT (European Research Council grant agreement No. 647038).

Data Availability Statement: The data used have been appropriately cited.

Acknowledgments: The authors are indebted to numerous research projects and and some good colleagues such as Sussane Schnabel, J.F. Lavado Contador, Vasilios Papanastadis or Juan Puigdefábregas.

References

1. Menke, J.; Bradford, G.E. Rangelands. *Agric. Ecosyst. Environ.* **1992**, *42*, 217–230. [CrossRef]
2. Cherlet, M.; Hutchinson, C.; Reynolds, J.; Hill, J.; Sommer, S.; Von Maltitz, G. (Eds.) *World Atlas of Desertification*; Publication Office of the European Union: Luxembourg, 2018; ISBN 978-92-79-75350-3.
3. Lund, G. Accounting for the World's Rangelands. *BioOne* **2007**, *29*, 3–10. [CrossRef]
4. Bedunah, D.J.; Angerer, J.P. Rangeland Degradation, Poverty, and Conflict: How Can Rangeland Scientists Contribute to Effective Responses and Solutions? *Rangel. Ecol. Manag.* **2012**, *65*, 606–612. [CrossRef]
5. Asner, G.P.; Elmore, A.J.; Olander, L.P.; Martin, R.E.; Harris, A.T. Grazing Systems, Ecosystem Responses, and Global Change. *Annu. Rev. Environ. Resour.* **2004**, *29*, 261–299. [CrossRef]
6. Maestre, F.T.; Le Bagousse-Pinguet, Y.; Delgado-Baquerizo, M.; Eldridge, D.J.; Saiz, H.; Berdugo, M.; Gozalo, B.; Ochoa, V.; Guirado, E. Grazing and Ecosystem Service Delivery in Global Drylands. *Science* **2022**, *378*, 915–920. [CrossRef]
7. Maestre, F.T.; Eldridge, D.J.; Soliveres, S.; Sonia, K.; Delgado-baquerizo, M.; Bowker, M.A.; Garc, P.; Gait, J.; Gallardo, A.; Roberto, L. Structure and Functioning of Dryland Ecosystems in a Changing Structure and Functioning of Dryland Ecosystems in a Changing World. *Annu. Rev. Ecol. Evol. Syst.* **2016**, *47*, 215–237. [CrossRef]
8. Reynolds, J.F.; Smith, D.M.S.; Lambin, E.F.; Turner II, B.L.; Mortimore, M.; Batterbury, S.P.J.; Downing, T.E.; Dowlatabadi, H.; Fernandez, R.J.; Herrick, J.E.; et al. Global Desertification: Building a Science for Dryland Development. *Science* **2007**, *316*, 847–851. [CrossRef]
9. Allington, G.R.H.; Li, W.; Brown, D.G. Modeling System Dynamics in Rangelands of the Mongolian Plateau. In Proceedings of the Trans-disciplinary Research Conference: Building Resilience of Mongolian Rangelands, Ulaanbaatar, Mongolia, 9–10 June 2015; pp. 216–221.
10. Fu, Q.; Feng, S. Responses of Terrestrial Aridity to Global Warming. *J. Geophys. Res. Atmos.* **2014**, *119*, 7863–7875. [CrossRef]
11. Huang, J.; Yu, H.; Guan, X.; Wang, G.; Guo, R. Accelerated Dryland Expansion under Climate Change. *Nat. Clim. Chang.* **2015**, *6*, 166–172. [CrossRef]

12. Lin, L.; Gettelman, A.; Fu, Q.; Xu, Y. Simulated Differences in 21st Century Aridity Due to Different Scenarios of Greenhouse Gases and Aerosols. *Clim. Chang.* **2018**, *146*, 407–422. [CrossRef]
13. Park, C.-E.; Jeong, S.-J.; Joshi, M.; Osborn, T.J.; Ho, C.-H.; Piao, S.; Chen, D.; Liu, J.; Yang, H.; Park, H.; et al. Keeping Global Warming within 1.5 °C Constrains Emergence of Aridification. *Nat. Clim. Chang.* **2018**, *8*, 70–74. [CrossRef]
14. Le Houérou, H.N. Impact of Man and His Animals on Mediterranean Vegetation. In *Mediterranean-Type Shrublands, Ecosystems of the World 11*; Castri, F., Goodall, D., Specht, R., Eds.; Elsevier Science Publishers Co.: New York, NY, USA, 1981; pp. 479–521.
15. Samaniego, L.; Thober, S.; Kumar, R.; Wanders, N.; Rakovec, O.; Pan, M.; Zink, M.; Sheffield, J.; Wood, E.F.; Marx, A. Anthropogenic Warming Exacerbates European Soil Moisture Droughts. *Nat. Clim. Chang.* **2018**, *8*, 421–426. [CrossRef]
16. Cramer, W.; Guiot, J.; Fader, M.; Garrabou, J.; Gattuso, J.P.; Iglesias, A.; Lange, M.A.; Lionello, P.; Llasat, M.C.; Paz, S.; et al. Climate Change and Interconnected Risks to Sustainable Development in the Mediterranean. *Nat. Clim. Chang.* **2018**, *8*, 972–980. [CrossRef]
17. Papanastasis, V.P.; Mansat, P. Grasslands and related forage resources in Mediterranean areas. In *Grasslands and Land Use Systems*; Parente, G., Frame, J., Orsi, S., Eds.; Grassland Science in Europe, ERSA: Gorizia, Italy, 1996; Volume 1, pp. 47–57.
18. Steffen, W.; Broadgate, W.; Deutsch, L.; Gaffney, O.; Ludwig, C. The Trajectory of the Anthropocene: The Great Acceleration. *Anthr. Rev.* **2015**, *2*, 81–98. [CrossRef]
19. Puigdefábregas, J.; Mendizabal, T. Perspectives on Desertification: Western Mediterranean. *J. Arid Environ.* **1998**, *39*, 209–224. [CrossRef]
20. Liao, C.; Agrawal, A.; Clark, P.E.; Levin, S.A.; Rubenstein, D.I. Landscape Sustainability Science in the Drylands: Mobility, Rangelands and Livelihoods. *Landsc. Ecol.* **2020**, *35*, 2433–2447. [CrossRef]
21. Nefzaoui, A.; Ketata, H.; El Mourid, M. Changes in North Africa Production Systems to Meet Climate Uncertainty and New Socio-Economic Scenarios with a Focus on Dryland Areas. In *New Approaches for Grassland Research in a Context of Climate and Socio-Economic Changes*; Acar, Z., López-Francos, A., Porqueddu, C., Eds.; Options Méditerranéennes: Série A. Séminaires Méditerranéens; CIHEAM: Zaragoza, Spain, 2012; Volume 102, pp. 403–421.
22. Soto, D.; Infante-Amate, J.; Guzmán, G.I.; Cid, A.; Aguilera, E.; García, R.; González de Molina, M. The Social Metabolism of Biomass in Spain, 1900–2008: From Food to Feed-Oriented Changes in the Agro-Ecosystems. *Ecol. Econ.* **2016**, *128*, 130–138. [CrossRef]
23. UN (United Nations) United Nations Convention to Combat Desertification in Countries Experiencing Serious Drought and/or Desertification, Particularly in Africa. Document A/AC. 241/27, 12. 09. 1994 with Annexe; United Nations, New York. 1994. Available online: https://catalogue.unccd.int/936_UNCCD_Convention_ENG.pdf (accessed on 1 January 2023).
24. Mensching, H.G. Desertification in Europe? A Critical Comment with Examples From Mediterranean Europe. In *Desertification in Europe*; Fantechi, R., Margaris, N.S., Eds.; Springer: Dordrecht, The Netherlands, 1986; pp. 3–8.
25. European Court of Auditors. *Combating Desertification in the EU: A Growing Threat in Need of More Action*; European Court: Brussels, Belgium, 2018.
26. MAGRAMA. *Programa de Acción Nacional Contra La Desertificación*. Madrid; Ministerio de Agricultura y Medio Ambiente: Madrid, Spain, 2008.
27. Martínez-Valderrama, J.; Ibáñez, J.; Del Barrio, G.; Sanjuán, M.E.; Alcalá, F.J.; Martínez-Vicente, S.; Ruiz, A.; Puigdefábregas, J. Present and Future of Desertification in Spain: Implementation of a Surveillance System to Prevent Land Degradation. *Sci. Total Environ.* **2016**, *563–564*, 169–178. [CrossRef]
28. MITERD. *Estrategia Nacional de Lucha Contra La Desertificación En España*; Ministerio Para La Transición Ecológica Y El Reto Demográfico MITERD: Madrid, Spain, 2022.
29. Martínez-Valderrama, J.; del Barrio, G.; Sanjuán, M.E.; Guirado, E.; Maestre, F.T. Desertification in Spain: A Sound Diagnosis without Solutions and New Scenarios. *Land* **2022**, *11*, 272. [CrossRef]
30. Manzano, P.; Burgas, D.; Cadahía, L.; Eronen, J.T.; Fernández-Llamazares, Á.; Bencherif, S.; Holand, Ø.; Seitsonen, O.; Byambaa, B.; Fortelius, M.; et al. Toward a Holistic Understanding of Pastoralism. *One Earth* **2021**, *4*, 651–665. [CrossRef]
31. Ayoub, A.T. Extent, Severity and Causative Factors of Land Degradation in the Sudan. *J. Arid Environ.* **1998**, *38*, 397–409. [CrossRef]
32. Schnabel, S. *Soil Erosion and Runoff Production in a Small Watershed under Silvo-Pastoral Landuse (Dehesas) in Extremadura, Spain*; Geoforma Ediciones: Logroño, Spain, 1997.
33. Teketay, D. Deforestation, Wood Famine, and Environmental Degradation in Ethiopia's Highland Ecosystems: Urgent Need for Action. *Northeast. Afr. Stud.* **2011**, *8*, 53–76. [CrossRef]
34. Wilcox, B.P.; Thurow, T.L. Emerging Issues in Rangeland Ecohydrology: Vegetation Change and the Water Cycle. *Rangel. Ecol. Manag.* **2006**, *59*, 220–224. [CrossRef]
35. Perevolotsky, A.; Seligman, N.G. Role of Grazing in Mediterranean Rangeland Ecosystems—Inversion of a Paradigm. *Bioscience* **1998**, *48*, 1007–1017. [CrossRef]
36. Moreira, F.; Rego, F.C.; Ferreira, P.G. Temporal (1958–1995) Pattern of Change in a Cultural Landscape of Northwestern Portugal: Implications for Fire Occurrence. *Landsc. Ecol.* **2001**, *16*, 557–567. [CrossRef]
37. Puigdefábregas, J. Erosión y Desertificación en España. *Campo* **1995**, *132*, 63–83.
38. Ruffault, J.; Moron, V.; Trigo, R.M.; Curt, T. Objective Identification of Multiple Large Fire Climatologies: An Application to a Mediterranean Ecosystem. *Environ. Res. Lett.* **2016**, *11*, 7. [CrossRef]

39. Turco, M.; Llasat, M.C.; Von Hardenberg, J.; Provenzale, A. Climate Change Impacts on Wildfires in a Mediterranean Environment. *Clim. Chang.* **2014**, *125*, 369–380. [CrossRef]
40. Tuel, A.; Eltahir, E.A.B. Why Is the Mediterranean a Climate Change Hot Spot? *J. Clim.* **2020**, *33*, 5829–5843. [CrossRef]
41. Syphard, A.; Radeloff, V.; Hawbaker, T.; Stewart, S. Conservation Threats Due to Human-Caused Increases in Fire Frequency in Mediterranean-Climate Ecosystems. *Conserv. Biol.* **2009**, *23*, 758–769. [CrossRef]
42. Zedler, P.H.; Gautier, C.R.; McMaster, G.S. Vegetation Change in Response to Extreme Events: The Effect of a Short Interval between Fires in California Chaparral and Coastal Scrub. *Ecology* **1983**, *64*, 809–818. [CrossRef]
43. Canadell, J.; López-Soria, L. Lignotuber Reserves Support Regrowth Following Clipping of Two Mediterranean Shrubs. *Funct. Ecol.* **1998**, *12*, 31–38. [CrossRef]
44. Arianoutsou, M.; Vilà, M. Fire and Invasive Plant Species in the Mediterranean Basin. *Isr. J. Ecol. Evol.* **2012**, *58*, 195–203.
45. Godde, C.M.; Garnett, T.; Thornton, P.K.; Ash, A.J.; Herrero, M. Grazing Systems Expansion and Intensification: Drivers, Dynamics, and Trade-Offs. *Glob. Food Sec.* **2018**, *16*, 93–105. [CrossRef]
46. Forrester, J.W. *Industrial Dynamics*; The MIT Press: Cambridge, MA, USA, 1961.
47. Moxnes, E. Overexploitation of Renewable Resources: The Role of Misperceptions. *J. Econ. Behav. Organ.* **1998**, *37*, 107–127. [CrossRef]
48. Badarch, D.; Ochirbat, B. On Sustainable Development of Cashmere Production and Goat Population in Mongolia. In *Mongolian Studies at CNEAS. CNEAS Monograph Series 6*; Oka, H., Ed.; Center for Northeast Asian Studies, Tohoku University: Sendai, Japan, 2002; pp. 187–200.
49. Noy-Meir, I. Grazing and Production in Seasonal Pastures: Analysis of a Simple Model. *J. Appl. Ecol.* **1978**, *15*, 809–835. [CrossRef]
50. Glasscock, S.N.; Grant, W.E.; Drawe, D.L. Simulation of Vegetation Dynamics and Management Strategies on South Texas, Semi-Arid Rangeland. *J. Environ. Manag.* **2005**, *75*, 379–397. [CrossRef]
51. Belcher, K.W.; Boehm, M. Evaluating Agroecosystem Sustainability Using an Integrated Model. In *Managing for Healthy Ecosystems*; Rapport, D.J., Lasley, B.L., Rolston, D.E., Ole Nielsen, N., Calvin, O., Qualset, A.B.D., Eds.; CRC Press (Taylor & Francis): Boca Raton, FL, USA, 2002; pp. 1209–1226, ISBN 978-0-42914-323-6.
52. Tedeschi, L.O.; Nicholson, C.F.; Rich, E. Using System Dynamics Modelling Approach to Develop Management Tools for Animal Production with Emphasis on Small Ruminants. *Small Rumin. Res.* **2011**, *98*, 102–110. [CrossRef]
53. Dace, E.; Muizniece, I.; Blumberga, A.; Kaczala, F. Searching for Solutions to Mitigate Greenhouse Gas Emissions by Agricultural Policy Decisions—Application of System Dynamics Modeling for the Case of Latvia. *Sci. Total Environ.* **2015**, *527–528*, 80–90. [CrossRef]
54. Moxnes, E.; Danell, Ö.; Gaare, E.; Kumpula, J. Optimal Strategies for the Use of Reindeer Rangelands. *Ecol. Modell.* **2001**, *145*, 225–241. [CrossRef]
55. Aderinto, R.F.; Ortega-s, J.A.; Anoruo, A.O.; Machen, R.; Turner, B.L. Can the Tragedy of the Commons Be Avoided in Common-Pool Forage Resource Systems? An Application to Small-Holder Herding in the Semi-Arid Grazing Lands of Nigeria. *Sustainability* **2020**, *12*, 5947. [CrossRef]
56. Hart, R.H.; Samuel, M.J.; Test, P.S.; Smith, M.A. Cattle, Vegetation, and Economic Responses to Grazing Systems and Grazing Pressure. *J. Range Manag.* **1988**, *41*, 282–286. [CrossRef]
57. Manley, W.A.; Hart, R.H.; Samuel, M.J.; Smith, M.A.; Waggoner, J.W.; Manley, J.T. Vegetation, Cattle, and Economic Responses to Grazing Strategies and Pressures. *J. Range Manag.* **1997**, *50*, 638–646. [CrossRef]
58. Ritten, J.P.; Bastian, C.T.; Frasier, W.M. Economically Optimal Stocking Rates: A Bioeconomic Grazing Model. *Rangel. Ecol. Manag.* **2010**, *63*, 407–414. [CrossRef]
59. Dunn, B.H.; Smart, A.J.; Gates, R.N.; Johnson, P.S.; Beutler, M.K.; Diersen, M.A.; Janssen, L.L. Long-Term Production and Profitability From Grazing Cattle in the Northern Mixed Grass Prairie. *Rangel. Ecol. Manag.* **2010**, *63*, 233–242. [CrossRef]
60. Turner, B.L.; Rhoades, R.D.; Tedeschi, L.O.; Hanagriff, R.D.; McCuistion, K.C.; Dunn, B.H. Analyzing Ranch Profitability from Varying Cow Sales and Heifer Replacement Rates for Beef Cow-Calf Production Using System Dynamics. *Agric. Syst.* **2013**, *114*, 6–14. [CrossRef]
61. Teague, W.R.; Kreuter, U.P.; Grant, W.E.; Diaz-Solis, H.; Kothmann, M.M. Economic Implications of Maintaining Rangeland Ecosystem Health in a Semi-Arid Savanna. *Ecol. Econ.* **2009**, *68*, 1417–1429. [CrossRef]
62. Torell, L.A.; Doll, J.P. Public Land Policy and the Value of Grazing Permits. *West. J. Agric. Econ.* **1991**, *16*, 174–184.
63. Torell, L.A.; Drummond, T.W. The Economic Impacts of Increased Grazing Fees on Gila National Forest Grazing Permittees. *J. Range Manag.* **1997**, *50*, 94–105. [CrossRef]
64. Baumgärtner, S.; Quaas, M.F. Ecological-Economic Viability as a Criterion of Strong Sustainability under Uncertainty. *Ecol. Econ.* **2009**, *68*, 2008–2020. [CrossRef]
65. Quaas, M.F.; Baumgärtner, S.; Becker, C.; Frank, K.; Müller, B. Uncertainty and Sustainability in the Management of Rangelands. *Ecol. Econ.* **2007**, *62*, 251–266. [CrossRef]
66. Freier, K.P.; Schneider, U.A.; Finckh, M. Dynamic Interactions between Vegetation and Land Use in Semi-Arid Morocco: Using a Markov Process for Modeling Rangelands under Climate Change. *Agric. Ecosyst. Environ.* **2011**, *140*, 462–472. [CrossRef]
67. Iglesias, E.; Báez, K.; Diaz-Ambrona, C.H. Assessing Drought Risk in Mediterranean Dehesa Grazing Lands. *Agric. Syst.* **2016**, *149*, 65–74. [CrossRef]

68. Gies, L.; Agusdinata, D.B.; Merwade, V. Drought Adaptation Policy Development and Assessment in East Africa Using Hydrologic and System Dynamics Modeling. *Nat. Hazards* **2014**, *74*, 789–813. [CrossRef]
69. Tinsley, T.L.; Chumbley, S.; Mathis, C.; Machen, R.; Turner, B.L. Managing Cow Herd Dynamics in Environments of Limited Forage Productivity and Livestock Marketing Channels: An Application to Semi-Arid Paci Fi c Island Beef Production Using System Dynamics. *Agric. Syst.* **2019**, *173*, 78–93. [CrossRef]
70. Oniki, S.; Shindo, K.; Yamasaki, S.; Toriyama, K. Simulation of Pastoral Management in Mongolia: An Integrated System Dynamics Model. *Rangel. Ecol. Manag.* **2018**, *71*, 370–381. [CrossRef]
71. Marandure, T.; Dzama, K.; Bennett, J.; Makombe, G.; Mapiye, C. Application of System Dynamics Modelling in Evaluating Sustainability of Low-Input Ruminant Farming Systems in Eastern Cape Province, South Africa. *Ecol. Modell.* **2020**, *438*, 109294. [CrossRef]
72. Herrero, M.; Thornton, P.K. Livestock and Global Change: Emerging Issues for Sustainable Food Systems. *Proc. Natl. Acad. Sci. USA* **2013**, *110*, 20878–20881. [CrossRef]
73. Briske, D.D.; Coppock, D.L.; Illius, A.W.; Fuhlendorf, S.D. Strategies for Global Rangeland Stewardship: Assessment through the Lens of the Equilibrium—Non-Equilibrium Debate. *J. Appl. Ecol.* **2020**, *57*, 1056–1067. [CrossRef]
74. Lubell, M.N.; Cutts, B.B.; Roche, L.M.; Hamilton, M.; Derner, J.D.; Kachergis, E.; Tate, K.W. Conservation Program Participation and Adaptive Rangeland Decision-Making. *Rangel. Ecol. Manag.* **2013**, *66*, 609–620. [CrossRef]
75. Marshall, N.A.; Stokes, C.J. Influencing Adaptation Processes on the Australian Rangelands for Social and Ecological Resilience. *Ecol. Soc.* **2014**, *19*, 14. [CrossRef]
76. Roche, L.M.; Schohr, T.K.; Derner, J.D.; Lubell, M.N.; Cutts, B.B.; Kachergis, E.; Eviner, V.T.; Tate, K.W. Sustaining Working Rangelands: Insights from Rancher Decision Making. *Rangel. Ecol. Manag.* **2015**, *68*, 383–389. [CrossRef]
77. Wilmer, H.; Fernández-Giménez, M.E. Rethinking Rancher Decision-Making: A Grounded Theory of Ranching Approaches to Drought and Succession Management. *Rangel. J.* **2015**, *37*, 517–528. [CrossRef]
78. Gea-Izquierdo, G.; Cañellas, I.; Montero, G. Acorn Production in Spanish Holm Oak Woodlands. *Investig. Agrar. For. Syst.* **2006**, *13*, 339–354. [CrossRef]
79. Liu, J.; Mooney, H.; Hull, V.; Davis, S.J.; Gaskell, J.; Hertel, T.; Lubchenco, J.; Seto, K.C.; Gleick, P.; Kremen, C.; et al. Systems Integration for Global Sustainability. *Science* **2015**, *347*, 1258832. [CrossRef] [PubMed]
80. Costanza, R. Ecological Economics: Reintegrating the Study of Humans and Nature. *Ecol. Appl.* **1996**, *6*, 978–990. [CrossRef]
81. Engler, J.O.; Von Wehrden, H. Global Assessment of the Non-Equilibrium Theory of Rangelands: Revisited and Refined. *Land Use Policy* **2018**, *70*, 479–484. [CrossRef]
82. Maestre, F.T.; Salguero-Gómez, R.; Quero, J.L. It Is Getting Hotter in Here: Determining and Projecting the Impacts of Global Environmental Change on Drylands. *Philos. Trans. R. Soc. B Biol. Sci.* **2012**, *367*, 3062–3075. [CrossRef] [PubMed]
83. Stafford Smith, D.M.; McKeon, G.M.; Watson, I.W.; Henry, B.K.; Stone, G.S.; Hall, W.B.; Howden, S.M. Learning from Episodes of Degradation and Recovery in Variable Australian Rangelands. *Proc. Natl. Acad. Sci. USA* **2007**, *104*, 20690–20695. [CrossRef]
84. Reynolds, J.F.; Stafford Smith, M. (Eds.) Do Humans Cause Deserts? In *Global Desertefication. Do Human Cause Deserts*; Dahlem University Press: Berlin, Germany, 2002; pp. 1–21.
85. Vetter, S. Rangelands at Equilibrium and Non-Equilibrium: Recent Developments in the Debate. *J. Arid Environ.* **2005**, *62*, 321–341. [CrossRef]
86. Liu, J.; Dietz, T.; Carpenter, S.R.; Alberti, M.; Folke, C.; Moran, E.; Pell, A.N.; Deadman, P.; Kratz, T.; Lubchenco, J.; et al. Complexity of Coupled Human and Natural Systems. *Science* **2007**, *317*, 1513–1516. [CrossRef]
87. Sterman, J.D. *Business Dynamics: Systems Thinking and Modeling for a Complex World*; Mc Graw Hill: New York, NY, USA, 2000; ISBN 0-07-231135-5.
88. Davies, E.G.R.; Simonovic, S.P. Global Water Resources Modeling with an Integrated Model of the Social-Economic-Environmental System. *Adv. Water Resour.* **2011**, *34*, 684–700. [CrossRef]
89. Kelly, R.A.; Jakeman, A.J.; Barreteau, O.; Borsuk, M.E.; ElSawah, S.; Hamilton, S.H.; Henriksen, H.J.; Kuikka, S.; Maier, H.R.; Rizzoli, A.E.; et al. Selecting among Five Common Modelling Approaches for Integrated Environmental Assessment and Management. *Environ. Model. Softw.* **2013**, *47*, 159–181. [CrossRef]
90. MEA (Millenium Ecosystem Assessment). *Ecosystems and Human Well-Being: Desertification Synthesis*; World Resources Institute: Washington, DC, USA, 2005.
91. Aracil, J. *Introducción a La Dinámica de Sistemas*; Alianza Editorial: Madrid, Spain, 1986.
92. Lotka, A.J. *Elements of Mathematical Biology*; Dover Publications: New York, NY, USA, 1956.
93. Volterra, V. Variations and Fluctuations of the Number of Individuals in Animal Species Living Together. In *Animal Ecology*; Chapman, B.M., Ed.; McGraw-Hill: New York, NY, USA, 1931; pp. 409–448.
94. Noy-Meir, I. Stability of Grazing Systems: An Application of Predator-Prey Graphs. *J. Ecol.* **1975**, *63*, 459–481. [CrossRef]
95. Thornes, J.B. The Ecology of Erosion. *Geography* **1985**, *70*, 222–235.
96. Thornes, J.B. Erosional Equilibria under Grazing. In *Conceptual Issues in Enviromental Archaelology*; Bintliff, J.L., Davidson, D.A., Grant, E.G., Eds.; Edinburgh University Press: Edinburgh, UK, 1988.
97. Ibáñez, J.; Martínez-Valderrama, J.; Puigdefábregas, J. Assessing Desertification Risk Using System Stability Condition Analysis. *Ecol. Modell.* **2008**, *213*, 180–190. [CrossRef]

98. Ibáñez, J.; Martínez-Valderrama, J.; Schnabel, S. Desertification Due to Overgrazing in a Dynamic Commercial Livestock-Grass-Soil System. *Ecol. Modell.* **2007**, *205*, 277–288. [CrossRef]

99. Martínez-Valderrama, J.; Ibáñez, J. Foundations for a Dynamic Model to Analyze Stability in Commercial Grazing Systems. In *Sustainability of Agrosilvopastoral Systems—Dehesas, Montados*; Schnabel, S., Ferreira, A., Eds.; Advances in Geoecology; International Union of Soil Sciences: Vienna, Austria, 2004; pp. 173–182, ISBN 3-923381-50-6.

100. Holling, C.S. The Components of Predation as Revealed by a Study of Small-Mammal Predation of the European Pine Sawfly. *Can. Entomol.* **1959**, *91*, 293–320. [CrossRef]

101. Band, L.E. Field Parameterization of an Empirical Sheetwash Transport Equation. *Catena* **1985**, *12*, 201–210. [CrossRef]

102. Kirkby, M.J. Soil Development Models as a Component of Slope Models. *Earth Surf. Process.* **1977**, *2*, 203–230. [CrossRef]

103. Thornes, J.B. The Interaction of Erosional and Vegetational Dynamics in Land Degradation: Spatial Outcomes. In *Vegetation and Erosion*; Thornes, J.B., Ed.; John Wiley & Sons: Chichester, UK, 1990; pp. 41–53.

104. Elwell, H.A.; Stocking, M. Vegetal Cover to Estimate Soil Erosion Hazard in Rhodesia. *Geoderma* **1976**, *15*, 61–70. [CrossRef]

105. Ibáñez, J.; Lavado Contador, J.F.; Schnabel, S.; Pulido Fernández, M.; Martínez-Valderrama, J. A Model-Based Integrated Assessment of Land Degradation by Water Erosion in a Valuable Spanish Rangeland. *Environ. Model. Softw.* **2014**, *55*, 201–213. [CrossRef]

106. Ibáñez, J.; Lavado Contador, J.F.; Schnabel, S.; Martínez-Valderrama, J. Evaluating the Influence of Physical, Economic and Managerial Factors on Sheet Erosion in Rangelands of SW Spain by Performing a Sensitivity Analysis on an Integrated Dynamic Model. *Sci. Total Environ.* **2016**, *544*, 439–449. [CrossRef] [PubMed]

107. Montgomery, D.R. Soil Erosion and Agricultural Sustainability. *Proc. Natl. Acad. Sci. USA* **2007**, *104*, 13268–13272. [CrossRef] [PubMed]

108. Tilman, D. *Resource Competition and Community Structure*; Princeton University Press: Princeton, NJ, USA, 1982.

109. Ibáñez, J.; Martinez-Valderrama, J.; Papanastasis, V.; Evangelou, C.; Puigdefabregas, J. A Multidisciplinary Model for Assessing Degradation in Mediterranean Rangelands. *Land Degrad. Dev.* **2014**, *25*, 468–482. [CrossRef]

110. Atanasova, N.; Džeroski, S.; Kompare, B.; Todorovski, L.; Gal, G. Automated Discovery of a Model for Dinoflagellate Dynamics. *Environ. Model. Softw.* **2011**, *26*, 658–668. [CrossRef]

111. Ventana Systems Inc. VensimDSS. 2019. Available online: https://www.ventanasystems.com/software/ (accessed on 1 January 2023).

112. Martinez-Valderrama, J.; Ibáñez, J.; Alcalá, F.J.; Martínez, S. SAT: A Software for Assessing the Risk of Desertification in Spain. *Sci. Program.* **2020**, *2020*, 7563928. [CrossRef]

113. Perry, G.L.W.; Millington, J.D.A. Spatial Modelling of Succession-Disturbance Dynamics in Forest Ecosystems: Concepts and Examples. *Perspect. Plant Ecol. Evol. Syst.* **2008**, *9*, 191–210. [CrossRef]

114. May, R.M. Thresholds and Breakpoints in Ecosystems with a Multiplicity of Stable States. *Nature* **1977**, *269*, 471–477. [CrossRef]

115. Frank, P.M. *Introduction to System Sensitivity Theory*; Academic Press: New York, NY, USA, 1978; ISBN 978-0-12265-650-7.

116. Swart, J. Sensitivity of a Hyrax-Lynx Mathematical Model to Parameter Uncertainty. *South Afr. J. Sci.* **1987**, *83*, 545–547.

117. Van Coller, L. Automated Techniques for the Qualitative Analysis of Ecological Models: Continuous Models. *Conserv. Ecol.* **1997**, *1*, 5. [CrossRef]

118. Martínez-Vicente, S.; Requena, A. *Dinámica de Sistemas. 1. Simulación Por Ordenador*; Alianza Editorial: Madrid, Spain, 1986.

119. Rosenzweig, M.L.; MacArthur, R.H. Graphical Representation and Stability Conditions of Predator-Prey Interactions. *Am. Nat.* **1963**, *97*, 209–223. [CrossRef]

120. Edelstein-Keshet, L. *Mathematical Models in Biology*; The Random House: New York, NY, USA, 1988.

121. Holling, C.S. Resilience and Stability of Ecological Systems. *Annu. Rev. Ecol. Syst.* **1973**, *4*, 1–23. [CrossRef]

122. Gotelli, N.J. *A Primer in Ecology*, 2nd ed.; Sinaver Associates Inc. Publishers: Sunderland, MA, USA, 1998.

123. Martínez-Valderrama, J. *Análisis de La Desertificación Por Sobrepastoreo Mediante Un Modelo de Simulación Dinámico*; Universidad Politécnica de Madrid: Madrid, Spain, 2006.

124. Ibáñez, J.; Martínez-Valderrama, J.; Martínez Vicente, S.; Martínez Ruiz, A.; Rojo Serrano, L. *Procedimientos de Alerta Temprana y Estimación de Riesgos de Desertificación Mediante Modelos de Dinámica de Sistemas*; Ministerio de Agricultura, Alimentación y Medio Ambiente: Madrid, Spain, 2015; ISBN 978-84-491-0074-1.

125. Xu, D.; Li, C.; Zhuang, D.; Pan, J. Assessment of the Relative Role of Climate Change and Human Activities in Desertification: A Review. *J. Geogr. Sci.* **2011**, *21*, 926–936. [CrossRef]

126. Plackett, R.L.; Burman, J.P. The Design of Optimum Multifactorial Experiments. *Biometrika* **1946**, *33*, 305–325. [CrossRef]

127. Beres, D.L.; Hawkins, D.M. Plackett–Burman Technique for Sensitivity Analysis of Many-Parametered Models. *Ecol. Modell.* **2001**, *141*, 171–183. [CrossRef]

128. Gan, Y.; Duan, Q.; Gong, W.; Tong, C.; Sun, Y.; Chu, W.; Ye, A.; Miao, C.; Di, Z. A Comprehensive Evaluation of Various Sensitivity Analysis Methods: A Case Study with a Hydrological Model. *Environ. Model. Softw.* **2014**, *51*, 269–285. [CrossRef]

129. Saltelli, A.; Ratto, M.; Andres, T.; Campolongo, F.; Cariboni, J.; Gatelli, D.; Saisana, M. *Global Sensitivity Analysis: The Primer*; John Wiley & Sons: Chichester, UK, 2008.

130. Sobol, I.M. Global Sensitivity Indices for Nonlinear Mathematical Models and Their Monte Carlo Estimates. *Math. Comput. Simul.* **2001**, *55*, 271–280. [CrossRef]

131. Ibáñez, J.; Martínez-Valderrama, J.; Contador, J.F.L.; Fernández, M.P. Exploring the Economic, Social and Environmental Prospects for Commercial Natural Annual Grasslands by Performing a Sensitivity Analysis on a Multidisciplinary Integrated Model. *Sci. Total Environ.* **2020**, *705*, 135860. [CrossRef]

132. Potkonjak, V.; Gardner, M.; Callaghan, V.; Mattila, P.; Guetl, C.; Petrović, V.M.; Jovanović, K. Virtual Laboratories for Education in Science, Technology, and Engineering: A Review. *Comput. Educ.* **2016**, *95*, 309–327. [CrossRef]

133. Martínez-Valderrama, J.; Ibáñez, J.; Ibáñez, M.A.; Alcalá, F.J.; Sanjuán, M.E.; Ruiz, A.; Del Barrio, G. Assessing the Sensitivity of a Mediterranean Commercial Rangeland to Droughts under Climate Change Scenarios by Means of a Multidisciplinary Integrated Model. *Agric. Syst.* **2021**, *187*, 103021. [CrossRef]

134. Martínez-Valderrama, J.; Ibáñez, J.; Del Barrio, G.; Alcalá, F.J.; Sanjuán, M.E.; Ruiz, A.; Hirche, A.; Puigdefábregas, J. Doomed to Collapse: Why Algerian Steppe Rangelands Are Overgrazed and Some Lessons to Help Land-Use Transitions. *Sci. Total Environ.* **2018**, *613–614*, 1489–1497. [CrossRef] [PubMed]

135. Hirche, A.; Salamani, M.; Abdellaoui, A.; Benhouhou, S.; Martínez-Valderrama, J. Landscape Changes of Desertification in Arid Areas: The Case of South-West Algeria. *Environ. Monit. Assess.* **2011**, *179*, 403–420. [CrossRef]

136. Slimani, H.; Aidoud, A. Desertification in the Maghreb: A Case Study of an Algerian High-Plain Steppe. In *Environmental Challenges in the Mediterranean 2000–2050*; Marquina, A., Ed.; Kluwer Academic Publishers: Amsterdam, The Netherlands, 2004; pp. 93–108.

137. Aidoud, A. Vegetation Changes and Land Use Changes in Algerian Steppe Rangelands. In *Restauración de la Cubierta Vegetal en Ecosistemas Mediterráneos*; Pastor-López, A., Seva-Román, E., Eds.; Instituto de Cultura Juan-Gil Albert: Alicante, Spain, 2002; pp. 53–80.

138. Martínez-Valderrama, J.; Guirado, E.; Maestre, F.T. Unraveling Misunderstandings about Desertification: The Paradoxical Case of the Tabernas-Sorbas Basin in Southeast Spain. *Land* **2020**, *9*, 269. [CrossRef]

139. Yu, Y.; Feng, K.; Hubacek, K. Tele-Connecting Local Consumption to Global Land Use. *Glob. Environ. Chang.* **2013**, *23*, 1178–1186. [CrossRef]

140. WWF (World Wildlife Fund). *The Growth of Soy: Impacts and Solutions*; WWF: Gland, Switzerland, 2014.

141. Martínez-Valderrama, J.; Sanjuán, M.E.; Barrio, G.; Guirado, E.; Ruiz, A.; Maestre, F.T. Mediterranean Landscape Re-Greening at the Expense of South American Agricultural Expansion. *Land* **2021**, *10*, 204. [CrossRef]

142. Ibáñez, J.; Martínez-Valderrama, J. Global Effectiveness of Group Decision-Making Strategies in Coping with Forage and Price Variabilities in Commercial Rangelands: A Modelling Assessment. *J. Environ. Manag.* **2018**, *217*, 531–541. [CrossRef] [PubMed]

143. Polasky, S.; Kling, C.L.; Levin, S.A.; Carpenter, S.R.; Daily, G.C.; Ehrlich, P.R.; Heal, G.M.; Lubchenco, J. Role of Economics in Analyzing the Environment and Sustainable Development. *Proc. Natl. Acad. Sci. USA* **2019**, *116*, 5233–5238. [CrossRef] [PubMed]

144. Herrero, M.; Thornton, P.K.; Mason-D'Croz, D.; Palmer, J.; Bodirsky, B.L.; Pradhan, P.; Barrett, C.D.; Benton, T.G.; Hall, A.; Pikaar, I.; et al. Articulating the Impact of Food Systems Innovation on the Sustainable Development Goals. *Lancet Planet. Health* **2020**, *5196*, 50–62.

145. Willett, W.; Rockström, J.; Loken, B.; Springmann, M.; Lang, T.; Vermeulen, S.; Garnett, T.; Tilman, D.; DeClerck, F.; Wood, A.; et al. Food in the Anthropocene: The EAT-Lancet Commission on Healthy Diets from Sustainable Food Systems. *Lancet* **2019**, *393*, 447–492. [CrossRef]

146. Berkes, F.; Folke, C. (Eds.) *Linking Social and Ecological Systems: Management Practices and Social Mechanisms for Building Resilience*; Cambridge University Press: Cambridge, UK, 1998.

147. Ringler, C.; Bhaduri, A.; Lawford, R. The Nexus across Water, Energy, Land and Food (WELF): Potential for Improved Resource Use Efficiency? *Curr. Opin. Environ. Sustain.* **2013**, *5*, 617–624. [CrossRef]

148. Sterman, J. All Models Are Wrong: Reflections on Becoming a Systems Scientist. *Syst. Dyn. Rev.* **2002**, *18*, 501–531. [CrossRef]

149. Wang, E.; Zhang, L.; Cresswell, H.; Hickel, K. Comparison of Top-Down and Bottom-Up Models for Simulation of Water Balance as Affected by Seasonality, Vegetation Type and Spatial Land Use. In Proceedings of the 3rd International Congress on Environmental Modelling and Software, Burlington, VT, USA, 6–9 July 2006.

150. Van Delden, H.; Seppelt, R.; White, R.; Jakeman, A.J. A Methodology for the Design and Development of Integrated Models for Policy Support. *Environ. Model. Softw.* **2011**, *26*, 266–279. [CrossRef]

151. D'Odorico, P.; Bhattachan, A.; Davis, K.; Ravi, S.; Runyan, C. Global Desertification: Drivers and Feedbacks. *Adv. Water Resour.* **2013**, *51*, 326–344. [CrossRef]

152. Oxley, T.; McIntosh, B.S.; Winder, N.; Mulligan, M.; Engelen, G. Integrated Modelling and Decision-Support Tools: A Mediterranean Example. *Environ. Model. Softw.* **2004**, *19*, 999–1010. [CrossRef]

 *land*

Article

Superabsorbent Polymer Use in Rangeland Restoration: Glasshouse Trials

Shannon V. Nelson [1], **Neil C. Hansen** [1], **Matthew D. Madsen** [1], **Val Jo Anderson** [1], **Dennis L. Eggett** [2] **and Bryan G. Hopkins** [1,*]

[1] Department of Plant and Wildlife Sciences (PWS), Brigham Young University, Provo, UT 84602, USA
[2] Department of Statistics (STAT), Brigham Young University, Provo, UT 84602, USA
* Correspondence: hopkins@byu.edu

Abstract: Post-disturbance rangeland restoration efforts are often thwarted due to soil moisture deficits. Superabsorbent polymers (SAPs) absorb hundreds of times their weight in water, increasing soil moisture when the SAP is mixed with soil. The objective of this study was to evaluate banded SAPs under the soil surface to increase plant available water and thus seedling establishment for perennial rangeland species during restoration efforts. Five glasshouse experiments with two rangeland perennial grass species, bottlebrush squirreltail (*Elymus elymoides*) or Siberian wheatgrass (*Agropyron fragile*), were conducted. Treatments varied, including SAP rates ranging from 11–3000 kg ha^{-1} with placement mostly banded at depths extending from the surface up to a 15 cm depth. Generally, SAPs increased soil moisture at all rates and depths for up to 49 days. However, rates $\geq$ 750 kg ha^{-1} caused the soil to swell and crack, potentially hastening soil drying later in the season. Seedling longevity was increased up to 12 days, especially at the high SAP band rate of 3000 kg ha^{-1} when the band was 8 or 15 cm deep. Further work is needed to verify banded SAP rates and placement depths in the field, ascertain conditions to reduce soil displacement, and evaluate benefits across species.

Keywords: superabsorbent polymer; SAP; hydrogel; rangeland; bottlebrush squirreltail; Siberian wheatgrass; banding; restoration; *Elymus elymoides*; *Agropyron fragile*

Citation: Nelson, S.V.; Hansen, N.C.; Madsen, M.D.; Anderson, V.J.; Eggett, D.L.; Hopkins, B.G. Superabsorbent Polymer Use in Rangeland Restoration: Glasshouse Trials. *Land* **2023**, *12*, 232. https://doi.org/10.3390/land12010232

Academic Editors: Michael Vrahnakis, Yannis (Ioannis) Kazoglou and Manuel Pulido Fernádez

Received: 30 November 2022
Revised: 6 January 2023
Accepted: 9 January 2023
Published: 11 January 2023

1. Introduction

Rangelands can be defined as "all lands, except for urban, agricultural, or densely forested lands, that support predominantly native or naturalized vegetation capable of sustaining native or domestic grazing and/or browsing ungulates, whether or not those animals are present" [1]. They provide economic benefits such as hunting, fishing, grazing, and mining as well as environmental and public service benefits such as recreation, habitat, water quality, and education [2]. In arid regions, the establishment of perennial species after a disturbance is key to restoration success and invasive species management [3–5]. Drought conditions make direct seeding efforts in rangelands notoriously challenging [6,7]. Seasonal and yearly variations in precipitation impact seedling emergence and establishment; exotic species add further stress to the system [8]. Their introduction and spread adversely affects landscapes [9,10] by changing the make-up of the local plant community [11–14]. This alters wildlife habitat and food supplies, increases erosion, and modifies wildfire characteristics [3,11].

Deep-rooted perennial grasses, forbs, and shrubs can reduce annual weed invasion, thereby minimizing erosion and fire danger while providing forage [9,15,16]. Utilizing water, nutrients, sunlight, space, and other resources, they inhibit the establishment of exotic species [17,18]. Fire, insects, disease, overgrazing, and other destructive forces that diminish these perennials free the resources that enable the establishment of invasive species [19–21]. Extensive root systems help surviving established plants to effectively compete for water and nutrients after a disturbance [16,22,23]. However, young perennial seedlings generally struggle to compete for water in arid and semi-arid systems where invasive plants use early season soil moisture [4,16,21,24–26].

Superabsorbent polymers (SAPs) may help tip the scales in favor of perennial species when banded directly below seedlings. This soil additive absorbs hundreds of times its weight in water and then releases it slowly for plant use [27]. The use of SAPs reduces soil compaction and water lost to deep percolation while increasing pore volume, water infiltration, and moisture retention [27–31].

In agriculture, SAPs have been shown to help increase the time between the need for irrigation, increase plant biomass, and improve fertilizer retention in the soil [32–35]. In the greenhouse industry, plant survival is increased when root plugs of plants grown for transplanting into areas with a water deficit are formed by mixing in SAPs [36]. When placed in a band under the soil surface, SAPs can act as a reservoir of water for young seedlings, which can help alleviate drought conditions with the onset of summer heat [34,35]. Though it diminishes over time, SAPs have the capacity to reabsorb water during precipitation events, increasing the duration of the soil water reservoir effect [37–40].

Although less commonly used than in crop production, SAPs have been used in native ecosystems to increase seedling longevity and control run-off and erosion [4,23,41–43]. El-Asmar et al. [35] showed increases in evapotranspiration, water use efficiency, and seedling growth parameters and survival time in irrigated agricultural conditions. However, to the best of our knowledge, the banding of SAPs under the soil surface has not been explored in native ecosystems. When facing competition for water from established invasive species, young seedlings of seeded species may not be able to persist until their root systems grow large enough to access moisture deeper in the soil profile. If banded SAPs can widen the window of persistence, seedlings may grow roots deep enough to access available water and increase the probability of survival until additional precipitation is received [24].

We hypothesize that dry SAPs placed in bands at or near seedling rooting depth will act as a localized soil water reservoir to increase soil moisture and seedling establishment. Five glasshouse studies, with bottlebrush squirreltail (*Elymus elymoides* (Raf.) Swezey) and/or Siberian wheatgrass (*Agropyron fragile* (Roth) Candargy), were conducted to serve as proof of concept and to evaluate various management strategies to best increase soil moisture and seedling establishment in preparation for field studies, with the following study objectives/justifications:

1. SAP Rate and Depth: Evaluate the effect of SAP rates (0, 1500, or 3000 kg ha^{-1}) and placement depths (0, 3, 8, or 15 cm depth bands, or mixed) to explore optimum management strategies.
2. Reduced Seeding Rate: Assess the effect of seeding rates (2, 4, 8, or 16 kg ha^{-1}) with SAP bands (0 or 3000 kg ha^{-1}) at 8 cm depth to determine if excessive inter-species competition for soil moisture occurs as a function of increased germination with SAP.
3. Low SAP Rate: Evaluate the efficacy of SAP rates (0, 11, 47, 190, 750, or 1500 kg ha^{-1}) at 8 cm depth to determine if relatively low rates of SAPs would sufficiently increase soil moisture to positively impact seedling health while keeping the soil surface intact, which was a problem observed at high rates in field conditions.
4. SAP Depth and Root Growth: Assess the effect of SAP rates (0 or 3000 kg ha^{-1}) and placement depths (0, 3, or 8 cm, or mixed) to measure seedling root growth response to SAPs.
5. Fertilizer and SAP Interaction: Evaluate the impact of fertilizer (with and without) used in conjunction with SAPs (0 or 3000 kg ha^{-1}) placed at 8 cm depth in addition to measuring SAPs' ability to reabsorb water.

2. Materials and Methods

2.1. Treatments

Five glasshouse experiments were conducted at Brigham Young University (BYU), Provo, UT, USA (40.2454, −111.6415, elevation 1391 m). Various rates of SAP (11–3000 kg ha^{-1})

were compared to an untreated control in a full factorial design with all combinations of various SAP placement depths, seeding rates, species, and/or fertilizer (Table 1) arranged in a randomized complete block design (RCBD). Generally, the SAP was applied in a concentrated band at or below the soil surface at various depths (Table 1). In Studies 1 and 4, there was a non-banded SAP treatment which consisted of mixing the SAP uniformly with the soil ("mixed")to a depth of 15 cm (Study 1) or 8 cm (Study 4).

Table 1. Five glasshouse study treatments, parameters, and measurements. Gray squares in the treatments section indicate a treatment parameter for that study.

Study	1	2	3	4	5
			Treatments		
SAP, kg ha^{-1}	0, 1500, 3000	0, 3000	0, 11, 47, 190, 750, 1500	0, 3000	0, 3000
SAP Depth, cm	0, 3, 8, 15, mixed (top 15)	8	8	3, 8, mixed (top 8)	8
Fertilizer	no	no	no	no	with/without
Species *	BB	BB and SW	BB and SW	BB and SW	BB and SW
Seeding Rate, kg ha^{-1}	24	2, 4, 8, 16	8	4	6
			Study Parameters		
Study length, d	107	76	78	70	76/133 **
Dates	14 February to 1 June 2017	7 September to 22 November 2018	8 February to 27 April 2019	19 July to 27 September 2019	16 December to 28 April 2018
Replicates	4	4	3	4	6
Dimensions, cm	10 (each side)	10 (each side)	30 × 21.5	10 (diameter)	10 (each side)
Depth, cm	23	23	15	25	10
Thinned, number@DAP	3@14 & 1@29	no	1@19	1@13	no
Saturation time, d	2	15	16	18	16
			Measurements		
Soil Moisture	3x/wk	3x/wk	3x/wk until 36 DAP then 57 & 78 DAP	weekly @4 depths	3x/wk
Seedling emergence and total alive ***	8, 25, 29, 70, 72, 76, 81, 84, 86, 105, and 107 DAP	3x/wk	~3x/wk	weekly	weekly
Seedling length and Blade number	76 DAP	weekly	no	weekly	weekly
Root length/branching	no	no	no	weekly	no
Root/shoot biomass	no	no	no	yes	no

* BB = bottlebrush squirreltail; SW = Siberian wheatgrass. ** all seedlings were dead by 76 d in Study 5, with the rewetting portion of the study beginning at that time and ending on 133 d. *** total seedlings alive was not measured in Study 1.

The SAP product used in these studies was Stockosorb® 660 micro (Evonik Industries AG; Essen, Germany). This polymer is made of crosslinked potassium (K$^+$) polyacrylate, which produces an absorptive capacity of 260 l water (H$_2$O) kg^{-1} [44]. It eventually, after 1–3 years in the soil, degrades to carbon dioxide (CO$_2$), H$_2$O, and K$^+$ [28].

A potential unwanted effect of the degradation of Stockosorb® 660 is the release of the essential plant nutrient K into the soil solution. However, the soil used in this study has high K concentration (Table 2). As the K was not limiting for plant growth and high soil K is not generally a specific ion toxicity concern, the K was not balanced across treatments.

The Stockosorb® 660 manufacturer recommends blending the product into the top 10 cm of soil or placing it in a band below the soil surface at rates up to 11 kg ha^{-1} at the time of seeding in irrigated agricultural systems [45,46]. Higher rates are suggested when the product is used under conditions of low rainfall and high temperatures [46]. Our study treatments included SAP placed at depths extending from the surface to 15 cm deep at rates up to 272 times higher than the recommended agricultural rate to ensure observable treatment effects in a non-irrigated, xeric system [47].

Study 5 included a fertilization treatment in addition to SAP rates. The fertilizer was applied at 4 N, 17 P_2O_5, 17 K_2O, 0.6 S, 0.6 Fe, 0.1 Zn, 0.1 Mn, 0.1 Cu, and 0.1 B (kg ha^{-1}). The micronutrients were all applied as the chelated form with ethylene diamine tetracetic acid (EDTA) at a 1:1 ratio.

2.2. Soil and SAP Placement

The soil used in these studies was collected from the site for the eventual field testing of SAP at Murray's Mesa on the Utah Test and Training Range (UTTR) located in the desert west of Salt Lake City, UT (41.040976, −112.982474; elevation 1392 m). The UTTR serves as a practice bombing and gunnery range for the United States Department of Defense by Hill Air Force Base, located in Layton, Utah, USA, and is an active revegetation research site. The soil in the area is generally classified as a Tooele Fine Sandy Loam soil [coarse-loamy, mixed (calcareous), mesic Typic Torriorthents] [48]. Although the specific soil collected for this study mostly fits this classification, its textural class is loam (Table 2). Soil analysis was performed at BYU's Environmental Analytical Laboratory (BYU-EAL, https://pws.byu.edu/eal (accessed on 13 May 2022)) and Servi-Tech Laboratories (www.servitechlabs.com (accessed on 13 May 2022)) (Table 2).

Table 2. Soil properties [49]. Analysis performed at BYU's Environmental Analytical Laboratory (BYU-EAL, https://pws.byu.edu/eal (accessed on 13 May 2022)) except where noted. Properties marked with "*" were analyzed at Servi-Tech Laboratories (www.servitechlabs.com (accessed on 24 May 2022)).

Properties		Nutrients, mg kg^{-1}	
pH *[a]	7.8	NO_3-N [d]	20
Salinity (dS m^{-1}) [a]	8.9	P [e]	12
Texture [b]	Loam	K [e]	1068
% Sand	30	Zn [f]	0.2
% Silt	49	Fe [f]	3.1
% Clay	21	Mn [f]	2.1
% Organic Matter [c]	1.3	Cu [f]	0.5
SAR [a]	2.6	Ca *[g]	3903
		Mg *[g]	345
		Na *[h]	2230

[a] Saturated Paste Analysis; [b] Hydrometer Method; [c] Loss-On-Ignition (LOI); [d] KCl extraction, analysis with RFA (rapid flow analysis), FIAlab Flow Injection Cadmium Reduction Nitrate (FIAlabs, Tacoma, WA); [e] Olsen bicarbonate, analysis with AAS (Atomic Absorption Spectroscopy), AAnalyst 200 Atomic Absorption Spectrometer (PerkinElmer, Seattle, WA); [f] DTPA (diethylenetriaminepentaacetic acid) with ICP-OES (Inductively Coupled Plasma-Optical Emission Spectrometry), iCAP 7400 ICP-OES Radial Analyzer (Thermo Scientific, Waltham, MA, USA); [g] Ammonium Acetate Analysis; [h] Mehlich 3 ICP.

Containers for growing plants (Table 1) consisted of tall square plastic pots (Studies 1 and 2), short square plastic pots (Study 5), rectangular wooden boxes (Study 3; Figure 1), or round clear plastic canisters (Study 4; B076 80 oz Clear PET Canister 110–400, Industrial Container and Supply, Salt Lake City, UT, USA). The boxes in Study 3 allowed for a relatively large surface area for a simulation of the planting furrows used in field reclamation projects (Figure 1). All containers had holes in the bottom to enable the free flow of water into and out of the containers. The bottom of each was lined to prevent soil loss. A single layer of commercial grade laboratory paper towel was used in Studies 1, 2, and 5. A single layer of Vigoro Weed Control Fabric (Nex Matrix, Wilmington, DE, USA) was used in Studies 3 and 4.

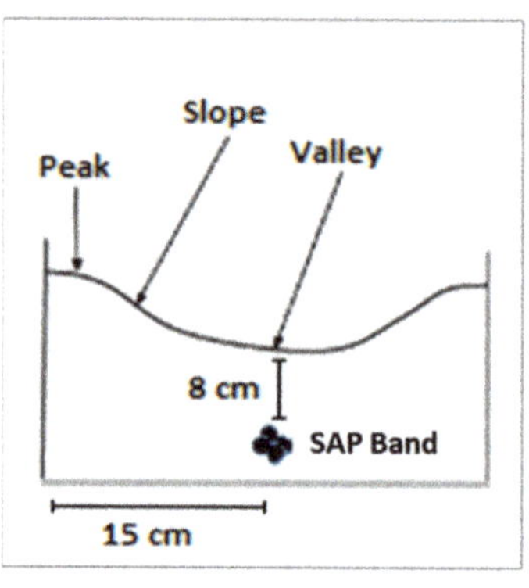

Figure 1. Frontal cross section diagram (**a**) of a grow box unit (**b**) containing soil with a superabsorbent polymer (SAP) band 8 cm below the soil surface of the furrow valley.

For Study 4, holes were drilled in the sides of each canister at 3, 5, 8, and 10 cm below the soil surface for volumetric soil moisture content measurements during the trial. The holes for the 3 and 8 cm depths and the 5 and 10 cm depths were directly above/below each other. The two groupings were offset by 8 cm. These holes were covered on the outside of the canister with Parafilm "M" Laboratory Film (Pechiney Plastic Packaging, Chicago, IL, USA), except during soil moisture measurements, to prevent soil and moisture loss. The canisters were slipped into opaque sleeves of Prodex AD5 Insulation (Prodex, El Coyol de Alajuela, Costa Rica) to prevent light reaching the roots. The covered canisters were maintained at an approximate 30° tilt from horizontal with the seeded edge down to encourage root growth along the side of the canister for ease in monitoring root growth.

For banded treatments, the soil was added to the desired depth, the SAP added, and then additional soil added to within 5–6 cm of the top of the pots/canisters (Studies 1–2, 4–5) or 1 cm from the top of the wood box (Study 3). In Studies 1, 2, and 5, the SAP was placed in a band across the middle of the pot. In Study 3, the SAP band ran front to back in the middle of the box under the furrow. In Study 4, the banded SAP treatments were added within 1 cm of the edge around the circumference of the canisters. Two studies included treatments with the SAP mixed thoroughly into the top 15 (Study 1) or 8 cm (Study 4) of soil. Soil was added to the containers to the correct depth then filled with the soil/SAP mixture. The control treatments were added to the same depths without any layering.

2.3. Irrigation

The soil, with or without SAP, was saturated with deionized (DI) water for several days (Table 1) immediately after (Study 1) or before (Studies 2–5) planting. This was done by partially submerging the pots/canisters for all but Study 3, which was not submerged due to the large size of the connected wooden box containers holding the soil (Table 1; Figure 1). Rather, paper towels were temporarily laid on the soil to reduce soil erosion while 4 l of DI water was added, followed by 0.4 l added on each of the next two days, and then 0.2 l were added three times per week for 14 d.

In all but Study 1, weed seedlings had sufficient time to germinate and be removed, and the soil was covered with black plastic sheeting during this time to reduce evaporation and surface salt accumulation. In all studies, gravitational water was allowed to drain for 1 d to reach field capacity, followed by the initiation of Study 1 or planting for Studies 2–5. The plastic covered the soil until the desired seedlings began to emerge. For Studies

1–3, the goal was to drought-stress the plants; thus, no additional water was added after initial saturation. In Study 4, the soil was rewet at 28–29 days after planting (DAP), as the objective was to see if healthy plant roots avoid the SAP band. In Study 5, the soil was rewet at 77–90 DAP after all seedlings had died to measure the ability of the SAP bands to reabsorb water after complete dry down.

2.4. Species

Bottlebrush squirreltail was used in all studies, and Siberian wheatgrass was used in all but Study 1 as model species. In studies with both species, they were seeded in separate pots/canisters or in separate rows in Study 3 where they were randomly assigned to two of four evenly spaced rows in each box. The rows ran perpendicular to the valley from one peak to the other along the 30 cm width to create three seed positions relative to the band location (valley, slope, and peak) (Figure 1).

Seeds were planted in dry soil for Study 1 and in moist soil in Studies 2–5. Seeding rates ranged 2–24 kg ha^{-1} (Table 1), including variable seeding rate treatments in Study 2. Seedlings were thinned in Studies 1, 3, and 4 (Table 1) (note: the recommended seeding rates for these species are 7 and 6 kg ha^{-1}, respectively [50,51]). Bottlebrush squirreltail is a native perennial rangeland grass in the Great Basin region. Siberian wheatgrass is an introduced perennial rangeland grass. Both species have been shown to compete well against cheatgrass (*Bromus tectorum* L.) and other invasive species in arid environments [48,49]. These were seeded at 1 cm depth immediately prior to saturation in Study 1 and 1 d after the saturation period ended, with the soil approximately at field capacity, for Studies 2–5.

Seedlings were grown without artificial lighting. Temperatures fluctuated from 13–28 °C, which is similar to the naturally occurring diurnal cycles for the region these soils were collected from.

2.5. Measurements

Soil moisture was measured gravimetrically (Studies 1, 2, and 5) or volumetrically (Studies 3 and 4) (Table 1). Gravimetric soil moisture was measured by weighing each container of soil to determine the added weight of water to soil. Volumetric water content was measured by an ML3 ThetaProbe Soil Moisture Sensor with HH2 Moisture Meter (Delta-T Devices; Cambridge, England). In Study 3, inserting the moisture probe into the soil created four openings in the soil (0.3 cm diameter × 6 cm deep and 2.5 cm apart in a triangular pattern with a fourth hole in the center of the pattern). The probe was inserted into the same locations each time to avoid excessive soil disturbance. It is logical that the soil would be slightly dryer near these holes due to aeration. This issue was reduced in Study 4 because the measurements were taken via holes drilled through the side of the canister. The holes were kept covered with Parafilm "M" Laboratory Film (Pechiney Plastic Packaging, Chicago, IL, USA) when measurements were not being taken.

For Study 3, the volumetric soil moisture was measured at the peak, slope, and valley for each row in each compartment. When seedlings were thinned, every attempt was made to leave them in locations that were not being used to measure soil moisture so as to minimize disturbance. At 78 DAP, after all seedlings had died, corresponding soil moisture measurements in undisturbed locations in each row were made to quantify the difference between disturbed and undisturbed locations.

Generally, seedling emergence, total seedlings alive (longevity), length, and blade count were measured (Table 1). In Study 2, the percent of planted seeds that emerged (persistence) was determined weekly. Root length and branching and root/shoot biomass were measured for Study 4 (Table 1). Seedling emergence is the number of seedlings alive on the day of measurement. Total seedlings alive is the cumulative seedlings that emerged during the study. Time to emergence was the amount of time it took for the seedlings to emerge. Longevity is the number of days a seedling lived (seedlings were considered dead if they snapped when the blade was bent and pinched at the base). Seedlings were thinned in studies 1, 3, and 4.

2.6. Statistical Analysis

The data from each trial were initially analyzed using a mixed model analysis in JMP (SAS Institute Inc., Cary, NC, USA). Across the studies, the fixed variables included SAP rate, SAP placement depth, seeding rate, date, species, and fertilizer depth. Random effects included block, pot or box, row, and seed number. Each effect was analyzed as appropriate for the given study. Random effects were removed when variance estimates were negative or not significant. As appropriate, post hoc mean separation by the Tukey–Kramer or Student's *t*-tests was performed on seedling and soil moisture variables and their interactions. In some instances, we were only interested in treatments compared to the control, not to each other. In those cases, mean separation by the Student's *t*-test with a pseudo-Bonferroni adjustment set at 0.005 was used to analyze the desired comparisons.

3. Results

3.1. SAP Rate and Depth

3.1.1. Soil Moisture

The three-way interaction of SAP Rate*Placement Depth*Time on soil moisture was not significant, but all other comparisons were highly significant (Table 3). The SAP increased soil moisture with an average of 7.9% compared to the untreated control at 6.9%. When averaged over the course the study, all SAP treatments but one had higher moisture than the control (Figure 2).

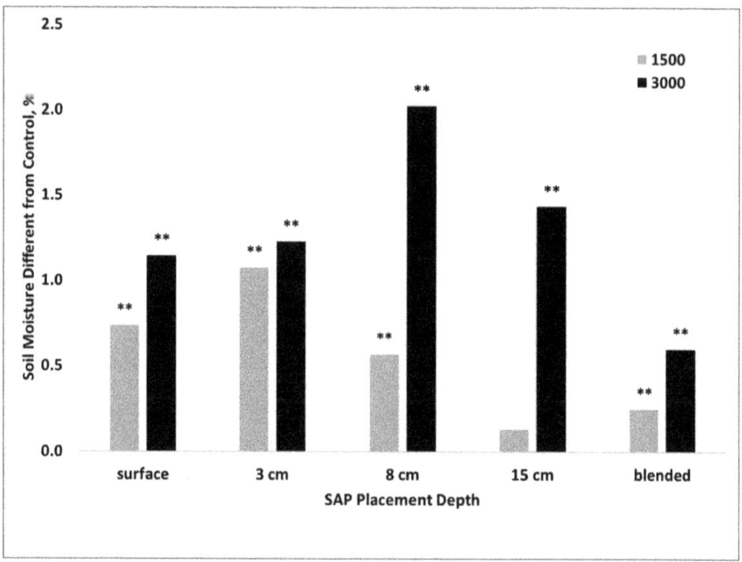

Figure 2. Soil moisture as a function of a superabsorbent polymer (SAP) rate and placement relative to an untreated control averaged across 108 d of a glasshouse study with bottlebrush squirreltail. The SAP (1500 or 3000 kg ha^{-1}) was blended with the soil to a depth of 15 cm or placed in a band at the surface or at 3, 8, or 15 cm deep directly below the seed. The soil was then saturated and allowed to dry down. Bars marked with "**" were highly significant at *p* < 0.0001.

Table 3. Statistical results of a glass house study evaluating the effect of SAP depth (D), rate (R), time (T), and their interactions on soil moisture, seedling longevity, shoot length, and blade count. Bolded numbers indicate $p < 0.05$.

Effect	DF	Soil Moisture		Longevity		Shoot Length		Blade Count	
		F Ratio	Prob > F	F Ratio	Prob > F	F Ratio	Prob > F	F Ratio	Prob. F
D	4	33.6114	**<0.0001**	1.8968	0.137	2.3625	0.076	1.7452	0.172
R	1	146.3037	**<0.0001**	4.3468	**0.046**	4.848	**0.036**	6.9231	**0.014**
T	41	21.6824	**<0.0001**						
D*R	4	28.3439	**<0.0001**	2.8533	**0.041**	3.5722	**0.017**	1.6961	0.182
D*T	164	2.9596	**<0.0001**						
R*T	41	4.977	**<0.0001**						
D*R*T	164	0.6947	0.998						

3.1.2. Seedling Growth Parameters

The two-way interaction of SAP Placement Depth*Rate on seedling longevity was significant (Table 2). The deepest placement depths, 8 and 15 cm, of 3000 kg ha^{-1} SAP bands resulted in increased seedling longevity of 21 and 18 d, respectively (Figure 3). There appears to be a trend toward increased seedling longevity for surface placement of both the 1500 and 3000 kg ha^{-1} SAP bands. It is noteworthy that there was no negative impact on seedling longevity at any depth or SAP rate.

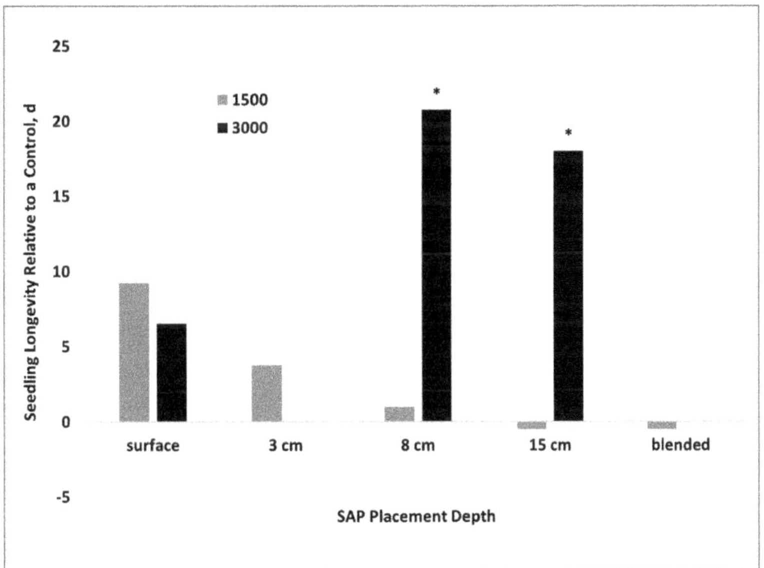

Figure 3. Seedling longevity, number of days alive, as a function of a superabsorbent polymer (SAP) rate and placement relative to an untreated control averaged across 108 d of a glasshouse study with bottlebrush squirreltail. The SAP (1500 or 3000 kg ha^{-1}) was blended with the soil to a depth of 15 cm or placed in a band at the surface or at 3, 8, or 15 cm deep directly below the seed. The soil was then saturated and allowed to dry down. Bars marked with "*" were significant at $p = 0.05$.

As with longevity, the interaction SAP Rate*Placement Depth had a significant impact on seedling length (Table 2). Although differences were measured, there were no clear trends for shoot length. The 3000 kg SAP ha^{-1} rate at 8 cm depth increased seedling length, while 1500 kg SAP ha^{-1} decreased it (Table 4).

Table 4. Bottlebrush squirreltail seedling length, relative to an untreated control, as a function of superabsorbent polymer (SAP) rate and placement for a glasshouse study. The SAP (1500 or 3000 kg ha^{-1}) was blended with the soil to a depth of 15 cm or placed in a band at the surface or at 3, 8, or 15 cm deep directly below the seed; the soil was saturated and allowed to dry down. Bolded lengths are significant.

| | Superabsorbent Polymer Rate | | | |
| | 1500 kg ha^{-1} | | 3000 kg ha^{-1} | |
SAP Placement Depth	Seedling Length Relative to Control (cm)	*p*-Value	Seedling Length Relative to Control (cm)	*p*-Value
surface	−0.8	0.722	2.8	0.232
3 cm	4.7	0.052	−0.6	0.787
8 cm	−3.3	0.155	**8.0**	**0.002**
15 cm	**−5.9**	**0.032**	−0.2	0.931
blended	−2.8	0.220	−2.1	0.362

In contrast to longevity and shoot length, the only significant effect on the number of blades was SAP Rate (Table 2). Both SAP rates had positive impacts on the number of blades per seedling. The control averaged 1.7 blades per plant. The 1500 kg SAP ha^{-1} rate increased the blade count over the control by 0.5, and the 3000 kg ha^{-1} rate increased it by 1.1 more blades per seedling. These are increases of 131 and 166% for each rate, respectively. This response, at least in part, could be due to the increased longevity associated with SAP treatments.

3.2. Reduced Seeding Rate

Soil Moisture and Seedling Growth Parameters

The four-way interaction of Species*Seeding Rate*SAP Rate*Time as well as all but one of the interactions involving SAP were significant for soil moisture (Table 5). The overall pattern was similar for both species, with the greatest difference in gravimetric water content (GWC) occuring over the first ~20 d and then dropping approximately linearly (Table A1, Figure 4). However, the magnitude of the increase in soil moisture for most treatments was slightly greater for bottlebrush squirreltail (Table A1).

Table 5. Statistical results of a glass house study evaluating the effect of species (S), seeding rate (R), SAP rate (SAP), time (T), and their interactions on soil moisture and persistence (percent of plants alive) each day. Bolded numbers indicate $p < 0.05$.

| | | Soil Moisture (Mixed Model) | | | Persistence (Anova) | | |
Effect	DF	DF Den	F Ratio	Prob > F	DF	F Ratio	Prob > F
S	1	48	0.0107	0.722	1	316.8774	**<0.0001**
R	3	48	5.5306	**0.012**	3	1.0705	0.361
SAP	1	48	211.2814	**<0.0001**	1	122.6885	**<0.0001**
T	22	1056	32,415.25	**<0.0001**	23	70.231	**<0.0001**
S*R	3	48	3.2271	**0.022**	3	12.3834	**<0.0001**
S*SAP	1	48	3.1131	0.139	1	60.0828	**<0.0001**
S*T	22	1056	0.4209	0.995	23	3.404	**<0.0001**
R*SAP	3	48	3.1285	**0.031**	3	5.3187	**0.001**
R*T	66	1056	1.4408	**0.005**	69	0.6621	0.985
SAP*T	22	1056	63.6763	**<0.0001**	23	2.7546	**<0.0001**
S*R*SAP	3	48	3.5213	**0.041**	3	8.2626	**<0.0001**
S*R*T	66	1056	2.5588	**<0.0001**	69	0.3022	1
S*SAP*T	22	1056	2.6144	**<0.0001**	23	1.0073	0.452
R*SAP*T	66	1056	2.21	**<0.0001**	69	0.2722	1
S*R*SAP*T	66	66	2.8337	**<0.0001**	69	0.4985	1

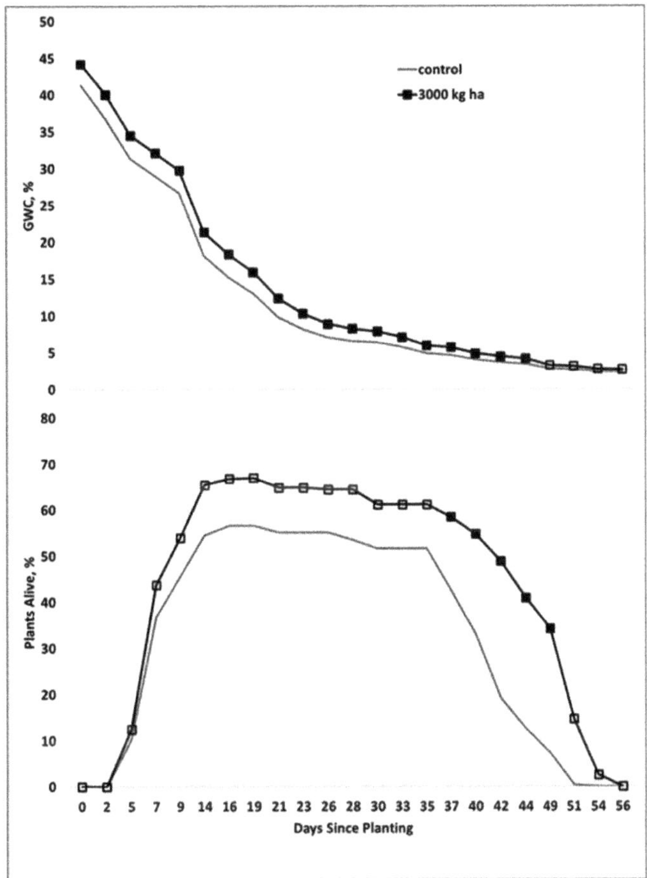

Figure 4. Increase in gravimetric water content (GWC) and viable seedlings as a function of superabsorbent polymer (SAP) application and time in a glasshouse study for bottlebrush squirreltail and Siberian wheatgrass. Data were averaged across all seeding rates of 2, 4, 8, and 16 kg ha^{-1}. The SAP rate of 3000 kg ha^{-1} was placed in a band at a depth of 8 cm directly below the seed. Following SAP treatment, the soil was saturated once and then dried down naturally over the time of the study. For the treated plots, with square markers, the solid (filled in) markers indicate highly significant ($p < 0.0001$) differences compared to the control for that day, and open (not filled in) markers are not significant.

In contrast to soil moisture, the four-way interaction and most of the three-way interactions were not significant for seedling persistence (Table 5). The only three-way interaction that was significant for seedling persistence was Species*Seeding Rate*SAP Rate (Table 5). All Siberian wheatgrass treatments had an increase in the percent of seedlings alive. An increase for bottlebrush squirreltail was only seen at the 8 kg ha^{-1} seeding rate (Figure 5). Interestingly, the 8 kg ha^{-1} seeding rate for bottlebrush squirreltail was the only treatment that did not have an increase in soil moisture, although it trended similarly.

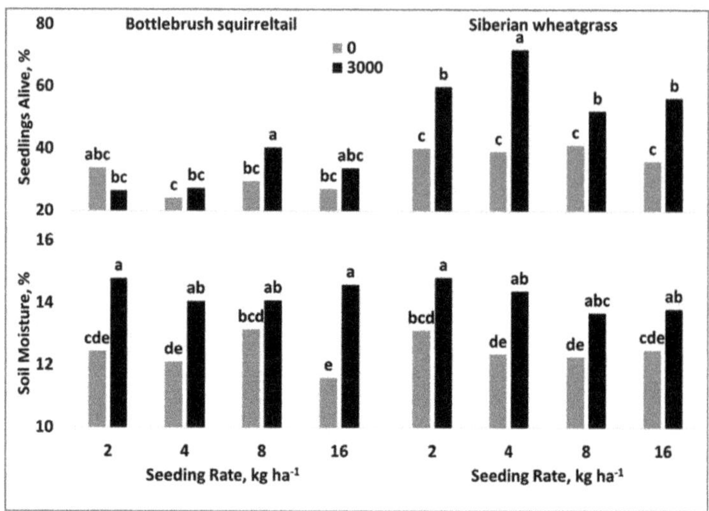

Figure 5. Percentage of live seedlings ($p < 0.0001$) and percent soil moisture ($p = 0.0218$) as a function of seeding rate and superabsorbent polymer (SAP) application averaged across 56 d of a glasshouse study for bottlebrush squirreltail (on the left) and Siberian wheatgrass (on the right). The SAP rate of 3000 kg ha^{-1} was placed in a band at a depth of 8 cm directly below the seed. Following SAP treatment, the soil was saturated once and then dried down naturally over the time of the study. Bars within a species for each variable that share the same letter are not statistically different from each other.

Though the three- and four-way interactions for soil moisture and the one three-way interaction for seedling persistence previously mentioned are statistically significant, the associated F-statistics are relatively low, indicating that they are significant but not practically important relative to other effects. Both soil moisture and persistence have multiple significant two-way interactions (Table 5). However, each has one with a relatively large F-statistic, indicating that this interaction is more likely to have mean differences representative of the population at large than the others or even than the three- or four-way interactions. As such, it is instructive and informative to analyze the effect of both the SAP*Date and the Species*SAP interactions.

For both soil moisture and persistence, the SAP*Time interaction is highly significant (Table 5). However, the F ratio for the impact of that interaction on soil moisture is an order of magnitude larger than all other interactions for that variable (Table 5). There was a difference in soil moisture between SAP treatments until day 44, remaining constant at about 3% greater for the first 19 d, then reducing steadily for the rest of the study (Figure 4). On average, the gravimetric water content (GWC) of 14.3% for SAP treatments throughout the study was higher than the control at 12.4% (Table A1). The GWC in SAP treatments ranged from 44.2% on the day of planting to 2.6% on the day that there were no more living seedlings. In contrast, the control ranged from 41.3% to 2.4% in the same period (Table A1).

The difference in the percent of viable seedlings (persistence) between SAP treatments was not significant but rose steadily for the first 14 d. It then stayed relatively constant at about 10%, until day 37 when it sharply increased (Figure 4). The 3000 kg ha^{-1} SAP band kept up to 30% more plants alive than the control for at least the last 7 d of increased soil moisture (day 37 to 44). By day 49, there was no longer a difference in soil moisture. By day 51, the difference in the relative percentage of viable seedlings disappeared. All seedlings had died by day 56 (Figure 4).

The interaction of Species*SAP is highly significant for persistence, with an F ratio an order of magnitude larger than all but one interaction (Table 5). This interaction was not significant for soil moisture. Treatment with SAP increased persistence of Siberian wheatgrass from 39 to 60% (Table A2, Figure 6). A trend in the same direction was observed for bottlebrush squirreltail, but the magnitude was much less, 29 to 32%, and it was not statistically significant.

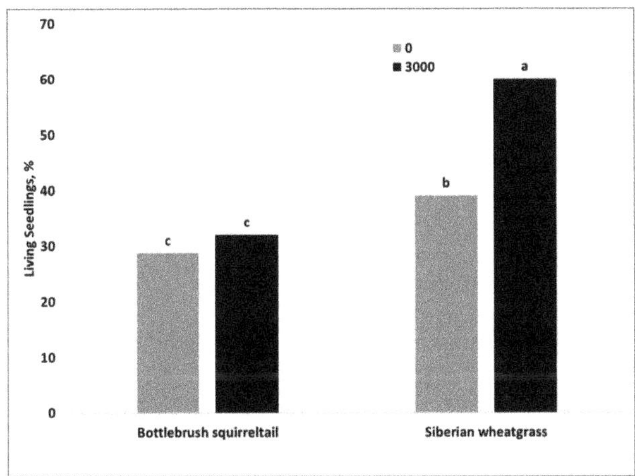

Figure 6. Percentage of live seedlings ($p < 0.0001$) as a function of a superabsorbent polymer (SAP) application averaged across 56 d of a glasshouse study and seeding rates of 2, 4, 8, or 16 kg ha^{-1} for bottlebrush squirreltail and Siberian wheatgrass. The SAP rate of 3000 kg ha^{-1} was placed in a band at a depth of 8 cm directly below the seed. Following SAP treatment, the soil was saturated once and then dried down naturally over the time of the study. Bars that share the same letter are not statistically different from each other.

Seeding Rate and SAP did not impact seedling time to emergence or emergence percentage (Table 6). However, there were differences across species for both (Table 6). Days to emergence were 9 and 8 d for bottlebrush squirreltail and Siberian wheatgrass, respectively. Percentage of seeds emerging were 50 and 80% for bottlebrush squirreltail and Siberian wheatgrass, respectively.

Table 6. Statistical results of a glass house study evaluating the effect of species (S), seeding rate (R), SAP rate (SAP), and their interactions on days to emergence and percent emergence. Bolded numbers indicate $p < 0.05$.

	Days to Emerge (Anova)			% Emerged (Anova)		
Effect	DF	F Ratio	Prob > F	DF	F Ratio	Prob > F
S	1	10.0268	**0.002**	1	24.2163	**<0.0001**
R	3	0.335	0.800	3	0.6647	0.578
SAP	1	0.0462	0.830	1	1.6028	0.212
S*R	3	0.3258	0.807	3	0.7814	0.510
S*SAP	1	1.7208	0.191	1	0.0098	0.922
R*SAP	3	0.2373	0.870	3	0.6838	0.566
S*R*SAP	3	0.4244	0.736	3	0.537	0.659

3.3. Low SAP Rate

3.3.1. Soil Moisture

Soil moisture in SAP treatments was nearly always numerically greater than the control regardless of SAP Rate or Time, although only significant for some rates and dates (Table 7, Figure 7). Regularly testing the soil with a probe affected soil moisture. Comparison on day 99 of the soil moisture of the regularly tested areas of each row with the comparable untested areas of the same row revealed measurements that were nearly double on the undisturbed side ($p < 0.0001$). The regularly tested areas had an average volumetric soil moisture of 0.8% on day 99, in contrast to previously undisturbed soil at 1.5% (data not shown).

Table 7. Statistical results of a glass house study evaluating the effect of species (S), planting location (L), SAP rate (R), time (T), and their interactions on soil moisture, as well as the effect of species and location on seedling longevity. Bolded numbers indicate $p < 0.05$.

Effect	DF	Soil Moisture		Longevity	
		F Ratio	Prob > F	F Ratio	Prob > F
L	2	18.8207	**<0.0001**		
S	1	0.2055	**0.658**	7.1565	**0.008**
R	5	0.9274	0.497	3.9492	**0.002**
T	16	7467.511	**<0.0001**		
L*S	2	7.9641	**0.0004**		
L*R	10	4.7621	**<0.0001**		
L*T	32	3.7136	**<0.0001**		
S*R	5	1.141	0.391	0.3532	0.880
R*T	80	4.7752	**<0.0001**		

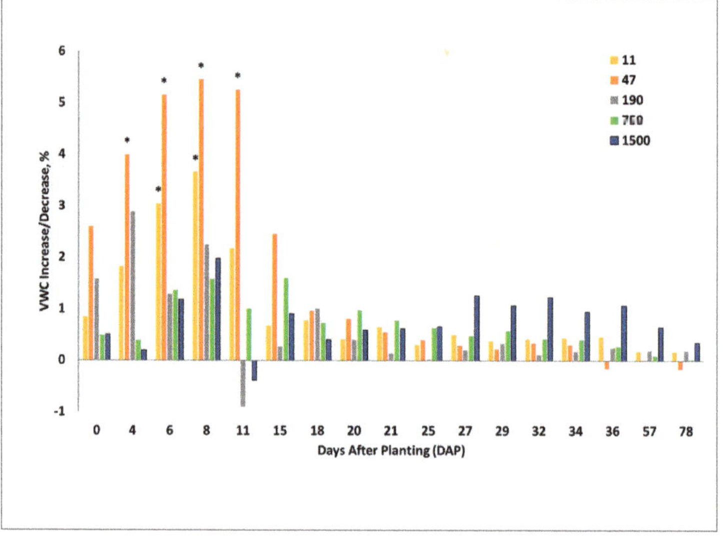

Figure 7. Volumetric Water Content (VWC) different from the control as a function of a superabsorbent polymer (SAP) rate and time relative to an untreated control in a glasshouse study with bottlebrush squirreltail and Siberian wheatgrass seeded at NRCS recommended rates. Data are averaged across both species. The SAP rates of 11, 47, 190, 750, or 1500 kg ha^{-1} were placed in a band at a depth of 8 cm perpendicular to the seed rows. Following the SAP treatment, the soil was saturated once and then dried down naturally over the time of the study. Bars with "*" indicate statistical significance relative to the control within that day ($p = 0.05$).

3.3.2. Seedling Growth Parameters

There were no differences in seedling emergence ($p = 0.539$). The effects of Species and SAP rate on seedling longevity were highly significant ($p = 0.008$ and $p = 0.002$, respectively), while their interaction was not ($p = 0.880$) (Table 7). The 1500 kg SAP ha^{-1} rate had greater seedling longevity than the next two lower rates and trended towards having greater longevity than the control. However, seedling longevity was not statistically different than the control for any treatment (data not shown).

3.4. SAP Depth and Root Growth

3.4.1. Soil Moisture

The interaction of SAP Placement Depth*Measurement Depth*Time on soil moisture was highly significant (Table 8, Table A3), but only when the soil was saturated. On 0 DAP at 10 cm measurement depth, the treatments with SAP bands at 3 and 8 cm depths held less soil moisture than the control (p <0.001). This difference disappeared over time. (Appendix Table A3). The two-way interactions of SAP Placement Depth*Time and Moisture Measurement Depth*Time were also highly significant; however, there were no differences between SAP treatments and the control. (Table 8, Table A3).

Table 8. Statistical results of a glass house study evaluating the effect of species (S), SAP placement depth (P), moisture measurement depth (M), time (T), and their interactions on soil moisture. Bolded numbers indicate $p < 0.05$.

Effect	DF	Soil Moisture DF Den	F Ratio	Prob > F
S	1	27	0.0082	0.929
P	3	27	0.9046	0.452
M	3	84	347.4975	**<0.0001**
T	7	189	1741.724	**<0.0001**
S*T	7	189	0.6996	0.672
P*M	9	84	0.7963	0.621
P*T	21	189	2.5507	**0.0004**
M*T	21	588	26.6123	**<0.0001**
P*M*T	63	588	1.864	**0.0001**

3.4.2. Seedling Growth Parameters

The interaction of SAP Placement Depth*Species on the total number of seedlings emerged was significant (Table 9). The same was true when evaluating the effect of the presence or absence of SAP on the total number emerged (Table 9). However, the difference was only between the controls (no SAP) of both species, where two bottlebrush squirreltail seedlings emerged, compared to one Siberian wheatgrass seedling. Neither species had SAP treatments that were different than their respective species control. Seedlings in most pots began emerging by 20 DAP. The exception was one bottlebrush squirreltail control, which did not have a seedling emerge until 34 DAP. Excluding that pot from the analysis, the Species*SAP Placement Depth interaction was highly significant for days to emerge (Table 9). Seedlings emerged in the bottlebrush squirreltail treatment with SAP placed at a depth of 3 cm 2.3 d faster than the control. Bottlebrush squirreltail with SAP placed at a depth of 8 cm trended in the same direction, but the difference was not significant. No other bottlebrush squirreltail and none of the Siberian wheatgrass treatments were different from their respective controls.

Table 9. Statistical results of a glass house study evaluating the effect of species (S), superabsorbent polymer (SAP) placement depth (D), and their interaction on seedling emergence, shoot and root biomass, shoot:root ratio, time for roots to reach the bottom of the canister (Time), and root length at 3 weeks. Mean comparisons were also made orthogonally evaluating species (S), SAP presence or absence (P), and the interaction of the two. Bolded numbers indicate $p < 0.05$.

Metric	DF	F Ratio	Prob > F
Total emergence	7	4.6899	**<0.0001**
S	1	4.9145	**0.028**

Table 9. *Cont.*

Metric	DF	F Ratio	Prob > F
D	3	5.4387	**0.001**
S*D	3	3.8661	**0.010**
Total emergence orthogonal	3	3.9670	**0.009**
S	1	10.2095	**0.002**
P	1	0.0604	0.806
S*P	1	7.3098	**0.008**
Days to emerge	7	6.4688	**<0.0001**
S	1	27.2552	**<0.0001**
D	3	3.4673	**0.018**
S*D	3	3.4673	**0.018**
Days to emerge orthogonal	3	8.9142	**<0.0001**
S	1	22.2726	**<0.0001**
P	1	1.9267	0.167
S*P	1	1.9267	0.167
Shoot biomass	7	1.3774	0.266
Shoot biomass orthogonal	3	1.8039	0.172
Root biomass	7	3.0537	**0.022**
S	1	7.5978	**0.012**
D	3	2.3923	0.097
S*D	3	1.8964	0.161
Root biomass orthogonal	3	2.5516	0.078
shoot:root ratio	7	0.3117	0.941
Shoot:Root ratio orthogonal	3	0.119	0.948
Time	7	1.7013	0.166
Root length 21 DAP	7	0.8454	0.563
Root length 21 DAP orthogonal	3	1.4936	0.241

A visual assessment suggests that root growth was not negatively impacted by SAP presence, as roots traveled into and through the SAP band with no signs of diversion in any of the pots (Figure 8). There were no impacts of SAP on shoot or root biomass, shoot:root ratio, time for roots to reach the bottom of the canister, or length of roots 21 DAP for either species (Table 9).

Figure 8. Seedling roots 5–12 cm below soil surface 51 d after planting: (**a**) bottlebrush squirreltail control, (**b**) bottlebrush squirreltail with 3000 kg ha^{-1} SAP band placed at 8 cm depth, (**c**) Siberian wheatgrass control, (**d**) Siberian wheatgrass with 3000 kg ha^{-1} SAP band placed at 8 cm depth.

3.5. Fertilizer and SAP Interaction

3.5.1. Soil Moisture

Three of the four three-way interactions for soil moisture were significant ($p < 0.0001$): Species*Fertilizer Depth*SAP, Species*SAP*Time, and Fertilizer Depth*SAP*Time (Table 10). The SAP treatments with no fertilizer or fertilizer placed on the soil surface both held a similar amount of gravimetric water throughout the study (Figure 9A). Both of these treatments held more soil moisture than their control (the same fertilizer placement without SAP), for the first 35 d for fertilizer placed on the soil surface and 40 d for no fertilizer. The treatment of fertilizer placed directly into the SAP band 8 cm below the surface held relatively more water than treatments with no fertilizer or fertilizer placed on the soil surface throughout the study. That treatment also held more water than its control (fertilizer placed at 8 cm depth with no SAP) for 49 d. The difference disappeared towards the end of the study. The GWC was significantly higher for the fertilizer placed directly into the SAP band at 8 cm depth than SAP treatment with no fertilizer 21, 24, and 26 DAP (Figure 9A).

Table 10. Statistical results of a glass house study evaluating the effect of Species (S), Fertilizer placement depth (F), SAP presence (SAP), Time (T), and their interactions on soil moisture each day. Bolded numbers indicate $p < 0.05$.

Effect	Nparm	DF Den	F Ratio	Prob > F
S	1	59.79	0.0309	0.861
F	2	59.79	0.1003	0.905
SAP	1	59.79	84.938	**<0.0001**
T	29	1715	12,268.02	**<0.0001**
S*F	2	59.79	501,227	**0.009**
S*SAP	1	59.79	1.2507	0.268
S*T	29	1715	2.325	**<0.0001**
F*SAP	2	59.79	2.2206	0.117
F*T	58	1715	4.0967	**<0.0001**
SAP*T	29	1715	78.2095	**<0.0001**
S*F*SAP	2	59.79	0.9599	0.389
S*F*T	58	1715	3.5676	**<0.0001**
S*SAP*T	29	1715	4.0605	**<0.0001**
F*SAP*T	58	1715	2.2783	**<0.0001**
S*F*SAP*T	58	1715	1.2506	0.100

The two-way interaction of SAP*Time was also highly significant and had an F-value an order of magnitude larger than any of the three-way or other two-way interactions (Table 10). This indicates that that interaction held the bulk of the influence over the changes in soil moisture compared to the influence of species in the interaction. Both species behaved very similarly in that SAP treatments held more soil moisture for the first ~49 d of the study (Figure 9B).

3.5.2. Seedling Growth Parameters

The number of days to emergence was not impacted by fertilizer or SAP application, species, nor any of their interactions ($p = 0.460$). However, the number of seedlings emerged was significantly impacted by the Species*SAP interaction and Fertilizer Placement Depth ($p = 0.011$ and $p = 0.013$, respectively). Bottlebrush squirreltail emergence was not impacted by SAP, but Siberian wheatgrass increased 1.6 times that of the control (Figure 10). Treatments with fertilizer did not increase the number of emerged seedlings, and deeper fertilizer placement had a detrimental effect (Figure 11). The remaining two-way interactions, Species*Fertilizer Depth and SAP*Fertilizer Depth, and the three-way interaction did not have an impact ($p = 0.720$, $p = 0.145$, and $p = 0.813$, respectively).

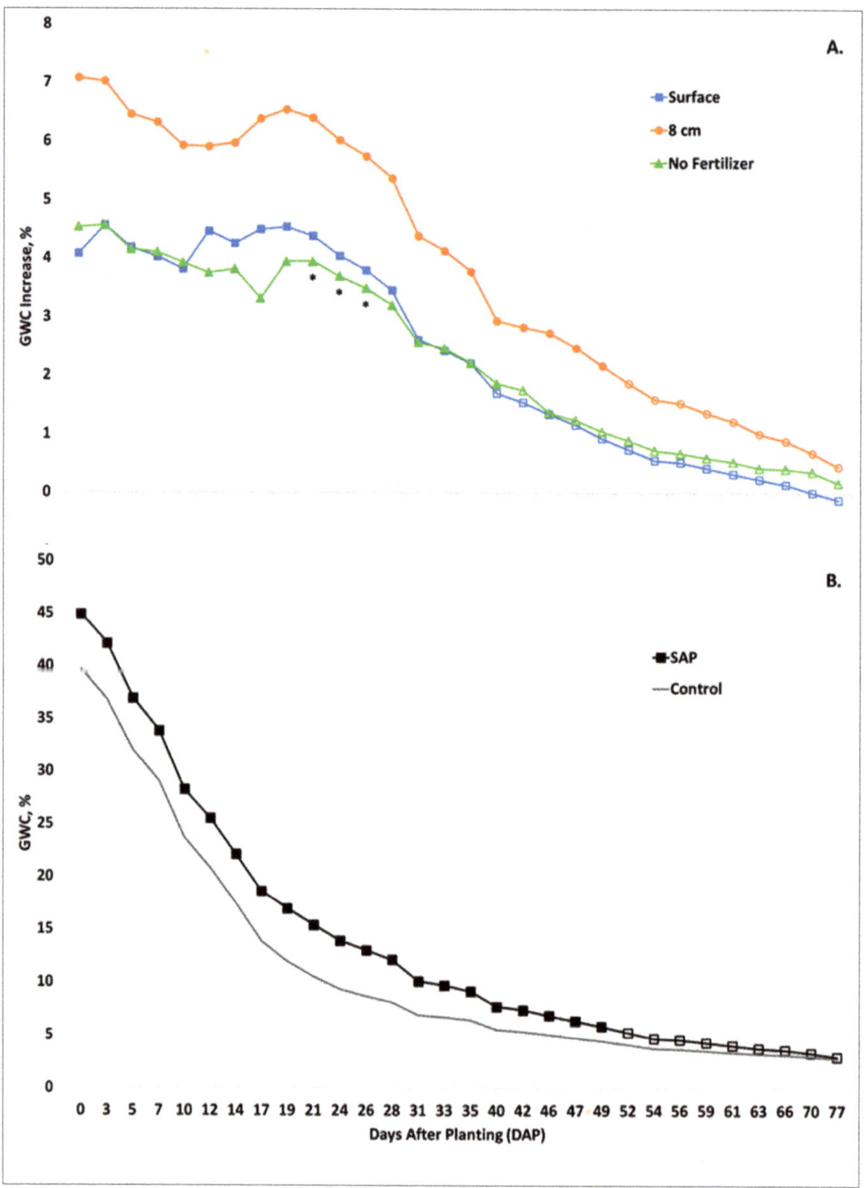

Figure 9. Gravimetric water content (GWC) for a superabsorbent polymer (SAP) fertilizer glasshouse study with bottlebrush squirreltail and Siberian wheatgrass (data averaged across species). (**A**) Increase in GWC with SAP (compared to the control without SAP) for fertilizer applied at the: Surface, 8 cm below the surface, or No Fertilizer. Filled markers indicate significance compared to the control (no SAP) for each fertilizer treatment ($p < 0.0001$). The symbol "*" indicates days where No Fertilizer was statistically less than with fertilizer applied at 8 cm depth ($p = 0.05$). (**B**) The GWC with or without SAP. Data are averaged across fertilizer treatments. Filled markers on the SAP line indicate significance compared to the control ($p < 0.0001$).

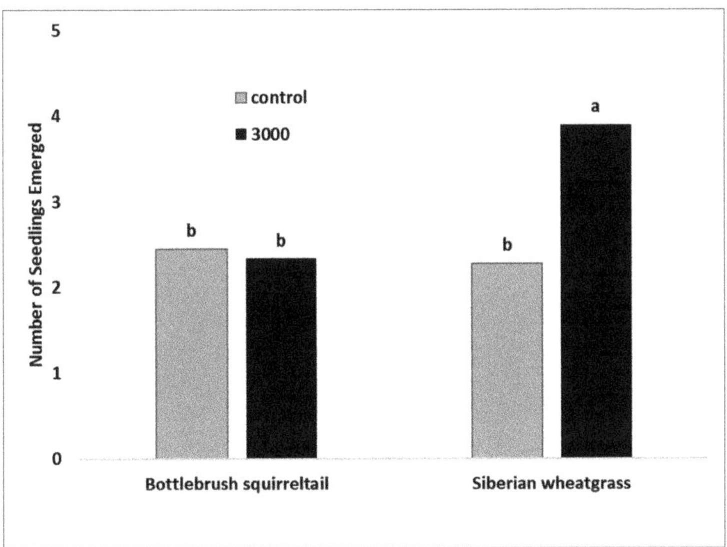

Figure 10. Number of seedlings emerged as a function of superabsorbent polymer (SAP) application averaged across applied fertilizer depths of 0 and 8 cm and 77 d of a glasshouse study with bottlebrush squirreltail and Siberian wheatgrass. The SAP rate of 3000 kg ha^{-1} was placed in a band at a depth of 8 cm directly below the seed. Following SAP treatment, the soil was saturated once and then dried down naturally over the time of the study. Bars with the same letters are not statistically different from each other. $p = 0.05$.

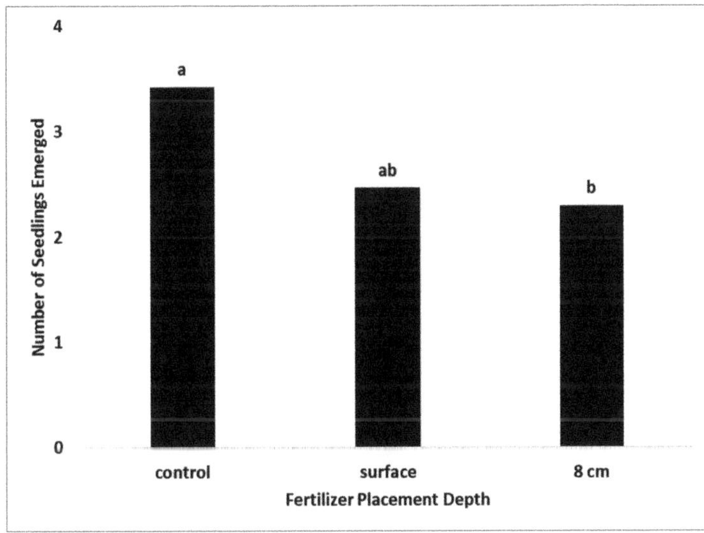

Figure 11. Number of seedlings emerged as a function of fertilizer placement depth compared to an untreated control in a glasshouse study with bottlebrush squirreltail and Siberian wheatgrass. Data were averaged over SAP presence at rates of 0 or 3000 kg ha^{-1} and across species. The soil was saturated once and then dried down naturally over the time of the study. The fertilizer was placed at a depth of 8 cm directly below the seed prior to saturation or at the soil surface at the time of planting. Bars with the same letters are not statistically different from each other. $p = 0.05$.

Seedling longevity was impacted by Species, SAP Application, and Fertilizer Placement Depth ($p = 0.008$, $p = 0.0002$, and $p = 0.058$, respectively). No interactions were significant. Seedlings in treatments with SAP at a rate of 3000 kg ha^{-1} lived 38 d, which was 11 d longer than the control (Figure 12). The fertilizer used in this study did not favorably impact seedling longevity and had a negative impact when placed at the surface (Figure 13).

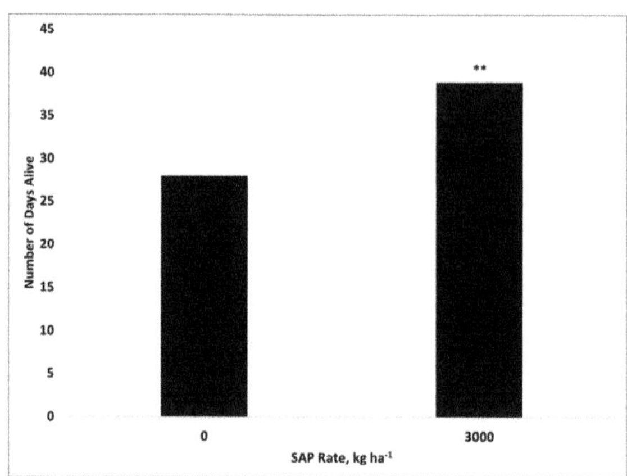

Figure 12. Seedling longevity, number of days alive, as a function of superabsorbent polymer (SAP) application averaged across 77 d of a glasshouse study with bottlebrush squirreltail and Siberian wheatgrass. Data were averaged across species and fertilizer placement depth (0 or 8 cm). The SAP rate of 3000 kg ha^{-1} was placed in a band at a depth of 8 cm directly below the seed. Following SAP treatment, the soil was saturated once and then dried down naturally over the time of the study. Bars marked with "**" are highly significant at $p = 0.0002$.

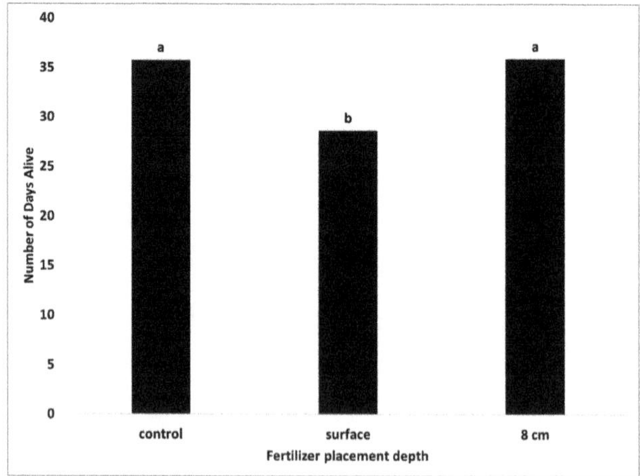

Figure 13. Seedling longevity, number of days alive, as a function of fertilizer placement depth in a glasshouse study with bottlebrush squirreltail and Siberian wheatgrass. Data were averaged across species and SAP rates of 0 and 3000 kg ha^{-1} applied in a band at a depth of 8 cm directly below the seed. Following SAP treatment, the soil was saturated once and then dried down naturally over the time of the study. The fertilizer was applied at a depth of 8 cm directly below the seed prior to saturation or at the soil surface at the time of planting. Bars with different letters are statistically different from each other $p = 0.0584$.

3.5.3. Soil Rewet

SAP application increased soil moisture content when rehydrated but with reduced effects after the initial saturation and dry down. An orthogonal comparison of gravimetric water content on day 0 and day 40 after initial saturation compared to the same days after rehydration as a result of SAP application was significant ($p < 0.0001$) (Figure 14). Treatments containing SAP held 11.2% more soil moisture on day 0 after their initial saturation in the fertilizer study compared to day 0 of their rehydration. At day 40 of the fertilizer study, treatments with SAP after the initial saturation held 4.3% more moisture compared to SAP treatments on day 40 after rehydration. Treatments with SAP held more moisture than the control at both day 0 and day 40 after initial saturation in the fertilizer study ($p < 0.0001$). That was also true for day 0 after rehydration. However, by day 40 after rehydration, there was no difference in soil moisture content between SAP treatment and the control (Figure 14).

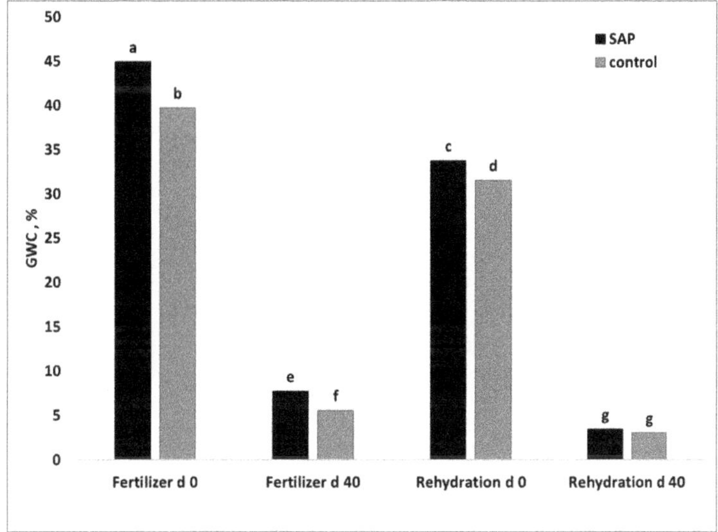

Figure 14. Comparison of gravimetric water content (GWC) as a function of superabsorbent polymer (SAP) application between original saturation for the fertilizer study and rehydration after dry-down period of 77 d. Data were averaged across species and fertilizer placement depth (0 or 8 cm). The SAP rate of 3000 kg ha^{-1} was placed in a band at a depth of 8 cm directly below the seed. Following SAP treatment, the soil was saturated once and then dried down naturally over the time of the fertilizer study then rehydrated. Bars with different letters are statistically different from each other. $p < 0.0001$.

Six bottlebrush squirreltail and eleven Siberian wheatgrass seedlings emerged after the soil was rehydrated. They were evenly spread across SAP and fertilizer treatments. As the number emerged represented only 4% (17/432) of the total number of seeds planted and only 7% (17/247) of the seeds that did not emerge after the initial saturation, no further analysis was done.

4. Discussion

4.1. Soil Moisture

In our perennial grass studies, soil moisture in a loam soil was increased significantly with Stockosorb 660 micro bands applied at a depth of 8 cm below the soil surface at rates of 1500 and 3000 kg ha^{-1} (0.1 and 0.2%). Soil moisture was increased over the control for up to seven weeks, with the largest differences occurring at the time of saturation and diminishing over time. The largest impact of soil moisture on seedling longevity occurred during approximately the last 12 days, when there was a significant soil moisture difference, despite the difference in soil moisture being near its lowest point.

Higher levels of soil moisture observed in this study as a function of the band-applied SAP are similar to other container studies with a variety of soil types, plant species, and SAP particle sizes, sources, and rates when mixed into the soil [35,40,42,52–56]. In general, such studies found SAP to increase soil moisture in sandy-textured soils at application rates between 0.04 to 1.0% (approximately 600–15,000 kg ha^{-1}). The effect decreased in finer-textured soils. In studies with both sandy and clayey soils, the sands had increased soil moisture, but the clays did not [35,52,54,55].

Our results were similar to El-Asmar et al. [35]. In a soil moisture characteristic study with clay (C) or sandy clay loam (SCL) soils, they found numerically increased water content with increasing SAP application rates of 0.1, 0.2, 0.3, and 0.4% (approximately 1700–7100 kg SAP ha^{-1}) when Stockosorb 660 was blended into the full container of soil. However, water holding capacity was only increased significantly at the 0.4% rate at matric potentials of 30 and 200 kPa and the 0.3% rate at a matric potential of 100 kPa in SCL. In a further study growing maize (*Zea mays* L.) in the same soils with the same SAP rates, they reported increases in evapotranspiration (ET), which is indicative of increased soil moisture [57]. Banded SAP at rates of 0.2, 0.3, and 0.4% (approximately 3400–7100 kg ha^{-1}) resulted in increased ET in SCL. Mixing the SAP with the soil resulted in increased ET only for the 0.3 and 0.4% SAP rates in SCL. There were no differences in water content or ET in their C soil. In our studies, when the SAP was mixed with the soil, we saw an increase in soil moisture at SAP rates of 1500 and 3000 kg SAP ha^{-1} (approximately the 0.1 and 0.2% rates used by El-Asmar et al. [35]). We did not test higher rates in our studies. This difference could be explained by the soil texture differences. Our loam soil had only 21% clay compared to 45% clay in the El-Asmar C soil and 37% clay in their SCL, giving us a much larger percentage of the soil particles in the larger sand and silt proportions. As discussed above, SAP effects on soil moisture decrease in finer-textured soils.

Additionally, our studies were conducted over a much wider range of matric potentials. El-Asmar et al. [35] measured water content at matric potentials no lower than 200 kPa with the lowest gravimetric water content (GWC) of 23%. A matric potential of 200 kPa is well within the range of plant available water, which is typically assumed to be between 30 kPa (field capacity) and 1500 kPa (permanent wilting point). We did not monitor matric potential in our studies. However, the soil in our studies was saturated and then allowed to dry down for 76–107 d—until all seedlings had wilted and died and GWC was measured as low as 2.2%. It may be that the SAP holds more water than the surrounding soil as the soil matric potential decreases, relatively increasing water content as the rest of the soil dries.

A third factor that may have increased soil moisture in our studies in treatments with the SAP mixed with the soil is the fact that the SAP was only blended into the top 8 or 15 cm of soil of 25 or 23 cm deep containers, respectively, not the full soil volume. When El-Asmar et al. [35] included treatments of SAP mixed into the soil, the SAP was blended into the full container of soil. They [35] suggest that banded SAP would allow water to diffuse into the soil below when it was saturated, but then act as a barrier to soil capillary rise as the soil dried. Concentrating the SAP into a portion of the container, even if not banded, could have a similar, though reduced, impact—slowing water movement from the soil below and increasing the overall moisture holding capacity.

Application of SAP increases plant available water not only by absorbing and holding water within the polymer, but it has also been shown to decrease bulk density and increase soil porosity [28,34,35,53,58,59]. Soil pores vary in size and shape and are categorized as macropores (>0.08 mm) and micropores (<0.08 mm). Macropores readily allow movement of air, water, and plant roots through the soil profile, while movement through micropores is more limited. Thus, an increase in soil macropore size can be beneficial. If, however, they are large enough to create large subsurface voids, moisture may be more easily lost due to drainage or evaporation [23]. This may result in a reduction in seedling longevity. Sarvaš et al. [60] found reduced transplant seedling survival with Stockosorb micro with Scotch pine (also known as "Scot's pine"; *Pinus sylvestris*). This was attributed to the swelling of the SAP that "pushed up" the seedlings, resulting in their demise, likely due to loss of soil moisture from large subsurface voids. No information is given on the size of the planting hole, but we speculate that their application rate (7 g tree^{-1}) was higher than ours. Depending on the transplant method, the planting holes could have ranged between approximately 8 and 25 cm in diameter. The applied rate of 7 g SAP mixed into each planting hole would be equivalent to approximately 1400 and 14,000 kg ha^{-1}.

Similarly, we frequently observed raised or cracked soil in our studies with SAP bands at $\geq$750 kg ha^{-1}. In general, the SAP broke through the soil surface, creating cracks, as it became saturated and swelled. Soil cracking did not occur with the deepest (15 cm) SAP placement. However, the entire soil surface in those pots was raised ~1 cm compared to the controls. Bakass et al. [61] found that SAPs dry out ahead of the surrounding soil. Since it is the absorption of water that causes the SAP to swell, it shrinks as it dries. It is plausible that as the banded SAP swelled, it also spread to form a layer. This could have resulted in a void below the surface as it dried, and the displaced soil retained its shape.

We found that the soil in the pots was divided into separate layers at the level of the dried SAP after complete dry-down. This was similar to El-Asmar et al. [35] who observed the formation of large air pores as individual SAP granules dried when mixed in the soil, as well as the formation of distinct soil layers after drying when SAP was applied as a band. The formation of large cracks and subsurface voids was likely exacerbated by both the placement in a concentrated band and the relatively high SAP rates in our studies. These cracks exposed the SAP and deeper layers of soil to air, light, and heat and could potentially lead to faster drying of the soil. However, this did not result in reduced soil moisture in these studies. As mentioned above, El-Asmar et al. [35] suggest that banded SAP, wet or dry, prevents evaporative loss of soil moisture from beneath it, even when exposed to the elements, thus maintaining higher soil moisture levels.

Reducing the SAP rates to reduce the threat of soil cracking always resulted in a reduced impact on soil moisture level and longevity, if any impact occurred at all. Similarly, Bandak et al. [62] found that the application of Super AB A-200 (Rahab Resin Co., Tehran, Iran or Iramont, Inc. Laval, Quebec, Canada) at rates of 1000 and 2000 kg ha^{-1} increased soil moisture. However, no effect was measured at the low rate of 500 kg ha^{-1}. Hüttermann et al. [42] suggest that there may be a minimum SAP concentration required to see an increase in soil moisture. Curiously, in our reduced rate study, the two lowest rates, 11 and 47 kg ha^{-1}, increased soil moisture for up to 11 d, but the 180 and 750 kg ha^{-1} rates did not, possibly due to drying due to greater soil swelling. The 1500 kg ha^{-1} rate also did not increase moisture in that study but did so in other studies. Increased soil moisture at the lowest rates could possibly be a result of the formation of relatively smaller air pockets created within the soil that still allowed for increased percolation and reduced evaporative loss without disturbing the soil surface.

The one exception to increased soil moisture with high-rate SAP bands was the SAP and Root Depth Study that showed no significant difference in soil moisture. Interestingly, unlike the other studies, there was no soil cracking, either. That study was conducted in round canisters with SAP placed in a band around the circumference of the canister, rather than square pots with the SAP band running across the middle. This may have been equivalent to reducing the amount of SAP at any one spot, thus reducing the soil cracking as the band swelled, but also reducing the positive impact on increased soil moisture.

Several other studies have also reported reduced SAP absorptive capacity when resaturated after complete dry-down of the soil [38–41,55,63–67]. Holliman et al. [39] reported that the largest decrease was observed in the first 18 months of use. Bai et al. [38] reported significant reductions of 73–99% in SAP water absorbency of four different SAPs after five rewet cycles over 5 months. Banedjschafie and Durner [40] found that water retention of Super AB A-200 at application rates of 0.3, 0.6, and 1.0% (approximately 3400–15,000 kg ha^{-1}) thoroughly mixed into sandy soil reduced significantly after the first wetting/drying cycle but was significantly increased over soil alone. They also found plant available water decreased by about 50% compared to the initial saturation after six months of repeated wetting and drying. Zhang et al. [37] found that the superabsorbent resin AG_{101} (Formosa Plastic Corporation, Kaohsiung City, Taiwan) absorbed and reabsorbed water 30–50 times before degrading. The reduction in absorbency has been attributed to degradation due to weathering, microbial action, and exposure to salts [27,34,38,40,43,53,55,56,63–66].

The gradual replacement of the original structural SAP cations (usually sodium (Na^+) or K^+) with calcium (Ca^{2+}) and/or magnesium (Mg^{2+}) is one possible source of the reduced absorption efficiency of SAPs [38,40,55,56,67]. Banedjschafie and Durner [40] attributed the reduction in SAP absorption to specific salts, especially Ca^{2+}, in the soil solution, while water-soluble phosphorus (P) and K had only a moderate effect on the reduction in water uptake. Yu et al. [56] attribute the reduction to the exchange in adsorbed cations within the SAP structure itself. They report that the strongest reduction is associated with bivalent cations, especially Ca^{2+} and Mg^{2+}, compared to monovalent cations, especially Na^+ and K^+. We saw indications of this effect in our studies (Figure 9). The UTTR soil used in this study was especially high in Ca^{2+} and Mg^{2+} (Table 2), but GWC was increased with the use of a fertilizer high in K^+ applied to the band. If the application of monovalent cations in or near a subsurface SAP band would improve its absorptive capacity, lower SAP rates could provide desired establishment or soil moisture results while reducing soil disturbance.

4.2. Seedling Growth Parameters

The use of banded SAP in these studies resulted in the same or better seedling longevity and/or persistence without any apparent negative impact on rooting. Significant longevity increases were measured at times, but only consistently at the highest rate of 3000 kg ha^{-1} and only at placement depths of 8 and 15 cm, where the highest soil moisture was also measured. Despite soil cracking and subsurface voids, there was no negative impact on longevity at either the 1500 or 3000 kg ha^{-1} rate at any placement depth. When it was measured, we found no effect on seedling emergence, longevity, shoot or root biomass, shoot:root ratio, or root growth.

There has been very little research done on the effects of placing SAP in a concentrated band in the soil on plant growth, but it may have a positive impact. El Asmar et al. [35] found similar longevity results which they attributed to the creation of soil water reservoirs in banded SAP. They applied SAP in clay (C) and sandy clay loam (SCL) soils at rates of 0.1, 0.2, 0.3, and 0.4% (approximately 1700–7100 kg SAP ha^{-1}), either mixed with the soil or placed in bands 25 cm below the soil surface to pine (*Pinus pinea*) seedlings. They found an increase in seedling survival time with SAP rates of 0.2 and 0.4% when placed in bands. There was no difference relative to the control when the SAP was mixed in the soil. They also found increased corn shoot fresh and dry weights when the same SAP rates were banded and placed at a depth of 15 cm, but not when the SAP was mixed into the soil. We found no clear influence of SAP band application on plant height but did see an increase in the number of blades per plant. The extreme drought conditions of our studies prevented the seedlings from developing into mature plants. However, the increased blade count indicates the possibility that an increase in above-ground biomass may have resulted if they had persisted.

Other works have demonstrated that SAPs, in general, do have a positive impact on plant growth [25,42,43,68,69]. Lucero et al. [43] used both starch- and acrylic-based SAPs and found a significant effect on leaf biomass and area in black grama grass (*Bouteloua eriopoda*), another long-lived, warm-season rangeland grass species. Additionally, Hütter-mann et al. [42] reported pronounced growth of Aleppo pine (*Pinus halepensis*) seedling shoots and roots under drought conditions with Stockosorb K 400 mixed into the soil at a concentration of 0.4% (*w/w* approximately 7000 kg ha^{-1}). Rezashateri et al. [25] found similar results in a containerized study of wormwood (*Artemisia sieberi*), a variety of sagebrush, with three different SAPs, including one from Stockosorb, at rates of 5 and 10 g kg^{-1} soil (or approximately 1800–3700 kg SAP ha^{-1}). Yang et al. [69] found that SAP rates between

30–45 kg ha^{-1} increased biomass, grain number, and yield in corn grown in low-rainfall conditions. Coello et al. [68] found that Aleppo pine growth generally increased with increased SAP application but noted a saturating effect.

Reducing the banded SAP rates below 1500 kg ha^{-1} in these studies did not impact seedling longevity compared to the control. This is in contrast to Johnston and Garbowski [70], who documented benefits to perennial grass establishment in field studies with in-season irrigation and application of two SAPs (Luquasorb 1280 RM and Tramfloc 1001) at rates of 310 and 450 kg ha^{-1} blended with the seed. This contrast could be due to different SAPs, higher soil moisture, field conditions, and direct seed placement in their study. The swelling of the dispersed SAP and any resulting soil cracking or subsurface void formation would also have been diffused across the planting area.

Root growth affects seedling establishment [25,26,71]. This may be especially true for seeded native species competing with invasive annuals in arid conditions [26]. We saw no SAP influence on any seedling root parameters for the first 7 weeks (approximately 20 cm) of growth of bottlebrush squirreltail and Siberian wheatgrass in the SAP Depth and Root Growth study. In contrast, Garbowski [72] saw higher root mass fraction in a field study using the same SAP as in our studies but at rates of 250 kg ha^{-1} and blended to a depth of 10 cm. The increase in root mass was dependent on other treatments such as ambient precipitation and increased cheatgrass presence. Rezashateri et al. [25] found increases in sagebrush root dry weight and root/shoot ratio with all tested SAPs and irrigation levels in sandy loam soil. Some treatments, including Stockosorb at 10 g kg^{-1} (4000 kg ha^{-1}) at 75% average irrigation rate, resulted in increased root length, perimeter, area, and volume. Bandak et al. [62] saw an increase in root length and weight of rain-fed wheat with the application of the superabsorbent polymer A200 SAP. Zhou et al. [66] found shorter roots with more surface area in an irrigated summer maize (*Zea mays*) using organic–inorganic composite superabsorbent polymers, which are SAPs incorporating inorganic materials such as clays. Hüttermann et al. [42] found that the growth of Aleppo pine root tips stopped during water stress, but in the presence of SAP, the adventitious and side roots were able to continue to grow.

Our contrasting results may be a function of the length and the severe drought conditions of our studies. The impact of SAPs on seedling survival is diminished over multiple growing seasons [4,36,72]. This could be a function of the degradation of the product [39] or the plant roots growing beyond the placement location. Our trials were too short to observe SAP degradation, but we did not find any difference in root growth to and through SAP bands compared to the control in our study. However, the use of SAP has been shown to increase soil moisture in this and other studies, which may positively impact seedling survival in the short term [35,42,43,68,70].

Some differences were observed between species in these studies, with Siberian wheatgrass responding more favorably to SAP than bottlebrush squirreltail. The number of days to emergence was not affected by SAP treatment in any study where it was measured. In most studies, the number of seedlings emerged was also not impacted. However, the number of Siberian wheatgrass seedlings that emerged was increased 1.6 times with SAP treatment in the Fertilizer study. Siberian wheatgrass persistence increased with SAP application in the Reduced Seeding Rate study. Bottlebrush squirreltail trended in the same direction, but this trend was not significant. This is similar to Minnick and Alward [4], who found that Aquasorb (Ark Enterprises, Warsaw, MO, USA), a cross-linked Na-polyacrylate SAP, increased survival of rubber rabbitbrush (*Ericameria nauseosa*) but not big sagebrush (*Artemisia tridentata*) or four wing saltbush (*Atriplex canescens*). The effects may be varied for different types of grasses as well. More research is needed to determine species that may benefit from banded SAP use more than others.

Though SAPs are reported to hold fertilizers in place to increase plant access and growth [27,73], we found the use of a low-N fertilizer, with or without SAP, had no positive impact on seedling emergence or longevity. In fact, application of the fertilizer at a depth of 8 cm negatively affected the number of seedlings that emerged, but not their longevity.

This was despite an increase in soil moisture when SAP was also present at that depth. Applying the fertilizer at the soil surface negatively impacted seedling longevity but did not impact soil moisture. The use of fertilizers has been shown to negatively impact seedling establishment in restoration efforts in arid regions [74,75]. It is possible that rangeland species that have evolved in harsh conditions without additional nutrition inputs could be negatively impacted if SAPs held not only water, but also increased nutrient concentrations near their roots [40,56].

Drought severity may impact the effectiveness of SAP. In a containerized study using Aleppo pine, Del Campo et al. [76] found that three SAPs (Aguaspon, Stockosorb, and Terracottem) mixed into the soil at rates of 0.01 and 0.1% (180–1800 kg ha^{-1}) were effective in moderate drought conditions, with effects diminished at suction tensions higher than 30 kPa. This would indicate that the effects of SAPs diminish beginning at about field capacity, much wetter than the dry soils in these studies or those found in the field during the summer. Garbowski et al. [72] reported no establishment effects on first-year field-grown rangeland seedlings under drought conditions, in contrast to positive effects in the deficit-irrigated agriculture condition. They suggest there may be a soil moisture threshold for positive effects on soil moisture content. Their SAP application rates were 1/6–1/12 those used in these studies, and higher rates could allow for higher moisture holding capacity. Rezashateri et al. [25] found that higher rates of SAP, 10 g kg^{-1} compared to 5 g kg^{-1} (or approximately 1800–3700 kg SAP ha^{-1}), had a greater impact on the growth of wormwood at 75% of the normal precipitation rate than 100 or 150% applied every 30 d to simulate natural precipitation patterns. In contrast, most of our studies were conducted under what should be considered severe drought conditions. It is entirely possible, though unlikely, that no further moisture inputs would occur during a growing season in the Great Basin. If the SAP was present and could increase soil moisture enough to allow seedlings to survive to the next precipitation event or allow the seedling roots to grow deep enough to find moisture lower in the soil profile, it could greatly impact seedling establishment rates. The benefits would need to be balanced against the need for higher SAP rates to capture and retain enough soil moisture as well as possible soil cracking and seedling death due soil displacement by swelling higher SAP rates.

Severe drought conditions may also explain why we did not see an impact on seedling longevity or other growth parameters when the SAP was mixed in the soil. The effects of mixing SAP into the soil were measured in only two of our studies. In the SAP Rate and Depth study, we observed a significant increase in soil moisture over the duration of the study but no increase in seedling longevity. This could be due to the trial conditions. SAPs have been used in agriculture when mixed in the soil to reduce the number of irrigations needed [32–34]. They have also been shown to increase seedling growth in rangeland applications when mixed into the soil [36]. In each of these situations, there are at least occasional additions of water through irrigation or precipitation. In our studies, the soil and SAP were saturated once and then allowed to dry down. No additional water inputs after initial saturation would have greatly increased the drought severity over the course of the studies relative to field conditions.

The utility of SAP use for seedling establishment will likely vary depending on location and project. Precipitation, temperature, terrain, and site size are just some of the factors that determine restoration success generally and SAP feasibility specifically. Priced between USD 5.00–6.18 kg^{-1} (USD 2.27–2.80 lb^{-1}) (Ken Aguilar, Global Plastic Sheeting, San Diego, CA, USA; personal communication (2017); Scott Mecom, Creasorb, Omaha, NE, USA; personal communication (2017)), the cost per kg of Stockosorb 660 micro is low relative to restoration seed mixes. Seed budgets for reseeding projects in the Great Basin run between USD 185–865 ha^{-1} (USD 75–350 acre^{-1}), with the average falling between USD 370–500 ha^{-1} (USD 150–200 acre^{-1}) (Josh Buck, Granite Seed, Lehi, UT, USA; personal communication (2018)). To increase the likelihood of seeing a response, SAP was used in these initial studies at rates ranging from 11 to 3000 kg ha^{-1}. At these rates and prices, the cost for Stockosorb 660 micro would range from USD 55–18,500 ha^{-1}.

There are several mitigating financial factors to consider, however. It may be possible to procure SAP at a lower price if purchased in very large quantities. The soil in these studies only received one saturating irrigation at the time of planting. Having the ability to reabsorb water during multiple precipitation events, lower amounts of SAP may be required to increase establishment in the field. If used in a band, the SAP must be buried in the ground, presumably with an attachment to a range drill. As such, it may only be used with species and in areas with terrain and slope that permit that seeding method or something similar, limiting the amount needed to be purchased. Improved establishment rates could reduce the amount of seed, and thus the cost, needed each year. If establishment rates are successfully improved, the need and cost to reseed the same area multiple times may also be reduced or eliminated. Rather than being used on the entire reseeding area, SAP could be used facilitate assisted succession. Bio-islands or green firebreaks of established perennial grasses could provide a free seed source, invasive species control, fire protection, and erosion reduction for the larger landscape, allowing other species to establish. Invasive weed and fire suppression provided by perennial grasses could also help reduce the overall restoration costs for the site by reducing the amount of money spent to manually, chemically or mechanically do the same. Once SAP effectiveness is demonstrated in the field, future studies should be done to evaluate the SAP effect on establishment in the field, determine best SAP and seeding rates, and determine species best suited to SAP use to improve the seedling establishment rate while lowering restoration costs.

5. Conclusions

We found that when mixed with the soil or placed in bands, moisture in the soil was increased at SAP rates of 1500 and 3000 kg ha^{-1} for up to 7 weeks under glasshouse conditions. We found seedling longevity improvements under the same conditions in two rangeland species, bottlebrush squirreltail and Siberian wheatgrass, when SAP was placed in bands at a rate of 3000 kg ha^{-1} at 8 and 15 cm depth. In some cases, we saw inconsistent trends or impacts at lower SAP rates. Soil moisture increased with SAP bands placed more shallowly, but seedling longevity was not impacted. However, at the high SAP rates, we observed soil cracking and the creation of subsurface voids which could impair soil moisture. Fertilizer placed in the SAP band increased soil moisture, but there was no benefit to seedling longevity. Generally, the two species responded differently to the presence of SAP, but there were some similarities. Further research should determine which species would benefit most. Because these are preliminary data from a glasshouse study, future work should evaluate banded SAP use under field conditions. The low cost of the SAP relative to the cost of seed makes it an appealing option in rangeland restoration. However, swelling SAP at high rates displaces soil, and lower rates do not seem to provide enough increase in water holding capacity within the soil to affect seedling establishment. Targeted applications where deeper band placement is possible could counteract the soil cracking. As such, banded SAP is potentially a better option to establish bio-islands or green firebreaks rather than as a general application.

Author Contributions: Conceptualization, B.G.H. and S.V.N.; methodology, B.G.H. and S.V.N.; validation, S.V.N.; formal analysis, S.V.N. and D.L.E.; investigation, S.V.N.; resources, B.G.H. and M.D.M.; data curation, S.V.N.; writing—original draft preparation, S.V.N.; writing—review and editing, S.V.N., B.G.H., M.D.M., V.J.A. and N.C.H.; visualization, S.V.N. and B.G.H.; supervision, S.V.N.; project administration, B.G.H. and S.V.N.; funding acquisition, B.G.H. All authors have read and agreed to the published version of the manuscript.

Funding: This research was funded by the United States Army Corps of Engineers, sponsor award number F19AC00962–03.

Data Availability Statement: Data available upon request from the corresponding author.

Conflicts of Interest: The authors declare no conflict of interest.

Appendix A

Table A1. Gravimetric water content (GWC) for bottlebrush squirreltail and Siberian wheatgrass grown in treatments with and without superabsorbent polymer (SAP) at seeding rates of 2, 4, 8, or 16 kg ha^{-1} (1/4, 1/2, 1, or 2 times the NRCS recommended seeding rate). All treatments were watered only once to saturation and allowed to dry down.

	Bottlebrush Squirreltail								Siberian Wheatgrass							
	Seeding Rate, kg ha^{-1}															
	—2—		—4—		—8—		—16—		—2—		—4—		—8—		—16—	
	ctrl	SAP	ctrl	SAP	ctrl	SAP	ctrl	SAP	ctrl	SAP	ctrl	SAP	ctrl	SAP	ctrl	SAP
Day	Gravimetric Water Content, %															
0	41.0	44.6	42.0	44.9	41.3	44.2	41.2	44.0	42.2	44.5	41.2	43.0	39.6	45.3	42.1	42.9
2	36.4	40.5	37.7	40.9	36.2	39.9	37.0	40.2	38.0	40.0	36.3	38.8	35.5	41.0	36.6	39.3
5	31.2	34.8	32.9	35.5	30.6	34.5	31.8	34.3	32.8	34.6	30.9	33.4	29.3	35.3	31.5	33.8
7	28.7	32.5	30.6	33.3	28.4	32.1	29.3	31.9	30.4	32.1	28.4	31.0	26.9	32.8	29.2	31.6
9	26.4	30.3	28.4	31.0	26.1	29.7	26.8	29.7	28.2	29.9	26.1	28.6	24.8	30.2	26.7	29.2
14	18.4	22.1	19.4	22.4	17.3	21.5	17.5	21.3	20.2	20.4	17.7	20.3	16.2	22.3	18.6	20.6
16	15.6	19.2	16.2	19.5	14.5	18.3	14.5	18.3	17.2	17.2	14.8	17.5	13.5	19.4	15.5	17.6
19	13.1	16.8	14.0	17.0	12.5	15.8	12.5	16.0	14.6	15.1	12.7	15.1	11.7	16.8	13.3	15.2
21	9.7	13.2	10.6	13.2	9.6	11.9	9.4	12.6	10.7	12.0	9.6	11.6	9.0	12.7	9.8	11.7
23	8.3	11.1	8.9	10.8	7.8	10.0	7.9	10.6	9.0	10.0	8.1	9.6	7.5	10.6	8.2	9.8
26	7.2	9.6	7.7	9.3	6.6	8.8	7.0	9.2	7.7	8.7	7.1	8.3	6.5	9.1	7.1	8.5
28	6.8	9.0	7.2	8.5	6.0	8.2	6.5	8.5	7.1	8.1	6.6	7.7	6.1	8.5	6.5	7.9
30	6.5	8.5	6.9	8.2	6.2	7.8	6.3	8.2	6.8	7.7	6.4	7.4	5.9	7.9	6.3	7.5
33	5.8	7.7	6.3	7.4	5.6	7.2	5.7	7.4	6.1	7.0	5.7	6.7	5.4	7.0	5.6	6.8
35	4.9	6.5	5.3	6.2	4.0	5.9	4.0	6.2	5.1	5.9	4.9	5.7	4.6	5.9	4.0	5.6
37	4.7	6.2	5.1	5.9	4.5	5.7	4.7	5.9	4.9	5.6	4.6	5.5	4.3	5.6	4.5	5.4
40	4.0	5.4	4.3	5.1	3.9	4.5	4.0	5.1	4.2	4.9	4.2	4.7	3.7	4.9	3.9	4.6
42	3.7	4.9	3.9	4.7	3.5	4.0	3.6	4.7	3.8	4.4	3.6	4.3	3.4	4.4	3.6	4.2
44	3.4	4.6	3.7	4.4	3.3	3.7	3.4	4.4	3.5	4.2	3.4	4.0	3.2	4.1	3.4	4.0
49	2.8	3.6	2.9	3.4	2.6	2.7	2.7	3.4	2.8	3.2	2.7	3.2	2.6	3.2	2.8	3.1
51	2.7	3.4	2.8	3.2	2.5	2.6	2.7	3.2	2.7	3.1	2.6	3.0	2.5	3.1	2.6	3.0
54	2.5	3.0	2.5	2.9	2.3	2.3	2.5	2.9	2.4	2.7	2.4	2.7	2.3	2.8	2.4	2.7
56	2.4	2.9	2.5	2.8	2.3	2.2	2.4	2.8	2.4	2.7	2.4	2.6	2.2	2.7	2.4	2.6

Table A2. Persistence (percent of seedlings alive) for bottlebrush squirreltail and Siberian wheatgrass grown in treatments with and without superabsorbent polymer (SAP) at seeding rates of 2, 4, 8, or 16 kg ha^{-1} (1/4, 1/2, 1, or 2 times the NRCS recommended seeding rates). All treatments were watered once to saturation and allowed to dry down.

	Bottlebrush Squirreltail								Siberian Wheatgrass							
	Seeding Rate, kg ha^{-1}															
	—2—		—4—		—8—		—16—		—2—		—4—		—8—		—16—	
	ctrl	SAP	ctrl	SAP	ctrl	SAP	ctrl	SAP	ctrl	SAP	ctrl	SAP	ctrl	SAP	ctrl	SAP
Day	Persistence, %															
4	0	0	0	0	0	0	0	0	0	0	6	0	3	0	0	0
5	0	0	0	0	0	0	2	0	13	25	25	31	28	19	16	25
7	13	0	19	13	22	28	34	25	50	75	56	75	50	69	50	66
9	50	38	25	38	31	38	36	31	63	75	56	75	50	69	53	69
12	50	38	38	44	31	53	41	36	63	75	69	81	63	72	53	70
14	50	38	38	50	47	59	45	52	63	75	69	94	69	75	56	81
16	50	38	38	56	50	63	48	50	63	75	75	94	69	78	61	81

Table A2. *Cont.*

Day	Bottlebrush Squirreltail								Siberian Wheatgrass							
	Seeding Rate, kg ha^{-1}															
	—-2—-		—-4—-		—-8—-		—-16—-		—-2—-		—-4—-		—-8—-		—-16—-	
	ctrl	SAP	ctrl	SAP	ctrl	SAP	ctrl	SAP	ctrl	SAP	ctrl	SAP	ctrl	SAP	ctrl	SAP
	Persistence, %															
19	50	38	38	56	50	63	48	52	63	75	75	94	69	78	61	81
21	50	38	38	44	50	59	47	52	50	75	75	94	69	78	63	80
23	50	38	38	44	50	59	47	52	50	75	75	94	69	78	63	80
26	50	38	38	44	50	56	47	52	50	75	75	94	69	78	63	80
28	50	38	38	44	50	56	47	52	50	75	63	94	69	78	63	80
30	50	38	38	38	50	53	47	50	50	75	56	94	69	66	53	77
33	50	38	38	38	50	53	47	50	50	75	56	94	69	66	53	77
35	50	38	38	38	50	53	47	50	50	75	56	94	69	66	53	77
37	50	38	38	38	44	50	42	50	50	75	31	88	44	56	42	73
40	50	38	38	25	34	50	17	42	50	75	13	88	28	56	34	64
42	38	25	19	19	16	47	6	36	38	75	6	81	19	50	13	58
44	25	25	13	13	13	41	2	28	38	63	0	75	6	41	5	42
47	25	25	13	13	13	41	2	27	38	63	0	75	6	38	5	42
49	13	25	6	6	6	38	0	22	25	50	0	63	3	31	5	39
51	0	13	0	0	3	9	0	3	0	38	0	38	0	9	0	8
54	0	0	0	0	0	0	0	0	0	0	0	19	0	0	0	2
56	0	0	0	0	0	0	0	0	0	0	0	0	0	0	0	0

Table A3. Soil moisture with associated *p*-values for individual dates × superabsorbent polymer (SAP) placement depth × measurement depth. Values within a grouping that are significant (*p*-value < 0.05) are in **bold-face type** and have means separated for statistical significance. Values not sharing the same letter(s) are statistically different from one another.

Days after Planting	0	4	13	20	27	34	41	50
SAP depth								
	3 cm measurement depth							
control	**36.0 g**	31.9	22	9.4	4.6	20.6	15.5	13.4
3 cm	**35.9 g**	30.4	18.9	10.2	6.3	23	18.2	13.3
8 cm	**36.6 fg**	31.7	21.4	12.4	4.2	19.8	14.8	13.5
mixed	**34.8 g**	29.7	19.1	10.5	4.8	19.4	16.8	13.8
	5 cm measurement depth							
control	**38.3 ef**	36.4	30.1	19.2	13.5	17.4	12.8	10.6
3 cm	**38.3 ef**	34.8	27.9	19.1	15.2	20.9	16.4	11.4
8 cm	**38.4 def**	35.3	28.9	19.1	11	20.8	15.6	10.5
mixed	**39.2 cde**	36.5	29.5	19	12.8	20.5	15.5	9.9
	8 cm measurement depth							
control	**39.9 cde**	37.7	31.54	23.4	16.8	18.3	14.1	11.7
3 cm	**39.2 cde**	37	32.6	23.7	18.7	22.6	18.3	12.5
8 cm	**40.1 bcd**	37.7	31.7	20.7	13.3	20.4	15.4	11.8
mixed	**39.5 cde**	37.5	30.2	22.6	16.7	21.9	17.4	12.5
	10 cm measurement depth							
control	**42.6 a**	39.6	33.8	25.7	19.1	19.1	15.1	10.2
3 cm	**39.5 cde**	37.5	34.6	25.9	21.1	21.9	14.5	10.5
8 cm	**40.6 bc**	38.6	33.4	24.5	19.9	21.8	15.3	10.9
mixed	**41.7 ab**	38.9	32.2	24.9	17.8	20.6	16.3	13
p-values	**<0.001**	0.514	0.219	0.221	0.128	0.159	0.491	0.586

References

1. Warren, S.D. Perceptions and History of Rangeland. In *Northeastern California Plateaus Bioregion Science Synthesis*; Gen. Tech. Rep RMRS-GTR-409; Dumroese, R., Moser, W., Eds.; US Department of Agriculture, Forest Service, Rocky Mountain Research Station: Fort Collins, CO, USA, 2020; pp. 45–47.
2. Kennedy, J.J.; Fox, B.L.; Osen, T.D. Changing Social Values and Images of Public Rangeland Management. *Rangelands* **1995**, *17*, 127–132.
3. Davies, K.W.; Johnson, D.D. Established Perennial Vegetation Provides High Resistance to Reinvasion by Exotic Annual Grasses. *Rangel. Ecol. Manag.* **2017**, *70*, 748–754. [CrossRef]
4. Minnick, T.J.; Alward, R.D. Soil Moisture Enhancement Techniques Aid Shrub Transplant Success in an Arid Shrubland Restoration. *Rangel. Ecol. Manag.* **2012**, *65*, 232–240. [CrossRef]
5. Ott, J.E.; Kilkenny, F.F.; Summers, D.D.; Thompson, T.W. Long-Term Vegetation Recovery and Invasive Annual Suppression in Native and Introduced Postfire Seeding Treatments. *Rangel. Ecol. Manag.* **2019**, *72*, 640–653. [CrossRef]
6. Brown, V.S.; Erickson, T.E.; Merritt, D.J.; Madsen, M.D.; Hobbs, R.J.; Ritchie, A.L. A Global Review of Seed Enhancement Technology Use to Inform Improved Applications in Restoration. *Sci. Total Environ.* **2021**, *798*, 149096. [CrossRef]
7. Madsen, M.D.; Davies, K.W.; Boyd, C.S.; Kerby, J.D.; Svejcar, T.J. Emerging Seed Enhancement Technologies for Overcoming Barriers to Restoration. *Restor. Ecol.* **2016**, *24*, S77–S84. [CrossRef]
8. Jessop, B.D.; Anderson, V.J. Cheatgrass Invasion in Salt Desert Shrublands: Benefits of Postfire Reclamation. *Rangel. Ecol. Manag.* **2007**, *60*, 235–243. [CrossRef]
9. Chambers, J.C. Invasive Plant Species and the Great Basin. In *USDA Forest Service—General Technical Report RMRS-GTR*; USDA Forest Service, Rocky Mountain Research Station: Reno, NV, USA, 2008; pp. 1–7.
10. Litt, A.R.; Pearson, D.E. Non-Native Plants and Wildlife in the Intermountain West. *Wildl. Soc. Bull.* **2013**, *37*, 517–526. [CrossRef]
11. Bradley, B.A.; Curtis, C.A.; Fusco, E.J.; Abatzoglou, J.T.; Balch, J.K.; Dadashi, S.; Tuanmu, M.N. Cheatgrass (*Bromus tectorum*) Distribution in the Intermountain Western United States and Its Relationship to Fire Frequency, Seasonality, and Ignitions. *Biol. Invasions* **2018**, *20*, 1493–1506. [CrossRef]
12. Garbowski, M.; Johnston, D.B.; Baker, D.V.; Brown, C.S. Invasive Annual Grass Interacts with Drought to Influence Plant Communities and Soil Moisture in Dryland Restoration. *Ecosphere* **2021**, *12*, e03417. [CrossRef]
13. Simberloff, D. The Role of Propagule Pressure in Biological Invasions. *Annu. Rev. Ecol. Evol. Syst.* **2009**, *40*, 81–102. [CrossRef]
14. Zouhar, K. Bromus Tectorum; Fire Effects Information System [Online]. U.S. Department of Agriculture, Forest Service, Rocky Mountain Research Station, Fire Sciences Laboratory (Porducer). 2003. Available online: https://www.fs.usda.gov/database/feis/plants/graminoid/brotec/all.html (accessed on 1 October 2020).
15. Hirsch-Schantz, M.C.; Monaco, T.A.; Call, C.A.; Sheley, R.L. Large-Scale Downy Brome Treatments Alter Plant-Soil Relationships and Promote Perennial Grasses in Salt Desert Shrublands. *Rangel. Ecol. Manag.* **2014**, *67*, 255–265. [CrossRef]
16. McGlone, C.M.; Sieg, C.H.; Kolb, T.E.; Nietupsky, T. Established Native Perennial Grasses Out-Compete an Invasive Annual Grass Regardless of Soil Water and Nutrient Availability. *Plant Ecol.* **2012**, *213*, 445–457. [CrossRef]
17. Rau, B.M.; Chambers, J.C.; Pyke, D.A.; Roundy, B.A.; Schupp, E.W.; Doescher, P.; Caldwell, T.G. Soil Resources Influence Vegetation and Response to Fire and Fire-Surrogate Treatments in Sagebrush-Steppe Ecosystems. *Rangel. Ecol. Manag.* **2014**, *67*, 506–521. [CrossRef]
18. Ryel, R.J.; Leffler, A.J.; Ivans, C.; Peek, M.S.; Caldwell, M.M. Functional Differences in Water-Use Patterns of Contrasting Life Forms in Great Basin Steppelands. *Vadose Zone J.* **2010**, *9*, 548–560. [CrossRef]
19. Chambers, J.C.; Miller, R.F.; Board, D.I.; Pyke, D.A.; Roundy, B.A.; Grace, J.B.; Schupp, E.W.; Tausch, R.J. Resilience and Resistance of Sagebrush Ecosystems: Implications for State and Transition Models and Management Treatments. *Rangel. Ecol. Manag.* **2014**, *67*, 440–454. [CrossRef]
20. Monaco, T.A.; Mangold, J.M.; Mealor, B.A.; Mealor, R.D.; Brown, C.S. Downy Brome Control and Impacts on Perennial Grass Abundance: A Systematic Review Spanning 64 Years. *Rangel. Ecol. Manag.* **2017**, *70*, 396–404. [CrossRef]
21. Roundy, B.A.; Hardegree, S.P.; Chambers, J.C.; Whittaker, A. Prediction of Cheatgrass Field Germination Potential Using Wet Thermal Accumulation. *Rangel. Ecol. Manag.* **2007**, *60*, 613–623. [CrossRef]
22. Chambers, J.C.; Roundy, B.A.; Blank, R.R.; Meyer, S.E.; Whittaker, A. What Makes Great Basin Sagebrush Ecosystems Invasible by *Bromus Tectorum*? *Ecol. Monogr.* **2007**, *77*, 117–145. [CrossRef]
23. Vallejo, V.R.; Smanis, A.; Chirino, E.; Fuentes, D.; Valdecantos, A.; Vilagrosa, A. Perspectives in Dryland Restoration: Approaches for Climate Change Adaptation. *New For.* **2012**, *43*, 561–579. [CrossRef]
24. Cline, N.; Roundy, B.; Gill, R.A.; Hopkins, B. *Wet—Thermal Time and Plant Available Water in the Seedbeds and Root Zones Across the Sagebrush Steppe Ecosystem of the Great Basin*; Brigham Young University: Provo, UT, USA, 2014.
25. Rezashateri, M.; Khajeddin, S.J.; Abedi-Koupai, J.; Majidi, M.M.; Matinkhah, S.H. Growth Characteristics of *Artemisia sieberi* Influenced by Super Absorbent Polymers in Texturally Different Soils under Water Stress Condition. *Arch. Agron. Soil. Sci.* **2017**, *63*, 984–997. [CrossRef]
26. Ferguson, S.D.; Leger, E.A.; Li, J.; Nowak, R.S. Natural Selection Favors Root Investment in Native Grasses during Restoration of Invaded Fields. *J. Arid. Environ.* **2015**, *116*, 11–17. [CrossRef]
27. Zohuriaan-Mehr, M.J.; Kabiri, K. Superabsorbent Polymer Materials: A Review. *Iran. Polym. J.* **2008**, *17*, 451–477.

28. Evonik Industries STOCKOSORB®660—Evonik Industries. Available online: https://www.break-thru.com/en/products/STOCKOSORB (accessed on 8 December 2021).

29. Abdallah, A.M. The Effect of Hydrogel Particle Size on Water Retention Properties and Availability under Water Stress. *Int. Soil Water Conserv. Res.* **2019**, *7*, 275–285. [CrossRef]

30. Singh, N.; Agarwal, S.; Jain, A.; Khan, S. 3-Dimensional Cross Linked Hydrophilic Polymeric Network "Hydrogels": An Agriculture Boom. *Agric. Water Manag.* **2021**, *253*, 106939. [CrossRef]

31. Abdallah, A.M.; Mashaheet, A.M.; Burkey, K.O. Super Absorbent Polymers Mitigate Drought Stress in Corn (*Zea mays* L.) Grown under Rainfed Conditions. *Agric. Water Manag.* **2021**, *254*, 106946. [CrossRef]

32. Varaprasad, K.; Raghavendra, G.M.; Jayaramudu, T.; Yallapu, M.M.; Sadiku, R. A Mini Review on Hydrogels Classification and Recent Developments in Miscellaneous Applications. *Mater. Sci. Eng. C* **2017**, *79*, 958–971. [CrossRef] [PubMed]

33. Evonik. *Stockosorb_660 Tech Data*; Evonik Industries: Krefeld, Germany, 2005.

34. Demitri, C.; Scalera, F.; Madaghiele, M.; Sannino, A.; Maffezzoli, A. Potential of Cellulose-Based Superabsorbent Hydrogels as Water Reservoir in Agriculture. *Int. J. Polym. Sci.* **2013**, *2013*, 435073. [CrossRef]

35. El-Asmar, J.; Jaafar, H.; Bashour, I.; Farran, M.T.; Saoud, I.P. Hydrogel Banding Improves Plant Growth, Survival, and Water Use Efficiency in Two Calcareous Soils. *CLEAN–Soil Air Water* **2017**, *45*, 1700251. [CrossRef]

36. Chirino, E.; Vilagrosa, A.; Vallejo, V.R. Using Hydrogel and Clay to Improve the Water Status of Seedlings for Dryland Restoration. *Plant Soil* **2011**, *344*, 99–110. [CrossRef]

37. Zhang, J.; Zhao, T.; Sun, B.; Song, S.; Guo, H.; Shen, H.; Wu, Y. Effects of Biofertilizers and Super Absorbent Polymers on Plant Growth and Soil Fertility in the Arid Mining Area of Inner Mongolia, China. *J. Mt. Sci.* **2018**, *15*, 1920–1935. [CrossRef]

38. Bai, W.; Song, J.; Zhang, H. Repeated Water Absorbency of Super-Absorbent Polymers in Agricultural Field Applications: A Simulation Study. *Acta Agric. Scand. B Soil Plant Sci.* **2013**, *63*, 433–441. [CrossRef]

39. Holliman, P.J.; Clark, J.A.; Williamson, J.C.; Jones, D.L. Model and Field Studies of the Degradation of Cross-Linked Polyacrylamide Gels Used during the Revegetation of Slate Waste. *Sci. Total Environ.* **2005**, *336*, 13–24. [CrossRef]

40. Banedjschafie, S.; Durner, W. Water Retention Properties of a Sandy Soil with Superabsorbent Polymers as Affected by Aging and Water Quality. *J. Plant Nutr. Soil Sci.* **2015**, *178*, 798–806. [CrossRef]

41. Abu-Zreig, M. Control of Rainfall-Induced Soil Erosion with Various Types of Polyacrylamide. *J. Soils Sediments* **2006**, *6*, 137–144. [CrossRef]

42. Hüttermann, A.; Zommorodi, M.; Reise, K. Addition of Hydrogels to Soil for Prolonging the Survival of *Pinus halepensis* Seedlings Subjected to Drought. *Soil Tillage Res.* **1999**, *50*, 295–304. [CrossRef]

43. Lucero, M.E.; Dreesen, D.R.; VanLeeuwen, D.M. Using Hydrogel Filled, Embedded Tubes to Sustain Grass Transplants for Arid Land Restoration. *J. Arid Environ.* **2010**, *74*, 987–990. [CrossRef]

44. Evonik. STOCKOSORB®—BREAK-THRU®—Technology for Agriculture. Available online: https://www.break-thru.com/product/break-thru/en/products/STOCKOSORB/ (accessed on 30 June 2020).

45. Evonik. STOCKOSORB® for Water and Soil Management in Agriculture and Horticulture. Evonik Industries: Krefeld, Germany, 2008.

46. Evonik. *Stockosorb 660 Label*; Evonik Corporation: Greensboro, NC, USA, 2015.

47. Whitlock, M.C.; Schluter, D. *The Analysis of Biological Data*; Roberts and Company Publishers: Greenwood Village, CO, USA, 2009; ISBN 978-0-9815194-0-1.

48. Soil Survey Staff Web Soil Survey. Available online: https://websoilsurvey.sc.egov.usda.gov/App/WebSoilSurvey.aspx (accessed on 30 January 2020).

49. Miller, R.O.; Gavlak, R.; Horneck, D. *Soil, Plant and Water Reference Methods for the Western Region, 4th Edition*; Colorado State University: Fort Collins, CO, USA, 2013.

50. Plumb, H. *NRCS Bottlebrush Squirreltail Elymus Elymoides Plant Fact Sheet*; USDA-National Resources Conservation Service, Upper Colorado Environmental Plant Center: Meeker, CO, USA, 2010.

51. USDA-NRCS-Aberdeen Plant Materials Center. *Release Brochure for Vavilov II Siberian Wheatgrass (Agropyron Fragile)*; USDA-National Resources Conservation Service, Aberdeen Plant Materials Center: Aberdeen, ID, USA, 2012.

52. Narjary, B.; Aggarwal, P.; Singh, A.; Chakraborty, D.; Singh, R. Water Availability in Different Soils in Relation to Hydrogel Application. *Geoderma* **2012**, *187–188*, 94–101. [CrossRef]

53. Smagin, A.V.; Sadovnikova, N.B.; Smagina, M.V. Biodestruction of Strongly Swelling Polymer Hydrogels and Its Effect on the Water Retention Capacity of Soils. *Eurasian Soil Sci.* **2014**, *47*, 591–597. [CrossRef]

54. Koupai, J.A.; Eslamian, S.S.; Kazemi, J.A. Enhancing the Available Water Content in Unsaturated Soil Zone Using Hydrogel, to Improve Plant Growth Indices. *Ecohydrol. Hydrobiol.* **2008**, *8*, 67–75. [CrossRef]

55. Saha, A.; Sekharan, S.; Manna, U. Superabsorbent Hydrogel (SAH) as a Soil Amendment for Drought Management: A Review. *Soil Tillage Res.* **2020**, *204*, 104736. [CrossRef]

56. Yu, J.; Shainberg, I.; Yan, Y.L.; Shi, J.G.; Levy, G.J.; Mamedov, A.I. Superabsorbents and Semiarid Soil Properties Affecting Water Absorption. *Soil Sci. Soc. Am. J.* **2011**, *75*, 2305–2313. [CrossRef]

57. Allen, R.G.; Pereira, L.S.; Raes, D.; Smith, M. Crop Evapotranspiration: Guidelines for Computing Crop Requirements. In *Irrigation and Drainage Paper No. 56*; FAO: Italy, Rome, 1998; Volume 300. [CrossRef]

58. Bai, W.; Zhang, H.; Liu, B.; Wu, Y.; Song, J.Q. Effects of Super-Absorbent Polymers on the Physical and Chemical Properties of Soil Following Different Wetting and Drying Cycles. *Soil Use Manag.* **2010**, *26*, 253–260. [CrossRef]
59. Han, Y.G.; Yang, P.L.; Luo, Y.P.; Ren, S.M.; Zhang, L.X.; Xu, L. Porosity Change Model for Watered Super Absorbent Polymer-Treated Soil. *Environ. Earth Sci.* **2010**, *61*, 1197–1205. [CrossRef]
60. Sarvaš, M.; Pavlenda, P.; Takácová, E. Effect of Hydrogel Application on Survival and Growth of Pine Seedlings in Reclamations. *J. For. Sci.* **2007**, *53*, 204–209. [CrossRef]
61. Bakass, M.; Mokhlisse, A.; Lallemant, M. Absorption and Desorption of Liquid Water by a Superabsorbent Polymer: Effect of Polymer in the Drying of the Soil and the Quality of Certain Plants. *J. Appl. Polym. Sci.* **2002**, *83*, 234–243. [CrossRef]
62. Bandak, S.; Naeini, S.A.R.M.; Zeinali, E.; Bandak, I. Effects of Superabsorbent Polymer A200 on Soil Characteristics and Rainfed Winter Wheat Growth (*Triticum aestivum* L.). *Arab. J. Geosci.* **2021**, *14*, 712. [CrossRef]
63. Do Nascimento, C.D.V.; Feitosa, J.P.D.A.; Simmons, R.; Dias, C.T.D.S.; do Nascimento, Í.V.; Mota, J.C.A.; Costa, M.C.G. Durability Indicatives of Hydrogel for Agricultural and Forestry Use in Saline Conditions. *J. Arid Environ.* **2021**, *195*, 104622. [CrossRef]
64. Evonik. *Stockosorb 660 Material Safety Data Sheet*; Evonik Corporation: Garyville, LA, USA, 2014.
65. Lentz, R.D. Long-Term Water Retention Increases in Degraded Soils Amended with Cross-Linked Polyacrylamide. *Agron. J.* **2020**, *112*, 2569–2580. [CrossRef]
66. Zhou, B.; Liao, R.; Li, Y.; Gu, T.; Yang, P.; Feng, J.; Xing, W.; Zou, Z. Water-Absorption Characteristics of Organic-Inorganic Composite Superabsorbent Polymers and Its Effect on Summer Maize Root Growth. *J. Appl. Polym. Sci.* **2012**, *126*, 423–435. [CrossRef]
67. He, M.; Lv, L.; Li, H.; Meng, W.; Zhao, N. Analysis on Soil Seed Bank Diversity Characteristics and Its Relation with Soil Physical and Chemical Properties after Substrate Addition. *PLoS ONE* **2016**, *11*, e0147439. [CrossRef]
68. Coello, J.; Ameztegui, A.; Rovira, P.; Fuentes, C.; Piqué, M. Innovative Soil Conditioners and Mulches for Forest Restoration in Semiarid Conditions in Northeast Spain. *Ecol. Eng.* **2018**, *118*, 52–65. [CrossRef]
69. Yang, W.; Guo, S.; Li, P.; Song, R.; Yu, J. Foliar Antitranspirant and Soil Superabsorbent Hydrogel Affect Photosynthetic Gas Exchange and Water Use Efficiency of Maize Grown under Low Rainfall Conditions. *J. Sci. Food Agric.* **2019**, *99*, 350–359. [CrossRef] [PubMed]
70. Johnston, D.B.; Garbowski, M. Responses of Native Plants and Downy Brome to a Water-Conserving Soil Amendment. *Rangel Ecol. Manag.* **2020**, *73*, 19–29. [CrossRef]
71. Ovalle, J.F.; Arellano, E.C.; Ginocchio, R. Trade-Offs between Drought Survival and Rooting Strategy of Two South American Mediterranean Tree Species: Implications for Dryland Forests Restoration. *Forests* **2015**, *6*, 3733–3747. [CrossRef]
72. Garbowski, M.; Brown, C.S.; Johnston, D.B. Soil Amendment Interacts with Invasive Grass and Drought to Uniquely Influence Aboveground vs. Belowground Biomass in Aridland Restoration. *Restor. Ecol.* **2019**, *28*, 1–11. [CrossRef]
73. Islam, M.R.; Ren, C.; Zeng, Z.; Jia, P.; Eneji, E.; Hu, Y. Fertilizer Use Efficiency of Drought-Stressed Oat (*Avena sativa* L.) Following Soil Amendment with a Water-Saving Superabsorbent Polymer. *Acta Agric. Scand. B Soil Plant Sci.* **2011**, *61*, 721–729. [CrossRef]
74. Jacobs, D.F.; Rose, R.; Haase, D.L.; Alzugaray, P.O. Fertilization at Planting Impairs Root System Development and Drought Avoidance of Douglas-Fir (*Pseudotsuga menziesii*) Seedlings. *Ann. For. Sci.* **2004**, *61*, 643–651. [CrossRef]
75. Fehmi, J.S.; Kong, T.M. Effects of Soil Type, Rainfall, Straw Mulch, and Fertilizer on Semi-Arid Vegetation Establishment, Growth and Diversity. *Ecol. Eng.* **2012**, *44*, 70–77. [CrossRef]
76. Del Campo, A.D.; Hermoso, J.; Flors, J.; Lidón, A.; Navarro-Cerrillo, R.M. Nursery Location and Potassium Enrichment in Aleppo Pine Stock 2. Performance under Real and Hydrogel-Mediated Drought Conditions. *Forestry* **2011**, *84*, 235–245. [CrossRef]

land

Article

Current and Potential Future Distribution of Endemic *Salvia ceratophylloides* Ard. (Lamiaceae)

Valentina Lucia Astrid Laface [1], Carmelo Maria Musarella [1,*], Gianmarco Tavilla [2], Agostino Sorgonà [1], Ana Cano-Ortiz [3], Ricardo Quinto Canas [4] and Giovanni Spampinato [1,*]

[1] Department of AGRARIA, Mediterranean University of Reggio Calabria, 89122 Reggio Calabria, Italy
[2] Department of Biological, Geological and Environmental Sciences, University of Catania, 95131 Catania, Italy
[3] Department of Didactics of Experimental Social Sciences and Mathematics, Section of Didactics of Experimental Sciences, Faculty of Education, Complutense University of Madrid, 28040 Madrid, Spain
[4] Faculty of Sciences and Technology, University of Algarve, Campus de Gambelas, 8005-139 Faro, Portugal
* Correspondence: carmelo.musarella@unirc.it (C.M.M.); gspampinato@unirc.it (G.S.)

Abstract: Human activities and climate change are the main factors causing habitat loss, jeopardising the survival of many species, especially those with limited range, such as endemic species. Recently, species distribution models (SDMs) have been used in conservation biology to assess their extinction risk, environmental dynamics, and potential distribution. This study analyses the potential, current and future distribution range of *Salvia ceratophylloides* Ard., an endemic perennial species of the Lamiaceae family that occurs exclusively in a limited suburban area of the city of Reggio Calabria (southern Italy). The MaxEnt model was employed to configure the current potential range of the species using bioclimatic and edaphic variables, and to predict the potential suitability of the habitat in relation to two future scenarios (SSP245 and SSP585) for the periods 2021–2040 and 2041–2060. The field survey, which spanned 5 years (2017–2021), involved 17 occurrence points. According to the results of the MaxEnt model, the current potential distribution is 237.321 km^2, which considering the preferred substrates of the species and land-use constraints is re-estimated to 41.392 km^2. The model obtained from the SSP245 future scenario shows a decrease in the area suitable for the species of 35% in the 2021–2040 period and 28% in the 2041–2060 period. The SSP585 scenario shows an increase in the range suitable for hosting the species of 167% in the 2021–2040 period and 171% in the 2041–2060 period. Assessing variation in the species distribution related to the impacts of climate change makes it possible to define priority areas for reintroduction and in situ conservation. Identifying areas presumably at risk or, on the contrary, suitable for hosting the species is of paramount importance for management and conservation plans for *Salvia ceratophylloides*.

Keywords: conservation; Calabria; climate changes; endangered species; Italy; MaxEnt; SSP245; SSP585; vascular plants

Citation: Laface, V.L.A.; Musarella, C.M.; Tavilla, G.; Sorgonà, A.; Cano-Ortiz, A.; Quinto Canas, R.; Spampinato, G. Current and Potential Future Distribution of Endemic *Salvia ceratophylloides* Ard. (Lamiaceae). *Land* **2023**, *12*, 247. https://doi.org/10.3390/land12010247

Academic Editors: Michael Vrahnakis, Yannis (Ioannis) Kazoglou and Manuel Pulido Fernádez

Received: 23 December 2022
Revised: 9 January 2023
Accepted: 10 January 2023
Published: 13 January 2023

1. Introduction

Human activities and climate change are the main causes of habitat and biodiversity loss [1–6], seriously threatening the survival of many species, especially those with limited distribution and, especially, endemic species [7–14].

The 20th century saw the strongest warming trend of the last millennium, with an increase in average temperatures of approximately 0.6 °C, compared to pre-industrial times (1850–1900) [15–19]. Estimates suggest that future temperature increases could exceed this value, with an increase of between 0.1 and 0.2 °C expected per decade [19]. In addition, climate change, combined with economic globalisation, rapid infrastructure development, and human activities, has favoured the spread of invasive alien species, which, by rapidly expanding their range, affect natural habitats and lead to the extinction of species, especially those with limited ranges [20–26].

The Mediterranean region is characterised by high plant biodiversity and a remarkable richness of endemic species, which is due to several factors acting simultaneously [27–30]. Several authors assessed the impact that climate change could have on the distribution of species, particularly species with limited distributions, such as endemic species, which are more sensitive than others to environmental change and are at greater risk of extinction [10,31–36]. To this end, the ecological variables that influence the natural distribution of endemic species must be studied to identify the areas where they occur or could occur [10,32,37,38]. Currently, one of the most widely used systems for determining the environmental limits of species is the MaxEnt prediction model (Maximum Entropy Species Distribution Modeling) [39], which uses bioclimatic data and species occurrence to predict species distributions based on the maximum entropy theory, estimating a probabilistic distribution that is as uniform as possible but subject to environmental constraints [40–47].

The MaxEnt model has been used extensively in the field of conservation biology: it allows the prediction of the current and future potential range of a species [48,49]. Compared to other prediction models, it is more stable and reliable and works quickly and easily in modelling rare species with restricted ranges and limited occurrence data [43,47,50–53].

Lamiaceae, one of the largest families of angiosperms, includes more than 7000 species distributed throughout the world, with several species characterised by essential oils [54–58]. In the Italian flora, among the endemic species of this family with an extremely limited range [59,60], whose existence may be threatened in the near future by climate change, there is *Salvia ceratophylloides* Ard. (Figure 1), a species growing exclusively in southern Italy in the hill belt of the suburb of Reggio Calabria. It is clearly distinguished from the other perennial sage species of the *Salvia pratensis* L. group, to which it belongs [61,62], mainly by its wrinkled, pinnatifid leaves with toothed lobes [63,64]. Its chromosome number is 2n = 6x = 54 [65]. *Salvia ceratophylloides* (Figure 1) is a perennial herbaceous plant (scapose hemicryptophyte), densely pubescent with both glandular and simple patent hairs, has a main flowering period in spring from April to June, and has a second flowering period in autumn from October to November. Pollination is entomophilous, mediated mainly by hymenoptera (*Eucera* sp., *Bombus* sp., *Apis* sp.). The fruiting occurs after some flowering weeks. Seed dispersal is mainly carried out by ants (myrmecochory) [64]. Seed germination takes place mostly in spring, seedlings reach reproductive maturity (small generative) within 4–5 months, while they tiller (Large Generative) in the following year [13].

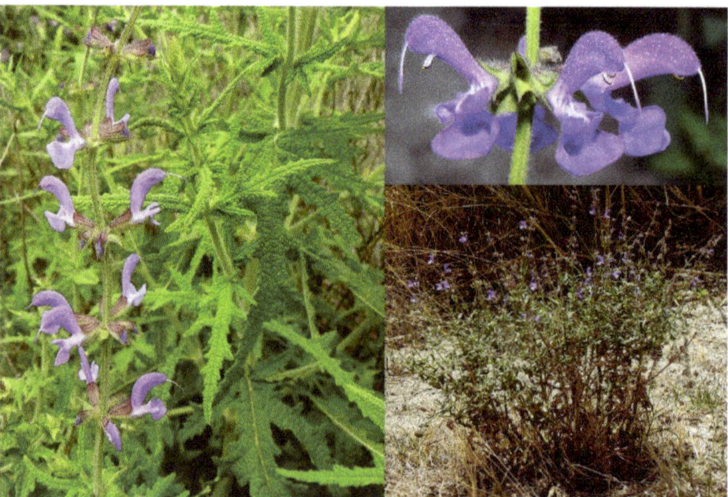

Figure 1. Details of the inflorescence, flowers, and habitat of *Salvia ceratophylloides* Ard. in its natural habitat (Ph. V.L.A. Laface).

The species was known only in a few nearby places, as can be seen from bibliographical references from 1800 [66,67] to the early 1900s [68,69], when, moreover, it was already very rare. Subsequently, despite the research of various botanists, the species was no longer found, having disappeared from the locations mentioned in the literature (Gallico Superiore, Terreti, Straorino, Ortì, Vito Superiore, Pietrastorta) [69]. For this reason, the species was considered extinct in 1997 and included in the "Libro rosso della flora d'Italia" (Red Book of the Flora of Italy) among the extinct species (EX) [70] and confirmed by Del Carratore and Garbari [71] and Scoppola and Spampinato [72].

Subsequent surveys in 2008 revealed four new occurrence points in the surroundings of Reggio Calabria at sites approximately 10 km from those for which the species was known in the literature of the early 1900s, each consisting of a few dozen individuals, totalling nearly 100 mature individuals [73–76].

Laface et al. [13] carried out field surveys between 2017 and 2021 and identified 17 occurrence points, always in the suburbs of Reggio Calabria, some of these with a small number of individuals. *Salvia ceratophylloides* covers an "Extent of Occurrence" (EOO) of 4.2 km^2 and an "Area of Occupancy" (AOO) of 7 km^2: this made it possible to assess the species as "Critically Endangered" (CR) [13,64,76] according to IUCN (International Union for Conservation of Nature) criteria and categories [77].

Salvia ceratophylloides grows spontaneously in the habitat of the EEC Directive 43/93: "5330 thermo Mediterranean and predesert scrub" subtype "32.23 Diss dominated garrigues". This habitat includes Mediterranean steppe, such as grasslands with *Ampelodesmos mauritanicus* (Poir.) Dur. & Schinz., sands vegetation with *Artemisia campestris* subsp. *variabilis* (Ten.) Greuter, and more rarely in garrigues, characterised by *Cistus creticus* L. subsp. *creticus* and *Thymbra capitata* (L.) Cav. The most frequently growing species with *S. ceratophylloides*, in addition to the aforementioned species, are some grasses (*Lagurus ovatus* L., *Avena barbata* Link, *Macrobriza maxima* (L.) Tzvelev, *Hyparrhenia hirta* (L.) Stapf., *Dasypyrum villosum* (L.) P. Candargy), several dwarf shrubs (*Micromeria graeca* (L.) Benth. ex Rchb., *Phlomis fruticosa* L.), and some shrubs (*Cytisus infestus* (C.Presl) Guss. subsp. *infestus*, *Spartium junceum* L.). Mostly, they are widespread species in the Mediterranean steppic grassland and garrigues [64,76]. The populations are located on hills at altitudes between 250 and 450 m a.s.l., characterised exclusively by layers of loose sands, alternating with banks of soft Pliocene calcarenites [78]. The species grows in a territory with average annual temperatures of 18 °C and an average annual rainfall of 600 mm, concentrated in the autumn, the months of November and December, and a summer dry period of approximately 5 months [13,64]. According to Pesaresi [79], the bioclimate is classified as oceanic pluviostagional Mediterranean, with upper thermo-Mediterranean thermotype and lower sub-humid ombrotype.

Numerous physiological studies have been carried out on *S. ceratophylloides*, and these have shown that the species has a very strong adaptive capacity to future climate change, and develops resilient forms of defence [80–82].

In order to safeguard the habitat of *S. ceratophylloides*, it is of fundamental importance, both theoretically and practically, to understand which areas are potentially suitable from a current and future climatic perspective. This, correlated with population dynamics [13], will make it possible to determine the most appropriate locations for effectively targeting conservation strategies aimed at protecting and reintroducing this critically endangered species. The aim of our study, therefore, is to analyse the species distribution patterns (SDM) of *S. ceratophylloides* by interpolating the occurrence points with environmental variables, and to model current and future scenarios to assess the current distribution and predict the habitat's conservation capacity in the context of climate change [19].

2. Materials and Methods

2.1. Species Occurrence Data

Information concerning the current distribution of *S. ceratophylloides* was obtained during fieldwork carried out between 2017 and 2021 [13], and also considering historical information reported in the literature by several authors [66–69,73–76] and verified in the field. For each point of occurrence, field coordinates were taken and the substrate and plant community recorded.

The collected data were analysed using QGIS 3.26.3® software (OSGeo, Beaverton, OR, USA) [83].

2.2. Environmental Variables

In order to model the potential habitat of *S. ceratophylloides*, based on its current occurrence, a total of 22 ecological variables were considered (Table 1); specifically, 19 bioclimatic and 3 topographic. This information was obtained from the WorldClim database [84,85] at a spatial resolution (expressed as minutes of a degree of longitude and latitude) of 30 s (approx. 1×1 km). The topographic variables were extracted using QGIS 3.26.3® software [83].

Table 1. Description of variables used in the prediction of the MaxEnt model. The variables in bold were selected through Pearson's correlation analysis and were used in the modelling.

Code	Description	Unit
Bio 1	**Annual Mean Temperature**	°C
Bio 2	Mean Diurnal Range (Mean of monthly (max temp–min temp))	°C
Bio 3	Isothermality (BIO2/BIO7) (*100)	%
Bio 4	**Temperature Seasonality (standard deviation *100)**	%
Bio 5	Max Temperature of Warmest Month	°C
Bio 6	Min Temperature of Coldest Month	°C
Bio 7	Temperature Annual Range (BIO5-BIO6)	°C
Bio 8	Mean Temperature of Wettest Quarter	°C
Bio 9	Mean Temperature of Driest Quarter	°C
Bio 10	Mean Temperature of Warmest Quarter	°C
Bio 11	Mean Temperature of Coldest Quarter	°C
Bio 12	Annual Precipitation	mm
Bio 13	**Precipitation of Wettest Month**	mm
Bio 14	**Precipitation of Driest Month**	mm
Bio 15	Precipitation Seasonality (Coefficient of Variation)	%
Bio 16	Precipitation of Wettest Quarter	mm
Bio 17	Precipitation of Driest Quarter	mm
Bio 18	Precipitation of Warmest Quarter	mm
Bio 19	**Precipitation of Coldest Quarter**	mm
Elev.	**Elevation**	**meter**
Slope	Slope	degree
Aspe.	Aspect	degree

Information on the environmental variables is an essential parameter for building a predictive model: however, overuse of the environmental variables may increase the spatial correlation between them, leading to overfitting and reducing the transferability of the model [86]. To avoid overfitting, it is necessary to calculate the correlation between all variables considered and exclude the highly correlated variables, which exponentially improves the predictive ability of the model [87]. For this purpose, Pearson's correlation analysis [88] was carried out using Past 4.1.4 © software (Hammer, Oslo, Norway) [89]. Environmental variables with correlation values falling in the following range were considered significant: $-0.8 \leq r \leq +0.8$. To assess the dominant environmental variables, i.e., those that defined the potential distribution of the species, the jackknife test [90] was performed. For the modelling of future scenarios, the Global Climate Model (GCM) BCC-CSM2-MR was used, with this model producing excellent results in many studies at the European and Mediterranean level [91,92]. For the scenarios reference for the IPCC's Sixth Assessment Report [19], where four Shared Socioeconomic Pathways (SSPs) are assumed:

- SSP585: with an additional radiative forcing of 8.5 W/m^2 by the year 2100;
- SSP370: with an additional radiative forcing of 7 W/m^2 by the year 2100;
- SSP245: with an additional radiative forcer of 4.5 W/m^2 by the year 2100;
- SSP126: with an additional radiative forcer of 2.6 W/m^2 by the year 2100.

To make the modelling more reliable and plausible, the scenarios SSP585 (most extreme) and SSP245 (intermediate) were chosen, for the periods 2021–2040 and 2041–2060.

Pearson's [88] correlation analysis made it possible to determine six ecological variables (out of 22) useful for modelling the distribution of the species. Five bioclimatic variables (Bio 1, Bio 4, Bio 13, Bio 14, Bio 19) and one topographic variable (Elev.) were found to be significant ($-0.8 \leq r \leq +0.8$) (Table 1, Figure 2). These variables were also used for modelling the future scenarios. Variables with values >0.8 and those <-0.8 were not considered in order to avoid overfitting.

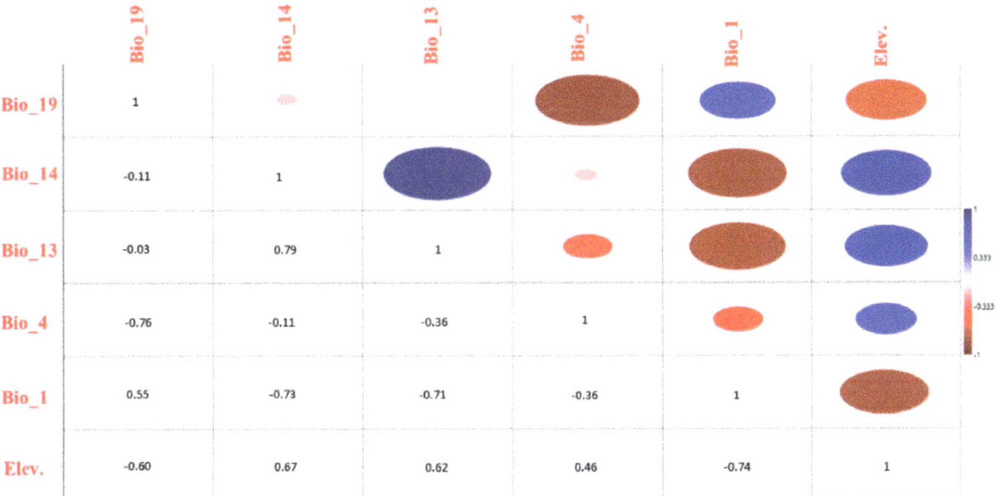

Figure 2. Pearson correlation analysis of significative environment variables for *Salvia ceratophylloides* Ard. ($-0.8 \leq r \leq +0.8$). Created with Past 4.1.4 © (Hammer, Oslo, Norway).

2.3. Model Construction

The distribution point data (species and geographical coordinates, saved in .csv format) and the resulting bioclimatic variable data were imported into MaxEnt 3.4.4® (American Museum of Natural History, New York, NY, USA) [93,94].

In the analysed models, 75% of the data were selected for model training (calibration), using a maximum number of iterations of 1000, and 25% as test data, for model validation [93,94], keeping the other values as defaults. The Bootstrap method was used, implemented with 10 repetitions and the multiplier value at 0.5. The output format is complementary log-log (cloglog).

The accuracy of the generated model was verified using the Receiver Operating Characteristic (ROC) curve analysis method. The ROC curve has as the ordinate the percentage of true positive values (the ratio that exists and is expected to exist) and as the abscissa the percentage of false positive values (the ratio that does not exist but is expected to exist) [95]. The AUC (Area under the Curve) value is the area enclosed between the abscissa and the ROC curve, and has a range between 0.5 and 1. The higher the AUC value, the greater the distance from the random distribution, the more relevant the correlation between the environmental variables and the geographic distribution of the species, and the more reliable the predictive power of this model.

Conversely, the predictive power of the model is not very reliable. The model's performance is classified as: inadequate with AUC values ranging from 0.5 to 0.6; poor with values ranging from 0.6 to 0.7; reasonable with values ranging from 0.7 to 0.8; good with values ranging from 0.8 to 0.9; and excellent with values ranging from 0.9 to 1. The necessary means of measuring the model performance is the AUC score, as it has a strong independence from threshold choices. The smallest difference between the training and test AUC data (AUCDiff) was also observed; a lower difference indicates less overfitting in the model [96].

2.4. Distribution Maps: Visualisation and Analysis

For the visualisation and investigation of the distribution areas of the species, the models created with the software MaxEnt (range 0–1) [39] were imported into the software QGIS 3.26.3 [83]. The areas found to be suitable for the species were grouped into 5 habitat potential classes (ranging from 0 to 1): highly unsuitable ($\leq$0.20); unsuitable (0.21–0.40); moderately suitable (0.41–0.60); highly suitable (0.61–0.80); very highly suitable ($\geq$0.80). For each model, the area for each selected class was calculated using QGIS [83].

To define the real distribution of the species, we interpolated the current and future models on the geological map of Calabria [78] and with the land use map of the Region of Calabria "Carta di Uso del Territorio" [97] using the software QGIS. In the first case, we considered the geological substrates on which the species grows, i.e., sands, calcarenites and conglomerates more or less cemented. In the land use map, which is divided into five macro-categories of land cover (1. Artificial surfaces; 2. Agricultural areas; 3. Forests and semi-natural areas; 4. Wetlands; 5. Water bodies), we considered land cover 2 and 3, because *S. ceratophylloides* grows in areas with a highly fragmented mosaic of agricultural and semi-natural habitats [13].

3. Results

3.1. Natural Distribution Data

A total of 23 occurrence points of *S. ceratophylloides* are known (Figure 3), of which 17 currently occur in the area (albeit with a small number of individuals for occurrence points) while 6 are extinct: occurrence point 13 became extinct in 2019 and had only one individual in the previous year; occurrence point 19 was reported in 2008 [73] and was not found in subsequent years during field surveys; occurrence points 20, 21, 22, and 23 are historical reports dating back to the early 1900s [68,69] and were not found in the second half of the last century [71].

3.2. Analysis and Evaluation of Environmental Variables

The calibration of the current potential distribution model for *S. ceratophylloides*, using the variables thus selected, was optimal (AUC mean = 0.986, $\pm$0.001; AUCDiff (0.09 $\pm$ 0.006).

From the results obtained with the jackknife test, we know that the distribution of *S. ceratophylloides* is mainly influenced by the precipitation of wettest month (Bio 13), the annual mean temperature (Bio 1), and the precipitation of the coldest quarter (Bio 19); these contributed 69.3%, 7.8%, and 11.4%, respectively, to the MaxEnt model (Figure 4). In addition, two other environmental variables (Bio 4, Bio 14) contributed a total of 8.3% to the habitat distribution model and 3.2% to the topographic variable (Elev.) (Figure 4, Table 2).

In view of the importance of the permutation, the precipitation of wettest month (Bio 13) had the greatest impact on the model with 66.9%, the annual mean temperature (Bio 1) with 13.3%, while the other variables contributed a smaller percentage, totalling 19.8%.

Considering the six bioclimatic variables previously selected, the mean annual temperature range (Bio 1) of *S. ceratophylloides* is 15.7–19.7 °C, and the temperature seasonality (Bio 4) is 549–576%. In addition, the average precipitation in the wettest month (Bio 13) is 108–111 mm, in the driest month (Bio 14) it is 10–15 mm, and in the coldest quarter (Bio 19) it is 256–297 mm, on average. The altitude ranges from 11 to 689 m a.s.l.

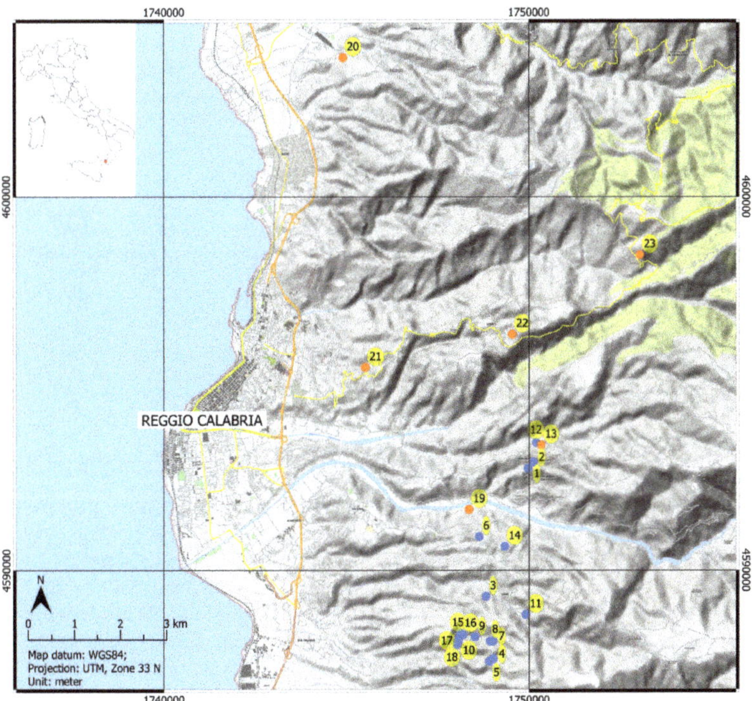

Figure 3. Occurrence points of *Salvia ceratophylloides* Ard. In blue are the micro-populations currently occurring, in orange the extinct ones. 1—Serro Ciugna, Mosorrofa; 2—Serro Ciugna, Mosorrofa; 3—Spilingari, Armo; 4—Contrada S. Todaro, Aretina; 5—Contrada S. Todaro, Aretina; 6—Serro dei Morti, Puzzi fraz. di Gallina; 7—Prai, Aretina; 8—Prai, Aretina; 9—Aretina; 10—Aretina; 11—Grotta di S. Arsenio, Armo; 12—Mosorrofa vecchio; 13—Mosorrofa vecchio; 14—Serro d'Angelo, Puzzi fraz. of Gallina; 15—Prai, Aretina; 16—Prai, Aretina; 17—Serro della Cattina, Aretina; 18—Serro della Cattina, Aretina; 19—Lutrà, Fiumara di Sant'Agata; 20—Galluzzi, Gallico Superiore; 21—Pietra Storta; 22—Croce Missionaria, Terreti; 23—Fontana Acqua Fresca, Straorino. In the top left-hand corner, the distribution area of the points of occurrence is highlighted in red.

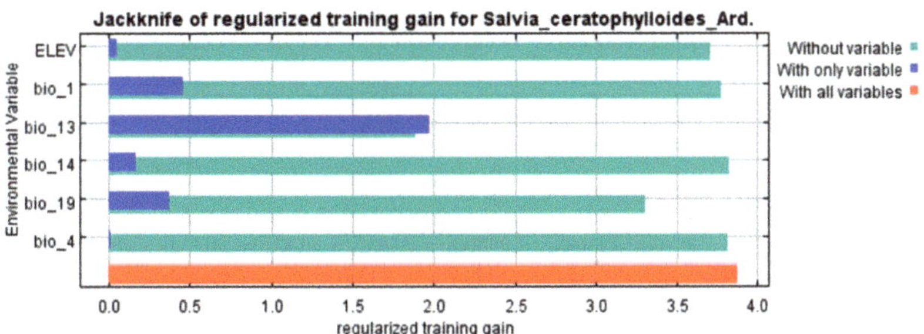

Figure 4. Relative predictive power of different environmental variables based on the jackknife of regularised training gain in MaxEnt models for *Salvia ceratophylloides* Ard.

Table 2. Percent contribution and permutation importance of environmental variables used to predict the MaxEnt model of *Salvia ceratophylloides* Ard. [SSPs- future scenarios (see text) Bio 1, Bio 4, Bio 13, Bio 14, and Bio 19. Elev (see Table 1)].

Time	SSPs	Variable	Bio 1	Bio 4	Bio 13	Bio 14	Bio 19	Elev.
Present time		Percent contribution (%)	7.8	7.3	69.3	1	11.4	3.2
		Permutation importance (%)	13.3	1	66.9	0.8	9	9
2021/2040	245	Percent contribution (%)	7	5.5	86.3	0.1	0.9	0.2
		Permutation importance (%)	0.1	1.1	97.8	0.2	0.6	0.
	585	Percent contribution (%)	24.4	15.4	40.7	0	3	16.5
		Permutation importance (%)	52.4	0	25	0.2	21.7	0.7
2041/2060	245	Percent contribution (%)	11.9	12.1	73.4	1	1.3	0.3
		Permutation importance (%)	13.7	0.6	78.7	0.8	5.4	0.7
	585	Percent contribution (%)	16.5	14.2	25.6	0.1	42.6	1
		Permutation importance (%)	15.4	1.5	76	0.1	0.8	6.2

3.3. Current Potential Distribution of Salvia ceratophylloides

The current estimated potential habitat for *S. ceratophylloides* is located exclusively in the south/west of the Italian peninsula and Calabria (Figure 5): this corresponds to a total area of 237.321 km^2, equal to 1.58% of the entire regional territory and 0.08% of the Italian territory. In relation to the probability of occurrence of the species, the area is distributed as follows: very highly suitable ($\geq$0.80) with an area of 30,440 km^2 (0.20%); highly suitable (0.61–0.80) with an area of 20,962 km^2 (0.14); moderately suitable (0.41–0.60) with a surface area of 59,434 km^2 (0.39%); and unsuitable (0.21–0.40) with a surface area of 126,485 km^2 (0.84%). The remaining territory (14,813,597 km^2, 98.42%) is unsuitable for the species (Table 3).

Table 3. Classification of habitat suitability in relation to the probability values for the presence of *Salvia ceratophylloides* Ard. (highly unsuitable ($\leq$0.20); unsuitable (0.21–0.40); moderately suitable (0.41–0.60); highly suitable (0.61–0.80); very highly suitable ($\geq$0.80); area in km^2, relative percentage (%) in relation to the entire regional territory, % decrease ($-$) or increase ($+$) of the area suitable for the species.

Time	SSP	Unit	Area tot.	$\geq$0.80	0.61–0.80	0.41–0.60	0.21–0.40	$\leq$0.20
Present time		km^2	237.321	30.440	20.962	59.434	126.485	14,813.597
		%	1.58	0.20	0.14	0.39	0.84	98.42
2021/2040	245	km^2	153.986	25.020	18.258	26.326	84.382	14,896.932
		%	1.02	0.17	0.12	0.17	0.56	98.98
		% inc./dec.	$-$35.11	$-$17.81	$-$12.90	$-$55.71	$-$33.29	$+$0.56
	585	km^2	633.513	129.708	94.667	171.816	237.322	14,417.405
		%	4.21	0.86	0.63	1.14	1.58	95.79
		% inc./dec.	$+$166.94	$+$326.11	$+$351.61	$+$189.09	$+$87.63	$-$2.67
2041/2060	245	km^2	171.414	29.071	22.984	41.241	78.118	14,879.504
		%	1.14	0.19	0.15	0.27	0.52	98.86
		% inc./dec.	$-$27.77	$-$4.50	$+$9.65	$-$30.61	$-$38.24	$+$0.44
	585	km^2	643.814	145.446	122.362	150.514	225.492	14,407.104
		%	4.28	0.97	0.81	1.00	1.50	95.72
		% inc./dec.	$+$171.28	$+$377.81	$+$483.73	$+$153.25	$+$78.28	$-$2.74

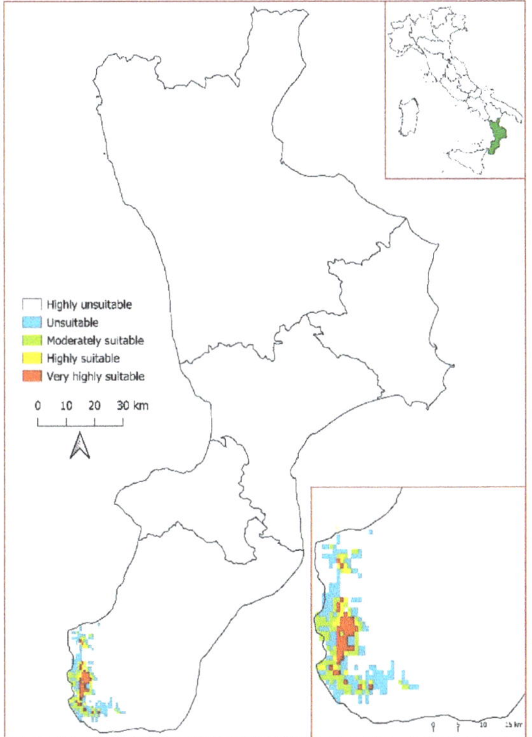

Figure 5. Prediction of the current potential distribution of *Salvia ceratophylloides* Ard. In white, highly unsuitable habitat (≤0.20); in blue, unsuitable (0.21–0.40); in green, moderately suitable (0.41–0.60); in yellow, highly suitable (0.61–0.80); in red, highly suitable (≥0.80). In the top left-hand corner, the Calabria region within the Italian territory is highlighted in green.

3.4. Future Potential Distribution of Salvia ceratophylloides

The jackknife test (Figure 6) reveals that the distribution of *S. ceratophylloides* with SSP 245 over the 2021–2040 period is mainly influenced by the precipitation of the wettest month (Bio 13) with 86.3%, an annual mean temperature (Bio 1) with 7%, and a temperature seasonality (Bio 4) with 5.5%; the remaining variables contributing a total of 1.2%. Regarding the importance of permutation, the most influential variable is Bio 13 with 97.8%. With SSP 245 in the 20-year period between 2041–2060, the variables that contribute the most are the precipitation of the wettest month (Bio 13) with 73.4%, the temperature seasonality (Bio 4) with 12.1%, and the annual mean temperature (Bio 1) with 11.9%; the remaining variables contribute a total of 2.6%.

Regarding the SSP585 scenario in the 20-year period between 2021–2040, the variables contributing most to the model are Bio 13 with 40.7%, Bio 1 with 24.4%, Bio 4 with 15.4%, and Elev. with 16.5%; the remaining variables contribute 3% (Table 2). By permutation importance, there are Bio 1 with 52.4%, Bio 13 with 25%, and Bio 19 with 21.7%; the remaining variables with 0.8%. For the 20-year period between 2041–2060, the variables contributing most to the model are Bio 19 with 42.6%, Bio 13 with 25.6%, Bio 1 with 16.5%, and Bio 4 with 14.2%; the other variables contribute 1.1%. With regard to the importance of permutation, the most influential variables are Bio 4 with 76%, Bio 1 with 15.4%, and Elev. with 6.2%; the other variables account for 2.4% (Table 2).

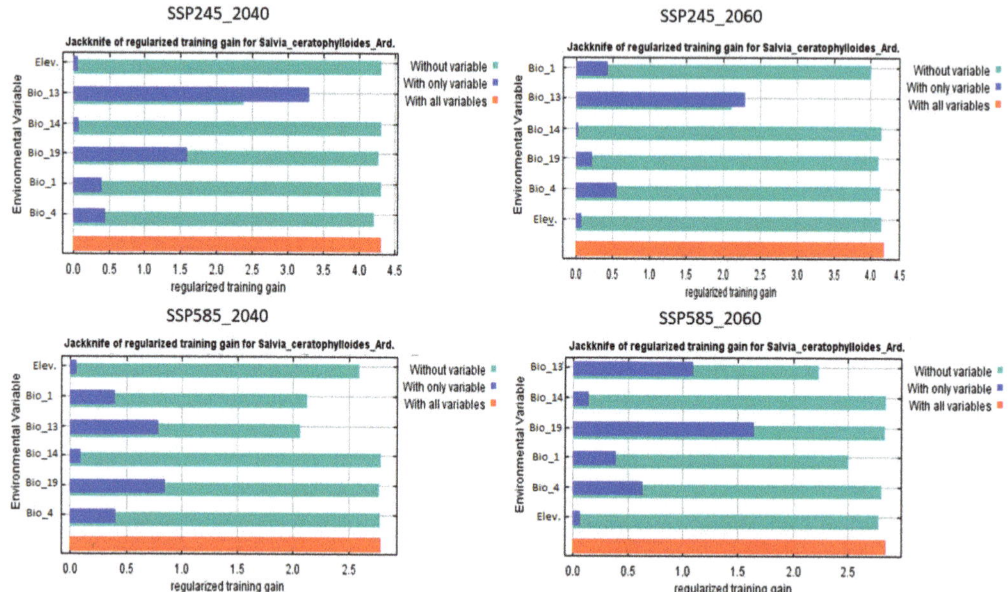

Figure 6. Relative predictive power of different environmental variables based on the jackknife of regularised training gain in MaxEnt models for *Salvia ceratophylloides* Ard.

The future potential distribution of *S. ceratophylloides*, estimated for two types of scenarios (SSP245 and SSP585), always occupies the south/west part of the Italian peninsula and Calabria (Figure 7), without expanding into other parts of the region. The habitat suitable for the species covers a total area of 153,321 km^2 (1.02%) in the SSP245 scenario, 2021–2040 and 171,414 km^2 (1.14%) in the SSP245 scenario of the following 20 years. It can be seen that, from the current distribution model, there is a decrease of 83,335 km^2 or 35% in the 20-year period between 2021–2040, and a decrease of 28% in the following 20-year period, with a loss of 65,907 km^2. In relation to the probability values for the presence of the species in the area of the SSP245 model, 2021–2040, is distributed as follows: very highly suitable ($\geq$0.80) with an area of 25,020 km^2 (0.17%); and highly suitable (0.61–0.80) with an area of 18,258 km^2 (0.12%) (Table 3, Figure 7). The area of the SSP245 model, 2041–2060 is distributed as follows: very highly suitable ($\geq$0.80) with an area of 29.071 km^2 (0.19%); and highly suitable (0.61–0.80) with an area of 22.984 km^2 (0.15%) (Table 3, Figure 7). It can be seen that the most significant decrease is in the optimal occurrence probability value of the species ($\geq$0.80) with 17.81%, or 5420 km^2, less in the SSP245 scenario 2021–2040, compared to the current scenario; in the SSP245 2041–2060 scenario, it is 4.5%, or 1369 km^2, less.

The distribution model with the SSP585 scenario, shows a total area, suitable to host the species, of 633.513 km^2 (4.21%) in the 20-year period between 2021–2040 and 643.814 km^2 (4.28%) in the 20-year period between 2041–2061. Compared to the modelling of the current potential distribution, we can see an increase in area of 396.192 km^2 or 167%, in the 20-year period between 2021–2040, and an increase of 171% in the following 20-year period, with an increase of 65.907 km^2. In relation to the probability values for the presence of the species, the area of the SSP585 model, 2021–2040 is distributed as follows: very highly suitable ($\geq$0.80) with an area of 129.708 km^2 (0.86%); and highly suitable (0.61–0.80) with an area of 94.667 km^2 (0. 63%) (Table 3, Figure 7). The potential area of the SSP585 model, 2041–2060 is distributed differently: very highly suitable ($\geq$0.80) with an area of 145.446 km^2 (0.97%); and highly suitable (0.61–0.80) with an area of 122.362 km^2 (0.81%) (Table 3, Figure 7). It can be seen that the most significant increase is in the probability value of optimal occurrence of the species ($\geq$0.80) in the SSP585 scenario 2021–2040 with

326.11%, or 99.268 km^2, more than the current scenario, in the SSP585, 2041–2060 scenario it is 377.81%, or 115.006 km^2, more.

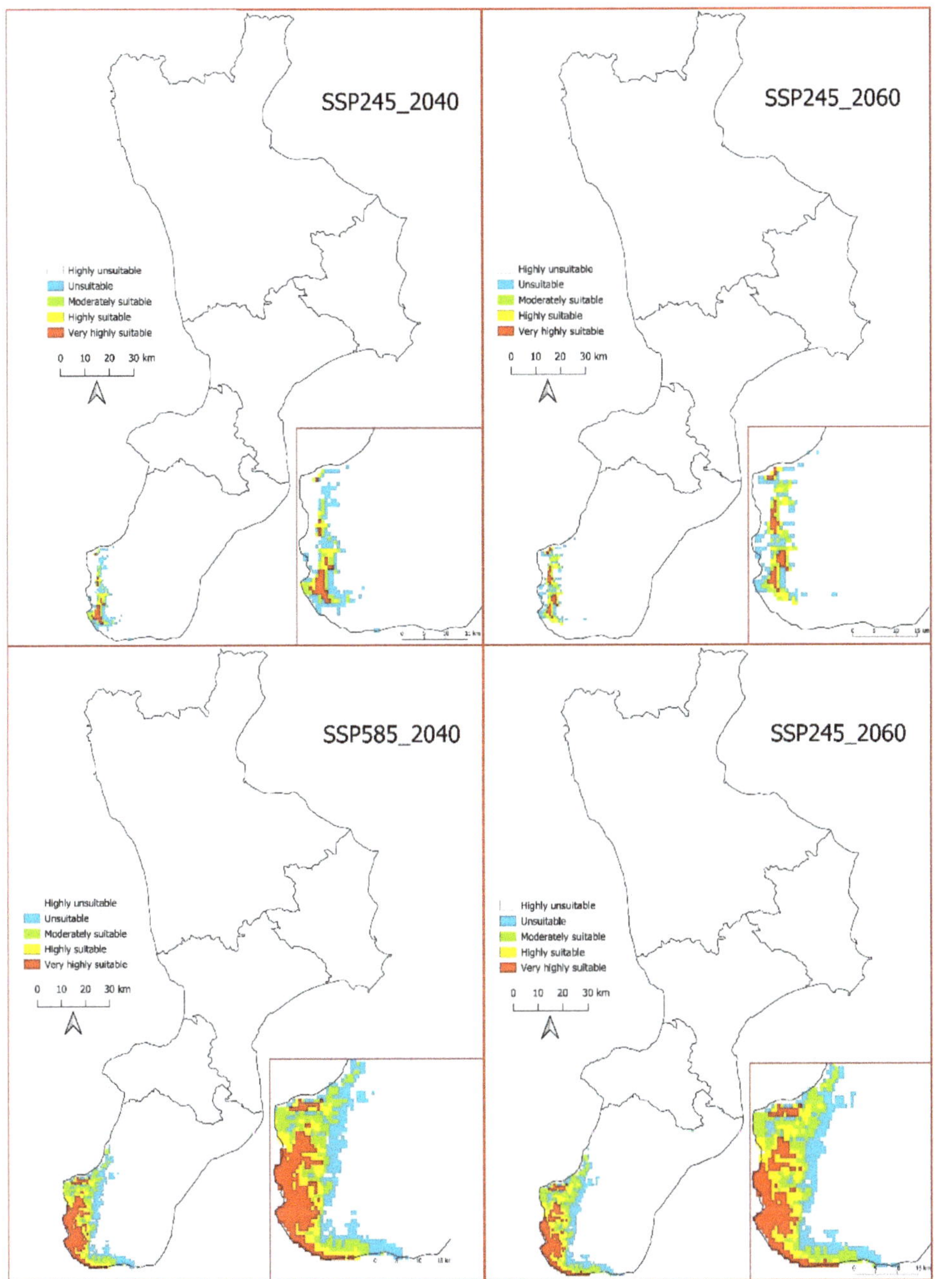

Figure 7. Prediction of the future potential distribution of *Salvia ceratophylloides* Ard. in two different scenarios SSP245 and SSP585 in two periods 2021–2040, 2041–2060. In white, highly unsuitable ($\leq$0.20); in blue, unsuitable (0.21–0.40); in green, moderately suitable (0.41–0.60); in yellow, highly suitable (0.61–0.80); and in red, very highly suitable ($\geq$0.80).

3.5. Real Distribution of Salvia ceratophylloides Analysed with Two Limiting Factors: Geology and Land Use

Field studies and bibliographical references [13,64,68,69,74,75] show that *S. ceratophylloides* grows, in nature, exclusively on loose, sandy, and calcarenite substrates of Pliocene and Pleistocene origin; in particular, the analysis of the geological map [78] shows that there are three types of sandy substrates in Calabria: sands and conglomerates (Pleistocene); sands and conglomerates (Pleistocene–Pliocene); sands and conglomerates (Yellow Sands)–Pliocene, widespread throughout the region.

Although the geological substratum suitable for the species occupies 2066.204 km^2 (14.14% of the regional territory), from the superimposition of the current potential distribution models of the species we can observe that: the habitat suitable for the species covers a total area of 62.427 km^2, or 2.93% of the area occupied in the region by the geological substratum, and 0.41% of the entire regional territory. The suitable area is subdivided as follows in relation to the probability of occurrence of the species: very highly suitable ($\geq$0.80) 16.285 km^2 (0.77% of the area occupied in the region by the substratum, 0.11% of the regional territory); and highly suitable (0.61–0.80) 4.071 km^2 (0.19% of the geological substratum, 0.03% of the regional territory (Table 4, Figure 8). This modelling shows a decrease of 74% compared to the current potential distribution model.

The model with SSP245 2021–2040 presents a total area of 31.892 km^2, equal to 1.50% of the geological substrate and 0.21% of the entire regional territory. In detail, the probability values for the presence of the species are distributed as follows: very highly suitable ($\geq$0.80) with an area of 5.428 km^2 (0.25% of the geological substratum, 0.04% of the regional territory); and highly suitable (0.61–0.80) with an area of 9.500 km^2 (0.45% of the geological substratum, 0.06% of the regional territory) (Table 4, Figure 8).

Table 4. Classification of habitat suitability related to the probability values for the presence of the *Salvia ceratophylloides* Ard. in areas with sandy substrate and conglomerates in the Calabria Region. [highly unsuitable ($\leq$0.20); unsuitable (0.21–0.40); moderately suitable (0.41–0.60); highly suitable (0.61–0.80); very highly suitable ($\geq$0.80); area (km^2), and the relative percentage in relation to the geological substratum (% sub.) and percentage in relation to the entire regional territory (% reg. ter.)].

Time	SSP	Unit	Area tot.	$\geq$0.80	0.61–0.80	0.41–0.60	0.21–0.40	$\leq$0.20
Present time		km^2	62.427	16.285	4.071	19.000	23.071	2066.204
		% sub.	2.93	0.77	0.19	0.89	1.08	97.07
		% reg. ter.	0.41	0.11	0.03	0.13	0.15	13.73
2021/2040	245	km^2	31.892	5.428	9.500	5.428	11.536	2096.739
		% sub.	1.50	0.25	0.45	0.25	0.54	98.50
		% reg. ter.	0.21	0.04	0.06	0.04	0.08	13.93
	585	km^2	187.960	56.320	17.642	37.321	76.677	1940.671
		% sub.	8.83	2.65	0.83	1.75	3.60	91.17
		% reg. ter.	1.25	0.37	0.12	0.25	0.51	12.89
2041/2060	245	km^2	50.213	4.750	6.107	10.857	28.499	2078.418
		% sub.	2.36	0.22	0.29	0.51	1.34	97.64
		% reg. ter.	0.33	0.03	0.04	0.07	0.19	13.81
	585	km^2	192.031	50.892	19.678	48.856	72.605	1936.600
		% sub.	9.02	2.39	0.92	2.30	3.41	90.98
		% reg. ter.	1.28	0.34	0.13	0.32	0.48	12.87

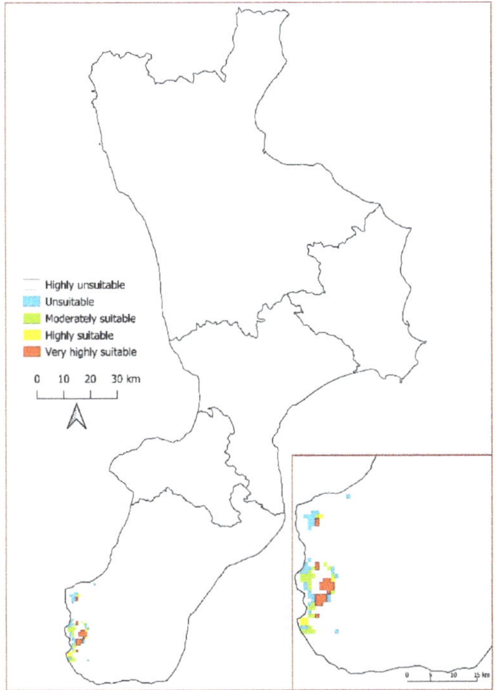

Figure 8. Prediction of the current potential distribution of *Salvia ceratophylloides* Ard. In relation to geological substrate. In white, highly unsuitable habitat ($\leq$0.20); in blue, unsuitable (0.21–0.40); in green, moderately suitable (0.41–0.60); in yellow-low, highly suitable (0.61–0.80); and in red, highly suitable ($\geq$0.80).

The SSP245 2041–2060 scenario presents a total area of 50.213 km^2, equal to 2.36% of the geological substratum and 0.33% of the regional territory. In relation to the probability values for the presence of the species, the area of model SSP245 2041–2060 is distributed as follows: very highly suitable ($\geq$0.80) with an area of 4.750 km^2 (0.22% of the substratum, 0.03% of the regional territory); and highly suitable (0.61–0.80) with an area of 6.107 km^2 (0.29% of the substratum, 0.04% of the regional territory) (Table 4, Figure 9). Compared to the modelling of the current potential distribution interpolated with geological substrate data, we can see an increase in area of 18.164 km^2 or 132% in the 2021–2040 period, and an increase of 266% in the following 20 years with an increase of 36.485 km^2. Considering, on the other hand, the distribution area with the highest probability of hosting the species ($\geq$0.80), overall, there is a decrease. In the 20-year period between 2021–2040, there is a decrease of 66.67% with a reduction in area of 10.857 km^2, and for the 20-year period between 2041–2060, there is a reduction of 70.83% and a loss of 11.535 km^2.

The distribution model with the SSP585 scenario shows a total area, suitable for hosting the species, of 187.960 km^2 in the 20-year period between 2021–2040, equal to 8.83% of the geological substratum considered and 1.25% of the entire Calabrian territory. In the following 20-year period (2041–2061), the area involved is 192.031 km^2, equal to 9.02% of the geological substratum and 1.28% of the regional territory. Considering the probability values for the presence of the species the area of the SSP585 model, in 2021–2040, is divided as follows: very highly suitable ($\geq$0.80) 56.320 km^2 (2.65% of the geological substratum, 0.37% of the regional territory); and highly suitable (0.61–0.80) 17.642 km^2 (0.83% of the geological substratum, 0.12% of the regional territory) (Table 4, Figure 9). The next 20 years (2041–2060) show a distribution area of the species divided as follows: very highly suitable ($\geq$0.80) 50.892 km^2 (2.39% of the geological substratum, 0.34% of the regional territory);

and highly suitable (0.61–0.80) 19.678 km^2 (0.92% of the geological substratum, 0.13% of the regional territory) (Table 4, Figure 9). Comparing the current potential distribution model interpolated with substrate data, with the SSP585 scenario, we find an increase of 174.232 km^2, or 1269%, for the 2021–2040 period, and an increase of 1299% in the following 20 years, with an increase of 178.303 km^2. Considering the distribution area with the highest probability of hosting the species, for the 20-year period between 2021–2040, we would have an increase of 245.84% with an increase of 40.035 km^2, for the 20-year period between 2041–2060, we would have an increase of 212.51% and a gain of 34.607 km^2. The future potential distribution models, interpolated with the geological substratum, show a decrease in area compared to the current potential distribution model. In the 20 years between 2021–2040, with scenario SSP245, there is a 79% decrease, while in the 20 years between 2041–2060 scenario SSP245, the decrease is 71%. The SSP585 scenario shows a decrease of 70% for both of the 20-year periods considered in the modelling.

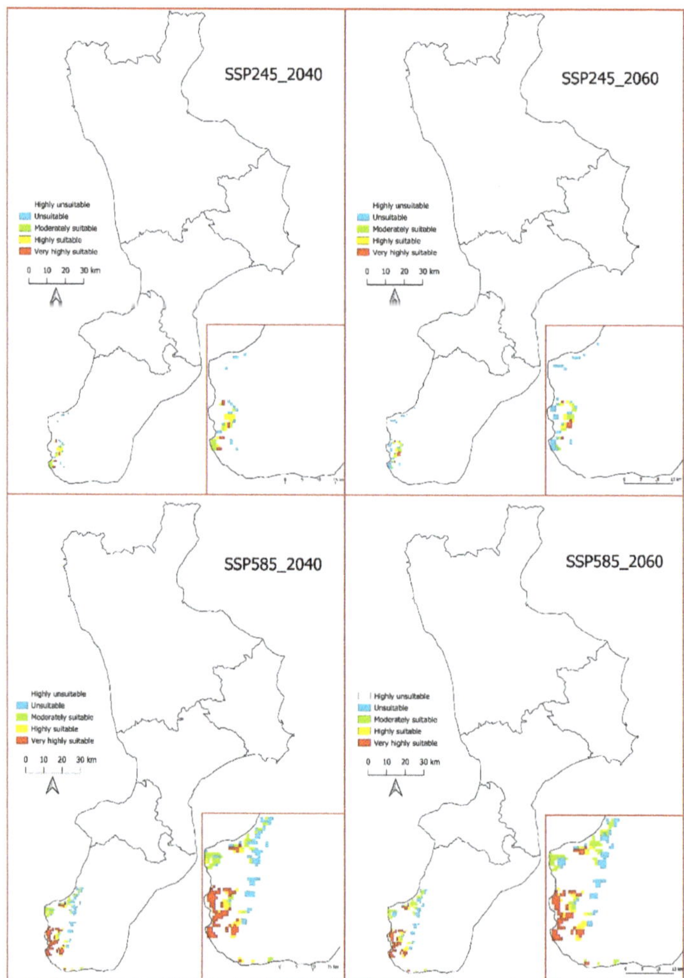

Figure 9. Prediction of the current potential distribution of *Salvia ceratophylloides* Ard. related to geological substrate, in two different scenarios SSP245 and SSP585 and two periods 2021–2040, 2041–2060. In white, highly unsuitable ($\leq$0.20); in blue, unsuitable (0.21–0.40); in green, moderately suitable (0.41–0.60); in yellow, highly suitable (0.61–0.80); and in red, very highly suitable ($\geq$0.80).

In accordance with the CORINE Land Cover system [98], class 2 (agricultural areas) and class 3 (forests and semi-natural areas) were considered in relation to the actual occurrence of the species; the second class was also considered because the Calabrian territory has fragmented agricultural areas that form a complex cultivation mosaic with the forests and semi-natural areas. All other land-use classes were omitted from the analyses.

Interpolating the current distribution model and the land use map shows that the area suitable for the species corresponds to 183.295 km², i.e., 1.22% of the entire regional territory; relating this to the current potential distribution model shows a decrease in the area suitable for the species of 54.026 km², i.e., 23% less (Table 5, Figure 10). The interpolation of the current distribution model with the exclusion of areas where there is no geological substrate suitable for the growth of the species, shows that the entire distribution area is 41.392 km², or 0.28%, of the entire regional territory, and in relation to the current potential distribution model, the area undergoes a decrease of 83%, or 195.929 km² (Figure 10). The table also shows the measures and percentages relating to the classification of habitat suitability in relation to the probability of occurrence values (Table 5).

Table 5. Ranking of habitat suitability of *Salvia ceratophylloides* Ard. Related to the values of probability of occurrence values in the different modelling obtained by interpolation with the current potential distribution. Very unsuitable (≤0.20); unsuitable (0.21–0.40); moderately suitable (0.41–0.60); very suitable (0.61–0.80); very suitable (≥0.80); and the area in km² and relative percentage (%) of decline compared to current potential modelling.

Time	Unit	Area tot.	≥0.80	0.61–0.80	0.41–0.60	0.21–0.40
Present time	km²	237.321	30.440	20.962	59.434	126.485
Present time–land use	km²	183.295	26.476	16.972	40.732	99.115
	%	23	13	19	31	22
Present time–geological substrate	km²	62.427	16.285	4.071	19.000	23.071
	%	74	47	81	68	82
Present time–geological substrate–land use	km²	41.392	14.928	2.714	10.857	12.893
	%	83	51	87	82	90

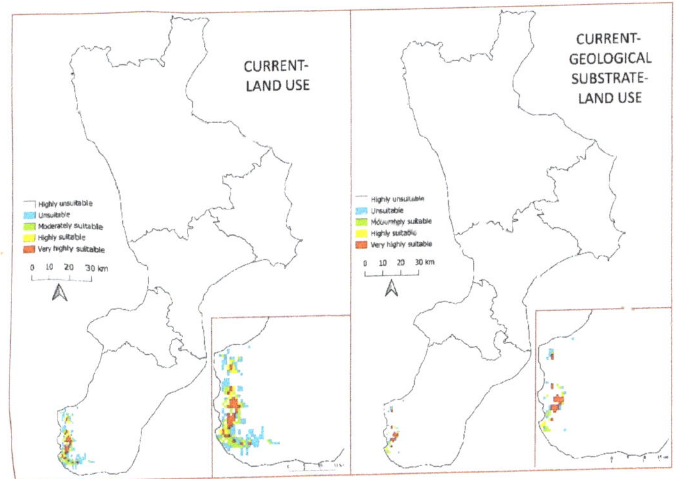

Figure 10. Prediction of the current potential distribution of *Salvia ceratophylloides* Ard. Only in relation to land use and land use with geological substrate. In white, highly unsuitable (≤0.20); in blue, unsuitable (0.21–0.40); in green, moderately suitable (0.41–0.60); in yellow, highly suitable (0.61–0.80); and in red, very highly suitable (≥0.80).

4. Discussion

The results of the current potential modelling show that the environmental suitability of *S. ceratophylloides* always falls within the same range as the observations made in the field in recent years, and are in accordance with the known distribution reported in the literature [13,64,66–69,73,74,76]. We can observe that the species does not extend its range; it is localised exclusively on the extreme southwestern side of the Italian Peninsula and the Calabria region, overlooking the Strait of Messina.

Further analyses of the current model suggest that the distribution of the species is strongly influenced by the same climatic conditions reported earlier in the literature [13,64]. The temperature and humidity variables that condition the reproductive biology of the species [13] proved important in defining the species' current potential distribution pattern. In particular, the average annual temperature parameters limit the distribution (7.8%) of this typically thermophilus species, as do the humidity parameters (precipitation of wettest month, precipitation of driest month, precipitation of coldest quarter) and temperature seasonality, which are closely linked to the species' ecological needs for germination and the release of young seedlings [13,64,80–82]. The elevation variable also has a range that does not differ from elevations measured at actual occurrence points [13,64]. The current potential distribution model includes (with a probability value of very highly suitable occurrence ≥0.80) areas where the species occurs as well as those where it is extinct [70,71]. Therefore, the extinction of the species in the latter areas is the result of severe environmental changes in the suburban area of Reggio Calabria, which is subject to extensive urbanisation and frequent devastating fires [99]. Future model projections for 2021–2040 and 2041–2060, obtained from the SSP245 and SSP585 scenarios, indicate that climate change will significantly influence the distribution of this species. The models with the SSP585 scenario show more significant impacts than the SSP245 scenario, which considers the same bioclimatic characteristics currently in place [19]. The SSP585 scenario shows that, the range suitable for the species will increase by 167% in the 2021–2040 period and 171% in the 2041–2060 period (Table 3, Figure 7). This trend can also be seen in other similar studies [100,101]. Furthermore, the SSP585 scenario predicts an extension of the optimal range to lower altitudes, down to sea level, and other authors also point to an altitudinal shift in the current potential distribution area of the examined species [43].

The SSP245 scenario shows a potential distribution of *S. ceratophylloides* similar to the current modelling, but with a decrease of 35% in the 2021–2040 period and 28% in the 2041–2060 period (Table 3, Figure 7). Similar decreases with the same scenario are also shown for other species [100].

Salvia ceratophylloides is a species with remarkable edaphic specialisation, as it grows exclusively on loose substrates characterised by Pliocene and/or Pleistocene sands and sandy conglomerates [13,64]; these substrates occupy 14% of the entire regional territory, but only 2.93% is occupied by the current potential distribution range of *S. ceratophylloides*. The geological substrate, in this case, becomes one of the limiting factors for the distribution of the species. Compared to the current distribution pattern, there is a decrease of 74%, with a loss of 174,894 km² of suitable area (Table 4, Figure 9), which considerably reduces the potential distribution range of the species. The model obtained by interpolating the SSP245 scenario with the geological substratum shows that the range suitable for the species will decrease by 79% in the period of 2021–2040 and by 71% in the period of 2041–2060 (Table 4, Figure 9), while the SSP585 scenario will show a decrease of 70% in both of the 20 years examined (2021–2040/2041–2060).

A further limiting factor is land use. The species' real range is entirely within a complex environmental mosaic, where agricultural areas (land use class 2) and natural and semi-natural habitats (land use class 3) are highly fragmented and interconnected. On the other hand, it is not present in urban areas (land use category 1), where it was probably present in the past before the expansion of the city, as bibliographic references attest [69]. Excluding the distribution of the species from urban areas, the potential distribution is reduced by 23%, with a loss of 54,026 km^2, which mainly affects the lower elevation band (Table 5, Figure 10).

Considering the constraints imposed by the combination of geologic substrate and land use, the area suitable for *S. ceratophylloides* is reduced by 83% with a total area of 41,392 km^2 (Table 5, Figure 10) compared to 237,321 km^2 (Table 5, Figure 10) in the current distribution model that considers only bioclimatic variables and elevation.

Modelling obtained by subtracting the two limiting factors (geological substrate and land use) from the current range shows that the very highly suitable habitat ($\geq$0.80), i.e., the one in which the probability of finding the species is very high, occupies 14.928 km^2; on the other hand, Laface et al. [13] show that the species has an Area of Occupancy (AOO) of 7 km^2. The modelled distribution is therefore greater than the observed AOO, and *S. ceratophylloides* could potentially be found in other areas where it has not yet been observed or where it is not present due to anthropogenic urbanisation [13] or other limitations, such as pests [102]. The current AOO and anthropogenic pressures justify the assessment of this species as Critically Endangered (CR) [13,64,75]. Without pressures and threats that currently limit the distribution of the species, the range will be 14.928 km^2 (very highly suitable $\geq$0.80), which would reassess the species as Endangered (EN).

Research on *S. ceratophylloides* confirms that the magnitude of change in the distribution of potential 2021–2040 and 2041–2060 niches is comparable. That is, the changes expected for the later period will occur approximately 20 years earlier than is commonly believed, as 2041–2060 is often overlooked in many studies, most of which are for 2061–2080. Hence, there is less time to develop strategies to mitigate the effects of climate change than is usually believed [103,104].

5. Conclusions

This study allowed us to develop very efficient models of the current and future potential distribution of *S. ceratophylloides*. These showed that habitat suitable for the species will decrease in 2021–2040 and 2041–2060 in the SSP245 scenario, and increase in the SSP585 scenario, but it should be noted that important constraints on the species' distribution are due to the geological substrate and land use, which significantly limit the current potential distribution.

The potential distribution model identifies areas of suitable habitat for the species occurrence to evaluate the presence of new occurrence points or to identify locations where there is a high probability of the species occurrence.

The assessment of changes in species distribution related to climate change impacts also made it possible to identify priority areas for reintroduction. Therefore, considering the results obtained, to reduce the risk of extinction of *S. ceratophylloides* in the wild, the reintroduction of the species in areas that are suitable according to modelling is an important in situ conservation measure

The model also gives us clear indications of where to focus conservation activities; for example, by establishing micro-reserves, small protected areas created to ensure the conservation, study, and monitoring of endemic endangered flora in the future, which can be entrusted to environmental associations or the landowner. Furthermore, this work may be useful for future actions to reintroduce and reinforce the existing *S. ceratophylloides* population in areas showed according to modelling. These activities should be accompanied by greater awareness raising among public opinion and political authorities to reduce the impact of human activities.

Author Contributions: Conceptualisation, V.L.A.L. and G.S.; methodology, V.L.A.L. and G.S.; validation, V.L.A.L., C.M.M. and G.S.; formal analysis, V.L.A.L., G.T. and G.S.; investigation, V.L.A.L., C.M.M., G.T. and G.S.; data curation, V.L.A.L., G.T., C.M.M. and G.S.; writing—original draft preparation, V.L.A.L., C.M.M. and G.S.; writing—review and editing, V.L.A.L., C.M.M., G.T., A.S., A.C.-O., R.Q.C. and G.S.; visualisation, V.L.A.L., C.M.M., G.T., A.S., A.C.-O., R.Q.C. and G.S.; supervision, C.M.M. and G.S.; project administration, G.S.; funding acquisition, G.S. All authors have read and agreed to the published version of the manuscript.

Funding: The present research work has been made possible thanks to the Research Project "PROGRAMMA OPERATIVO CALABRIA FESR-FSE 2014/2020-ASSE VI-AZIONE 6.5.A.1-Sub 1-2 (scientific manager Giovanni Spampinato)".

Data Availability Statement: Not applicable.

Conflicts of Interest: The authors declare no conflict of interest.

References

1. Walther, G.R.; Post, E.; Convey, P.; Menzel, A.; Parmesan, C.; Beebee, T.J.; Fromentin, J.M.; Hoegh-Guldberg, O.; Bairlein, F. Ecological responses to recent climate change. *Nature* **2002**, *416*, 389–395. [CrossRef] [PubMed]
2. Thuiller, W.; Lavorel, S.; Araújo, M.B.; Sykes, M.T.; Prentice, I.C. Climate change threats to plant diversity in Europe. *Proc. Natl. Acad. Sci. USA* **2005**, *102*, 8245–8250. [CrossRef]
3. Moreno-Saiz, J.C.; Martínez García, F.; Gavilán, R.G. Plant Conservation in Spain: Strategies to halt the loss of plant diversity. *Mediterr. Bot.* **2018**, *39*, 65–66. [CrossRef]
4. Wood, A.; Stedman-Edwards, P.; Mang, J. *The Root Causes of Biodiversity Loss*; Routledge: London, UK, 2000; p. 416.
5. Del Río, S.; Canas, R.; Cano, E.; Cano-Ortiz, A.; Musarella, C.; Pinto-Gomes, C.; Penas, A. Modelling the impacts of climate change on habitat suitability and vulnerability in deciduous forests in Spain. *Ecol. Indic.* **2021**, *131*, 108202. [CrossRef]
6. Dineva, S. Applying Artificial Intelligence (AI) for Mitigation Climate Change Consequences of the Natural Disasters. *Res. J. Ecol. Environ. Sci.* **2022**, *1*, 1–8. [CrossRef]
7. Içik, K. Rare and endemic species: Why are they prone to extinction? *Turk. J. Bot.* **2011**, *35*, 411–417.
8. Signorino, G.; Cannavò, S.; Crisafulli, A.; Musarella, C.M.; Spampinato, G. Fagonia cretica L. *Inf. Bot. Ita.* **2011**, *43*, 397–399.
9. Spampinato, G.; Musarella, C.M.; Cano-Ortiz, A.; Signorino, G. Habitat, occurrence and conservation status of the Saharo-Macaronesian and Southern-Mediterranean element *Fagonia cretica* L. (Zygophyllaceae) in Italy. *J. Arid. Land* **2018**, *10*, 140–151. [CrossRef]
10. Abdelaal, M.; Fois, M.; Fenu, G.; Bacchetta, G. Using MaxEnt modeling to predict the potential distribution of the endemic plant Rosa arabica Crép. in Egypt. *Ecol. Inform.* **2019**, *50*, 68–75. [CrossRef]
11. Carmona, E.C.; Ortiz, A.C.; Musarella, C.M. Introductory Chapter: Endemism as a Basic Element for the Conservation of Species and Habitats. In *Endemic Species [Internet]*; Carmona, E.C., Musarella, C.M., Ortiz, A.C., Eds.; IntechOpen: London, UK, 2019; Available online: https://www.intechopen.com/chapters/65963 (accessed on 16 October 2022).
12. Orsenigo, S.; Bernardo, L.; Cambria, S.; Gargano, D.; Laface, V.L.A.; Musarella, C.M.; Passalacqua, N.G.; Spampinato, G.; Tavilla, G.; Fenu, G. Global and Regional IUCN Red List Assessments: 9. *Ital. Botanist.* **2020**, *9*, 111–123. [CrossRef]
13. Laface, V.L.A.; Musarella, C.M.; Sorgonà, A.; Spampinato, G. Analysis of the population structure and dynamic of endemic *Salvia ceratophylloides* Ard. (Lamiaceae). *Sustainability* **2022**, *14*, 10295. [CrossRef]
14. Caruso, G. Anthyllis hermanniae L. subsp. brutia Brullo & Giusso (Fabaceae): Population survey and conservation tasks. *Res. J. Ecol. Environ. Sci.* **2022**, *2*, 92–102. [CrossRef]
15. Jones, P.D.; Osborn, T.J.; Briffa, K.R. The evolution of climate over the last millennium. *Science* **2001**, *292*, 662–667. [CrossRef] [PubMed]
16. Xoplaki, E.; Gonzales-Rouco, F.J.; Luterbacher, J.; Wanner, H. Mediterranean summer air temperature variability and its connection to the large-scale atmospheric circulation and SSTs. *Clim. Dyn.* **2003**, *20*, 723–739. [CrossRef]
17. Solomon, S.; Qin, D.; Manning, M.; Alley, R.B.; Berntsen, T.; Bindoff, N.L.; Chen, Z.; Chidthaisong, A.; Gregory, J.M.; Hegerl, G.C.; et al. Technical summary. In *Climate Change 2007: The Physical Science Basis. Contribution of Working Group I to the Fourth Assessment Report of the Intergovernmental Panel on Climate Change*; Solomon, S., Qin, D., Manning, M., Chen, Z., Marquis, M., Averyt, K., Tignor, M., Miller, H., Eds.; Cambridge University Press: Cambridge, UK; New York, NY, USA, 2007.
18. Mariotti, A.; Pan, Y.; Zeng, N.; Alessandri, A. Long-term climate change in the Mediterranean region in the midst of decadal variability. *Clim. Dyn.* **2015**, *44*, 1437–1456. [CrossRef]
19. Intergovernmental Panel on Climate Change (IPCC). Climate Change 2022: Mitigation of Climate Change. In *Contribution of Working Group III to the Sixth Assessment Report of the Intergovernmental Panel on Climate Change*; Shukla, P.R., Skea, J., Slade, R., al Khourdajie, A., van Diemen, R., McCollum, D., Pathak, M., Some, S., Vyas, P., Fradera, R., et al., Eds.; Cambridge University Press: Cambridge, UK; New York, NY, USA, 2022.
20. Laface, V.L.A.; Musarella, C.M.; Ortiz, A.C.; Canas, R.Q.; Cannavò, S.; Spampinato, G. Three New Alien Taxa for Europe and a Chorological Update on the Alien Vascular Flora of Calabria (Southern Italy). *Plants* **2020**, *9*, 1181. [CrossRef]
21. Musarella, C.M.; Stinca, A.; Cano-Ortíz, A.; Laface, V.L.A.; Petrilli, R.; Esposito, A.; Spampinato, G. New data on the alien vascular flora of Calabria (Southern Italy). *Ann. Bot.* **2020**, *10*, 55–66. [CrossRef]
22. Musarella, C.M. Solanum torvum Sw. (Solanaceae): A new alien species for Europe. *Genet. Resour. Crop Evol.* **2020**, *67*, 515–522. [CrossRef]
23. Raposo, M.A.M.; Pinto Gomes, C.J.; Nunes, L.J.R. Evaluation of Species Invasiveness: A Case Study with *Acacia dealbata* Link. on the Slopes of Cabeça (Seia-Portugal). *Sustainability* **2021**, *13*, 11233. [CrossRef]
24. Zhang, Y.; Tang, J.; Ren, G.; Zhao, K.; Wang, X. Global potential distribution prediction of *Xanthium italicum* based on Maxent model. *Sci. Rep.* **2021**, *11*, 16545. [CrossRef]

25. de Carvalho, C.A.; Raposo, M.; Pinto-Gomes, C.; Matos, R. Native or Exotic: A Bibliographical Review of the Debate on Ecological Science Methodologies: Valuable Lessons for Urban Green Space Design. *Land* **2022**, *11*, 1201. [CrossRef]
26. Spampinato, G.; Laface, V.L.A.; Posillipo, G.; Cano-Ortiz, A.; Quinto-Canas, R.; Musarella, C.M. Alien flora in Calabria (Southern Italy): An updated checklist. *Biol. Invasions* **2022**, *24*, 2323–2334. [CrossRef]
27. Ighbareyeh, J.M.H.; Suliemieh, A.A.-R.A.; Sheqwarah, M.; Cano-Ortiz, A.; Carmona, E.C. Flora and Phytosociological of Plant in Al-Dawaimah of Palestine. *Res. J. Ecol. Environ. Sci.* **2022**, *2*, 58–91. [CrossRef]
28. Molina-Venegas, R.; Aparicio, A.; Lavergne, S.; Arroyo, J. Climatic and topographical correlates of plant palaeo- and neoendemism in a Mediterranean biodiversity hotspot. *Ann. Bot.* **2017**, *119*, 229–238. [CrossRef] [PubMed]
29. Rundel, P.W.; Arroyo, M.T.K.; Cowling, R.M.; Keeley, J.E.; Lamont, B.B.; Vargas, P. Mediterranean biomes: Evolution of their vegetation, floras, and climate. *Annu. Rev. Ecol. Evol. Syst.* **2016**, *47*, 383–407. [CrossRef]
30. Musarella, C.M.; Tripodi, G. La flora della rupe e dei ruderi di Pentidattilo (Reggio Calabria). *Inform. Bot. Ital.* **2004**, *36*, 3–12.
31. Pecl, G.T.; Araújo, M.B.; Bell, J.D.; Blanchard, J.; Bonebrake, T.C.; Chen, I.C.; Clark, T.D.; Colwell, R.K.; Danielsen, F.; Evengård, B.; et al. Biodiversity redistribution under climate change: Impacts on ecosystems and human well-being. *Science* **2017**, *355*, eaai9214. [CrossRef]
32. Qin, A.; Liu, B.; Guo, Q.; Bussmann, R.W.; Ma, F.; Jian, Z.; Xu, G.; Pei, S. Maxent modeling for predicting impacts of climate change on the potential distribution of *Thuja sutchuenensis* Franch., an extremely endangered conifer from southwestern China. *Glob. Ecol. Conserv.* **2017**, *10*, 139–146. [CrossRef]
33. del Río, S.; Álvarez-Esteban, R.; Cano, E.; Pinto-Gomes, C.; Penas, Á. Potential Impacts of Climate Change on Habitat Suitability of *Fagus sylvatica* L. For. Spain *Plant Biosyst.* **2018**, *152*, 1205–1213. [CrossRef]
34. Rus, J.D.; Ramírez-Rodríguez, R.; Amich, F.; Melendo-Luque, M. Habitat Distribution Modelling, under the Present Climatic Scenario, of the Threatened Endemic Iberian *Delphinium fissum* subsp. *sordidum* (Ranunculaceae) and Implications for Its Conservation. *Plant Biosyst.* **2018**, *152*, 891–900. [CrossRef]
35. Anand, V.; Oinam, B.; Singh, H. Predicting the current and future potential spatial distribution of endangered *Rucervus eldii eldii* (Sangai) using MaxEnt model. *Environ. Monit. Assess.* **2021**, *193*, 147. [CrossRef]
36. Rojo, J.; Fernández-González, F.; Lara, B.; Bouso, V.; Crespo, G.; Hernández-Palacios, G.; Rodríguez-Rojo, M.P.; Rodríguez-Torres, A.; Smith, M.; Pérez-Badia, R. The effects of climate change on the flowering phenology of alder trees in Southwestern Europe. *Mediterr. Bot.* **2021**, *42*, e67360. [CrossRef]
37. Wilson, C.D.; Roberts, D.; Reid, N. Applying species distribution modelling to identify areas of high conservation value for endangered species: A case study using *Margaritifera margaritifera* (L.). *Biol. Conserv.* **2011**, *144*, 821–829. [CrossRef]
38. Reich, D.; Flatscher, R.; Pellegrino, G.; Hülber, K.; Wessely, J.; Gattringer, A.; Greimler, J. Biogeography of amphi-adriatic *Gentianella crispata* (Gentianaceae): A northern refugium and recent trans-adriatic migration. *Plant Biosyst.* **2022**, *156*, 754–768. [CrossRef]
39. Phillips, S.J.; Anderson, R.P.; Schapire, R.E. Maximum entropy modeling of species geographic distributions. *Ecol. Model.* **2006**, *190*, 231–259. [CrossRef]
40. Elith, J.; Phillips, S.J.; Hastie, T.; Dudík, M.; Chee, Y.E.; Yates, C.J. A statistical explanation of maxent for ecologists. *Divers. Distrib.* **2011**, *17*, 43–57. [CrossRef]
41. Soilhi, Z.; Sayari, N.; Benalouache, N.; Mekki, M. Predicting current and future distributions of *Mentha pulegium* L. in Tunisia under climate change conditions, using the MaxEnt model. *Ecol. Inform.* **2022**, *68*, 101533. [CrossRef]
42. Thakur, S.; Rai, I.D.; Singh, B.; Dutt, H.C.; Musarella, C.M. Predicting the Suitable Habitats of *Elwendia persica* in Indian Himalayan Region (IHR). *Plant Biosyst.* 2023, *in press.*
43. Gebrewahid, Y.; Abrehe, S.; Meresa, E.; Eyasu, G.; Abay, K.; Gebreab, G.; Kidanemariam, K.; Adissu, G.; Abreha, G.; Darcha, G. Current and future predicting potential areas of *Oxytenanthera abyssinica* (A. Richard) using MaxEnt model under climate change in Northern Ethiopia. *Ecol. Process.* **2020**, *9*, 6. [CrossRef]
44. Bhandari, M.S.; Shankhwar, R.; Maikhuri, S.; Pandey, S.; Meena, R.K.; Ginwal, H.S.; Kant, R.; Rawat, P.S.; Martins-Ferreira, M.A.; Silveira, L.H. Prediction of ecological and geological niches of *Salvadora oleoides* in arid zones of India: Causes and consequences of global warming. *Arab. J. Geosci.* **2021**, *14*, 524. [CrossRef]
45. Dai, X.; Wu, W.; Ji, L.; Tian, S.; Yang, B.; Guan, B.; Wu, D. MaxEnt model-based prediction of potential distributions of *Parnassia wightiana* (Celastraceae) in China. *Biodivers. Data J.* **2022**, *10*, e81073. [CrossRef]
46. Khan, A.M.; Li, Q.; Saqib, Z.; Khan, N.; Habib, T.; Khalid, N.; Majeed, M.; Tariq, A. MaxEnt Modelling and Impact of Climate Change on Habitat Suitability Variations of Economically Important Chilgoza Pine (*Pinus gerardiana* Wall.) in South Asia. *Forests* **2022**, *13*, 715. [CrossRef]
47. Zhang, L.; Zhu, L.; Li, Y.; Zhu, W.; Chen, Y. Maxent Modelling Predicts a Shift in Suitable Habitats of a Subtropical Evergreen Tree (*Cyclobalanopsis glauca* (Thunberg) Oersted) under Climate Change Scenarios in China. *Forests* **2022**, *13*, 126. [CrossRef]
48. Parveen, S.; Kaur, S.; Baishya, R.; Goel, S. Predicting the potential suitable habitats of genus *Nymphaea* in India using MaxEnt modeling. *Environ. Monit. Assess.* **2022**, *194*, 853. [CrossRef]
49. Khajoei Nasab, F.; Mehrabian, A.; Mostafavi, H.; Neemati, A. The influence of climate change on the suitable habitats of Allium species endemic to Iran. *Environ. Monit. Assess.* **2022**, *194*, 169. [CrossRef] [PubMed]
50. Kumar, S.; Stohlgren, T.J. Maxent modeling for predicting suitable habitat for threatened and endangered tree *Canacomyrica monticola* in New Caledonia. *J. Ecol. Nat. Environ.* **2009**, *1*, 94–98.

51. Marcer, A.; Sáez, L.; Molowny-Horas, R.; Pons, X.; Pino, J. Using species distribution modelling to disentangle realised versus potential distributions for rare species conservation. *Biol. Conserv.* **2013**, *166*, 221–230. [CrossRef]
52. Shcheglovitova, M.; Anderson, R.P. Estimating optimal complexity for ecological niche models: A jackknife approach for species with small sample sizes. *Ecol. Model.* **2013**, *269*, 9–17. [CrossRef]
53. Booth, T.H. Why understanding the pioneering and continuing contributions of BIOCLIM to species distribution modelling is important. *Austral Ecol.* **2018**, *43*, 852–860. [CrossRef]
54. Heywood, V.H.; Richardson, I.B.K. Labiatae. In *Flora Europaea*; Tutin, T.G., Heywood, V.H., Burges, N.A., Moore, D.M., Valentine, D.H., Walters, S.M., Webb, D.A., Eds.; Cambridge at the University Press: Cambridge, UK, 1972; Volume 3, pp. 126–192.
55. Kadereit, J.W. *The Families and Genera of Vascular Plants*; Lamiales: Berlin, Germany, 2004; Volume VII.
56. Perrino, E.V.; Valerio, F.; Gannouchi, A.; Trani, A.; Mezzapesa, G. Ecological and Plant Community Implication on Essential Oils Composition in Useful Wild Officinal Species: A Pilot Case Study in Apulia (Italy). *Plants* **2021**, *10*, 574. [CrossRef]
57. Valerio, F.; Mezzapesa, G.N.; Ghannouchi, A.; Mondelli, D.; Logrieco, A.F.; Perrino, E.V. Characterization and Antimicrobial Properties of Essential Oils from Four Wild Taxa of *Lamiaceae* Family Growing in Apulia. *Agronomy* **2021**, *11*, 1431. [CrossRef]
58. Cianfaglione, K.; Bartolucci, F.; Ciaschetti, G.; Conti, F.; Pirone, G. Characterization of *Thymus vulgaris* subsp. vulgaris Community by Using a Multidisciplinary Approach: A Case Study from Central Italy. *Sustainability* **2022**, *14*, 3981. [CrossRef]
59. Peruzzi, L.; Conti, F.; Bartolucci, F. An inventory of vascular plants endemic to Italy. *Phytotaxa* **2014**, *168*, 1–75. [CrossRef]
60. Bartolucci, F.; Peruzzi, L.; Galasso, G.; Albano, A.; Alessandrini, A.; Ardenghi, N.M.G.; Astuti, G.; Bacchetta, G.; Ballelli, S.; Banfi, E.; et al. An updated checklist of the vascular flora native to Italy. *Plant Biosyst.* **2018**, *152*, 179–303. [CrossRef]
61. Pignatti, S. *Flora d'Italia*; Edagricole: Bologna, Italy, 1982; Volume 2, pp. 502–507.
62. Pignatti, S. *Flora d'Italia*, 2nd ed.; Edagricole: Bologna, Italy, 2018; Volume 3, pp. 301–310.
63. Spampinato, G. *Guida alla flora dell'Aspromonte*, 2nd ed.; Laruffa Editore: Reggio Calabria, Italy, 2014; pp. 38, 244–245.
64. Spampinato, G.; Laface, V.L.A.; Ortiz, A.C.; Canas, R.Q.; Musarella, C.M. *Salvia ceratophylloides* Ard. (Lamiaceae): A rare endemic species of Calabria (Southern Italy). In *Endemic Species*; Cano Carmona, E., Musarella, C.M., Cano Ortiz, A., Eds.; IntechOpen: London, UK, 2019.
65. Salmeri, C. Karyological data of some plant species native to South Italy. *Flora Mediterr.* **2019**, *29*, 334–340. [CrossRef]
66. Tenore, M. *Sylloge Plantarum Vascularium Florae Neapolitanae Hucusque Detectarum*; Ex Typographia Fibreni: Neapoli, Greece, 1831; p. 1.
67. Macchiati, C. Catalogo delle piante raccolte nei dintorni di Reggio Calabria dal settembre 1881 al febbraio 1883. *Nuovo Giorn. Bot. Ital.* **1884**, *16*, 59–100.
68. Lacaita, C. Addenda et emendanda ad floram italicam. *Bull. Soc. Bot. Ital.* **1921**, *28*, 18–19.
69. Lacaita, C. Piante italiane critiche o rare: 67. *Salvia ceratophylloides* Arduino. *Nuovo Giorn. Bot. Ital.* **1921**, *28*, 144–147.
70. Conti, F.; Manzi, A.; Pedrotti, F. *Liste Rosse Regionali delle Piante d'Italia*; WWF Italia, Società Botanica Italiana: Camerino, Italy, 1997.
71. Del Carratore, F.; Garbari, F. Il Gen. *Salvia* Sect. *Plethiosphace* (Lamiaceae) in Italia. *Arch. Geobot.* **2001**, *7*, 41–62.
72. Scoppola, A.; Spampinato, G. Atlante delle specie a rischio di estinzione. Versione 1.0. CD-Rom enclosed to the volume. In *Stato Delle Conoscenze Sulla Flora Vascolare d'Italia*; Scoppola, A., Blasi, C., Eds.; Palombi Editori: Roma, Italy, 2005.
73. Spampinato, G.; Crisafulli, A. Struttura delle popolazioni e sinecologia di *Salvia ceratophylloides* (Lamiaceae) specie endemica minacciata di estinzione, 56. In *Book of Abstracts of 103° S.B.I. Congress*; Università Mediterranea di Reggio Calabria: Reggio Calabria, Italy, 2008; pp. 17–19.
74. Crisafulli, A.; Cannavò, S.; Maiorca, G.; Musarella, C.M.; Signorino, G.; Spampinato, G. Aggiornamenti floristici per la Calabria. *Inform. Bot. Ital.* **2010**, *42*, 431–442.
75. Spampinato, G.; Crisafulli, A.; Marino, A.; Signorino, G. *Salvia ceratophylloides* Ard. *Inf. Bot. Ital.* **2011**, *43*, 381–458.
76. Laface, V.L.A.; Musarella, C.M.; Spampinato, G. Conservation status of the Aspromontana flora: Monitoring and new stations of *Salvia ceratophylloides* Ard. (Lamiaceae) endemic species in Reggio Calabria (southern Italy). In *Abstracts Book of 113° Congresso della Società Botanica Italiana (V International Plant Science Conference (IPSC)*; Società Botanica Italiana: Fisciano, Italy, 2018; pp. 12–15, ISBN 978-88-85915-22-0.
77. International Union for Conservation of Nature (IUCN). *IUCN Red List Categories and Criteria: Version 3.1*, 2nd ed.; IUCN: Gland, Switzerland; Cambridge, UK, 2012; pp. iv + 32. ISBN 978-2-8317-1435-6.
78. *Cassa per il Mezzogiorno—Cassa per Opere Straordinarie di Pubblico Interesse nell'Italia Meridionale, Carta Geologica della Calabria*; Cassa per il Mezzogiorno: Rome, Italy, 1967–1972.
79. Pesaresi, S.; Galdenzi, D.; Biondi, E.; Casavecchia, S. Bioclimate of Italy: Application of the worldwide bioclimatic classification system. *J. Maps* **2014**, *10*, 538–553. [CrossRef]
80. Abate, E.; Azzarà, M.; Trifilò, P. When Water Availability Is Low, Two Mediterranean *Salvia* Species Rely on Root Hydraulics. *Plants* **2021**, *10*, 1888. [CrossRef] [PubMed]
81. Abate, E.; Nardini, A.; Petruzzellis, F.; Trifilò, P. Too dry to survive: Leaf hydraulic failure in two *Salvia* species can be predicted on the basis of water content. *Plant Physiol. Biochem.* **2021**, *166*, 215–224. [CrossRef] [PubMed]
82. Vescio, R.; Abenavoli, M.R.; Araniti, F.; Musarella, C.M.; Sofo, A.; Laface, V.L.A.; Spampinato, G.; Sorgonà, A. The Assessment and the Within-Plant Variation of the Morpho-Physiological Traits and VOCs Profile in Endemic and Rare *Salvia ceratophylloides* Ard. (Lamiaceae). *Plants* **2021**, *10*, 474. [CrossRef]

83. QGIS 2022. QGIS Geographic Information System. Open Source Geospatial Foundation Project. Available online: http://qgis.osgeo.org (accessed on 10 October 2022).

84. WordClim. Available online: https://www.worldclim.org/ (accessed on 10 October 2022).

85. Fick, S.E.; Hijmans, R.J. WorldClim 2: New 1km spatial resolution climate surfaces for global land areas. *Int. J. Climatol.* **2017**, *37*, 4302–4315. [CrossRef]

86. Dormann, C.F.; Elith, J.; Bacher, S.; Buchmann, C.; Carl, G.; Carré, G.; Marquéz, J.R.G.; Gruber, B.; Lafourcade, B.; Leitão, P.J.; et al. Collinearity: A review of methods to deal with it and a simulation study evaluating their performance. *Ecography* **2013**, *36*, 27–46. [CrossRef]

87. Saupe, E.E.; Barve, V.; Myers, C.E.; Soberón, J.; Barve, N.; Hensz, C.M.; Peterson, A.T.; Owens, H.L.; Lira-Noriega, A. Variation in niche and distribution model performance: The need for a priori assessment of key causal factors. *Ecol. Model.* **2012**, *237–238*, 11–22. [CrossRef]

88. Pearson, K. Notes on regression and inheritance in the case of two parents. *Proc. R. Soc. Lond.* **1895**, *58*, 240–242.

89. Hammer, Ø. Past Software. In *Natural History Museum*; University of Oslo: Oslo, Norway, 2022.

90. Quenouille, M.H. Approximate Tests of Correlation in Time Series. *J. R. Stat. Soc.* **1949**, *11*, 68–84.

91. Castellana, S.; Martin, M.Á.; Solla, A.; Alcaide, F.; Villani, F.; Cherubini, M.; Neale, D.; Mattioni, C. Signatures of local adaptation to climate in natural populations of sweet chestnut (*Castanea sativa* Mill.) from southern Europe. *Ann. For. Sci.* **2021**, *78*, 1–21. [CrossRef]

92. Abou-Shaara, H.F.; Darwish, A.A. Expected prevalence of the facultative parasitoid Megaselia scalaris of honey bees in Africa and the Mediterranean region under climate change conditions. *Int. J. Trop. Insect Sci.* **2021**, *41*, 3137–3145. [CrossRef]

93. Phillips, S.J. Transferability, sample selection bias and background data in presence-only modeling: A response to Peterson et al. and (2007). *Ecography* **2008**, *31*, 272–278. [CrossRef]

94. Phillips, S.J.; Dudik, M.; Schapire, R.E. Maxent Software for Modeling Species Niches and Distributions (Version 3.4.4). Available online: http://biodiversityinformatics.amnh.org/open_source/maxent/ (accessed on 11 October 2022).

95. Fielding, A.H.; Bell, J.F. A review of methods for the assessment of prediction errors in conservation presence/absence models. *Environ. Conserv.* **1997**, *24*, 38–49. [CrossRef]

96. Warren, D.L.; Seifert, S.N. Ecological niche modeling in Maxent: The importance of model complexity and the performance of model selection criteria. *Ecol. Appl.* **2001**, *21*, 335–342. [CrossRef]

97. Carta di Uso del Territorio della Regione Calabria. ARSAC. Available online: http://93.51.147.138/corine_land_cover.html (accessed on 11 October 2022).

98. Bossard, M.; Feranec, J.; Otahel, J. *CORINE Land Cover Technical Guide: Addendum 2000*; European Environment Agency: Copenhagen, Denmark, 2000; Volume 40.

99. European Forest Fire Informationon System (EFFIS). Available online: https://effis.jrc.ec.europa.eu/ (accessed on 14 October 2022).

100. Canturk, U.; Kulaç, Ş. The effects of climate change scenarios on *Tilia* ssp. in Turkey. *Environ. Monit. Assess.* **2021**, *193*, 771. [CrossRef]

101. Ma, B.; Sun, J. Predicting the distribution of *Stipa purpurea* across the Tibetan Plateau via the MaxEnt model. *BMC Ecol.* **2018**, *18*, 10. [CrossRef]

102. Bonsignore, C.P.; Laface, V.L.A.; Vono, G.; Marullo, R.; Musarella, C.M.; Spampinato, G. Threats Posed to the Rediscovered and Rare *Salvia ceratophylloides* Ard. (Lamiaceae) by Borer and Seed Feeder Insect Species. *Diversity* **2021**, *13*, 33. [CrossRef]

103. Olszewski, P.; Dyderski, M.K.; Dylewski, Ł.; Bogusch, P.; Schmid-Egger, C.; Ljubomirov, T.; Zimmermann, D.; Le Divelec, R.; Wiśniowski, B.; Twerd, L.; et al. European beewolf (*Philanthus triangulum*) will expand its geographic range as a result of climate warming. *Reg. Environ. Chang.* **2022**, *22*, 129. [CrossRef]

104. Puchałka, R.; Dyderski, M.K.; Vítková, M.; Sádlo, J.; Klisz, M.; Netsvetov, M.; Prokopuk, Y.; Matisons, R.; Mionskowski, M.; Wojda, T.; et al. Black locust (*Robinia pseudoacacia* L.) range contraction and expansion in Europe under changing climate. *Glob. Chang. Biol.* **2021**, *27*, 1587–1600. [CrossRef]

 *land*

Article

How Much Complexity Is Required for Modelling Grassland Production at Regional Scales?

Iris Vogeler [1,2,*], Christof Kluß [1], Tammo Peters [1] and Friedhelm Taube [1,3]

[1] Grass and Forage Science/Organic Agriculture, Christian Albrechts University, 24118 Kiel, Germany
[2] Department of Agroecology, Aarhus University, 8830 Tjele, Denmark
[3] Grass Based Dairy Systems, Animal Production Systems Group, Wageningen University, 6708 WD Wageningen, The Netherlands
* Correspondence: ivogeler@gfo.uni-kiel.de

Abstract: Studies evaluating the complexity of models, which are suitable to simulate grass growth at regional scales in intensive grassland production systems are scarce. Therefore, two different grass growth models (GrasProg1.0 and APSIM) with different complexity and input requirements were compared against long-term observations from variety trials with perennial ryegrass (*Lolium perenne*) in Germany and Denmark. The trial sites covered a large range of environmental conditions, with annual average temperatures ranging from 5.9 to 10.3 °C, and annual rainfall from 536 to 1154 mm. The sites also varied regarding soil type, which were for modelling categorised into three different groups according to their plant available water (PAW) content: light soils with a PAW of 60 mm, medium soils with a PAW of 80 mm, and heavy soils with a PAW of 100 mm. The objective was to investigate whether the simple model performed equally well with the given low number of inputs, namely climate and PAW group. Evaluation statistics showed that both models provided satisfactory results, with root mean square errors for individual cuts ranging from 0.59 to 1.28 t dry matter ha^{-1}. The model efficiency (Nash–Sutcliffe efficiency) for the separate cuts were also good for both models, with 81% of the sites having a positive Nash–Sutcliffe efficiency value with GrasProg1.0, and 72% with APSIM. These results reveal that without detailed site-specific descriptions, the less complex GrasProg1.0 model can be incorporated into a simple decision support tool for optimising grassland management in intensive livestock production systems.

Keywords: GrasProg1.0; APSIM; perennial ryegrass; North-West Europe

Citation: Vogeler, I.; Kluß, C.; Peters, T.; Taube, F. How Much Complexity Is Required for Modelling Grassland Production at Regional Scales? *Land* **2023**, *12*, 327. https://doi.org/10.3390/land12020327

Academic Editors: Michael Vrahnakis and Yannis (Ioannis) Kazoglou

Received: 20 December 2022
Revised: 19 January 2023
Accepted: 23 January 2023
Published: 25 January 2023

1. Introduction

Despite the importance of grasslands in sustaining ruminant livestock farming, information about grassland productivity and its response to changing climatic conditions, with increasing frequency and severity of extreme events, is scarce [1–3]. Simulation models constitute a key tool to understanding and predicting the effects of climate variations and management strategies on biophysical systems. Various models have been developed and used for predicting grass growth. Modelling approaches vary from simple empirical to complex mechanistic models, and operate on different hierarchical levels, from the individual plant [4], to plant communities based on plant functional types [5,6], and to the field [7–10] or even global scale [11,12].

Complex process-based models at the individual plant level include numerous plant-physiological functions, which are very parameter intensive and data demanding [6,13]. For modelling at higher hierarchical levels, simple physiological and morphological plant traits as well as statistical functions, which represent dynamic plant growth processes, have been integrated into mechanistic models [7,10,14]. Some of these simpler dynamic and mechanistic modelling approaches have also been integrated into decision support tools for practical grassland management [15]. The compromise between model complexity and input data requirement has been addressed in several studies, and the selection of the model for a given application depends, among other factors, on the expected performance, data availability as well as the users' familiarity with the model [16,17].

In Europe, grasslands amount to about 34% of the agricultural area [18], with a similar share in Germany of 28% [19]. Grasslands provide a wide range of ecosystem services including carbon sequestration, water filtering, and the provision of habitats for wildlife [20–22]. Apart from delivering substantial ecosystem services, grasslands are a low-cost feed source for ruminants. Perennial ryegrass (*Lolium perenne*) is the most important forage grass in temperate climates due to its high dry matter (DM) productivity potential in combination with a high forage digestibility and nutritive value throughout the grazing season [23]. However, temperature-limited herbage growth in spring and autumn and moisture-limited growth in summer can result in feed deficits in intensively managed systems. Thus, future grass growth and thus feed supply is highly uncertain within and between seasons and locations. Extreme drought periods have been shown to prolong the start of growth after rewetting, which has been referred to as the legacy effect. This can particularly occur in shallow-rooted grasses such as perennial ryegrass, which further influences the annual yield variability [24].

Due to climatic conditions, grass growth rates are highly variable both in time (within and between seasons at one location), and in space (between locations). For example, average annual dry matter (DM) yields of perennial ryegrass in Germany show substantial inter-annual variations as well as large variations within the various states of Germany, with annual DM yields in the last decade ranging from <3 t ha^{-1} to >9 t ha^{-1} (Figure 1). This is due to differences in the soils (including availability of water and nutrients) as well as the high temporal variability in weather conditions. For example, the extended summer drought all over North-West Europe in 2018 is reflected in a substantial drop in DM yield, with reductions ranging from 60 to 93% compared with the average of the last 10 years. This high variability has direct impacts on the levels of forage produced on farms, and thus the feeding management.

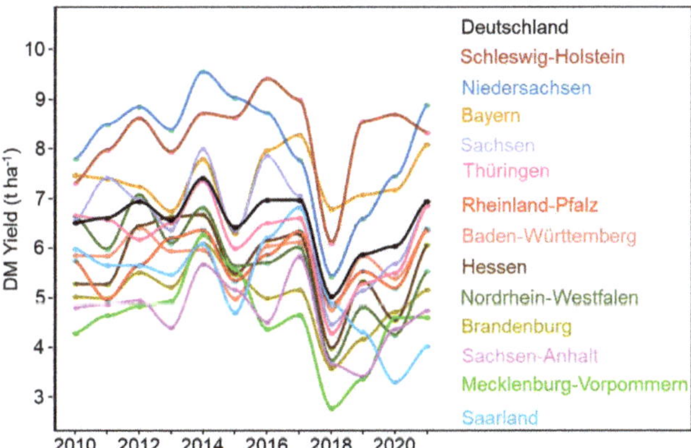

Figure 1. Annual dry matter (DM) yields for the different states of Germany obtained from the Statistisches Bundesamt (https://www.destatis.de/DE/Themen/Branchen-Unternehmen/Landwirtschaft-Forstwirtschaft-Fischerei/Publikationen/Bodennutzung/landwirtschaftliche-nutzflaeche-2030312217004.pdf; accessed 11 January 2021).

Many model comparison studies have been conducted for cropping systems, and the use of model outputs of model ensembles is gaining attention [25,26]. Only a few comparisons of grass growth models with different complexities (deterministic and empirical) have been conducted [27–29]. While these few studies suggest that empirical models are often comparable to more complex deterministic models, further evaluation is required, especially with studies covering more diverse climatic conditions. Moreover, evaluating the model complexity (i.e., the number of variables) is of importance as the addition of new variables may increase the probability of additional errors and less accurate simulations [30,31].

In mechanistic and process-oriented modelling approaches such as the Agricultural Production Systems Simulator (APSIM; [32]), pools and fluxes of carbon, nitrogen, and water are represented, with sub-models for soil, water, carbon, and nitrogen, and the plant. Plant growth is calculated based on physiological processes at the plant scale (i.e., considering the leaf photosynthetic rate and carbon accumulation based on incoming radiation, carbon dioxide, temperature, water, nitrogen, plant respiration, fertility, and tissue turnover including senescence and detachment of dead material) [33]. In more simple semi-mechanistic models, detailed physiological processes are not considered in detail. In the GrasProg model, for example, the simulations are based on functional growth equations that consider more general physiological principles of perennial grasses. Here, the relative growth rate and development of the leaf area index as a function of time are integral parts of the model to account for the photosynthetic efficiency of the grassland canopy with subsequent considerations of the growth limiting factors for temperature, radiation, and precipitation [34].

The objective of the current study was to compare two models with different complexity, GrasProg1.0, an updated version of GrasProg [34] and APSIM, for predicting grass growth across Northern Europe using only basic soil information and typical fertilisation rates. The evaluation was based on long-term variety trials with perennial ryegrass from Germany and Denmark.

2. Materials and Methods

2.1. Trial Sites

For the comparison of the two models, the grass growth data from Germany and Denmark were used (Figure 2). In Germany, data were obtained from the states' variety testing trials (Landessortenversuche; sourced from http://www.landwirtschaftskamme rn.de/ accessed 12 April 2021), and in Denmark from the national trials recorded in the Nordic Field Trial System (NFTS; https://nfts.dlbr.dk/Forms/Forside.aspx; assessed 14 June 2021). These testing trials run over a period of three years, after which another set of new varieties is started. To represent the perennial character of permanent grassland and avoid the effects of poor grassland establishment in the first production year, data were limited to the second and third production/trial year. Additionally, only perennial ryegrass varieties from the medium maturity group (including reference varieties) were selected. These resulted in 28 sites for Germany and four for Denmark, spanning different soil types and a range of climatic conditions, with mean temperatures ranging from 5.9 to 10.3 °C and a mean annual rainfall from 536 to 1154 mm (Table 1).

According to the protocol in the trials from Germany, nitrogen (N) fertilisation ranged between 300 and 360 kg N ha^{-1}, of which 80–100 kg ha^{-1} was applied in early spring for the first cut, which, according to the prescribed management protocols [37], should be carried out at BBCH51 (Biologische Bundesanstalt für Land- und Forstwirtschaft, Bundessortenamt und CHemische Industrie). For further details, see [38]. Depending on the site, cuts were taken in an area of 10 to 12 m^2, with four replications and cut to a height of 5–6 cm above ground. The DM content of the herbage was determined after oven drying the subsamples at 60 °C for 48 h. In Denmark, N fertilisation was applied according to the Danish Plant Directorates standards of 340 kg N ha^{-1} yr^{-1} for a pure grass, with 40% applied in early spring, 30% after the first harvest, 20% after the second harvest, and 10% after the third harvest. The plot size was 18 m^2, and the cutting height was 5 cm. Dry matter was determined by oven drying the subsamples at 60 °C for 40 h.

For the modelling, the soils were categorised into three different groups based on their plant available water content (PAW) in the rootzone, namely 'low' with 60 mm, 'medium' with 80 mm, and 'high' with 100 mm (with a rootzone depth for ryegrass set as 500 mm).

For Germany, the PAW for the soils was derived from the Ackerzahl, which is based on
the German land appraisal system (Reichsbodenschätzung), which was initiated in 1932 to
rank soils according to their potential productivity [35]. The Ackerzahl is scaled from 1 to
100 (for highest productivity), and takes the soil type, formation, topography, and climatic
conditions into account. Soils with an Ackerzahl up to 25 were classified into the 'low'
PAW soil group, those with an Ackerzahl 25 and 65 to the 'medium' soil group, and those
with an Ackerzahl larger than 65 to the 'high' soil group. For the Danish sites, soils were
grouped according to the Danish soil classification scheme [36], with JB1 (coarse sandy)
and JB2 (fine sandy soil) in the 'low' soil group, and JB5 and JB6 (sandy clay) in the 'high'
soil group (Table 1).

Daily weather data were gathered from meteorological sites close by the trial sites. For
Germany, these were obtained from the Deutschen Wetterdienst (Germany's National Meteorological Service, DWD; https://www.dwd.de/; accessed 12 April 2021). For Denmark,
they were obtained from the online database (http://agro-web01t.uni.au.dk/KlimaDB/;
accessed 14 June 2021) managed by Aarhus University.

Figure 2. Locations of the national ryegrass variety trials in Germany and Denmark, used for the
comparison of two models, GrasProg1.0 and APSIM, for the prediction of grass growth.

Table 1. Site descriptions and meteorological stations (Met Station) used for the modelling. DWD (=Germany's National Meteorological Service, Der Deutsche Wetterdienst) was used for the modelling. Lon = longitude, Lat = latitude in degrees, Alt = altitude in meters above sea level, PAW = plant available water (mm) in 500 mm depth, T = mean annual temperature (°C), mean annual RF = mean annual rainfall (mm), Lon Met = longitude in degrees of the meteorological station, Lat Met = latitude in degrees of the meteorological station, * proximate values. Soil classification (Soil) in Germany (DE) was based on the Ackerzahl [35] and on the Danish soil classification scheme [36] in Denmark (DK).

Site	Lon	Lat	Alt	PAW	Soil	Met Station	Lon Met	Lat Met	T	RF
Aulendorf, DE	9.66	47.94	570	80	56	Weingarten	9.62	47.81	9.3	926
Burkersdorf, DE	11.88	50.65	594	80	36	Schleiz	11.80	50.57	8.2	652
Dasselsbruch, DE	10.02	52.56	35 *	60	20	Celle	10.03	52.60	10.0	679
Eichhof, DE	9.68	50.85	200	80	57	Bad Hersfeld	9.74	50.85	9.1	658
Eslohe, DE	8.17	51.25	370 *	80	40	Eslohe	8.16	51.25	8.5	1086
Forchheim 2, DE	13.27	50.71	565	80	33	Marienberg	13.15	50.65	7.3	890
Haufeld, DE	11.28	50.80	430	80	56	Jena	11.58	50.93	10.3	594
Hayn-Schwenda, DE	11.08	51.57	441	80	40	Harzgerode	11.14	51.65	8.0	582
Heßberg, DE	10.78	50.42	380	80	45	Lautertal	10.97	50.31	9.1	739
Hjerm, DK	8.65	56.43	30 *	100	JB5/6	Vemb	8.22	56.71	8.6	796
Hohenschulen, DE	9.99	54.32	30	80	50	Kiel-Holtenau	10.14	54.38	9.4	759
Iden, DE	11.90	52.78	18	100	67	Seehausen	11.73	52.89	9.6	565
Kalteneber, DE	10.14	51.32	450 *	80	45	Leinefelde	10.31	51.39	8.6	700
Kißlegg, DE	9.89	47.79	700 *	80	58	Weingarten	9.62	47.81	9.3	926
Kleve, DE	6.17	51.79	15	80	56	Kleve	6.10	51.76	10.3	837
Kranichfeld, DE	11.20	50.86	330 *	80	46	Erfurt-Weimar	10.96	50.98	9.0	536
Kyllburgweiler, DE	6.62	50.07	529	80	34	Manderscheid	6.80	50.10	8.6	887
Malchow/Poel, DE	11.47	53.99	10 *	80	34	Boltenhagen	11.19	54.00	9.3	597
Nørager, DK	9.63	56.75	40 *	60	JB2	Aars	9.51	56.76	8.3	706
Obershagen, DE	10.06	52.50	40 *	80	45	Celle	10.03	52.60	10.0	679
Oberstaudhausen, DE	11.95	47.86	500 *	80		Rosenheim	12.13	47.88	9.2	1068
Oberweißbach, DE	11.14	50.58	660	60	23	Neuhaus	11.13	50.50	5.9	1154
Osterseeon, DE	11.93	48.07	560	80	45	Ebersberg	11.99	48.10	8.7	1036
Ovelgönne, DE	8.42	53.34	0 *	100	88	Bremerhaven	8.58	53.53	10.1	753
Paulinenaue, DE	12.71	52.67	30 *	60	30	Neuruppin	12.85	52.94	9.6	620
Scharnhorst, DE	9.52	52.53	38	80	50	Wunstorf	9.43	52.46	10.3	650
Schoonorth-Otterham, DE	7.22	53.50	−0.3	100	85	Emden	7.23	53.39	9.4	823
Schuby, DE	9.45	54.52	42.7	60	22	Schleswig	9.55	54.53	8.6	885
Skælskør, DK	11.31	55.24	6 *	100	JB6	Flakkebjerg	11.39	55.31	8.9	581
Sophienhof, DE	9.06	49.81	453	100	72	Michelstadt-Vielbrunn	9.10	49.72	8.5	1031
Steinach, DE	12.61	48.98	508 *	80	56	Straubing	12.56	48.83	9.2	691
Vemb, DK	8.38	56.35	6 *	60	JB1/2	Vemb	8.22	56.71	8.6	796

2.2. Model Descriptions

2.2.1. GrasProg1.0

The GrasProg model is a semi-mechanistic model for simulating grass growth for intensively managed ryegrass (*Lolium perenne*) dominated swards with typical non limiting N fertilisation rates. The model only requires a few input parameters, and aside from proxies for the number of generative tillers and the tiller density, only the soil's plant available water (PAW) and meteorological factors (global radiation, mean daily temperature, precipitation, and evaporation) are necessary. The model has previously been calibrated for intensively managed ryegrass dominated grass swards with typical non limiting N fertilisation rates in North-West Germany (GrasProg; [34]). Now included in the updated version, GrasProg1.0 is a drought legacy factor that accounts for a period of unusually dry weather. Such extreme and long drought events can, aside from an immediate reduction in canopy photosynthesis, have longer-lasting legacy effects on vegetation growth [39,40]. The drought legacy factor is assumed to start after a drought period of seven days, after which the start of the grass growth is delayed by 7 days, where a drought is defined as the soil having a water content $\leq$30% PAW.

The model was set up for the trial sites described above using the meteorological data from the climate stations nearby (Table 1) and the site relevant soil PAW, either low (PAW = 60 mm), medium (PAW = 80 mm), or high (PAW = 100 mm).

2.2.2. APSIM

APSIM is a modular process-oriented simulation framework maintained by the APSIM Initiative (www.apsim.info; accessed 14 June 2021). APSIM is climate-driven and comprises a range of submodels including SoilWat for simulating water movement, SoilNitrogen for simulating N cycling, AgPasture for pasture growth and N uptake, and the Micromet module [41] for computing evapotranspiration using the Penman–Monteith equation. AgPasture is based on the physiological model of Thornley and Johnson [42], which has been shown to simulate growth patterns and seasonal yields well [43,44]. In brief, grass growth is modelled with a daily time-step calculation based on intercepted global solar radiation, radiation use efficiency, and growth modifiers for temperature, soil water, and N supply. APSIM with the AgPasture model has been used successfully for simulating grass growth under a range of climatic conditions in New Zealand, mainly binary mixtures of ryegrass/white clover [45], but also for diverse pastures [46,47] and for annual and perennial ryegrass in Australia [48]. The model has also been tested for predicting seasonal grass growth rates under different climatic conditions for New Zealand and using generic soils with PAWs estimated from the land use capability classes [49].

The model was set up with a pure perennial ryegrass (*Lolium perenne* L.), a rooting depth of 500 mm, and three different soil profiles: light (PAW = 60 mm), medium (PAW = 80 mm), and heavy (PAW = 100 mm). The soil organic carbon in the top 100 mm was set according to averages for grassland and different soil types across Germany [50], with 3.8% for sandy soils (used for the light soils), 3.9 for loamy soils (used for the medium soils), and 2.9% for clay soils (used for the heavy soils). The grass was cut according to the trial management, and fertiliser was applied via a manager script, with the amounts and timings as described above. Meteorological data required by APSIM are daily values of rainfall, minimum and maximum daily temperature, and radiation.

2.2.3. Data Analysis and Statistical Analysis

Grass growth data were screened for outliers using the linear regression of pasture production of the first cut vs. global radiation sum and temperature sum from the beginning of the growing season (taken after a temperature sum of 250 °C with a base temperature of 3 °C) to the date of the first cut (Figure 3). Cook's distance, which measures the change in fitted response for all observations with and without the presence of observation i, was then used to identify outliers. Observations that have a Cook's distance >4 times the mean were classified as outliers.

The performance of GrasProg1.0 and APSIM were evaluated based on common measures including the coefficient of determination (R^2), Nash–Sutcliffe efficiency (NSE), root mean square error (RMSE), and percent bias. For these, the R package hydroGOF [51] was used. Additionally, a paired *t*-test was conducted using the R function: *t*-test (x, y, paired = TRUE, alternative = "two.sided").

These statistics were calculated for both the entire dataset and for the individual sites using data from each individual cut. For the evaluation of the two models, the biomass of individual cuts as well as the annual amounts were used.

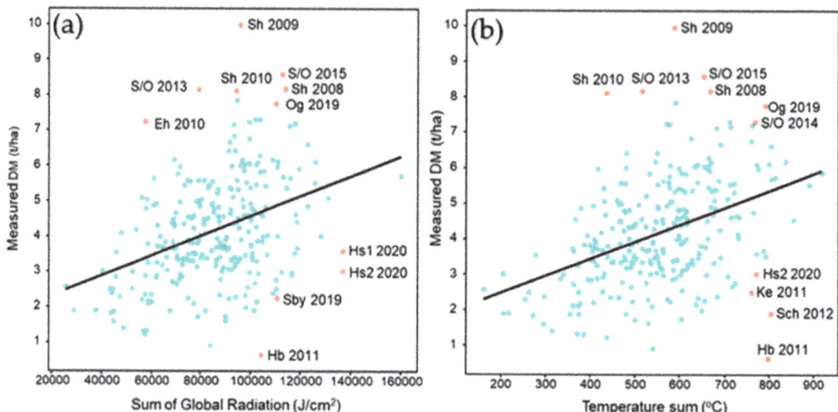

Figure 3. Measured grass dry matter (DM) yield of the first cut vs. temperature sum (**a**) and global radiation sum (**b**) after a temperature sum of 250 °C with a base temperature of 3 °C. Blue symbols indicated data that were included in the analysis, and red symbols those that were excluded for the years provided. Eh = Eichhof. Hb = Heßberg; Hs = Hohenschulen; Ke = Kalteneber; Og = Ovelgönne; Sby = Schuby, Sch = Scharnhorst; Sh = Sophienhof; S/O = Schoonorth/Otterham.

3. Results

3.1. Inclusion of a Legacy Effect

The improvement in GrasProg1.0 with the legacy effect can be seen in some of the data collected in 2018, which had a prolonged summer drought (Figure 4). While both versions of the model predicted the first cut in Kyllburgweiler well, the second and third were overestimated without the legacy effect. For Osterseeon, including the legacy effect reduced the grass growth during June too much, but in August, the simulations were much closer to the measurements. This shows that the inclusion of a legacy effect improved the model, but better parametrisation and/or its description in the model is required.

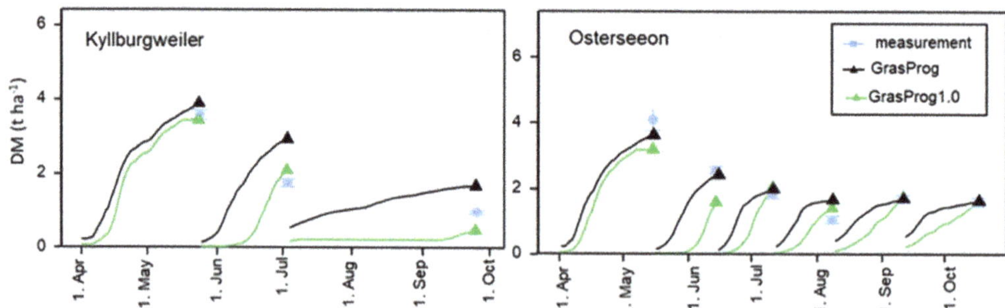

Figure 4. GrasProg1.0 predictions of the grass growth (dry matter; DM) in 2018 for Kyllburgweiler and Osterseeon, either without (black lines) or with the legacy (green lines) effect.

3.2. Measured and Predicted Dry Matter Production—Individual Cuts

Predictions of seasonal DM production (individual cuts) by GrasProg1.0 for four selected sites from varying geographical areas, altitudes, and with different meteorological conditions (Aulendorf, Kleve, Osterseeon and Schuby with measurements over 11 to 12 years) showed generally good agreement with measurements (Figures 5–9). For GrasProg1.0, the RMSE ranged from 0.59 to 0.77 t ha^{-1} and NSE from 0.55 to 0.71 for these four sites (Table 2; Figure 9). APSIM showed a slightly less good prediction for these four sites, with RMSE ranging from 0.59 to 0.91 t DM ha^{-1} and NSE from 0.22 to 0.54. In some instances, GrasProg1.0 slightly underpredicted the first cuts while APSIM at times overpredicted these. The underestimation may be because GrasProg1.0 was calibrated on a dataset, which was more intensively defoliated (8 cuts yr^{-1}) compared with the data used for evaluation in the present study (4–5 cuts yr^{-1}). The defoliation frequency influences various plant traits such as tiller density, which greatly influence grass growth [52], which might explain the disparities between the measurements and simulations.

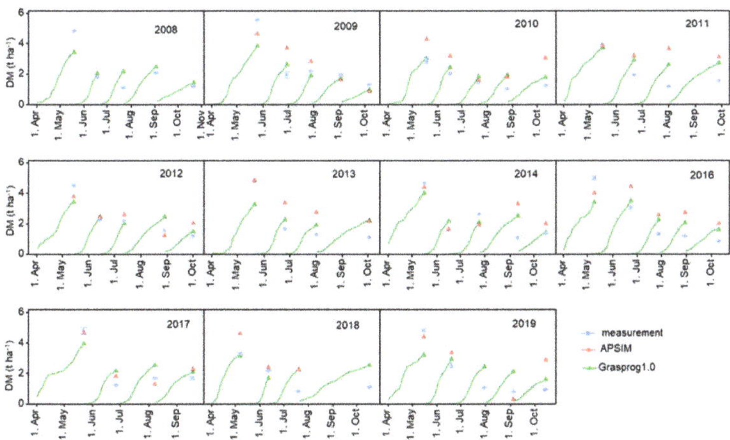

Figure 5. Measured grass dry matter (DM) for different cuts and years for Aulendorf, Germany. The predictions by GrasProg1.0 and APSIM are also shown.

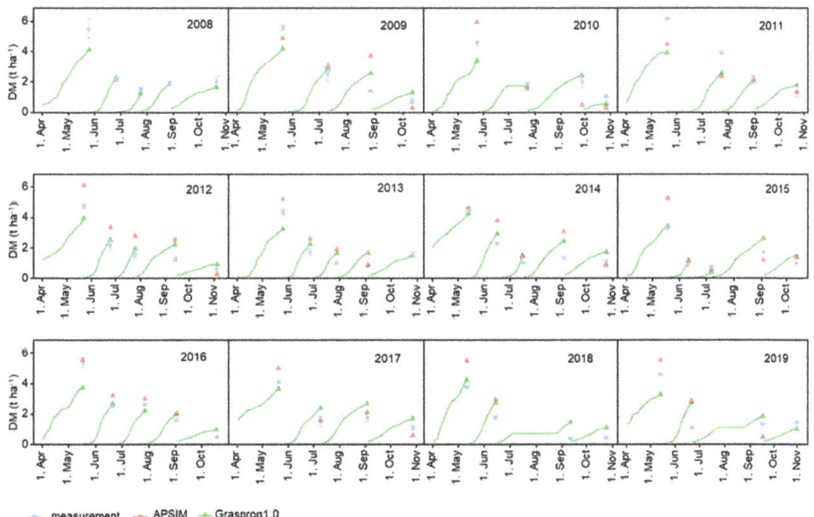

Figure 6. Measured grass dry matter (DM) for different cuts and years for Kleve, Germany. The predictions by GrasProg1.0 and APSIM are shown.

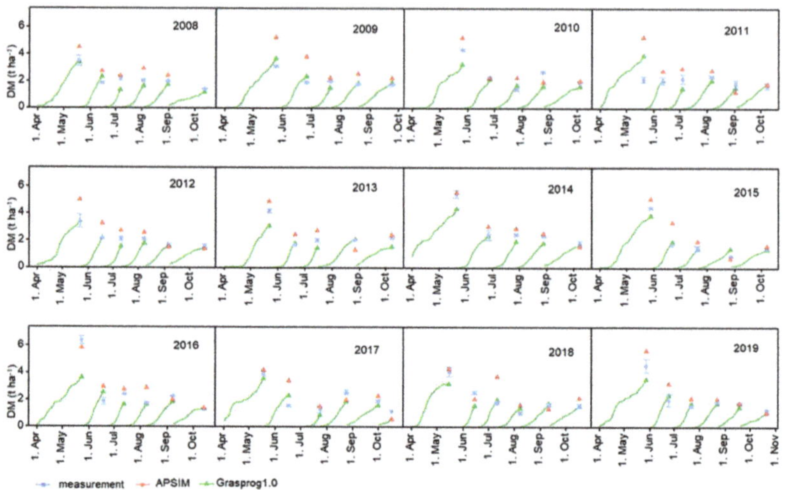

Figure 7. Measured grass dry matter (DM) for different cuts and years for Osterseeon, Germany. The predictions by GrasProg1.0 and APSIM are also shown.

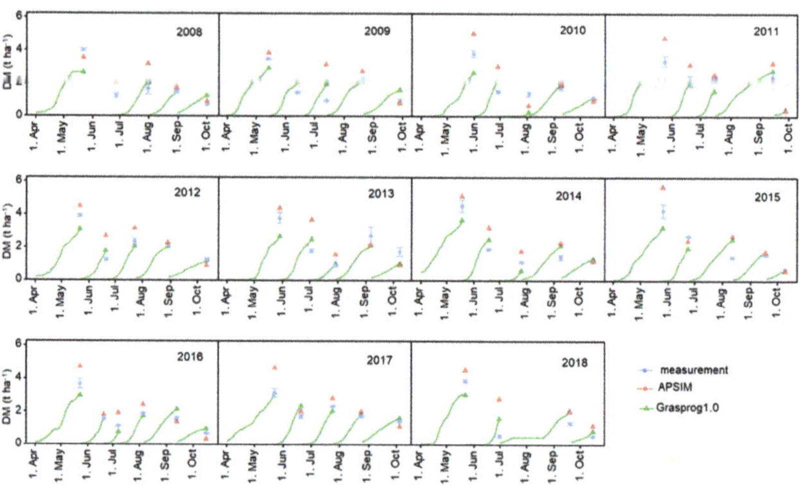

Figure 8. Measured grass dry matter (DM) for different cuts and years for Schuby, Germany. The predictions by GrasProg1.0 and APSIM are also shown.

For the entire datasets with a total of 32 different sites and measurement periods ranging 1 to 13 years, the RMSE values were acceptable, ranging from 0.59 to 1.28 t DM ha^{-1} for GrasProg1.0 and 0.56 to 1.26 t ha^{-1} for APSIM (Table 2). Accurately predicting grass growth with its high seasonal and interannual variation is not an easy task [28,53]. These RMSE values are lower to those reported with ranges from 0.7 to 2.1 t DM ha^{-1}, using three different models for predicting the first two cuts of timothy grass under northern European conditions [29]. They concluded that there is a need for a better understanding of the processes involved and how they are described in models. However, for some sites, the individual cuts were vastly over- or underpredicted, and the NSE values are close to zero or negative.

While there are no explicit standards for evaluating model performance, we used the suggested thresholds for monthly values to judge if the model results were satisfactory, namely NSE > 0.3, R^2 and p-value > 0.025 [54]. Out of the 32 datasets, GrasProg1.0 predicted grass growth for individual cuts satisfactorily for 11 datasets according to these criteria, and APSIM only slightly more with 13 datasets. Closer inspections also showed that GrasProg1.0 did not predict grass growth for the sites in Denmark satisfactorily, with high underestimation (high Pbias). This is not astonishing, as the model has not been calibrated for high latitude sites with long day-lengths. The APSIM model seemed to capture this slightly better, with two of the sites being satisfactorily simulated. Many of the datasets that were not satisfactorily predicted also had very short observation periods of $\geq$5 years.

Looking at the NSE values, GrasProg1.0 predicted the pattern of grass growth (individual cuts) on 81% of the sites better than just using the average values. When considering only observations $\geq$5 years, 85% of the sites were better predicted than using the averages.

Over the entire datasets, the individual cuts were reasonably predicted with GrasProg1.0, with a RMSE of 0.94 t DM ha^{-1} and NSE of 0.43, and an overprediction of 12.9% (Figure 10). The performance of APSIM was slightly worse, with a RMSE of 0.99 t DM ha^{-1}, a NSE of 0.29, and an underprediction of 12.8%.

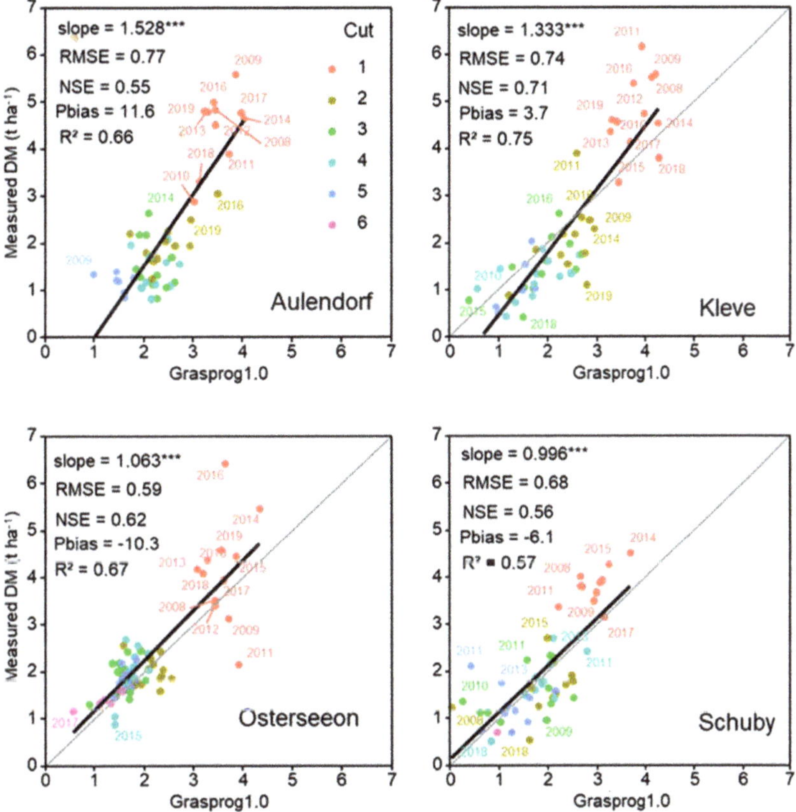

Figure 9. Measured grass dry matter (DM) vs. predictions by GrasProg1.0 for different cuts and years for Aulendorf, Kleve, Osterseeon, and Schuby, Germany. *** indicates significance $p < 0.001$.

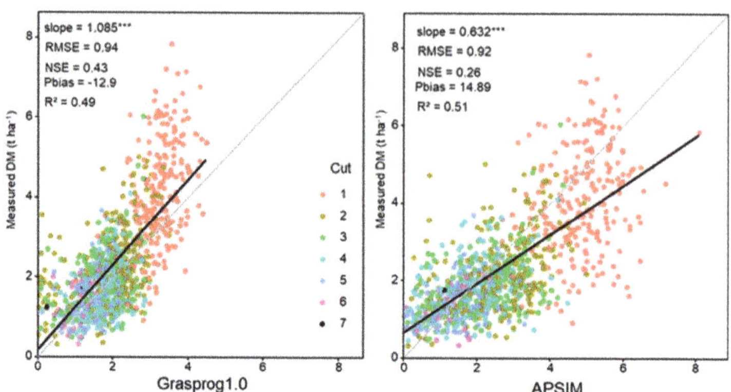

Figure 10. Measured grass dry matter (DM) vs. the predictions by GrasProg1.0 and APSIM for different cuts and years for the entire datasets from Germany and Denmark, with a total of 33 locations. *** indicates significance *p* < 0.001.

3.3. Measured and Predicted Dry Matter Production—Annual

Considering that the models were not tuned to any of the data at each site and the weather data were not obtained directly on the sites, both models performed well with a RMSE ranging from 0.12 to 2.85 t DM ha^{-1} for GrasProg1.0 and from 0.06 to 2.55 t DM ha^{-1} for APSIM. The inter-annual variability in annual yield was well-reflected (Figure 11). Furthermore, GrasProg was calibrated with a dataset from permanent grasslands, in which perennial ryegrass was the dominant species, but in which other grasses were also present. For accurate predictions, it has been suggested that the genetic variability between cultivars should be accounted for [29]. In our study, data from the national trials were restricted to the medium maturity group but comprised different cultivars. Furthermore, to increase the accuracy of the model for simulating growth in spring, the soil temperature should be considered, rather than the air temperature [34].

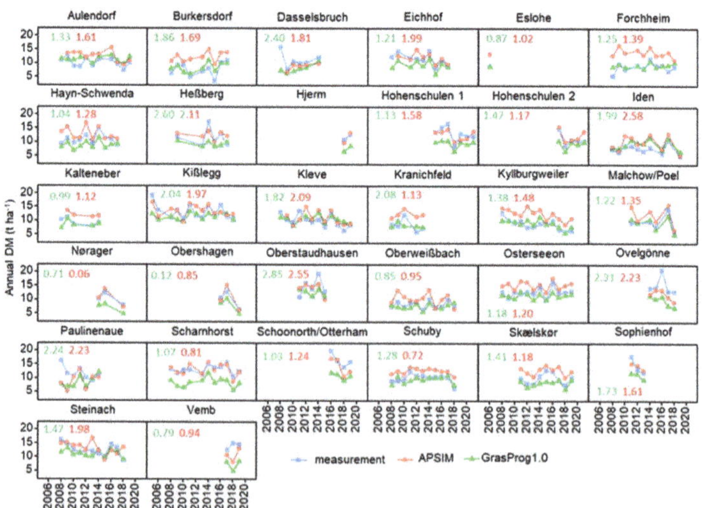

Figure 11. Annual measured grass dry matter (DM) and predictions by GrasProg1.0 and APSIM for different years for different locations in Germany and Denmark. The numbers show the RMSE for the two models (GrasProg1.0 green, APSIM red).

Table 2. Model performance statistics GrasProg1.0 and APSIM for individual cuts with data from variety trials from national trials in Germany and Denmark, spanning different numbers of years. Bold numbers indicate good model performance (NSE > 0.3, R^2 > 0.5, and p > 0.025), and x indicates if the model satisfies all three criteria.

Site	RMSE		R^2		NSE		P (Paired t-Test)		Pbias		Slope		Years	Evaluation	
	GrasProg	APSIM	GrasProg	APSIM	GrasProg	APSIM	GrasProg	APSIM	GrasProg	APSIM	GrasProg	APSIM		GrasProg	APSIM
Iden	0.72	0.49	**0.68**	**0.76**	**0.64**	0.10	**0.094**	**0.555**	12.40	30.60	1.10	0.57	14	x	
Kißlegg	0.69	0.70	**0.55**	**0.55**	**0.43**	**0.40**	0.009	**0.649**	−14.00	2.00	0.98	0.65	13		x
Hayn-Schwenda	0.92	0.76	0.42	**0.60**	0.32	−0.24	0.000	0.001	−18.50	19.70	0.99	0.48	12		
Kleve	0.74	0.79	**0.75**	**0.72**	**0.71**	**0.54**	**0.509**	**0.984**	3.70	11.90	1.33	0.69	12	x	x
Kyllburgweiler	0.68	0.73	**0.56**	0.49	**0.56**	−0.45	**0.600**	0.000	2.60	43.10	1.01	0.60	12	x	
Osterseeon	0.59	0.59	**0.67**	**0.67**	**0.62**	0.29	0.004	0.000	−10.30	21.00	1.06	0.67	12		
Aulendorf	0.77	0.91	**0.66**	**0.52**	**0.55**	**0.31**	0.019	**0.374**	11.60	24.80	1.53	0.77	11		x
Oberweißbach	0.80	0.70	0.06	0.28	−0.10	−1.86	0.015	0.006	−10.20	23.50	0.46	0.29	11		
Scharnhorst	1.01	0.85	**0.67**	**0.77**	**0.36**	**0.77**	0.000	**0.779**	−35.50	−0.80	1.35	1.06	11		x
Schuby	0.68	0.60	**0.57**	**0.66**	**0.56**	0.22	**0.187**	0.000	−6.10	25.50	1.00	0.64	11	x	
Steinach	0.80	0.95	0.38	0.13	0.29	−0.57	0.021	**0.300**	−11.80	6.70	0.77	0.31	11		
Forchheim	0.60	0.56	0.37	0.44	**0.36**	−2.65	**0.349**	0.000	4.50	59.50	0.86	0.43	10		
Burkersdorf	1.01	0.75	**0.53**	**0.74**	**0.49**	0.01	**0.382**	0.000	7.90	58.10	1.38	0.89	9	x	
Eichhof	0.98	0.73	**0.59**	**0.73**	**0.48**	**0.63**	0.000	**0.538**	−20.70	3.90	1.22	0.73	9		x
Skælskør	0.96	0.82	0.39	**0.56**	0.29	0.11	0.007	0.000	−18.10	27.80	0.97	0.64	9		
Hohenschulen 1	0.85	0.81	**0.66**	**0.69**	0.25	**0.69**	0.001	**0.513**	−27.40	−3.50	1.55	0.95	7		x
Paulinenaue	1.30	1.26	0.10	0.15	−0.15	−0.30	**0.042**	**0.088**	−26.50	−23.40	0.58	0.43	7		
Dasselsbruch	1.21	0.93	0.16	0.21	−0.25	−0.19	**0.032**	**0.173**	−29.40	−14.10	0.63	0.47	6		
Eslohe	0.68	0.57	**0.53**	**0.66**	**0.54**	0.15	**0.378**	0.002	−4.20	28.00	1.08	0.77	6	x	
Heßberg	0.85	0.82	**0.53**	**0.57**	**0.38**	−0.10	**0.100**	**0.459**	−20.40	6.90	0.97	0.48	6	x	
Malchow/Poel	0.92	0.69	**0.51**	**0.73**	**0.46**	**0.49**	**0.479**	0.023	−9.50	26.80	1.41	0.78	6	x	
Hohenschulen 2	0.95	0.74	**0.60**	**0.76**	**0.48**	**0.73**	**0.064**	**0.087**	−16.80	12.50	1.42	1.02	5	x	x
Kalteneber	0.90	0.81	**0.56**	**0.58**	**0.55**	0.09	**0.196**	**0.912**	−9.40	32.30	1.16	0.68	5	x	
Kranichfeld	0.81	0.80	**0.52**	**0.52**	**0.49**	−0.38	**0.879**	0.004	−9.60	44.70	0.81	0.55	5	x	

Table 2. *Cont.*

Site	RMSE		R^2		NSE		P (Paired *t*-Test)		Pbias		Slope		Years	Evaluation	
	GrasProg	APSIM	GrasProg	APSIM	GrasProg	APSIM	GrasProg	APSIM	GrasProg	APSIM	GrasProg	APSIM		GrasProg	APSIM
Oberstaudhausen	0.70	0.75	**0.43**	0.36	0.24	0.10	**0.216**	**0.638**	−17.40	−3.30	0.91	0.54	5		
Ovelgönne	0.95	0.94	**0.66**	**0.67**	−0.08	**0.46**	0.008	**0.047**	−37.30	−21.00	1.38	0.96	5		x
Schoonorth	0.74	0.72	**0.75**	**0.76**	0.01	**0.66**	0.005	**0.087**	−34.90	−14.40	1.63	1.01	4		x
Nørager	0.73	0.45	**0.56**	**0.83**	0.05	**0.53**	0.024	**0.593**	−31.70	4.00	0.98	0.62	3		x
Obershagen	0.39	0.52	**0.83**	**0.71**	**0.60**	**0.33**	0.012	**0.701**	−20.20	4.60	1.00	0.58	3		x
Sophienhof	0.97	0.82	**0.71**	**0.79**	0.40	**0.78**	**0.119**	**0.964**	−24.30	−0.50	2.03	1.23	3		x
Vemb	1.01	1.02	0.00	0.00	−3.34	−1.89	**0.050**	**0.176**	−48.70	-23.90	−0.33	0.15	3		
Hjerm	1.28	1.10	0.33	**0.51**	−0.04	**0.56**	0.000	**0.821**	−36.70	-2.10	1.60	0.84	2		x

4. Discussion

In practical grassland farming, information about yield often remains unknown or is only roughly estimated because management in recent decades has been animal-focused, rather than on grassland management [55]. The good prediction of grass growth by GrasProg1.0 means that the model can be used to aid farm management, especially by providing information on the best cutting or grazing times dependent on the climatic conditions. The model can also be used to evaluate likely changes due to climate change such as how increasing temperatures and temperature sums affect the cutting regimes and grassland productivity. The inclusion of the legacy factor in Grasprog1.0 means that the model can be employed to aid in the development of farm-level adaptations according to changes in the productivity and seasonality of grasslands resulting from the expected increases in drought and heat extremes [56]. Furthermore, model-based knowledge of annual yields can help to assist in optimising fertilisation strategies and avoid the risks of negative environmental effects due to N losses [57]. Knowledge about annual yields is also crucial, because according to the new German fertiliser ordinance [58], the permitted amount of N fertilisation needs to account for the yield of the grassland in previous years.

Trade-offs between model complexity and validation have been discussed [59] and include a more complete entity representation by complex models at the cost for the need of a greater requirement for validation and meta data. In contrast, simple models require less validation data, and model parameters are more generalised, with a greater probability of a large difference between the observed and estimated values. The APSIM model was not tuned to any of the data at each site, and general data were used rather than site-specific values such as soil hydraulic properties and organic carbon. The importance of accurate soil parameterisation when using a complex model such as APSIM for simulating soil water and nitrogen dynamics, and pasture production has also been emphasised by Craig et al. [60].

The similar fit between measurements and predictions by the two models means that GrasProg1.0 is a suitable grass growth model for North-West European conditions, especially where site-specific parameters are not available. This finding is in line with other studies where Hurtado-Uria et al. [28] found that an empirical model performed equally well as a complex model for predicting grass growth across Ireland, and Skinner et al. [27] also found no difference in the ability to simulate the grass forage yield in Pennsylvania, USA. The authors highlighted the need for better validation datasets for a robust comparison and parameterisation of the models. In contrast to other grass growth models such as GrazeGro [15], STICS [9], LINGRA [61], and even the simpler MoSt GG model [8], GrasProg only requires a few input parameters (namely temperature, radiation, rainfall, evaporation, and PAW soil group), and thus can be incorporated into a simple decision support tool for use by farmers and advisers.

When evaluating models, uncertainties in the observed data and the exact management of the sites should also be considered. The high spatial heterogeneity of botanical composition, nutrient availability, and defoliation strategy influence the forage biomass and quality, especially in permanent grasslands [62,63]. Additionally, due to the extreme variability of individual plants even at a small scale, the determination of grass biomass is very difficult [64], and the methodology of measurements influences the data of biomass and quality [65,66].

Furthermore, although the data used for model evaluation in this study were from national trials with prescribed management protocols, the specific management varied slightly between sites and years including differences in the amount of N applied and cutting regimes. However, the defoliation frequency can influence various plant traits such as the tiller density, which greatly influence grass growth [52] and may lead to additional disparities between the measurements and simulations. One limitation of GrasProg1.0 is that it does not account for N fertilisation management and currently does not include pasture quality indicators. This is mainly due to a lack of sufficient forage quality data across regions with different climatic conditions, covering seasonal pasture growth data under a range of fertiliser treatments. However, due to its generic structure, GrasProg1.0

has been shown to be suitable to deliver the information of biomass yields for intensively managed and perennial ryegrass dominated grasslands in Germany and northern Europe.

5. Conclusions

The hierarchical (plant, leaf, molecular) and spatial level (field, farm, landscape) at which grass growth is simulated is strongly dependent on the subsequent practical implementation of a model. For implementation as a decision support tool for grassland management, the simple semi-mechanistic model GrasProg1.0 is highly suitable and showed similar results to the more complex and process-oriented model APSIM. Such complex models are very data rich and require site-specific input parameters, which are often not known. In contrast, GrasProg1.0 only requires a few input parameters including meteorological data and the classification of the soil into a PAW class.

Author Contributions: Conceptualisation, F.T. and T.P.; Methodology, F.T. and T.P.; Software, C.K.; Validation, C.K., T.P. and I.V.; Formal analysis, C.K., T.P. and I.V.; Writing—original draft preparation, T.P. and I.V.; Writing—review and editing, I.V., T.P. and F.T.; Visualisation, C.K.; Supervision, F.T.; Funding acquisition, F.T. All authors have read and agreed to the published version of the manuscript.

Funding: This research was funded by the arismo GmbH within the project "GrasProg (Ertragsmodell Grünlandversicherung)".

Institutional Review Board Statement: Not applicable.

Informed Consent Statement: Not applicable.

Data Availability Statement: Data will be made available upon request.

Acknowledgments: We would like to thank Ingrid Nöhles and Regina Rehermann (Vereinigte Hagelversicherung, VVaG) and the arismo GmbH within the project "GrasProg (Ertragsmodell Grünlandversicherung)". We further would like to thank Mathias Herbst and Cathleen Frühauf (Deutscher Wetterdienst; DWD—Zentrum für Agrarmeteorologische Forschung Braunschweig; ZAMF) for weather data provision, and the institute managers of the "Landesdienststellen für Landwirtschaft" (state offices of agriculture) for the access to the national variety trials data, and special thanks to Christine Kalzendorf and Hubert Kivelitz for valuable discussion about the experimental setup and the analysis of modelling results.

Conflicts of Interest: The authors declare no conflict of interest.

References

1. Chang, J.; Viovy, N.; Vuichard, N.; Ciais, P.; Campioli, M.; Klumpp, K.; Martin, R.; Leip, A.; Soussana, J.-F. Modeled Changes in Potential Grassland Productivity and in Grass-Fed Ruminant Livestock Density in Europe over 1961–2010. *PLoS ONE* **2015**, *10*, e0127554. [CrossRef] [PubMed]
2. Olesen, J.E.; Trnka, M.; Kersebaum, K.C.; Skjelvåg, A.O.; Seguin, B.; Peltonen-Sainio, P.; Rossi, F.; Kozyra, J.; Micale, F. Impacts and adaptation of European crop production systems to climate change. *Eur. J. Agron.* **2011**, *34*, 96–112. [CrossRef]
3. Taube, F.; Gierus, M.; Hermann, A.; Loges, R.; Schonbach, P. Grassland and globalization—Challenges for north-west European grass and forage research. *Grass Forage Sci.* **2014**, *69*, 2–16. [CrossRef]
4. Gauthier, M.; Barillot, R.; Schneider, A.; Chambon, C.; Fournier, C.; Pradal, C.; Robert, C.; Andrieu, B. A functional structural model of grass development based on metabolic regulation and coordination rules. *J. Exp. Bot.* **2020**, *71*, 5454–5468. [CrossRef]
5. Jouven, M.; Carrère, P.; Baumont, R. Model predicting dynamics of biomass, structure and digestibility of herbage in managed permanent pastures. 1. Model description. *Grass Forage Sci.* **2006**, *61*, 112–124. [CrossRef]
6. Duru, M.; Adam, M.; Cruz, P.; Martin, G.; Ansquer, P.; Ducourtieux, C.; Jouany, C.; Theau, J.; Viegas, J. Modelling above-ground herbage mass for a wide range of grassland community types. *Ecol. Model.* **2009**, *220*, 209–225. [CrossRef]
7. Topp, C.F.; Doyle, C.J. Simulating the impact of global warming on milk and forage production in Scotland: 1. The effects on dry-matter yield of grass and grass-white clover swards. *Agric. Syst.* **1996**, *52*, 213–242. [CrossRef]
8. Ruelle, E.; Hennessy, D.; Delaby, L. Development of the Moorepark St Gilles grass growth model (MoSt GG model): A predictive model for grass growth for pasture based systems. *Eur. J. Agron.* **2018**, *99*, 80–91. [CrossRef]
9. Brisson, N.; Gary, C.; Justes, E.; Roche, R.; Mary, B.; Ripoche, D.; Zimmer, D.; Sierra, J.; Bertuzzi, P.; Burger, P.; et al. An overview of the crop model stics. *Eur. J. Agron.* **2003**, *18*, 309–332. [CrossRef]
10. Bouman, B.A.M.; Schapendonk, A.H.C.M.; Stol, W.; van Kraalingen, D.W.G. *Description of the Growth Model LINGRA as Implemented in CGMS*; Quantitative Approaches in Systems Analysis No. 7; AB-DLO: Wageningen/Haren, The Netherlands; PE: Wageningen, The Netherlands, 1996; Volume 56, p. 32.

11. Rolinski, S.; Müller, C.; Heinke, J.; Weindl, I.; Biewald, A.; Bodirsky, B.L.; Bondeau, A.; Boons-Prins, E.R.; Bouwman, A.F.; Leffelaar, P.A.; et al. Modeling vegetation and carbon dynamics of managed grasslands at the global scale with LPJmL 3.6. *Geosci. Model Dev.* **2018**, *11*, 429–451. [CrossRef]

12. Sun, Y.; Feng, Y.; Wang, Y.; Zhao, X.; Yang, Y.; Tang, Z.; Wang, S.; Su, H.; Zhu, J.; Chang, J.; et al. Field-Based Estimation of Net Primary Productivity and Its Above- and Belowground Partitioning in Global Grasslands. *J. Geophys. Res. Biogeosci.* **2021**, *126*, e2021JG006472. [CrossRef]

13. Hetzer, J.; Huth, A.; Taubert, F. The importance of plant trait variability in grasslands: A modelling study. *Ecol. Model.* **2021**, *453*, 109606. [CrossRef]

14. Sinclair, T.R.; Seligman, N. Criteria for publishing papers on crop modeling. *Field Crop. Res.* **2000**, *68*, 165–172. [CrossRef]

15. Barrett, P.; Laidlaw, A.; Mayne, C. GrazeGro: A European herbage growth model to predict pasture production in perennial ryegrass swards for decision support. *Eur. J. Agron.* **2005**, *23*, 37–56. [CrossRef]

16. Avanzi, F.; De Michele, C.; Morin, S.; Carmagnola, C.M.; Ghezzi, A.; Lejeune, Y. Model complexity and data requirements in snow hydrology: Seeking a balance in practical applications. *Hydrol. Process.* **2016**, *30*, 2106–2118. [CrossRef]

17. Albanito, F.; McBey, D.; Harrison, M.; Smith, P.; Ehrhardt, F.; Bhatia, A.; Bellocchi, G.; Brilli, L.; Carozzi, M.; Christie, K.; et al. How Modelers Model: The Overlooked Social and Human Dimensions in Model Intercomparison Studies. *Environ. Sci. Technol.* **2022**, *56*, 13485–13498. [CrossRef]

18. Eurostat. Share of Main Land Types in Utilised Agricultural Area (UAA) by NUTS 2 Regions. 2020. Available online: https://ec.europa.eu/eurostat/web/products-datasets/-/tai05 (accessed on 7 February 2022).

19. Destatis, S.B. Bodennutzung der Betriebe—Landwirtschaftlich Genutzte Flächen. 2021. Available online: https://www.destatis.de/DE/Themen/Branchen-Unternehmen/Landwirtschaft-Forstwirtschaft-Fischerei/Publikationen/Bodennutzung/landwirtschaftliche-nutzflaeche-2030312217004.pdf;jsessionid=5B5577CA66935CE90997CD2A7F253CBB.live742?__blob=publicationFile (accessed on 14 February 2022).

20. Schils, R.L.; Bufe, C.; Rhymer, C.M.; Francksen, R.M.; Klaus, V.H.; Abdalla, M.; Milazzo, F.; Lellei-Kovács, E.; Berge, H.T.; Bertora, C.; et al. Permanent grasslands in Europe: Land use change and intensification decrease their multifunctionality. *Agric. Ecosyst. Environ.* **2022**, *330*, 107891. [CrossRef]

21. Soussana, J.; Tallec, T.; Blanfort, V. Mitigating the greenhouse gas balance of ruminant production systems through carbon sequestration in grasslands. *Animal* **2010**, *4*, 334–350. [CrossRef] [PubMed]

22. Otto, C.R.; Zheng, H.; Hovick, T.; van der Burg, M.P.; Geaumont, B. Grassland conservation supports migratory birds and produces economic benefits for the commercial beekeeping industry in the U.S. Great Plains. *Ecol. Econ.* **2022**, *197*, 107450. [CrossRef]

23. Finneran, E.; Crosson, P.; O'kiely, P.; Shalloo, L.; Forristal, D.; Wallace, M. Simulation modelling of the cost of producing and utilising feeds for ruminants on Irish farms. *J. Farm Manag.* **2010**, *14*, 95–116.

24. Hofer, D.; Suter, M.; Haughey, E.; Finn, J.A.; Hoekstra, N.J.; Buchmann, N.; Lüscher, A. Yield of temperate forage grassland species is either largely resistant or resilient to experimental summer drought. *J. Appl. Ecol.* **2016**, *53*, 1023–1034. [CrossRef]

25. Groh, J.; Diamantopoulos, E.; Duan, X.; Ewert, F.; Heinlein, F.; Herbst, M.; Holbak, M.; Kamali, B.; Kersebaum, K.; Kuhnert, M.; et al. Same soil, different climate: Crop model intercomparison on translocated lysimeters. *Vadose Zone J.* **2022**, *21*, 20202. [CrossRef]

26. Pirttioja, N.; Carter, T.; Fronzek, S.; Bindi, M.; Hoffmann, H.; Palosuo, T.; Ruiz-Ramos, M.; Tao, F.; Trnka, M.; Acutis, M.; et al. Temperature and precipitation effects on wheat yield across a European transect: A crop model ensemble analysis using impact response surfaces. *Clim. Res.* **2015**, *65*, 87–105. [CrossRef]

27. Skinner, R.H.; Corson, M.S.; Rotz, C.A. Comparison of two pasture growth models of differing complexity. *Agric. Syst.* **2008**, *99*, 35–43. [CrossRef]

28. Hurtado-Uria, C.; Hennessy, D.; Shalloo, L.; Schulte, R.P.O.; Delaby, L.; O'Connor, D. Evaluation of three grass growth models to predict grass growth in Ireland. *J. Agric. Sci.* **2013**, *151*, 91–104. [CrossRef]

29. Korhonen, P.; Palosuo, T.; Persson, T.; Höglind, M.; Jégo, G.; Van Oijen, M.; Gustavsson, A.-M.; Bélanger, G.; Virkajärvi, P. Modelling grass yields in northern climates—A comparison of three growth models for timothy. *Field Crop. Res.* **2018**, *224*, 37–47. [CrossRef]

30. Wallach, D.; Makowski, D.; Jones, J.W.; Brun, F. *Working with Dynamic Crop Models: Methods, Tools and Examples for Agriculture and Environment*; Academic Press—Elsevier: Cambridge, MA, USA, 2018; 613p.

31. Yu, L.; Lai, K.K.; Wang, S.; Huang, W. A Bias-Variance-Complexity Trade-Off Framework for Complex System Modeling. In *Computational Science and Its Applications—ICCSA 2006*; Lecture Notes in Computer Science; Springer: Berlin/Heidelberg, Germany, 2006; Volume 3980, pp. 518–527. [CrossRef]

32. Holzworth, D.P.; Snow, V.; Janssen, S.; Athanasiadis, I.N.; Donatelli, M.; Hoogenboom, G.; White, J.W.; Thorburn, P. Agricultural production systems modelling and software: Current status and future prospects. *Environ. Model. Softw.* **2015**, *72*, 276–286. [CrossRef]

33. Andreucci, M.P.; Snow, V.; Cichoty, R. The APSIM AgPasture Model. 2022. Available online: https://apsimdev.apsim.info/ApsimX/Documents/AgPastureScience.pdf (accessed on 12 September 2022).

34. Peters, T.; Kluß, C.; Vogeler, I.; Loges, R.; Fenger, F.; Taube, F. GrasProg: Pasture Model for Predicting Daily Pasture Growth in Intensive Grassland Production Systems in Northwest Europe. *Agronomy* **2022**, *12*, 1667. [CrossRef]

35. Rothkegel, W. *Geschichtliche Entwicklung der Bodenbonitierungen und Wesen und Bedeutung der Deutschen Bodenschätzung*; AGRIS: Stuttgart, Germany, 1950.
36. Greve, M.H.; Breuning-Madsen, H. *Soil Mapping in Denmark*; European Soil Bureau—Research Report No. 9; European Soil Bureau: Ispra, Italy, 1999.
37. BSA. Amendments to the Guidelines for Conducting Agricultural VCU Testing and Variety Testing 2000. Chapter 4.18 Grass and Clover Species, Including Lucerne. 30627 Hannover, Germany. 2008. Available online: https://www.bundessortenamt.de/bsa/media/Files/RILI_4_18_Graeser_Klee_200804.pdf (accessed on 6 September 2021).
38. Lancashire, P.D.; Bleiholder, H.; Van Den Boom, T.; Langelüddeke, P.; Stauss, R.; Weber, E.; Witzenberger, A. A uniform decimal code for growth stages of crops and weeds. *Ann. Appl. Biol.* **1991**, *119*, 561–601. [CrossRef]
39. Wu, X.; Liu, H.; Li, X.; Ciais, P.; Babst, F.; Guo, W.; Zhang, C.; Magliulo, V.; Pavelka, M.; Liu, S.; et al. Differentiating drought legacy effects on vegetation growth over the temperate Northern Hemisphere. *Glob. Change Biol.* **2018**, *24*, 504–516. [CrossRef]
40. Hahn, C.; Lüscher, A.; Ernst-Hasler, S.; Suter, M.; Kahmen, A. Timing of drought in the growing season and strong legacy effects determine the annual productivity of temperate grasses in a changing climate. *Biogeosciences* **2021**, *18*, 585–604. [CrossRef]
41. Snow, V.O.; Huth, N.I. The APSIM MICROMET Module. 2004, HortResearch. Internal Report No. 2004/12848. HortResearch, Auckland, p. 18. Available online: www.apsim.info/wiki/public/Attachments/Module-Documentation/Micromet.pdf (accessed on 25 February 2012).
42. Thornley, J.H.M.; Johnson, I.R. *Plant and Crop Modelling—A Mathematical Approach to Plant and Crop Physiology*; The Blackburn Press: Caldwell, NJ, USA, 2000.
43. White, T.A.; Johnson, I.R.; Snow, V.O. Comparison of outputs of a biophysical simulation model for pasture growth and composition with measured data under dryland and irrigated conditions in New Zealand. *Grass Forage Sci.* **2008**, *63*, 339–349. [CrossRef]
44. Cullen, B.R.; Eckard, R.J.; Callow, M.N.; Johnson, I.R.; Chapman, D.F.; Rawnsley, R.P.; Garcia, S.C.; White, T.; Snow, V.O. Simulating pasture growth rates in Australian and New Zealand grazing systems. *Aust. J. Agric. Res.* **2008**, *59*, 761–768. [CrossRef]
45. Li, F.Y.; Snow, V.O.; Holzworth, D.P. Modelling seasonal and geographical pattern of pasture production in New Zealand—Validating a pasture model in APSIM. *N. Z. J. Agric. Res.* **2011**, *54*, 331–352. [CrossRef]
46. Vogeler, I.; Vibart, R.; Cichota, R. Potential benefits of diverse pasture swards for sheep and beef farming. *Agric. Syst.* **2017**, *154*, 78–89. [CrossRef]
47. Cichota, R.; Snow, V. Simulating plant growth in diverse pastures with new forage models in APSIM. *Agron. N. Z.* **2018**, *48*, 77–89.
48. Harrison, M.T.; De Antoni Migliorati, M.; Rowlings, D.; Dougherty, W.; Grace, P.; Eckard, R.J. Modelling biomass, soil water content and mineral nitrogen in dairy pastures: A comparison of DairyMod and APSIM. In Proceedings of the 2018 Australasian Dairy Science Symposium, Palmerston North, New Zealand, 21–23 November 2018.
49. Vogeler, I.; Cichota, R.; Beautrais, J. Linking Land Use Capability classes and APSIM to estimate pasture growth for regional land use planning. *Soil Res.* **2016**, *54*, 94–110. [CrossRef]
50. Düwel, O.; Siebner, C.S.; Utermann, J.; Krone, F. BGR Gehalte an organischer Substanz in Oberböden Deutschlands—Bericht über länderübergreifende Auswertungen von Punktinformationen im FISBo BGR. In Rohstoffe; B.B.f.G.u., Editor. 2008; Archiv-Nr.: 0126616. Available online: https://www.bgr.bund.de/DE/Themen/Boden/Produkte/Schriften/Downloads/Humusgehalte_Bericht.pdf?__blob=publicationFile (accessed on 7 February 2022).
51. Zambrano-Bigiarini, M. hydroGOF (04-1). Goodness-of-Fit Functions for Comparison of Simulated and Observed Hydrological Time Series. 2020. [R Package HydroGOF Version 0.4-0]. Available online: https://cran.r-project.org/web/packages/hydroGOF/index.html (accessed on 7 February 2022).
52. Peters, T.; Taube, F.; Kluß, C.; Reinsch, T.; Loges, R.; Fenger, F. How does nitrogen application rate affect plant functional traits and crop growth rate of perennial ryegrass-dominated permanent pastures? *Agronomy* **2021**, *11*, 2499. [CrossRef]
53. McDonnell, J.; Brophy, C.; Ruelle, E.; Shalloo, L.; Lambkin, K.; Hennessy, D. Weather forecasts to enhance an Irish grass growth model. *Eur. J. Agron.* **2019**, *105*, 168–175. [CrossRef]
54. Chung, S.W.; Gassman, P.W.; Huggins, D.R.; Randall, G.W. Evaluation of EPIC for Three Minnesota Cropping Systems. *Am. J. Agric. Econ.* **2000**, *45*, 1135–1146.
55. van den Pol-van Dasselaar, A.; Hennessy, D.; Isselstein, J. Grazing of Dairy Cows in Europe—An In-Depth Analysis Based on the Perception of Grassland Experts. *Sustainability* **2020**, *12*, 1098. [CrossRef]
56. Ciais, P.; Reichstein, M.; Viovy, N.; Granier, A.; Ogée, J.; Allard, V.; Aubinet, M.; Buchmann, N.; Bernhofer, C.; Carrara, A.; et al. Europe-wide reduction in primary productivity caused by the heat and drought in 2003. *Nature* **2005**, *437*, 529–533. [CrossRef] [PubMed]
57. Clay, N.; Garnett, T.; Lorimer, J. Dairy intensification: Drivers, impacts and alternatives. *Ambio* **2020**, *49*, 35–48. [CrossRef] [PubMed]
58. Düngeverordnung, Düngeverordnung vom 26. Mai 2017 (BGBl. I S. 1305), die zuletzt durch Artikel 97 des Gesetzes vom 10. August 2021 (BGBl. I S. 3436) geändert worden ist. 2021. Düngeverordnung (DüV): Landwirtschaftskammer Niedersachsen (Duengebehoerde-Niedersachsen.de). Available online: https://www.duengebehoerde-niedersachsen.de/duengebehoerde/news/38985_Duengeverordnung_DueV (accessed on 7 February 2022).
59. Bellocchi, G.; Rivington, M.; Donatelli, M.; Matthews, K. Validation of biophysical models: Issues and methodologies. A review. *Agron. Sustain. Dev.* **2010**, *30*, 109–130. [CrossRef]

60. Craig, P.R.; Badgery, W.; Millar, G.; Moore, A. Achieving modelling of pasture-cropping systems with APSIM and GRAZPLAN. In Proceedings of the 17th ASA Conference-Building Productive, Diverse and Sustainable Landscapes, Hobart, Australia, 20–24 September 2015.

61. Schapendonk, A.; Stol, W.; van Kraalingen, D.; Bouman, B. LINGRA, a sink/source model to simulate grassland productivity in Europe. *Eur. J. Agron.* **1998**, *9*, 87–100. [CrossRef]

62. Trott, H.; Wachendorf, M.; Ingwersen, B.; Taube, F. Performance and environmental effects of forage production on sandy soils. I. Impact of defoliation system and nitrogen input on performance and N balance of grassland. *Grass Forage Sci.* **2004**, *59*, 41–55. [CrossRef]

63. Bloor, J.M.G.; Tardif, A.; Pottier, J. Spatial Heterogeneity of Vegetation Structure, Plant N Pools and Soil N Content in Relation to Grassland Management. *Agronomy* **2020**, *10*, 716. [CrossRef]

64. Rueda-Ayala, V.P.; Peña, J.M.; Höglind, M.; Bengochea-Guevara, J.M.; Andújar, D. Comparing UAV-Based Technologies and RGB-D Reconstruction Methods for Plant Height and Biomass Monitoring on Grass Ley. *Sensors* **2019**, *19*, 535. [CrossRef]

65. Binnie, R.C.; Chestnutt, D.M.B. Effect of regrowth interval on the productivity of swards defoliated by cutting and grazing. *Grass Forage Sci.* **1991**, *46*, 343–350. [CrossRef]

66. Calder, F.W.; Nicholson, J.W.G.; Carson, R.B. Effect of actual versus simulated grazing on pasture productivity and chemical composition of forage. *Can. J. Anim. Sci.* **1970**, *50*, 475–482. [CrossRef]

Article

Opportunities for Adaptation to Climate Change of Extensively Grazed Pastures in the Central Apennines (Italy)

Edoardo Bellini [1], Raphaël Martin [2], Giovanni Argenti [1,*], Nicolina Staglianò [1], Sergi Costafreda-Aumedes [3], Camilla Dibari [1], Marco Moriondo [3] and Gianni Bellocchi [2]

[1] Department of Agriculture, Food, Environment and Forestry (DAGRI), University of Florence, 50144 Florence, Italy

[2] Unité Mixte de Recherche sur l'Écosystème Prairial (UREP), Université Clermont Auvergne, INRAE, VetAgro Sup, UREP, 63000 Clermont-Ferrand, France

[3] Institute of BioEconomy, Italian National Research Council (IBE-CNR), 50019 Sesto Fiorentino, Italy

* Correspondence: giovanni.argenti@unifi.it; Tel.: +39-055-2755747

Abstract: Future climate change is expected to significantly alter the growth of vegetation in grassland systems, in terms of length of the growing season, forage production, and climate-altering gas emissions. The main objective of this work was, therefore, to simulate the future impacts of foreseen climate change in the context of two pastoral systems in the central Italian Apennines and test different adaptation strategies to cope with these changes. The PaSim simulation model was, therefore, used for this purpose. After calibration by comparison with observed data of aboveground biomass (AGB) and leaf area index (LAI), simulations were able to produce various future outputs, such as length of growing season, AGB, and greenhouse gas (GHG) emissions, for two time windows (i.e., 2011–2040 and 2041–2070) using 14 global climate models (GCMs) for the generation of future climate data, according to RCP (Representative Concentration Pathways) 4.5 and 8.5 scenarios under business-as-usual management (BaU). As a result of increasing temperatures, the fertilizing effect of CO_2, and a similar trend in water content between present and future, simulations showed a lengthening of the season (i.e., mean increase: +8.5 and 14 days under RCP4.5 and RCP8.5, respectively, for the period 2011–2040, +19 and 31.5 days under RCP4.5 and RCP8.5, respectively, for the period 2041–2070) and a rise in forage production (i.e., mean biomass peak increase of the two test sites under BaU: +53.7% and 62.75% for RCP4.5. and RCP8.5, respectively, in the 2011–2040 period, +115.3% and 176.9% in RCP4.5 and RCP8.5 in 2041–2070, respectively,). Subsequently, three different alternative management strategies were tested: a 20% rise in animal stocking rate (+20 GI), a 15% increase in grazing length (+15 GL), and a combination of these two management factors (+20 GI × 15 GL). Simulation results on alternative management strategies suggest that the favorable conditions for forage production could support the increase in animal stocking rate and grazing length of alternative management strategies (i.e., +20 GI, +15 GL, +20 GI × 15 GL). Under future projections, net ecosystem exchange (NEE) and nitrogen oxide (N_2O) emissions decreased, whereas methane (CH_4) rose. The simulated GHG future changes varied in magnitude according to the different adaptation strategies tested. The development and assessment of adaptation strategies for extensive pastures of the Central Apennines provide a basis for appropriate agricultural policy and optimal land management in response to the ongoing climate change.

Keywords: grasslands; modeling; PaSim; climatic scenarios; aboveground biomass

Citation: Bellini, E.; Martin, R.; Argenti, G.; Staglianò, N.; Costafreda-Aumedes, S.; Dibari, C.; Moriondo, M.; Bellocchi, G. Opportunities for Adaptation to Climate Change of Extensively Grazed Pastures in the Central Apennines (Italy). *Land* **2023**, *12*, 351. https://doi.org/10.3390/land12020351

Academic Editors: Michael Vrahnakis, Yannis (Ioannis) Kazoglou and Pulido Fernádez Manuel

Received: 29 December 2022
Revised: 21 January 2023
Accepted: 26 January 2023
Published: 28 January 2023

1. Introduction

With an herbage production potential up to ~15 t DM ha^{-1} [1], grasslands contribute significantly to global food security by providing fodder for ruminants used in the production of protein-rich foods, such as meat and milk [2,3]. In Italy, grassland areas (i.e., permanent meadows and pastures) cover approximately 3.6 Mha [4], roughly 12% of the entire Italian territory, and are located mainly along the Alpine and Apennine mountain ranges and on the islands [5]. Differing in climate and land use, factors that influence productivity and botanical composition, Italian grasslands can be divided into three different biogeographic regions: Alpine, Apennine, and Mediterranean [6]. They are mostly large-scale rainfed pastoral systems, with permanent pastures dominant in the mountains and hilly areas and fodder crops also dominant in the Mediterranean region. Generally, these systems provide forage for only short periods of time during spring and summer, exhibiting great inter-annual variability in production [7,8]. With regard to mountain areas (i.e., Alps and Apennines), grasslands are often located in areas with nutrient-poor soils and/or extreme climate conditions that make vegetation growth, and consequently forage production, reliant on seasonal dynamics [9]. Focusing specifically on Apennine mountain pastures, forage quality is generally lower than in Alpine pasturelands [10], due mainly to the great variability in pedo-climatic conditions that can be found along the latitudinal gradient of Italy [11].

In addition to forage production, grasslands provide several other ecosystem services important for human well-being, such as water and nutrient regulation and protection from soil erosion [12–16]. Particularly important is the role that these systems can play in climate-changing emissions [17], as they can stock/emit carbon dioxide CO_2 [18,19] and emit non-CO_2 greenhouse gases, such as methane (CH_4) and nitrous oxide (N_2O) [20]. According to Guillaume et al. [21], soil organic C stock measured from surface to 50 cm depth in permanent grasslands is approximately 7 kg C m^{-2}, and evidence from European grasslands shows that soil C sequestration rates can reach $0.77 \text{ g C m}^{-2} \text{ yr}^{-1}$ [22]. Compared with other ecosystems, grasslands are, in fact, an important store of C [23], and management (grazing in particular) is an important regulator of C and N fluxes [24]. Grasslands have the advantage of potentially acting as C and N sinks, compared with croplands, and can mitigate GHG emissions in livestock production systems, as C and N sequestration can offset GHG emissions [17,25,26].

Pastoral resources in the Apennines during the last decades have shown fragility in the face of changes induced by recent global warming. There was a shift in air temperature distribution towards warmer values in all seasons (especially for minimum temperature, while maximum temperature shows a more intense warming and a pronounced peak in summer) since the 1980s, with an acceleration in the 2000s [27], and it is projected to increase in the future [28]. In view of the expected increase in temperatures associated with a decrease in precipitation during the summer period, forage production is assumed to change in terms of quantity and quality [29,30]. Moreover, evolution of the distribution of species in herbaceous communities and changes in the botanical composition of semi-natural grasslands are highlighted [31]. In fact, rising temperatures and summer droughts tend to promote the predominance of thermophilic communities or species more adapted to xeric environments, which now grow in environments at lower altitudes, as was already observed in the Alps [32] and Apennines [33,34].

In this view, simulation models, through the reproduction of system biophysical processes, can help stakeholders in decision-making by assessing the impacts of climate change and/or testing different management strategies under current [35,36] or future scenarios [37–41]. In this context, appropriate management (e.g., stocking rate and grazing period) can preserve grassland biodiversity, maintain socio-ecological systems, and counteract the effects of climate change. On the basis of assessment of the previous literature, it can be said that a very small number of modeling exercises have examined the effect of foreseen climate changes on pasture production characteristics in the Apennine area [6], as almost all works have analyzed the effects on vegetation features and biodiversity, e.g., [42–44]. Therefore, the present research aims to analyze the expected effect of climatic changes mainly from an agronomic perspective, providing an approach that can be repeated in other contexts and that is aimed at evaluating the impacts on productive features of forage resources and the possible adaptation strategies of some of the main pasture management characteristics.

This perspective forms the basis for the design and implementation of this study initiated in 2020 on two pastoral farms in the Apennines territory of central Italy, based on field observations and model-based simulations. Modeling the performance of pastoral systems is helpful in defining management strategies that maximize pastoral production and minimize environmental impacts [45]. Field data support the modeling exercises by providing detailed on-farm information on the spatial and temporal variation of important canopy state variables, which are often difficult to obtain [46]. Simulation results under future climate change scenarios were the key tools for the design and assessment of the analytical framework concerning climate change adaptation strategies, pivotal factors for the conservation of grassland resources [47]. Based on the hypothesis that future climate change will significantly affect extensive grazing systems of the Central Apennines, the specific objectives of this study were: (1) to inform the modeling via calibration with field data; (2) to use the calibrated models to project the impacts of climate change; and (3) to assess a set of adaptation options for pastoral management identified locally.

2. Materials and Methods

The study was initially conducted by calibrating the grassland simulation model PaSim [48] with observed data collected on two specific farms in the Italian Apennines (Suite 1). The parameterization obtained was subsequently used, together with the climate models, to simulate the impacts of climate change on grasslands (Suite 2). In parallel, a sensitivity analysis was performed with specific attention to biomass production (Suite 3). Finally, on the basis of the results obtained in the impact analysis, possible adaptation strategies were identified and tested (Suite 4). A general outline of the methodology used can be seen in Figure 1.

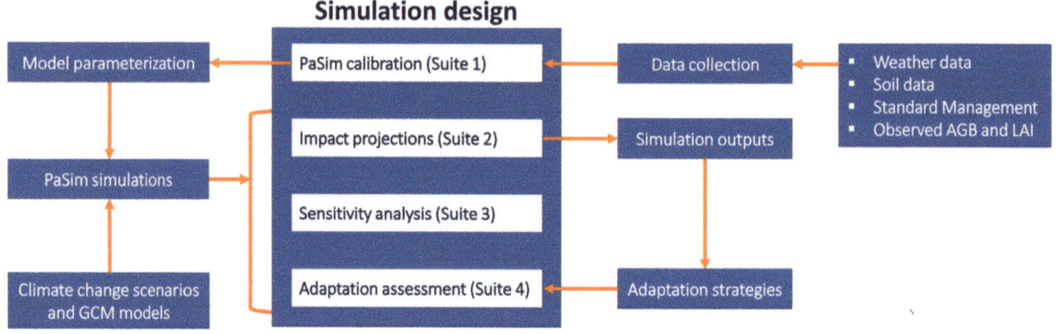

Figure 1. Workflow of the methodology applied in this study. PaSim is the grassland simulation model used for the analysis.

2.1. Study Sites, Experimental Layout, and Data Collection

The study considered two pastoral farms (Figure 2) located at different altitudes in the Tuscan Apennines (Table 1), both managed under continuous grazing system of Limousin cattle (Table 2).

The Marradi study site (M) covers more than 5 ha of upland sown pasture that tends towards a re-naturalization, usually grazed from May to July. The Borgo San Lorenzo study site (B) covers 30 ha of lowland sown pasture. For the purpose of the trial, Site B was divided in 2020 into two differently managed sub-areas, B1 (approx. 10 ha) and B2 (approx. 20 ha). Specifically, in sub-area B1 the pasture was grazed by Limousin cattle from April until the end of October, while sub-area B2 was managed under a mixed utilization: mowed in May and grazed from June until the end of October.

Table 1. Description of the study sites.

Description	Unit	Site M (Marradi)	Site B (Borgo San Lorenzo)
Location			
Latitude (WGS84)	degree N	44.08°	43.95°
Longitude (WGS84)	degree E	11.63°	11.35°
Elevation	m a.s.l.	600	200
Climate			
Mean annual temperature [1]	°C	12.4	13.4
Mean annual precipitation [2]	mm	1330	990
Soil [3]			
Depth	m	1	1
Clay	%	37	37
Silt	%	42	36
Sand	%	21	27
Total organic carbon	g kg^{-1}	33.6	23.5
Total nitrogen	g kg^{-1}	3.0	2.5
Soil pH	-	6.6	7.4
Bulk density	g cm^{-3}	1.29	1.44
Saturated soil water content	m^3 m^{-3}	0.52	0.51
Field capacity	m^3 m^{-3}	0.36	0.35
Wilting point	m^3 m^{-3}	0.21	0.21
Dominant vegetation	-	*Dactylis glomerata, Lolium* sp., *Festuca arundinacea, Phleum pratense,* and *Onobrychis viciifolia,* with other minor forbs and a large presence in some sectors of shrubs, such as *Rubus ulmifolius.*	*Lolium* sp., *Dactylis glomerata, Trifolium pratense, Trifolium repens, Lotus corniculatus,* and *Festuca arundinacea,* with other minor forbs.

[1] Site M: mean of 2016, 2017, and 2020; Site B: mean of 1951–2020. [2] Site M: mean of 2001–2020; Site B: mean of 2001–2020. Data collected from regional weather stations of Tuscany Region (SIR, Servizio Idrologico Regionale, https://www.sir.toscana.it/index.php, accessed on 20 January 2023). Distance from sites <10 km. [3] 1 m soil profile mean.

Table 2. Management of the two study sites. Livestock Standard Unit (LSU) refers to a dairy cow producing 3000 kg of milk per year, without additional concentrated feed (EC, 2008).

Management	Unit	Site M (Marradi)		Site B (Borgo San Lorenzo)			
				B1		B2	
		2020	2021	2020	2021	2020	2021
Surface	ha	5.4		10		20	10
Cut	day of year	-	-		-	125	-
Grazing period	days of year (start, end)	139–244 [a]; 244–267 [b]	135–176 [a]; 176–276 [b]	100–180 [a]; 186–300 [b]	100–145 [a]; 145–306 [b]	180–186 [a]; 186–300 [b]	110–145 [a]; 145–306 [b]
Stocking rate	LSU ha^{-1} d^{-1}	4.0 [a]; 3.4 [b]	3.3 [a]; 2.0 [b]	2.9 [a]; 1.0 [b]		1.5 [a]; 1.0 [b]	0.9; 1.2 [b]

[a] and [b] represent two distinctive grazing periods during the season in terms of stocking rate.

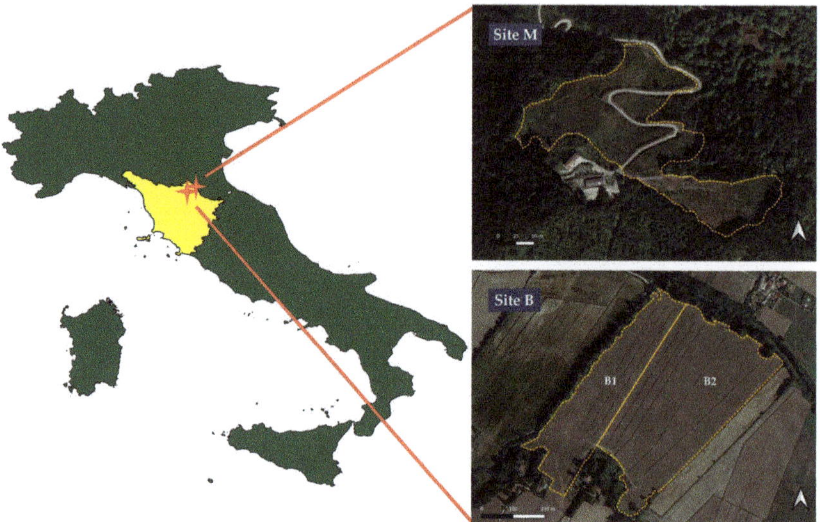

Figure 2. Aerial view of the study sites of Marradi (M, left) and Borgo San Lorenzo (B, right). Satellite images of the sites were obtained from Google Earth.

Samples of aboveground dry matter (DM) biomass (AGB, kg DM m^{-2}) and measurements of leaf area index (LAI, m^2 m^{-2}) were collected during field surveys conducted in spring/summer (2020 and 2021) at both sites and used for the modeling work (Table S1). Field data were collected in 16 randomly arranged samples in an area of 1 m^2 each (eight in M, four in B1, and four in B2). The sampling position was changed from time to time, taking care to choose areas that represented the general situation. The AccuPAR PAR/LAI Ceptometer Model LP-80 (Decagon Devices, 2017) was used to measure LAI in each plot.

2.2. Climate Scenarios and Models

Daily-downscaled (bias-corrected) weather data were selected to map a broad range of climate outputs for impact modeling [49] (Table S2).

In order to take into account the uncertainties of the different climate models in the projected simulations [50], the outputs of an ensemble of models were considered for the modeling exercise under the future scenarios RCP4.5 (intermediate scenario) and RCP8.5 (extreme scenario). The climate change scenario ensemble included 14 members deriving from the combination of 14 Global Climate Models (GCMs) downscaled to six high-resolution (~0.12°) Regional Climate Models (RCMs) in the framework of the Med-CORDEX project [51]. Daily climate outputs (minimum and maximum temperatures and cumulative rainfall) obtained from the 14 GCMs (available at https://www.medcordex.eu/index.php/, accessed on 20 January 2023) were then bias-corrected over the study sites according to Cornes et al. (2018) and Lange (2019) [52,53] in order to drive the relevant simulations in future periods. Daily global radiation and relative humidity were retrieved from daily temperature according to Bristow-Campbell [54] and the FAO Irrigation and Drainage paper [55], respectively. CO_2 annual concentrations (ppm) for past, current, and future projections were calculated from the IPCC report [56].

2.3. The Grassland Model

The Pasture Simulation model (PaSim) was chosen for this study because it can describe in detail the dynamic biogeochemical responses of a grassland system under altered climate and management. Originally developed by Riedo et al. [48], PaSim simulates the cycling of water, C, and N in grassland systems at a sub-daily time step (1/50th of a day) or, as in this work, at a daily time step. Microclimate, soil biophysics, vegetation, herbivores, and management practices are interacting modules. The simulations are not spatially resolved (e.g., inhomogeneity is not taken into account) and input/output data are assumed to be representative of the entire field. The assimilated photosynthetic C is dynamically allocated to a root and three shoot compartments (each composed of four age classes) or lost through animal metabolism (ecosystem respiration). Accumulated aboveground biomass is cut, grazed, or relocated to the litter pool. Management includes the application of organic and mineral N fertilizers, mowing, and grazing. Details on the model processes are provided in published articles [57–61], which have contributed to the recognition of PaSim as a suitable tool to reproduce biophysical and biogeochemical processes in managed grasslands and its inclusion in international modeling exercises [17,62].

2.4. Simulation Design

The modeling work was performed in four simulation suites: Suite 1 with observational data (calibration), Suite 2 with projected climate change scenarios with CO_2 fertilization effect (impact projections), Suite 3 with projected climate change scenarios without CO_2 fertilization effect (sensitivity), and Suite 4 with modified management under projected climate change scenarios with CO_2 fertilization effect (adaptation assessment).

For Suite 1, the simulations setup included weather, soil, vegetation variables and management implementation in the studied years (2020 and 2021). The weather variables included daily minimum and maximum air temperatures, precipitation, and solar radiation. Temperature, precipitation, and wind speed data for 2020 and 2021 were collected from the regional weather stations of Tuscany Region (SIR, Servizio Idrologico Regionale, https://www.sir.toscana.it/index.php, accessed on 20 January 2023) located near the study sites. Daily global solar radiation data were generated from the R package "sirad", developed by Bojanowski et al. [63], based on the model of Bristow and Campbell [54]. The soil data were extracted from the SoilgridsTM dataset (https://soilgrids.org, accessed on 20 January 2023), described in Poggio et al. [64]. The actual management practices (grazing intensity and periods) are described in Table 2.

Model calibration was not applied separately to each site. The model was calibrated on all datasets to obtain more realistic and robust parameter values for application on a larger scale, as in Ma et al. [59]. The availability of detailed LAI and AGB data from two grassland sites offered the possibility of a genuine (multi-location and multi-output) calibration of the model, on the assumption that a unique calibration across sites is appropriate under these conditions. We assumed that a common set of eco-physiological model parameters can be established to simulate C3 grasslands (including grass, forb, and legume species) under contrasting climatic and management regimes (e.g., Site M represents hill situations, and Site B represents plain situations), while site-specific climatic and management conditions provide the local drivers of actual grassland biomass and foliage production.

In particular, PaSim calibration (Suite 1) was performed against LAI and AGB data collected in the years 2020 and 2021 by modifying the values of a set of parameters (Table S3) to which model sensitivity was determined in previous studies [58–61]. Parameter values were modified (with the generation of 1000 sets of values using the random Latin hypercube method) within their plausible ranges [48] to ensure satisfactory performance, which is a realistic representation of both outputs. The sets of parameter values resulting from the model calibration were used to compare the PaSim outputs (AGB and LAI) with the observations in each study site. The agreement between simulated and observed AGB and LAI was assessed by inspection of time-series plots (fluctuations of output variables over time) and numerically, through two performance metrics commonly used in model evaluation [65]: relative root mean square error (best, $0 \leq$ RRMSE $< +\infty$, worst) and coefficient of determination (worst, $0 \leq R^2 \leq 1$, best).

For Suites 2, 3, and 4, simulated pastoral outputs were obtained by forcing the calibrated PaSim with the downscaled (bias-corrected) daily weather data described in Section 2.2, Climate Scenarios and Models. Projected PaSim responses to climate change forcing options were calculated on changes in a set of agro-ecosystem outputs related to growing season length, fodder production, water cycle, and C-N fluxes (Table 3). At both sites, we assessed the sensitivity of the grassland model to climate change (RCP4.5 and RCP8.5 for the ongoing and mid-future periods) under business-as-usual (BaU) management (Suites 2 and 3) and alternative management scenarios (Suite 4).

For Suite 2 (impact projections) and Suite 4 (adaptation assessment), grassland modeling results were obtained with a climate forcing based on atmospheric CO_2 concentration set at 363 ppm, on average, for the baseline scenario (near past: 1981–2010). In this way, the year 2010 was taken as the end of the time horizon used in this study to emulate the near-past climate, i.e., 30-year time span until the late 2000s, which includes the limit of the historical period (1765–2005) of the atmospheric observations used to drive the climate models [66]. Then, mean atmospheric CO_2 concentrations were prescribed according to the selected RCPs (middle impact: 4.5; extreme impact: 8.5) and timeframes (ongoing: 2011–2040; mid-future: 2041–2070): 431 (ongoing) and 523 (mid-future) mean ppm under RCP4.5; and 438 (ongoing) and 613 (mid-future) mean ppm under RCP8.5. The results related to the pasture system obtained in Suite 2 were then used in the choice of the possible future adaptation strategies (e.g., increase or decrease in animal load and/or length of grazing season).

For Suite 3 (sensitivity), any fertilization effect from the additional CO_2 emitted during the period from 2011 to 2070 was eliminated. What has been carried out here is, in effect, a test of the sensitivity of PaSim to alterations in weather inputs, this exercise being ultimately focused on understanding the grassland modeling process (not on assessing impacts of climate change and elevated CO_2).

Table 3. Climate change impact metrics.

Type	Output	Acronym	Unit	Description
Date	Growing season start	GSs	day of year (doy)	Day after seven consecutive days with a mean air temperature ≥ 8 °C from 1 January onwards [67]
	Growing season end	GSe		Day after seven consecutive days with a mean air temperature <8 °C from 1 July onwards [67]
	Biomass peak date	BPd		Day of the year with the highest value of aboveground biomass
Count	Growing season length	GS	days	Number of days between the GSs and GSe
Amount	Biomass peak	BP	kg DM m^{-2}	Aboveground biomass value at the peak date
	Aboveground biomass	AGB	kg DM m^{-2}	Aboveground biomass values
	Net ecosystem exchange	NEE	kg C m^{-2} yr^{-1}	C-N fluxes (annual balance) (These include emissions from ecosystem respiration, RECO = plant + soil + animal respiration, as well as estimates of the plant production of organic compounds from atmospheric CO_2 (GPP: gross primary production) and other system variables: NEE = RECO - GPP, enteric emissions of CH_4 from grazing animals and N_2O emissions from the N cycle)
	Methane	CH$_4$	kg C m^{-2} yr^{-1}	
	Nitrous oxide	N$_2$O	kg N m^{-2} yr^{-1}	
	Soil water content	SWC	m^3 m^{-3}	Annual mean of daily soil water content values (0.35-m topsoil). In Supplementary Materials.

3. Results

3.1. Climate Analysis

The monthly distribution of air temperatures at the two study sites (Figure 3), averaged from the outputs of 14 climate models, showed an overall increase in temperature towards the mid-future, similar for both sites, with the highest increases in summer (roughly +2.6 °C at both sites under the warmest scenario) and the lowest in autumn–winter (roughly +2.1 °C at both sites under the warmest scenario).

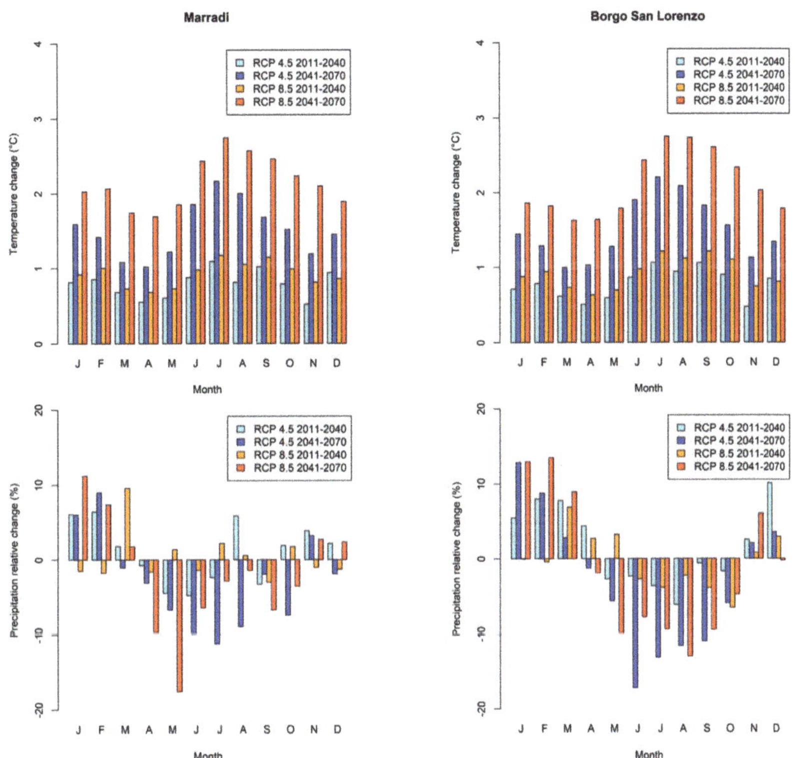

Figure 3. Absolute change (°C) in monthly mean air temperature (top graphs) and relative change (%) of monthly cumulated rainfall (bottom graphs) generated in the two study sites with the RCM ensemble (14 models) for two climate scenarios (RCP4.5, RCP8.5) and two periods–2011–2040 (ongoing) and 2041–2070 (mid-future)—over the baseline period 1981–2010 (near past).

Analysis of simulated rainfall data (Figure 3) showed increases in the November–March period relative to the baseline in both scenarios and sites (Site M: +3.1% and +5.1%; Site B: +6.0% and +8.2%, for RCP4.5 and RCP8.5, respectively), while between April and October there was a sharp decrease in rainfall at both sites (−7.0% and −8.9% at M and −9.4% and −8.9% at B for RCP4.5 and RCP8.5, respectively).

3.2. Suite 1 of Simulations: Evaluation of the Model against Observed Data

AGB simulations (Figure 4, Table 4) indicate that estimates substantially reflect patterns of vegetation dynamics (R^2~0.70) although some departures from observed data are noted. The RRMSE values (<15%), in particular, suggest that the model has strong predictive ability for biomass production. This was also obtained with the LAI, with $R^2 < 0.50$ only in sub-area B1 of Site B, where the RRMSE of ~25% was acceptable.

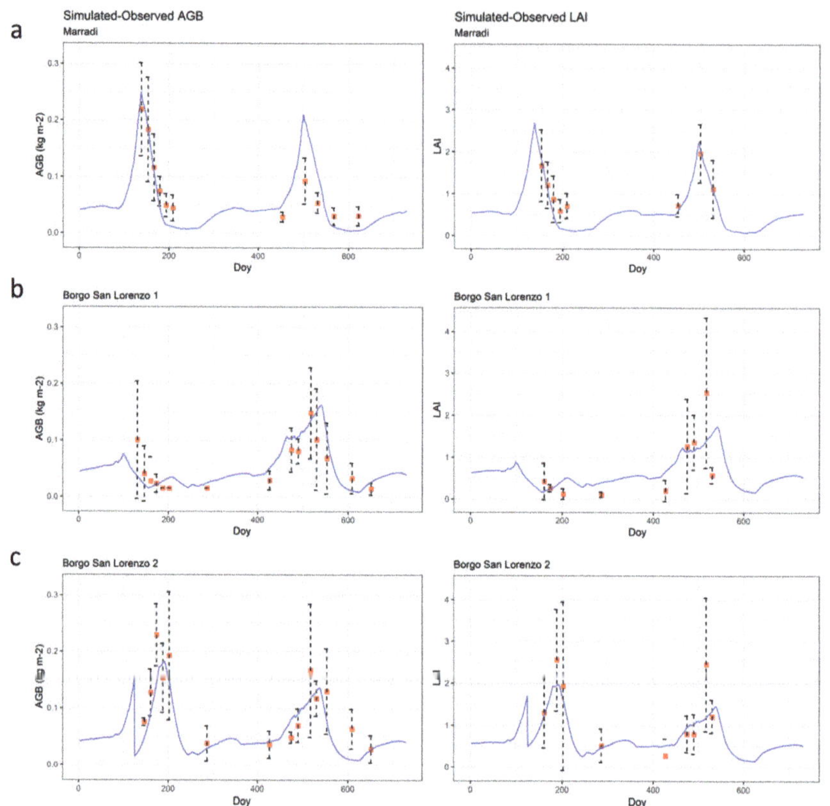

Figure 4. Simulated (blue line) and observed (red square dots) patterns of aboveground biomass (AGB) and leaf area index (LAI) at Sites M (**a**), B1 (**b**), and B2 (**c**) for the period 2020–2021.

Table 4. Model performance for the two study sites (M: Marradi; B: Borgo San Lorenzo, sub-areas B1 and B2) based on two performance metrics: R^2, coefficient of determination of the linear regression between estimates and observations; and RRMSE (%), Relative Root Mean Square Error. AGB: aboveground biomass; LAI: Leaf Area Index.

Output	Site M		Site B			
			B_1		B_2	
	R^2	RRMSE	R^2	RRMSE	R^2	RRMSE
AGB	0.76	14.9	0.66	13.5	0.68	10.0
LAI	0.96	9.6	0.47	24.5	0.71	12.6

3.3. Suites 2, 3, and 4 of Simulations: Impacts of Future Scenarios, Sensitivity to Weather Inputs, and Adaptation Strategies

For both sites, we assessed the response of the grassland model to climate change (RCP4.5 and RCP8.5 for the ongoing and mid-future periods) with business-as-usual (BaU) management (Suite 2) and to different management options (Suite 4). Multi-year mean responses for growing season length (GS), biomass production (AGB), and biogeochemical (C-N fluxes) were calculated. Sensitivity analysis was performed without the CO_2 fertilization (Suite 3) effect by observing future AGB trends over the season for the different RCPs and time periods.

3.4. Growing Season

Under the climate change scenarios, the estimated length of the growing season increases at both sites because optimal thermal conditions for vegetation growth occur earlier and later in the season. This leads to an earlier onset (GSs) and later end (GSe) of the growing season (GS) in both sites, especially in the mid-future (i.e., 2041–2070) (Figure 5). Specifically, for RCP4.5, GSs was advanced by 4 and 8 days, on average, in Site M and by 6 and 12 days in Site B for the periods 2011–2040 and 2041–2070, respectively. In addition, GSe was delayed by 3 and 9 days, on average, for the periods 2011–2040 and 2041–200, respectively, at Site M and by 4 and 9 days, on average, at Site B for the periods 2011–2040 and 2041–2070, respectively. The most pronounced differences from the baseline are visible for the RCP8.5 scenario. Earlier onsets of 4 and 17 days for Site M and 11 and 15 days for Site B under the periods 2011–2040 and 2041–2070, respectively, are accompanied by delays in GSe (5 and 18 days for Site M and 8 and 13 days for site B under the periods 2011–2040 and 2041–2070, respectively).

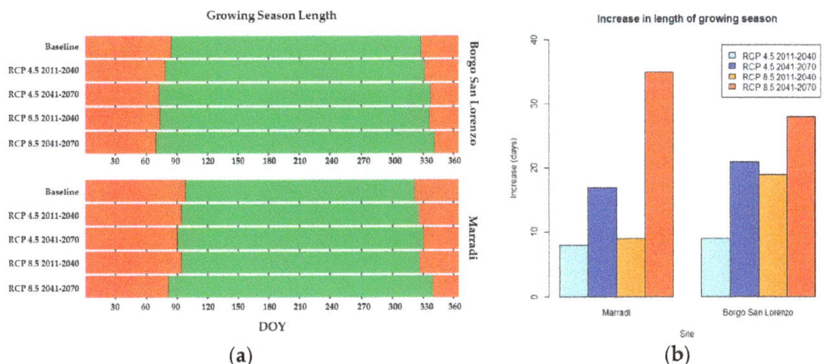

Figure 5. Estimated durations (30-year mean values) of vegetation growing seasons (green bars) for baseline and climate change scenarios under business-as-usual management at both study sites (**a**). On the right, increases of growing season length compared with the baseline (**b**).

3.5. Aboveground Biomass

Figure 6 shows the AGB production patterns under BaU management in both sites for the baseline and future projections, while the AGB patterns obtained with all alternative management options can be found in the Supplementary Material (Figures S2–S5).

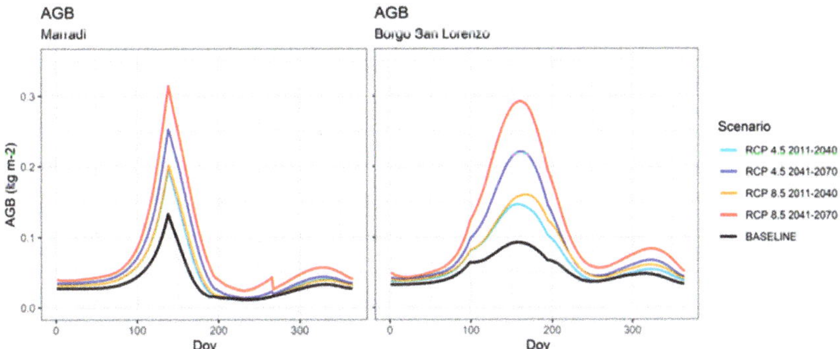

Figure 6. Daily simulation (30-year mean) of aboveground biomass (AGB) with PaSim for baseline and climate change scenarios under business-as-usual management at both study sites.

The main differences in AGB patterns among alternative management and climate scenarios were assessed from changes in peak biomass dates (BPd) and corresponding AGB values (BP), which strongly influence stakeholders' and farmers' decisions in choosing the most suitable periods for grazing.

With the baseline climate scenarios, PaSim reported peak biomass on days 138 (Site M) and 157 (Site B). With the future climate scenarios, the model indicated the same BPd at Site M (day 138) with both scenarios and time slices, as grazing starts on day 139, while Site B showed a general delay in BPd, specifically 1 to 5 days in RCP4.5 and 3 to 10 days in RCP8.5.

In the baseline scenarios, the peak biomass production (BP) is 0.13 ($\pm$0.03 standard deviation) kg DM m^{-2} at Site M and 0.09 ($\pm$0.02 standard deviation) kg DM m^{-2} at Site B. With the climate change patterns, PaSim estimated higher BP values with both RCP4.5 (by 48.4 and 90.8% at Site M and 58.9 and 139.7% at Site B, for 2011–2040 and 2041–2070, respectively) and RCP8.5 (by 52.1 and 136.9% at Site M and 73.4 and 216.8% at Site B, respectively), mainly due to the fertilizing role of CO_2 in the selected emission scenarios and the absence of sensible water deficits simulated by PaSim (Figure S1). With respect to SWC, in fact, although the simulated patterns suggest that, with drier summer conditions, grassland growth may be limited by some water stress in the future, differences between the baseline and climate change scenarios are limited at both sites. In particular, no significant changes in SWC are evident during the spring period, when plant growth activity is the greatest.

To assess the effect of CO_2 fertilization (Suite 3), we tested the same climate change scenarios using the mean baseline CO_2 concentration (i.e., 363 ppm recorded, on average, during 1981–2010), showing that BP values under the baseline CO_2 concentration did not increase to the same extent as observed for the future scenarios with higher CO_2 concentration (Figure 7). Specifically, compared with the baseline, the BP increased by 24.8 and 29.5% at Site M and 10.5 and 16.5% at Site B for RCP4.5 (for 2011–2040 and 2041–2070, respectively) and by 25.2 and 50.0% at Site M and 15.4 and 27.0% at Site B for RCP8.5 (for 2011–2040 and 2041–2070, respectively).

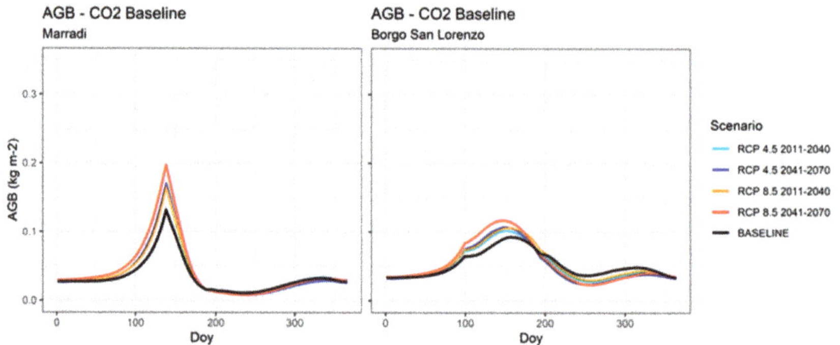

Figure 7. Daily simulation (30-year mean) of aboveground biomass (AGB) with PaSim for baseline and climate change scenarios (no CO_2 fertilization) under business-as-usual management at both study sites.

Considering the results of Suite 2, alternative management practices (Suite 4) included: (1) livestock grazing intensity increased by 20% (i.e., +20 GI); (2) extension of the grazing period length by 15% (i.e., +15 GL), specifically 7 days earlier start and 7 days later end at Marradi, 16 days earlier start and 16 days later end at Borgo San Lorenzo; (3) combination of (1) and (2) (i.e., +20GI x 15GL). For the impact of adaptation strategies, the value of the peak biomass obtained with alternative management practices (i.e., BaU and adaptation management options) was compared with the peak biomass from business-as-usual (BaU) management under the projected scenarios (Figure 8).

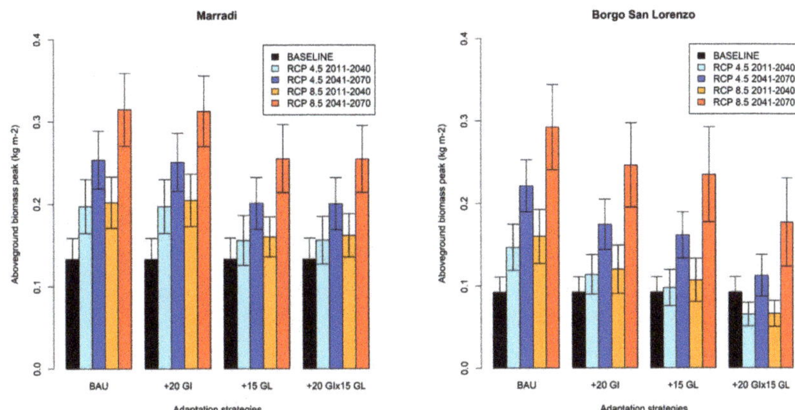

Figure 8. Changes in peak aboveground biomass (kg DM m^{-2}) among business-as-usual management (BaU) under the baseline climate (black histogram) and all alternative management options under RCP4.5 (cyan and blue histograms) and RCP8.5 (orange and red histograms) in both sites as provided by PaSim. Vertical bars are standard deviations.

According to model's outputs, the aboveground peak (Figure 8) and the trends over the season (Figures S2–S5), obtained using the different adaptation strategies, show that future biomass availability will reach higher values when compared with the baseline, even by increasing the animal stocking rate (i.e., +20 GI) and/or the number of grazing days (i.e., +15 GL or +20 GI × 15 GL).

3.6. Carbon–nitrogen Fluxes

Under current climate and management conditions, PaSim shows limited non-CO$_2$ emissions at both sites, i.e., ~2 g C m^{-2} yr^{-1} for CH$_4$ and 4.6–4.7 g N m^{-2} yr^{-1} for N$_2$O emissions, while the C exchanges reflect that both sites are sources of C (NEE $\geq$ 350 g C m^{-2} yr^{-1}, Table 5).

Table 5. C-N emissions (NEE: net ecosystem CO$_2$ exchange; CH$_4$: methane; and N$_2$O: nitrous oxide) from the two study sites (baseline climate), estimated (30-year mean with standard deviation) using PaSim. The estimated components of the C budget (GPP: gross primary production; RECO: ecosystem respiration) can be found in Supplementary Material (Table S4).

Site	NEE	CH$_4$	N$_2$O
	g C m^{-2} yr^{-1}		g N m^{-2} yr^{-1}
Site M	381.3 ± 245.6	2.2 ± 0.3	4.6 ± 3.4
Site B	350.1 ± 236.1	1.8 ± 0.2	4.7 ± 3.2

Heatmaps of the % differences between current conditions (i.e., baseline climate and BaU management) and combinations of alternative climate and management scenarios allow the impact of altered climate and management changes on gas emissions at the two study sites to be assessed (Figure 9). For NEE, in particular, the PaSim heatmaps show overall trends towards C uptake (more negative NEE values) in both study sites by moving towards extreme climate conditions (i.e., RCP8.5 and time-frame 2041–2070), with all management options. This reflects the AGB pattern (Figure 6) resulting from a higher photosynthetic plant production from atmospheric CO$_2$, even with increased animal respiration under the option of increased livestock density (GPP and RECO values in Table S4).

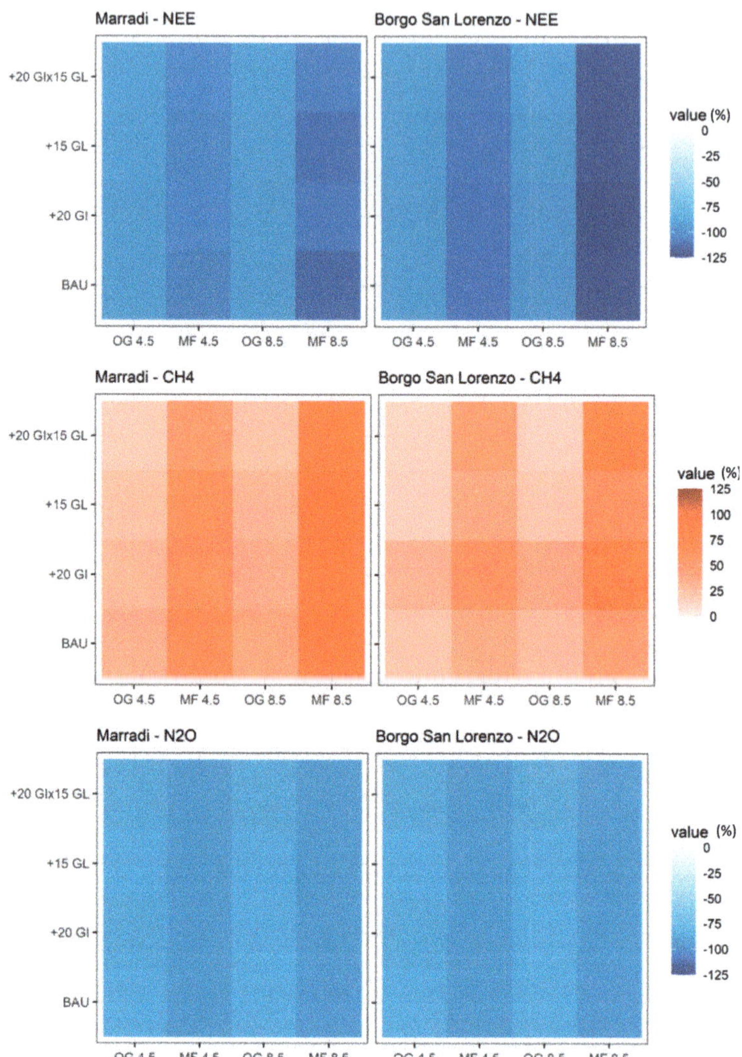

Figure 9. Heatmap visualization of the relative differences (%) of the three main greenhouse gas emissions (NEE: net ecosystem exchange; CH_4: methane; and N_2O: nitrous oxide), estimated using PaSim, for alternative management and climate change scenarios compared with current climate and management in the two study sites. OG: ongoing period, MF: mid-future period, 4.5: RCP4.5, 8.5: RCP8.5, GI: grazing intensity; and GL: grazing length, 20: +20%, 15: +15%.

As for CH_4 emissions, the PaSim heatmap indicates that emissions are higher with the warmest scenario and as livestock density increases (up to <100%). Finally, the N_2O emissions estimated by PaSim tend to be lower under future climate and alternative management scenarios.

4. Discussion

4.1. Model Parameterisation

The great deal of fundamental research incorporated into the mechanistic PaSim model has ensured satisfactory estimates, which are also comparable to published grassland modeling studies [68,69]. This is relevant considering that simulations for grasslands are generally less accurate compared with arable crops [70] since large uncertainties in biomass and LAI measurements cause simulation of grassland vegetation dynamics to be difficult to perform [67,71].

This was obtained with calibrated parameter values (Table S3) that do not deviate substantially from those obtained in previous studies on continental and Mediterranean grasslands. For instance, the maximum specific leaf area, slam = 27.2 m^2 kg^{-1}, is similar to slam = 29 m^2 kg^{-1} obtained in the Europe-wide calibration of Ma et al. [59]. Light-saturated leaf photosynthetic rates for reproductive (pmco2rep = 12.88 µmol C m^{-2} s^{-1}) and vegetative (pmco2veg = 9.49 µmol C m^{-2} s^{-1}) stages are similar to the values obtained for Mediterranean grasslands (pmco2rep = 14.0 µmol C m^{-2} s^{-1} and pmco2veg = 10.0 µmol C m^{-2} s^{-1}) from Pulina et al. [60]. The root and shoot turnover rates at 20 °C, kturnrt20 = 0.0155 d^{-1} and kturnsh20 = 0.0468 d^{-1}, respectively, exceed those estimated by Pulina et al. [60] for grasslands dominated by annual self-seeding plant species: 0.0144 d^{-1} and 0.0250 d^{-1}, respectively. With the obtained calibration, the shoot turnover parameter dwindled to approximately 21 days (1/0.0144 d^{-1}), which is lower than 40 days (1/0.0250 d^{-1}), as in Pulina et al. [60]. In fact, perennial plants tend to invest mainly in long-lived and competitive adult individuals, and, consequently, shoot turnover tends to be faster in perennial plants than in annual species, as the former allocate more resources for new leaf growth to maximize photosynthetic efficiency [72].

4.2. Uncertainties in Climate Change Impacts and Adaptation Strategies

The adopted impact model was widely applied in various contexts [59,60,73,74], dealing with multifaceted territorial and vegetation structures and extreme weather conditions, which are often difficult to parameterize [75] due to the complex response of the vegetation growth with respect to critical thresholds (e.g., air temperatures, water requirements, and radiation use efficiency) for mixed plant communities [17]. In this study, PaSim represented the effects of climate change and management options on the timing and extent of the growing season and C-N fluxes, together with biomass production and peaks. The longer growing season length was due to the extension of the potential growing season in both spring and autumn, as already observed in grasslands during the last decades [76,77]. The mean plant growth trend simulated with the model (30-year means) mirrors the observed pattern of vegetation growth during the growing season, indicating that the overall pattern of response to elevated atmospheric CO_2 concentration significantly stimulates leaf photosynthesis [78,79]. Sensitivity analysis performed in Suite 2 highlighted this fertilization effect of increased CO_2 concentration simulated by PaSim; nevertheless, it must be underlined that similar trends of increased aboveground biomass in both future climate change scenarios and time periods are visible also with steady CO_2 concentration (i.e., baseline concentration, 363 ppm), albeit to a lesser extent. In addition, although a down-regulation strategy can be useful to limit the effect of increased CO_2 concentration on plant growth [80], it is worth emphasizing that the production increases projected for the mid-future (2041–2070) resulted in being particularly high when compared with a baseline that reflects a situation of the near past (period 1981–2010). When compared, instead, with the ongoing period (2011–2040), which reflects average aboveground biomass values similar to the present and to the calibration period, the increases are smaller, comparable to those found in other studies [81,82]. The CO_2 positive effect is reflected in the higher C uptake estimated by PaSim as a result of increased productivity, also with higher stocking rates (i.e., higher C losses due to higher animal respiration), which confirms the increased worldwide productivity of grasslands exposed to increased CO_2 [83].

PaSim estimated increasing CH_4 emissions and decreasing N_2O emissions with climate scenarios. The former logically reflects evidence that grasslands emit more CH_4 at higher temperatures [84]. Although the latter does not reflect the direct effect of temperature on the enzymatic processes involved in N_2O production, N_2O emissions are controlled mainly by soil properties and current soil N levels [85], which may have been reduced with increased plant demand due to higher biomass production under climate scenarios. This increase in future biomass production, driven by the higher average annual GPP (gross primary production), also led to a consequent decrease in simulated NEE over the years.

4.3. Consequences for Grassland Sustainability

Herders depend on pasture and water resources for their livestock and are among the groups most vulnerable to climate change impacts in dry regions [39,86–89]. Although there are reasons to be concerned, some impacts of climate change are expected to be positive. Foreseen climate variability can be an opportunity for effective management, as actions could be timed to the most effective conditions, and climate change could be a motivation to develop a broader and more responsive and collaborative management paradigm. We showed that increases in plant productivity and longer growing seasons in central Italy may support more livestock and increase economic benefits. Rising air temperatures simulated by climatic models, combined with increasing concentrations of CO_2 in the atmosphere in RCP4.5 and RCP8.5 scenarios, are expected to offer important opportunities in terms of forage production for livestock systems in central Italy. This is possible if future water availability is not a limiting factor, as stressed by various research studies on grassland potential production [90]. Indeed, as seen from the results of climatic models, precipitations are expected to decrease in the future, mostly in summer months but not particularly in spring. The availability of water in the soil, therefore, does not vary significantly over time and future climate change scenarios, as is visible from the soil water content simulated by PaSim (Figure S1). These trends on future pasture productivity are consistent with other studies, originating also from different geographical sectors. Already in the understanding of Rounsevell et al. [91], it seemed unlikely that climate change would have a negative impact on grasslands in England and Wales, while Riedo et al. [48] predicted a positive effect on grassland productivity in central Europe. Additionally, in the case of grasslands in the United States, pasture production is generally expected to increase under projected climate scenarios [92]. Moreover, Morales et al. [93] predicted an increase in grassland productivity in Europe, albeit with significant regional variability. In this regard, it should be emphasised that the impacts of climate change on grazing systems may be region-specific [94].

Adaptation strategies must face different and opposite effects on rangeland productivity, as already previously pointed out [95,96], and in some cases, it is foreseen that climate change can produce a positive effect, being able to support greater livestock numbers [97] and to lengthen the duration of the grazing season due to a higher herbage availability early in the year [98]. In our study, we provided clues for increasing stocking rates and extending grazing periods (mainly by putting animals out to pasture earlier) to take advantage of the change in seasonality and increased forage production compared with the baseline (1981–2010), especially in the mid-future (i.e., 2041–2070). The possibility of having an earlier vegetative recovery that prolongs the duration of the grazing season allows, along with the higher productivity assumed, an increase in animal density, and, in this way, a biomass intake more consistent with the forage availability. Consequently, these conditions allow a more efficient management of the resource [99] with less waste and a more adequate stocking rate, a factor that ensures less degradation of the pasture itself [100]. Results confirm these opportunities also comparing mid-future aboveground biomass under adaptation strategies (peak and trend, Figures 8 and S2–S5) with those of the ongoing period under BaU (i.e., 2011–2040), which is the condition most similar to the one of calibration. In this view, it is, however, important to emphasize that in the simulation of adaptation strategies, the model does not specifically consider the role of increased animal stocking rate and/or duration of the grazing season on soil compaction, a condition that may disadvantage forage quality, vegetation regrowth, and biodiversity [101,102]. In addition, warming and altered rainfall patterns may reduce the forage quality and palatability of Italian grasslands [6]. Indeed, climatic changes, as well as land-use changes, have already strongly modified the botanical composition, species distribution, and size of grasslands in the central Italian massifs since the 1950s [43]. The observed floral, ecological, and structural variations confirm that grassland ecosystems in mountainous environments in Italy have undergone a process of thermophilization, with an evolutionary trend towards more nutrient-demanding vegetation [34,42]. Variations in vegetation composition in response to increased competition for environmental factors indicate, at higher altitudes, less displacement of plant species from higher slopes as well as dispersal of species from south-facing to north-facing slopes, with greater presence of grass- and shrub-dominated communities replacing rare and cold-tolerant species [103]. This reflects the narrower thermal niche of mountain plant species, which makes short-term adaptation/acclimation more difficult [104]. As a narrow thermal niche prevents plant species from adapting quickly to high altitudes, site elevation explains the response of species richness to warming [105]. Indeed, although changes in species cover and plant community composition indicate an accelerated transformation to more heat-demanding vegetation, this colonization process may occur at a slower rate than the continued decline of cryophilic species, favoring periods of accelerated species decline [106].

The analyses performed in this study identified the possible impacts of climate change on a typical grazing system of the Apennines in Central Italy, highlighting future trends of different system characteristics, such as length of the growing season, pasture productivity, soil water conditions, and gas emissions, as well as possible alternative management strategies in a context of future climate change. In fact, the results obtained in this study highlight the potential of employing specific models for simulating the behavior of pastoral resources under actual utilization and different future scenarios (i.e., RCP4.5 and RCP8.5), testing adaptation management options. In this sense, the study has produced a significant step forward compared with previous studies that analyzed climate change impacts on Apennine grasslands, mainly with regard to the botanical evolution of the plant communities, by providing insights on future agronomic conditions and possible adaptation strategies. The modeling approach used has, thus, been demonstrated to be a useful tool to support the management decisions that breeders will have to make in the near future.

5. Conclusions

The results of this study represent a step forward in the knowledge of the impacts of future climate change on a typical pasture system in the central Apennines. Specifically, this study fills a lack of information on future grassland development, as well as providing detailed information on the length of the growing season, GHG emissions, water conditions, and the effectiveness of different adaptation strategies in response to the increase in forage production simulated by PaSim in future scenarios. In particular, the analysis of adaptation strategies investigated possible management changes to cope with climate change impacts, providing useful indications to stakeholders and policy-makers for appropriate agricultural policy and optimal land management strategies for ongoing climate change.

However, while modeling approaches capture distinct aspects of the adaptive process, they have done so in relative isolation from the use of other technological supports (e.g., remote sensing and precision farming) and participatory approaches, without producing improved unified representations. As well, management options to sustain grassland ecosystems under global change are many and need to be tested for their ability to maintain or enhance resource values in the future. Social impact assessment studies are, thus, needed to examine how the impacts, i.e., the effects of climatic anomalies on the performance of Apennine pastures, propagate through the socio-economic and political systems. This type of integrated approach, which would include the potential for adaptation and adjustment to climate pressure, would reflect the reality of pastoral communities much better than the modeling used and raises fruitful research questions regarding the vulnerability of Apennine territories and their adaptive capacity.

Supplementary Materials: The following supporting information can be downloaded at: https://www.mdpi.com/article/10.3390/land12020351/s1, Table S1: Aboveground dry matter biomass (AGB) and leaf area index (LAI) collected in 2020 and 2021 in the two study-sites (sample mean and standard deviation of eight sub-samples in Marradi and four sub-samples in Borgo San Lorenzo); Table S2: Climate models used in this study, an indication of their origin (institute), version, realisation and frequency. The suffixes i and p of each realisation (r) indicate the initialisation and physics indices, respectively; Table S3: Summary of the PaSim parameters considered for the calibration; Table S4: Simulated flux components (30-year mean) from the two study-sites for the baseline (1981–2010) and climate scenarios (RCP4.5 and RCP8.5) under different management options, estimated using PaSim (GPP: gross primary production; RECO: ecosystem respiration; NEE: net ecosystem exchange). +20 GI represent a 20% rise in animal stocking rate, 15 GL a 15% increase in grazing length and +20 GI × 15 GL a combination of these two management factors. RCP4.5 and 8.5 are the different Representatives Concentration Pathways used in the simulations; Figure S1: Daily simulation (30-year mean) of 0.35-m soil water content (SWC) with PaSim for baseline and climate-change scenarios under business-as-usual management at both study-sites. RCP4.5 and 8.5 are the different Representatives Concentration Pathways used in the simulations; Figure S2: Daily simulation (30-year mean) of aboveground biomass (AGB) with PaSim for climate-change scenarios under different adaptation strategies at Marradi site for 2011–2040 period. +20 GI represent a 20% rise in animal stocking rate, 15 GL a 15% increase in grazing length and +20 GI × 15 GL a combination of these two management factors. RCP4.5 and 8.5 are the different Representatives Concentration Pathways used in the simulations; Figure S3: Daily simulation (30-year mean) of aboveground biomass (AGB) with PaSim for climate-change scenarios under different adaptation strategies at Marradi site for 2041–2070 period. +20 GI represent a 20% rise in animal stocking rate, 15 GL a 15% increase in grazing length and +20 GI × 15 GL a combination of these two management factors. RCP4.5 and 8.5 are the different Representatives Concentration Pathways used in the simulations; Figure S4: Daily simulation (30-year mean) of aboveground biomass (AGB) with PaSim for climate-change scenarios under different adaptation strategies at Borgo San Lorenzo site for 2011–2040 period. +20 GI represent a 20% rise in animal stocking rate, 15 GL a 15% increase in grazing length and +20 GI × 15 GL a combination of these two management factors. RCP4.5 and 8.5 are the different Representatives Concentration Pathways used in the simulations; Figure S5: Daily simulation (30-year mean) of aboveground biomass (AGB) with PaSim for climate-change scenarios under different adaptation strategies at Borgo San Lorenzo site for 2041–2070 period. +20 GI represent a 20% rise in animal stocking rate, 15 GL a 15% increase in grazing length and +20 GI × 15 GL a combination of these two management factors. RCP4.5 and 8.5 are the different Representatives Concentration Pathways used in the simulations.

Author Contributions: Conceptualization, G.B. and E.B.; methodology, G.B., E.B., R.M. and S.C.-A.; software, R.M. and G.B.; validation, E.B., G.A., M.M. and C.D.; formal analysis, R.M. and E.B.; investigation, G.A., C.D., N.S., E.B., S.C.-A. and M.M.; resources, G.A. and N.S.; data curation, C.D., G.A. and N.S.; writing—original draft preparation, G.B. and E.B.; writing—review and editing, G.A., M.M., C.D., S.C.-A. and N.S.; visualization, E.B. and G.B.; supervision, R.M., G.A., G.B. and M.M.; project administration, C.D. and G.B.; funding acquisition, G.A., C.D. and G.B. All authors have read and agreed to the published version of the manuscript.

Funding: The study was developed in the framework of E.B.'s Ph.D. thesis, supported by the VISTOCK project, funded by GAL-START Mugello (Tuscany Region), grant number 853175. It falls within the thematic area of the French government IDEX-ISITE initiative (reference: 16-IDEX-0001; project CAP 20-25).

Data Availability Statement: Not applicable.

Acknowledgments: We thank the farms involved in the study for their practical support.

Conflicts of Interest: The authors declare no conflict of interest.

References

1. Dillon, P. The Evolution of Grassland in the European Union in Terms of Utilisation, Productivity, Food Security and the Importance of Adoption of Technical Innovations in Increasing Sustainability of Pasture-Based Ruminant Production Systems. *Grassl. Sci. Eur.* **2018**, *23*, 3–15.
2. Mara, F.P.O. The Role of Grasslands in Food Security and Climate Change. *Ann. Bot.* **2012**, *210*, 1263–1270. [CrossRef]
3. Barbour, R.; Young, R.H.; Wilkinson, J.M. Production of Meat and Milk from Grass in the United Kingdom. *Agronomy* **2022**, *12*, 914. [CrossRef]
4. ISTAT. Available online: Http://Dati.Istat.It/Index.Aspx?Queryid=33704 (accessed on 5 December 2022).
5. Burrascano, S.; Caccianiga, M.; Gigante, D. *Dry Grasslands Habitat Types in Italy*. 2010. Bulletin of European Dry Grassland Group. Available online: https://air.unimi.it/handle/2434/155321 (accessed on 20 January 2023).
6. Dibari, C.; Pulina, A.; Argenti, G.; Aglietti, C.; Bindi, M.; Moriondo, M.; Mula, L.; Pasqui, M.; Seddaiu, G.; Roggero, P.P. Climate Change Impacts on the Alpine, Continental and Mediterranean Grassland Systems of Italy: A Review. *Ital. J. Agron.* **2021**, *16*, 1843. [CrossRef]
7. Cavallero, A.; Aceto, P.; Gorlier, A.; Lombardi, G.; Lonati, M.; Martinasso, B.; Tagliatori, C. *I Tipi Pastorali Delle Alpi Piemontesi*; Alberto Perdisa Editore: Bologna, Italy, 2007. (In Italian)
8. Argenti, G.; Bottai, L.; Chiesi, M.; Maselli, F.; Staglianò, N.; Targetti, S. Analisi e Valutazione Di Pascoli Montani Attraverso l'integrazione Di Dati Multispettrali e Ausiliari. *Riv. Ital. Telerilevamento* **2011**, *43*, 45–57. (In Italian)
9. Orlandi, S.; Probo, M.; Sitzia, T.; Trentanovi, G.; Garbarino, M.; Lombardi, G.; Lonati, M. Environmental and Land Use Determinants of Grassland Patch Diversity in the Western and Eastern Alps under Agro-Pastoral Abandonment. *Biodivers. Conserv.* **2016**, *25*, 275–293. [CrossRef]
10. Targetti, S.; Messeri, A.; Staglianò, N.; Argenti, G. Leaf Functional Traits for the Assessment of Succession Following Management in Semi-Natural Grasslands: A Case Study in the North Apennines, Italy. *Appl. Veg. Sci.* **2013**, *16*, 325–332. [CrossRef]
11. Metzger, M.J.; Bunce, R.G.H.; Jongman, R.H.G.; Mücher, C.A.; Watkins, J.W. A Climatic Stratification of the Environment of Europe. *Glob. Ecol. Biogeogr.* **2005**, *14*, 549–563. [CrossRef]
12. Bengtsson, J.; Bullock, J.M.; Egoh, B.; Everson, C.; Everson, T.; O'Connor, T.; O'Farrell, P.J.; Smith, H.G.; Lindborg, R. Grasslands—More Important for Ecosystem Services than You Might Think. *Ecosphere* **2019**, *10*, e02582. [CrossRef]
13. Hao, R.; Yu, D.; Liu, Y.; Liu, Y.; Qiao, J.; Wang, X.; Du, J. Impacts of Changes in Climate and Landscape Pattern on Ecosystem Services. *Sci. Total Environ.* **2017**, *579*, 718–728. [CrossRef]
14. Ponzetta, M.P.; Cervasio, F.; Crocetti, C.; Messeri, A.; Argenti, G. Habitat Improvements with Wildlife Purposes in a Grazed Area on the Apennine Mountains. *Ital. J. Agron.* **2010**, *5*, 233–238. [CrossRef]
15. Tamburini, G.; Aguilera, G.; Öckinger, E. Grasslands Enhance Ecosystem Service Multifunctionality above and Below-Ground in Agricultural Landscapes. *J. Appl. Ecol.* **2022**, 3061–3071. [CrossRef]
16. Wepking, C.; Mackin, H.C.; Raff, Z.; Shrestha, D.; Orfanou, A.; Booth, E.G.; Kucharik, C.J.; Gratton, C.; Jackson, R.D. Perennial Grassland Agriculture Restores Critical Ecosystem Functions in the U.S. Upper Midwest. *Front. Sustain. Food Syst.* **2022**, *6*, 1010280. [CrossRef]
17. Sándor, R.; Ehrhardt, F.; Brilli, L.; Carozzi, M.; Recous, S.; Smith, P.; Snow, V.; Soussana, J.F.; Dorich, C.D.; Fuchs, K.; et al. The Use of Biogeochemical Models to Evaluate Mitigation of Greenhouse Gas Emissions from Managed Grasslands. *Sci. Total Environ.* **2018**, *642*, 292–306. [CrossRef]
18. Smith, P.; Martino, D.; Cai, Z.; Gwary, D.; Janzen, H.; Kumar, P.; McCarl, B.; Ogle, S.; O'Mara, F.; Rice, C.; et al. Greenhouse Gas Mitigation in Agriculture. *Philos. Trans. R. Soc. B Biol. Sci.* **2008**, *363*, 789–813. [CrossRef]
19. Oates, L.G.; Jackson, R.D. Livestock Management Strategy Affects Net Ecosystem Carbon Balance of Subhumid Pasture. *Rangel. Ecol. Manag.* **2014**, *67*, 19–29. [CrossRef]
20. Franzluebbers, A.J. *Cattle Grazing Effects on the Environment: Greenhouse Gas Emissions and Carbon Footprint*, Elsevier Inc.: Cambridge, MA, USA, 2020; ISBN 9780128144749.
21. Guillaume, T.; Makowski, D.; Libohova, Z.; Elfouki, S.; Fontana, M.; Leifeld, J.; Bragazza, L.; Sinaj, S. Geoderma Carbon Storage in Agricultural Topsoils and Subsoils Is Promoted by Including Temporary Grasslands into the Crop Rotation. *Geoderma* **2022**, *422*, 115937. [CrossRef]
22. Soussana, J.F.; Tallec, T.; Blanfort, V. Mitigating the Greenhouse Gas Balance of Ruminant Production Systems through Carbon Sequestration in Grasslands. *Anim. Int. J. Anim. Biosci.* **2010**, *4*, 334–350. [CrossRef]
23. Dass, P.; Houlton, B.Z.; Wang, Y.; Warlind, D. Grasslands May Be More Reliable Carbon Sinks than Forests in California. *Environ. Res. Lett.* **2018**, *13*, 074027. [CrossRef]
24. Steinfeld, H.; Wassenaar, T. The Role of Livestock Production in Carbon and Nitrogen Cycles. *Annu. Rev. Environ. Resour.* **2007**, *32*, 271–296. [CrossRef]
25. Hortnagl, L.; Barthel, M.; Buchmann, N.; Eugster, W.; Butterbach-bahl, K.; Eugenio, D.; Zeeman, M.; Lu, H.; Kiese, R.; Bahn, M.; et al. Greenhouse Gas Fluxes over Managed Grasslands in Central Europe S. *Glob. Chang. Biol.* **2018**, *24*, 1843–1872. [CrossRef] [PubMed]
26. Soussana, J.F.; Allard, V.; Pilegaard, K.; Ambus, P.; Amman, C.; Campbell, C.; Raschi, A.; Baronti, S.; Rees, R.M.; Skiba, U.; et al. Full Accounting of the Greenhouse Gas (CO_2, N_2O, CH_4) Budget of Nine European Grassland Sites. *Agric. Ecosyst. Environ.* **2007**, *121*, 121–134. [CrossRef]

27. Toreti, A.; Desiato, F. Temperature Trend over Italy from 1961 to 2004. *Theor. Appl. Climatol.* **2008**, *91*, 51–58. [CrossRef]

28. Tomozeiu, R.; Pasqui, M.; Quaresima, S. Future Changes of Air Temperature over Italian Agricultural Areas: A Statistical Downscaling Technique Applied to 2021–2050 and 2071–2100 Periods. *Meteorol. Atmos. Phys.* **2018**, *130*, 543–563. [CrossRef]

29. Scocco, P.; Piermarteri, K.; Malfatti, A.; Tardella, F.M.; Catorci, A. Increase of Drought Stress Negatively Affects the Sustainability of Extensive Sheep Farming in Sub-Mediterranean Climate. *J. Arid Environ.* **2016**, *128*, 50–58. [CrossRef]

30. Chelli, S.; Wellstein, C.; Campetella, G.; Canullo, R.; Tonin, R.; Zerbe, S.; Gerdol, R. Climate Change Response of Vegetation across Climatic Zones in Italy. *Clim. Res.* **2017**, *71*, 249–262. [CrossRef]

31. Petriccione, B.; Bricca, A. Thirty Years of Ecological Research at the Gran Sasso d'Italia LTER Site: Climate Change in Action. *Nat. Conserv.* **2019**, *34*, 9–39. [CrossRef]

32. Dibari, C.; Costafreda-Aumedes, S.; Argenti, G.; Bindi, M.; Carotenuto, F.; Moriondo, M.; Padovan, G.; Pardini, A.; Staglianò, N.; Vagnoli, C.; et al. Expected Changes to Alpine Pastures in Extent and Composition under Future Climate Conditions. *Agronomy* **2020**, *10*, 926. [CrossRef]

33. Stanisci, A.; Frate, L.; Morra Di Cella, U.; Pelino, G.; Petey, M.; Siniscalco, C.; Carranza, M.L. Short-Term Signals of Climate Change in Italian Summit Vegetation: Observations at Two GLORIA Sites. *Plant Biosyst.* **2016**, *150*, 227–235. [CrossRef]

34. Evangelista, A.; Frate, L.; Carranza, M.L.; Attorre, F.; Pelino, G.; Stanisci, A. Changes in Composition, Ecology and Structure of High-Mountain Vegetation: A Re-Visitation Study over 42 Years. *AoB Plants* **2016**, *8*, 1–11. [CrossRef]

35. Bebeley, J.F.; Kamara, A.Y.; Jibrin, J.M.; Akinseye, F.M.; Tofa, A.I.; Adam, A.M. Evaluation and Application of the CROPGRO-Soybean Model for Determining Optimum Sowing Windows of Soybean in the Nigeria Savannas. *Sci. Rep.* **2022**, *12*, 6747. [CrossRef] [PubMed]

36. Kamilaris, C.; Dewhurst, R.J.; Sykes, A.J.; Alexander, P. Modelling Alternative Management Scenarios of Economic and Environmental Sustainability of Beef Fi Nishing Systems. *J. Clean. Prod.* **2020**, *253*, 119888. [CrossRef]

37. Fullman, T.J.; Bunting, E.L.; Kiker, G.A.; Southworth, J. Predicting Shifts in Large Herbivore Distributions under Climate Change and Management Using a Spatially-Explicit Ecosystem Model. *Ecol. Modell.* **2017**, *352*, 1–18. [CrossRef]

38. Kalaugher, E.; Beukes, P.; Bornman, J.F.; Clark, A.; Campbell, D.I. Modelling Farm-Level Adaptation of Temperate, Pasture-Based Dairy Farms to Climate Change. *Agric. Syst.* **2017**, *153*, 53–68. [CrossRef]

39. Moore, A.D.; Ghahramani, A. Climate Change and Broadacre Livestock Production across Southern Australia. 3. Adaptation Options via Livestock Genetic Improvement. *Anim. Prod. Sci.* **2014**, *54*, 111–124. [CrossRef]

40. Snow, V.O.; Rotz, C.A.; Moore, A.D.; Martin-Clouaire, R.; Johnson, I.R.; Hutchings, N.J.; Eckard, R.J. The Challenges–and Some Solutions–to Process-Based Modelling of Grazed Agricultural Systems. *Environ. Model. Softw.* **2014**, *62*, 420–436. [CrossRef]

41. Ma, L.; Derner, J.D.; Harmel, R.D.; Tatarko, J.; Moore, A.D.; Rotz, C.A.; Augustine, D.J.; Boone, R.B.; Coughenour, M.B.; Beukes, P.C.; et al. Application of Grazing Land Models in Ecosystem Management: Current Status and next Frontiers. *Adv. Agron.* **2019**, *158*, 173–215. [CrossRef]

42. Ferrarini, A.; Alatalo, J.M.; Gervasoni, D.; Foggi, B. Exploring the Compass of Potential Changes Induced by Climate Warming in Plant Communities. *Ecol. Complex.* **2017**, *29*, 1–9. [CrossRef]

43. Frate, L.; Carranza, M.L.; Evangelista, A.; Stinca, A.; Schaminée, J.H.J.; Stanisci, A. Climate and Land Use Change Impacts on Mediterranean High-Mountain Vegetation in the Apennines since the 1950s. *Plant Ecol. Divers.* **2018**, *11*, 85–96. [CrossRef]

44. Dibari, C.; Argenti, G.; Catolfi, F.; Moriondo, M.; Staglianò, N.; Bindi, M. Pastoral Suitability Driven by Future Climate Change along the Apennines. *Ital. J. Agron.* **2015**, *10*, 109–116. [CrossRef]

45. Vigan, A.; Lasseur, J.; Benoit, M.; Mouillot, F.; Eugène, M.; Mansard, L.; Vigne, M.; Lecomte, P.; Dutilly, C. Evaluating Livestock Mobility as a Strategy for Climate Change Mitigation: Combining Models to Address the Specificities of Pastoral Systems. *Agric. Ecosyst. Environ.* **2017**, *242*, 89–101. [CrossRef]

46. Insua, J.R.; Utsumi, S.A.; Basso, B. Estimation of Spatial and Temporal Variability of Pasture Growth and Digestibility in Grazing Rotations Coupling Unmanned Aerial Vehicle (UAV) with Crop Simulation Models. *PLoS ONE* **2019**, *14*, e0212773. [CrossRef]

47. Gao, Q.Z.; Li, Y.; Xu, H.M.; Wan, Y.F.; Jiangcun, W. zha Adaptation Strategies of Climate Variability Impacts on Alpine Grassland Ecosystems in Tibetan Plateau. *Mitig. Adapt. Strateg. Glob. Chang.* **2014**, *19*, 199–209. [CrossRef]

48. Riedo, M.; Grub, A.; Rosset, M. A Pasture Simulation Model for Dry Matter Production, and Fluxes of Carbon, Nitrogen, Water and Energy. *Ecol. Model.* **1998**, *105*, 141–183. [CrossRef]

49. Wilcke, R.A.I.; Bärring, L. Selecting Regional Climate Scenarios for Impact Modelling Studies. *Environ. Model. Softw.* **2016**, *78*, 191–201. [CrossRef]

50. Pierce, D.W.; Barnett, T.P.; Santer, B.D.; Gleckler, P.J. Selecting Global Climate Models for Regional Climate Change Studies. *Proc. Natl. Acad. Sci. USA* **2009**, *106*. [CrossRef]

51. Ruti, P.M.; Somot, S.; Giorgi, F.; Dubois, C.; Flaounas, E.; Obermann, A.; Dell'Aquila, A.; Pisacane, G.; Harzallah, A.; Lombardi, E.; et al. Med-CORDEX Initiative for Mediterranean Climate Studies. *Bull. Am. Meteorol. Soc.* **2016**, *97*, 1187–1208. [CrossRef]

52. Cornes, R.C.; van der Schrier, G.; van den Besselaar, E.J.M.; Jones, P.D. An Ensemble Version of the E-OBS Temperature and Precipitation Data Sets. *J. Geophys. Res. Atmos.* **2018**, *123*, 9391–9409. [CrossRef]

53. Lange, S. Trend-Preserving Bias Adjustment and Statistical Downscaling with ISIMIP3BASD (v1.0). *Geosci. Model Dev.* **2019**, *12*, 3055–3070. [CrossRef]

54. Bristow, K.L.; Campbell, G.S. On the Relationship between Incoming Solar Radiation and Daily Maximum and Minimum Temperature. *Agric. For. Meteorol.* **1984**, *31*, 159–166. [CrossRef]

55. Allen, R.G.; Pereira, L.S.; Raes, D.; Smith, M. FAO Irrigation and Drainage Paper No. 56 - Crop Evapotranspiration. *Rome Food Agric. Organ. United Nations* **1998**, *56*, e156.
56. IPCC Annex III: Tables of Historical and Projected Well-Mixed Greenhouse Gas Mixing Ratios and Effective Radiative Forcing of All Climate Forcers. In *Climate Change 2021: The Physical Science Basis. Contribution of Wor*; Dentener, F.J.; Hall, B.; Smith, C. (Eds.) Cambridge University Press: Cambridge, UK; New York, NY, USA, 2021; pp. 2139–2152.
57. Graux, A.I.; Bellocchi, G.; Lardy, R.; Soussana, J.F. Ensemble Modelling of Climate Change Risks and Opportunities for Managed Grasslands in France. *Agric. For. Meteorol.* **2013**, *170*, 114–131. [CrossRef]
58. Touhami, H.B.; Lardy, R.; Barra, V.; Bellocchi, G. Screening Parameters in the Pasture Simulation Model Using the Morris Method. *Ecol. Modell.* **2013**, *266*, 42–57. [CrossRef]
59. Ma, S.; Lardy, R.; Graux, A.I.; Ben Touhami, H.; Klumpp, K.; Martin, R.; Bellocchi, G. Regional-Scale Analysis of Carbon and Water Cycles on Managed Grassland Systems. *Environ. Model. Softw.* **2015**, *72*, 356–371. [CrossRef]
60. Pulina, A.; Lai, R.; Salis, L.; Seddaiu, G.; Roggero, P.P.; Bellocchi, G. Modelling Pasture Production and Soil Temperature, Water and Carbon Fluxes in Mediterranean Grassland Systems with the Pasture Simulation Model. *Grass Forage Sci.* **2018**, *73*, 272–283. [CrossRef]
61. Sándor, R.; Picon-Cochard, C.; Martin, R.; Louault, F.; Klumpp, K.; Borras, D.; Bellocchi, G. Plant Acclimation to Temperature: Developments in the Pasture Simulation Model. *Field Crop. Res.* **2018**, *222*, 238–255. [CrossRef]
62. Ehrhardt, F.; Soussana, J.F.; Bellocchi, G.; Grace, P.; McAuliffe, R.; Recous, S.; Sándor, R.; Smith, P.; Snow, V.; de Antoni Migliorati, M.; et al. Assessing Uncertainties in Crop and Pasture Ensemble Model Simulations of Productivity and N_2O Emissions. *Glob. Chang. Biol.* **2018**, *24*, e603–e616. [CrossRef]
63. Bojanowski, J.S.; Donatelli, M.; Skidmore, A.K.; Vrieling, A. An Auto-Calibration Procedure for Empirical Solar Radiation Models. *Environ. Model. Softw.* **2013**, *49*, 118–128. [CrossRef]
64. Poggio, L.; De Sousa, L.M.; Batjes, N.H.; Heuvelink, G.B.M.; Kempen, B.; Ribeiro, E.; Rossiter, D. SoilGrids 2.0: Producing Soil Information for the Globe with Quantified Spatial Uncertainty. *Soil* **2021**, *7*, 217–240. [CrossRef]
65. Richter, K.; Atzberger, C.; Hank, T.B.; Mauser, W. Derivation of Biophysical Variables from Earth Observation Data: Validation and Statistical Measures. *J. Appl. Remote Sens.* **2012**, *6*, 063557-1. [CrossRef]
66. Meinshausen, M.; Smith, S.J.; Calvin, K.; Daniel, J.S.; Kainuma, M.L.T.; Lamarque, J.; Matsumoto, K.; Montzka, S.A.; Raper, S.C.B.; Riahi, K.; et al. The RCP Greenhouse Gas Concentrations and Their Extensions from 1765 to 2300. *Clim. Chang.* **2011**, *109*, 213–241. [CrossRef]
67. Movedi, E.; Bellocchi, G.; Argenti, G.; Paleari, L.; Vesely, F.; Staglianò, N.; Dibari, C.; Confalonieri, R. Development of Generic Crop Models for Simulation of Multi-Species Plant Communities in Mown Grasslands. *Ecol. Modell.* **2019**, *401*, 111–128. [CrossRef]
68. Sándor, R.; Barcza, Z.; Acutis, M.; Doro, L.; Hidy, D.; Köchy, M.; Minet, J.; Lellei-Kovács, E.; Ma, S.; Perego, A.; et al. Multi-Model Simulation of Soil Temperature, Soil Water Content and Biomass in Euro-Mediterranean Grasslands: Uncertainties and Ensemble Performance. *Eur. J. Agron.* **2017**, *88*, 22–40. [CrossRef]
69. Schucknecht, A.; Seo, B.; Krämer, A.; Asam, S.; Atzberger, C.; Kiese, R. Estimating Dry Biomass and Plant Nitrogen Concentration in Pre-Alpine Grasslands with Low-Cost UAS-Borne Multispectral Data-a Comparison of Sensors, Algorithms, and Predictor Sets. *Biogeosciences* **2022**, *19*, 2699–2727. [CrossRef]
70. Kollas, C.; Kersebaum, K.C.; Nendel, C.; Manevski, K.; Müller, C.; Palosuo, T.; Armas-Herrera, C.M.; Beaudoin, N.; Bindi, M.; Charfeddine, M.; et al. Crop Rotation Modelling-A European Model Intercomparison. *Eur. J. Agron.* **2015**, *70*, 98–111. [CrossRef]
71. Vuichard, N.; Soussana, J.F.; Ciais, P.; Viovy, N.; Ammann, C.; Calanca, P.; Clifton-Brown, J.; Fuhrer, J.; Jones, M.; Martin, C. Estimating the Greenhouse Gas Fluxes of European Grasslands with a Process-Based Model: 1. Model Evaluation from in Situ Measurements. *Glob. Biogeochem. Cycles* **2007**, *21*. [CrossRef]
72. Schippers, P.; Van Groenendael, J.M.; Vleeshouwers, L.M.; Hunt, R. Herbaceous Plant Strategies in Disturbed Habitats. *Oikos* **2001**, *95*, 198–210. [CrossRef]
73. Fuchs, K.; Merbold, L.; Buchmann, N.; Bellocchi, G.; Bindi, M.; Brilli, L.; Conant, R.T.; Dorich, C.D.; Ehrhardt, F.; Fitton, N.; et al. Evaluating the Potential of Legumes to Mitigate N_2O Emissions From Permanent Grassland Using Process-Based Models. *Glob. Biogeochem. Cycles* **2020**, *34*, e2020GB006561. [CrossRef]
74. Vital, J.A.; Gaurut, M.; Lardy, R.; Viovy, N.; Soussana, J.F.; Bellocchi, G.; Martin, R. High-Performance Computing for Climate Change Impact Studies with the Pasture Simulation Model. *Comput. Electron. Agric.* **2013**, *98*, 131–135. [CrossRef]
75. Wang, Z.; Ma, Y.; Zhang, Y.; Shang, J. Review of Remote Sensing Applications in Grassland Monitoring. *Remote Sens.* **2022**, *14*, 2903. [CrossRef]
76. Bellini, E.; Moriondo, M.; Dibari, C.; Leolini, L.; Staglianò, N.; Stendardi, L.; Filippa, G.; Galvagno, M.; Argenti, G. Impacts of Climate Change on European Grassland Phenology: A 20-Year Analysis of MODIS Satellite Data. *Remote Sens.* **2023**, *15*, 218. [CrossRef]
77. Ren, S.; Vitasse, Y.; Chen, X.; Peichl, M.; An, S. Assessing the Relative Importance of Sunshine, Temperature, Precipitation, and Spring Phenology in Regulating Leaf Senescence Timing of Herbaceous Species in China. *Agric. For. Meteorol.* **2022**, *313*, 108770. [CrossRef]
78. Ellsworth, D.S.; Reich, P.B.; Naumburg, E.S.; Koch, G.W.; Kubiske, M.E.; Smith, S.D. Photosynthesis, Carboxylation and Leaf Nitrogen Responses of 16 Species to Elevated PCO_2 across Four Free-Air CO2 Enrichment Experiments in Forest, Grassland and Desert. *Glob. Chang. Biol.* **2004**, *10*, 2121–2138. [CrossRef]

79. Ainsworth, E.A.; Long, S.P. What Have We Learned from 15 Years of Free-Air CO2 Enrichment (FACE)? A Meta-Analytic Review of the Responses of Photosynthesis, Canopy Properties and Plant Production to Rising CO_2. *New Phytol.* **2005**, *165*, 351–372. [CrossRef] [PubMed]

80. Ainsworth, E.A.; Rogers, A. The Response of Photosynthesis and Stomatal Conductance to Rising [CO2]: Mechanisms and Environmental Interactions. *Plant Cell Environ.* **2007**, *30*, 258–270. [CrossRef]

81. Shrestha, S.; Abdalla, M.; Hennessy, T.; Forristal, D.; Jones, M.B. Irish Farms under Climate Change - Is There a Regional Variation on Farm Responses? *J. Agric. Sci.* **2015**, *153*, 385–398. [CrossRef]

82. Zarrineh, N.; Abbaspour, K.C.; Holzkämper, A. Integrated Assessment of Climate Change Impacts on Multiple Ecosystem Services in Western Switzerland. *Sci. Total Environ.* **2020**, *708*, 135212. [CrossRef] [PubMed]

83. Chang, J.; Ciais, P.; Gasser, T.; Smith, P.; Herrero, M.; Havlík, P.; Obersteiner, M.; Guenet, B.; Goll, D.S.; Li, W.; et al. Climate Warming from Managed Grasslands Cancels the Cooling Effect of Carbon Sinks in Sparsely Grazed and Natural Grasslands. *Nat. Commun.* **2021**, *12*, 118. [CrossRef] [PubMed]

84. Zhu, Y.; Purdy, K.J.; Eyice, Ö.; Shen, L.; Harpenslager, S.F.; Yvon-Durocher, G.; Dumbrell, A.J.; Trimmer, M. Disproportionate Increase in Freshwater Methane Emissions Induced by Experimental Warming. *Nat. Clim. Chang.* **2020**, *10*, 685–690. [CrossRef]

85. Butterbach-Bahl, K.; Baggs, E.M.; Dannenmann, M.; Kiese, R.; Zechmeister-Boltenstern, S. Nitrous Oxide Emissions from Soils: How Well Do We Understand the Processes and Their Controls? *Philos. Trans. R. Soc. B Biol. Sci.* **2013**, *368*, 20130122. [CrossRef]

86. Cullen, B.R.; Johnson, I.R.; Eckard, R.J.; Lodge, G.M.; Walker, R.J.; Rawnsley, R.P.; McCaskillF, M.R. Climate Change Effects on Pasture Systems in South-Eastern Australia. *Crop Pasture Sci.* **2009**, *60*, 933–942. [CrossRef]

87. Parton, W.J.; Ojima, D.S.; Cole, C.V.; Schimel, D.S. A General Model for Soil Organic Matter Dynamics: Sensitivity to Litter Chemistry, Texture and Management. *Quant. Model. Soil Form. Process. Proc. Symp. Minneapolis* **1994**, *39*, 147–167. [CrossRef]

88. Murphy, K.L.; Burke, I.C.; Vinton, M.A.; Lauenroth, W.K.; Aguiar, M.R.; Wedin, D.A.; Virginia, R.A.; Lowe, P.N. Regional Analysis of Litter Quality in the Central Grassland Region of North America. *J. Veg. Sci.* **2002**, *13*, 395–402. [CrossRef]

89. Ouled Belgacem, A.; Louhaichi, M. The Vulnerability of Native Rangeland Plant Species to Global Climate Change in the West Asia and North African Regions. *Clim. Chang.* **2013**, *119*, 451–463. [CrossRef]

90. He, P.; Ma, X.; Sun, Z.; Han, Z.; Ma, S.; Xiaoyu, M. Compound Drought Constrains Gross Primary Productivity in Chinese Grasslands. *Environ. Res. Lett.* **2022**, *17*, 104054. [CrossRef]

91. Rounsevell, M.D.A.; Brignall, A.P.; Siddons, P.A. Potential Climate Change Effects on the Distribution of Agricultural Grassland in England and Wales. *Soil Use Manag.* **1996**, *12*, 44–51. [CrossRef]

92. Edmonds, J.A.; Rosenberg, N.J. Climate Change Impacts for the Conterminous USA: An Integrated Assessment Summary. *Clim. Chang. Impacts Conterminous USA Integr. Assess.* **2005**, 151–162. [CrossRef]

93. Morales, P.; Hickler, T.; Rowell, D.P.; Smith, B.; Sykes, M.T. Changes in European Ecosystem Productivity and Carbon Balance Driven by Regional Climate Model Output. *Glob. Chang. Biol.* **2007**, *13*, 108–122. [CrossRef]

94. Harrison, M.T.; Cullen, B.R.; Armstrong, D. Management Options for Dairy Farms under Climate Change: Effects of Intensification, Adaptation and Simplification on Pastures, Milk Production and Profitability. *Agric. Syst* **2017**, *155*, 19–32. [CrossRef]

95. Joyce, L.A.; Briske, D.D.; Brown, J.R.; Polley, H.W.; McCarl, B.A.; Bailey, D.W. Climate Change and North American Rangelands: Assessment of Mitigation and Adaptation Strategies. *Rangel. Ecol. Manag.* **2013**, *66*, 512–528. [CrossRef]

96. Cheng, M.; Mccarl, B.; Fei, C. Climate Change and Livestock Production: A Literature Review. *Atmosphere* **2022**, *13*, 140. [CrossRef]

97. Briske, D.D.; Joyce, L.A.; Polley, H.W.; Brown, J.R.; Wolter, K.; Morgan, J.A.; McCarl, B.A.; Bailey, D.W. Climate-Change Adaptation on Rangelands: Linking Regional Exposure with Diverse Adaptive Capacity. *Front. Ecol. Environ.* **2015**, *13*, 249–256. [CrossRef] [PubMed]

98. Hristov, A.N.; Degaetano, A.T.; Rotz, C.A.; Hoberg, E.; Skinner, R.H.; Felix, T.; Li, H.; Patterson, P.H.; Roth, G.; Hall, M.; et al. Climate Change Effects on Livestock in the Northeast US and Strategies for Adaptation. *Clim. Chang.* **2018**, *146*, 33–45. [CrossRef]

99. Xu, C.; Liu, H.; Williams, A.P.; Yin, Y.; Wu, X. Trends toward an Earlier Peak of the Growing Season in Northern Hemisphere Mid-Latitudes. *Glob. Chang. Biol.* **2016**, *22*, 2852–2860. [CrossRef] [PubMed]

100. Allen, V.G.; Batello, C.; Berretta, E.J.; Hodgson, J.; Kothmann, M.; Li, X.; McIvor, J.; Milne, J.; Morris, C.; Peeters, A.; et al. An International Terminology for Grazing Lands and Grazing Animals. *Grass Forage Sci.* **2011**, *66*, 2–28. [CrossRef]

101. Li, W.; Cao, W.; Wang, J.; Li, X.; Xu, C.; Shi, S. Effects of Grazing Regime on Vegetation Structure, Productivity, Soil Quality, Carbon and Nitrogen Storage of Alpine Meadow on the Qinghai-Tibetan Plateau. *Ecol. Eng.* **2017**, *98*, 123–133. [CrossRef]

102. Liu, J.; Isbell, F.; Ma, Q.; Chen, Y.; Xing, F.; Sun, W.; Wang, L.; Li, J.; Wang, Y.; Hou, F.; et al. Overgrazing, Not Haying, Decreases Grassland Topsoil Organic Carbon by Decreasing Plant Species Richness along an Aridity Gradient in Northern China. *Agric. Ecosyst. Environ.* **2022**, *332*, 107935. [CrossRef]

103. Porro, F.; Tomaselli, M.; Abeli, T.; Gandini, M.; Gualmini, M.; Orsenigo, S.; Petraglia, A.; Rossi, G.; Carbognani, M. Could Plant Diversity Metrics Explain Climate-Driven Vegetation Changes on Mountain Summits of the GLORIA Network? *Biodivers. Conserv.* **2019**, *28*, 3575–3596. [CrossRef]

104. Löffler, J.; Pape, R. Thermal Niche Predictors of Alpine Plant Species. *Ecology* **2020**, *101*, e02891. [CrossRef]

105. Piseddu, F.; Bellocchi, G.; Picon-Cochard, C. Mowing and Warming Effects on Grassland Species Richness and Harvested Biomass: Meta-Analyses. *Agron. Sustain. Dev.* **2021**, *41*, 74. [CrossRef]
106. Lamprecht, A.; Semenchuk, P.R.; Steinbauer, K.; Winkler, M.; Pauli, H. Climate Change Leads to Accelerated Transformation of High-Elevation Vegetation in the Central Alps. *New Phytol.* **2018**, *220*, 447–459. [CrossRef] [PubMed]

Article

Early Evidence That Soil Dryness Causes Widespread Decline in Grassland Productivity in China

Panxing He [1,2], Yiyan Zeng [3], Ningfei Wang [4], Zhiming Han [5], Xiaoyu Meng [6], Tong Dong [2], Xiaoliang Ma [4], Shangqian Ma [7], Jun Ma [3] and Zongjiu Sun [2,*]

[1] Henan Normal University, Xinxiang 453007, China
[2] Ministry of Education Key Laboratory for Western Arid Region Grassland Resources and Ecology, Xinjiang Agricultural University, Urumqi 830000, China
[3] Ministry of Education Key Laboratory for Biodiversity Science and Ecological Engineering, Fudan University, Shanghai 200438, China
[4] Lanzhou University, Lanzhou 730020, China
[5] College of Resources and Environment, Northwest A&F University, Yangling 712100, China
[6] Key Research Insititute of Yellow River Civilization and Sustainable Development Collaborative Innovation Center on Yellow River Civilization, Henan University, Kaifeng 475001, China
[7] College of Resources and Environmental Sciences, China Agricultural University, Beijing 100193, China
* Correspondence: szj@xjau.edu.com

Citation: He, P.; Zeng, Y.; Wang, N.; Han, Z.; Meng, X.; Dong, T.; Ma, X.; Ma, S.; Ma, J.; Sun, Z. Early Evidence That Soil Dryness Causes Widespread Decline in Grassland Productivity in China. *Land* 2023, *12*, 484. https://doi.org/10.3390/land12020484

Academic Editors: Michael Vrahnakis, Yannis (Ioannis) Kazoglou and Manuel Pulido Fernádez

Received: 9 January 2023
Revised: 1 February 2023
Accepted: 2 February 2023
Published: 15 February 2023

Abstract: The burning of fossil fuels by humans emits large amounts of CO_2 into the atmosphere and strongly affects the Earth's carbon balance, with grassland ecosystems changing from weak carbon sinks that were previously close to equilibrium to core carbon sinks. Chinese grasslands are located in typical arid–semi-arid and semi-arid climatic regions, and drought events in the soil and atmosphere can have strong and irreversible consequences on the function and structure of Chinese grassland ecosystems. Based on this, we investigated the response of the gross primary production (GPP) of Chinese grasslands to land–atmosphere moisture constraints, using GPP data simulated through four terrestrial ecosystem models and introduced copula functions and Bayesian equations. The main results were as follows: (1) Soil moisture trends were not significant, and changes were dominated by interannual variability. The detrended warm-season SM correlated with GPP at 0.48 and 0.63 for the historical and future periods, respectively; thus, soil moisture is the critical water stress that regulates interannual variability in Chinese grassland GPP. (2) The positive correlation between shallow SM (0–50 cm) and GPP was higher (r = 0.62). Shallow-soil moisture is the main soil layer that constrains GPP, and the soil moisture decrease in shallow layers is much more likely to cause GPP decline in Chinese grasslands than that in deep-soil water. (3) The probability of GPP decline in Chinese grasslands caused by drought in shallow soils of 0–20 and 20–50 cm is 32.49% and 27.64%, respectively, which is much higher than the probability of GPP decline in deeper soils. In particular, soil drought was more detrimental to grassland GPP in Xinjiang and the Loess Plateau. (4) The probability of soil drought causing GPP decline was higher than that of atmospheric drought during the historical period (1.78–8.19%), but the probability of an atmospheric drought-induced GPP deficit increases significantly in the future and becomes a key factor inhibiting GPP accumulation in some regions (e.g., the Loess Plateau). Our study highlighted the response of grassland ecosystems after the occurrence of soil drought, especially for the shallow-soil-water indicator, which provides important theoretical references for grassland drought disaster emergency prevention and policy formulation.

Keywords: gross primary productivity; water constraint; Chinese grasslands; soil drought; probabilistic framework

1. Introduction

Global climate change has fundamentally altered the inherent patterns of variability of weather phenomena, such as precipitation and temperature, leading to the increasing frequency of extreme weather events such as droughts, heat waves, and cold waves [1]. Extreme events severely affect the composition and function of terrestrial ecosystems, thus affecting the terrestrial carbon cycle and its feedback to the climate system. Extreme droughts, in particular, are highly anticipated natural disasters because they occur with the highest frequency and exert the greatest influence [2,3]. The intensity and frequency of droughts are the most limiting factors affecting terrestrial vegetation growth and carbon cycling [4], and droughts reduce the water available to vegetation directly [5], with extensive and profound effects on carbon uptake in terrestrial ecosystems [6].

Recent research pointed out that terrestrial carbon sinks are often strongly influenced by interannual fluctuations in terrestrial water storage. Terrestrial water constraints are also an important limiting factor for the accumulation of ecosystem gross primary productivity (GPP) [7], that is, the largest carbon flux [8]. However, the terrestrial water storage referred to by previous studies includes all water types, such as soil water, groundwater, surface water, and canopy water. In contrast, there is a lack of exploration of the possible effects of water shortage from the soil on ecosystem productivity, as the terrestrial water variability is considerably sensitive [9]. Soil moisture accounts for a relatively large proportion of terrestrial water storage [10,11]. However, soil moisture is a generalized concept, and it is generally loosely considered as all the water within 3 m below the ground surface. Considering the differences in terrestrial water adaptation of different plants and the strong link between water availability to grass plants and root depth [12], it is necessary to explore the varying effects of soil water constraints at different soil depths on ecosystem GPP.

In addition to soil drought, atmospheric drought has been recently reported as one of the hydraulic processes that affect ecosystem productivity [13]. Plants initially draw water from the soil for photosynthesis and simultaneously dissipate it in the atmosphere through transpiration from plant leaves [14]. Both supply and depletion stresses can significantly reduce terrestrial carbon uptake and crop yields [6] and cause widespread vegetation mortality. However, current research on the primacy of soil moisture (SM) and atmospheric moisture (vapor pressure difference, VPD) on ecosystem productivity response is controversial. Novick et al. (2016) pointed out that VPD has a greater constraining effect on stomatal conductance and evapotranspiration than SM, which is more important for carbon accumulation. Moreover, as atmospheric VPD will continue to rise in the future, the negative effect of VPD on stomatal conductance during the plant growing season will increase significantly, as will its dominance on carbon fixation [15]. Liu et al. (2020) proposed the opposite conclusion and claimed that soil drought is the main stress that threatens GPP accumulation in more than 70% of regional ecosystems worldwide, and by decoupling SM–VPD, found that the impact of VPD on terrestrial ecosystems is much smaller than that of soil moisture [16]. Given the complexity of the coupling between VPD and SM, it is highly desirable to use novel methods (e.g., probabilistic) to compare the dominance of VPD and SM on ecosystem GPP.

Chinese grasslands are mainly distributed in arid and semi-arid regions (typical temperate grasslands and montane grasslands) and the Tibetan Plateau (typical alpine grasslands), where the ecological environment is fragile [17,18], providing a natural barrier to ecological security in Central and Western China. Due to the unique arid–semi-arid environment and distinct vertical zonality, the carbon cycling processes in Chinese grassland ecosystems are very sensitive to climate change and human disturbances [19–21]. In this study, Chinese grasslands were the study area, and copula functions and Bayesian equations were introduced to explore the conditional probability of GPP decline in Chinese grassland ecosystems when atmospheric drought (high-VPD event) and soil drought (low-SM event) occurred, respectively. We aim to address the following three issues.

(1) To analyze the interannual variability of SM and GPP in Chinese grasslands and to explore the mechanisms of SM regulation of GPP in Chinese grassland ecosystems during historical and future periods.

(2) To compare the correlation between the effects of SM on ecosystem GPP in different soil layers, and analyze the conditional probability of drought in soils of different soil layers causing a decline in GPP in Chinese grasslands.

(3) To calculate the difference between the probability of ecosystem loss due to soil drought minus the probability of ecosystem loss due to atmospheric drought and determine the key moisture constraints controlling GPP in Chinese grasslands.

2. Materials and Methods

2.1. Materials

2.1.1. GPP Datasets

We used the model outputs from four global vegetation dynamics models (CARAIB, LPJ GUESS, LPJmL, ORCHIDEE DGVM) as GPP (monthly, 0.5°) products. The model GPP products are all derived from the harmonized general circulation model IPSL-CM5A-LR as meteorological forcing data, considering historical (1901–2005) and further (2006–2099) scenarios. The results of the model simulations of GPP are all implemented strictly in accordance with the ISIMIP 2b standard protocol, so differences between model outputs are only related to the complexity of the model. Considering that a single model only provides valuable and usable insights at the regional scale, to effectively eliminate intra-model variability and reduce uncertainty in GPP products, we conducted pooled averaging of multi-model GPPs for further processing. Since only LPJmL of the four models of ISIMIP 2b provides mainstream high-emission RCP8.5 climate projections, the future scenarios consider only medium to high GHG emission scenarios (RCP6.0).

2.1.2. SM Datasets

SM data were used to characterize the degree of soil drought. The previous comparison of individual simulated values and pooled averages shows that the amplitudes of the pooled averages are much smaller than those of individual simulated values [22]. In other words, the ensemble averaging method can effectively eliminate the effect of variability within the model, which can effectively reduce the uncertainty between different models. Therefore, we selected the pooled average soil moisture products from the four model outputs.

However, it should be additionally noted that the different models differ in their simulation of land atmospheric exchange fluxes and carbon- and water-cycle stocks in natural and agro-ecosystems, so each model provides different soil moisture soil thicknesses. Here, to match soil layer thickness and reduce data errors, the soil moisture counted was limited to the sum of the moisture in all soil layers within 3 m below ground.

In addition, to explore the binding effect of soil moisture in different soil layers on ecosystem GPP, the LPJmL model with explicit soil stratification thickness was extracted for further analysis. The LPJmL model output included soil moisture at different soil depths of 0–20 cm, 20–50 cm, 50–100 cm, 100–200 cm, and 200–300 cm.

2.1.3. VPD Datasets

We also obtained standard-corrected model input parameters (temperature and relative humidity) from ISIMIP to calculate VPD, which refers to the difference between saturated vapor pressure and actual vapor pressure (AVP) at a given temperature and is a direct measure of the intensity of atmospheric drought [23].

$$VPD = SVP - AVP$$

Saturated vapor pressure is a non-linear function of atmospheric temperature and can be obtained directly from atmospheric temperature calculations with the following empirical formula:

$$SVP = 6.112 \times fa \times e^{\frac{17.67T_a}{T_a + 243.5}}$$

$$f_a = 1 + 7 \times 10^{-4} + 3.46 \times 10^{-6} P_{mst}$$

$$P_{mst} = P_{msl}\left(\frac{(T_a + 273.16)}{(T_a + 273.16) + 0.0065 \times Z}\right)$$

where T_a is the atmospheric surface temperature (°C); Z is the altitude (m); P_{msl} is the atmospheric pressure at mean sea level ($\approx$1013.25 hPa); and P_{mst} is the atmospheric pressure (hPa).

$$\text{AVP} = \frac{RH}{100} \times SVP$$

The above equation shows that warming significantly increases the amount of water vapor held by the atmosphere at saturation (saturation vapor pressure), while the actual vapor pressure of the atmosphere (depending on the relative humidity) remains relatively constant. Consequently, the warming is followed by a non-linear increase in VPD.

2.1.4. Definition of Warm Season and Screening for Warm-Season GPP, SM, and VPD

The warm season in the grid cell was defined as the hottest three month average (one value per year) given that the warm season coincides with the main growing season of plants [24].

We adopted the previous idea of averaging (one value per year) GPP, SM, and VPD for the three months with larger mean temperature values, as carbon loss due to moisture shortage is often most intense in the hottest three months. Therefore, the location of the three months with the largest temperature on the grid unit was extracted by counting the monthly mean temperatures over the period 1901–2005, and the mean GPP, SM, and VPD for the three months with the largest temperature on the metric scale were then filtered from the time series for analysis.

2.2. Methods

2.2.1. Interannual Correlation Measures

For extreme value theory, the correlation coefficient is a good indicator and is commonly used to measure the effect of dependence on the likelihood of binary extreme values [24]. We calculated the interannual correlation between the SM and GPP model dataset per pixel. To remove the influence of climate change signals on long-term trends, we performed a linear detrending of the bivariate prior to calculating correlations.

Commonly used correlations include Pearson and Spearman [25]. Pearson correlation coefficients are only applicable to correlation analyses where the two variables are linear; however, the effect of SM on GPP is often non-linear. Therefore, we chose the Spearman's rank correlation coefficient to analyze the correlation between SM and GPP, with a stronger negative correlation between the two variables indicating more significant negative feedback.

2.2.2. Bivariate Linkage to Calculate the Probability of Conditions under Soil (or Atmospheric) Drought Conditions

Based on the copula function and Bayesian equation, we drew on the novel probabilistic assessment framework constructed by Wang et al. (2021) and He et al. (2022) to calculate the conditional probability of a simultaneous soil drought (or atmospheric drought) scenario. This consists of the following three steps:

(1) Fitting of marginal distributions

Bivariate frequency analysis requires that the distributions of the random variables U and V be determined, so determining the marginal distribution of the bivariate is a prerequisite for constructing the joint probability distribution. We used a non-parametric method for fitting the marginal distributions, because non-parametric estimation methods do not require prior estimation or assumptions about all parameters of the copula function of the dependence structure between the bivariate variables, and can be directly estimated to obtain the fitted values at any point [26]. Non-parametric kernel density estimation is therefore more widely used in practice than conventional parameter estimation (e.g., normal, gamma), and can effectively eliminate errors in the fitted joint probability distribution due to partial singular values. Kernel density estimation is the most widely used test in the field of non-parametric estimation [27], where the kernel distribution produces a non-parametric probability density estimate that adapts itself to the data, rather than selecting a probability density estimate with a particular parameter. Here, we used kernel density estimation to derive marginal distribution fits for the bivariate.

The kernel function density estimation method is described by the following equation:

$$f_n(x) = \frac{1}{nh}\sum\nolimits_{i=1}^{n} K\left(\frac{x - x_i}{h}\right)$$

$f_n(x)$ represents the kernel density value; n represents the number of samples in the bandwidth range; h is the window or bandwidth, representing a reasonable smoothing parameter and $h > 0$; K is the kernel smoothing function; $(x - x_i)$ is the distance between points to x_i; and $x_1, x_2, \ldots x_n$ are random samples from an unknown distribution.

The kernel density estimate of the cumulative distribution function is described as follows:

$$F_n(x) = \int_{-\infty}^{x} f_n(t)dt = \frac{1}{n}\sum\nolimits_{i=1}^{n} G\left(\frac{x - x_i}{h}\right), G(x) = \int_{-\infty}^{x} K(t)dt$$

(2) Fitting and optimization of joint probability distributions

Copulas are multivariate distribution functions defined in the domain of [0, 1] and are used to describe correlations between multiple variables. The bivariate copula function is commonly used to describe the dependence structure between two sets of random variables and to count the joint probability of an event (such as a compound drought) occurring. Sklar confirms that the copula function is unique in that if $F(\cdot,\cdot)$ is a joint distribution function and $X(\cdot)$ and $Y(\cdot)$ are marginal distribution functions of independent variables, then there must be a copula function $C(\cdot,\cdot)$, satisfying

$$F(x,y) = C(X,Y)$$

When $X(\cdot)$ and $Y(\cdot)$ are continuous, there must be a uniquely determined C. Conversely, $X(\cdot)$ and $Y(\cdot)$ are only one-dimensional distribution functions.

According to Sklar's theorem, the joint probability distribution function $F_{X,Y}(x, y)$ for variables X and Y can be expressed as:

$$F_{X,Y}(x,y) = C(F_X(x), F_Y(y)) = P(X \leq x, Y \leq y)$$

where $Fx(x) = P(X \leq x)$ and $Fy(y) = P(Y \leq y)$ are the cumulative distribution functions of the variables X (e.g., SM) and Y (e.g., GPP). C is the joint distribution function of $U = Fx(x)$ and $V = Fy(y)$ after marginal fitting, and the new sequence U and V after the marginal distribution fitting transformation has the characteristics of a uniform distribution.

In addition, a binary copula family connection function is required to calculate the joint probability of an event, and we chose the joint distribution functions Clayton, Frank, Gumbel, t, and Gaussian copula to describe the possible dependency structure of the two variables. The function expression is described in Table 1.

Table 1. Expressions of marginal distribution function.

Copula	Expression of Distribution Function $C(u,v)$	Range of θ Values
Clayton	$max\left(\left[u^{-\theta} + v^{-\theta} - 1^{-1/\theta}\right], 0\right)$	$(0, +\infty)$
Frank	$-\frac{1}{\theta}ln\left(1 + \frac{(e^{-\theta u} - 1)(e^{-\theta v} - 1)}{e^{-\theta} - 1}\right)$	$(-\infty, 0) \cap (0, +\infty)$

Table 1. *Cont.*

Copula	Expression of Distribution Function $C(u,v)$	Range of θ Values
Gumbel	$\exp\left(-\left[(-\ln u)^{\theta} + -\ln v^{\theta}\right]^{1/\theta}\right)$	$(1, +\infty)$
t	$\int_{-\infty}^{t_k^{-1}(u)} \int_{-\infty}^{t_k^{-1}(v)} \frac{1}{2\pi\sqrt{1-\theta^2}} exp\left[1 + \frac{s^2 - 2\theta st + t^2}{k(1-\theta^2)}\right]^{-\frac{k+2}{2}} dsdt$	$(-1, 1), k \neq 0$
Gaussian	$\int_{-\infty}^{\varnothing^{-1}(u)} \int_{-\infty}^{\varnothing^{-1}(v)} \frac{1}{2\pi\sqrt{1-\theta^2}} exp\left[1 + \frac{s^2 - 2\theta st + t^2}{k(1-\theta^2)}\right]^{-\frac{k+2}{2}} dsdt$	$(-1, 1), k \neq 0$

The copula family linkage function was used to reconstruct the dependence structure of the bivariate. To more accurately describe the dependence structure of the bivariate, the best-fitting distribution function was then selected from the five mentioned above for further analysis. The goodness-of-fit test is based on the minimized squared Euclidean distance (*SED*) and is described as follows:

$$SED = (\mathrm{CUV}/\mathrm{C})^2$$

where CUV is the empirical value of the individual binary copula function fit and C is the theoretical value.

(3) Bayesian formula modeling conditional probabilities

Bayesian formulas are causal imputations related to a priori probabilities and phenomenal probabilities (observed objective probabilities) and are a deformation of conditional probabilities. The descriptive formula is:

$$P(A|B) = \frac{P(A)P(B|A)}{P(B)} = \frac{P(AB)}{P(B)}, P(B) \neq 0$$

where $P(B)$ is the probability of the occurrence of event B (a priori probability); $P(AB)$ is derived from step 2 and represents the probability of simultaneous occurrence of event A and event B. By simple reasoning, we know that to determine the conditional probability of the occurrence of loss of ecosystem GPP under drought stress conditions, we need to calculate the probability of the occurrence of drought and the probability of the co-occurrence of increased drought and decreased productivity, where $P(AB)$ can be interpreted as the compound probability of the simultaneous occurrence of drought and loss of vegetation productivity.

$P(A \mid B)$ represents the probability of event A occurring if event B is known to occur (conditional probability); in other words, $P(-\mathrm{GPP} > 90\% \mid \mathrm{VPD} > 90\%)$ can be interpreted as the probability of occurrence of $-\mathrm{GPP}$ above 90% when VPD is above 90%. The specific formula is described as follows:

$$P(-\mathrm{GPP} > 90\%|\mathrm{VPD} > 90\%) = \frac{1 - P(-\mathrm{GPP} > 90\%) - P(\mathrm{VPD} > 90\%) + C(-\mathrm{GPP} > 90\%, \mathrm{VPD} > 90\%)}{1 - P(\mathrm{VPD} > 90\%)}$$

Note that it takes a threshold above a certain value of atmospheric VPD to initiate a decline in ecosystem GPP, so we used $-\mathrm{GPP}$ to calculate the conditional probability in the bivariate relationship. Here, the conditional probability of $P(-\mathrm{GPP} > 90\% \mid \mathrm{VPD} > 90\%)$ is identical to that of $P(\mathrm{GPP} < 10\% \mid \mathrm{VPD} > 90\%)$. Therefore, above equation can be interpreted as the conditional probability of an event occurring with a GPP below 10% in the case of VPD above 90%. It is expressed as follows:

$$P(\mathrm{GPP} < 10\%|\mathrm{VPD} > 90\%) = \frac{1 - P(\mathrm{GPP} < 10\%) - P(\mathrm{VPD} > 90\%) + C(\mathrm{GPP} < 10\%, \mathrm{VPD} > 90\%)}{1 - P(\mathrm{VPD} > 90\%)}$$

We used percentages to define the degree of atmospheric drought and the degree of ecosystem GPP deficit, defining all samples with VPD values above 90%, 70%, and 50% as severe, moderate, and mild atmospheric drought and all samples with GPP values below 10%, 30%, and 50% as severe, moderate, and mild vegetation deficit.

Unlike atmospheric drought, a soil drought event can only be recognized as occurring when soil moisture falls below a certain threshold. We focused on the joint probability of extreme SM (below 10%, 30%, 50%) and GPP (below 10%, 30%, 50%). Thus, the conditional probability of a decline in ecosystem GPP in the presence of an extreme soil drought event can be derived by combining the Bayes' equation described as:

$$P(\text{GPP} < 10\%|\text{SM} < 10\%) = \frac{C(\text{GPP} < 10\%, \text{SM} < 10\%)}{P(\text{SM} < 10\%)}$$

Similarly to atmospheric drought, we defined severe, moderate, and mild soil drought for all samples with SM values below 10%, 30%, and 50%.

3. Results

3.1. Characteristics of Long-Term Changes in Chinese Grassland SM and Its Constraints on Ecosystem GPP

Th long-term trend of soil moisture simulation in Chinese grasslands is not insignificant over the past 100 years. The annual SM in the historical period had an insignificant soil-degenerating aridity trend, and future projections were largely constant (Figure 1a). The warm-season SM showed a similar trend, but the moisture anomaly domain was significantly higher in the warm season than in the annual (Figure 1b). Compared to the trends, the interannual variability of soil moisture in Chinese grasslands is highly significant, as can be seen in periods such as 1930–1950 and 2050–2060 when soil moisture is significantly higher. In conclusion, the long-term change of soil moisture is mainly controlled by its interannual fluctuations. Therefore, the next analysis will focus on the regulation of the interannual variability of soil moisture on ecosystem GPP.

Figure 2 shows that the interannual variability of SM and GPP is similar, i.e., years with low SM correspond to low GPP, and the two are strongly positively correlated. The correlations between SM_{IAV} and GPP_{IAV} are 0.171 and 0.477 for the whole year in the historical and future periods, respectively, and the fluctuation patterns of both are found to have been basically consistent since 1950. The degree of correlation was significantly stronger in the warm season than in the whole year, with correlations of 0.48 and 0.63 between them year by year for the historical and future periods, respectively, and both passed the significance level ($p < 0.05$) test. We concluded that soil moisture is an important factor that strongly regulates ecosystem GPP, and when soil moisture is below a certain threshold, it significantly constrains the accumulation of ecosystem GPP, because an insufficient soil moisture supply significantly forces a decrease in plant photosynthetic capacity, which in turn leads to a decrease in ecosystem GPP. Moreover, this effect more strongly restricts plant growth during the growing season.

Next, the correlations of SM_{IAV} and GPP_{IAV} were calculated using a sliding window of 41 years for the whole year and warm season from 1901 to 2099 (Figure 3). The bivariate correlations were found to be consistently positive in most periods and regions of Chinese grasslands, and the correlation values were higher for the warm season than for the whole year. The correlations between SM_{IAV} and GPP_{IAV} in Mongolia, the Loess Plateau, and Xinjiang all exceeded 0.4 and passed the significance test, while the correlation in the Tibetan Plateau grasslands gradually increased to 0.2 from year to year. This reflects that the regulation of GPP by soil moisture is always positive, i.e., it suppresses plant growth in water-scarce years and promotes it in water-abundant years. In comparison, the positive correlations in the warm season are higher in all regions and stages than in the whole year, and the correlations in Chinese grasslands are always around 0.4 throughout the year, while the correlations in the warm season have been over 0.5 since 1970. In the Xinjiang and Loess Plateau regions in particular, the correlations of the warm-season bivariate exceeded 0.6. This indicated that the regulation of GPP by soil moisture is higher in the warm season than in the whole year; in other words, the warm-season soil moisture deficit can trigger more serious ecosystem

GPP loss events in most cases. Given that our study discusses the effect of soil moisture constraint events on ecosystem GPP, the choice of warm-season bivariate better reflects the limiting effect of extreme moisture constraints on carbon uptake in grasslands, and, therefore, the SM and GPP in the warm season were chosen for further analysis.

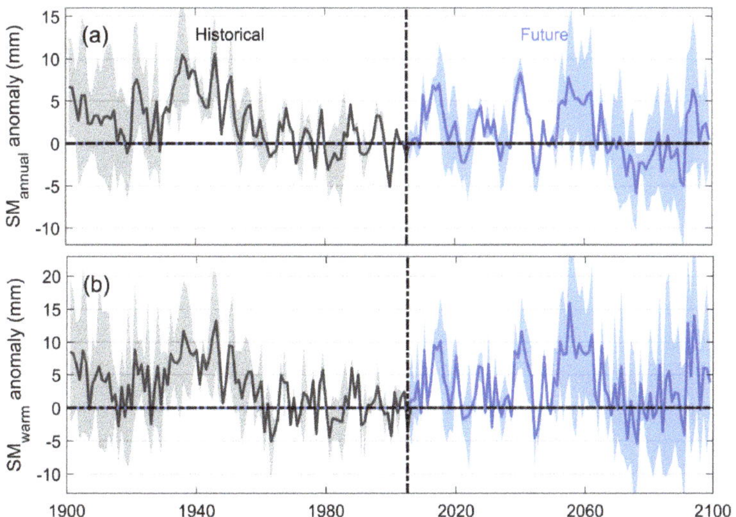

Figure 1. Long-term variation in soil moisture anomalies based on process-model simulations. (**a**) Annual SM; (**b**) warm-season SM. SM anomalies are relative to the distance to the mean during 2006–2015. The shaded line is the doubled standard deviation between the four models. The warm seasons mentioned are the three months with the largest annual mean temperatures in 1901–2005.

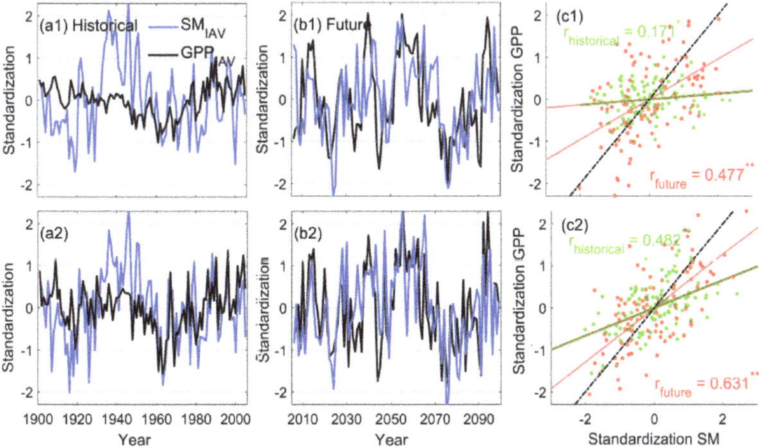

Figure 2. Standardized interannual variability characteristics of soil moisture (SM_{IAV}) and gross primary productivity (GPP_{IAV}) in Chinese grasslands. The upper side shows the annual time series, and the lower side shows the warm-season time series. (**a**) historical period; (**b**) future period; (**c**) correlation between SM_{IAV} and GPP_{IAV} per year, * represents passing 0.1 significance test, and ** represents passing 0.05 significance test.

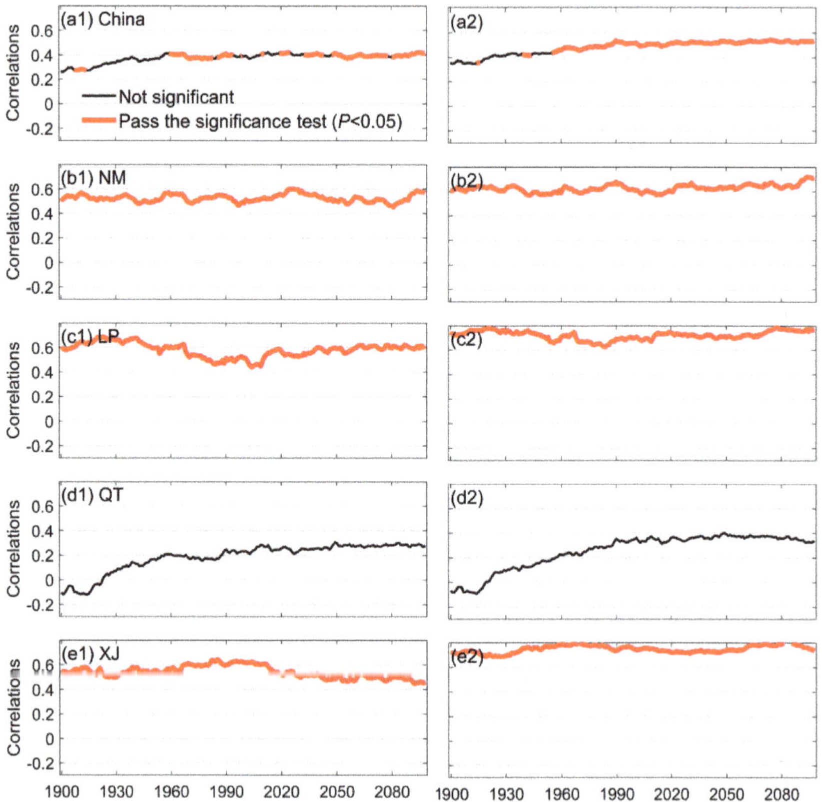

Figure 3. Correlation between GPP$_{IAV}$ and SM$_{IAV}$ for Chinese grasslands during the period 1901–2099. (**a1,b1,c1,d1,e1**) is annual, and (**a2,b2,c2,d2,e2**) is the warm season. The correlation between SM$_{IAV}$ and GPP$_{IAV}$ per year in the warm season was calculated with a 41-year sliding window, i.e., by calculating the correlation between the two variables from 1861–1901 as the result in 1901, and so on, to find the correlation from 1901–2099. The red line in the figure shows the years in which the correlations passed the significance test ($p < 0.05$).

3.2. Comparison of the Regulation of GPP by Different Soil Layers

The impact on the GPP of the ecosystem varies across soil layers due to differences in rainwater recharge and evapotranspiration at different depths. The warm-season SM and GPP were correlated at the meta-scale to compare the bivariate structure between SM and GPP at different soil depths (Figure 4). The bivariate correlations for shallow soils (0–20 and 20–50 cm) were found to show significant interannual agreement, with highly significant positive correlations in both historical and future periods. Deeper soils (100–200 and 200–300 cm) with higher possible difficulty in water use for plants, did not show a significant consistent interannual correlation in the historical period, and the interannual relationship between the two variables tended to be consistent in the future period. As shown in Figure 4c, most of the pixel correlations between shallow-soil water and GPP are greater than 0, implying a strong positive coupling between the two, with shallow-soil water deficit significantly constraining the accumulation of GPP in the ecosystem. In contrast, about half of the pixels of the deep-soil water are on both sides of the 0 value line, showing very high uncertainty in the regulation of GPP. Overall, the correlation between ecosystem GPP and the SM of different soil layers showed a higher correlation the closer to the surface the soil layer was. Comparing different soil layers, it can be seen that soil moisture at different depths has an inconsistent regulating effect on ecosystem GPP, and uncertainty has a non-negligible effect on GPP constraint.

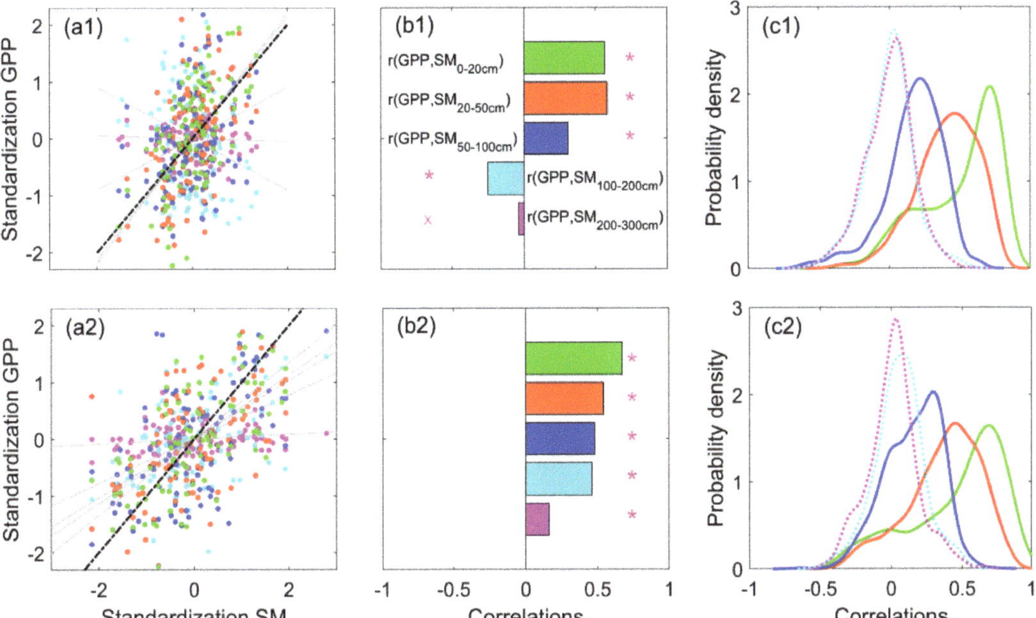

Figure 4. Correlation between warm-season SM and GPP for different soil depths. The upper column is the historical period and the lower column is the future period. (**a1,a2**) Scatter plots of SM and GPP per year. Red, green, blue, cyan, and pink represent soil moisture at 0–20 cm, 20–50 cm, 50–100 mm, 100–200 mm, and 200–300 mm soil depths, respectively; (**b1,b2**) correlation between SM and GPP regional averages; (**c1,c2**) probability density plot of correlation between SM and GPP.

Figure 5 shows the spatial distribution of the probability of different degrees of ecosystem GPP loss under the precondition of extreme drought in soils with different soil moisture levels, and found that the probabilities show great differences between different soil levels. The probability of severe GPP loss induced by drought events at the 0–20 cm soil depth was as high as 32.49%, decreasing to 27.64%, 16.79%, 16.43%, 16.43%, and 12.61% with the increase in soil depth. We found that after drought in shallow soils, the ecosystem GPP tended to be severely deficient due to water limitations, while the probability of drought-induced vegetation deficit in deep soils was lower. In addition, we also emphasized that the probability of severe shallow-soil drought induced moderate and mild ecosystem GPP loss, while the ecosystem GPP was significantly higher than that of deep-soil drought.

The locations of the spatial distribution of ecosystem GPP loss probabilities for severe, moderate, and mild were basically the same, indicating that soil drought events of different soil depths had weaker effects on the changes in probability at different locations. The spatial heterogeneity of ecosystem GPP loss induced by shallow-soil drought events was stronger. Xinjiang and the Loess Plateau are the most susceptible grasslands to soil drought, followed by Inner Mongolia grasslands and Tibetan Plateau grasslands being the least affected. The probabilities associated with deep-soil drought were more uniform and consistent in space, with less variation among regions (Figure 5(a6)).

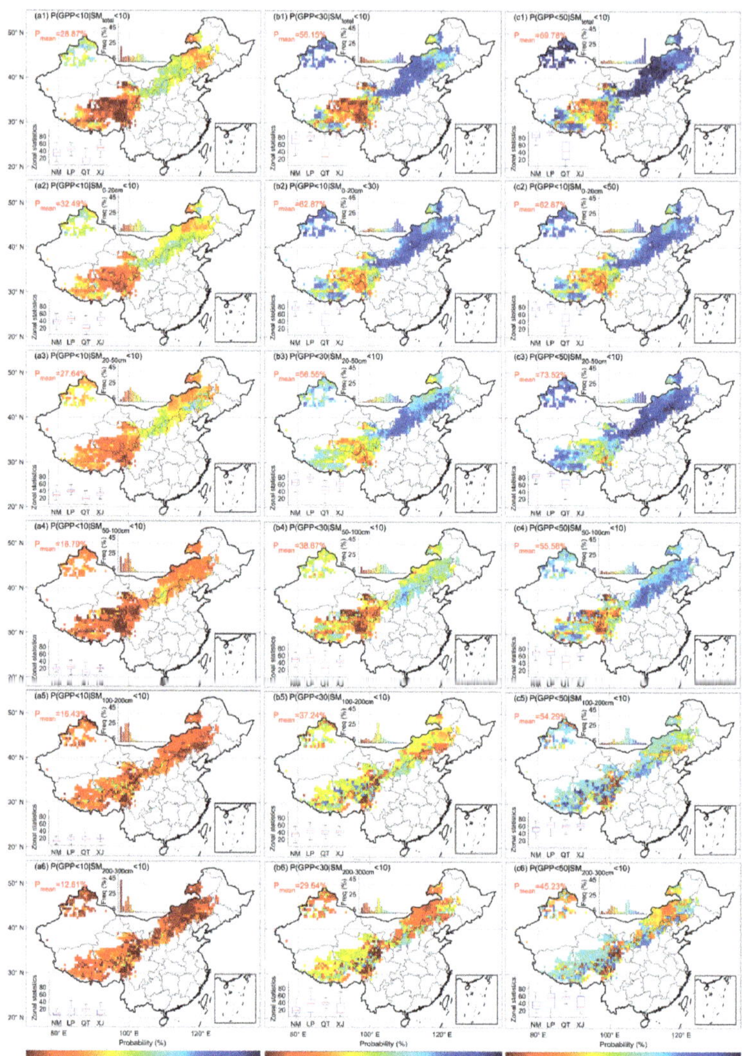

Figure 5. Conditional probabilities of ecosystem GPP loss under severe soil drought conditions (i.e., probability when soil moisture is below 10% threshold in the 1901–2005 time series) at different soil depths of Chinese grasslands during the historical period. (**a1–a6,b1–b6,c1–c6**) The probability of severe ecosystem GPP loss under severe soil drought conditions; 1–6 represent the soil depths of 0–300 mm, 0–20 mm, 20–50 mm, 50–100 mm, 100–200 mm, and 200–300 mm, respectively. The regional average trend values are counted as box plots in the lower left corner. NM, LP, QT, and XJ represent grasslands in Inner Mongolia, the Loess Plateau, the Tibetan Plateau, and Xinjiang, respectively, and red dots represent anomalies. The histogram is the frequency statistics.

3.3. Comparison of the Probability of High-VPD and Low-SM Events Leading to Ecosystem GPP Deficits

The difference was obtained by subtracting the probability of GPP decline due to soil drought from the probability of GPP decline due to atmospheric drought, which can be used to compare the primary and secondary effects of soil drought and atmospheric drought on the constraint of ecosystem GPP. A positive value represents a stronger constraint of soil drought on GPP, while a negative value represents a stronger constraint of atmospheric drought. For the historical period, we found that soil drought in Chinese grasslands causes a higher probability of ecosystem GPP loss than atmospheric drought, but the results vary widely with different threshold settings (Figure 6). The difference between the two was relatively large in the most severe drought scenario, with 8.16%, 10.41%, and 6.93% differences for severe, moderate, and mild ecosystem GPP losses, respectively (Figure 6a–c). The differences also showed different degrees of reduction with decreasing drought levels. From the probability perspective, we concluded that the overall impact of soil drought on Chinese grassland ecosystems is significantly higher than that of atmospheric drought, and the probability of GPP deficit caused by soil drought is on average 1–10% higher than atmospheric drought. However, the difference was found to be spatially heterogeneous, with soil drought in Inner Mongolia, the Tibetan Plateau, and the Loess Plateau apparently causing a higher probability of ecosystem GPP loss, which is the main constraint limiting GPP accumulation. In contrast, the probability of atmospheric drought causing ecosystem GPP deficit was higher in the grassland of the Tibetan Plateau, which represents a stronger constraint of atmospheric drought on the GPP of the Tibetan Plateau.

For the future period, soil drought remains the main water constraint limiting the accumulation of GPP in the ecosystem (Figure 7). However, compared to the historical period, the spatial differences changed significantly. The difference in the Loess Plateau tends to be negative from the positive value in the historical period, indicating that atmospheric drought will become the most important moisture constraint in the future. Inner Mongolia and Xinjiang, although still positive, have decreased compared with the historical period, indicating that the moisture constraint from the atmosphere has increased and the soil moisture constraint has weakened. The Tibetan Plateau is very unique, and soil drought will become the main constraint on GPP accumulation in the Tibetan Plateau ecosystem in the future, and its probability difference is more than 10%.

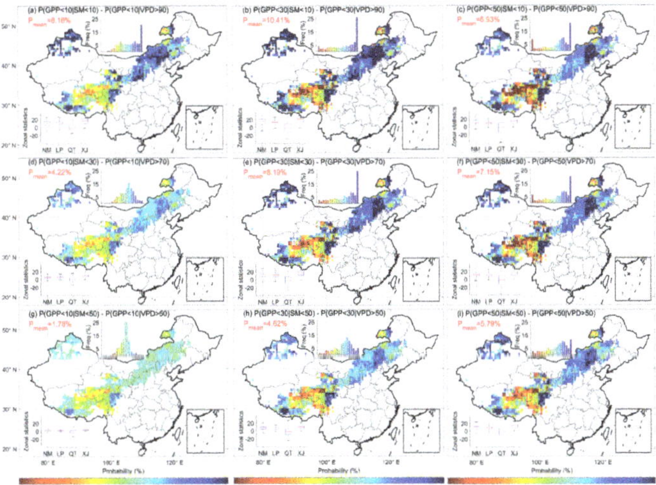

Figure 6. Difference between the probability of ecosystem GPP loss under soil drought conditions minus the probability of ecosystem GPP loss in the atmospheric drought scenario for the historical period. (**a–c**) The difference between the probability of severe, moderate, and mild ecosystem GPP loss corresponding to severe soil drought and atmospheric drought; (**d–f**) the difference between the probability of severe, moderate, and mild ecosystem GPP loss corresponding to moderate soil drought and atmospheric drought; (**g–i**) the difference between the probability of severe, moderate, and mild ecosystem GPP loss corresponding to mild soil drought and atmospheric drought; (**g–i**) the difference in the probability of severe, moderate, and mild ecosystem GPP loss corresponding to mild soil and atmospheric drought.

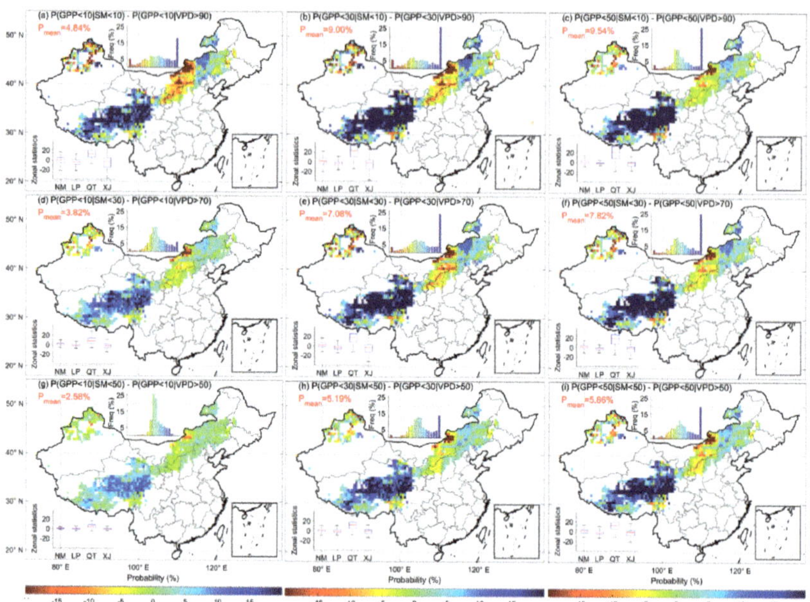

Figure 7. Difference between the probability of ecosystem GPP loss under soil drought conditions minus the probability of ecosystem GPP loss in the atmospheric drought scenario for the future period. (**a–c**) The difference between the probability of severe, moderate, and mild ecosystem GPP loss corresponding to severe soil drought and atmospheric drought; (**d–f**) the difference between the probability of severe, moderate, and mild ecosystem GPP loss corresponding to moderate soil drought and atmospheric drought; (**g–i**) the difference between the probability of severe, moderate, and mild ecosystem GPP loss corresponding to mild soil drought and atmospheric drought; (**g–i**) the difference in the probability of severe, moderate, and mild ecosystem GPP loss corresponding to mild soil and atmospheric drought..

4. Discussion

4.1. Soil Moisture More Strongly Regulates Carbon Balance Than Atmospheric Indicators in Chinese Grasslands

Terrestrial ecosystem models can obtain high-resolution long time series of soil moisture products, bridging the gap between remote sensing and station observations [22] and providing important information for exploring long-term soil drought evolution. We confirmed that soil moisture deficit can severely weaken the accumulation of ecosystem GPP and is an important aspect in regulating the interannual dynamics of GPP. Previous studies have shown that CO_2 growth rates are sensitive to observed changes in terrestrial moisture, i.e., drought years are associated with rapid increases in atmospheric CO_2 [7]; higher water availability promotes ecosystem productivity fixation, and water loss attenuates it [28,29]. Our results are similar and showed that low SM corresponds to low GPP, reflecting a reduced potential for ecosystem CO_2 uptake in years with low soil water. Thus, a lack of soil water supply would affect the division and expansion of individual plant cells and greatly reduce the ability to obtain carbon from the atmosphere, which in turn would result in a decrease in ecosystem GPP.

Considering that soil water content is closely related to precipitation, most of the precipitation infiltrates into the soil, except for a small portion that evaporates and is trapped by the canopy [30]. Most of the water available for plant uptake originates from precipitation, which affects the photosynthetic capacity of plants and is mainly regulated by the intra-annual distribution of precipitation and differences in precipitation intensity [31]. Changes in the intra-annual distribution of precipitation can lead to a mismatch between water availability and plant growth requirements [32]. For example, precipitation events that occur during the germination period do not match actual water demand and supply. As a result, much of the water is likely to be inefficient and 'wasted' in the evapotranspiration process [33]. It has been suggested that increasing precipitation early in the growing season can have a positive impact on semi-arid grassland productivity [34]. This may be due to a number of factors, including the fact that the earlier the precipitation, the higher the soil moisture, and the fact that more precipitation early in the growing season promotes plant root growth. In contrast, more precipitation events later in the growing season do not significantly improve photosynthetic capacity or carbon accumulation [35]. In addition, changes in the magnitude or intensity of precipitation events can alter the vertical distribution of soil moisture, which is closely linked to the underground activity of plants [36]. Small precipitation events tend to increase the shallow-soil water content and stimulate shallow-rooted plant activity (photosynthesis and autotrophic respiration) and soil microbial activity. Large precipitation events are more effective at replenishing deep-soil water content, which may be more effective in triggering deep plant activity [37].

We keenly captured the extremely close coupling between Chinese grassland productivity and soil moisture, with soil moisture surplus and deficit directly regulating the direction of ecosystem carbon revenues and expenditures. Precipitation is the most direct source of replenishment for soils [38], and precipitation surplus and deficit can be considered to regulate productivity in Chinese grasslands. However, given that soil moisture is the main source of water directly available to plants, this study innovatively used soil moisture as the main water constraint to analyze its relative influence on GPP.

4.2. Soil Moisture Is a Key Water Constraint Controlling the Grassland Productivity in China

Previous studies have found that soil drought is the dominant constraint on drought stress in most ecosystems worldwide [16]. We have confirmed that soil drought is indeed the main water constraint threatening GPP in Chinese grassland ecosystems from a probabilistic perspective. However, we noted that the probability difference between soil drought and atmospheric drought causing an ecosystem GPP deficit gradually decreases or reverses to a negative value in most regions in the future. This represents a clear increase in the extent to which atmospheric drought will affect ecosystems in the future and a decrease in the importance of soil drought in dominating vegetation deficits. This is likely due to the fact that atmospheric VPD has increased rapidly over the past century and will remain the growing trend in the future (Supplementary Figure S1), while soil moisture trends have remained stable over time (Figure 1). Continued atmospheric constraints force ecosystems to adapt to water stress by closing stomata or stopping plant growth due to the inability to coordinate water–carbon fluxes [39]. We found that the future scenario for the Tibetan Plateau is one in which the temperature limitation of alpine ecosystems is lifted by rising temperatures, possibly due to increased atmospheric pressure that promotes vegetation growth rather than inhibiting it, thus causing soil drought to remain the main moisture constraint limiting GPP accumulation in the Tibetan Plateau in the future. It is important that identifying the primary and secondary moisture constraints from the soil and atmosphere remains a challenge [40], and a probabilistic perspective provides only a possibility rather than a definitive conclusion. In the future, there is a need for further clarification of this long-standing and complex issue to open up new avenues for improved modeling and better management of drought risk.

However, the interannual regulation of ecosystem GPP was significantly different for soil moisture at different soil depths. We found that shallow-soil water (0–50 cm) was the dominant soil layer regulating interannual variability in GPP, probably because the most accessible water to grassland plants comes from shallow soils and is the direct source [41]. In water-scarce weather, most plants preferentially draw water from the soil [12], and roots easily take up residual water from shallow soils in a variety of ways, including mass flow, diffusion, and interception [42]. Only when shallow-soil water is exhausted can deep-rooted plants draw water from deeper soils [43]. We also confirmed that the probability of ecosystem GPP loss induced by water deficit in shallow soils is much higher than that of deep-soil water, representing a greater dependence on shallow-soil water for plant growth, and that negative anomalies in shallow-soil water can cause substantial ecosystem carbon loss.

The innovation of our study is to not only point out that soil drought outweighs atmospheric drought as the key to the restriction of ecosystem GPP but also that shallow-soil water is critical to controlling ecosystem GPP. This addresses the fact that previous studies have only detailed the effects of soil drought on ecosystems in general, without distinguishing the relative contribution of soil moisture at different soil depths [16,44], which also provides precedent for future validation on a global scale.

5. Conclusions

We used simulation results from a terrestrial ecosystem model to analyze the moderating effect of soil drought on GPP changes in Chinese grasslands to explore the conditional probability of ecosystem GPP loss due to soil drought and to analyze the dominance of terrestrial–atmosphere moisture constraints on ecosystem GPP. The main conclusions are as follows:

(1) No significant trends were found for soil moisture in the historical or future periods, and its long-term change was mainly reflected through interannual fluctuations. Soil moisture showed a highly significant positive correlation with ecosystem GPP in the time series, indicating that when soil water decreases, it causes a decrease in ecosystem GPP. Moreover, the correlation between SM and GPP was higher in the warm season than annually, and higher in the future period than in the historical period, representing a stronger constraint on GPP in Chinese grasslands in the warm season and a deeper constraint in the future period than in the historical period.

(2) Using the LPJmL model's soil moisture data at different soil depths and analyzing their relationship with ecosystem GPP, it was found that the correlation between shallow-soil moisture (0–50 cm) and GPP was clearly higher than that of deeper soils, and the probability of an ecosystem GPP deficit due to a shortage of soil water in the shallow layer was much higher than that of soil water in the middle and deep layers.

(3) In probabilistic terms, soil drought has a higher probability of initiating the loss of ecosystem GPP than atmospheric drought, with moisture scarcity originating from the soil becoming the main aspect that constrains ecosystem GPP. In the future, with the rapid rise of global VPD, the probability of ecosystem GPP loss induced by atmospheric drought increases and overtakes soil drought as the main water constraint in some regions.

Supplementary Materials: The following supporting information can be downloaded at: https://www.mdpi.com/article/10.3390/land12020484/s1, Figure S1.

Author Contributions: Conceptualization, Z.H., Z.S. and P.H.; methodology, X.M. (Xiaoyu Meng) and P.H.; software, T.D., P.H. and N.W.; validation, P.H. and X.M. (Xiaoyu Meng); formal analysis, P.H.; investigation, P.H. and X.M. (Xiaoyu Meng); resources, P.H.; data curation, P.H.; writing—original draft preparation, X.M. (Xiaoliang Ma), Z.H., S.M. and P.H.; writing—review and editing, X.M. (Xiaoliang Ma), X.M. (Xiaoyu Meng), J.M., Y.Z. and P.H.; visualization, P.H.; supervision, P.H.; project administration, Z.S. and P.H.; funding acquisition, Z.S. and P.H. All authors have read and agreed to the published version of the manuscript.

Funding: This research was funded by Foundation items: National Natural Science Foundation of China, No. 32060408; National Basic Resources Survey Special, No. 2017FY100200; the Scientific Innovation Project of Postgraduates of Xinjiang Uygur Autonomous Region, China, No. XJ2021G169.

Institutional Review Board Statement: Not applicable.

Informed Consent Statement: Not applicable.

Data Availability Statement: Data are already published and publicly available, datasets utilized for this research are as follows: The ISIMIP dataset: https://esg.pik-potsdam.de/projects/isimip/, accessed on 6 December 2022.

Conflicts of Interest: The authors declare no conflict of interest.

References

1. Ciais, P.; Reichstein, M.; Viovy, N.; Granier, A.; Ogee, J.; Allard, V.; Aubinet, M.; Buchmann, N.; Bernhofer, C.; Carrara, A.; et al. Europe-wide reduction in primary productivity caused by the heat and drought in 2003. *Nature* **2005**, *437*, 529–533. [CrossRef]
2. Zhao, M.; Running, S. Drought-induced reduction in global terrestrial net primary production from 2000 through 2009. *Science* **2010**, *329*, 940–943. [CrossRef]
3. Peng, L.; Xia, J.; Li, Z.; Fang, C.; Deng, X. Spatio-temporal dynamics of water-related disaster risk in the Yangtze River Economic Belt from 2000 to 2015. *Resour. Conserv. Recycl.* **2020**, *161*, 104851. [CrossRef]
4. Xiao, J.; Sun, G.; Chen, J.; Chen, H.; Chen, S.; Dong, G.; Gao, S.; Guo, H.; Guo, J.; Han, S.; et al. Carbon fluxes, evapotranspiration, and water use efficiency of terrestrial ecosystems in China. *Agric. For. Meteorol.* **2013**, *182*, 76–90. [CrossRef]
5. Yang, Y.; Guan, H.; Batelaan, O.; McVicar, T.; Long, D.; Piao, S.; Liang, W.; Liu, B.; Jin, Z.; Simmons, C. Contrasting response of water use efficiency to drought across global terrestrial ecosystems. *Sci. Rep.* **2016**, *6*, 23283. [CrossRef]
6. Zhang, Y.; Xiao, X.; Zhou, S.; Ciais, P.; McCarthy, H.; Luo, Y. Canopy and physiological control of GPP during drought and heatwave. *Geophys. Res. Lett.* **2016**, *43*, 3325–3333. [CrossRef]
7. Humphrey, V.; Zscheischler, J.; Ciais, P.; Gudmundsson, L.; Sitch, S.; Seneviratne, S.I. Sensitivity of atmospheric CO2 growth rate to observed changes in terrestrial water storage. *Nature* **2018**, *560*, 628–631. [CrossRef]
8. Zhang, J.; Wang, X.; Ren, J. Simulation of Gross Primary Productivity Using Multiple Light Use Efficiency Models. *Land* **2021**, *10*, 329. [CrossRef]
9. Liancourt, P.; Sharkhuu, A.; Ariuntsetseg, L.; Boldgiv, B.; Helliker, B.R.; Plante, A.F.; Petraitis, P.S.; Casper, B.B. Temporal and spatial variation in how vegetation alters the soil moisture response to climate manipulation. *Plant and Soil* **2011**, *351*, 249–261. [CrossRef]
10. Han, Z.; Huang, Q.; Huang, S.; Leng, G.; Bai, Q.; Liang, H.; Wang, L.; Zhao, J.; Fang, W. Spatial-temporal dynamics of agricultural drought in the Loess Plateau under a changing environment: Characteristics and potential influencing factors. *Agric. Water Manag.* **2021**, *244*, 106540. [CrossRef]
11. He, P.; Sun, Z.; Han, Z.; Ma, X.; Zhao, P.; Liu, Y.F.; Ma, J. Divergent trends of water storage observed via gravity satellite across distinct areas in China. *Water* **2020**, *12*, 2862. [CrossRef]
12. Sehgal, V.; Gaur, N.; Mohanty, B. Global surface soil moisture drydown patterns. *Water Resour. Res.* **2021**, *57*, e2020WR027588. [CrossRef]
13. Yuan, W.; Zheng, Y.; Piao, S.; Ciais, P.; Lombardozzi, D.; Wang, Y.; Ryu, Y.; Chen, G.; Dong, W.; Hu, Z.; et al. Increased atmospheric vapor pressure deficit reduces global vegetation growth. *Sci. Adv.* **2019**, *5*, eaax1396. [CrossRef]
14. Shen, Q.; Gao, G.; Fu, B.; Lu, Y. Responses of shelterbelt stand transpiration to drought and groundwater variations in an arid inland river basin of Northwest China. *J. Hydrol.* **2015**, *531*, 738–748. [CrossRef]
15. Novick, K.; Ficklin, D.; Stoy, P.; Williams, C.A.; Bohrer, G.; Oishi, A.C.; Papuga, S.A.; Blanken, P.D.; Noormets, A.; Sulman, B.N.; et al. The increasing importance of atmospheric demand for ecosystem water and carbon fluxes. *Nat. Clim. Change* **2016**, *6*, 1023–1027. [CrossRef]
16. Liu, L.; Gudmundsson, L.; Hauser, M.; Qin, D.; Li, S.; Seneviratne, S.I. Soil moisture dominates dryness stress on ecosystem production globally. *Nat. Commun.* **2020**, *11*, 4892. [CrossRef]
17. Kang, L.; Han, X.; Zhang, Z.; Sun, O.J. Grassland ecosystems in China: Review of current knowledge and research advancement. *Philos. Trans. R. Soc. London. Ser. B Biol. Sci.* **2007**, *362*, 997–1008. [CrossRef]
18. Jia, W.; Liu, M.; Wang, D.; He, H.; Shi, P.; Li, Y.; Wang, Y. Uncertainty in simulating regional gross primary productivity from satellite-based models over northern China grassland. *Ecol. Indic.* **2018**, *88*, 134–143. [CrossRef]
19. Liu, Y.; Lei, H. Responses of natural vegetation dynamics to climate drivers in China from 1982 to 2011. *Remote Sens.* **2015**, *7*, 10243–10268. [CrossRef]
20. Rui, Z.; Zhao, X.; Zuo, X.; Degen, A.; Yulin, L.; Liu, X.; Luo, Y.; Qu, H.; Lian, J.; Wang, R. Drought-induced shift from a carbon sink to a carbon source in the grasslands of Inner Mongolia, China. *Catena* **2020**, *195*, 104845. [CrossRef]
21. He, P.; Ma, X.; Sun, Z. Interannual variability in summer climate change controls GPP long-term changes. *Environ. Res.* **2022**, *212*, 113409. [CrossRef]
22. Cheng, S.; Huang, J.; Ji, F.; Lin, L. Uncertainties of soil moisture in historical simulations and future projections: Uncertainties of Soil Moisture. *J. Geophys. Res. Atmos.* **2017**, *122*, 2239–2253. [CrossRef]
23. Buck, A. New equations for computing vapor pressure and enhancement Factor. *J. Appl. Meteorol.* **1981**, *20*, 1527–1532. [CrossRef]
24. Zscheischler, J.; Seneviratne, S. Dependence of drivers affects risks associated with compound events. *Sci. Adv.* **2017**, *3*, e1700263. [CrossRef]
25. Bishara, A.; Hittner, J. Testing the significance of a correlation with nonnormal data: Comparison of pearson, spearman, transformation, and resampling approaches. *Psychol. Methods* **2012**, *17*, 399–417. [CrossRef]

26. Wang, L.; Huang, S.; Huang, Q.; Leng, G.; Han, Z.; Zhao, J.; Guo, Y. Vegetation vulnerability and resistance to hydrometeorological stresses in water- and energy-limited watersheds based on a Bayesian framework. *Catena* **2021**, *196*, 104879. [CrossRef]
27. Jones, M.; Hardle, W. Applied nonparametric regression. *Economica* **1992**, *58*, 538. [CrossRef]
28. Wu, Z.; Dijkstra, P.; Koch, G.; Penuelas, J.; Hungate, B. Responses of terrestrial ecosystems to temperature and precipitation change: A meta-analysis of experimental manipulation. *Glob. Change Biol.* **2011**, *17*, 927–942. [CrossRef]
29. Shao, C.; Chen, J.; Chu, H.; Lafortezza, R.; Dong, G.; Abraha, M.; Ochirbat, B.; John, R.; Ouyang, Z.; Zhang, Y.; et al. Grassland productivity and carbon sequestration in Mongolian grasslands: The underlying mechanisms and nomadic implications. *Environ. Res.* **2017**, *159*, 124–134. [CrossRef]
30. Arca, V.; Power, S.A.; Delgado-Baquerizo, M.; Pendall, E.; Ochoa-Hueso, R. Seasonal effects of altered precipitation regimes on ecosystem-level CO2 fluxes and their drivers in a grassland from Eastern Australia. *Plant Soil* **2021**, *460*, 435–451. [CrossRef]
31. Ren, H.; Xu, Z.; Huang, J.; Lü, X.; Zeng, D.-H.; Yuan, Z.; Han, X.; Fang, Y. Increased precipitation induces a positive plant-soil feedback in a semi-arid grassland. *Plant Soil* **2014**, *389*, 211–223. [CrossRef]
32. Suttle, K.; Thomsen, M.; Power, M. Species interactions reverse grassland responses to changing climate. *Science* **2007**, *315*, 640–642. [CrossRef]
33. Deng, Y.; Wang, X.; Lu, T.; Du, H.; Ciais, P.; Lin, X. Divergent seasonal responses of carbon fluxes to extreme droughts over China. *Agric. For. Meteorol.* **2023**, *328*, 109253. [CrossRef]
34. Ru, J.; Zhou, Y.; Hui, D.; Zheng, M.; Wan, S. Shifts of growing-season precipitation peaks decrease soil respiration in a semiarid grassland. *Glob. Change Biol.* **2017**, *24*, 1001–1011. [CrossRef]
35. Huxman, T.; Snyder, K.; Tissue, D.; Leffler, A.J.; Ogle, K.; Pockman, W.; Sandquist, D.; Potts, D.; Schwinning, S. Precipitation pulses and carbon fluxes in semiarid and arid ecosystems. *Oecologia* **2004**, *141*, 254–268. [CrossRef]
36. Nielsen, U.; Ball, B. Impacts of altered precipitation regimes on soil communities and biogeochemistry in arid and semi-arid ecosystems. *Glob. Change Biol.* **2014**, *21*, 1407–1421. [CrossRef]
37. Parton, W.; Morgan, J.; Smith, D.; Del Grosso, S.; Prihodko, L.; Lecain, D.; Kelly, R.; Lutz, S. Impact of precipitation dynamics on net ecosystem productivity. *Glob. Change Biol.* **2012**, *18*, 915–927. [CrossRef]
38. Parker, N.; Patrignani, A. Reconstructing Precipitation Events Using Co-located Soil Moisture Information. *J. Hydrometeorol.* **2021**, *22*, 3275–3290. [CrossRef]
39. Grossiord, C.; Buckley, T.N.; Cernusak, L.A.; Novick, K.A.; Poulter, B.; Siegwolf, R.T.W.; Sperry, J.S.; McDowell, N.G. Plant responses to rising vapor pressure deficit. *New Phytol.* **2020**, *226*, 1550–1566. [CrossRef]
40. He, P.; Ma, X.; Zongjiu, S.; Han, Z.; Ma, S.; Meng, X. Compound drought constrains gross primary productivity in Chinese grasslands. *Environ. Res. Lett.* **2022**, *17*, 104054. [CrossRef]
41. Allen-Diaz, B.H. Water Table and Plant Species Relationships in Sierra Nevada Meadows. *Am. Midl. Nat.* **1991**, *126*, 30–43. [CrossRef]
42. Darrouzet-Nardi, A.; D'Antonio, C.M.; Dawson, T.E. Depth of water acquisition by invading shrubs and resident herbs in a Sierra Nevada meadow. *Plant Soil* **2006**, *285*, 31–43. [CrossRef]
43. Hoekstra, N.; Finn, J.; Hofer, D.; Luscher, A. The effect of drought and interspecific interactions on depth of water uptake in deep- and shallow-rooting grassland species as determined by delta O-18 natural abundance. *Biogeosciences* **2014**, *11*, 4493–4506. [CrossRef]
44. Yongming, C.; Liu, L.; Cheng, L.; Fa, K.; Liu, X.; Huo, Z.; Huang, G. A shift in the dominant role of atmospheric vapor pressure deficit and soil moisture on vegetation greening in China. *J. Hydrol.* **2023**, *615*, 128680. [CrossRef]

 *land*

Article

Species Enriched Grass–Clover Pastures Show Distinct Carabid Assemblages and Enhance Endangered Species of Carabid Beetles (Coleoptera: Carabidae) Compared to Continuous Maize

Henriette Beye [1,*], Friedhelm Taube [2,3], Tobias W. Donath [1], Jan Schulz [1], Mario Hasler [4] and Tim Diekötter [1,*]

1   Institute of Natural Resource Conservation, Landscape Ecology, Kiel University, 24118 Kiel, Germany
2   Institute of Plant Production and Plant Breeding, Grass and Forage Science/Organic Agriculture, Kiel University, 24118 Kiel, Germany
3   Grass Based Dairy Systems, Animal Production Systems Group, Wageningen University (WUR), 6700 HB Wageningen, The Netherlands
4   Department of Statistics, Kiel University, 24118 Kiel, Germany
*   Correspondence: hbeye@ecology.uni-kiel.de (H.B.); tdiekoetter@ecology.uni-kiel.de (T.D.)

Abstract: There is an urgent global need for the ecological intensification of agricultural systems to reduce negative impacts on the environment while meeting the rising demand for agricultural products. Enriching grasslands with floral species is a tool to promote diversity and the associated services at higher trophic levels, and ultimately, to enhance the agricultural landscape matrix. Here, we studied an organic pastures-based dairy production system with plant species enhanced grass–clover pastures with respect to the effect on the activity density, functional traits, carabid assemblages, and species richness of carabid beetles. To understand the effect of land management on carabid beetles, we studied two types of grass–clover pastures with low and relatively high plant diversities in an integrated crop–livestock rotational grazing system (ICLS). As a comparison, organic permanent grasslands and conventionally managed maize were studied. We installed pitfall traps for three weeks in early summer, and for two weeks in autumn. In total, 11,347 carabid beetles of 66 species were caught. Grass–clover pastures did not differ in activity density, functional traits, habitat guilds, or species richness, but conventional maize did show a higher activity density in autumn and a higher proportion of eurytopic species and mobile species compared to grass–clover pastures. On grass–clover pastures, we found more endangered species, *Carabus* beetles, and a distinct carabid assemblage compared to maize. However, we attribute the lack of an effect of increased plant diversity of the grass–clover pastures on carabid species richness and functional traits to the intensive grazing regime, which resulted in the compositional and structural homogeneity of vegetation. Still, the presence of specialized and endangered species indicated the potential for organically managed grass–clover pastures to promote dispersal through an otherwise depleted and fragmented agricultural landscape. By increasing crop diversity in ICLS, more resources for foraging and nesting are created; therefore, organically managed grass–clover pastures add to the multi-functionality of agricultural landscapes.

Keywords: multi-species mixtures; agrobiodiversity; multifunctionality; carabid beetles; Carabidae; ecological intensification; grazing; dairy systems; ley grassland

Citation: Beye, H.; Taube, F.; Donath, T.W.; Schulz, J.; Hasler, M.; Diekötter, T. Species Enriched Grass–Clover Pastures Show Distinct Carabid Assemblages and Enhance Endangered Species of Carabid Beetles (Coleoptera: Carabidae) Compared to Continuous Maize. *Land* 2023, *12*, 736. https://doi.org/10.3390/land12040736

Academic Editors: Michael Vrahnakis, Yannis (Ioannis) Kazoglou and Manuel Pulido Fernádez

Received: 1 February 2023
Revised: 15 March 2023
Accepted: 16 March 2023
Published: 24 March 2023

1. Introduction

The rising demand of food worldwide has led to intensified land use, and the spatial and temporal homogenization of agricultural landscapes [1]. As a consequence of the associated habitat fragmentation and resource degradation, insect abundance and species richness is in decline [2,3]. In order to restore resource availability and diversity, and thereby biodiversity in agroecosystems, the (re)introduction of flowering plants [4,5], grazing cattle [6,7], and organic production systems [8–10] are discussed as potential measures to enhance land management. Here, we tested if plant species-enriched grass–clover pastures in organic dairy production can promote carabid beetle activity density, species richness, carabid assemblage, and functional traits.

Dairy systems range from intensive confinement systems to full-grazing systems, which vary, not only in access to pastures by cattle, but also in fertilizer input and the addition of supplementary food for the cattle [11]. Conventional confinement systems can operate without grazing, and incorporate intensively used grasslands and maize crops for fodder. As a result, monocultures of maize, for example, dominate in these systems until now. In contrast, ley systems include temporary grassland in a crop rotation and thereby enhance crop diversity. Ley systems support ecological intensification [11] by improving soil structure and health [12], N-cycling [13], and weed abundance [14]. Further, the type of production system and management strongly affects the species richness of many taxa, among them, insects such as bees [15] or carabids [16–18].

Carabid beetles are predators of soil insect pests and weed seeds; therefore, they provide biological control as a key ecosystem service in agroecosystems [19]. In addition, carabid beetles are themselves a food resource for higher trophic levels such as birds, and are an essential link in food webs [20,21]. Carabids respond to habitat changes with shifts in their community structure [22,23]. The composition of carabid beetles has been shown to shift towards medium-sized herbivorous species such as *Harpalus affinis* once arable land is converted to flowering fields, whereas small carnivorous species, such as *Bembidion*, decrease [22]. Herbivorous carabids also increased in organic winter spelt [24] and under organic management in wheat and meadows [25]. The management type may also affect mobility-related traits in carabids. Thus, intensively managed and disturbed habitats, such as maize fields, and intensively managed grasslands are often colonized by high-mobile species that are able to fly [17,26], while less mobile flightless species may be better supported by extensive, less disturbed land-use systems [26,27]. Due to these system- or management-type specific assemblages of carabid beetles, increased heterogeneity in agro-ecosystems has been shown to positively affect carabid richness [28,29].

Ley systems can increase carabid beetle species richness in comparison to cereal fields and pastures [30], although crop diversity in the landscape might be a precondition for the size of this effect [29]. Adding ley grasslands into arable crop rotations is one option to increase the heterogeneity of land-use types compared to specialized systems. To enhance the effect of crop diversification with ley systems, a rotational grazing system with cattle will increase environmental heterogeneity via selective grazing, trampling, and the release of dung [31,32]. While some studies found no effect of grazing on carabid beetle species [20,33] or trait diversity [17], others have shown beneficial effects of low (0.2 LSU/ha/year) grazing intensities [18,34,35]. While plant species richness was greater where grazing occurred, no effect of plant species richness was found in carabid assemblage [18], abundance, biomass, or species richness [20]. Plant community type within the semi-natural grasslands, however, was an effective predictor of carabid assemblage [36].

Here, we tested whether, in addition to climate- [37,38], water- [39], and other biodiversity-related [15] benefits, species-enriched grass–clover pastures enhance carabid beetle species richness in an organic ley farming system. We measured carabid beetle activity density, species richness, assemblage composition, and functional traits in (a) conventionally managed maize (CM), (b) organic grass–clover pastures (GC) with grazed and ungrazed management, (c) organic grass–clover pastures with herbs (GCH) with grazed and ungrazed management, and (d) organically managed permanent grassland (PG). We hypothesized that (i) GC and GCH support higher carabid beetle species richness, activity, and functional traits; (ii) they show a different carabid assemblage than CM or PG; (iii) higher plant species richness sown in GCH increases carabid beetle activity, and species richness and functional trait diversity; and (iv) ungrazed strips of GC and GCH support less mobile, flightless carabid beetles than the grazed pastures. We do, however, expect (v) higher carabid abundance and species richness on ungrazed strips as a result of increased plant species richness [15] compared to grazed pastures.

2. Materials and Methods

2.1. Study Region and Design

The study took place on the Lindhof experimental farm of Kiel University, Germany (54°27′ N; 9°57′ E) between 6 May and 16 September 2019. The mean annual temperature in the study area is 10.24 °C, and the mean annual precipitation is 745 mm. A crop rotation system has been in place since 2015, where in spring, grass–clover was sown and was used as pasture for 2–3 years, followed by successive annual cultures of oat (*Avena sativa*), potato, and winter wheat. In winter wheat, grass–clover is re-established to start the rotation. The organic grass–clover swards were sown in two mixtures: the binary grass–clover mixture (GC, n = 3) containing perennial ryegrass *Lolium perenne* and white clover *Trifolium repens*, and the grass–clover mixture with herbs (GCH, n = 3) containing perennial ryegrass *L. perenne*, white clover *T. repens*, red clover *Trifolium pratense*, birdsfoot trefoil *Lotus corniculatus*, chicory *Cichorium intybus*, plantain *Plantago lanceolata*, caraway *Carum carvi*, and sheep's burnet *Sanguisorba minor*. We assume that grass–clover pastures displayed a higher plant species richness, based on the initial seed mixture. The organic grass–clover pastures were rotationally stocked with Jersey cattle from April to September every 3–4 weeks for 1–3 days, with a stocking rate of 2.0 livestock units per hectare. The grass–clover fields were present in their first, second, and third years of usage. Because no replicates for the year of usage were present within mixtures, year was omitted as a factor in our study design, and the three sites per mixture were considered a random sample for this land-use type. To investigate the full potential of grass–clover swards without grazing, an area of 0.042 ha was excluded from grazing for each pasture. These ungrazed grass–clover strips (n = 3 for each of the two mixtures) were cut once on 20 August 2019. As an alternative to ley-pastures in dairy production, this study included organic permanent grasslands (PG, n = 3) at the Lindhof, with one cut per year and a less intense stocking rate of 1.2 livestock units per hectare, which are 20 years in age. In addition, conventional maize (CM) for fodder production for cattle in confinement systems was included in the study. CM (n = 3) of conventional farms in spatial proximity to the Lindhof were investigated. CM was fertilized with cow slurry at 40 m³/ha, Yara Mila NP 20/20 1.5 dt/ha, 40er potassium 1.5 dt/ha during seed drill, nitrogen 180 kg/ha, phosphorus 30 kg/ha, and potassium 170 kg/ha, and treated with herbicides (MaisTer Powder 0.9 L/ha, Aspect 0.9 L/ha). Harvest took place at the end of September.

2.2. Carabid Beetle Sampling

The grass–clover pastures (GC and GCH) were present in their first, second, and third years of production, totaling three sites per mixture. We installed three traps (triple) at three locations (nine traps in total) on each of the sites. The triples had a minimum distance of 350 m to each other, and the pitfall traps within the triple had a distance of 15 m to each other. On the ungrazed strips of the grass–clover pastures, three pitfall traps were installed. On PG and CM, we installed three pitfall traps per site. The pitfall traps were clear cups with a diameter of 10 cm and a volume of 500 mL [40], and they were filled with 50 mL of vinegar solution and a drop of unscented detergent. We used vinegar instead of ethylene glycol, as the study sites were frequently grazed and we wanted to prevent harm to the cattle. A wire mesh with a mesh size of 31 × 31 mm in the upper part of the traps prevented vertebrates from falling into the traps [40]. The traps were emptied once a week for three weeks beginning in May (6 May–9 July 2019), and for two weeks in September (30 August–17 September 2019). Species determination was performed [41] and supervised by an expert in the field. Carabid beetles were stored in ethanol at Kiel University. We identified endangered species according to the red list of Schleswig Holstein [42]. Habitat guilds were defined according to the catalogue supplied by "Gesellschaft für Angewandte Carabidologie" [43]. For our analyses, we selected eurytopic beetles, open-habitat beetles, and agrotopic beetles as the most typical for the studied habitat types. Eurytopic beetles occur across multiple habitat types including shaded areas, whereas open-habitat beetles occur in multiple habitat types without shading. Agrotopic beetles occur on croplands,

Table 1. *Cont.*

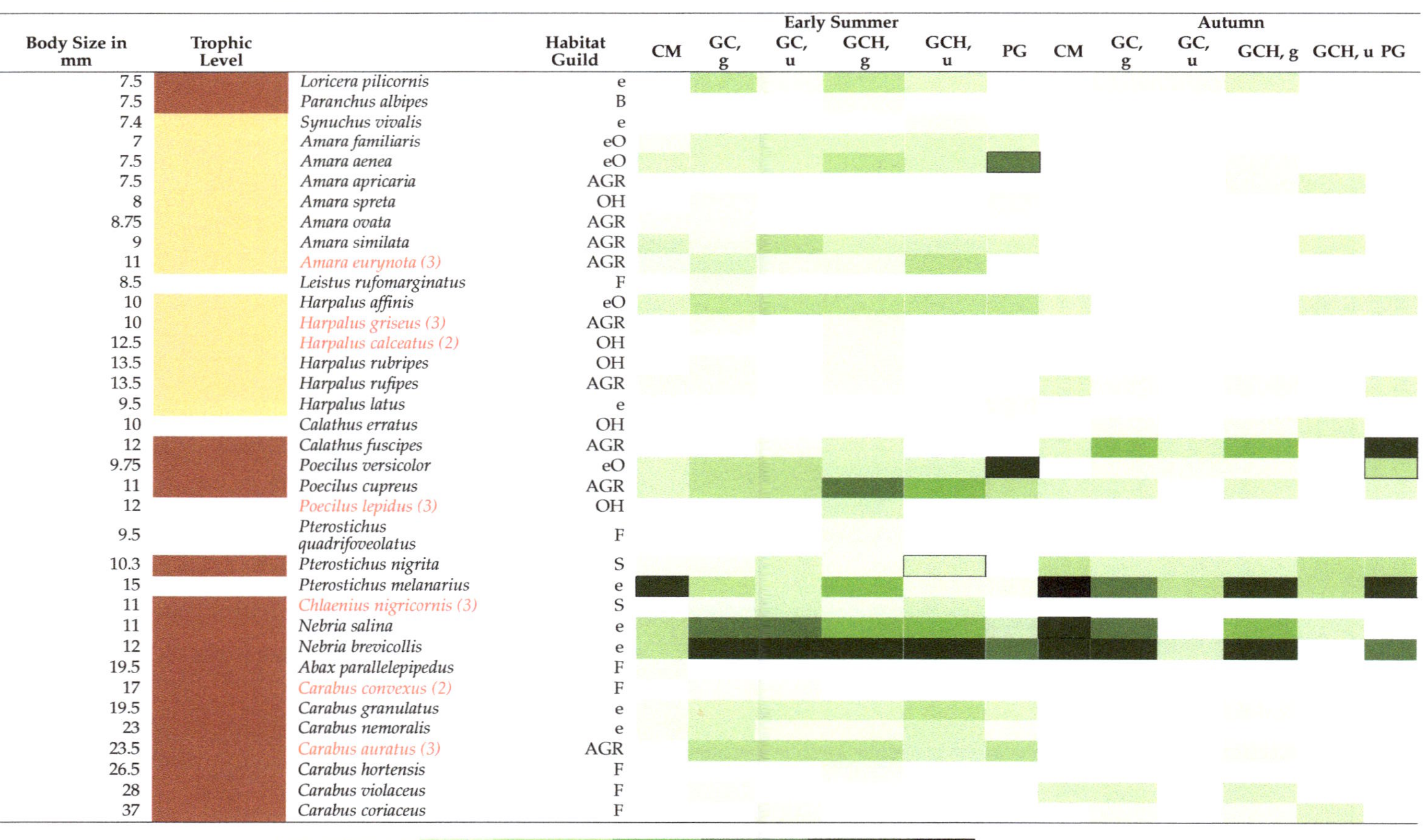

Body Size in mm	Trophic Level		Habitat Guild	Early Summer						Autumn					
				CM	GC, g	GC, u	GCH, g	GCH, u	PG	CM	GC, g	GC, u	GCH, g	GCH, u	PG
7.5		*Loricera pilicornis*	e												
7.5		*Paranchus albipes*	B												
7.4		*Synuchus vivalis*	e												
7		*Amara familiaris*	eO												
7.5		*Amara aenea*	eO												
7.5		*Amara apricaria*	AGR												
8		*Amara spreta*	OH												
8.75		*Amara ovata*	AGR												
9		*Amara similata*	AGR												
11		*Amara eurynota (3)*	AGR												
8.5		*Leistus rufomarginatus*	F												
10		*Harpalus affinis*	eO												
10		*Harpalus griseus (3)*	AGR												
12.5		*Harpalus calceatus (2)*	OH												
13.5		*Harpalus rubripes*	OH												
13.5		*Harpalus rufipes*	AGR												
9.5		*Harpalus latus*	e												
10		*Calathus erratus*	OH												
12		*Calathus fuscipes*	AGR												
9.75		*Poecilus versicolor*	eO												
11		*Poecilus cupreus*	AGR												
12		*Poecilus lepidus (3)*	OH												
9.5		*Pterostichus quadrifoveolatus*	F												
10.3		*Pterostichus nigrita*	S												
15		*Pterostichus melanarius*	e												
11		*Chlaenius nigricornis (3)*	S												
11		*Nebria salina*	e												
12		*Nebria brevicollis*	e												
19.5		*Abax parallelepipedus*	F												
17		*Carabus convexus (2)*	F												
19.5		*Carabus granulatus*	e												
23		*Carabus nemoralis*	e												
23.5		*Carabus auratus (3)*	AGR												
26.5		*Carabus hortensis*	F												
28		*Carabus violaceus*	F												
37		*Carabus coriaceus*	F												

Legend: >0–5 | >5–25 | >25–75 | >75–150 | >150–250 | >250–400 | >400

Grass–clover (GC) and grass–clover herbs (GCH) contained all seven threatened species, while with *Amara eurynota*, only one threatened species was found in conventionally managed maize (CM) (Table 1). Furthermore, GC and GCH promoted species typical for oligotrophic grasslands/heathland; these were not found on CM. All of the indicator species for CM were either eurytopic or open-habitat species (*Bembidion quadrimaculatum*, *Trechus quadristriatus*, *Bembidion tetracolum*, *Clivina fossor*, *Pterostichus melanarius*, and *Nebria salina*) while indicator species for GC and GCH were agrotopic or swampland species (*Acupalpus meridianus*, *Notiophilus substriatus*, and *Pterostichus nigrita*, Table 1 and Tables A1 and A2. All species found in CM were present on GC and GCH, except for *Abax parallelepipedus* (Table 1). Yet, the NMDS for early summer showed a distinct species composition of CM compared to GC and GCH, with no overlap. There are correlations with flying species, activity density, endangered species, and species richness. According to the NMDS, activity density was a strong predictor for CM, whereas GC and GCH were predicted by endangered species and flightless beetles. The final NMDS for the data in early summer had three dimensions, with a stress value of 14.116. The explanatory power was highest for the second (24.3%) and third (42.9%) axis compared to the first axis (19.7%). All of the grass–clover plots showed a larger overlap, indicating similar species composition (Figure 1). The NMDS revealed a distinction between grazed and ungrazed plots of GC and GCH in autumn, but a larger overlap of the grazed pastures and the CM fields compared to the NMDS in summer (Figure 2). The MRPP results verified this pattern, as it showed significant differences between GC and CM ($p < 0.001$), and GCH and CM ($p < 0.001$, Table 2). The final ordination for autumn had two dimensions, with an explanatory value of 32.1% for the first axis, 28.6% explanatory power of the second axis, and a stress value of 31.156. Similar to the results of the NMDS in summer, the activity density was a strong predictor of the CM fields. According to the MRPP, there were significant differences comparing GC to CM ($p < 0.001$) and PG ($p < 0.001$), as well as GCH to CM ($p < 0.001$) and PG ($p = 0.003$). We also found significant differences comparing the management; the grazed pastures of GCH were significantly different from the ungrazed stripes of GCH ($p < 0.001$, Table 3).

Table 2. Results of the multi-response permutation procedure (MRPP) for carabid assemblages in early summer. T is the test statistic calculating the difference between observed and expected delta, while A is the chance-corrected within-group agreement.

	T	A	*p*-Value
GC–GCH	0.018	−0.0001	0.375
GC–CM	−11.372	0.127	<0.001
GC–PG	−2.494	0.031	0.029
GCH–CM	−8.711	0.097	<0.001
GCH–PG	−1.826	0.022	0.059
GC grazed–GC ungrazed	−0.094	0.001	0.344
GCH grazed–GCH ungrazed	−0.846	0.001	0.160

Table 3. Results of the multi-response permutation procedure (MRPP) for carabid assemblage in fall, revealing significant differences comparing the grass–clover pastures GC and GCH to CM and PG. T is the test statistic calculating the difference between observed and expected delta, while A is the chance-corrected within-group agreement.

	T	A	*p*-Value
GC–GCH	−0.946	0.007	0.155
GC–CM	−8.941	0.133	<0.001
GC–PG	−7.746	0.085	<0.001
CM–GCH	−6.201	0.094	<0.001
PG–GCH	−4.054	0.049	0.003
GC grazed–GC ungrazed	−3.364	0.032	0.011
GCH grazed–GCH ungrazed	−5.365	0.067	<0.001

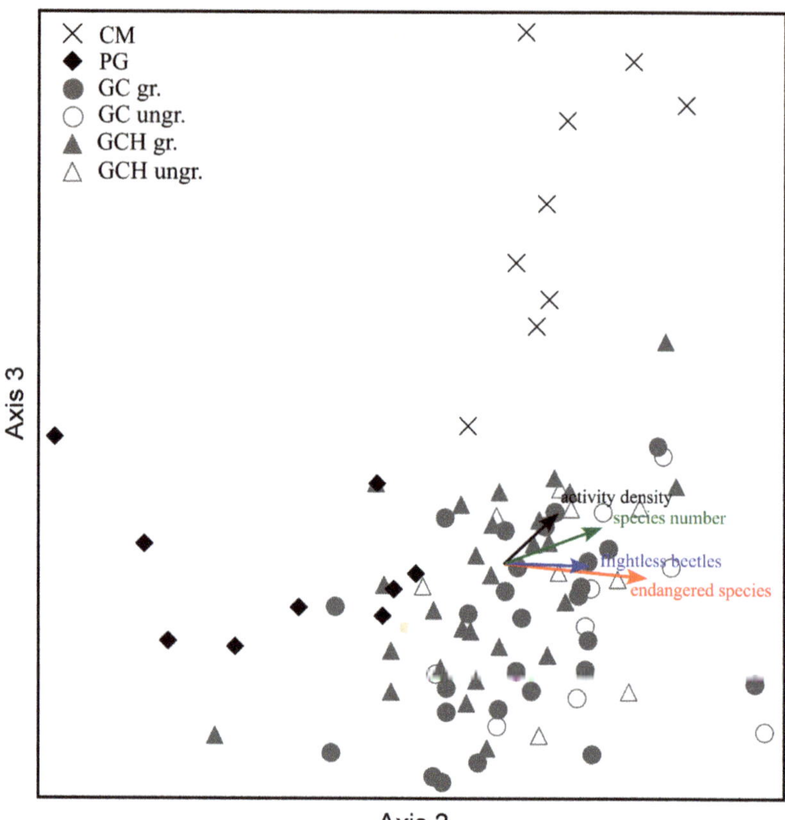

Figure 1. Non-metric multidimensional scaling (NMDS) of carabid assemblages in the different land-use systems (conventional maize CM, organic grass–clover GC, organic grass–clover herbs GCH, organic permanent grasslands PG) and management types (grazed gr., ungrazed ungr.) in early summer. We found correlations with beetle activity density, species richness, endangered species, and flightless species. Carabid assemblages of grass–clover pastures, irrespective of mixture and management, were distinct from the assemblages of CM.

There was no difference in the activity densities of the carabid beetles comparing CM to GC and GCH, in May and June. In CM, the overall activity density was, however, significantly higher ($p < 0.05$) compared to GC in autumn (Figure 3, Table A4). CM also showed a significantly higher activity density of eurytopic species compared to GC ($p < 0.05$) and GCH ($p < 0.05$, Table A5) in autumn, whereas in early summer, GCH showed a higher activity of open-habitat beetles compared to CM ($p < 0.05$, Figure 4, Table A6). Yet, CM showed a significantly higher species richness of open-habitat species compared to GC (early summer $p < 0.05$, autumn $p < 0.05$) and to GCH (early summer $p < 0.05$, Table A7). None of the treatments differed in activity density and species richness of agrotopic species (Tables A8 and A9), or in the species richness of eurytopic beetles (Table A10). On organic permanent grassland (PG), significantly more herbivorous beetles were found compared to GC ($p < 0.05$, Table A11), and on PG, significantly more flightless beetle species were present compared to GC ($p < 0.05$) and GCH ($p < 0.05$) in autumn (Figure 5, Table A12). In CM, we found a significantly higher activity density of flying carabid beetles compared to GC ($p < 0.05$) and GCH ($p < 0.05$, Table A13). Comparing GC and GCH, we did not find significant differences in activity density, carabid beetle biomass, habitat preferences,

Chao diversity index, the activity density of endangered carabids, or feeding behavior (herbivorous and carnivorous species, Tables A14–A17.

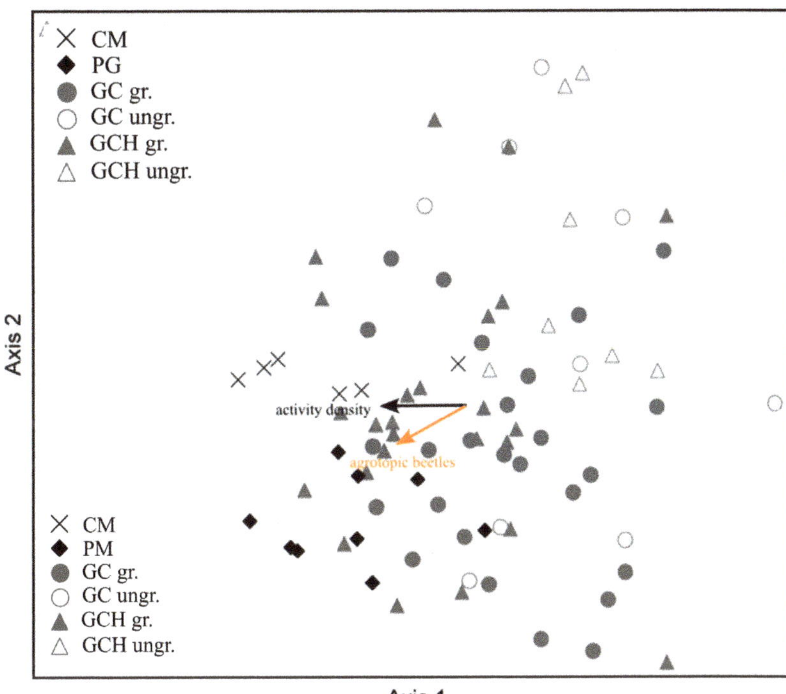

Figure 2. Non-metric multidimensional scaling (NMDS) of carabid beetle assemblages in the different land-use types (conventional maize CM, organic grass–clover GC, organic grass–clover herbs GCH, organic permanent grasslands PG) and management types (grazed gr., ungrazed ungr.) in autumn. We found correlations with activity density and agrotopic beetle activity density. Compared to the NMDS in early summer, there was less distinction between the carabid assemblages of CM and grass–clover pastures.

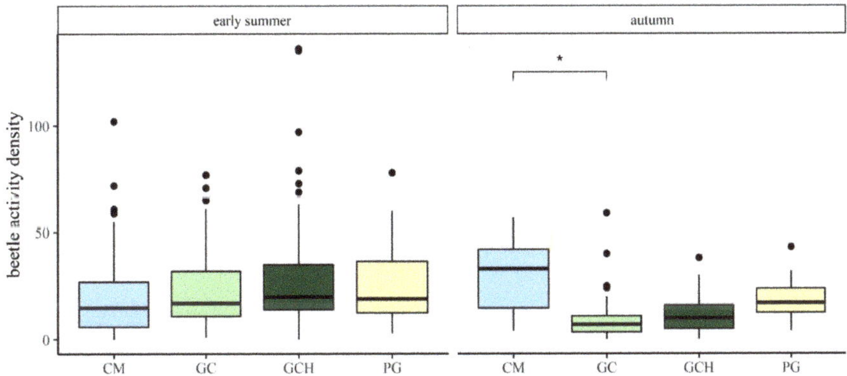

Figure 3. Beetle activity density of organic grass–clover (GC) and organic grass–clover herbs (GCH) in comparison to conventional maize (CM) and organic permanent grasslands (PG) in early summer and in autumn. *p*-values are indicated as * $p < 0.05$.

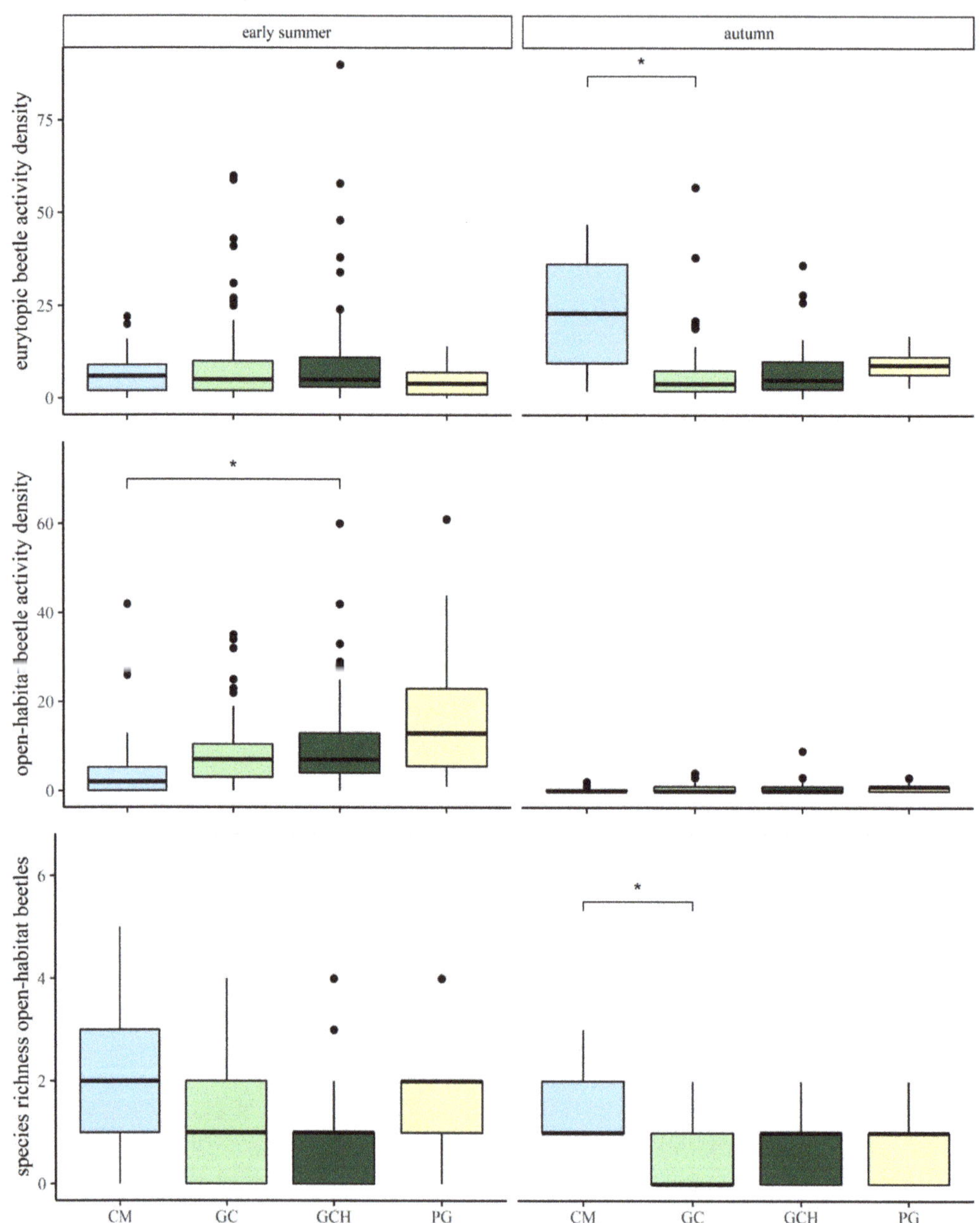

Figure 4. Beetle activity densities of eurytopic and open-habitat beetles, and species richness of open-habitat beetles of organic grass–clover (GC) and organic grass–clover herbs (GCH) in comparison to conventional maize (CM) and organic permanent grasslands (PG) in early summer and in autumn. *p*-values are indicated as * $p < 0.05$.

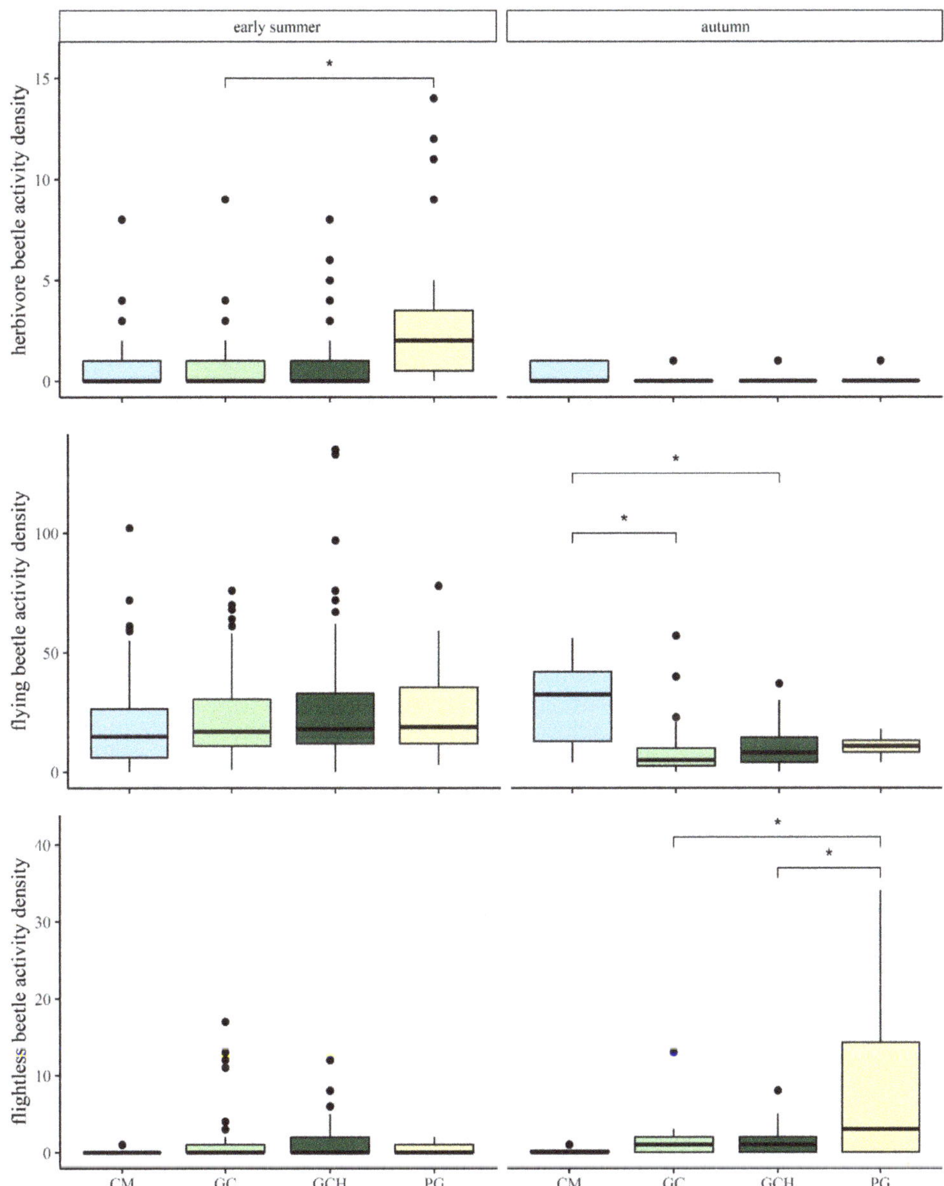

Figure 5. Trait-specific activity densities of herbivorous carabid beetles, and carabid beetles with a high mobility or low mobility of organic grass–clover (GC) and organic grass–clover herbs (GCH), in comparison to conventional maize (CM) and organic permanent grasslands (PG), in early summer and in autumn. *p*-values are indicated as * $p < 0.05$.

On the grazed pastures of GCH, significantly more eurytopic species were observed, compared to the ungrazed strips of GCH in autumn ($p < 0.05$, Figure 6, Table A18). We found no significant differences between grazed and ungrazed management in carabid beetle activity density, biomass, habitat guilds, endangered species, Chao diversity index, or functional traits (Tables A19–A31).

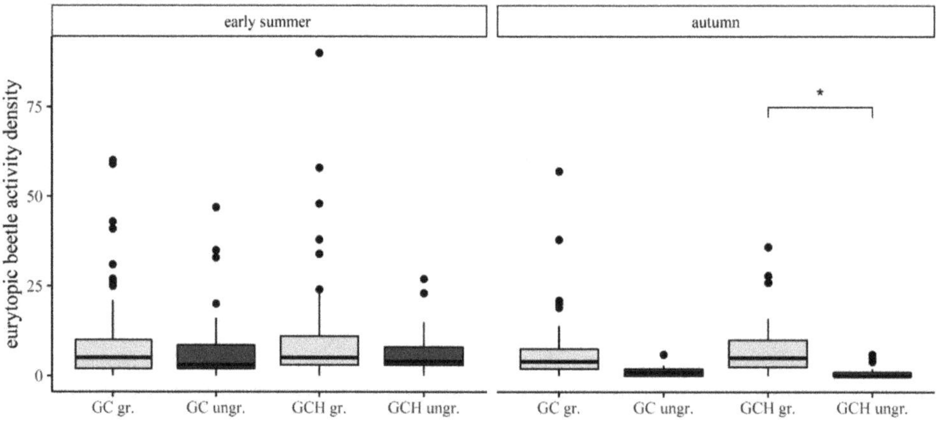

Figure 6. Eurytopic beetle activity densities of the different managements (grazed and ungrazed) and mixtures (grass–clover GC and grass–clover herbs GCH) in early summer and autumn. *p*-values are indicated as * $p < 0.05$.

4. Discussion

As part of studying the eco-efficiency of pasture-based milk production [37,39,53], we found organic grass–clover pastures to support more endangered carabid beetle species, and a seasonally distinct species composition on both grass–clover pastures in early summer compared to conventional maize (CM). CM fields offered a suitable habitat for eurytopic species and species with high mobility, especially in autumn, when vegetation cover was high. In maize, however, only one endangered species was found. Open-habitat species showed highest richness in CM, and some species were identified indicators of this habitat type, while their activity density was highest in grass–clover (GC) and grass–clover herbs (GCH) in early summer. While these results for open-habitat beetles are equivocal, the NMDS did show a distinct carabid assemblage in GC and GCH compared to CM, especially in early summer.

Although statistical analysis did not show a significantly higher activity density of endangered carabid beetles in GC and GCH, GC supported several endangered carabid beetle species in contrast to CM, with many frequently abundant eurytopic species [54]. Among the endangered species in GC and GCH, *Carabus auratus* is considered to be an indicator for organic agriculture. In Schleswig-Holstein, the species has almost exclusively been found in organic farms, and was shown to exponentially colonize organic crops after its conversion from conventional to organic management [55]. *Carabus* beetles in general, because of their low mobility, prefer stable vegetation structures and extensive grazing [18,26], explaining their generally higher activity density in the organic GC and GCH in this study. The CM fields are a less stable habitat, as vegetation cover is only present for a limited amount of time and they were harvested in late autumn, whereas GC and GCH were present for three subsequent years. Similarly, the carnivorous species *Poecilus lepidus* shows a low dispersal ability, and as an stenoecious xerophilic open-habitat species, it may thus prefer GC and GCH over meadows or maize. In contrast to *Carabus*, the two endangered species, *Harpalus griseus* and *Harpalus calceatus*, are herbivorous, feeding on plant seeds. With a sown grass–clover mixture of a maximum of seven plant species in GCH, including frequently utilized species such as *Cichorium intybus* [56] or *Plantago lanceolata* [57], it is likely that the diversity and availability of seeds were higher in GC and GCH than in CM fields [58]. *Chlaenius nigricornis* was most likely trapped on grass–clover pastures while it was dispersing to reach other habitats, as the species prefers wet habitat conditions. This was likely also the case for *Amara eurynota*, which was found on grass–clover fields and was the only endangered species found in CM as well.

GC and GCH showed the highest numbers of endangered species. No difference, though, was found between GC and GCH with respect to carabid activity density or species richness. This is in accordance with a previous study investigating a similar plant species mixture, which found no effect of a high diversity seed mixture on carabid beetle activity density, species richness, or biomass, compared to a low diversity seed mixture [20]. The lack of an effect of plant species richness, which was generally highest in the GCH in our study, may be due to the intensive grazing regime in these ley-grasslands. As a result, not many plant species set flowers or established well, and so the plant community was probably less heterogeneous than anticipated, a situation that was also observed in other studies [32,59]. As the permanent grasslands (PG) in this study were not rotationally stocked, the disturbance regime may actually have been lower than that of GC and GCH, as indicated by a higher abundance of flightless beetles, which agrees with other studies [26,27]. Even though GC and GCH were more permanent habitats than CM, their management changes every few years with the crop rotation, and no increase in less mobile species was found.

The effects of grazing on carabid beetles vary, as studies have shown that carabid abundance is increased on grazed sites [20] and systems grazed by sheep [34], as well as showing no effect of grazing to carabid abundance [20]. This may be attributed to differences in the grazing intensity and the studies' environmental contexts [60]. Most studies agree that moderate grazing benefits carabid richness [18] and the activity density of herbivorous [60] or less immobile (flightless) species [26], while heavy grazing reduces carabid richness [61], possible as a result of a more open and permeable vegetation structures on pastures. In our study, the high grazing intensity of 2.0 livestock units per hectare and per year most likely prevented the general benefits to carabid beetle species richness. Benefits to carabid richness have been previously observed at lower grazing intensities (0.2 LSU/ha/year). Unexpectedly, in this regard, activity density, species richness, or functional traits were similar in the ungrazed strips as compared to the pasture, despite a higher flower cover that benefited bumblebees [15]. Possibly, the dense vegetation cover in ungrazed strips restricted the movements of carabid species, many of which prefer bare ground due to a higher permeability [25].

Higher permeability may also be the reason for the higher carabid beetle activity density in CM in autumn. During our surveys in autumn, CM was still standing and offering vegetation cover for carabid beetles, while bare ground was also present as rows were separated by approximately 0.7 m. In addition to just bare ground, the higher looseness of ploughed soil and favorable microclimatic conditions in CM can increase beetle abundance [62], and soil temperature has possibly risen as a result of decreasing leaf area index, which favors, e.g., *Poecilus* and *Amara* [63]. Particularly mobile carabid beetles with a high colonization rate [17] potentially shifted from other habitats to maize in autumn, when habitat conditions became favorable [64]. Despite this attractiveness of maize to some carabid beetle species or during specific times of the year, several studies suggest that the habitat quality of maize fields is low and therefore lacking in carabid beetle species richness [65,66], and similar to our findings, this shows a low proportion of endangered species [65]. In future experiments, carabid activity density could be measured after harvest in autumn, as we suppose that the lack of vegetation cover might decrease activity density, proving further that maize production is a less favorable dairy system in comparison to crop–livestock systems, in terms of the promotion of carabid beetles.

The NMDS showed a shift in the carabid assemblage of GC and GCH compared to maize; however, the analysis of the habitat guilds that could clarify the direction of this shift was equivocal. The direction of this shift may have been unclear, because carabid assemblages with distinct functional traits develop over long periods, suggesting that with GC and GCH being present only for three subsequent years, this may not be long enough to develop an even more distinct assemblage. Changes in guild structure were previously found only after 10 years of grazing [67]. Grass–clover pastures in this study did not increase the species richness of carabid beetles or act as a key habitat structure. Implementing a regime with moderate grazing intensity (e.g., 0.2–1.4 livestock units per hectare) in the investigated system may be a suitable management for enhancing plant resource diversity, and consequently, carabid beetles. Nonetheless, in comparison to conventional confinement systems, integrated crop–livestock systems (ICLS) with crop rotation, rotational grazing, and multi-species pastures such as GCH diversify the agricultural landscape matrix, promote endangered carabid species, and indicate a shifted carabid assemblage. Despite the observed lack of a significant increase in carabid species richness in intensively grazed ley-systems, the presence of more specialized and endangered species may indicate their potential in promoting dispersal through fragmented landscapes. Further, crop diversity increases in ICLS, and more edge habitats are created, which provide habitat niches for nesting, foraging, and overwintering, and this has been shown to enhance carabid trait diversity [24] and Shannon diversity in landscapes that are rich in semi-natural habitats [28]. Improving the matrix quality in agricultural landscapes is essential for allowing species dispersal [68,69], which may be achieved with crop–livestock integrated grass–clover pastures. To solely focus on nature conservation efforts for protected habitats bears the risk of creating isolated habitat patches in an otherwise depleted landscape [69], which limits gene flow in carabid beetles [70], and therefore the long-term resilience of carabid populations. Therefore, ICLS with species enriched grass–clover pastures can help to support biodiversity in agricultural production systems, and in addition, it may also buffer protected areas from being isolated.

5. Conclusions

Reintroducing plant diversity and grazing in ICLS offers a new solution for dairy production that joins agricultural production with benefits for biodiversity, greenhouse gas emissions, and soil properties. Considering the large proportion of intensively used grasslands occupied worldwide, enhancing their plant diversity may have large-scale positive effects. An increased number of endangered species of carabid beetles, and a compositional shift in their assemblages in the species-enriched grass–clover pastures as compared to conventional maize indicates their potential for the promotion of heterogeneity and biodiversity in agricultural landscapes. Yet, in order to express their full potential to increase the species richness of carabid beetles, and particularly immobile and herbivorous species, a moderately reduced grazing regime would benefit plant diversity and flower cover, and thus, biodiversity in general. The less intense management regime may also promote solitary wild bees, as our previous study on the same grass–clover pastures found [15]. In addition to hosting more biodiversity itself, species-enriched grass–clover pastures also enhance the quality of the agricultural matrix, thereby promoting species dispersal and the associated ecosystem services in multifunctional agricultural landscapes. Instead of solely focusing nature conservation to a limited amount of protected areas, ICLS with species enriched grass–clover can counteract landscape fragmentation and facilitate carabid beetle exchange between protected habitats.

Author Contributions: Conceptualization, H.B., F.T. and T.D.; Methodology, H.B. and T.D.; Field work, H.B. and J.S.; Lab analysis, H.B. and J.S.; Data analysis, M.H., H.B., T.W.D. and T.D.; Writing—original draft preparation, H.B.; Writing—review and editing, T.W.D., T.D. and F.T.; Supervision, T.D. and F.T.; Funding acquisition, H.B. and F.T. All authors have read and agreed to the published version of the manuscript.

Funding: H.B. is supported by the Evangelisches Studienwerk Villigst foundation, under the research program: "Third Ways of Feeding the World" providing funding in the form of a doctoral scholarship. Scholarship number: 851227.

Data Availability Statement: The data presented in this study are available on request from the corresponding author.

Acknowledgments: We want to thank Alice Mockford and Christof Kluß for their help in editing the manuscript.

Conflicts of Interest: The authors declare no conflict of interest.

Appendix A

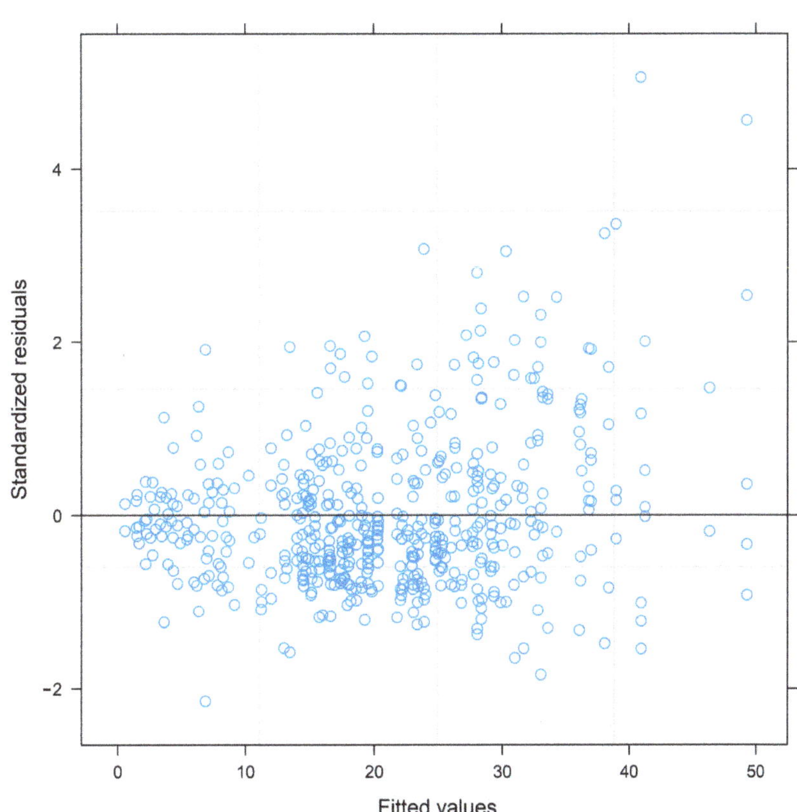

Figure A1. Residuals graph of the linear mixed effect model analyzing the carabid beetle activity density.

Table A1. Full species list of all caught ground beetles in total, and separated by the season they were caught in (May/June and September).

Carabidae	Carabid Individuals, Total	Carabid Individuals May–June	Carabid Individuals September
Abax parallelepipedus (Pill. and Mitt., 1783)	1	1	0
Acupalpus meridianus (L., 1761)	184	181	3
Agonum emarginatum (Gyll., 1827)	2	2	0
Agonum fuliginosum (Panzer, 1809)	1	1	0
Agonum muelleri (Herbst, 1784)	830	823	7
Agonum viduum (Panzer, 1796)	1	1	0
Amara aenea (De Geer, 1774)	147	146	1
Amara apricaria (Payk., 1790)	3	0	3
Amara eurynota (Panzer, 1796)	24	24	0
Amara familiaris (Duft., 1812)	27	27	0
Amara ovata (F., 1792)	3	3	0

Table A1. *Cont.*

Carabidae	Carabid Individuals, Total	Carabid Individuals May–June	Carabid Individuals September
Amara similata (Gyll., 1810)	56	55	1
Amara spreta Dejean, 1831	2	2	0
Anchomenus dorsalis (Pont., 1763)	389	388	1
Badister bullatus (Schrank, 1798)	9	9	0
Bembidion aeneum Germar, 1824	25	25	0
Bembidion guttula (Fabricius, 1792)	3	3	0
Bembidion lampros (Herbst, 1784)	2860	2774	86
Bembidion lunatum (Duftschmid, 1812)	17	388	1
Bembidion mannerheimii Sahlb., 1827	17	17	0
Bembidion obtusum Aud.-Serv., 1821	56	54	2
Bembidion properans (Steph., 1828)	238	226	12
Bembidion quadrimaculatum (L., 1761)	30	30	0
Bembidion tetracolum Say, 1823	731	724	7
Blemus discus (F., 1792)	10	2	8
Calathus erratus (Sahlb., 1827)	4	0	4
Calathus fuscipes (Goeze, 1777)	257	13	244
Carabus convexus F., 1775	2	2	0
Carabus coriaceus L., 1758	2	1	4
Carabus auratus L., 1761	197	196	1
Carabus granulatus L., 1758	71	70	1
Carabus hortensis L., 1758	2	2	0
Carabus nemoralis Müller, 1764	19	19	0
Carabus violaceus L., 1758	14	1	13
Chlaenius nigricornis (F., 1787)	14	14	0
Clivina fossor (L., 1758)	474	459	15
Demetrias atricapillus (L., 1758)	5	5	0
Dyschirius globosus (Herbst, 1784)	1	1	0
Epaphius secalis (Paykull, 1790)	1	1	0
Harpalus affinis (Schrank, 1781)	131	127	4
Harpalus calceatus (Duft. 1812)	3	3	0
Harpalus griseus (Panzer, 1796)	3	3	0
Harpalus rubripes (Duft., 1812)	3	3	0
Harpalus rufipes (De Geer, 1774)	10	5	5
Harpalus marginellus (Gyllenhal, 1827)	1	1	0
Leistus rufomarginatus (Duft., 1812)	1	1	0
Loricera pilicornis (F., 1775)	90	76	14
Microlestes minutulus (Goeze, 1777)	1	1	0
Nebria brevicollis (F., 1792)	1812	1422	390
Nebria salina (Fairm. and Lab., 1854)	708	511	197
Notiophilus biguttatus (F., 1779)	5	4	1
Notiophilus substriatus (G. R. Waterhouse, 1833)	66	65	1
Paranchus albipes (Fabricius, 1796)	1	1	0
Paradromius linearis (Ol., 1795)	1	1	0
Poecilus cupreus (L., 1758)	373	361	12
Poecilus lepidus (Leske, 1785)	7	7	0
Poecilus versicolor (Sturm, 1824)	165	156	9
Pterostichus melanarius (Ill., 1798)	888	404	484
Pterostichus nigrita (Payk., 1790)	53	13	40
Pterustichus strenuus (Panzer, 1796)	23	23	0
Pterostichus quadrifoveolatus (Letzner, 1852)	1	1	0
Pterostichus vernalis (Panzer, 1796)	13	11	2
Stomis pumicatus (Panzer, 1796)	8	8	0
Synuchus vivalis (Ill., 1798)	1	1	0
Trechoblemus micros (Herbst, 1784)	17	17	0
Trechus quadristriatus (Schrank, 1781)	230	16	214
Total	11.347	9.561	1.786

Table A2. Results of the indicator species analysis in May and June (CM: conventional maize, GC: grass–clover, GCH: grass–clover herbs, PG: permanent grasslands).

Species	Group	Indicator Value	*p*
Abax parallelepipedus	CM	2.0	0.4009
Acumenus meridianus	GC, ungrazed	15.9	0.0034
Agonum emarginatum	GCH, ungrazed	2.6	0.1696
Agonum fuliginosum	GCH, grazed	0.9	1.000
Agonum muelleri	PG	21.5	0.0030
Agonum viduum	GCH, ungrazed	2.6	0.1712
Amara aenea	PG	44.8	0.0002
Amara eurynota	GCH, ungrazed	7.3	0.0122
Amara familiaris	PG	4.9	0.1154
Amara ovata	CM	1.6	0.5709
Amara similata	GC, ungrazed	6.2	0.1136
Amara spreta	PG	3.0	0.1294
Anchomenus dorsalis	GC, ungrazed	20.1	0.0022
Badister bullatus	GC, grazed	1.6	0.5853
Bembidion aeneum	GCH, ungrazed	6.9	0.0250
Bembidion guttula	GCH, grazed	1.1	0.7572
Bembidion lampros	PG	18.7	0.2517
Bembidion lunatum	CM	3.3	0.1464
Bembidion mannerheimii	GCH, grazed	3.0	0.2731
Bembidion obtusum	GCH, ungrazed	6.8	0.0706
Bembidion properans	PG	29.4	0.0002
Bembidion quadrimaculatum	CM	30.9	0.0002
Bembidion tetracolum	CM	40.1	0.0002
Blemus discus	GCH, ungrazed	1.9	0.3037
Calathus fuscipes	GCH, grazed	4.6	0.0536
Carabus convexus	GC, ungrazed	1.7	0.4227
Carabus coriaceus	GC, ungrazed	2.4	0.2701
Carabus auratus	GC, grazed	7.4	0.2478
Carabus granulatus	GCH, ungrazed	8.0	0.0750
Carabus hortensis	GCH, grazed	1.7	0.5859
Carabus nemoralis	GCH, ungrazed	2.7	0.4153
Carabus violaceus	GC, grazed	0.9	0.6991
Chlaenius nigricornis	GCH, ungrazed	2.6	0.3369
Clivina fossor	CM	20.9	0.0022
Demetrias atricapillus	GCH, ungrazed	0.8	0.9628
Dyschirius globosus	GC, grazed	0.9	0.6891
Epaphius secalis	GCH, grazed	0.9	1.0000
Harpalus affinis	GC, ungrazed	5.9	0.7057
Harpalus calceatus	GCH, grazed	0.9	1.0000
Harpalus griseus	GCH, grazed	1.1	0.7590
Harpalus rubripes	GC, grazed	0.6	0.9162
Harpalus rufipes	CM	2.4	0.2318
Harpalus marginatus	PG	3.7	0.0678
Leistus rufomarginatus	GC, grazed	0.9	0.7019
Loricera pilicornis	GC, grazed	5.7	0.2134
Micros minutulus	GC, ungrazed	2.4	0.2757
Nebria brevicollis	GC, ungrazed	17.4	0.2360
Nebria salina	GC, grazed	14.0	0.1676
Notiophilus bigutattus	GC, ungrazed	0.9	0.8370
Notiophilus substratius	GC, grazed	9.8	0.0200
Paranchus albipes	GCH, grazed	0.9	1.0000
Paradromius linearis	GCH, grazed	0.9	1.0000
Poecilus cupreus	GCH, grazed	14.5	0.0180
Poecilus versicolor	PG	49.7	0.0002
Poecilus Lepidus	GCH, grazed	1.7	0.5449
Pterostichus melanarius	CM	33.4	0.0002
Pterostichus nigrita	GCH, ungrazed	6.5	0.0182

Table A2. *Cont.*

Species	Group	Indicator Value	*p*
Pterostichus strenuus	PG	10.7	0.0026
Pterostichus quadrifoveolatus	GCH, grazed	0.9	1.0000
Pterostichus vernalis	GCH, ungrazed	4.7	0.0724
Stomis pumicollis	GC, grazed	0.6	0.9844
Synuchus vivalis	GCH, ungrazed	2.6	0.1620
Trechus micros	PG	3.5	0.1856
Trechus quadristriatus	CM	11.1	0.0008

Table A3. Results of indicator species analysis in September (CM: conventional maize, GC: grass–clover, GCH: grass–clover herbs, PG: permanent grasslands).

Species	Group	Indicator Value	*p*
Acupalpus meridianus	CM	16.7	0.0064
Agonum muelleri	GCH, ungrazed	12.2	0.0404
Amara aenea	GCH, grazed	2.2	1.0000
Amara apricaria	GCH, ungrazed	11.5	0.0274
Amara similata	GCH, ungrazed	6.7	0.1688
Anchomenus dorsalis	CM	8.3	0.0808
Bembidion lampros	GCH, ungrazed	24.7	0.0102
Bembidion obtusum	CM	16.7	0.0048
Bembidion properans	GCH, ungrazed	6.7	0.2763
Bembidion tetracolum	CM	13.4	0.0210
Blemus discus	GCH, grazed	13.0	0.0296
Calathus [illegible]	GCH, ungrazed	5.0	0.2334
Calathus fuscipes	PG	51.2	0.0002
Carabus coriaceus	GCH, ungrazed	10.0	0.0582
Carabus auratus	GCH, grazed	2.2	1.0000
Carabus granulatus	GCH, grazed	2.2	1.0000
Carabus violaceus	GC, grazed	6.8	0.3025
Clivina fossor	GCH, ungrazed	7.8	0.2418
Harpalus affinis	CM	9.4	0.0638
Harpalus rufipes	CM	10.2	0.0566
Loricera pilicornis	GCH, grazed	12.1	0.0606
Nebria brevicollis	CM	29.5	0.0144
Nebria salina	CM	46.5	0.0002
Notiophilus bigutattus	GC, ungrazed	6.2	0.3963
Notiophilus substratius	GCH, grazed	2.2	1.0000
Poecilus cupreus	CM	5.9	0.3547
Poecilus versicolor	PG	17.7	0.0054
Pterostichus melanarius	CM	49.0	0.0002
Pterostichus nigrita	CM	10.7	0.1972
Pterostichus vernalis	PG	4.6	0.4217
Trechus quadristriatus	CM	49.1	0.0002

Table A4. Differences in the activity densities of carabid beetles, comparing grass–clover (GC) and grass–clover herbs (GCH) with each other, and in comparison to conventional maize (CM) and permanent grasslands (PG) in May/June and September.

	Estimate	SE	z-Value	*p*-Value
GCH–GC (September)	2.437	5.045	0.483	0.995
CM–GC (September)	20.577	6.213	3.312	0.041
PG–GC (September)	9.852	5.37	1.688	0.497
CM–GCH (September)	18.14	6.208	2.922	0.08
PG–GCH (September)	7.415	5.832	1.271	0.744
GCH–GC (May)	4.03	5.374	0.750	0.963

Table A4. *Cont.*

	Estimate	SE	z-Value	*p*-Value
CM–GC (May)	−0.27	6.032	−0.045	1.0
PG–GC (May)	3.437	6.278	0.547	0.991
CM–GCH (May)	−4.303	6.112	−0.704	0.972
PG–GCH (May)	−0.594	6.355	−0.093	1.0

Table A5. Differences in the activity densities of eurytopic carabid beetles, comparing grass–clover (GC) and grass–clover herbs (GCH) with each other, and in comparison to conventional maize (CM) and permanent grasslands (PG) in May/June and September. *p*-values are indicated as * *p* < 0.05.

	Estimate	SE	z-Value	*p*-Value
GCH–GC (September)	1.11	2.657	0.418	0.998
CM–GC (September)	16.71	4.850	3.445	0.033 *
PG–GC (September)	2.086	2.730	0.758	0.964
CM–GCH (September)	15.6	4.732	3.297	0.043 *
PG–GCH (September)	0.958	2.515	0.381	1.0
GCH–GC (May)	0.63	2.327	0.271	1.0
CM–GC (May)	−1.4489	2.299	−0.630	0.985
PG–GC (May)	−3.653	2.23	−1.638	0.535
CM–GCH (May)	−2.079	2.323	−0.895	0.928
PG–GCH (May)	−4.283	2.255	−1.899	0.39

Table A6. Differences in the activity densities of open-habitat carabid beetles, comparing grass–clover (GC) and grass–clover herbs (GCH) with each other, and in comparison to conventional maize (CM) and permanent grasslands (PG) in May/June and September. *p*-values are indicated as * *p* < 0.05.

	Estimate	SE	z-Value	*p*-Value
GCH–GC (September)	0.123	0.535	0.229	1.0
CM–GC (September)	−0.216	0.558	−0.386	0.999
PG–GC (September)	0.395	0.557	0.71	0.98
CM–GCH (September)	−0.338	0.577	−0.587	0.992
PG–GCH (September)	0.273	0.575	0.474	0.998
GCH–GC (May)	1.449	1.182	1.225	0.805
CM–GC (May)	−3.853	1.289	−2.989	0.076
PG–GC (May)	8.526	3.005	2.837	0.099
CM–GCH (May)	−5.301	1.393	−3.805	0.019 *
PG–GCH (May)	7.077	3.051	2.319	0.227

Table A7. Differences in the species richness of open-habitat carabid beetles, comparing grass–clover (GC) and grass–clover herbs (GCH) with each other, and in comparison to conventional maize (CM) and permanent grasslands (PG) in May/June and September. *p*-values are indicated as * *p* < 0.05.

	Estimate	SE	z-Value	*p*-Value
GCH–GC (September)	0.163	0.159	1.027	0.903
CM–GC (September)	0.897	0.239	3.745	0.021 *
PG–GC (September)	0.157	0.208	0.753	0.975
CM–GCH (September)	0.734	0.239	3.074	0.068
PG–GCH (September)	−0.007	0.207	−0.032	1.0
GCH–GC (May)	0.008	0.155	0.054	1.0
CM–GC (May)	0.944	0.238	3.964	0.014 *
PG–GC (May)	0.532	0.212	2.508	0.173
CM–GCH (May)	0.936	0.236	3.973	0.014 *
PG–GCH (May)	0.524	0.209	2.504	0.174

Table A8. Differences in activity densities of agrotopic carabid beetles, comparing grass–clover (GC) and grass–clover herbs (GCH) with each other, and in comparison to conventional maize (CM) and permanent grasslands (PG) in May/June and September.

	Estimate	SE	z-Value	p-Value
GCH–GC (September)	0.499	0.672	0.742	0.974
CM–GC (September)	−0.123	0.781	−0.157	1.0
PG–GC (September)	7.161	2.748	2.606	0.141
CM–GCH (September)	−0.622	0.766	−0.811	0.96
PG–GCH (September)	6.662	2.744	2.428	0.188
GCH–GC (May)	1.751	0.896	1.955	0.375
CM–GC (May)	0.017	1.542	0.011	1.0
PG–GC (May)	−0.489	1.068	−0.458	0.998
CM–GCH (May)	−1.734	1.579	−1.098	0.863
PG–GCH (May)	−2.24	1.122	−1.996	0.355

Table A9. Differences in the species richness of agrotopic carabid beetles, comparing grass–clover (GC) and grass–clover herbs (GCH) with each other, and in comparison to conventional maize (CM) and permanent grasslands (PG) in May/June and September.

	Estimate	SE	z-Value	p-Value
GCH–GC (September)	0.136	0.42	0.323	1.0
CM–GC (September)	0.359	0.482	0.745	0.958
PG–GC (September)	0.195	0.461	0.422	0.997
CM–GCH (September)	0.223	0.485	0.460	0.996
PG–GCH (September)	0.059	0.465	0.127	1.0
GCH–GC (May)	0.076	0.100	0.194	1.0
CM–GC (May)	−1.102	0.455	−2.425	0.173
PG–GC (May)	0.354	0.458	0.773	0.951
CM–GCH (May)	−1.178	0.456	−2.581	0.135
PG–GCH (May)	0.278	0.460	0.605	0.983

Table A10. Differences in the species richness of eurytopic carabid beetles, comparing grass–clover (GC) and grass–clover herbs (GCH) with each other, and in comparison to conventional maize (CM) and permanent grasslands (PG) in May/June and September.

	Estimate	SE	z-Value	p-Value
GCH–GC (September)	0.126	0.459	0.275	1.0
CM–GC (September)	1.097	0.485	2.262	0.216
PG–GC (September)	−0.008	0.473	−0.016	1.0
CM–GCH (September)	0.97	0.486	1.997	0.317
PG–GCH (September)	−0.134	0.475	−0.283	1.0
GCH–GC (May)	0.119	0.437	0.273	1.0
CM–GC (May)	−0.301	0.452	−0.666	0.97
PG–GC (May)	−0.177	0.484	−0.365	0.998
CM–GCH (May)	−0.42	0.452	−0.931	0.891
PG–GCH (May)	−0.296	0.484	−0.612	0.98

Table A11. Differences in activity densities of herbivore carabid beetles, comparing grass–clover (GC) and grass–clover herbs (GCH) with each other, and in comparison to conventional maize (CM) and permanent grasslands (PG) in May/June and September. p-values are indicated as * $p < 0.05$.

	Estimate	SE	z-Value	p-Value
GCH–GC (September)	−0.0001	0.445	0.000	1.0
CM–GC (September)	0.214	0.595	0.36	0.999
PG–GC (September)	0.117	0.856	0.136	1.0
CM–GCH (September)	0.215	0.601	0.357	0.999

Table A11. *Cont.*

	Estimate	SE	z-Value	p-Value
PG–GCH (September)	0.117	0.86	0.136	1.0
GCH–GC (May)	0.267	0.411	0.65	0.982
CM–GC (May)	0.11	0.452	0.243	1.0
PG–GC (May)	2.472	0.698	3.543	0.028 *
CM–GCH (May)	−0.157	0.454	−0.347	0.999
PG–GCH (May)	2.205	0.699	3.153	0.055

Table A12. Differences in activity densities of flightless/immobile carabid beetles, comparing grass–clover (GC) and grass–clover herbs (GCH) with each other, and in comparison to conventional maize (CM) and permanent grasslands (PG) in May/June and September. *p*-values are indicated as * $p < 0.05$.

	Estimate	SE	z-Value	p-Value
GCH–GC (September)	0.234	0.570	0.41	0.998
CM–GC (September)	−0.999	0.517	−1.931	0.37
PG–GC (September)	6.837	1.682	4.065	0.011 *
CM–GCH (September)	−1.232	0.47	−2.623	0.13
PG–GCH (September)	6.603	1.668	3.959	0.013 *
GCH–GC (May)	0.069	0.467	0.148	1.0
CM–GC (May)	−0.964	0.439	−2.194	0.255
PG–GC (May)	−0.496	1.309	−0.379	0.999
CM–GCH (May)	−1.033	0.416	−2.482	0.163
PG–GCH (May)	−0.595	1.301	−0.434	0.998

Table A13. Differences in activity densities of flying/mobile carabid beetles, comparing grass–clover (GC) and grass–clover herbs (GCH) with each other, and in comparison to conventional maize (CM) and permanent grasslands (PG) in May/June and September. *p*-values are indicated as * $p < 0.05$.

	Estimate	SE	z-Value	p-Value
GCH–GC (September)	2.327	4.677	0.5	0.995
CM–GC (September)	21.521	5.906	3.644	0.023 *
PG–GC (September)	3.356	5.296	0.634	0.983
CM–GCH (September)	19.194	5.901	3.252	0.046 *
PG–GCH (September)	1.029	5.291	0.194	1.0
GCH–GC (May)	3.784	5.020	0.754	0.963
CM–GC (May)	0.509	5.681	0.09	1.0
PG–GC (May)	3.85	5.954	0.647	0.982
CM–GCH (May)	−3.275	5.781	−0.566	0.99
PG–GCH (May)	0.066	6.049	−0.011	1.0

Table A14. Differences in the biomasses of carabid beetles, comparing grass–clover (GC) and grass–clover herbs (GCH) with each other, and in comparison to conventional maize (CM) and permanent grasslands (PG) in May/June and September.

	Estimate	SE	z-Value	p-Value
GCH–GC (September)	19.1	85.34	0.224	1.0
CM–GC (September)	407.78	128.9	3.163	0.054
PG–GC (September)	150.94	92.7	1.628	0.538
CM–GCH (September)	388.67	126.76	3.066	0.064
PG–GCH (September)	131.84	89.69	1.47	0.633
GCH–GC (May)	29.8	86.74	0.344	0.999
CM–GC (May)	−139.31	86.29	−1.615	0.546
PG–GC (May)	−95.7	87.86	−1.089	0.848
CM–GCH (May)	−169.12	81.22	−2.082	0.303
PG–GCH (May)	−125.5	82.89	−1.514	0.606

Table A15. Differences in the Chao diversity index of carabid beetles, comparing grass–clover (GC) and grass–clover herbs (GCH) with each other, and in comparison to conventional maize (CM) and permanent grasslands (PG) in May/June and September.

	Estimate	SE	z-Value	*p*-Value
GCH–GC (September)	−0.337	0.460	−0.733	0.978
CM–GC (September)	1.031	1.203	0.857	0.955
PG–GC (September)	−0.803	0.476	−1.685	0.541
CM–GCH (September)	1.369	1.18	1.16	0.846
PG–GCH (September)	−0.465	0.416	−1.117	0.866
GCH–GC (May)	0.401	0.452	0.887	0.948
CM–GC (May)	−0.533	0.46	−1.159	0.846
PG–GC (May)	0.275	0.697	0.395	0.999
CM–GCH (May)	−0.934	0.501	−1.863	0.438
PG–GCH (May)	−0.125	0.725	−0.173	1.0

Table A16. Differences in activity densities of red list carabid beetles, comparing grass–clover (GC) and grass–clover herbs (GCH) with each other, and in comparison to conventional maize (CM) and permanent grasslands (PG) in May/June and September.

	Estimate	SE	z-Value	*p*-Value
GCH–GC (September)	−0.065	0.586	−0.111	1.0
CM–GC (September)	0.017	0.553	0.03	1.0
PG–GC (September)	−0.033	0.563	−0.059	1.0
CM–GCH (September)	0.081	0.515	0.158	1.0
PG–GCH (September)	0.032	0.526	0.06	1.0
GCH–GC (May)	−0.048	0.52	−0.092	1.0
CM–GC (May)	−0.823	0.506	−1.627	0.485
PG–GC (May)	−0.526	0.512	−1.027	0.829
CM–GCH (May)	−0.775	0.489	−1.586	0.508
PG–GCH (May)	−0.478	0.495	−0.966	0.858

Table A17. Differences in activity densities of carnivore carabid beetles, comparing grass–clover (GC) and grass–clover herbs (GCH) with each other, and in comparison to conventional maize (CM) and permanent grasslands (PG) in May/June and September.

	Estimate	SE	z-Value	*p*-Value
GCH–GC (September)	0.176	4.136	0.042	1.0
CM–GC (September)	9.871	5.105	1.934	0.377
PG–GC (September)	4.609	5.09	0.905	0.927
CM–GCH (September)	9.695	5.08	1.907	0.39
PG–GCH (September)	4.433	5.07	0.874	0.936
GCH–GC (May)	3.152	4.549	0.693	0.977
CM–GC (May)	−2.343	5.067	−0.462	0.997
PG–GC (May)	1.503	5.487	0.274	1.0
CM–GCH (May)	−5.494	5.178	−1.061	0.866
PG–GCH (May)	−1.648	5.589	−0.295	1.0

Table A18. Differences in activity densities of eurytopic carabid beetles, comparing the grazed pastures and ungrazed strips of the grass–clover (GC) and the grass–clover herbs (GCH) in May/June and September. *p*-values are indicated as * $p < 0.05$.

	Estimate	SE	z-Value	*p*-Value
GC ungr.–GC gr. (September)	−5.675	2.632	−2.157	0.156
GCH ungr.–GCH gr. (September)	−6.8	2.407	−2.825	0.0495 *
GC ungr.–GC gr. (May)	−0.824	2.776	−0.297	0.994
GCH ungr.–GCH gr. (May)	−2.484	2.456	−1.013	0.715

Table A19. Differences in beetle activity densities, comparing the grazed pastures and ungrazed strips of the grass–clover (GC) and the grass–clover herbs (GCH) in May/June and September.

	Estimate	SE	z-Value	*p*-Value
GC ungr.–GC gr. (September)	−6.017	5.656	−1.064	0.677
GCH ungr.–GCH gr. (September)	−7.619	5.674	−1.343	0.502
GC ungr.–GC gr. (May)	1.941	6.054	0.321	0.991
GCH ungr.–GCH gr. (May)	−4.186	6.076	−0.689	0.891

Table A20. Differences in biomasses, comparing the grazed pastures and ungrazed strips of the grass–clover (GC) and the grass–clover herbs (GCH) in May/June and September.

	Estimate	SE	z-Value	*p*-Value
GC ungr.–GC gr. (September)	−193.86	84.52	−2.294	0.126
GCH ungr.–GCH gr. (September)	−155.67	90.12	−1.727	0.308
GC ungr.–GC gr. (May)	−45.96	106.03	−0.433	0.977
GCH ungr.–GCH gr. (May)	−158.67	84.67	−1.874	0.248

Table A21. Differences in activity densities of open-habitat carabid beetles, comparing the grazed pastures and ungrazed strips of the grass–clover (GC) and the grass–clover herbs (GCH) in May/June and September.

	Estimate	SE	z-Value	*p*-Value
GC ungr.–GC gr. (September)	0.468	0.603	0.776	0.889
GCH ungr.–GCH gr. (September)	0.738	0.586	1.259	0.612
GC ungr.–GC gr. (May)	1.375	1.521	0.904	0.827
GCH ungr.–GCH gr. (May)	−1.392	1.385	−1.005	0.77

Table A22. Differences in activity densities of agrotopic carabid beetles, comparing the grazed pastures and ungrazed strips of the grass–clover (GC) and the grass–clover herbs (GCH) in May/June and September.

	Estimate	SE	z-Value	*p*-Value
GC ungr.–GC gr. (September)	−0.971	0.73	−1.330	0.543
GCH ungr.–GCH gr. (September)	−1.306	0.717	−1.821	0.282
GC ungr.–GC gr. (May)	1.135	1.098	1.034	0.73
GCH ungr.–GCH gr. (May)	−1.313	1.161	−1.131	0.669

Table A23. Differences in species richness of eurytopic carabid beetles, comparing the grazed pastures and ungrazed strips of the grass–clover (GC) and the grass–clover herbs (GCH) in May/June and September.

	Estimate	SE	z-Value	*p*-Value
GC ungr.–GC gr. (September)	1.128	0.481	−2.343	0.11
GCH ungr.–GCH gr. (September)	−1.253	0.484	−2.59	0.072
GC ungr.–GC gr. (May)	−0.534	0.472	−1.131	0.621
GCH ungr.–GCH gr. (May)	−0.064	0.488	−0.131	1.0

Table A24. Differences in species richness of open-habitat carabid beetles, comparing the grazed pastures and ungrazed strips of the grass–clover (GC) and the grass–clover herbs (GCH) in May/June and September.

	Estimate	SE	z-Value	*p*-Value
GC ungr.–GC gr. (September)	0.042	0.228	0.186	1.0
GCH ungr.–GCH gr. (September)	0.264	0.243	1.085	0.728
GC ungr.–GC gr. (May)	0.007	0.197	0.035	1.0
GCH ungr.–GCH gr. (May)	−0.022	0.176	−0.126	1.0

Table A25. Differences in species richness of agrotopic carabid beetles, comparing the grazed pastures and ungrazed strips of the grass–clover (GC) and the grass–clover herbs (GCH) in May/June and September.

	Estimate	SE	z-Value	*p*-Value
GC ungr.–GC gr. (September)	0.132	0.441	0.298	0.994
GCH ungr.–GCH gr. (September)	0.388	0.456	0.851	0.81
GC ungr.–GC gr. (May)	0.107	0.472	0.226	0.998
GCH ungr.–GCH gr. (May)	−0.172	0.469	−0.366	0.987

Table A26. Differences in Chao diversity index, comparing the grazed pastures and ungrazed strips of the grass–clover (GC) and the grass–clover herbs (GCH) in May/June and September.

	Estimate	SE	z-Value	*p*-Value
GC ungr.–GC gr. (September)	[illegible]	[illegible]	[illegible]	[illegible]
GCH ungr.–GCH gr. (September)	0.2566	0.7578	0.339	0.994
GC ungr.–GC gr. (May)	−0.4892	0.5385	−0.908	0.832
GCH ungr.–GCH gr. (May)	−0.3111	0.6663	−0.467	0.981

Table A27. Differences in activity densities of red list carabid beetles, comparing the grazed pastures and ungrazed strips of the grass–clover (GC) and the grass–clover herbs (GCH) in May/June and September.

	Estimate	SE	z-Value	*p*-Value
GC ungr.–GC gr. (September)	0.002	0.667	0.004	1.0
GCH ungr.–GCH gr. (September)	0.047	0.568	0.083	1.0
GC ungr.–GC gr. (May)	−0.148	0.557	−0.265	0.996
GCH ungr.–GCH gr. (May)	−0.217	0.517	−0.420	0.979

Table A28. Differences in activity densities of herbivore carabid beetles, comparing the grazed pastures and ungrazed strips of the grass–clover (GC) and the grass–clover herbs (GCH) in May/June and September.

	Estimate	SE	z-Value	*p*-Value
GC ungr.–GC gr. (September)	−0.043	0.538	−0.079	1.0
GCH ungr.–GCH gr. (September)	0.183	0.508	0.359	0.989
GC ungr.–GC gr. (May)	0.314	0.465	0.676	0.904
GCH ungr.–GCH gr. (May)	−0.045	0.458	−0.098	1.0

Table A29. Differences in activity densities of herbivore carabid beetles, comparing the grazed pastures and ungrazed strips of the grass–clover (GC) and the grass–clover herbs (GCH) in May/June and September.

	Estimate	SE	z-Value	p-Value
GC ungr.–GC gr. (September)	−5.179	4.710	−1.1	0.672
GCH ungr.–GCH gr. (September)	−4.742	4.711	−1.006	0.731
GC ungr.–GC gr. (May)	2.302	5.191	0.443	0.977
GCH ungr.–GCH gr. (May)	−3.073	5.233	−0.587	0.94

Table A30. Differences in activity densities of flightless/immobile carabid beetles, comparing the grazed pastures and ungrazed strips of the grass–clover (GC) and the grass–clover herbs (GCH) in May/June and September.

	Estimate	SE	z-Value	p-Value
GC ungr.–GC gr. (September)	−1.037	0.652	−1.591	0.387
GCH ungr.–GCH gr. (September)	−1.359	0.503	−2.701	0.065
GC ungr.–GC gr. (May)	−0.215	0.505	−0.425	0.982
GCH ungr.–GCH gr. (May)	−0.467	0.436	−1.069	0.703

Table A31. Differences in activity densities of flying/mobile carabid beetles, comparing the grazed pastures and ungrazed strips of the grass–clover (GC) and the grass–clover herbs (GCH) in May/June and September.

	Estimate	SE	z-Value	p-Value
GC ungr.–GC gr. (September)	−4.83	5.275	−0.196	0.773
GCH ungr.–GCH gr. (September)	−6.375	5.275	−1.208	0.589
GC ungr.–GC gr. (May)	1.9	5.657	0.336	0.99
GCH ungr.–GCH gr. (May)	−3.834	5.708	−0.672	0.901

References

1. Hooftman, D.A.P.; Bullock, J.M. Mapping to Inform Conservation. *Biol. Conserv.* **2012**, *145*, 30–38. [CrossRef]
2. Hallmann, C.A.; Sorg, M.; Jongejans, E.; Siepel, H.; Hofland, N.; Schwan, H.; Stenmans, W.; Müller, A.; Sumser, H.; Hörren, T. More than 75 Percent Decline over 27 Years in Total Flying Insect Biomass in Protected Areas. *PLoS ONE* **2017**, *12*, e0185809. [CrossRef] [PubMed]
3. Homburg, K.; Drees, C.; Boutaud, E.; Nolte, D.; Schuett, W.; Zumstein, P.; von Ruschkowski, E.; Assmann, T. Where Have All the Beetles Gone? Long-term Study Reveals Carabid Species Decline in a Nature Reserve in Northern Germany. *Insect Conserv. Divers.* **2019**, *12*, 268–277. [CrossRef]
4. Bommarco, R.; Lindborg, R.; Marini, L.; Öckinger, E. Extinction Debt for Plants and Flower-visiting Insects in Landscapes with Contrasting Land Use History. *Divers. Distrib.* **2014**, *20*, 591–599. [CrossRef]
5. Isbell, F.; Adler, P.R.; Eisenhauer, N.; Fornara, D.; Kimmel, K.; Kremen, C.; Letourneau, D.K.; Liebman, M.; Polley, H.W.; Quijas, S. Benefits of Increasing Plant Diversity in Sustainable Agroecosystems. *J. Ecol.* **2017**, *105*, 871–879. [CrossRef]
6. Enri, S.R.; Probo, M.; Farruggia, A.; Lanore, L.; Blanchetete, A.; Dumont, B. A Biodiversity-Friendly Rotational Grazing System Enhancing Flower-Visiting Insect Assemblages While Maintaining Animal and Grassland Productivity. *Agric. Ecosyst. Environ.* **2017**, *241*, 1–10. [CrossRef]
7. Teague, R.; Barnes, M. Grazing Management That Regenerates Ecosystem Function and Grazingland Livelihoods. *Afr. J. Range Forage Sci.* **2017**, *34*, 77–86. [CrossRef]
8. Bengtsson, J.; Ahnström, J.; Weibull, A. The Effects of Organic Agriculture on Biodiversity and Abundance: A Meta-analysis. *J. Appl. Ecol.* **2005**, *42*, 261–269. [CrossRef]
9. Hole, D.G.; Perkins, A.J.; Wilson, J.D.; Alexander, I.H.; Grice, P.V.; Evans, A.D. Does Organic Farming Benefit Biodiversity? *Biol. Conserv.* **2005**, *122*, 113–130. [CrossRef]
10. Rundlöf, M.; Smith, H.G.; Birkhofer, K. Effects of Organic Farming on Biodiversity. *ELS* **2016**, *243*, 19–26.
11. Rockström, J.; Steffen, W.; Noone, K.; Persson, Å.; Chapin III, F.S.; Lambin, E.; Lenton, T.M.; Scheffer, M.; Folke, C.; Schellnhuber, H.J. Planetary Boundaries: Exploring the Safe Operating Space for Humanity. *Ecol. Soc.* **2009**, *14*, 32. [CrossRef]
12. Martin, G.; Durand, J.-L.; Duru, M.; Gastal, F.; Julier, B.; Litrico, I.; Louarn, G.; Mediene, S.; Moreau, D.; Valentin-Morison, M. Role of Ley Pastures in Tomorrow's Cropping Systems. A Review. *Agron. Sustain. Dev.* **2020**, *40*, 17. [CrossRef]

13. Reinsch, T.; Malisch, C.; Loges, R.; Taube, F. Nitrous Oxide Emissions from Grass–Clover Swards as Influenced by Sward Age and Biological Nitrogen Fixation. *Grass Forage Sci.* **2020**, *75*, 372–384. [CrossRef]
14. Schuster, M.Z.; Gastal, F.; Doisy, D.; Charrier, X.; De Moraes, A.; Médiène, S.; Barbu, C.M. Weed Regulation by Crop and Grassland Competition: Critical Biomass Level and Persistence Rate. *Eur. J. Agron.* **2020**, *113*, 125963. [CrossRef]
15. Beye, H.; Taube, F.; Lange, K.; Hasler, M.; Kluß, C.; Loges, R.; Diekötter, T. Species-Enriched Grass-Clover Mixtures Can Promote Bumblebee Abundance Compared with Intensively Managed Conventional Pastures. *Agronomy* **2022**, *12*, 1080. [CrossRef]
16. Gallé, R.; Happe, A.; Baillod, A.B.; Tscharntke, T.; Batáry, P. Landscape Configuration, Organic Management, and Within-field Position Drive Functional Diversity of Spiders and Carabids. *J. Appl. Ecol.* **2019**, *56*, 63–72. [CrossRef]
17. Gobbi, M.; Fontaneto, D.; Bragalanti, N.; Pedrotti, L.; Lencioni, V. *Carabid Beetle (Coleoptera: Carabidae) Richness and Functional Traits in Relation to Differently Managed Grasslands in the Alps*; Taylor & Francis: Abingdon, UK, 2015; Volume 51, pp. 52–59.
18. Lyons, A.; Ashton, P.A.; Powell, I.; Oxbrough, A. Impacts of Contrasting Conservation Grazing Management on Plants and Carabid Beetles in Upland Calcareous Grasslands. *Agric. Ecosyst. Environ.* **2017**, *244*, 22–31. [CrossRef]
19. Letourneau, D.K.; Jedlicka, J.A.; Bothwell, S.G.; Moreno, C.R. Effects of Natural Enemy Biodiversity on the Suppression of Arthropod Herbivores in Terrestrial Ecosystems. *Annu. Rev. Ecol. Evol. Syst.* **2009**, *40*, 573. [CrossRef]
20. Waite, E.S.; Houseman, G.R.; Jensen, W.E.; Reichenborn, M.M.; Jameson, M.L. Ground Beetle (Coleoptera: Carabidae) Responses to Cattle Grazing, Grassland Restoration, and Habitat across a Precipitation Gradient. *Insects* **2022**, *13*, 696. [CrossRef]
21. Larochelle, A.; Larivière, M.-C. *A Natural History of the Ground-Beetles (Coleoptera: Carabidae) of America North of Mexico*; Pensoft: Sofia, Bulgaria, 2003; ISBN 954-642-165-0.
22. Baulechner, D.; Diekötter, T.; Wolters, V.; Jauker, F. Converting Arable Land into Flowering Fields Changes Functional and Phylogenetic Community Structure in Ground Beetles. *Biol. Conserv.* **2019**, *231*, 51–58. [CrossRef]
23. Hoffmann, H.; Peter, F.; Donath, T.W.; Diekötter, T. Landscape-and Time-Dependent Benefits of Wildflower Areas to Ground-Dwelling Arthropods. *Basic Appl. Ecol.* **2022**, *59*, 44–58. [CrossRef]
24. Gayer, C.; Lövei, G.L.; Magura, T.; Dieterich, M.; Batáry, P. Carabid Functional Diversity Is Enhanced by Conventional Flowering Fields, Organic Winter Cereals and Edge Habitats. *Agric. Ecosyst. Environ.* **2019**, *284*, 106579. [CrossRef]
25. Batáry, P.; Holzschuh, A.; Orci, K.M.; Samu, F.; Tscharntke, T. Responses of Plant, Insect and Spider Biodiversity to Local and Landscape Scale Management Intensity in Cereal Crops and Grasslands. *Agric. Ecosyst. Environ.* **2012**, *146*, 130–136. [CrossRef]
26. Cole, L.; Pollock, M.; Robertson, D.; Holland, J.; McCracken, D. Carabid (Coleoptera) Assemblages in the Scottish Uplands: The Influence of Sheep Grazing on Ecological Structure. *Entomol. Fenn.* **2006**, *17*, 229–240. [CrossRef]
27. Ribera, I.; Dolédec, S.; Downie, I.S.; Foster, G.N. Effect of Land Disturbance and Stress on Species Traits of Ground Beetle Assemblages. *Ecology* **2001**, *82*, 1112–1129. [CrossRef]
28. Aguilar, J.; Gramig, G.G.; Hendrickson, J.R.; Archer, D.W.; Forcella, F.; Liebig, M.A.; Hart, J.P. Crop Species Diversity Changes in the United States. *PLoS ONE* **2015**, *10*, e0136580. [CrossRef]
29. Palmu, E.; Ekroos, J.; Hanson, H.I.; Smith, H.G.; Hedlund, K. Landscape-Scale Crop Diversity Interacts with Local Management to Determine Ground Beetle Diversity. *Basic Appl. Ecol.* **2014**, *15*, 241–249. [CrossRef]
30. Weibull, A.-C.; Östman, Ö.; Granqvist, Å. Species Richness in Agroecosystems: The Effect of Landscape, Habitat and Farm Management. *Biodivers. Conserv.* **2003**, *12*, 1335–1355. [CrossRef]
31. Adler, P.; Raff, D.; Lauenroth, W. The Effect of Grazing on the Spatial Heterogeneity of Vegetation. *Oecologia* **2001**, *128*, 465–479. [CrossRef]
32. Vickery, J.; Tallowin, J.; Feber, R.; Asteraki, E.; Atkinson, P.; Fuller, R.; Brown, V. The Management of Lowland Neutral Grasslands in Britain: Effects of Agricultural Practices on Birds and Their Food Resources. *J. Appl. Ecol.* **2001**, *38*, 647–664. [CrossRef]
33. Bassett, E.R.; Fraser, L.H. Effects of Cattle on the Abundance and Composition of Carabid Beetles in Temperate Grasslands. *J. Agric. Stud.* **2015**, *3*, 36–47. [CrossRef]
34. Pozsgai, G.; Quinzo-Ortega, L.; Littlewood, N.A. Grazing Impacts on Ground Beetle (Coleoptera: Carabidae) Abundance and Diversity on Semi-natural Grassland. *Insect Conserv. Divers.* **2022**, *15*, 36–47. [CrossRef]
35. Littlewood, N.A.; Pakeman, R.J.; Pozsgai, G. Grazing Impacts on Auchenorrhyncha Diversity and Abundance on a Scottish Upland Estate: Auchenorrhyncha Responses to Grazing. *Insect Conserv. Divers.* **2012**, *5*, 67–74. [CrossRef]
36. Schaffers, A.P.; Raemakers, I.P.; Sýkora, K.V.; Ter Braak, C.J. Arthropod Assemblages Are Best Predicted by Plant Species Composition. *Ecology* **2008**, *89*, 782–794. [CrossRef]
37. Loza, C.; Reinsch, T.; Loges, R.; Taube, F.; Gere, J.I.; Kluß, C.; Hasler, M.; Malisch, C.S. Methane Emission and Milk Production from Jersey Cows Grazing Perennial Ryegrass–White Clover and Multispecies Forage Mixtures. *Agriculture* **2021**, *11*, 175. [CrossRef]
38. Nyameasem, J.K.; Malisch, C.S.; Loges, R.; Taube, F.; Kluß, C.; Vogeler, I.; Reinsch, T. Nitrous Oxide Emission from Grazing Is Low across a Gradient of Plant Functional Diversity and Soil Conditions. *Atmosphere* **2021**, *12*, 223. [CrossRef]
39. Smit, H.P.J.; Reinsch, T.; Kluß, C.; Loges, R.; Taube, F. Very Low Nitrogen Leaching in Grazed Ley-Arable-Systems in Northwest Europe. *Agronomy* **2021**, *11*, 2155. [CrossRef]
40. Hoffmann, H.; Peter, F.; Herrmann, J.D.; Donath, T.W.; Diekötter, T. Benefits of Wildflower Areas as Overwintering Habitats for Ground-Dwelling Arthropods Depend on Landscape Structural Complexity. *Agric. Ecosyst. Environ.* **2021**, *314*, 107421. [CrossRef]
41. Freude, H.; Harde, K.; Lohse, G.; Klausnitzer, B. *Die Käfer Mitteleuropas. Bd: 2. Carabidae*; Springer: Berlin/Heidelberg, Germany, 2004.

42. Gürlich, S.; Suikat, R.; Ziegler, W. *Die Käfer Schleswig-Holsteins-Rote Liste*; Landesamt für Landwirtschaft, Umwelt und ländliche Räume Schleswig-Holstein: Flintbek, Germany, 2011; ISBN 3-937937-54-4.
43. Bräunicke, M.; Trautner, J. Lebensraumpräferenzen Der Laufkäfer Deutschlands–Wissensbasierter Katalog. *Angew. Carabidol. Suppl.* **2009**, *1*, 45.
44. Homburg, K.; Homburg, N.; Schuldt, A.; Assmann, T. Carabids. Org–a Dynamic Online Database of Ground Beetle Species Traits (Coleoptera, Carabidae). *Insect Conserv. Divers.* **2014**, *7*, 195–205. [CrossRef]
45. McCune, B.; Grace, J.B. *Analysis of Ecological Communities*; MjM Software Design: Lincoln County, OR, USA, 2002; ISBN 0-9721290-0-6.
46. Legendre, P.; Legendre, L. *Numerical Ecology*; Elsevier: Amsterdam, The Netherlands, 2012; ISBN 0-444-53869-0.
47. Pinheiro, J.; Bates, D. *Mixed-Effects Models in S and S-PLUS*; Springer Science & Business Media: New York, NY, USA, 2006; ISBN 0-387-22747-4.
48. Schaarschmidt, F.; Vaas, L. Analysis of Trials with Complex Treatment Structure Using Multiple Contrast Tests. *Hort. Sci* **2009**, *44*, 188–195. [CrossRef]
49. Nakagawa, S.; Schielzeth, H.; O'Hara, R.B. A General and Simple Method for Obtaining R2 from Generalized Linear Mixed-Effects Models. *Methods Ecol. Evol.* **2013**, *4*, 133–142. [CrossRef]
50. Bretz, F.; Hothorn, T.; Westfall, P. *Multiple Comparisons Using R*; Chapman and Hall/CRC: Boca Raton, FL, USA, 2016; ISBN 978-1-4200-1090-9.
51. Hothorn, T.; Bretz, F.; Westfall, P. Simultaneous Inference in General Parametric Models. *Biom. J.* **2008**, *50*, 346–363. [CrossRef] [PubMed]
52. R Core Team. *R: A Language and Environment for Statistical Computing*; R Foundation for Statistical Computing: Vienna, Austria, 2021.
53. Reinsch, T.; Loza, C.; Malisch, C.S.; Vogeler, I.; Kluß, C.; Loges, R.; Taube, F. Toward Specialized or Integrated Systems in Northwest Europe: On-Farm Eco-Efficiency of Dairy Farming in Germany. *Front. Sustain. Food Syst.* **2021**, *5*, 167. [CrossRef]
54. da Silva, P.M.; Aguiar, C.A.; Niemelä, J.; Sousa, J.P.; Serrano, A.R. Diversity Patterns of Ground-Beetles (Coleoptera: Carabidae) along a Gradient of Land-Use Disturbance. *Agric. Ecosyst. Environ.* **2008**, *124*, 270–274. [CrossRef]
55. Irmler, U. *Veränderung Der Laufkäfergemeinschaften (Carabidae) in 15 Jahren Sukzession Nach Der Umstellung Vom Konventionellen Auf Ökologischen Landbau Auf Hof Ritzerau*; Kiel University Publishing: Kiel, Germany, 2020; ISBN 3-928794-30-2.
56. Honek, A.; Martinkova, Z.; Jarosik, V. Ground Beetles (Carabidae) as Seed Predators. *EJE* **2013**, *100*, 531–544. [CrossRef]
57. Rusch, A.; Binet, D.; Delbac, L.; Thiéry, D. Local and Landscape Effects of Agricultural Intensification on Carabid Community Structure and Weed Seed Predation in a Perennial Cropping System. *Landsc. Ecol.* **2016**, *31*, 2163–2174. [CrossRef]
58. Wietzke, A.; van Waveren, C.-S.; Bergmeier, E.; Meyer, S.; Leuschner, C. Current State and Drivers of Arable Plant Diversity in Conventionally Managed Farmland in Northwest Germany. *Diversity* **2020**, *12*, 469. [CrossRef]
59. Kruess, A.; Tscharntke, T. Grazing Intensity and the Diversity of Grasshoppers, Butterflies, and Trap-nesting Bees and Wasps. *Conserv. Biol.* **2002**, *16*, 1570–1580. [CrossRef]
60. Barton, P.S.; Manning, A.D.; Gibb, H.; Wood, J.T.; Lindenmayer, D.B.; Cunningham, S.A. Experimental Reduction of Native Vertebrate Grazing and Addition of Logs Benefit Beetle Diversity at Multiple Scales. *J. Appl. Ecol.* **2011**, *48*, 943–951. [CrossRef]
61. Cagnolo, L.; Molina, S.I.; Valladares, G.R. Diversity and Guild Structure of Insect Assemblages under Grazing and Exclusion Regimes in a Montane Grassland from Central Argentina. *Biodivers. Conserv.* **2002**, *11*, 407–420. [CrossRef]
62. Baguette, M.; Hance, T. Carabid Beetles and Agricultural Practices: Influence of Soil Ploughing. *Biol. Agric. Hortic.* **1997**, *15*, 185–190. [CrossRef]
63. De Heij, S.E.; Benaragama, D.; Willenborg, C.J. Carabid Activity-Density and Community Composition, and Their Impact on Seed Predation in Pulse Crops. *Agric. Ecosyst. Environ.* **2022**, *326*, 107807. [CrossRef]
64. Triquet, C.; Roume, A.; Wezel, A.; Tolon, V.; Ferrer, A. In-field Cover Crop Strips Support Carabid Communities and Shape the Ecological Trait Repartition in Maize Fields. *Agri. For. Entomol.* **2022**, *25*, 152–163. [CrossRef]
65. Hagge, J.; Seibold, S.; Gruppe, A. Beetle Biodiversity in Anthropogenic Landscapes with a Focus on Spruce Plantations, Christmas Tree Plantations and Maize Fields. *J. Insect Conserv.* **2019**, *23*, 565–572. [CrossRef]
66. Hüber, C.; Zettl, F.; Hartung, J.; Müller-Lindenlauf, M. The Impact of Maize-Bean Intercropping on Insect Biodiversity. *Basic Appl. Ecol.* **2022**, *61*, 1–9. [CrossRef]
67. Woodcock, B.; Pywell, R.; Roy, D.; Rose, R.; Bell, D. Grazing Management of Calcareous Grasslands and Its Implications for the Conservation of Beetle Communities. *Biol. Conserv.* **2005**, *125*, 193–202. [CrossRef]
68. Jauker, F.; Diekoetter, T.; Schwarzbach, F.; Wolters, V. Pollinator Dispersal in an Agricultural Matrix: Opposing Responses of Wild Bees and Hoverflies to Landscape Structure and Distance from Main Habitat. *Landsc. Ecol.* **2009**, *24*, 547–555. [CrossRef]
69. Kremen, C.; Merenlender, A.M. Landscapes That Work for Biodiversity and People. *Science* **2018**, *362*, 51. [CrossRef]
70. Keller, I.; Largiader, C.R. Recent Habitat Fragmentation Caused by Major Roads Leads to Reduction of Gene Flow and Loss of Genetic Variability in Ground Beetles. *Proc. R. Soc. Lond. Ser. B Biol. Sci.* **2003**, *270*, 417–423. [CrossRef]

 *land*

Article ·

Enabling Regenerative Agriculture Using Remote Sensing and Machine Learning

Michael Gbenga Ogungbuyi [1], Juan P. Guerschman [2], Andrew M. Fischer [3], Richard Azu Crabbe [4], Caroline Mohammed [1], Peter Scarth [2], Phil Tickle [2], Jason Whitehead [5] and Matthew Tom Harrison [1,*]

[1] Tasmanian Institute of Agriculture, University of Tasmania, Launceston, TAS 7248, Australia; michael.ogungbuyi@utas.edu.au (M.G.O.)
[2] Cibo Labs Pty Ltd., 15 Andrew St., Point Arkwright, QLD 4573, Australia
[3] Institute for Marine and Antarctic Studies, University of Tasmania, Launceston, TAS 7248, Australia
[4] Gulbali Institute, Charles Sturt University, Albury, NSW 2640, Australia
[5] Cape Herbert Pty Ltd., Blackstone Heights, TAS 7250, Australia
* Correspondence: matthew.harrison@utas.edu.au

Citation: Ogungbuyi, M.G.; Guerschman, J.P.; Fischer, A.M.; Crabbe, R.A.; Mohammed, C.; Scarth, P.; Tickle, P.; Whitehead, J.; Harrison, M.T. Enabling Regenerative Agriculture Using Remote Sensing and Machine Learning. *Land* **2023**, *12*, 1142. https://doi.org/10.3390/land12061142

Academic Editor: Dailiang Peng

Received: 26 April 2023
Revised: 15 May 2023
Accepted: 23 May 2023
Published: 29 May 2023

Abstract: The emergence of cloud computing, big data analytics, and machine learning has catalysed the use of remote sensing technologies to enable more timely management of sustainability indicators, given the uncertainty of future climate conditions. Here, we examine the potential of "regenerative agriculture", as an adaptive grazing management strategy to minimise bare ground exposure while improving pasture productivity. High-intensity sheep grazing treatments were conducted in small fields (less than 1 ha) for short durations (typically less than 1 day). Paddocks were subsequently spelled to allow pasture biomass recovery (treatments comprising 3, 6, 9, 12, and 15 months), with each compared with controls characterised by lighter stocking rates for longer periods (2000 DSE/ha). Pastures were composed of wallaby grass (*Austrodanthonia species*), kangaroo grass (*Themeda triandra*), Phalaris (*Phalaris aquatica*), and cocksfoot (*Dactylis glomerata*), and were destructively sampled to estimate total standing dry matter (TSDM), standing green biomass, standing dry biomass and trampled biomass. We invoked a machine learning model forced with Sentinel-2 imagery to quantify TSDM, standing green and dry biomass. Faced with La Nina conditions, regenerative grazing did not significantly impact pasture productivity, with all treatments showing similar TSDM, green biomass and recovery. However, regenerative treatments significantly impacted litterfall and trampled material, with high-intensity grazing treatments trampling more biomass, increasing litter, enhancing surface organic matter and decomposition rates thereof. Pasture digestibility and sward uniformity were greatest for treatments with minimal spelling (3 months), whereas both standing senescent and trampled material were greater for the 15-month spelling treatment. TSDM prognostics from machine learning were lower than measured TSDM, although predictions from the machine learning approach closely matched observed spatiotemporal variability within and across treatments. The root mean square error between the measured and modelled TSDM was 903 kg DM/ha, which was less than the variability measured in the field. We conclude that regenerative grazing with short recovery periods (3–6 months) was more conducive to increasing pasture production under high rainfall conditions, and we speculate that – in this environment - high-intensity grazing with 3-month spelling is likely to improve soil organic carbon through increased litterfall and trampling. Our study paves the way for using machine learning with satellite imagery to quantify pasture biomass at small scales, enabling the management of pastures within small fields from afar.

Keywords: machine learning; satellite imagery; regenerative grazing; grassland biomass; total standing dry matter; digital agriculture; grassland management; climate change; land degradation; long-term monitoring

1. Introduction

Grasslands comprise key terrestrial ecosystems, providing feed and habitat for domesticated livestock and wildlife globally [1–3]. Grasslands allow significant carbon sequestration [4,5] in addition to existing carbon stocks they prevent from entering the atmosphere [6,7]. The resilience of grasslands to extreme drought and future climate requires an innovative agroecosystem approach that promotes functional biological drivers (such as soil microbial activities) and adaptive grazing management [8,9]. One such adaptive technique is using regenerative grazing principles [8,10] to stimulate ecosystem functions through short, intense grazing, adjustable stocking rate, and multi-paddock-system at the farm level (1–100 ha) with long rest periods allowing pasture biomass and land to recover. Residual biomass from trampling effects associated with regenerative grazing plays a significant role in reducing bare ground, enabling soil health (through soil microbial functionality), litter conversion, soil aggregation and porosity, and carbon sequestration [8,11]. Stimulation of organic microbial activities through residual biomass and trampling effects of grazing livestock contrasts with conventional farming systems in developed nations (through the use of irrigation, synthetic fertilizers, etc.) [8]. In practice, evidence of regenerative grazing impacts on pasture biomass, litterfall, and decomposition tend to be based on anecdotal rather than quantitative evidence [11–13]. Since the current information is not experimentally driven, available monitoring tools have not been tested to understand their usefulness to end-users. Due to large land areas and the dynamic and spatially variable nature of grazing [14,15], physical monitoring of grassland conditions is often cumbersome, particularly where land areas are remote, large, and/or geographically challenging. The rise of satellite imagery, cloud computing, big data analytics, and machine learning have paved the way for innovative opportunities for land managers to remotely monitor crop, pasture, or grassland biomass from afar [16].

Conventional methods for monitoring pasture biomass and livestock utilisation (i.e., ground-based measurement and proximal sensing) are limited in terms of scope, and both spatial and temporal extent [17]. Previous research in Australia [18], the United Kingdom [19], New Zealand [20], and the United States [21] has reported limitations of ground sampling approaches (i.e., visual, rising plate meter, and destructive method by clipping) in quantifying the spatial variability of pasture biomass. By contrast, remote sensing provides timely spatiotemporal information that can predict the availability of feed prior to grazing [19], allowing for feed budgeting. However, in most cases, remote sensing of pasture biomass is not process-driven (i.e., based on vegetation indices); often the use of such reflectance indices at small field scales (e.g., less than 50 ha) is constrained by the resolution of the satellite imagery [19,22] and accurate calibration [23]. Remote sensing that considers process-based retrieval of pasture biomass and other biophysical variables may invoke site-specific modelling and machine-learning techniques [24]. Although some successes have been reported, physical-based techniques such as radiative transfer modelling and light use efficiency modelling can be prohibitive as they may require a set of parametric rules for different study locations [25–27]. However, machine learning techniques including artificial neural networks (ANN) [16], random forest (RF) [28], and support vector machine (SVM) [21] are not site-specific and can be used to retrieve pasture biomass estimates [22]. ANN [16] was used to estimate pasture biomass leveraging multitemporal Sentinel-2 data collected over dairy farms in Tasmania [16]. The study showed that the accuracy of ANN improved when meteorological variables were included in the model; indeed, much process-based modelling is based primarily on longitudinal measurements of climate at a given site [2,23,29]. However, process-based applications are required as an operational service to support farm management—what is often known as a decision support system (DSS) [16,17,30,31]—and are often limited by the accuracy of site-specific soil characterisation [32,33].

Previous estimates of pasture biomass at the field (paddock) scale with machine learning algorithms have used standing green vegetation as a proxy to quantify the actual biomass from the normalised difference vegetation index (NDVI) [21,28,34,35]. Information derived from NDVI can provide sufficient information about active photosynthetic [36] vegetation, whereas non-green senescent pasture species or dormant vegetation are often much more difficult to quantify due to their low reflectance in the near-infrared [37]. To successfully realise improved land-use sustainability through more timely, accurate biologically-intelligent monitoring of pasture sustainability indicators, more robust approaches are urgently needed [30,38–40]. This would also allow livestock farmers to better predict feed on offer (for total green and non-green forage) enabling planning of their stocking rate to maximise liveweight production while maximising environmental stewardship [32,33]. While a range of commercial technologies exists, outputs from many of these applications are site-specific and others have not been validated. This raises questions as to how well such applications predict pasture biomass outside their zone of calibration.

The launch of the European Space Agency's Sentinel-2 satellites has enhanced the development of "agricultural technology" or "Ag-tech" companies offering products aimed at quantifying land surface conditions. One such company—"Cibo Labs" (https://www.cibolabs.com.au/; accessed on 1 December 2022)—uses a predictive time series machine learning approach to derive spectral information from Sentinel-2 data about local properties at the field scale. Cibo Labs uses pasture cuts to train and validate the total standing dry matter (TSDM) model. Several thousand fields from farms across Australia are used to train a deep neural network (DNN). Cibo Labs uses the dropout regularisation method to reduce overfitting and computational costs, hence improving the generalisation of the DNN [41]. This is achieved by randomly dropping units (i.e., hidden and visible layers) to improve the neural network's performance during training. Hitherto the present study, Cibo Labs validated total standing dry matter (TSDM) estimates using 2000 field measured samples collected over two years from across eastern and northern Australia. Thirty-three percent of field sites were used to train a three-layer, multilayer perceptron regression model (MPRM) using a 50% dropout and a maximum norm constraint [42–44]. The remainder of the field samples were used for validation. The model was trained with 100 iterations (~16,000 epochs) before reaching a termination criterion characterised by a median prediction error of 295 ± 8 kg DM/ha.

While such predictive accuracy was within the variability of measured data, the study was primarily conducted using measurements taken from low-latitude environments (the Northern part of Australia). Additionally, previous investigations of Cibo Labs' utility did not consider regenerative grazing principles implemented at the farm level. Therefore, it remains to be seen how well Cibo Labs performs in mid-latitude environments such as the island state of Tasmania, where cloud cover in winter and spring is frequent [45], as well as examine if the tool can support regenerative grazing at the farm level. Clouds reduce spatial and temporal coverage by reducing target clarity and increasing the time between clear useable images [16,46]. In the present study, we used a destructive sampling method to measure the total standing dry matter (kg DM/ha), equivalent to standing green and standing dry before and after grazing, with 3, 6, 9, 12, and 15 months of biomass regrowth. We applied regenerative grazing to the smaller plots of similar size (<1 ha), while three plots of size 10–50 ha were used as controls (i.e., business-as-usual grazing). Our hypothesis was that the treatment plots or disturbance caused by the high stocking density would account for the TSDM variability. The key aim was to examine the effects of regenerative grazing on TSDM productivity in the plots and whether Sentinel-2 imagery and the Cibo Labs model could estimate the TSDM at the plot level. This was conducted by comparing Cibo Labs estimates of TSDM with destructively sampled pasture biomass for a site in south-eastern Tasmania subject to sheep grazing treatments.

Our objectives were to thus provide insight into: (1) the effects of regenerative grazing on TSDM productivity, consumption, and trampling and (2) the usefulness of Sentinel-2 imagery and accuracy of the Cibo Labs model to estimate TSDM on effects of regenerative grazing at the farm level.

2. Materials and Methods

2.1. Study Site

The location for this study was south-eastern Tasmania, Australia. We worked on a case study farm (42°30′ S, 147°59′ E) north of the town of Triabunna called 'Okehampton'. The average annual rainfall at this location is 648 mm while the average annual minimum and maximum temperature are 7 °C and 17 °C, respectively [47]. Okehampton consists of 52 paddocks of sizes ranging from 1–138 ha covering an estimated area of 1446 ha (Figure 1). The botanical composition of fields comprises a mixture of native and sown pastures with mostly annual and perennial ryegrass (wallaby grass (*Austrodanthonia* species), kangaroo grass (*Themeda triandra*), Phalaris (*Phalaris aquatic*) and cocksfoot spear grass (*Dactylis glomerata*) [48]. The absence of irrigation and synthetic fertilizers on this site and the goal to stimulate pasture growth to improve livestock production, demand that agronomic systems implemented be sustainable, profitable, inclusive, and enduring—especially given the uncertainty of future climate conditions in this region [29,49]. The farm has a history of sheep grazing but the field layouts have evolved over time to accommodate inclusive, intensive grazing management, conservation of biodiversity, and environmental stewardship, including protection of endangered grass species and implementation of cultural burning practices informed by the local indigenous people (Pakana Services).

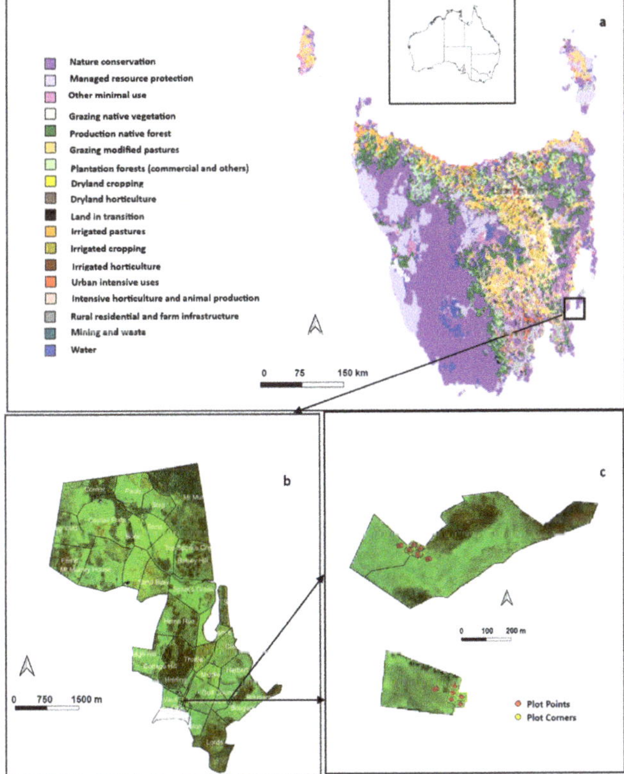

Figure 1. Study site (**a**) land use for Tasmania, (**b**) farm property, comprising 52 paddocks, and (**c**) subplots used for field sampling, [three larger plots (10 ha, 14 ha and 54 ha) were used as controls, while treatment plots had sizes of 0.2–0.4 ha]. The first six plots were located on a paddock called "Bougainville" on a hill. Land use data in (**a**) was obtained from the Australian Government, Department of Agriculture, Fisheries and Forestry, land use and management (accessed 10 October 2022).

2.2. Regenerative Grazing Data Collection

Twelve paddocks nominated by the case study farmer were used for the field sampling campaign. Biomass samples were collected by a local consultant from December 2021 through November 2022. Grazing was conducted for 1 day in the treatment plots (early morning to late evening). The control has "business-as-usual" grazing (Table 1). Pasture biomass fractions were generally quantified before grazing. Three plots [Vault control (VC), lower Bougainville (LB), and upper Bougainville (UB)] were used as controls following grazing regimes that were business-as-usual. These plots were grazed for longer periods (weeks) at lower stocking rates (2000 DSE/ha) than the intensive treatments (i.e., the other seven paddocks) and allowed less time between subsequent grazing compared with intensively grazed paddocks. Control paddocks were larger in size compared with treatment plots. Treatment plots were stocked at the same rate while following adjusted stocking density (Table 1) and grazed for one day on consecutive days within the same week to minimise potential confounding effects of weather impacts on pasture growth, then rested for three, six, nine or twelve months before re-grazing. Treatment plots were conducted based on 'regenerative' principles that conduct short, intense grazing, with long rest periods allowing pastures to recover [50]. In contrast, control paddocks were grazed at lighter stocking rates (Equation (1)), for longer durations, and allowed less time to recover (Table 1). Henceforth, the business-as-usual plots would be called BAU.

$$\text{Stocking rate (DSE/ha)} = \text{grazing area per dry sheep equivalent for a nominated period} \tag{1}$$

From Equation (1), if the stocking rate of BAU plots is 2000 (DSE/ha), then the stocking rate for the treatment plots is 1/4 of ha = 8000 DSE/ha.

Pasture biomass was harvested to the ground level from five locations (quadrats) that were predetermined within each plot (from plot points with red layouts in Figure 1) using a battery-operated shearing handpiece and a 0.25 m^2 quadrat (a square of 0.5 × 0.5 m). Standing biomass (green and dry) was cut prior to grazing while standing residuals (green and dry) and trampled biomass (green and dry) were taken post grazing, in a location immediately adjacent to the pre-grazing biomass harvest. Biomass was quickly placed in sealed, labelled plastic bags and transported to a 4 °C room in the laboratory where each bag was weighed after dung was excluded. The biomass was mixed, and using a quartering method, subsampled for separation and drying. Sub-samples of green and dry biomass were separated and then dried in a 60 °C oven for at least 48 h, before being weighed using a Mettler scale. This process was repeated for post-grazing biomass in some of the paddocks that were grazed. To account for the high volume of trampled biomass (i.e., biomass lying on the surface disturbed by the high density of sheep) this component was measured separately from the standing biomass (Table 1). Total standing dry matter (TSDM) was computed by the summation of green and dry biomass without trampled components. To determine actual biomass utilised during a post-grazing event, we used Equations (2) and (3) for total trampled dry matter (TTDM), as shown in Figure 2.

$$\text{total standing dry matter (TSDM)} - \text{trampled residual} = \text{Biomass consumed} \tag{2}$$

$$\text{trampled green dry matter} + \text{trampled senesced dry matter} = \text{Total trampled dry matter (TTDM)} \tag{3}$$

Since the sampled biomass collected from the five locations was completed only once in each plot following a predetermined layout (plot points in Figure 1), we computed the mean for these locations to account for sampling error and tested if the treatment plots and grazing days have a significant effect on biomass using statistical analysis (ANOVA and general linear model). We developed a time series analysis for the treatment plots (including BAU) and compared them with statistical outcomes.

The experiment was for twelve months, from December 2021 to November 2022, where the effects of short, intense grazing compared with the conventional grazing (control) on plot treatments for their drought resilience were observed. Hence, the experiment covered the four seasonal variations (summer, winter, autumn, and spring) in the study area. Summer is from December to February; autumn is from April to May; winter is from June to August; spring is from September to November [46].

Table 1. Experimental treatments and business-as-usual plots (controls). All plots were sampled and grazed in phase 1. Trampled residual was collected only post-grazing. Bougainville plots 1, 2, 3, and 4 were conducted with different treatments to Vault treatments. At the outset, Bougainville plots 2 and 4 were subjected to intense grazing, similar to the Vault treatments, whereas Bougainville 1 and 3 plots were grazed in accordance with BAU. After phase 1, all four plots were closed grazed no further. Asterisk (*) DSE represents "dry sheep equivalent", a standardised grazing unit representing one dry, non-lactating 45 kg castrated male (wether) consuming 7.6 MJ/day.

Treatments	Plot	Size (ha)	Phase 1 December 2021 & January 2022		Phase 2 April 2022	Phase 3 July 2022	Phase 4 November 2022
	Grazing		Pre	Post	Pre	Pre	Pre
	Trampled after post-grazing			✔			V4
	Stocking rate (DSE/ha) *		8000		6000	8800	8800
BAU & Regenerative	Bougainville (B1)	0.4	✔	✔			
Regenerative	Bougainville (B2)	0.2	✔	✔			
BAU & Regenerative	Bougainville (B3)	0.3	✔	✔			
Regenerative	Bougainville (B4)	0.2	✔	✔			
Control	Upper Bougainville (UB)	54	Business as usual				
Control	Lower Bougainville (LB)	10	Business as usual				
Regenerative	Vault 1 [12 months]	0.3	✔	✔			✔
	Vault 2 [9 months]	0.3	✔	✔		✔	
	Vault 3 [6 months]	0.3	✔	✔	✔		
	Vault 4 [3 months]	0.3	✔	✔	✔	✔	✔
	Vault 5 [15 months]	0.3	✔	✔			
Control	Vault Control	14	Business as usual				

The method was developed by the authors.

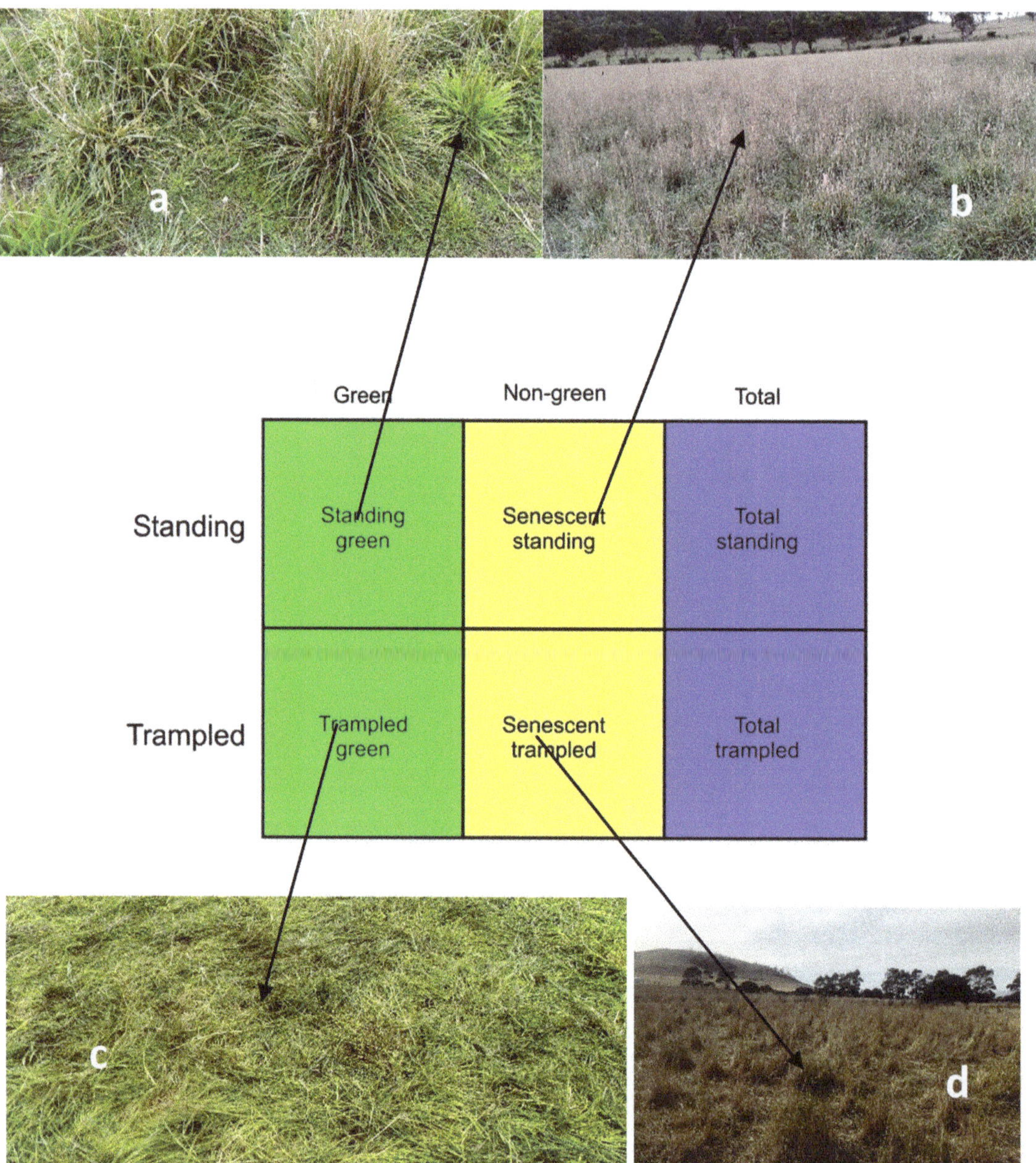

Figure 2. Pasture biomass categories enumerated using destructive harvests at Okehampton, Tasmania, Australia. We measured (**a**) standing green biomass and (**b**) standing dry biomass prior to grazing; post-grazing we also measured (**c**) trampled green biomass and (**d**) trampled dry biomass. Photographs (**a**) and (**b**) were taken in autumn, (**c**) in winter and (**d**) in summer. We refer to destructive sampling data herein as 'measured' data. Total standing dry matter (TSDM) was computed as the summation of green and dry standing biomass.

2.3. Remote Sensing

Estimates of TSDM were derived using the 'PastureKey' app within Cibo Labs, which is produced using 10 m resolution Sentinel-2 imagery provided by the European Space Agency (ESA). Only cloud-free pixels of Sentinel-2 imagery are used by Cibo Labs, and the application produces TSDM estimates for cloud-free paddocks every 5 days (Sentinel 2 revisit time). Cloudy pixels are detected and masked with the 'Fmask' algorithm [51]. Ten bands (b2, b3, b4, b5, b6, b7, b8, b8A, b11 and b12) of Sentinel-2 imagery were used to derive TSDM products. Using a predictive machine learning approach driven by deep neural networks (DNN), measured data are trained to predict TSDM within and across the paddock for every satellite revisit across a property. Cibo Labs uses Sentinel-2 bands from several thousand paddocks and dates of satellite imagery acquisition to train a three-layer, multilayer perceptron neural network regression model using a 20–50% dropout regularisation method. The dropout regularisation method addresses the problem of overfitting [41].

Pasture estimates in near real-time are available from the PastureKey application within Cibo Labs. Hereafter, the PastureKey application would be referred to as Cibo Labs for convenience. The multilayer perceptron model can learn in real-time, complementing the delivery of products to end-users in cloud optimised GeoTiff (COG) format. Estimates of pasture biomass are available on demand or in a batch mode through a high-performance computing (HPC) environment.

2.4. Comparing Measured Pasture Biomass with Satellite Estimates

On the account that Sentinel-2 could retrieve total standing green and dry matter from the plot with a size less than 1 ha, pasture estimates from Cibo Labs were evaluated by comparison with corresponding measured values for each time point (in each case using the most proximal Sentinel-2 imagery). Comparisons of measured against estimated data were assessed using, time series trendline and error bar, root mean square error, and R^2 following [15,29,52].

3. Results

3.1. The Effects of Regenerative Grazing on Pasture Productivity, Consumption and Trampling

In all treatments and BAU plots, pasture biomass removal through intensive and conventional (control) grazing typically shows biomass loss between pre-grazing (December 2021–January 2022) and post-grazing (January–February) in phase 1 (Figure 3). This is also observed in phase 4, where Vault 2 and Vault 4 plots went through a post-grazing regime (Figure 3). The actual biomass consumed in the one-day grazing for all treatments is shown in Figure 3. As shown in Figure 3, the actual biomass utilised (see Equation (2)) through grazing is low compared to the trampled biomass (see Equation (3)) for all treatment plots (BAU inclusive). Therefore, trampling has a more significant effect on the TSDM than the actual grazing (i.e., consumption).

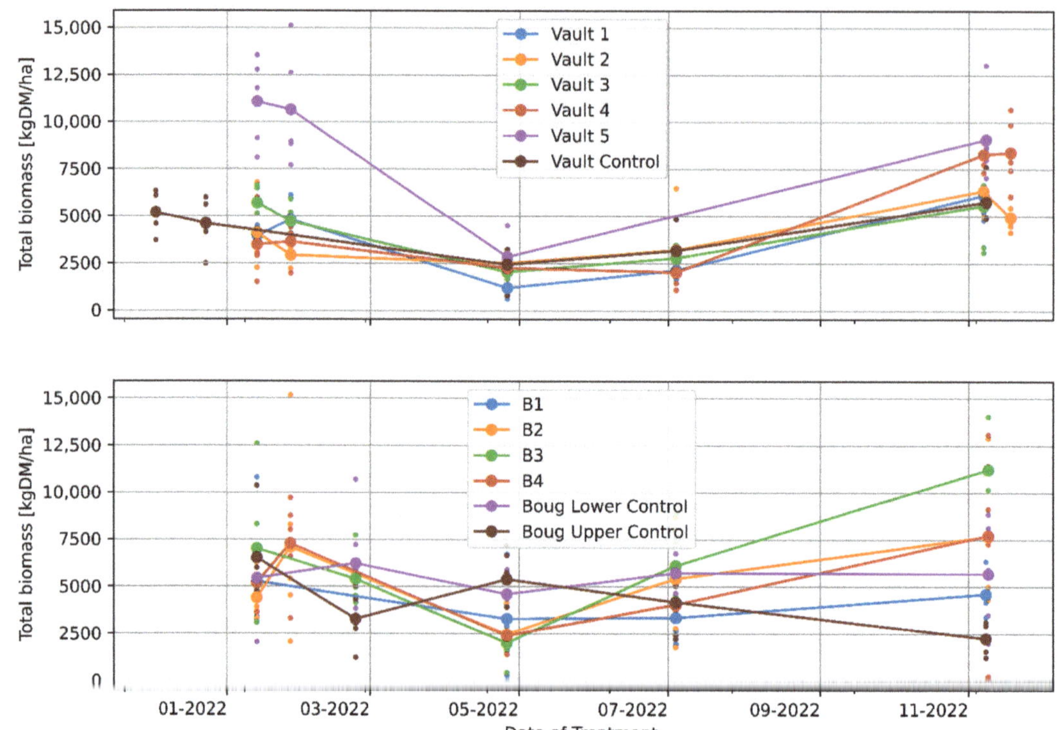

Figure 3. Effects of grazing treatment on total standing dry matter (TSDM), computed as the sum of standing green DM, standing dry DM and trampled residual. Small dot points show measurements obtained from five quadrats in each treatment plot; large dots show means for each plot.

All treatments, including BAU plots, show similar temporal variability and trends of total standing biomass (Figure 4). A similar trend is observed in standing green DM and standing dry DM (Figures 5 and 6). This indicates that the grazing intervals and the resting periods (3, 6, 9, 12, and 15 months) did not significantly influence biomass recovery or productivity in the experiment. For instance, after the first three months of rest in phase 1, where all treatment plots were grazed, pastures did not recover to the biomass level at the start of the experiment, likely due to seasonal variations in rainfall and temperature (Figure 5). However, following the consecutive increase in rainfall and temperature in winter through spring, total standing biomass increased, as observed in the Vaults and Bougainville 3 and 4 treatments (Figure 5). For example, the treatment plot (Vault 1) grazed only once (i.e., 12 months of rest) was similar to the Vault 4 plot, which was grazed every three months. In the same way, the Vault 5 plot that has not been grazed (i.e., 15 months of rest) is similar to the plot that was grazed every three months (Vault 4). In similar manner, the Vault 4 treatment is similar to the Bougainville 2–4 plots that were left ungrazed after phase 1 (Figures 3 and 4). Only Bougainville 3 plot exceeded Vault 4 treatment in the TSDM during spring by 3000 kg DM/ha. Therefore, biomass removal and recovery through regenerative grazing or conventional method does not influence Vaults and Bougainville treatments (Figure 3).

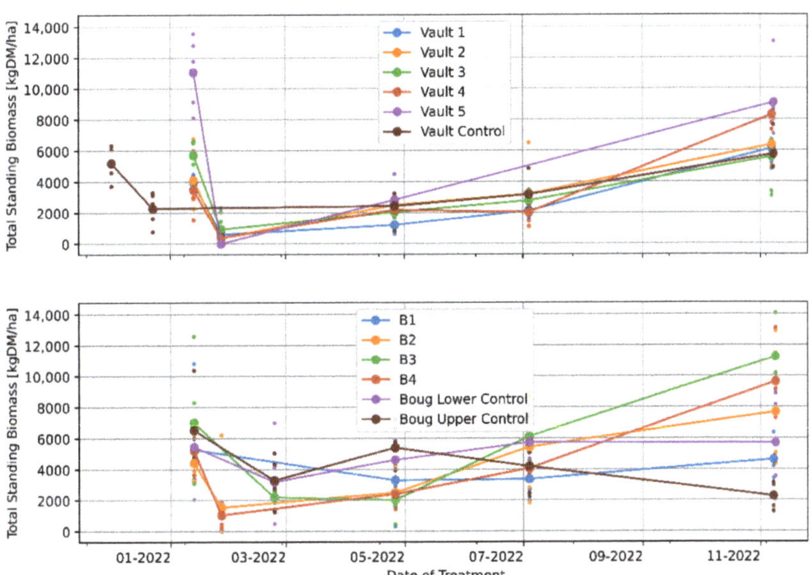

Figure 4. Effects of treatment on the total standing dry matter. TSDM was computed as the sum of standing green DM and standing dry DM, excluding trampled residual.

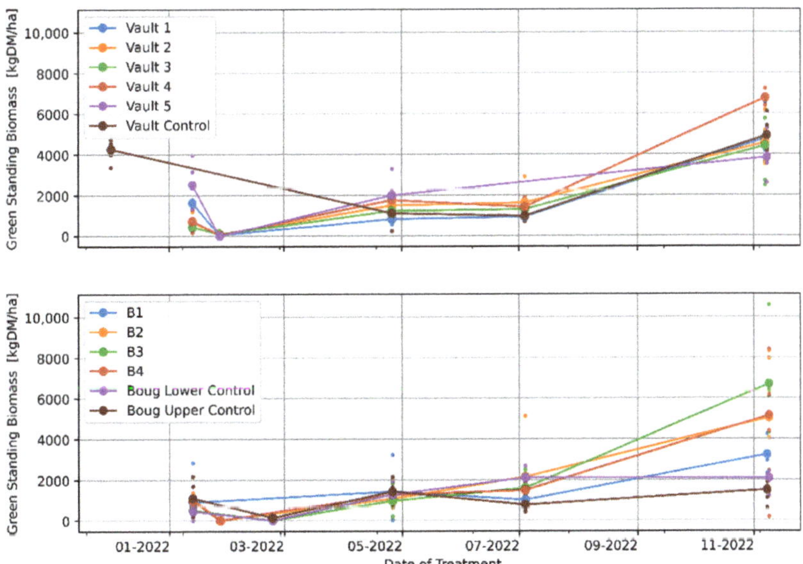

Figure 5. Effects of treatment on standing green biomass.

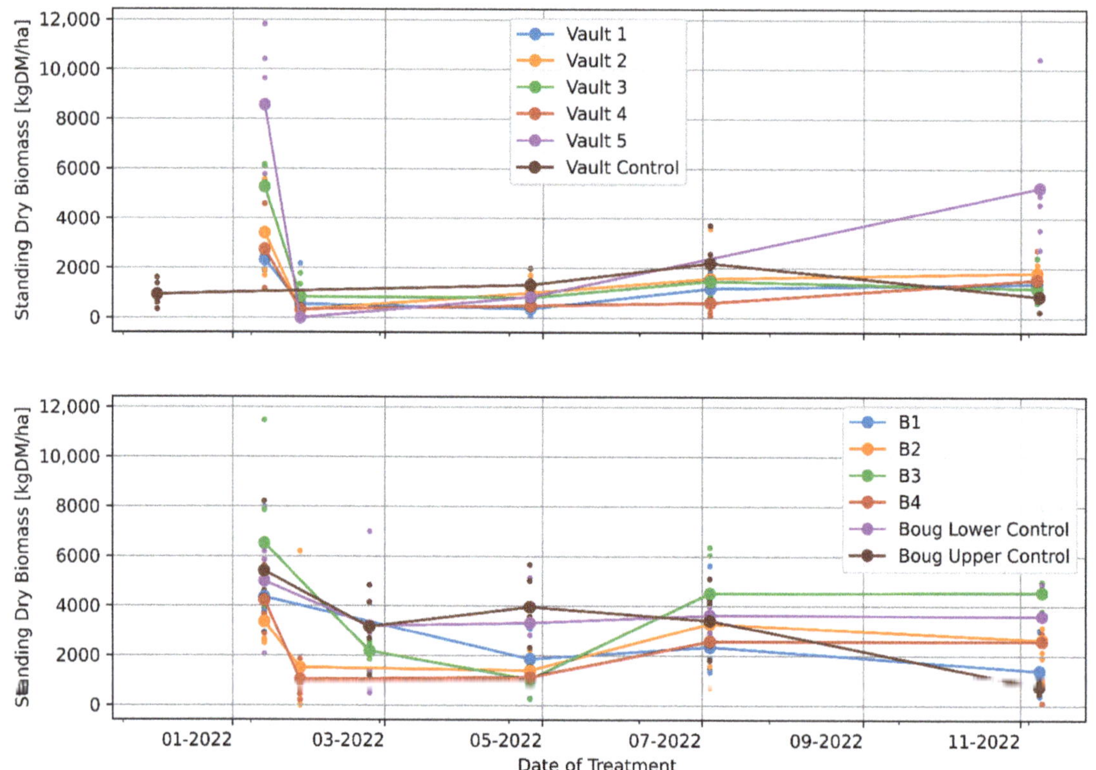

Figure 6. Effects of treatment on standing dry biomass.

During spring, the TSDM in the Bougainville 1–4 plots varied due to rainfall, as shown in Figures 3 and 4. The Bougainville 3 plot had more available TSDM than the Vault 5 treatment, as seen in Figure 4. Since the Bougainville plots are situated on a sloping hill, it is uncertain if their location played a role in the significant biomass growth observed in Bougainville 2–4 during spring, as depicted in Figure 4. The Bougainville 1 plot had the lowest TSDM volume.

ANOVA and generalized linear models showed no significant association between plots and pasture biomass productivity (TSDM). However, there were significant differences when the date of grazing was used as an effect of treatment. Generally, the ANOVA test shows the effect of grazing date is statistically significant ($p < 0.001$) to the TSDM, while the post hoc Dunnett test does not show the level of interaction. Analysing the effects of dates of treatments and TSDM further with interaction using the GML model shows strong evidence of significant difference ($p < 0.05$) on 27 January 2022 by an estimated $-11,076$ kg DM/ha compared to other grazing dates. The treatment plots associated with the 27 January grazing event include Bougainville 2, Bougainville 4, Vault 1, Vault 2, Vault 3, Vault 4, and Vault 5. In contrast, there is no statistical evidence that the BAU (Lower Bougainville, Upper Bougainville, and Vault Control) is significantly different ($p > 0.1$) from variability in biomass. Therefore, we conclude that the effect of regenerative treatments (i.e., short, intense grazing, and rest periods) in the plots did not affect TSDM productivity and consumption. Only the trampling effect (surface disturbed by the high density of sheep) associated with the 27 January 2022 post-grazing event in phase 1 for Vaults (1, 2, 3, 4, and 5) and Bougainville 2 and 4 plots explained the variability in biomass. We conclude that regenerative grazing did not have an effect on pasture biomass productivity in the wet year of 2022. All treatment plots have similar results.

In summary, the Vault 4 plot with three months grazing interval has the highest volume of standing green DM compared with Vault 5 with 15 months of resting interval (Figure 5). The Vault 5 plot with 15 months of resting interval has the highest standing dry matter compared to other treatment plots as there was no grazing in this period (Figure 6).

3.2. Satellite Estimate of Pasture Biomass

Cibo Labs (PastureKey application) utilises Sentinel-2 imagery to estimate TSDM, standing green DM, and standing dry DM in all the treatment plots, pre- and post-grazing (Figures 7 and 8). The matchup of Sentinel-2 imagery with the measured biomass measurements ranges from 2 to 40 days. In the summer (5 December 2021 to 13 January 2022), six plots (Upper Bougainville, Vault 1 to 5) had a lag of 40 days between Sentinel-2 imagery and the measured data. A two-day difference was experienced in autumn (between 3 and 5 July 2022).

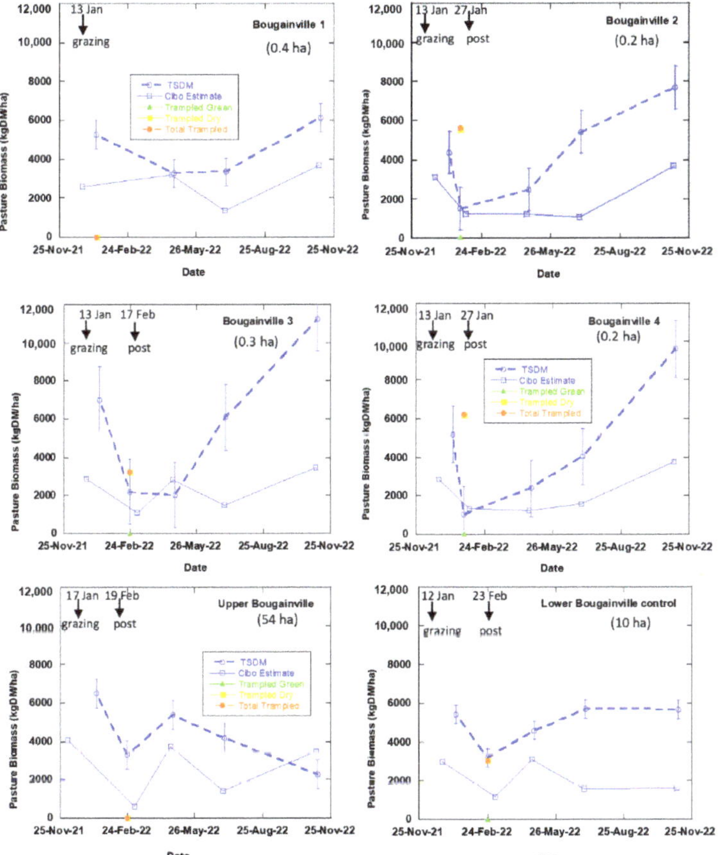

Figure 7. *Cont.*

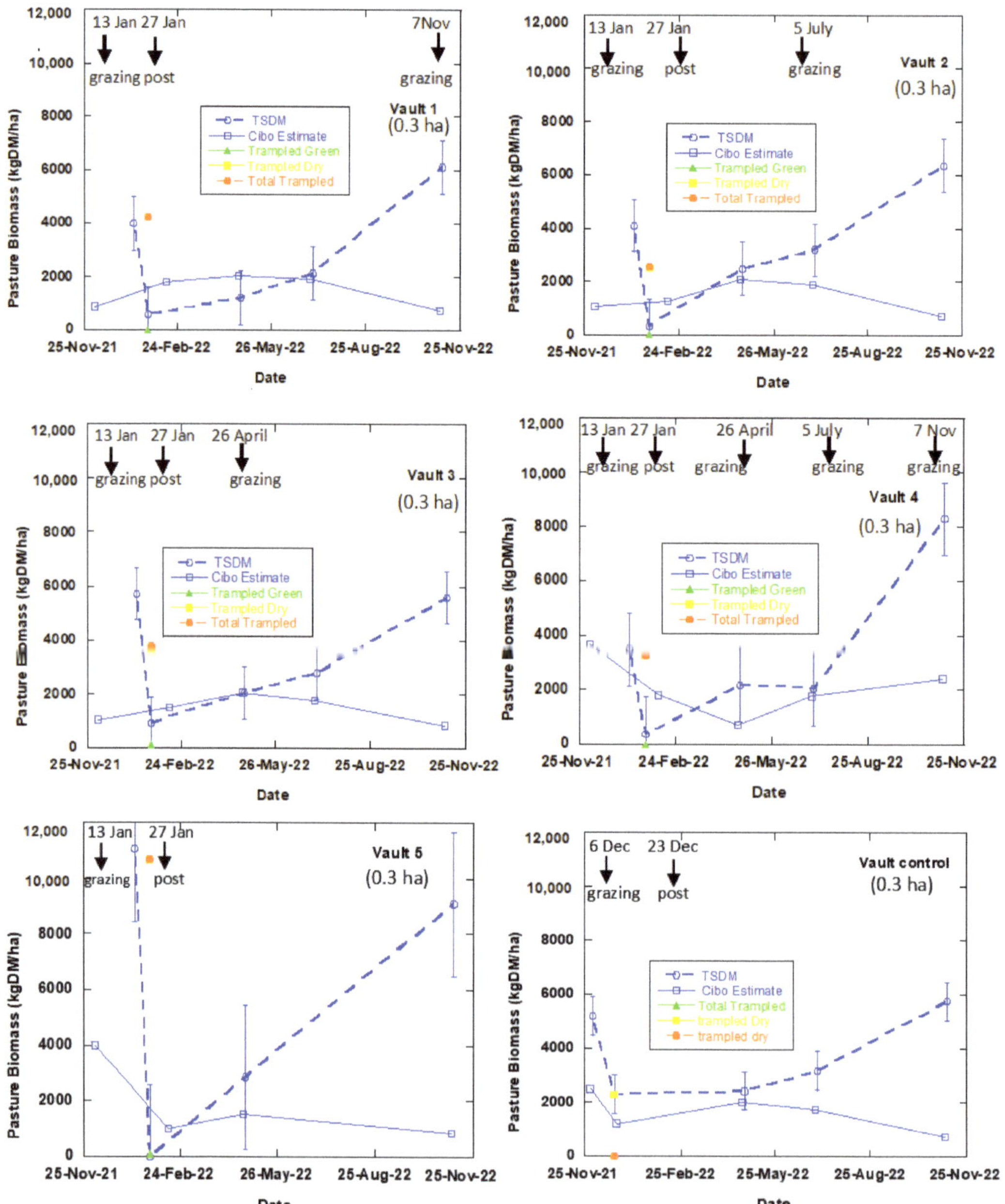

Figure 7. Measured and modelled TSDM data at Okehampton, Triabunna, Tasmania. Trampled material is vegetation pushed against the ground surface by grazing that was measured in phase 1 (Table 1) post-grazing. Broken lines represent measured TSDM; blue solid line represents Cibo Labs modelled TSDM. Bougainville 1 and 3 treatment plots were grazed as BAU at the start of the experiment before subsequently being closed to grazing. Error bars represent standard error of the mean.

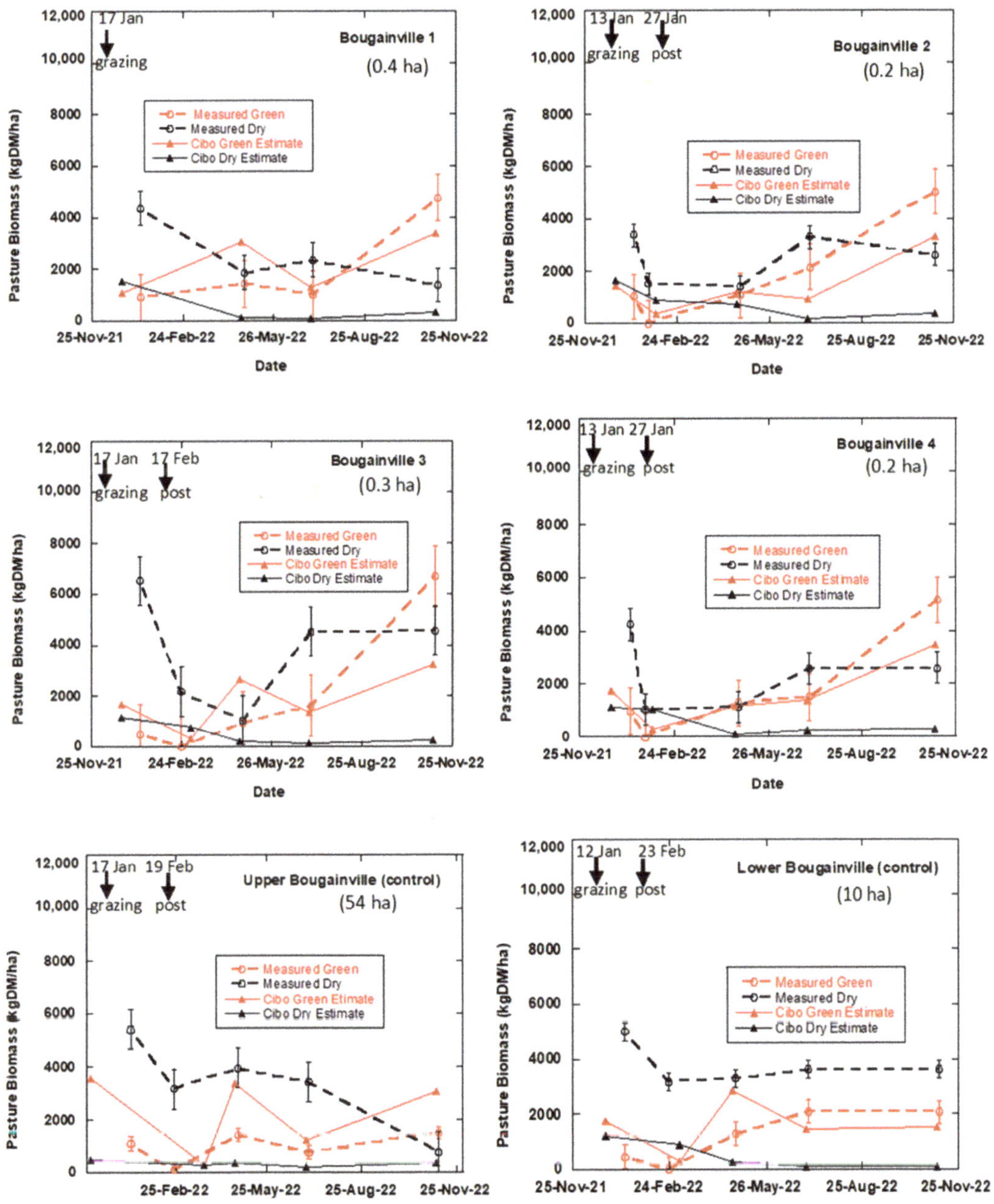

Figure 8. *Cont.*

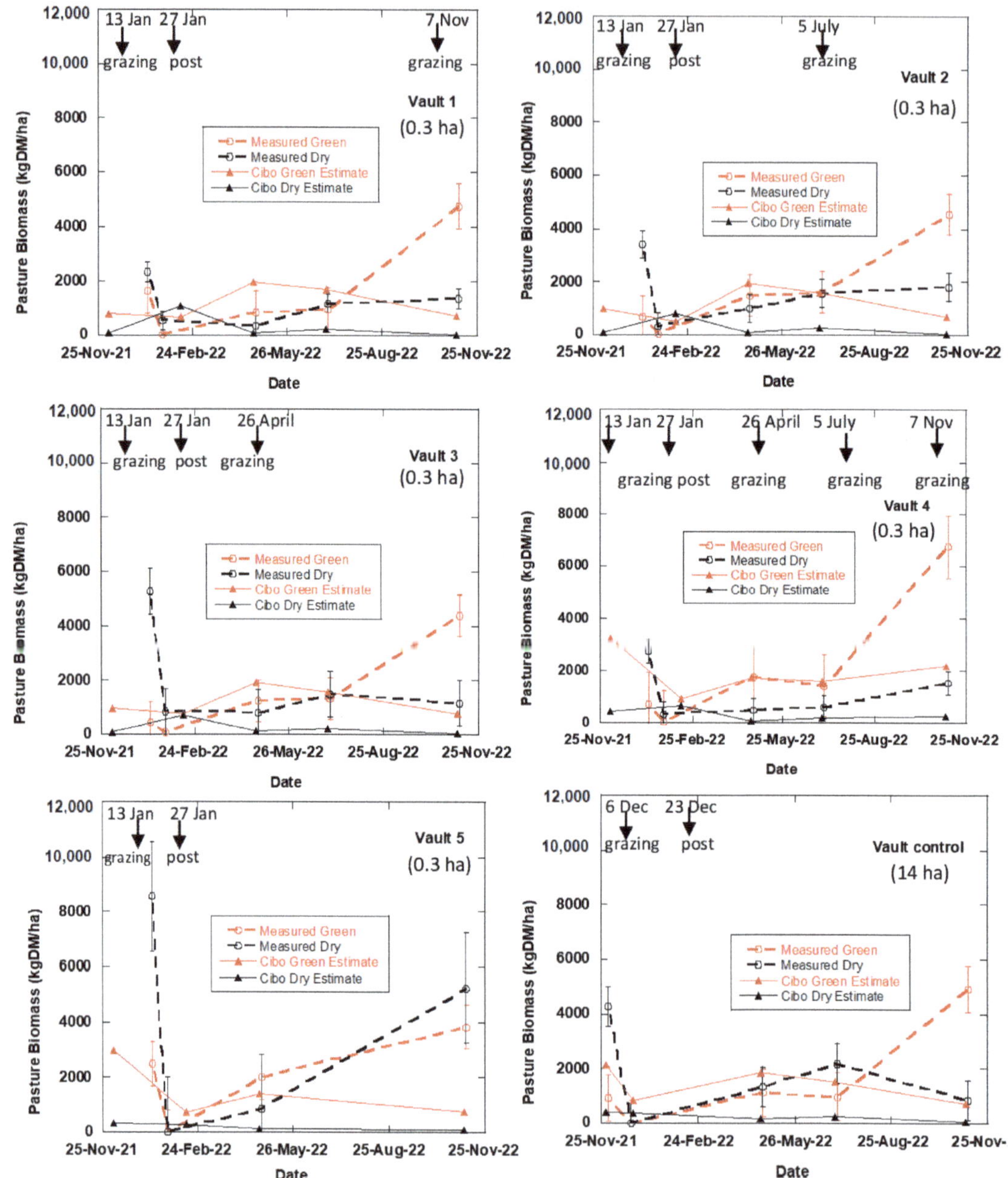

Figure 8. Measured standing green and dry pasture biomass compared with the Cibo Labs simulated values. Broken lines represent measured green DM and dry DM, while blue solid line represents Cibo Labs estimated green DM and dry DM. Error bars represent standard error of the mean.

Cibo Labs accounted for the variability in the TSDM in the treatment plots (between and within) but underestimated this value, compared with the measured TSDM (Figure 9). There are instances (phases 1–4) where the satellite estimated TSDM values closely (Bougainville 1, 2, and 4 and Vaults 1, 3, and 4) matched the measured points.

For all the treatment plots excluding Upper Bougainville and Lower Bougainville which went through conventional grazing (control), the Cibo Labs passes through one or more error bars, indicating it is within an acceptable variability of the measured biomass. The plot (Vault 4) which went through repeated grazing treatment every three months is more closely associated with the measured variability than Vault 1 which was grazed only once, or Vault 5 with 15 months of rest (Figure 7). The Vault 5 treatment plot has the highest variance.

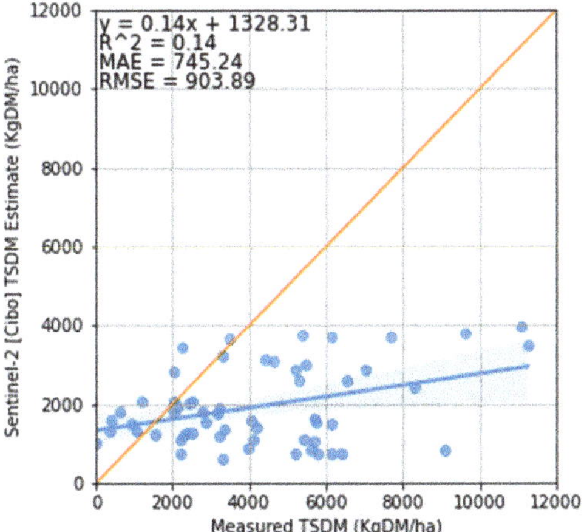

Figure 9. Relationship between measured total standing dry matter and Cibo Labs modelled values. Orange line is 1:1 and blue regression is line of best fit. Mean absolute error was 745 kg DM/ha and root mean square error was 904 kg DM/ha.

The measured TSDM collected for the post-grazing event in Vault 5 on 27 January 2022 shows the total trampled residual (trampled green DM and trampled dry DM) was zero which implies that biomass in this plot was lying on the surface due to the effect of the high-density of sheep (Figure 7). In this treatment, the biomass utilised was 355 kg DM/ha for one day of grazing [Total standing dry matter (TSDM) before grazing − Total trampled dry matter (TTDM) after grazing, (11,076–10,721 = 355 kg DM/ha]. The unutilised trampled residual (green and dry) that was measured, 10,721 kg DM/ha has a corresponding estimate of 1004 kg DM/ha from the Cibo Labs. It, therefore, implies that although Cibo Labs underestimates the TSDM, it can account for the trampled residual that is of high volume.

Sentinel-2 imagery integrated with the Cibo labs model has a better capability of estimating standing green DM than standing dry DM (Figure 8). Although estimates are within the variability of the measured data points, in phase 1 (except Vaults 1 and 2 and Bougainville 1 and 2) and phase 2 of the experiment, Cibo Lab overestimated standing green DM (Figure 8). No clear relationship exists between the measured and estimates (Figure 7) for the 7 November 2022 (phase 4), similar to Figure 7. There is no correlation between the measured standing dry DM and Cibo Labs estimates. Cibo Labs underestimated standing dry DM—estimates are barely above ground level (Figure 8).

The correlation between the measured standing biomass and Cibo Labs estimates and their respective linear regression plots with R^2, mean absolute error (MAE) and root mean squared error (RMSE) are shown in Figures 9–12.

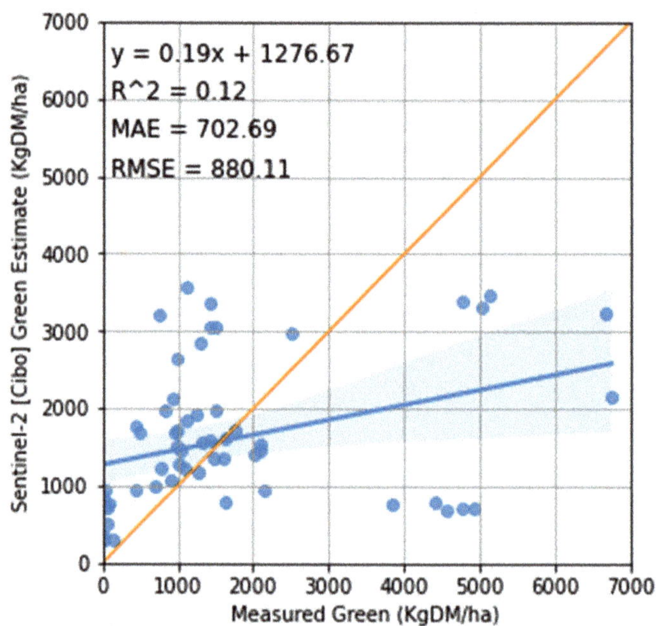

Figure 10. Relationship between the measured standing green and Cibo Labs estimate. Orange line is 1:1 and blue regression is line of best fit. Mean absolute error was 703 kg DM/ha and root mean square error was 880 kg DM/ha.

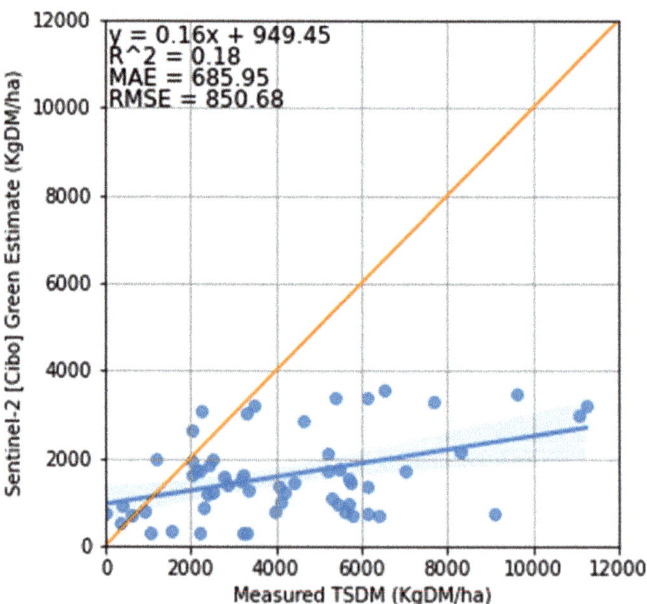

Figure 11. Relationship between measured total standing dry matter and Cibo Labs standing green estimates. Orange line is 1:1 and blue regression is line of best fit. Mean absolute error was 686 kg DM/ha and root mean square error was 851 kg DM/ha.

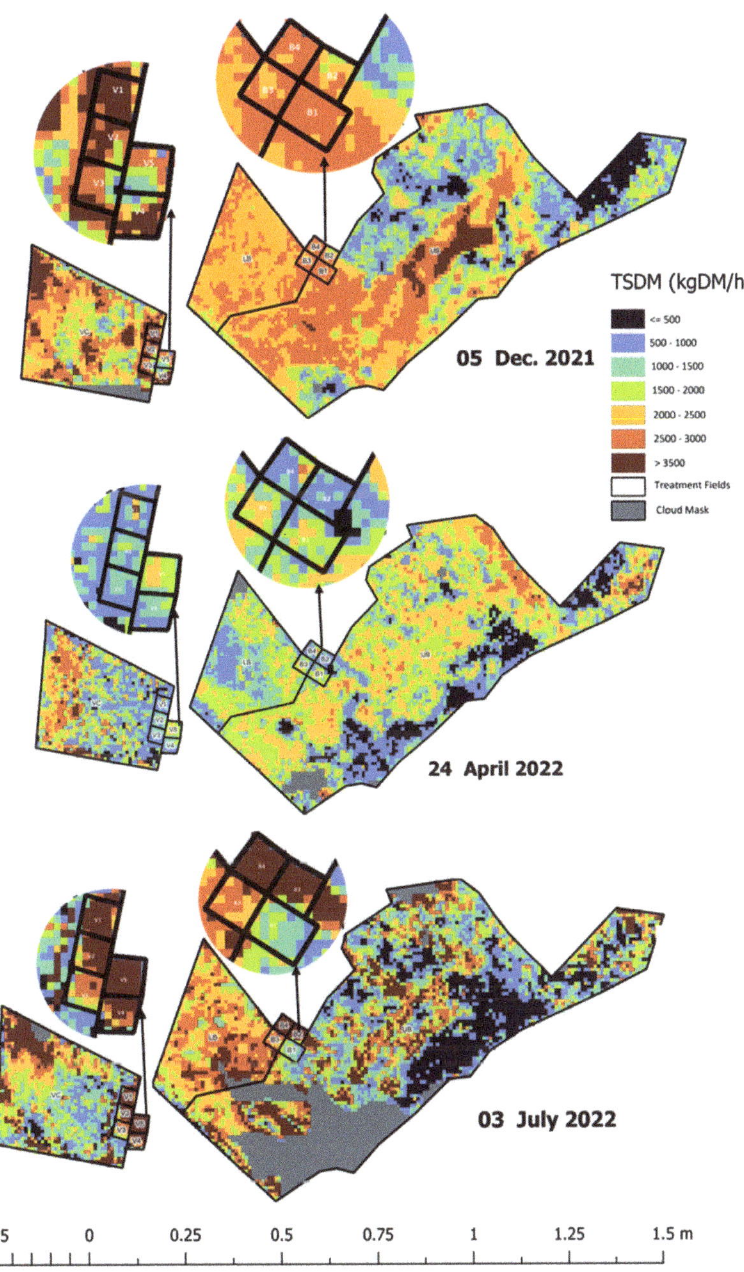

Figure 12. Spatiotemporal variability in pasture biomass across Okehampton. Smaller plots (expanded) represent regenerative grazing treatments, while the larger plots were conventional (business-as-usual) grazing treatments. Vaults plots 1 to 5 are shown in the lower-left expanded view, while Bougainville plots 1 to 4 plots are at the upper-right expanded view.

3.3. Spatial Maps Derived from Sentinel—2 Imagery and Cibo Labs Model

Cibo Labs derived Sentinel-2 maps for the treatment plots at Okehampton, Triabunna, Tasmania show spatiotemporal changes and the variability (within and across) in pasture biomass levels in all the treatment plots including control for pre-grazing and post-grazing activities such as the time series plot discussed in Figure 7. Cloud-free Sentinel-2 imagery to quantify the available pasture biomass against the ground measurement collected on 13 January 2022 before grazing the fields (paddocks) was on 5 December 2021. This makes a lag of 40 days between the available cloud-free satellite imagery and the ground measurement. All treatment plots started with more pasture biomass before grazing. All fields were grazed in phase 1 and left to rest for three months. All fields were grazed in phase 1 and left to rest for three months. After rest, the ground measurement collected on 26 April indicates the treatment plots have not recovered in autumn (proximal Sentinel-2 imagery available on 24 July). However, satellite imagery available on 3rd July against the measured pasture biomass collected on 5th July shows the plots (Vault and Bougainville) show increasing TSDM during winter. The map indicates Bougainville 1 (bottom) is the least-performing treatment plot with reference to phase 3 of the experiment. The maps correspond to Figures 3 and 4 and the modelling time series in Figure 7. As shown earlier (Figures 3 and 4), the map confirmed that Bougainville 1 is the least-performing plot.

4. Discussion

4.1. The Effects of Regenerative Grazing on Pasture Biomass Productivity, Consumption and Trampling

This study examined the effect of regenerative grazing treatments (i.e., short, intense grazing and rest periods) with smaller plots (less than 1 ha) on pasture productivity, consumption, and trampling. In the treatment plots examined, regenerative grazing did not influence pasture biomass productivity in the wet year of 2022. All treatment plots, including the ones used for conventional grazing (control), have similar results (Figures 3–6). ANOVA and generalized linear models (GLM) showed no significant association between treatment plots and pasture biomass productivity (TSDM). However, there were significant differences when the date of grazing was used as an effect of treatment (Section 3.1). GLM model shows a strong statistical significance exists only with the treatment plots (i.e., Vault 1, Vault 2, Vault 3, Vault 4, Vault 5, Bougainville 2, and Bougainville 4) associated with the post-grazing event of 27 January 2022 (Section 3.1). Therefore, this study concluded that the variability in the TSDM can only be explained by the treatment plots associated with the post-grazing regime in phase 1 of the experiment. The time series charts in Figures 3–6 confirm that although all treatment plots exhibited similar results, Vaults 4 and 5 showed significant variability with pasture biomass productivity. Similarly, the Bougainville 2, 3, and 4 plots benefited from rainfall to produce more biomass in the spring [53].

The effect of resting interval (3, 6, 9, 12, and 15 months) for TSDM to recover in the plots did not contribute to biomass variability (Figures 3 and 4). The main effect of treatment in the plots is associated with the high stocking rate, which resulted in a high volume of trampling residual (i.e., 27 January 2022). This implies that the actual biomass utilised (i.e., TSDM minus trampling residual) for grazing in the treatment plots was significantly low (Figure 3). In all treatment plots (including the BAU), the recovery or productivity of TSDM from summer through spring due to increasing rainfall followed a similar pattern (Figure 4). This showed that the influence of weather contributed to biomass recovery in a similar way, thereby confounding the effect of other treatments. For example, there was no significant difference between the Vault 4 treatment, which was grazed every three months, and Vault 5 with 15 months of rest. Similarly, there was no significant difference between Vault 1 treatment with 12 months grazing plan and the Vault 4 plot (Figures 3 and 4).

The present study has shown that although grazing through an intensive or conventional approach reduces pasture biomass [13], intensively grazed paddocks/fields through a regenerative strategy provide pastures with adaptive management for quick biomass recovery and reduction in bare ground. The plot (Vault 4) subjected to three months of resting interval utilised residual biomass from the trampling effects of grazing and optimum weather conditions to produce the highest volume of standing green DM over other treatment plots (Figure 5). Therefore, we conclude that Vault 4 is the treatment plot with the best pasture biomass productivity. In contrast, the Vault 5 treatment plot with 15 months of resting interval produced the highest standing dry DM compared to other plots (Figure 6). The pasture biomass produced is actively senescing from a lack of utilisation.

We emphasise that the impact of favourable weather confounded the effect of treatments on pasture biomass variability or biomass recovery. Hence, the resilience of pasture biomass to drought could not be established. A longer resting interval is not recommended in a situation such as this with good weather conditions. An earlier study under a simulated environment of rainfall and other treatment variables considered a 30-day resting period insufficient to recover soil samples from trampling caused by intensive grazing rotation [54]. Although this and few other studies approached regenerative grazing in the sense of soil recovery [54–57], the same principle as the strategy employed here (pasture biomass utilisation) is used to stimulate microbial activities and soil functions. However, this study is the first to use an approach where the experiment conditions followed natural processes with no farm inputs (fertilizer, irrigation etc.) and a simulated environment. Our results indicate that post-grazing data provides an incentive to determine the effect of trampling, which according to the analysis in this study, is limited. Trampling residual data provides information about the actual biomass utilised by the grazing livestock, which in turn gives insight into liveweight gain [54]. Furthermore, the actual biomass utilised for grazing is negligible compared to the trampled residual. Therefore, to minimise biomass wastage through trampling while achieving regenerative grazing sustainability [8], future work will focus on adjusting the stocking density to accommodate more grazing days (3 to 5 days). This is because, in practice, one day of grazing may be infeasible [54] with limited land resources and logistical constraints. In addition, having 3 to 5 days of adjustable stocking rate instead of 1-day grazing would support a more effective intensive rotational grazing regime within a multi-paddock system. Future research opportunities exist in understanding the resting period that will be sustainable to recover pasture from the trampling effect.

4.2. Satellite Estimates of Pasture Biomass

In this study, we examined the usefulness and accuracy of PastureKey, an application from the Cibo Labs, and derived from 10 m resolution Sentinel-2 imagery estimates of total standing dry matter to support regenerative grazing at the farm level. The usefulness of the tool was examined with respect to capturing TSDM (standing green DM and standing dry DM) variability in the treatment and business-usual plots, similar to the one obtained by the destructive sampling approach. The accuracy of the Cibo Labs (used instead of PastureKey for convenience) was then examined by performing regression analysis on the interacting variables (standing green DM, standing dry DM, and total standing dry matter) with the sampled biomass.

Satellite estimates derived from the Cibo Labs model are within the sampled biomass's variability for all treatment plots except the control (Figure 7). There is a closer correlation and high variability in TSDM with the Vault 4 plot, which has three months of resting interval and grazing treatment than other treatment plots. The standard error bars show that the measure of variability with the sampled biomass (Figure 7) correlates with the post-grazing event of 27 January 2022 for treatment plots Vault 3, Vault 4, Vault 5, Bougainville 2, and Bougainville 4, similar to the statistical (GML model) result obtained in Section 3.1. Therefore, Cibo Labs derived from Sentinel-2 imagery can monitor the spatiotemporal variability associated with TSDM for all post-grazing events and the plot (Vault 4) with a regular regenerative grazing plan at the farm level. The Vault 5 plot with the 15-month resting interval has the highest degree of uncertainty compared with other plots. In addition, our findings reveal that Sentinel-2 imagery can account for the trampled residual as in Vault 5, where the TSDM is zero against the trampled biomass (high volume of lying biomass) for the post-grazing event on 27 January (Figure 7). The total trampled residual in this plot was 11,072 kg DM/ha compared to 1004 kg DM/ha of Cibo Labs estimated as TSDM. While the Cibo Labs Sentinel-2-derived model could provide useful information about regenerative grazing for the treatment plots the plots used for conventional grazing (BAU) (Lower Bougainville, Upper Bougainville, and Vault Control) are challenging to estimate (Figure 7).

Regarding the accuracy of the Cibo Labs estimates, the model underestimated the total TSDM in all treatment plots with MAE of 745 kg DM/ha and RMSE of 903 kg DM/ha (Figures 7 and 9) and overestimated the standing green DM (Figures 8 and 10). In addition, the model significantly underestimated the standing dry DM. (Figures 8, 11 and 12). The overestimation of the standing green DM and underestimation of the standing dry DM by the Cibo Lab model reveals that the model calibration is too sensitive to green vegetation and less to dry vegetation. In spring, when biomass growth reached optimum, the model underestimated TSDM in all plots but performed better in the Bougainville plots. The performance of the Cibo Lab model in Bougainville 2, 3, and 4 plots in spring is associated with the slopy hill, which influences the vegetation growth, distribution, and variations in biomass and productivity [58]. In the same way, the underestimation of TSDM in all plots in spring was caused by environmental conditions [16,59] (excess rainfall and soil type), which were not considered during model calibration. In general, the confounding influence of rainfall discussed in Sections 3.1 and 4.1 hardly substantiated any variability in the treatment plots [60]. The fact that there was no statistical interaction between the treatment plots themselves and TSDM except with the grazing dates where we found strong evidence of significant difference with 27 January 2022, shows that there would have been a better correlation between satellite estimates and measured biomass with more post-grazing events. However, the time series charts (Figures 7 and 8) show Sentinel-2 imagery and predictive machine learning model can provide estimates of pasture biomass in monitoring regenerative grazing at the farm level. Such estimates are available as a spatial map providing management decisions per plot as an indication of available pasture biomass. Previous work has demonstrated the capability of machine learning to derive pasture estimates from Sentinel-2 imagery at the farm level [16,28], though not applied to regenerative grazing schemes.

4.3. The Feasibility of Using Sentinel-2 Imagery to Estimate Total Standing Dry Matter

Here, we demonstrated that Sentinel-2 imagery could be used to retrieve TSDM, standing green DM, and standing dry DM through a simple but powerful predictive machine learning to support regenerative grazing that considers a regular grazing and recovery period. To the best of our knowledge, this study is the first to examine this approach at the farm level. However, cloud constraints are one of the major limitations to model performance accuracy in this study. Tasmania is considered a medium-high latitude environment [45]. Despite using smaller fields of less than 1 ha, clouds over Tasmania hindered the consistent availability of Sentinel-2 imagery to feed the predictive machine-learning model used in this study. The lag effect between the available Sentinel-2 imagery and the sampling date ranges from 2 to 40 days (Figure 12). Earlier work supports the argument that time lag effects between field sampling and data from the satellite are a potential source of error to model performance [61–63].

Although the model could retrieve TSDM through inherent Sentinel-2 SWIR band inclusion [19], the association between Cibo Labs TSDM and standing dry DM shows a weak relationship (Figures 11 and 12). Summer is characterised by a high concentration of senescence (Figures 7, 8 and 12) intermixed with green vegetation. Hence, it is challenging to distinguish senesced from healthy vegetation despite the high spatial resolution of Sentinel-2 satellite [64].

A future optimisation study on estimating TSDM with a similar predictive machine learning model will consider more robust ground/field samples that will complement observation satellite data to improve accuracy. In the present study, the number of data available was limited to the summer of 2021 through the spring of 2022 on one farm. The review of [22] suggests that the accuracy of machine learning approaches in estimating aboveground biomass depends on the data source, the number of ground/field samples, pasture species composition, and addressing the errors associated with the algorithms. The improvement of pasture biomass prediction with the ANN algorithm from a similar study on five farms in Tasmania was based on the inclusion of more input parameters (meteorological data) to achieve 0.60 [16]. Sentinel-1 imagery, a synthetic aperture radar, may help address cloud constraints and saturation of optical instruments in cases where limited field datasets are available to estimate pasture biomass [65]. In addition, the frequency of Sentinel-2 imagery can be enhanced by interpolating a daily revisit high resolution of Planet Lab to account for missing data [64,66].

5. Conclusions

We conclude that regenerative grazing with short recovery periods (3–6 months) was most conducive to increasing pasture production under high rainfall conditions. In the one-day grazing treatment, sheep could not exploit selective grazing, but rather the trampling of pasture biomass, which is caused by the disturbance from the high stocking density in the treatment plots. The trampled residual from the post-grazing event was found to be statistically significant, thus, providing an insight into the source of variability in the treatment plots. In the one-day grazing, an insignificant biomass volume was utilised. Therefore, being one of the pioneering studies in this field, there is an opportunity for future research to understand the effect of regenerative grazing in drought or in a year with moderate rainfall. More work is needed to understand the effects of more grazing days (3 to 5) to make regenerative grazing sustainable. Additionally, more robust data on post-grazing should be considered since it is the main effect in the current study.

This study demonstrated that a predictive machine learning model could be developed using Sentinel-2 time-series imagery to estimate TSDM, standing green DM, and standing dry DM to support regenerative grazing at the farm scale. Although the model underestimated TSDM in all the plots, it is within the variability of the measured biomass. Specifically, the model could explain the variability in biomass for the plot (Vault 4) with a regular grazing and recovery period. Furthermore, the model could show the treatment plot (Vault 5) with the highest level of variance. Our subsequent study will use more timely imagery (PlanetScope) with radar imagery (Sentinel-1) with the aim of overcoming some of the limitations associated with the present study, including less frequent satellite pass-overs, as well as a lack of cloud-free images.

We conclude regenerative grazing with shorter recovery periods in wet seasons is more likely to improve grassland productivity, however, this result remains to be seen in drier seasons (e.g., El Nino). We also showed promise in machine learning with satellite imagery at very small field sizes, and we encourage further research into this area.

Author Contributions: Conceptualization, Methodology, Formal Analysis, Investigation, Writing-Original Draft Preparation, M.G.O.; Visualisation, M.G.O., M.T.H. and J.P.G.; Review and Editing, J.P.G.; Review and Editing, A.M.F.; Review and Editing, R.A.C.; Review and Editing, C.M.; Review and Editing; P.S.; Review and Editing, P.T.; Sampling design and field data collection, J.W. and M.T.H.; Review and Editing, M.T.H. All authors have read and agreed to the published version of the manuscript.

Funding: Funding for this work was provided from the Australian Government's Future Drought Fund and the University of Tasmania (Project Number 4-G37I1PA).

Data Availability Statement: All data is available upon request from the corresponding author.

Acknowledgments: Funding support for the Tasmanian Institute of Agriculture was provided through funding from the Australian Government's Future Drought Fund and the University of Tasmania. The authors acknowledge project delivery and infrastructure from the Cape Herbert Pty Ltd., Australia. Special acknowledgment to Peter Ball, who collected the measured data and made it available.

Conflicts of Interest: The authors declare no conflict of interest.

References

1. Harrison, M.T.; Cullen, B.R.; Mayberry, D.E.; Cowie, A.L.; Bilotto, F.; Badgery, W.B.; Liu, K.; Davison, T.; Christie, K.M.; Muleke, A.; et al. Carbon myopia: The urgent need for integrated social, economic and environmental action in the livestock sector. *Glob. Chang. Biol.* **2021**, *27*, 5726–5761. [CrossRef] [PubMed]
2. Christie, K.M.; Smith, A.P.; Rawnsley, R.P.; Harrison, M.T.; Eckard, R.J. Simulated seasonal responses of grazed dairy pastures to nitrogen fertilizer in SE Australia: N loss and recovery. *Agric. Syst.* **2020**, *182*, 102847. [CrossRef]
3. Thornton, P.K. Livestock production: Recent trends, future prospects. *Philos. Trans. R. Soc. London. Ser. B Biol. Sci.* **2010**, *365*, 2853–2867. [CrossRef]
4. Abberton, M. Grassland carbon sequestration: Management. In *Proceedings of the Workshop on the Role of Grassland Carbon Sequestration in the Mitigation Of Climate Change*; Food and Agriculture Organisation: Rome, Italy, 2010; Volume 11.
5. Franzluebbers, A.J. Soil organic carbon in managed pastures of the southeastern United States of America. In *Grassland Carbon Sequestration: Management, Policy and Economics. Integrated Crop Manage*; Food and Agriculture Organisation: Rome, Italy, 2010; pp. 163–175.
6. Henry, B.; Dalal, R.; Harrison, M.T.; Keating, B. Creating frameworks to foster soil carbon sequestration. In *Burleigh Dodds Series in Agricultural Science*; Burleigh Dodds Science Publishing: Cambridge, UK, 2022.
7. Sándor, R.; Ehrhardt, F.; Grace, P.; Recous, S.; Smith, P.; Snow, V.; Soussana, J.-F.; Basso, B.; Bhatia, A.; Brilli, L.; et al. Ensemble modelling of carbon fluxes in grasslands and croplands. *F. Crop. Res.* **2020**, *252*, 107791. [CrossRef]
8. Teague, R.; Kreuter, U. Managing Grazing to Restore Soil Health, Ecosystem Function, and Ecosystem Services. *Front. Sustain. Food Syst.* **2020**, *4*, 534187. [CrossRef]
9. Rawnsley, R.P.; Smith, A.P.; Christie, K.M.; Harrison, M.T.; Eckard, R.J. Current and future direction of nitrogen fertiliser use in Australian grazing systems. *Crop Pasture Sci.* **2019**, *70*, 1034–1043. [CrossRef]
10. Díaz de Otálora, X.; Epelde, L.; Arranz, J.; Garbisu, C.; Ruiz, R.; Mandaluniz, N. Regenerative rotational grazing management of dairy sheep increases springtime grass production and topsoil carbon storage. *Ecol. Indic.* **2021**, *125*, 107484. [CrossRef]
11. Teague, R.; Barnes, M. Grazing management that regenerates ecosystem function and grazingland livelihoods. *Afr. J. Range Forage Sci.* **2017**, *34*, 77–86. [CrossRef]
12. Spratt, E.; Jordan, J.; Winsten, J.; Huff, P.; van Schaik, C.; Jewett, J.G.; Filbert, M.; Luhman, J.; Meier, E.; Paine, L. Accelerating regenerative grazing to tackle farm, environmental, and societal challenges in the upper Midwest. *J. Soil Water Conserv.* **2021**, *76*, 15A–23A. [CrossRef]
13. Öllerer, K.; Varga, A.; Kirby, K.; Demeter, L.; Biró, M.; Bölöni, J.; Molnár, Z. Beyond the obvious impact of domestic livestock grazing on temperate forest vegetation—A global review. *Biol. Conserv.* **2019**, *237*, 209–219. [CrossRef]
14. Harrison, M.; Evans, J.; Moore, A.D. Using a mathematical framework to examine physiological changes in winter wheat after livestock grazing 2. Model validation and effects of grazing management. *Field Crop. Res.* **2012**, *136*, 127–137. [CrossRef]
15. Harrison, M.T.; Evans, J.R.; Moore, A.D. Using a mathematical framework to examine physiological changes in winter wheat after livestock grazing 1. Model derivation and coefficient calibration. *Field Crop. Res.* **2012**, *136*, 116–126. [CrossRef]
16. Chen, Y.; Guerschman, J.; Shendryk, Y.; Henry, D.; Harrison, M.T. Estimating pasture biomass using sentinel-2 imagery and machine learning. *Remote Sens.* **2021**, *13*, 603. [CrossRef]
17. Hudson, T.D.; Reeves, M.C.; Hall, S.A.; Yorgey, G.G.; Neibergs, J.S. Big landscapes meet big data: Informing grazing management in a variable and changing world. *Rangelands* **2021**, *43*, 17–28. [CrossRef]
18. Trotter, M.G.; Lamb, D.W.; Donald, G.E.; Schneider, D.A. Evaluating an active optical sensor for quantifying and mapping green herbage mass and growth in a perennial grass pasture. *Crop Pasture Sci.* **2010**, *61*, 389–398. [CrossRef]
19. Punalekar, S.M.; Verhoef, A.; Quaife, T.L.; Humphries, D.; Bermingham, L.; Reynolds, C.K. Application of Sentinel-2A data for pasture biomass monitoring using a physically based radiative transfer model. *Remote Sens. Environ.* **2018**, *218*, 207–220. [CrossRef]
20. Edirisinghe, A.; Clark, D.; Waugh, D. Spatio-temporal modelling of biomass of intensively grazed perennial dairy pastures using multispectral remote sensing. *Int. J. Appl. Earth Obs. Geoinf.* **2012**, *16*, 5–16. [CrossRef]
21. Wang, J.; Xiao, X.; Bajgain, R.; Starks, P.; Steiner, J.; Doughty, R.B.; Chang, Q. Estimating leaf area index and aboveground biomass of grazing pastures using Sentinel-1, Sentinel-2 and Landsat images. *ISPRS J. Photogramm. Remote Sens.* **2019**, *154*, 189–201. [CrossRef]
22. Morais, T.G.; Teixeira, R.F.M.; Figueiredo, M.; Domingos, T. The use of machine learning methods to estimate aboveground biomass of grasslands: A review. *Ecol. Indic.* **2021**, *130*, 108081. [CrossRef]
23. Harrison, M.T.; Roggero, P.P.; Zavattaro, L. Simple, efficient and robust techniques for automatic multi-objective function parameterisation: Case studies of local and global optimisation using APSIM. *Environ. Model. Softw.* **2019**, *117*, 109–133. [CrossRef]
24. Ali, I.; Greifeneder, F.; Stamenkovic, J.; Neumann, M.; Notarnicola, C. Review of Machine Learning Approaches for Biomass and Soil Moisture Retrievals from Remote Sensing Data. *Remote Sens.* **2015**, *7*, 16398–16421. [CrossRef]
25. Gitelson, A.A.; Gamon, J.A. The need for a common basis for defining light-use efficiency: Implications for productivity estimation. *Remote Sens. Environ.* **2015**, *156*, 196–201. [CrossRef]

26. Delegido, J.; Verrelst, J.; Rivera, J.P.; Ruiz-Verdú, A.; Moreno, J. Brown and green LAI mapping through spectral indices. *Int. J. Appl. Earth Obs. Geoinf.* **2015**, *35*, 350–358. [CrossRef]

27. Bsaibes, A.; Courault, D.; Baret, F.; Weiss, M.; Olioso, A.; Jacob, F.; Hagolle, O.; Marloie, O.; Bertrand, N.; Desfond, V.; et al. Albedo and LAI estimates from FORMOSAT-2 data for crop monitoring. *Remote Sens. Environ.* **2009**, *113*, 716–729. [CrossRef]

28. De Rosa, D.; Basso, B.; Fasiolo, M.; Friedl, J.; Fulkerson, B.; Grace, P.R.; Rowlings, D.W. Predicting pasture biomass using a statistical model and machine learning algorithm implemented with remotely sensed imagery. *Comput. Electron. Agric.* **2021**, *180*, 105880. [CrossRef]

29. Ibrahim, A.; Harrison, M.T.; Meinke, H.; Zhou, M. Examining the yield potential of barley near-isogenic lines using a genotype by environment by management analysis. *Eur. J. Agron.* **2019**, *105*, 41–51. [CrossRef]

30. Zhai, Z.; Martínez, J.F.; Beltran, V.; Martínez, N.L. Decision support systems for agriculture 4.0: Survey and challenges. *Comput. Electron. Agric.* **2020**, *170*, 105256. [CrossRef]

31. Ara, I.; Turner, L.; Harrison, M.T.; Monjardino, M.; deVoil, P.; Rodriguez, D. Application, adoption and opportunities for improving decision support systems in irrigated agriculture: A review. *Agric. Water Manag.* **2021**, *257*, 107161. [CrossRef]

32. Ge, Y.; Thomasson, J.A.; Sui, R. Remote sensing of soil properties in precision agriculture: A review. *Front. Earth Sci.* **2011**, *5*, 229–238. [CrossRef]

33. Bilotto, F.; Harrison, M.T.; Migliorati, M.D.A.; Christie, K.M.; Rowlings, D.W.; Grace, P.R.; Smith, A.P.; Rawnsley, R.P.; Thorburn, P.J.; Eckard, R.J. Can seasonal soil N mineralisation trends be leveraged to enhance pasture growth? *Sci. Total Environ.* **2021**, *772*, 145031. [CrossRef]

34. Ali, I.; Cawkwell, F.; Dwyer, E.; Barrett, B.; Green, S. Satellite remote sensing of grasslands: From observation to management. *J. Plant Ecol.* **2016**, *9*, 649–671. [CrossRef]

35. Smith, R.C.G.; Adams, M.; Gittins, S.; Gherardi, S.; Wood, D.; Maier, S.; Stovold, R.; Donald, G.; Khohkar, S.; Allen, A. Near real-time Feed On Offer (FOO) from MODIS for early season grazing management of Mediterranean annual pastures. *Int. J. Remote Sens.* **2011**, *32*, 4445–4460. [CrossRef]

36. Dingaan, M.N.V.; Tsubo, M. Improved assessment of pasture availability in semi-arid grassland of South Africa. *Environ. Monit. Assess.* **2019**, *191*, 1–12. [CrossRef] [PubMed]

37. Guerschman, J.P.; Hill, M.J.; Renzullo, L.J.; Barrett, D.J.; Marks, A.S.; Botha, E.J. Estimating fractional cover of photosynthetic vegetation, non-photosynthetic vegetation and bare soil in the Australian tropical savanna region upscaling the EO-1 Hyperion and MODIS sensors. *Remote Sens. Environ.* **2009**, *113*, 928–945. [CrossRef]

38. Zhang, B.; Carter, J. FORAGE—An online system for generating and delivering property-scale decision support information for grazing land and environmental management. *Comput. Electron. Agric.* **2018**, *150*, 302–311. [CrossRef]

39. Dong, S.; Shang, Z.; Gao, J.; Boone, R.B. Enhancing sustainability of grassland ecosystems through ecological restoration and grazing management in an era of climate change on Qinghai-Tibetan Plateau. *Agric. Ecosyst. Environ.* **2020**, *287*, 106684. [CrossRef]

40. Donnelly, J.R.; Moore, A.D.; Freer, M. GRAZPLAN: Decision support systems for Australian grazing enterprises—I. Overview of the GRAZPLAN project, and a description of the MetAccess and LambAlive DSS. *Agric. Syst.* **1997**, *54*, 57–76. [CrossRef]

41. Srivastava, N.; Hinton, G.; Krizhevsky, A.; Sutskever, I.; Salakhutdinov, R. Dropout: A simple way to prevent neural networks from overfitting. *J. Mach. Learn. Res.* **2014**, *15*, 1929–1958.

42. Verrelst, J.; Malenovský, Z.; Van der Tol, C.; Camps-Valls, G.; Gastellu-Etchegorry, J.-P.; Lewis, P.; North, P.; Moreno, J. Quantifying Vegetation Biophysical Variables from Imaging Spectroscopy Data: A Review on Retrieval Methods. *Surv. Geophys.* **2019**, *40*, 589–629. [CrossRef]

43. *Earth Observation: Data, Processing and Applications. Volume 1A: Data—Basics and Acquisition*; CRCSI: Melbourne, VIC, Australia, 2018.

44. Scarth, P.; Armston, J.; Flood, N.; Denham, R.; Collett, L.; Watson, F.; Trevithick, B.; Muir, J.; Goodwin, N.; Tindalla, D.; et al. Operation application of the Landsat tineseries to address large area landcover understanding. *Int. Arch. Photogramm. Remote Sens. Spat. Inf. Sci.* **2015**, *XL 3 W3*, 571 575. [CrossRef]

45. Roberts, D.; Mueller, N.; Mcintyre, A. High-Dimensional Pixel Composites From Earth Observation Time Series. *IEEE Trans. Geosci. Remote Sens.* **2017**, *55*, 6254–6264. [CrossRef]

46. Ara, I.; Harrison, M.T.; Whitehead, J.; Waldner, F.; Bridle, K.; Gilfedder, L.; Marques Da Silva, J.; Marques, F.; Rawnsley, R. Modelling seasonal pasture growth and botanical composition at the paddock scale with satellite imagery. *Silico Plants* **2021**, *3*, 1–15. [CrossRef]

47. BoM Australia Government. Bureau of Meteorology. Available online: http://www.bom.gov.au/climate/averages/tables/cw_092027.shtml (accessed on 25 October 2022).

48. Franklin, M. *Okehampton—Optimising Management of Production and Biodiversity Assets*; Devonport TAS; University of Tasmania: Tasmania, Australia, 2019.

49. Phelan, D.C.; Harrison, M.T.; McLean, G.; Cox, H.; Pembleton, K.G.; Dean, G.J.; Parsons, D.; do Amaral Richter, M.E.; Pengilley, G.; Hinton, S.J.; et al. Advancing a farmer decision support tool for agronomic decisions on rainfed and irrigated wheat cropping in Tasmania. *Agric. Syst.* **2018**, *167*, 113–124. [CrossRef]

50. Langworthy, A.D.; Rawnsley, R.P.; Freeman, M.J.; Pembleton, K.G.; Corkrey, R.; Harrison, M.T.; Lane, P.A.; Henry, D.A. Potential of summer-active temperate (C_3) perennial forages to mitigate the detrimental effects of supraoptimal temperatures on summer home-grown feed production in south-eastern Australian dairying regions. *Crop Pasture Sci.* **2018**, *69*, 808–820. [CrossRef]

51. Zhu, Z.; Wang, S.; Woodcock, C.E. Improvement and expansion of the Fmask algorithm: Cloud, cloud shadow, and snow detection for Landsats 4–7, 8, and Sentinel 2 images. *Remote Sens. Environ.* **2015**, *159*, 269–277. [CrossRef]
52. Piñeiro, G.; Perelman, S.; Guerschman, J.P.; Paruelo, J.M. How to evaluate models: Observed vs. predicted or predicted vs. observed? *Ecol. Modell.* **2008**, *216*, 316–322. [CrossRef]
53. Zhu, Q.; Lin, H. Influences of soil, terrain, and crop growth on soil moisture variation from transect to farm scales. *Geoderma* **2011**, *163*, 45–54. [CrossRef]
54. Warren, S.D.; Thurow, T.L.; Blackburn, W.H.; Garza, N.E. The influence of livestock trampling under intensive rotation grazing on soil hydrologic characteristics. *Rangel. Ecol. Manag. Range Manag. Arch.* **1986**, *39*, 491–495. [CrossRef]
55. Khatri-Chhetri, U.; Thompson, K.A.; Quideau, S.A.; Boyce, M.S.; Chang, S.X.; Kaliaskar, D.; Bork, E.W.; Carlyle, C.N. Adaptive multi-paddock grazing increases soil nutrient availability and bacteria to fungi ratio in grassland soils. *Appl. Soil Ecol.* **2022**, *179*, 104590. [CrossRef]
56. van Eekeren, N.; Jongejans, E.; van Agtmaal, M.; Guo, Y.; van der Velden, M.; Versteeg, C.; Siepel, H. Microarthropod communities and their ecosystem services restore when permanent grassland with mowing or low-intensity grazing is installed. *Agric. Ecosyst. Environ.* **2022**, *323*, 107682. [CrossRef]
57. Zwerts, J.A.; Prins, H.H.T.; Bomhoff, D.; Verhagen, I.; Swart, J.M.; de Boer, W.F. Competition between a Lawn-Forming Cynodon dactylon and a Tufted Grass Species Hyparrhenia hirta on a South-African Dystrophic Savanna. *PLoS ONE* **2015**, *10*, e0140789. [CrossRef] [PubMed]
58. Ivanov, V.Y.; Bras, R.L.; Vivoni, E.R. Vegetation-hydrology dynamics in complex terrain of semiarid areas: 2. Energy-water controls of vegetation spatiotemporal dynamics and topographic niches of favorability. *Water Resour. Res.* **2008**, *44*, 5595. [CrossRef]
59. Barrachina, M.; Cristóbal, J.; Tulla, A.F. Estimating above-ground biomass on mountain meadows and pastures through remote sensing. *Int. J. Appl. Earth Obs. Geoinf.* **2015**, *38*, 184–192. [CrossRef]
60. Andresen, J.A.; Alagarswamy, G.; Rotz, C.A.; Ritchie, J.T.; LeBaron, A.W. Weather Impacts on Maize, Soybean, and Alfalfa Production in the Great Lakes Region, 1895–1996. *Agron. J.* **2001**, *93*, 1059–1070. [CrossRef]
61. Edirisinghe, A.; Hill, M.J.; Donald, G.E.; Hyder, M. Quantitative mapping of pasture biomass using satellite imagery. *Int. J. Remote Sens.* **2011**, *32*, 2699–2724. [CrossRef]
62. Myrgiotis, V.; Harris, P.; Revill, A.; Sint, H.; Williams, M. Inferring management and predicting sub-field scale C dynamics in UK grasslands using biogeochemical modelling and satellite-derived leaf area data. *Agric. For. Meteorol.* **2021**, *307*, 108466. [CrossRef]
63. Moore, C.E.; Beringer, J.; Donohue, R.J.; Evans, B.; Exbrayat, J.-F.; Hutley, L.B.; Tapper, N.J. Seasonal, interannual and decadal drivers of tree and grass productivity in an Australian tropical savanna. *Glob. Chang. Biol.* **2018**, *24*, 2530–2544. [CrossRef]
64. Segarra, J.; Buchaillot, M.L.; Araus, J.L.; Kefauver, S.C. Remote Sensing for Precision Agriculture: Sentinel-2 Improved Features and Applications. *Agronomy* **2020**, *10*, 50641. [CrossRef]
65. Crabbe, R.A.; Lamb, D.W.; Edwards, C.; Andersson, K.; Schneider, D. A Preliminary Investigation of the Potential of Sentinel-1 Radar to Estimate Pasture Biomass in a Grazed Pasture Landscape. *Remote Sens.* **2019**, *11*, 70872. [CrossRef]
66. Sadeh, Y.; Zhu, X.; Dunkerley, D.; Walker, J.P.; Zhang, Y.; Rozenstein, O.; Manivasagam, V.S.; Chenu, K. Fusion of Sentinel-2 and PlanetScope time-series data into daily 3 m surface reflectance and wheat LAI monitoring. *Int. J. Appl. Earth Obs. Geoinf.* **2021**, *96*, 102260. [CrossRef]

 *land*

Review

Disentangling the Belowground Web of Biotic Interactions in Temperate Coastal Grasslands: From Fundamental Knowledge to Novel Applications

Gederts Ievinsh

Department of Plant Physiology, Faculty of Biology, University of Latvia, 1 Jelgavas Str., LV-1004 Rīga, Latvia; gederts.ievins@lu.lv

Abstract: Grasslands represent an essential part of terrestrial ecosystems. In particular, coastal grasslands are dominated by the influence of environmental factors resulting from sea–land interaction. Therefore, coastal grasslands are extremely heterogeneous both spatially and temporally. In this review, recent knowledge in the field of biotic interactions in coastal grassland soil is summarized. A detailed analysis of arbuscular mycorrhiza symbiosis, rhizobial symbiosis, plant–parasitic plant interactions, and plant–plant interactions is performed. The role of particular biotic interactions in the functioning of a coastal grassland ecosystem is characterized. Special emphasis is placed on future directions and development of practical applications for sustainable agriculture and environmental restoration. It is concluded that plant biotic interactions in soil are omnipresent and important constituents in different ecosystem services provided by coastal grasslands.

Keywords: arbuscular mycorrhiza; ecosystem services; coastal grasslands; parasitic plants; plant–plant interaction; resilience; rhizobial symbiosis; signal transfer

Citation: Ievinsh, G. Disentangling the Belowground Web of Biotic Interactions in Temperate Coastal Grasslands: From Fundamental Knowledge to Novel Applications. *Land* **2023**, *12*, 1209. https://doi.org/10.3390/land12061209

Academic Editors: Michael Vrahnakis, Yannis (Ioannis) Kazoglou and Manuel Pulido Fernádez

Received: 25 May 2023
Revised: 8 June 2023
Accepted: 9 June 2023
Published: 10 June 2023

1. Introduction

Grasslands form a large and essential part of terrestrial ecosystems in terms of occupied area as well as for both biodiversity maintenance and functional importance. Thirty years of progress in grassland ecosystem research has been reviewed recently, and it was revealed that grassland ecology and grassland ecosystem services were among the two most productive directions of research [1].

There is no doubt that grasslands are hotspots of plant diversity and important components of ecosystem service. In particular, carbon sequestration potential in grassland soils has been recently addressed and its global role has been emphasized [2]. Most importantly, interactions between plant and microbial diversity were recognized as the main driving force in carbon storage. Recently, there have been more attempts to make functional generalizations of grassland existence related to changes or gradients of environmental factors. For example, the belowground characteristics of plants—presence of clonal growth organs, vegetative buds, fine root spread—have been related to the degree of water availability in grasslands [3]. There has been an increasing awareness of the fact that precisely functional interactions involving plants and their diversity are important drivers of plant distribution and multiple ecosystem services in grasslands [4,5].

Among different grassland types, coastal grasslands are unique in that they are habitats where the influence of marine and terrestrial determining factors combine. On the other hand, both coastal areas and grasslands are important in their own right from the point of view of global ecosystem functioning. Conservation of coastal habitats in Europe has been an object of continuous scientific attention in the recent decades [6]. In a broader context, coastal habitats in general and coastal grasslands in particular play an important role in providing ecosystem services. Much attention has been focused on analysis of ecosystem services in the context of grasslands [7]. However, coastal grasslands have often remained underrecognized in the analysis of European grasslands [8].

Biotic interactions in soil have been a relatively understudied aspect of functioning of coastal grasslands, but considerable evidence on adaptive importance of these interactions has accumulated from studies performed in other grassland types as well as in controlled conditions. For example, microbial symbioses with plants have been emphasized for their importance both in vegetation establishment and resilience [9]. Moreover, the contribution of biotic interactions to ecosystem services in general is an aspect that is not widely recognized. However, several aspects of changes in ecosystem services provided by coastal grasslands have been analyzed, including abandonment of grazing [10].

Within the framework of this review, an attempt will be undertaken to clarify whether there is a reason to believe that biotic interactions in the soil are determinants of the diversity and resilience of coastal grasslands and if these interactions can make a significant contribution to ecosystem services. In particular, answers will be sought for the following questions: (i) what types of biotic interactions with possible effects on vegetation composition, productivity, and resilience in coastal grasslands exist; (ii) why biotic interactions in coastal grasslands are important for ecosystem services; (iii) what future studies are necessary, and what are the perspective practical applications?

2. Heterogeneity of Environmental Conditions in Coastal Grasslands

In this section, it will be briefly analyzed how coastal grasslands differ from the other types of grasslands and why these differences are important for biotic interactions in soil. Except location in the immediate vicinity of the seashore and presence of grass and legume species, the definition of "coastal grasslands" might seem like an artificial construct, mostly because of the high heterogeneity of environmental factors leading to extreme diversity and fragmentation of coastal grassland habitats. Primarily, both spatial and temporal variability in soil edaphic conditions have been studied in coastal habitats, and they also clearly affect biotic interactions in the soil [11,12]. On the other hand, it is likely that it is the heterogeneity of soil conditions that accounts for the remarkable diversity in plant species that generally characterizes European temperate grasslands [13]. In addition, it has been suggested that grasslands with higher richness in plant species can buffer the negative effects of environmental heterogeneity on productivity [14].

Unlike other types of grasslands, coastal grasslands are dominated by the influence of environmental factors resulting from sea–land interaction. Analysis of these factors is beyond the scope of this review, but it is necessary to understand how the main types of coastal grasslands are formed due to differences in prevailing conditions. In general, substantial differences in water regime related to geomorphological and littoral processes determine the formation of two main types of coastal grasslands. Dune systems are formed on sand-accumulating active coasts, whereas grasslands form as a continuation of fixed dunes, and the most characteristic dominant environmental factor is drought. On less active non-accumulating shores, grasslands form as a continuation of salt marshes or may already be found in the beach area.

As a result, dominating environmental conditions in fixed-dune-associated and salt-marsh-associated coastal grasslands are fundamentally different. To a large extent, this applies to the soil moisture regime (drought vs. flooding) and potential exposure to salinity (occasional surface spray with seawater vs. periodic inflow of saltwater). In any case, coastal grasslands are subject to sharp fluctuations in environmental conditions over time and large spatial variations even over short distances, and in general form a highly dynamic and heterogeneous system. From the perspective of functional analysis of coastal grasslands, such differences explain the fact that conventionally dry (associated with dunes) and relatively wet (associated with beach or salt marshes) grasslands are considered separately.

A European-scale study has confirmed that dune perennial grasslands are significantly affected by local climatic conditions, resulting in differences in plant species composition and distribution [15]. Mean annual temperature and mean annual precipitation appeared to be the main climate variables affecting floristic variability and community structure. Seven groups of grasslands in different geographical areas have been identified from north to south along an increasing temperature gradient, namely Baltic, North Sea, Atlantic, North Adriatic, Black Sea, South Atlantic, and Mediterranean–Atlantic. It appears that dry coastal grasslands associated with fixed dunes are more vulnerable to global climate change as rising temperature and changes in precipitation patterns can significantly affect species distribution, composition, and abundance [15]. For example, air-borne nitrogen deposition in stable dune grasslands results in a drastic decline in herbaceous species at the expense of dominance of fast-growing grass species [16].

There is not much research on the relationship between environmental heterogeneity in coastal habitats and plant taxonomic and functional diversity [17], but recent attention has at least focused on general aspects of environmental heterogeneity in terms of ecosystem services [18].

3. Diversity in Biological Interactions in Coastal Grasslands

Diversity in microbial interactions in coastal soils and important functional aspects for vegetation establishment and maintenance have been reviewed recently [19]. It is important to note that the main dichotomy between dune- and wetland-associated grasslands in the coastal zone is also reflected by major differences in microbial processes. While sea-water-affected wetland grasslands are characterized by high microbial activity and a high rate of mineralization of organic matter [20], microbial processes in dry grasslands are less active [21]. The complexity of the ongoing microbial processes and their dependence on the heterogeneity of conditions are characterized by ambiguous changes under the influence of complex environmental factors. For example, functional aspects of the nitrogen cycle in coastal habitats are especially affected by the influence of saltwater inflow. While flooding itself results in higher denitrification activity [20], an increase in salinity in freshwater-adapted wetlands leads to a decrease in denitrification rates [22]. However, in saltwater-adapted wetlands, the opposite effect may be observed, namely that intermediate salinity results in an increase in denitrification activity while freshwater intrusion results in almost complete loss of denitrification capacity [23]. These results point to the existence of special adaptation of a consortium of denitrifying microorganisms to a specific salinity level and indicate that hypersaline soils can be used as a source of such resistant strains for practical purposes.

It seems that decomposition of organic matter in coastal grasslands is similarly exposed to the effects of moisture regime and salinity. A community of saprotrophic fungi is a main decomposer of organic matter in grasslands, but bacterial decomposers participate mainly in degradation of relatively labile compounds [24]. Most importantly, microbial communities dominated by fungi shift to bacterial dominance as a result of increased salinity [25]. Many microorganisms involved in decomposition of organic matter produce biologically active substances with beneficial effects on plant growth as hormone-like substances or elicitors of defense responses [26]. There is no doubt that free-living microorganisms are important both for establishment of vegetation in coastal habitats as well as in adaptation of plants to heterogeneous conditions [27], but the main focus of the present review is on symbiotic interactions between plants and microorganisms and on plant–plant relationships, including the ones with parasitic plants.

Aside from mycorrhizal fungi, microbial endophytes represent another group of organisms forming intimate relationships with plants. In particular, fungal endophytes are considered to be important in adaptive responses, including abiotic stress tolerance [28]. However, in order not to complicate the picture and also considering the fact that endophytes are found in all parts of the plant, not only in the soil-bound roots, this aspect of the biotic interaction will not be further analyzed.

4. Mycorrhizal Symbiosis in Coastal Grasslands

4.1. Mycorrhizal vs. Non-Mycorrhizal Plants

For relevant information regarding classification of mycorrhizal associations, as well as characterization of functional aspects of mycorrhizal symbiosis, readers are invited to consult recent articles [29–31]. It needs to be mentioned, however, that, in grassland habitats, arbuscular mycorrhizal symbiosis is of primary importance, but there are also other types of mycorrhiza present, such as orchid mycorrhiza.

Most importantly, in order to understand the importance of mycorrhizal symbiosis, we should first determine whether mycorrhizal symbiosis is a widespread phenomenon in coastal grasslands. On the one hand, abundance, frequency, and anatomical diversity in mycorrhizal associations have been assessed in different coastal habitats, including grasslands [32]. On the other hand, a number of studies show the importance of mycorrhizae in the adaptation of plants to environmental factors, which are characteristic of coastal grasslands. For example, there is a considerable amount of research on the importance of mycorrhizal symbiosis in halophytes and its potential to increase salt tolerance in glycophytes. Interestingly, the first observations on mycorrhizal symbiosis in wild plants were from studies in coastal salt marshes and included several halophytic species [33].

One of the problems in assessing the functional importance of mycorrhizae in coastal grasslands is related to the limited universal nature of mycorrhizal symbiosis. This manifests as a low mycorrhizal intensity in different situations or even the appearance of non-mycorrhizal plant taxa. It needs to be emphasized that it is generally accepted that a relatively low degree of root mycorrhization does not automatically mean little functional importance, and, since the intensity of symbiosis is a highly variable quantity, it is easy to overlook the situation when individuals of a given plant species show significant signs of mycorrhization. Thus, even species described as non-mycorrhizal, such as *Triglochin maritima*, showed mycorrhizal structures in roots, such as vesicles and arbuscules, suggesting the presence of functionally active symbiosis but with relatively low intensity [32]. However, this species did not show any signs of mycorrhizal colonization in the previous studies [34], and genus *Triglochin* has been considered to be non-mycorrhizal [35]. Seasonal changes in mycorrhizal colonization showed that, in roots of *Triglochin maritima*, plants' intensity of mycorrhizal symbiosis increased from less than 5% in May to 25% in July, but the presence of arbuscules was very low in May and June but increased to 15% in July, reaching 25% in September [11]. Therefore, it can be hypothesized that the two dominant environmental factors are associated with the low intensity of mycorrhizal symbiosis on plants in relatively moist coastal grasslands, namely high salinity and soil flooding. These aspects will be analyzed further.

Given the fact that several plant families with significant halophyte occurrence (Ateraceae, Brassicaceae, Chenopodiaceae, Carophyllaceae) have been reported as essentially mostly non-mycorrhizal [35], the question of the general importance of mycorrhizal symbiosis in salt-affected habitats remains open.

4.2. Mycorrhizal Fungal Community Structure

Assessment of genetic diversity in arbuscular-mycorrhiza-forming fungal communities is an important aspect of microbial ecology. Because mycorrhizal fungi are associated with particular plant species and are subject to seasonality and environmental conditions, both the diversity and occurrence of mycorrhizal fungi are highly variable. Ecosystem-level comparison of different studies on the community structure has revealed that some Glomeromycota taxa are found globally, while others can be found only in certain ecosystems [36]. Ecological aspects of arbuscular mycorrhizal symbiosis in halophytic plant species have been reviewed, and it has been suggested that diversity in mycorrhiza-forming fungi seems to be more complex than usually assumed [37].

Only some studies so far have addressed community structure of mycorrhizal fungi in coastal grasslands. Using mycorrhizal roots of a single common plant species, *Hieracium pilosella*, high spatial diversity in fungal phylotypes was found in a sandy coastal grassland [38]. However, even a single root fragment from an individual plant contained almost all genetic variation found within the whole area. Most importantly, it was shown that there is a possibility that a single individual non-sporulating mycelium might cover an area at least 10 m in length. In another study, it was tested if the dominant mycorrhizal fungal strains found in roots of *Hieracium pilosella* can colonize individuals of other abundant plant species, *Hypochaeris radicata*, *Thymus serphyllum*, *Artemisia campestris*, and *Armeria maritima* [39]. As was expected, the dominating strains were found in root fragments of all five plant species but with spatial differences in intensity of occurrence. Therefore, it was concluded that presence of dominant fungal strain is an indication of presence of interconnecting mycelial mycorrhizal network in a coastal grassland.

The effect of changes in various environmental factors on the structure of mycorrhizal fungal communities in coastal grasslands has been studied very little. It should be assumed that the heterogeneity of dominant factors significantly affects this structure. Thus, changes in community structure of AM fungi have been assessed in respect to reclamation of saline coastal lands, and it was shown that vegetation succession following reclamation results in a decrease in overall fungal diversity and a shift from dominance of Acaulosporaceae and Gigasporaceae to Glomeraceae [40]. In salt-affected coastal plains, the community structure of arbuscular mycorrhizal fungi was strongly affected both by soil salinity and pH [41]. In a coastal dune ecosystem, diversity in mycorrhizal fungi was clearly segregated between the seaward (wind-disturbed) and landward (stabilized) slopes of dunes [42]. Therefore, it was concluded that zonal distribution of both abiotic and biotic (including host plant species) factors are determinants of the fungal community structure.

In addition, apart from genetical diversity, functional variability of mycorrhizal fungi seems to be important for the outcome of the effectiveness of the symbiotic relationship. Thus, it was shown that even communities of arbuscular mycorrhizal fungi with relatively low diversity may have significant functional heterogeneity [43]. Such characteristic features refer both to the nature of hyphal growth pattern and the intensity of mineral uptake, and they could also be important in maintaining the diversity in coastal grasslands, especially in mineral-poor soils, such as dune-associated grasslands.

4.3. Mycorrhizal Symbiosis in Resource Acquisition

It is generally accepted that typical mycorrhizal plants provide fungal partners with sugars and vitamins, receiving in return water and minerals, mostly N, P, and K [44,45]. However, terrestrial orchid species have so-called mixotrophic type of nutrition and are dependent on the mycorrhizal partner at certain stages of development and receiving sugars and vitamins from it. Nutrition of mycoheterotophic achlorophyllous plants occurs in the same way. Due to the potentially beneficial effect of the interaction on both partners, mycorrhizal symbiosis is designated as mutualistic and therefore essentially positive.

However, it must be remembered that mycorrhizal symbiosis is not always entirely mutualistic. While usually both partners benefit from the interaction, a continuum of mutualism–parasitism exists in nature [46]. The relationship can be shifted towards parasitism due to the genetic specificity of the particular plant–fungus interaction, or it can be induced by plant developmental stage or environmental factors. Some studies performed with grassland species in controlled conditions indicate that the nature of the mycorrhizal interaction may change differentially depending on the specific situation. Typical grassland species differentially responded to mycorrhizal colonization depending on their relative abundance: dominant species *Taraxacum officinale* and *Agrostis capillaris* were more negatively affected by parasitic-oriented strain of *Glomus intraradices*, but less abundant (subordinate) species *Prunella vulgaris* and *Achillea millefolium* were not negatively affected by the fungus [47]. Thus, plant hierarchy in grasslands can be significantly affected by the presence of particular taxa of mycorrhizal fungi. Moreover, mycorrhizal association of *Hieracium pilosella* was clearly beneficial, while it was parasitic for *Corynephorus canescens*, especially under species competition [48]. Both high available P concentration in soil and shade shifted mycorrhizal interaction from mutualistic to parasitic, showing that the costs of C sent to symbiont exceeded the benefits from increased mineral nutrient availability in these conditions. In this respect, it would be important to determine how the saltwater inundation characteristic of wet coastal grasslands affects the mycorrhizal dependence of different plant species relative to the mutualism–parasitism continuum, especially given the differences in salinity tolerance of various plant species.

4.4. Common Mycorrhizal Networks

The concept of common mycorrhizal networks (CMNs) has gained much scientific interest within recent decades [44,45,49–53]. However, due to obvious technical difficulties, experimental evidence for existence of CMNs in nature is still scant. Instead, studies of varying degrees of complexity are conducted in different model systems under controlled conditions.

Historically, insights into mycorrhizal hyphal associations between multiple plants began with observations of the specific type of nutrition of achlorophyllous parasitic plants, mycoheterotrophy, where reduced carbon substances are received from a mycorrhizal partner, associated with an autotrophic plant [54]. Later, with the discovery of the presence of long-branched extraradical hyphae present in both ectomycorrhizal and arbuscular mycorrhizal associations, the understanding of the potential globality and functional importance of the mycorrhizal network expanded significantly.

The importance of non-mycorrhizal plants (non-host plants) in CMNs has been recently reviewed [55]. In particular, it was concluded that only fungal hyphae from already established mycorrhizal symbiosis can penetrate roots of non-host plants, without formation of any characteristic symbiotic structures. Usually, non-host plants are negatively affected by this type of interaction, and effects on systemic resistance are highly likely.

In contrast to unidirectional movement of mineral nutrients towards host plants, water transport in CMNs is bidirectional and changes during the day [45]. Similarly, it is proposed that movement in signaling substances by means of CMNs can occur in different directions [45]. Carbon cycling has been shown to occur in ectomycorrhizal CMNs but is still controversial in respect to AM networks [50,56]. Therefore, it seems that, apart from some specific situations (as in the case of mycoheterotophic and mixotrophic associations), CMNs are less important as a mechanism for sharing mineral resources between symbiotic plants but rather act as means for information exchange between plants. Thus far, most evidence on the importance of mycorrhizal networks in plant communication is associated with studies on ectomycorrhiza-dominated ecosystems [51]. It has been proposed that stress-associated signals are transmitted more quickly through CMNs if compared to transfer of resources [56]. However, the chemical nature of the signals is far from clear. Initially, it was proposed that plant hormones salicylic acid and jasmonic acid are involved in the transfer of information through CMNs [57]. Recently, scientific information has begun to accumulate that small RNAs can be involved in important aspects of mycorrhizal symbiosis [58–60], but evidence for their role in signaling through CMNs is still lacking.

Moreover, recently, a term "hyposphere" was coined to describe a zone of soil around mycorrhizal hyphae where release of hyphal exudates results in establishment of specific abiotic and biotic conditions [61], forming similar differences from the bulk soil as in the case with the rhizosphere. Similar to exudates from plant roots, exudates from fungal hyphae also have an impact on bacterial diversity and abundance [61].

Continuum of specificity of mycorrhizal fungi is an important aspect to consider in respect to development and function of CMNs [52]. Overlap of host plant compatibility for a particular fungal strain is a critical characteristic for formation of functional CMNs, and, usually, this feature is found for dominant fungal taxa. Therefore, it is logical to assume that dynamics of plant communities are strongly dependent on functioning of CMNs, but empirical evidence for coastal grasslands is still not available. Potential mycorrhiza-related plant interactions in coastal grasslands are shown in Figure 1.

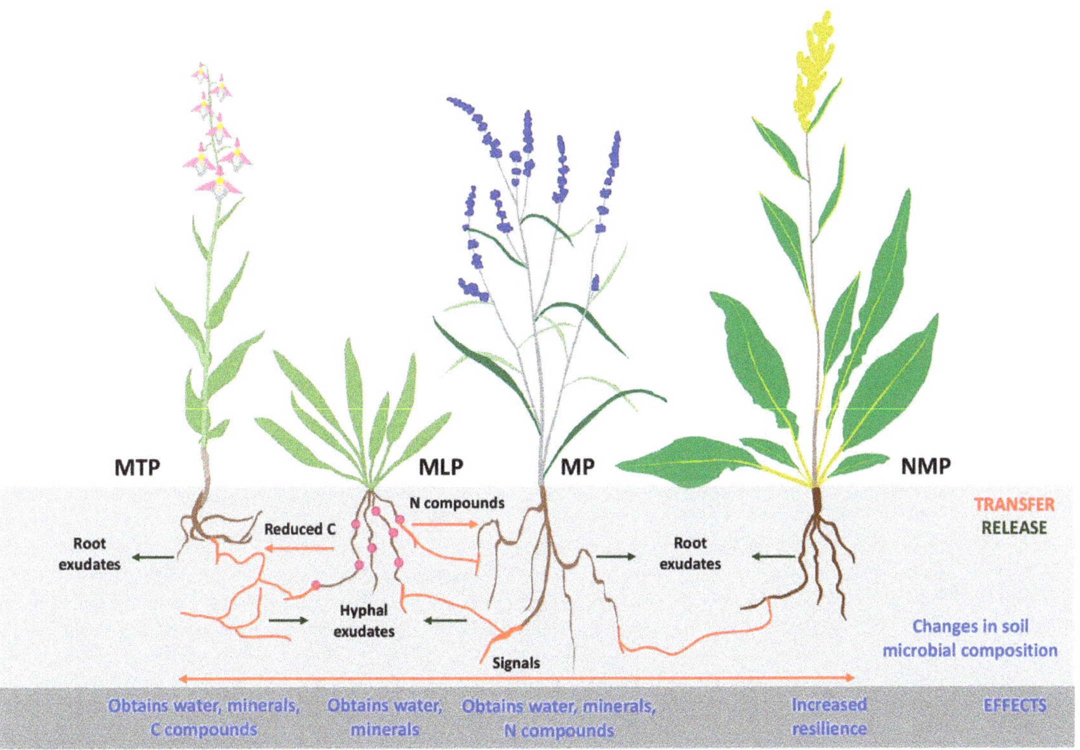

Figure 1. Mycorrhiza-related interactions in coastal grasslands. MTP, mixotrophic plant species; MLP, mycorrhizal legume plant species; MP, mycorrhizal plant species; NMP, non-mycorrhizal plant species. Mycorrhizal root fragments and mycorrhizal hyphae are shown in red.

4.5. Mycorrhizal Symbiosis in Environmental Resilience

The issue of plant adaptation to heterogeneous environmental conditions is particularly important in grassland systems that are subject to sharp fluctuations in environmental conditions, such as coastal grasslands. The presence of mycorrhizal symbiosis in coastal plants has led to the idea of the importance of this type of symbiosis in plant adaptation. However, there has been a scientific debate on which partner of mycorrhizal symbiosis is more vulnerable to environmental constraints, plant host or fungal symbiont? Because there is a concept that mycorrhizal symbiosis is important for the plant to overcome environmental extremes, one might think that the fungal partner is the stronger side in this respect. There is no doubt that, similar to plant species specificity in tolerance to particular environmental factors, mycorrhizal fungal species and strains also differ in their ability to tolerate unfavorable conditions. Indeed, different mycorrhizal fungal species and strains show a wide range of tolerance to one of the dominant environmental constraints in coastal grasslands, soil salinity [62]. More specifically, spore germination was delayed in the presence of NaCl, and, in some cases, the spores did not germinate at all in saline conditions, but, in others, they reached a maximum germination in the presence of 300 mM salt. Similarly, the rate of hyphal extension of some fungal taxa was even stimulated in the presence of 150 mM NaCl, but, in general, salinity inhibited hyphal growth to a varying extent.

Field studies of mycorrhizal symbiosis associated with halophytic plant species in highly saline habitats and often extremely variable moisture levels have usually revealed the presence of functionally active mycorrhizal structures in their roots, however with variable colonization intensity. Examples of such studies are from Tabriz Plain in Iran [63], Central European salt marshes [34] (Hildebrandt et al. 2001), Sečovlje solar salterns in Slovenia [64], and saline soils in Hungary [65]. Seasonal trends in mycorrhizal colonization have been assessed for halophytes in Hungarian steppe habitats [66] and in saltwater-affected wet grassland [11]. The results of these studies suggest that intensity of root colonization is indeed negatively affected by increasing soil salinity, but number of fungal spores does not depend on the level of salinity. It also seems that environmental factors that have an impact on host plant physiology also affect fungal symbionts.

Both diversity and composition of mycorrhizal fungi only marginally differed between halophyte and non-halophyte species growing on salt-affected coastal plains [41]. In a typical halophyte species well-adapted to intermediate and high salinity, such as *Tripolium pannonicum* (syn. *Aster tripolium*), early stages of symbiotic interaction were more negatively affected by salinity in comparison to expansion of root colonization [67]. On the other hand, seasonal changes in intensity of mycorrhizal colonization in roots of several halophyte species growing in salt-affected grassland (*Aster tripolium, Glaux maritima, Plantago maritima, Trifolium fragiferum, Triglochin maritima*) showed negative dependence on fluctuations of soil salinity [11].

A number of entirely practically oriented studies on improvement in salinity tolerance in glycophytic crop species as a result of mycorrhizal fungi application are available, indicating that the use of salt-tolerant fungal strains has great practical potential. Such strains could be isolated from the rhizosphere of salt-affected coastal grassland soils. In particular, mycorrhizal inoculation increased plant growth under saline conditions for *Gossypium arboreum* [68], *Pennisetum glaucum* [69], *Zea mays* [70], *Triticum aestivum* [71], *Lactuca sativa* [72], *Lens culinaris* [73], and *Ocimum basilicum* [74]. In some studies, it has been stressed that the fungal strains used were isolated from saline habitats [70]. In one case, effect of two isolates of *Glomus mosseae*, either form non-saline or saline soil, were compared, and it was shown that, contrary to what was initially expected, the isolate from non-saline soil had a higher capacity to alleviate negative effects of salinity [68].

Similar experiments in controlled conditions have been performed also with some halophyte species. Thus, it was shown that mycorrhizal plants of grass species *Puccinellia distans* had better growth potential under saline conditions due to enhanced photosynthesis and improved ion homeostasis [75]. In addition, mycorrhizal symbiosis affected accumulation of osmotically active mineral elements, allowing to avoid uptake of Na [76]. In particular, for *Trifolium alexandrinum*, increased phosphorus uptake in mycorrhizal plants was associated with their better growth in saline conditions [77]. Nine psammophilic species native to coastal sand dunes and evidently adapted only to salt spray were tested for their ability to recover after repeated seawater treatment in controlled conditions when inoculated with mycorrhizal fungus *Glomus intraradices* [78]. The intensity of the survival-promoting effect of mycorrhizal colonization was a distinctly species-specific feature. On the positive side, mycorrhizal *Ammophila arenaria* plants showed less than 20% mortality after fourth application of 100% seawater concentration, while all non-mycorrhizal plants died after the third application of 100% seawater. From the worst side, mycorrhization of *Dorycinum pentaphyllum* plants only marginally improved their survival under diluted seawater treatment. However, both species showed similarly high mycorrhizal dependency and high intensity of root colonization by mycorrhizal fungus.

Soil flooding, either with fresh water or seawater, represents another common environmental factor in low-lying coastal grasslands. A number of practically oriented studies on flooding tolerance of mycorrhizal plants have been performed, including seedlings of *Citrus sinensis* [79], *Prunus persica* [80,81], *Pterocarpus officinalis* [82], and showing the overall beneficial effect of mycorrhization on flooding tolerance. Improved mineral nutrition, proline production, and suppression of ethanol production in roots during anoxic conditions were among the mechanisms responsible for growth improvement due to mycorrhizal symbiosis in flooded conditions. However, in a study with *Oryza sativa*, it was shown that, while symbiosis activated the phosphorus uptake pathway in a fungal partner, it suppressed phosphorus uptake of the host plant [83]. As a result, mycorrhizal colonization decreased shoot phosphorus content in flooded conditions, and the effect of symbiosis was negative.

Closer to the topic of this review, two grass species (*Panicum hemitomon* and *Leersia hexandra*) native to nutrient-poor depressional wetlands in the southeastern USA coastal plain were used in a wetland mesocosm experiment to determine if controlled water regimes affect mycorrhizal colonization as well as if colonization affects plant growth [84]. It appears that intensity of mycorrhizal colonization decreases with increasing water levels even for species well adapted even to semi-aquatic conditions. However, mycorrhizal viability was not negatively affected, and symbiotic plants had higher phosphorus uptake even under flooded soil conditions in comparison to non-mycorrhizal plants. Mycorrhizal colonization of the same two grass species was assessed in field conditions along a hydrological gradient, and it was evident that the degree of root colonization decreased with water depth, but this did not affect number of mycorrhizal propagules in soil [85]. However, even plants growing in permanently flooded soil retained active mycorrhizal symbiosis in roots. Similarly, the number of hyphae and spores in soil with *Zea mays* plants was not affected by extended flooding in controlled conditions [86]. Other studies also supported the idea that flooding negatively influences root colonization with arbuscular mycorrhiza, but basic symbiotic functionality is not affected [67,87]. Similarly, in halophyte *Aster tripolium* plants, better tolerance of mycorrhizal plants to flooding was associated with improved osmotic balance and nitrogen uptake [88].

A field study along a tidal gradient in a mangrove swamp indicated that duration of flooding period mainly affected the community structure of arbuscular mycorrhizal fungi and resulted in increased intensity of mycorrhizal colonization [89]. When aquatic species *Polygonum hydropiper* and semiaquatic species *Panicum repens* grown under different hydrological regimes were compared in respect to mycorrhizal colonization and mycorrhizal community structure in natural conditions, it appeared that high flooding intensity led to a decrease in both mycorrhizal intensity and diversity level in both species [90]. However, moderate flooding resulted in an increase in mycorrhizal colonization and fungal species richness only in aquatic species *Polygonum hydropiper*. Recently, the role of arbuscular mycorrhizal symbiosis in wetland plants has been reviewed, and it was concluded that survival and development of these plants in native conditions is highly dependent on mycorrhizal colonization [91].

It is difficult to generalize a potential role of mycorrhiza in respect to plant adaptation to soil moisture regime in coastal grasslands due to extreme variability in this factor across different coastal grassland habitats. However, it is evident that soil moisture regime is a significant determinant of both mycorrhizal community structure as well as intensity of symbiosis and its functional properties [92]. On the other hand, there is no doubt that mycorrhizal symbiosis modulates morphological and biochemical adaptations of drought-stressed plants, as summarized in the recent reviews [93,94]. As mycorrhizal colonization usually results in formation of induced systemic resistance of host plants [95], further studies of specific responses to drought in mycorrhizal vs. non-mycorrhizal coastal grassland plants are needed.

5. Rhizobial Symbiosis in Coastal Grasslands

Legume plant species (Fabaceae) are of special importance both in natural as well as agroecosystems due to symbiosis with N_2-fixing bacteria. Rhizobial symbiosis in wild legume plants provides an important contribution to the nitrogen cycle on Earth, being a part of the biological nitrogen fixation process. From an ecological point of view, shortage in plant-available nitrogen is one of the factors limiting plant growth in heterogeneous habitats, such as coastal dunes and dune grasslands [96], and both plant community structure and productivity are affected by symbiotic rhizobia in these habitats [97]. Additionally, rhizobia–legume symbiosis is the major type of N acquisition into soil of arid ecosystems [98]. From the perspective of sustainable agriculture, inclusion of legumes in crop sequences allows additional fixed nitrogen to accumulate in the soil, increasing plant-available N pool and in general benefiting non-legumes cultivated in subsequent years and allowing to decrease application of N-based chemical fertilizers [99]. This allows for efficient use of low-input agricultural systems. In addition, factors not related to N are also important for soil sustainability, possibly related to nodule-emitted hydrogen, with further effects on soil microbial diversity [100]. It is also becoming clear that tolerance of legume crop species to adverse environmental conditions can be positively affected by rhizobial symbiosis [101]. This aspect is especially important due to global climate change and its negative impact on agricultural productivity. Possible rhizobial-symbiosis-related interactions in coastal grasslands are shown in Figure 2.

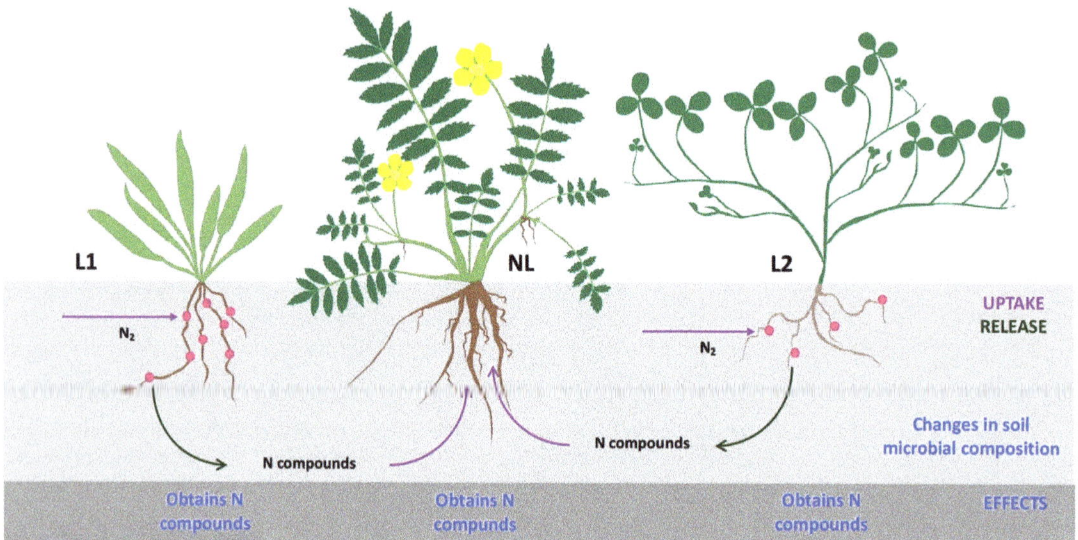

Figure 2. Rhizobial-symbiosis-related interactions in coastal grasslands. L1, L2, symbiotic legume plant species; NL, non-legume species.

Rhizobial bacteria (Pseudomonadota: α-proteobacteria and β-proteobacteria) are Gram-negative soil bacteria. As facultative symbionts, rhizobia are freely living in soil (resident rhizobia) but are able to benefit from forming symbiosis with legume species as N_2 fixation occurs only in symbiotic rhizobia. Competition among soil-resident rhizobia for nodule formation can lead to formation of completely or partially inefficient N_2 fixation as rhizobial strains with high competitive ability might have low N_2 fixation efficiency. Therefore, it is always necessary to distinguish between nodulation specificity (an ability to infect a legume to form a nodule) and effectiveness of N_2 fixation (an ability of formed nodules to fix N_2). Nodulation specificity seems to be associated with modulation of plant immunity as a result of developing interaction between the partners [102]. By this mechanism, hosts can restrict nodulation by even potentially efficient symbionts, resulting in nodule formation with inefficient symbionts leading to parasitic type of interaction [103].

It appears that nodulation specificity and intensity are determined mainly by a host plant [104]. Host plants are able to select microorganisms from bulk soils both at the taxonomic and functional level [105]. Some rhizobia have very high host specificity, such as the ones forming nodules only on *Cicer arietinum*, having highly conserved genes involved in both nodulation and N_2 fixation [106]. Some legume hosts (such as *Glycine max* and *Sophora flavescens*) have very low specificity for rhizobia, being nodulated by various rhizobial species possessing diverse symbiosis-associated genes [107]. In addition, soil factors (edaphic factors) can further modulate the outcome of the established symbiosis and also can have an effect on nodule microbiome, such as, for example, in the case of soil pH and *Trifolium* species [108]. Moreover, the importance of plant community effects cannot be ruled out [109]. On the other hand, metabolically more versatile rhizobial strains, being capable to use a wide range of energy-providing substrates, are usually more competitive in contrast to metabolic specialist strains [110]. For *Trifolium* and other legume genera, strains with effective N_2 fixation have been shown to be more competitive for nodule occupancy [111,112].

Critical soil conditions (such as low soil moisture, salinization, soil waterlogging, etc.), whose likelihood of occurrence continues to increase due to global climate changes and overall anthropogenic pressure, are likely to negatively affect symbiotic N_2 fixation in legume crops [101]. Many rhizobial strains native to local soils or commercially produced rhizobial products have low efficiency or even low viability in unfavorable conditions; therefore, they will not provide efficient contribution to the soil N pool necessary for sustainable agricultural production or to increase soil fertility. Isolation of resilient rhizobial strains will allow to develop new bacterial products suitable for problematic soils and highly heterogeneous environmental conditions. Use of local bacterial isolates adapted to particular (local) environmental and soil conditions is especially desirable for this purpose. Such an approach can possibly prevent so called 'rhizobial competition problem', when rhizobial strains effective in controlled conditions fail to be successful in field conditions, being outcompeted by highly nodulating but inefficient indigenous soil bacteria better adapted to local conditions [113,114].

In the ecological context of ecosystem functioning, there is a reason to believe that rhizobial symbiosis in coastal grasslands acts as an important determinant factor in interactions both between plant species as well as between plant species and their environment. While principal empirical evidence exists for the critical role of rhizobial symbiosis in determining both plant productivity and community structure in dune-associated coastal grassland derived from microcosm studies [97], similar information is sparse for salt-affected wetland grassland systems. However, separate studies have shown the ability of rhizobial symbiosis to promote growth of legume plants under salinity conditions. Thus, dual inoculation with mycorrhizal fungi and rhizobium stimulated growth and improved mineral nutrition of salt-stressed *Lathyrus sativus* plants [115]. Moreover, extreme tolerance of grassland species *Lotus tenuis* to drought, waterlogging, and salinity has been associated with its ability to form early associations both with rhizobia as well as arbuscular mycorrhizal fungi [116].

Recently, two experiments in controlled conditions have been performed with two coastal legume species, *Trifolium fragiferum* from salt-affected wet grassland [117] and *Anthyllis maritima* from dry dune grassland [118]. Rhizobial symbiosis was a significant factor, which determined the nature of the interaction between *Trifolium fragiferum* and *Trifolium repens* [117]. In particular, plant growth was affected by interaction between the origin of bacterial isolate, NaCl treatment, and species coexistence. It was also concluded that, in conditions when one legume species has established symbiosis with more efficient N_2-fixing bacteria in comparison to that of other species, the species with less efficient symbiosis can benefit from this interaction. This mechanism is similar to that described for interaction between symbiotic legume species and non-legume species [119]. For *Anthyllis maritima*, rhizobial symbiosis differentially affected growth and physiological performance of plants through interaction of salinity and burial with sand [118]. Symbiotic conditions positively affected photosynthesis-related traits, but the effect was negative for growth and tissue integrity indices.

Salinity tolerance vs. susceptibility of symbiotic nitrogen fixation in legume species has been reviewed, with an emphasis on practical use of legumes in saline agriculture [120]. In particular, the list of salt-tolerant nitrogen-fixing plant species was included, showing that more than 40 legume species have an important potential in this respect. Within the present review, 11 legume species characteristic for coastal grasslands of the Baltic Sea have been identified (Table 1). Clover species (*Trifolium*) from coastal grasslands are especially promising targets for assessing both genetical and functional diversity in rhizobial symbiosis. *T. fragiferum* is one of the most resistant clover species with high potential for practical use. While not economically used in Europe, it is exploited in the USA, Australia, and New Zealand as a resilient component of temperate perennial grasslands. The species has shown great tolerance against different unfavorable environmental and anthropogenic factors, and it can be classified as a crop wild relative, with a potential for use in breeding of tolerant forage crops. Recently, physiological and genetic diversity in *T. fragiferum* accessions from Latvia were comprehensively characterized and it was shown that geographically isolated wild populations of *T. fragiferum* from the Baltic Sea region are important as a source of abiotic-stress-tolerance related genes [121,122]. Especially interesting is the fact that, in Northern Europe, the species is naturally associated with coastal habitats. *T. fragiferum* micropopulations are geographically isolated and can be found in sites with relatively high soil salinity [123], and their salinity tolerance has been confirmed in controlled conditions [124]. Most importantly, dependence of *T. fragiferum* plants belonging to different accessions on their native symbiotic rhizobia was experimentally characterized [125]. Particular host plant–rhizobia combinations showed significant differences in plant growth stimulation and N acquisition, pointing to existence of genetic variation in N_2-fixing ability within the bacterial population in the Baltic Sea region. As several of the studied *T. fragiferum* accessions are especially tolerant to saline and waterlogging conditions, it is highly possible that their associated native rhizobia have pronounced tolerance to these conditions. While rhizobial diversity in other clover species, such as *Trifolium repens* or *Trifolium pratense*, has been relatively well characterized (for example, [126]), there are no previous functional studies involving rhizobial isolates from *T. fragiferum* plants. Therefore, further characterization and selection of salt- and soil-waterlogging-tolerant rhizobial strains from root nodules of *T. fragiferum* plants in the Baltic Sea region is a promising direction of future studies.

Table 1. Legume plant species from coastal grasslands.

Species	Presence in Coastal Habitats [1]	Salinity Tolerance [1]	Presence in eHALOPH Database (Life Form) [2]
Anthyllis vulneraria subsp. *maritima* (Hagen) Corb. (syn. *Anthyllis maritima* Schweigg. ex K.G.Hagen)	0	1	–
Lathyrus palustris L.	3	3	hydrohalophyte
Lotus maritimus L.	6	3	–
Lotus tenuis Waldst. and Kit. ex Willd.	7	4	hydrohalophyte
Melilotus albus Medik.	1	2	annual
Melilotus altissimus Thuill.	1	3	–
Melilotus dentatus (Waldts. and Kit.) Pers.	4	4	annual
Ononis spinosa L.	4	2	–
Trifolium fragiferum L.	7	3	herbaceous perennial
Trifolium pratense L.	1	2	–
Trifolium repens L.	1	2	–

[1] Tyler et al. [127]; [2] eHALOPH database (V4.65, https://ehaloph.uc.pt, accessed on 15 May 2023).

6. Plant–Parasitic Plant Interactions in Coastal Grasslands

Several groups of plants have evolved parasitic lifestyles and are benefiting from direct interaction with common plant species acting as their hosts [128]. In contrast to mycoheterotrophic parasites, obtaining resources from host plants indirectly through their symbiotic mycorrhizal partners, haustorial parasites feed directly on host tissues through modified root homologous structure, haustorium [128]. Differences in the degree of dependence of the parasite on the host plant determine their further classification, which is associated with significant functional differences. Parasites requiring attachment to their hosts for completing the life cycle are known as obligate, while facultative parasites are being able to reproduce without attachment. In relation to photosynthesis, parasitic plants are either hemiparasites (being able to photosynthesize) or holoparasites (lacking photosynthesis). Regarding the place of attachment of the haustoria to host plant, root versus stem parasites have been recognized. However, it was recently discussed that, since some hemiparasitic *Cuscuta* species possessing chlorophyll derive 99% of organic carbon from their hosts, the type of functional connection (either to xylem or phloem) is more important for classification regardless of photosynthetic ability [129]. Therefore, an alternative classification system of parasitic plants has been proposed based on particular functional characteristics in the life cycle: euphytoid parasites, mistletoes, parasitic vines, obligate root parasites, and endoparasites [129].

Interactions between parasitic plants and their hosts have gained recent scientific interest mainly for several practical reasons. First, several parasitic plants are economically important weeds to crops, such as species of *Orobanche* and *Striga* [130]. Second, hemiparasitic plants of genus *Rhinanthus* and some other genera are recognized as ecosystem engineers, significantly affecting species diversity and abundance in grassland habitats [131]. However, the relationship between parasitic plants and their hosts could also have a wider meaning, both in a fundamentally biological sense and in ensuring resilience of ecosystems.

Orobanchaceae is the largest parasitic plant family, with over 2100 species [128]. Facultative hemiparasitic plants (or euphytoid parasites according to the recent system of classification) of the genus *Rhinanthus* are photosynthetically active and at least partially autotrophic but benefit from haustorial contact with host plants as means for uptake of xylem water together with inorganic nutrients and organic substances. Species of the genus are widely distributed in grassland habitats. While having only xylem connectivity, even *Rhinanthus* spp. are able to obtain a significant part of carbohydrates from their host plants [132]. *Rhinanthus* spp. have low host specificity and can use several plant species as hosts simultaneously [133]. However, plant species differ in their ability to resist parasitic interactions, which could be dependent on host gene silencing [134]. Aside from the effects of a parasite on host plants, host plant functional characteristics seem to be important determinants of the relationship as the morphology of both *R. minor* and *R. angustifolius* plants is shown to be affected by the host species [135,136].

Recently, a role of *Rhinanthus* species in grassland biodiversity at multiple trophic levels has been reviewed [137]. It was found that the most common effect is decrease in abundance and/or biomass of grass species, but the effect on plant species diversity is either neutral or positive. As grasses are better hosts for *Rhinanthus* species, an increase in density of *Rhinanthus* plants usually linearly decreases both biomass of grasses as well as cumulative cover of legumes [138]. Due to differences in plant susceptibility to parasitic plants, competitive ability of hosts is decreased while that of non-host species concomitantly increases [139]. Therefore, different species of *Rhinanthus* have been used for restoration of grasslands in Europe [140]. Other hemiparasitic plant species, such as *Pedicularis canadensis* and *Comandra umbellata*, can be used for grassland restoration in different parts of the world [141].

Thus far, communications between parasitic plants and their hosts have been analyzed mostly from the perspective of bidirectional exchange of chemical factors during establishment of the relationship [142]. Existence of other effects of parasitic plant–host plant interactions beyond resource transfer are highly possible. A relatively early review discussed the possibility that transfer of mRNAs from host plant to parasite can affect the fate of their interaction [143]. In other parasite–host interactions besides *Rhinanthus* spp., exchange of proteins and RNAs is a factor contributing to the development of the interactions and their outcome. Thus, during interaction between *Cuscuta* spp. and their hosts, it was found that mRNAs move bidirectionally [144,145]. Most importantly, host-derived mRNAs are translated to protein in the parasite [146]. In addition, novel parasite-derived miRNAs target host plant mRNAs [147].

The idea that parasitic plants, similar to these of *Rhinanthus* spp., parasitizing multiple hosts simultaneously, might act as founders of common root networks similar to these made by mycorrhizal hyphae has been expressed recently [148]. In contrast to *Cuscuta* spp. plant parasites [149], these types of relationships have not been explored in *Rhinanthus*–host plant associations, and no information is available on three-way interactions between a parasite and two host plants belonging to different species. It can be proposed that, in addition to resources, there is an intensive exchange of signals, including small RNAs, between the parasitic *Rhinanthus* spp. and its host, which can influence their response to the action of other environmental factors. *Rhinanthus* species are commonly found also in coastal habitats [150], where periodic flooding with seawater is one of the crucial determining abiotic factors for species coexistence. It can be hypothesized that both species diversity and resilience in plant associations where hemiparasitic plants are present are positively affected through exchange of signals by means of a parasite–host network. In particular, salinity tolerance of non-halophytic plant species can be boosted by presence of halophytic plant species.

Numerous previous studies have shown the importance of miRNAs in post-transcriptional regulation of plant responses and tolerance to salinity both in halophytes [151,152] and glycophytes [153–155], and were reviewed recently [156,157]. In general, it was concluded that one of the most important miRNA target groups are transcription factors, in turn having control functions over salinity responses. Therefore, it is highly likely that, in conditions of salt-affected habitats, including coastal grasslands, transfer of miRNAs by parasitic plant–host network affects salinity tolerance of individual plants involved in it.

Several parasitic plant species have been described as able to make associations with halophytic plant species, including obligate holoparasite *Cynomorium coccineum* [158,159], *Cuscuta salina* [160], and *Cuscuta campestris* [161]. Interestingly, transmission of Na and Cl ions from host plant to mistletoe parasite *Plicosepalus acaciae* under increased salinity has been shown, and it was concluded that the parasite can be classified as euhalophyte [162]. In this respect, no information is available on putative salinity tolerance and ion accumulation potential of *Rhinanthus* spp., but it can be expected that *Rhinanthus serotinus* accessions found in salt-affected grasslands will have considerable salinity tolerance, at least when parasitizing on halophytic hosts.

In temperate coastal grassland plant communities, several hemiparasitic plant species of family Orobanchaceae are relatively common (Table 2, Figure 3). There is no information available if obligate plant species can be found in coastal grasslands, but these are frequently found in coastal salt marshes [160,163,164]. Interestingly, potential hemiparasitic plant species from coastal grasslands appear to be non-mycorrhizal (Table 2). Only plants from genus *Pedicularis* have been reported as facultatively mycorrhizal [35]. Potential plant–parasitic plant interactions in coastal grasslands are shown in Figure 4.

Table 2. Hemiparasitic plant species of family Orobanchaceae from coastal grasslands.

Species	Presence in Coastal Habitats [1]	Salinity Tolerance [1]	Presence in Coastal Habitats [1]	Mycorrhizal Status [2]
Euphrasia nemorosa (Pers.) Wettst.	1	2	–	NM
Euphrasia stricta J.P.Wolff ex J.F.Lehm.	1	2	–	NM
Melampyrum arvense L.	0	1	–	NM
Odontites litoralis Fr.	10	4	parasite	NM
Odontites vernus (Bellardi) Dumort.	3	2	–	NM
Odontites vulgaris Moench	2	2	–	NM
Pedicularis palustris L.	3	2	–	NM-AM
Rhinanthus minor L.	1	2	–	NM
Rhinanthus serotinus (Schön) Oborny (syn. *R. angustifolius* C.C.Gmel.)	1	2	–	NM

NM, non-mycorrhizal; AM, arbuscular mycorrhiza. [1] Tyler et al. [127]; [2] Soudzilovskaia et al., 2020 [35].

Figure 3. Hemiparasitic plant species found in coastal grasslands. (**A**) *Rhinanthus serotinus* together with *Agrostis stolonifera*, *Centaurea jacea*, *Phragmites australis*, *Trifolium fragiferum*, *Trifolium pratense* in salt-affected wet coastal grassland on island of Kihnu, Estonia. (**B**) *Melampyrum pratense* in coastal grassland on island of Saaremaa, Estonia. (**C**) *Odontites vulgaris* together with *Agrostis stolonifera*, *Centaurea jacea*, *Phragmites australis*, *Trifolium fragiferum*, *Trifolium pratense* in salt-affected wet coastal grassland on island of Kihnu, Estonia. (**D**) *Euphrasia nemorosa* in coastal dune grassland on Pape, Latvia.

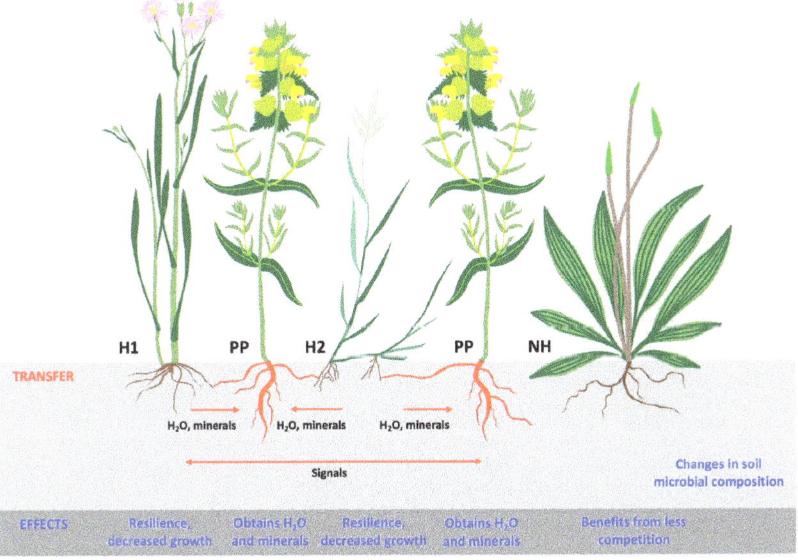

Figure 4. Parasitic-plant-related interactions in coastal grasslands. PP, parasitic plant species; H1, H2, different host plant species; NH, non-host plant species.

7. Plant–Plant Interactions in Coastal Grasslands

The problem of interactions between plants and their role in ecosystem functioning is not a very often studied problem. However, it has been assessed both experimentally as well as using synthetic approaches. In general, both competition (leading to detrimental effects) and facilitation (leading to beneficial effects) are considered as the main general principles of species interactions [165,166]. In respect to competition for resources, nutrients, water, and light are considered to have main importance [167]. Interaction between plants beyond resource acquisition can result from release of chemical substances into the environment as a result of a process known as allelopathy [168]. The definition of allelopathy includes direct effects of compounds released by one plant (or plant remains in soil) on other plants, but it is clearly evident that indirect effects are most common, such as the effect of root exudates on soil microbial diversity. A recent meta-analysis on allelopathic effects on plants has revealed that coexistence of taxonomically related species as well as dominance of single species can be facilitated by means of allelopathy [169]. Moreover, root-emitted volatile organic compounds are important clues in plant–plant interactions and can significantly affect plant defense responses [170]. A potential role of allelopathy in grasslands has been reviewed relatively recently [171,172], and it is evident that allelopathic effects in coastal grasslands cannot be ignored. On top of that, kin recognition acts as a mechanism controlling both plant communication and defense [173,174], but this aspect will not be further analyzed because there are not many studies specifically using coastal-grassland-related model systems.

It has been hypothesized that positive Interactions between plant species are more common in less favorable environmental conditions, while competition prevails under conditions approaching optimum [175,176]. There is some reason to believe that, similarly, interspecies competition will decrease under more heterogeneous conditions compared to less heterogeneous ones. Usually, only competition between individuals belonging to different species is assessed experimentally, but it is evident that both intraspecific and interspecific interaction need to be considered [177]. Detailed analysis of conceptual approaches in respect to facilitation, including differences between interspecific and intraspecific relationships, has been performed [178]. An additional problem is related to differences between pairwise vs. multi-species designs in species competition experiments, clearly indicating that plant interactions in complex plant communities show both additive and non-additive effects [179]. Diversity in plant–plant interactions in coastal grasslands is shown in Figure 5.

Grasslands in general have been studied in terms of plant–plant relationships. In particular, competition between grassland species has been assessed in respect to drought and heavy rainfall [180] and soil moisture gradient in alpine grasslands [181]. Information on plant–plant interactions in coastal grasslands is rather limited. More data are available for associated coastal habitats, salt marshes, and sand dunes. In salt-affected habitats, as in coastal salt marshes, plant competition is an important mechanism, which determines the distribution of species along the salinity gradient depending on the salt tolerance of the plants [182]. While the majority of typical halophyte species are able to grow and reproduce efficiently in non-saline conditions, they are not able to compete successfully with less-salt-tolerant species in low-salinity conditions. Therefore, species distribution in habitats with pronounced salinity gradients reflects their relative salinity tolerance. However, non-tidal salt marshes and salt-affected wet coastal grasslands are characterized by large spatial and temporal variation in soil salinity, not allowing for establishment of clear vegetation patterns [11,183]. Usually, in such habitats, individuals of halophyte species grow next to individuals of less-salt-tolerant species, suggesting that there are other types of interactions between these species besides competition.

Facilitation has been assessed in coastal communities, as in the case of established *Honckenya peploides* plants forming favorable conditions for germination and emergence of trapped seeds of *Leymus mollis* [184]. In addition, an increase in the intensity of sand accretion for dune-adapted plant species, while it reduced plant biomass, promoted facilitation between them [185].

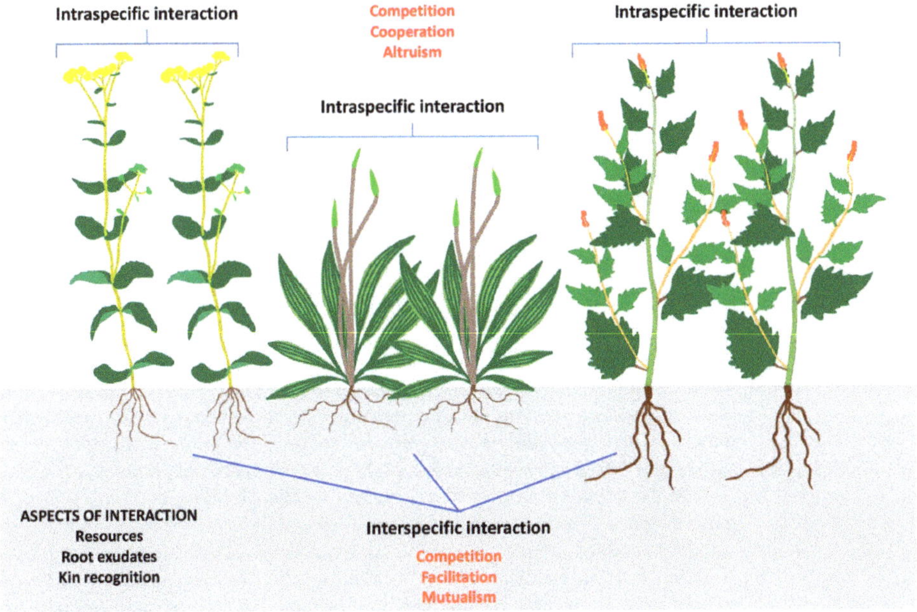

Intraspecific interaction

Competition
Cooperation
Altruism

Intraspecific interaction

ASPECTS OF INTERACTION
Resources
Root exudates
Kin recognition

Interspecific interaction

Competition
Facilitation
Mutualism

Figure 5. Diversity in plant–plant interactions in coastal grasslands.

Plant–plant interactions in coastal dunes have been reviewed, and it was concluded that understanding of this type of interactions is especially important for conservation and restoration [186]. There is no doubt that the mutual influence of different plant species at the functional level is also an important aspect for understanding the operation and resilience of the coastal grassland ecosystem in general.

Plant–plant interactions with possible importance in coastal grasslands are summarized in Figure 5. In addition, all types of biotic interactions in soil analyzed in this review evidently affect plant–plant interactions, including mycorrhizal [187] and rhizobial symbiosis [117,118]. Interactions between parasitic plants and their hosts include also effects on non-host species and need to be taken into the account when total plant interactions in a habitat are considered.

It is clear that one of the results of plant–plant interaction is the appearance of species associations. However, the formation and existence of these associations are usually examined only from the side of changes in environmental factors, but functional interactions are seldom analyzed. Only relatively recently, the conceptual basis of molecular aspects of plant–plant interactions has started to take shape [188]. This section is not mandatory but may be added if there are patents resulting from the work reported in this manuscript.

8. Conclusions and Perspectives

The performed analysis of information clearly shows that plant biotic interactions in soil are omnipresent and important constituents in different ecosystem services provided by coastal grasslands. Not only are supporting and regulating services strongly dependent on these interactions as affecting primary production, nutrient cycling, invasion resistance, etc., but provisioning services can also greatly benefit, for example, from discovery of resistant symbiotic microorganisms that could be used in provision of agricultural resilience. It is no less important that coastal grasslands can serve as a source of empirical knowledge about the impact of environmental heterogeneity on ecosystem functioning and the importance of plant interactions in it.

It seems that the coexistence of species in grassland habitats as well as environmental resilience of these plant assemblages are more directly affected by biotic interactions in soil than previously thought. Interactive effects of mycorrhiza and rhizobial symbiosis of legume plants have been assessed, showing that plants benefit more from dual interactions [189]. However, many potentially mycorrhizal legume plants can act as hosts of hemiparasitic plants. A study of such three-way interactions would be particularly challenging for salt-affected coastal grassland plants.

Based on analysis of biotic interactions in soil of coastal grasslands, several lines of research seem to be especially promising, both for the design of experimental systems as well as choice regarding model plant species. Such general possible research directions could include the following: role of clonal plants in environmental resilience of coastal grasslands, as recently analyzed in respect to the role of clonal growth in halophyte resistance to heterogenous salinity conditions [190]; transfer of hormonal signals and small RNAs between individual organisms by mycorrhizal and parasitic plant networks and their regulative effect on plant growth and responses to environmental constraints, especially, salinity and flooding; role of processes of epigenetic memory as mechanisms for fine-tuning plant adaptation to relatively short-term but persistent changes in environmental conditions in coastal grassland habitats; functional role of symbiotic interactions in adaptation to highly heterogeneous availability of plant nutrients in coastal grasslands; and many others.

At the level of mechanisms of interaction between plants and their symbiotic microorganisms, as well as between parasitic plants and their hosts, inoculation experiments in highly controlled conditions can be successfully applied. Simplified experimental systems, such as aseptically cultivated seedlings or root cultures, seem to be particularly promising as they enable eliminating undesirable effects of soil-related factors [191–195]. The exchange of chemical and molecular signals between the involved partners and the functional results can be monitored by means of various molecular biology approaches. High-throughput sequencing can be used to efficiently sequence transcriptome and small RNA libraries. Comparison of sequence data from different variants of experimentally manipulated plant–symbiont as well as host plant–parasitic plant combinations will allow for identification of differentially expressed genes as well as for possible movement of RNA molecules between interacting partner organisms. Combined with modern methods of data analysis, such as statistical network analysis [196,197], such studies will provide an opportunity to critically evaluate the importance of plant biological interactions in functioning of ecosystems and the services they provide, including these of coastal grasslands.

From the point of view of practical innovations and developments, based on an understanding of biotic relationships in coastal grasslands, resilient symbiotic microorganisms, both arbuscular mycorrhizal fungi and rhizobial bacteria, need to be isolated and identified. Detailed genetical and functional characterization of the isolated fungal and bacterial strains could lead to development of new plant fertilizers and growth stimulants for the promotion of sustainable agriculture or urban greening measures, especially useful in marginal or degraded lands.

Funding: This research received no external funding.

Data Availability Statement: All data are taken from published sources. All photographs are taken by the author.

Conflicts of Interest: The author declares no conflict of interest.

References

1. Zhu, X.; Zheng, J.; An, Y.; Xin, X.; Xu, D.; Yan, R.; Xu, L.; Shen, B.; Hou, L. Grassland Ecosystem Progress: A Review and Bibliometric Analysis Based on Research Publication over the Last Three Decades. *Agronomy* **2023**, *13*, 614. [CrossRef]
2. Bai, J.; Wang, X.; Jia, J.; Zhang, G.; Wang, Y.; Zhang, S. Denitrification of soil nitrogen in coastal and inland salt marshes with different flooding frequencies. *Phys. Chem. Earth Parts A/B/C* **2017**, *97*, 31–36. [CrossRef]

3. Klimešová, J.; Martínková, J.; Bartušková, A.; Ott, J.P. Belowground plant traits and their ecosystem functions along aridity gradients in grasslands. *Plant Soil* **2023**, 1–10. [CrossRef]

4. Hanisch, M.; Schweiger, O.; Cord, A.F.; Volk, M.; Knapp, S. Plant functional traits shape multiple ecosystem services, their trade-offs and synergies in grasslands. *J. Appl. Ecol.* **2020**, *57*, 1535–1550. [CrossRef]

5. In 't Zandt, D.; Hoekstra, N.J.; de Caluwe, H.; Cruijsen, P.M.J.M.; Visser, E.J.W.; de Kroon, H. Plant life-history traits rather than soil legacies determine colonization of soil patches in a multi-species grassland. *J. Ecol.* **2022**, *110*, 889–901. [CrossRef]

6. Delbosc, P.; Lagrange, I.; Rozo, C.; Bensettiti, F.; Bouzillé, J.-B.; Evans, D.; Lalanne, A.; Rapinel, S.; Bioret, F. Assessing the conservation status of coastal habitats under Article 17 of the EU Habitats Directive. *Biol. Conserv.* **2021**, *254*, 108935. [CrossRef]

7. Bengtsson, J.; Bullock, J.M.; Egoh, B.; Everson, C.; Everson, T.; O'Connor, T.; O'Farrell, P.J.; Smith, H.G.; Lindborg, R. Grasslands-more important for ecosystem services than you might think. *Ecosphere* **2019**, *10*, e02582. [CrossRef]

8. Feurdean, A.; Ruprecht, E.; Molnár, Z.; Hutchinson, S.M.; Hickler, T. Biodiversity-rich European grasslands: Ancient, forgotten ecosystems. *Biol. Conserv.* **2018**, *228*, 224–232. [CrossRef]

9. Revillini, D.; Gehring, C.A.; Johnson, N.C. The role of locally adapted mycorrhizas and rhizobacteria in plant–soil feedback systems. *Funct. Ecol.* **2016**, *30*, 1086–1098. [CrossRef]

10. Ford, H.; Garbutt, A.; Jones, D.L.; Jones, L. Impacts of grazing abandonment on ecosystem service provision: Coastal grassland as a model system. *Agric. Ecosyst. Environ.* **2012**, *162*, 108–115. [CrossRef]

11. Karlsons, A.; Druva-Lusite, I.; Osvalde, A.; Necajeva, J.; Andersone-Ozola, U.; Samsone, I.; Ievinsh, G. Adaptation strategies of rare plant species to heterogeneous soil conditions on a coast of a lagoon lake as revealed by analysis of mycorrhizal symbiosis and mineral constituent dynamics. *Environ. Exp. Biol.* **2017**, *15*, 113–126.

12. Karlsons, A.; Osvalde, A.; Ievinsh, G. Growth and mineral nutrition of two Triglochin species from saline wetlands: Adaptation strategies to conditions of heterogeneous mineral supply. *Environ. Exp. Biol.* **2011**, *9*, 83–90.

13. Pärtel, M.; Bruun, H.H.; Sammul, M. Biodiversity in temperate European grasslands: Origin and conservation. In *Integrating Efficient Grassland Farming and Biodiversity, Proceedings of the 13th International Occasional Symposium of the European Grassland Federation; Grassland Science in Europe, Tartu, Estonia, 29–31 August 2005*; Lillak, R., Viiralt, R., Linke, A., Geherman, V., Eds.; European Grassland Federation: Tartu, Estonia, 2005; Volume 10, pp. 1–14.

14. Daleo, P.; Alberti, J.; Chaneton, E.J.; Iribarne, O.; Tognetti, P.M.; Bakker, J.D.; Borer, E.T.; Bruschetti, M.; MacDougall, A.S.; Pascual, J.; et al. Environmental heterogeneity modulates the effect of plant diversity on the spatial variability of grassland biomes. *Nat. Commun.* **2023**, *14*, 1809. [CrossRef]

15. Del Vecchio, S.; Fantinato, E.; Janssen, J.; Bioret, F.; Acosta, A.; Prisco, I.; Tzonev, R.; Marcenò, C.; Rodwell, J.; Buffa, G. Biogeographic variability of coastal perennial grasslands at the European scale. *Appl. Veg. Sci.* **2018**, *21*, 312–321. [CrossRef]

16. van den Berg, L.J.; Tomassen, H.B.; Roelofs, J.G.; Bobbink, R. Effects of nitrogen enrichment on coastal dune grassland: A mesocosm study. *Environ. Pollut.* **2005**, *138*, 77–85. [CrossRef] [PubMed]

17. Ievinsh, G. Biological basis of biological diversity: Physiological adaptations of plants to heterogeneous habitats along a sea coast. *Acta Univ. Latv.* **2006**, *710*, 53–79.

18. Dronova, I. Environmental heterogeneity as a bridge between ecosystem service and visual quality objectives in management, planning and design. *Landsc. Urban Plan.* **2017**, *163*, 90–106. [CrossRef]

19. Ievinsh, G. Where land meets sea: Biology of coastal soils. In *Structure and Functions of Pedosphere*; Giri, B., Kapoor, R., Wu, Q.-S., Varma, A., Eds.; Springer: Singapore, 2022; pp. 151–173.

20. Bai, Y.; Cotfruto, M.F. Grassland soil carbon sequestration: Current understanding, challenges, and solutions. *Science* **2022**, *377*, 603–608. [CrossRef]

21. Jones, M.L.; Sowerby, A.; Williams, D.L.; Jones, R.E. Factors controlling soil development in sand dunes: Evidence from a coastal dune soil chronosequence. *Plant Soil* **2008**, *307*, 219–234. [CrossRef]

22. Huang, L.; Zhang, G.; Bai, J.; Xia, Z.; Wang, W.; Jia, J.; Wang, X.; Liu, X.; Cui, B. Desalinization via freshwater restoration highly improved microbial diversity, co-occurrence patterns and functions in coastal wetland soils. *Sci. Total. Environ.* **2021**, *765*, 142769. [CrossRef]

23. Marks, B.M.; Chambers, L.; White, J.R. Effect of Fluctuating Salinity on Potential Denitrification in Coastal Wetland Soil and Sediments. *Soil Sci. Soc. Am. J.* **2016**, *80*, 516–526. [CrossRef]

24. Garcia-Pausas, J.; Paterson, E. Microbial community abundance and structure are determinants of soil organic matter mineralisation in the presence of labile carbon. *Soil Biol. Biochem.* **2011**, *43*, 1705–1713. [CrossRef]

25. Chen, J.; Wang, H.; Hu, G.; Li, X.; Dong, Y.; Zhuge, Y.; He, H.; Zhang, X. Distinct accumulation of bacterial and fungal residues along a salinity gradient in coastal salt-affected soils. *Soil Biol. Biochem.* **2021**, *158*, 108266. [CrossRef]

26. Pii, Y.; Mimmo, T.; Tomasi, N.; Terzano, R.; Cesco, S.; Crecchio, C. Microbial interactions in the rhizosphere: Beneficial influences of plant growth-promoting rhizobacteria on nutrient acquisition process. A review. *Biol. Fertil. Soils* **2015**, *51*, 403–415. [CrossRef]

27. Otlewska, A.; Migliore, M.; Dybka-Stępień, K.; Manfredini, A.; Struszczyk-Świta, K.; Napoli, R.; Białkowska, A.; Canfora, L.; Pinzari, F. When Salt Meddles Between Plant, Soil, and Microorganisms. *Front. Plant Sci.* **2020**, *11*, 553087. [CrossRef]

28. Qin, Y.; Druzhinina, I.S.; Pan, X.; Yuan, Z. Microbially Mediated Plant Salt Tolerance and Microbiome-based Solutions for Saline Agriculture. *Biotechnol. Adv.* **2016**, *34*, 1245–1259. [CrossRef] [PubMed]

29. Wipf, D.; Krajinski, F.; van Tuinen, D.; Recorbet, G.; Courty, P. Trading on the arbuscular mycorrhiza market: From arbuscules to common mycorrhizal networks. *New Phytol.* **2019**, *223*, 1127–1142. [CrossRef]

30. Berger, F.; Gutjahr, C. Factors affecting plant responsiveness to arbuscular mycorrhiza. *Curr. Opin. Plant Biol.* **2021**, *59*, 101994. [CrossRef] [PubMed]

31. Sportes, A.; Hériché, M.; Boussageon, R.; Noceto, P.-A.; van Tuinen, D.; Wipf, D.; Courty, P.E. A historical perspective on mycorrhizal mutualism emphasizing arbuscular mycorrhizas and their emerging challenges. *Mycorrhiza* **2021**, *31*, 637–653. [CrossRef]

32. Druva-Lusite, I.; Ievinsh, G. Diversity of arbuscular mycorrhizal symbiosis in plants from coastal habitats. *Environ. Exp. Biol.* **2010**, *8*, 17–34.

33. Mason, E. Note on the presence of mycorrhiza in the roots of salt marsh plants. *New Phytol.* **1928**, *27*, 193–195. [CrossRef]

34. Hildebrandt, U.; Janetta, K.; Ouziad, F.; Renne, B.; Nawrath, K.; Bothe, H. Arbuscular mycorrhizal colonization of halophytes in Central European salt marshes. *Mycorrhiza* **2001**, *10*, 175–183. [CrossRef]

35. Soudzilovskaia, N.A.; Vaessen, S.; Barcelo, M.; He, J.; Rahimlou, S.; Abarenkov, K.; Brundrett, M.C.; Gomes, S.I.; Merckx, V.; Tedersoo, L. FungalRoot: Global online database of plant mycorrhizal associations. *New Phytol.* **2020**, *227*, 955–966. [CrossRef]

36. Öpik, M.; Moora, M.; Liira, J.; Zobel, M. Composition of root-colonizing arbuscular mycorrhizal fungal communities in different ecosystems around the globe. *J. Ecol.* **2006**, *94*, 778–790. [CrossRef]

37. Bothe, H. Arbuscular mycorrhiza and salt tolerance of plants. *Symbiosis* **2012**, *58*, 7–16. [CrossRef]

38. Rosendahl, S.; Stukenbrock, E.H. Community structure of arbsicualr mycorrhizal fungi in undisturbed vegetation revealed by analyses of LSU rDNA sequences. *Mol. Ecol.* **2004**, *13*, 3179–3186. [CrossRef]

39. Stukenbrock, E.H.; Rosendahl, S. Distribution of dominant arbuscular mycorrhizal fungi among five plant speccies in undisturbed vegetation of a coastal grassland. *Mycorrhiza* **2005**, *15*, 497–503. [CrossRef]

40. Cui, X.; Hu, J.; Wang, J.; Yang, J.; Lin, X. Reclamation negatively influences arbuscular mycorrhizal fungal community structure and diversity in coastal saline-alkaline land in Eastern China as revealed by Illumina sequencing. *Appl. Soil Ecol.* **2016**, *98*, 140–149. [CrossRef]

41. Guo, X.; Gong, J. Differential effects of abiotic factors and host plant traits on diversity and community composition of root-colonizing arbuscular mycorrhizal fungi in a salt-stressed ecosystem. *Mycorrhiza* **2014**, *24*, 79–94. [CrossRef] [PubMed]

42. Kawahara, A.; Ezawa, T. Characterization of arbuscular mycorrhizal fungal communities with respect to zonal vegetation in a coastal dune ecosystem. *Oecologia* **2013**, *173*, 533–543. [CrossRef] [PubMed]

43. Munkvold, L.; Kjøller, R.; Vestberg, M.; Rosendahl, S.; Jakobsen, I. High functional diversity within species of arbuscular mycorrhizal fungi. *New Phytol.* **2004**, *164*, 357–364. [CrossRef]

44. Selosse, M.-A.; Richard, F.; He, X.; Simard, S.W. Mycorrhizal networks: Des liaisons dangereuses? *Trends Ecol. Evol.* **2006**, *21*, 621–628. [CrossRef] [PubMed]

45. Barto, E.K.; Weidenhamer, J.D.; Cipollini, D.; Rillig, M.C. Fungal superhighways: Do common mycorrhizal networks enhance below ground communication? *Trends Plant Sci.* **2012**, *17*, 633–637. [CrossRef]

46. Johnson, N.C.; Graham, J.H.; Smith, F.A. Functioning of mycorrhizal associations along the mutualism-parasitism continuum. *New Phytol.* **1997**, *135*, 575–585. [CrossRef]

47. Mariotte, P.; Meugnier, C.; Johnson, D.; Thébault, A.; Spiegelberger, T.; Buttler, A. Arbuscular mycorrhizal fungi reduce the differences in competitiveness between dominant and subordinate plant species. *Mycorrhiza* **2013**, *23*, 267–277. [CrossRef]

48. Friede, M.; Unger, S.; Hellmann, C.; Beyschlag, W. Conditions Promoting Mycorrhizal Parasitism Are of Minor Importance for Competitive Interactions in Two Differentially Mycotrophic Species. *Front. Plant Sci.* **2016**, *7*, 1465. [CrossRef]

49. Simard, S.W.; Durall, D.M. Mycorrhizal networks: A review of their extent, function, and importance. *Can. J. Bot.* **2004**, *82*, 1140–1165. [CrossRef]

50. Simard, S.W.; Beiler, K.J.; Bingham, M.A.; Deslippe, J.R.; Philip, L.J.; Teste, F.P. Mycorrhizal networks: Mechanisms, ecology and modelling. *Fungal Biol. Rev.* **2012**, *26*, 39–60. [CrossRef]

51. Gorzelak, M.A.; Asay, A.K.; Pickles, B.J.; Simard, S.W. Inter-plant communication through mycorrhizal networks mediates complex adaptive behaviour in plant communities. *AoB Plants* **2015**, *7*, plv050. [CrossRef] [PubMed]

52. Molina, R.; Horton, T.R. Mycorrhiza specificity: Its role in the development and function of common mycelial networks. In *Mycorrhizal Networks*; Horton, T.R., Ed.; Springer Science + Business Media: Dordrecht, The Netherlands, 2015; pp. 1–39.

53. Gilbert, L.; Johnson, D. Plant–plant communication through common mycorrhizal networks. In *Advances in Botanical Research*; Becard, G., Ed.; How Plants Communicate with their Biotic Environment; Academic Press: London, UK, 2017; Volume 82, pp. 83–97.

54. Bidartondo, M.I. The evolutionary ecology of myco-heterotrophy. *New Phytol.* **2005**, *167*, 335–352. [CrossRef] [PubMed]

55. Wang, Y.; He, X.; Yu, F. Non-host plants: Are they mycorrhizal networks players? *Plant Divers.* **2022**, *44*, 127–134. [CrossRef] [PubMed]

56. Boyno, G.; Demir, S. Plant-mycorrhizal communication and mycorrhizae in inter-plant communication. *Symbiosis* **2022**, *86*, 155–168. [CrossRef]

57. Song, Y.; Zeng, R.S.; Xu, J.F.; Li, J.; Shen, X.; Yihdego, W.G. Interplant Communication of Tomato Plants through Underground Common Mycorrhizal Networks. *PLoS ONE* **2010**, *5*, e13324. [CrossRef] [PubMed]

58. Wu, P.; Wu, Y.; Liu, C.-C.; Liu, L.-W.; Ma, F.-F.; Wu, X.-Y.; Wu, M.; Hang, Y.-Y.; Chen, J.-Q.; Shao, Z.-Q.; et al. Identification of Arbuscular Mycorrhiza (AM)-Responsive microRNAs in Tomato. *Front. Plant Sci.* **2016**, *7*, 429. [CrossRef]

59. Couzigou, J.-M.; Lauressergues, D.; André, O.; Gutjahr, C.; Bruno Guillotin, B.; Bécard, G.; Combier, J.-P. Positive gene regulation by a natural protective miRNA enables arbuscular mycorrhizal symbiosis. *Cell Host Microbe* **2017**, *21*, 106–112. [CrossRef]

60. Šečić, E.; Kogel, K.-H.; Ladera-Carmona, M.J. Biotic stress-associated microRNA families in plants. *J. Plant Physiol.* **2021**, *263*, 153451. [CrossRef] [PubMed]

61. Zhang, L.; Zou, J.; George, T.S.; Limpens, E.; Feng, G. Arbuscular mycorrhizal fungi conducting the hyposphere bacterial orchestra. *Trends Plant Sci.* **2022**, *27*, 402–411. [CrossRef]

62. Juniper, S.; Abbott, L.K. Soil salinity delays germination and limits growth of hyphae from propagules of arbuscular mycorrhizal fungi. *Mycorrhiza* **2006**, *16*, 371–379. [CrossRef]

63. Aliasgharzadeh, N.; Rastin, S.N.; Towfighi, H.; Alizadeh, A. Occurrence of arbuscular mycorrhizal fungi in saline soils of the Tabriz Plain of Iran in relation to some physical and chemical properties of soil. *Mycorrhiza* **2001**, *11*, 119–122. [CrossRef]

64. Sonjak, S.; Udovič, M.; Wraber, T.; Likar, M.; Regvar, M. Diversity of halophytes and identification of arbuscular mycorrhizal fungi colonising their roots in an abandoned and sustained part of Sečovlje salterns. *Soil Biol. Biochem.* **2009**, *41*, 1847–1856. [CrossRef]

65. Füzy, A.; Biró, B.; Tóth, T. Effect of saline soil parameters on endomycorrhizal colonisation of dominant halophytes in four Hungarian sites. *Span. J. Agric. Res.* **2010**, *8*, 144. [CrossRef]

66. Füzy, A.; Biró, B.; Tóth, T.; Hildebrandt, U.; Bothe, H. Drought, but not salinity, determines the apparent effectiveness of halophytes colonized by arbuscular mycorrhizal fungi. *J. Plant Physiol.* **2008**, *165*, 1181–1192. [CrossRef]

67. Carvalho, L.M.; Correia, P.M.; Caçador, I.; Martins-Loução, M.A. Effects of salinity and flooding on the infectivity of salt marsh arbuscular mycorrhizal fungi in *Aster tripolium* L. *Biol. Fertil. Soils* **2003**, *38*, 137–143. [CrossRef]

68. Tian, C.; Feng, G.; Li, X.; Zhang, F. Different effects of arbuscular mycorrhizal fungal isolates from saline or non-saline soil on salinity tolerance of plants. *Appl. Soil Ecol.* **2004**, *26*, 143–148. [CrossRef]

69. Borde, M.; Dudhane, M.; Jite, P. Growth photosynthetic activity and antioxidant responses of mycorrhizal and non-mycorrhizal bajra (Pennisetum glaucum) crop under salinity stress condition. *Crop. Prot.* **2011**, *30*, 265–271. [CrossRef]

70. Estrada, B.; Aroca, R.; Barea, J.M.; Ruiz-Lozano, J.M. Native arbuscular mycorrhizal fungi isolated from a saline habitat improved maize antioxidant systems and plant tolerance to salinity. *Plant Sci.* **2013**, *201–202*, 42–51. [CrossRef]

71. Talaat, N.B.; Shawky, B.T. Protective effects of arbuscular mycorrizal fungi on wheat (*Triticum aestivum* L.) plants exposed to salinity. *Environ. Exp. Bot.* **2014**, *98*, 20–31. [CrossRef]

72. Hammer, E.C.; Forstreuter, M.; Rillig, M.C.; Kohler, J. Biochar increases arbuscular mycorrhizal plant growth enhancement and ameliorates salinity stress. *Appl. Soil Ecol.* **2015**, *96*, 114–121. [CrossRef]

73. Rahman, M.; Ali, M.; Alam, F.; Banu, M.; Anik, M.; Bhuiyan, M. Effect of Arbuscular Mycorrhizal Fungi on Germination, Nodulation and Sporulation of Lentil (*Lens culinaris*) at Different NaCl Levels. *Bangladesh J. Microbiol.* **2017**, *34*, 73–81. [CrossRef]

74. Elhindi, K.M.; El-Din, A.S.; Elgorban, A. The impact of arbuscular mycorrhizal fungi in mitigating salt-induced adverse effects in sweet basil (*Ocimum basilicum* L.). *Saudi J. Biol. Sci.* **2017**, *24*, 170–179. [CrossRef] [PubMed]

75. Dashtebani, F.; Hajiboland, R.; Aliasgharzad, N. Characterization of salt-tolerance mechanisms in mycorrhizal (*Claroideoglomus etunicatum*) halophytic grass, Puccinellia distans. *Acta Physiol. Plant.* **2014**, *36*, 1713–1726. [CrossRef]

76. Hammer, E.C.; Nasr, H.; Pallon, J.; Olsson, P.A.; Wallander, H. Elemental composition of arbuscular mycorrhizal fungi at high salinity. *Mycorrhiza* **2011**, *21*, 117–129. [CrossRef]

77. Shokri, S.; Maadi, B. Effects of Arbuscular Mycorrhizal Fungus on the Mineral Nutrition and Yield of Trifolium alexandrinum Plants under Salinity Stress. *J. Agron.* **2009**, *8*, 79–83. [CrossRef]

78. Camprubí, A.; Abril, M.; Estaún, V.; Calvet, C. Contribution of arbuscular mycorrhizal symbiosis to the survival of psammophilic plants after sea water flooding. *Plant Soil* **2012**, *351*, 97–105. [CrossRef]

79. Hartmond, U.; Schaesberg, N.V.; Graham, J.H.; Syvertsen, J.P. Salinity and flooding stress effects on mycorrhizal and non-mycorrhizal citrus rootstock seedlings. *Plant Soil* **1987**, *104*, 37–43. [CrossRef]

80. Rutto, K.L.; Mizutani, F.; Kadoya, K. Effect of root zone flooding on mycorrhizal and non-mycorrhizal peach (*Prunus persica* Batsch) seedlings. *Sci. Hortic.* **2002**, *94*, 285–295. [CrossRef]

81. Zheng, F.-L.; Liang, S.-M.; Chu, X.-N.; Yang, Y.-L.; Wu, Q.-S. Mycorrhizal fungi enhance flooding tolerance of peach through inducing proline accumulation and improving root architecture. *Plant Soil Environ.* **2020**, *66*, 624–631. [CrossRef]

82. Fougnies, L.; Renciot, S.; Müller, F.; Plenchette, C.; Prin, Y.; De Faria, S.M.; Bouvet, J.M.; Sylla, S.N.; Dreyfus, B.; Bâ, A.M. Arbuscular mycorrhizal colonization and nodulation improve flooding tolerance in Pterocarpus officinalis Jacq. seedlings. *Mycorrhiza* **2007**, *17*, 159–166. [CrossRef]

83. Wang, Y.; Bao, X.; Li, S. Effects of Arbuscular Mycorrhizal Fungi on Rice Growth Under Different Flooding and Shading Regimes. *Front. Microbiol.* **2021**, *12*, 756752. [CrossRef]

84. Miller, S.P.; Sharitz, R.R. Manipulation of flooding and arbuscular mycorrhiza formation influences growth and nutrition of two semiaquatic grass species. *Funct. Ecol.* **2000**, *14*, 738–748. [CrossRef]

85. Miller, S.P. Arbuscular mycorrhizal colonization of semi-aquatic grasses along a wide hydrologic gradient. *New Phytol.* **2000**, *145*, 145–155. [CrossRef]

86. Ellis, J.R. Post Flood Syndrome and Vesicular-Arbuscular Mycorrhizal Fungi. *J. Prod. Agric.* **1998**, *11*, 200–204. [CrossRef]

87. Vallino, M.; Fiorilli, V.; Bonfante, P. Rice flooding negatively impacts root branching and arbuscular mycorrhizal colonization, but not fungal viability. *Plant Cell Environ.* **2014**, *37*, 557–572. [CrossRef]

88. Neto, D.; Carvalho, L.M.; Cruz, C.; Martins-Loução, M.A. How do mycorrhizas affect C and N relationships in flooded *Aster tripolium* plants? *Plant Soil* **2006**, *279*, 51–63. [CrossRef]
89. Wang, Y.; Huang, Y.; Qiu, Q.; Xin, G.; Yang, Z.; Shi, S. Flooding Greatly Affects the Diversity of Arbuscular Mycorrhizal Fungi Communities in the Roots of Wetland Plants. *PLoS ONE* **2011**, *6*, e24512. [CrossRef]
90. Wang, Y.; Li, Y.; Bao, X.; Björn, L.O.; Li, S.; Olsson, P.A. Response differences of arbuscular mycorrhizal fungi communities in the roots of an aquatic and a semiaquatic species to various flooding regimes. *Plant Soil* **2016**, *403*, 361–373. [CrossRef]
91. Huang, G.-M.; Srivastava, A.K.; Zou, Y.-N.; Wu, Q.-S.; Kuča, K. Exploring arbuscular mycorrhizal symbiosis in wetland plants with a focus on human impacts. *Symbiosis* **2021**, *84*, 311–320. [CrossRef]
92. Deepika, S.; Kothamasi, D. Soil moisture—A regulator of arbuscular mycorrhizal fungal community assembly and symbiotic phosphorus uptake. *Mycorrhiza* **2015**, *25*, 67–75. [CrossRef] [PubMed]
93. Cheng, S.; Zou, Y.-N.; Kuča, K.; Hashem, A.; Abd_Allah, E.F.; Wu, Q.-S. Elucidating the Mechanisms Underlying Enhanced Drought Tolerance in Plants Mediated by Arbuscular Mycorrhizal Fungi. *Front. Microbiol.* **2021**, *12*, 809473. [CrossRef]
94. Wang, Y.; Zou, Y.-N.; Shu, B.; Wu, Q.-S. Deciphering molecular mechanisms regarding enhanced drought tolerance in plants by arbuscular mycorrhizal fungi. *Sci. Hortic.* **2023**, *308*, 111591. [CrossRef]
95. Pozo, M.J.; Azcón-Aguilar, C. Unraveling mycorrhiza-induced resistance. *Curr. Opin. Plant Biol.* **2007**, *10*, 393–398. [CrossRef] [PubMed]
96. Skiba, V.; Wainwright, M. Nitrogen transformation in coastal sands and dune soils. *J. Arid. Environ.* **1984**, *7*, 1–8. [CrossRef]
97. Van Der Heijden, M.G.; Bakker, R.; Verwaal, J.; Scheublin, T.R.; Rutten, M.; Van Logtestijn, R.; Staehelin, C. Symbiotic bacteria as a determinant of plant community structure and plant productivity in dune grassland. *FEMS Microbiol. Ecol.* **2006**, *56*, 178–187. [CrossRef] [PubMed]
98. Zahran, H.H. Rhizobia from wild legumes: Diversity, taxonomy, ecology, nitrogen fixation and biotechnology. *J. Biotechnol.* **2001**, *91*, 143–153. [CrossRef]
99. Peoples, M.B.; Brockwell, J.; Herridge, D.F.; Rochester, I.J.; Alves, B.J.R.; Urquiaga, S.; Boddey, R.M.; Dakora, F.D.; Bhattarai, S.; Maskey, S.L.; et al. The contributions of nitrogen-fixing crop legumes to the productivity of agricultural systems. *Symbiosis* **2009**, *48*, 1–17. [CrossRef]
100. Kirkegaard, J.; Christen, O.; Krupinsky, J.; Layzell, D. Break crop benefits in temperate wheat production. *Field Crop. Res.* **2008**, *107*, 185–195. [CrossRef]
101. Bindhu, B.; Dahiya, A.; Uera, R.; Bindhu, B.B. Mitigation of abiotic stress in legume-nodulating rhizobia for sustainable crop production. *Agric. Res.* **2020**, *9*, 444–459. [CrossRef]
102. Benezech, C.; Doudement, M.; Gourion, B. Legumes tolerance to rhizobia is not always observed and not always deserved. *Cell. Microbiol.* **2020**, *22*, e13124. [CrossRef]
103. Sugawara, M.; Takahashi, S.; Umehara, Y.; Iwano, H.; Tsurumaru, H.; Odake, H.; Suzuki, Y.; Kondo, H.; Konno, Y.; Yamakawa, T.; et al. Variation in bradyrhizobial NopP effector determines symbiotic incompatibility with Rj2-soybeans via effector-triggered immunity. *Nat. Commun.* **2018**, *9*, 3139. [CrossRef]
104. Ferguson, B.J.; Mens, C.; Hastwell, A.H.; Zhang, M.; Su, H.; Jones, C.H.; Chu, X.; Gresshoff, P.M. Legume nodulation: The host controls the party. *Plant Cell Environ.* **2019**, *42*, 41–51. [CrossRef]
105. Zhong, Y.; Yang, Y.; Liu, P.; Xu, R.; Rensing, C.; Fu, X.; Liao, H. Genotype and rhizobium inoculation modulate the assembly of soybean rhizobacterial communities. *Plant Cell Environ.* **2019**, *42*, 2028–2044. [CrossRef]
106. Zhang, J.; Peng, S.; Shang, Y.; Brunel, B.; Li, S.; Zhao, Y.; Liu, Y.; Chen, W.; Wang, E.; Singh, R.P.; et al. Genomic diversity of chickpea-nodulating rhizobia in Ningxia (north central China) and gene flow within symbiotic Mesorhizobium muleiense populations. *Syst. Appl. Microbiol.* **2020**, *43*, 126089. [CrossRef]
107. Chen, W.F.; Wang, E.T.; Ji, Z.J.; Zhang, J.J. Recent development and new insight of diversification and symbiosis specificity of legume rhizobia: Mechanism and application. *J. Appl. Microbiol.* **2020**, *131*, 553–563. [CrossRef] [PubMed]
108. Shah, A.S.; Wakelin, S.A.; Moot, D.J.; Blond, C.; Laugraud, A.; Ridgway, H.J. Trifolium repens and T. subterraneum modify their nodule microbiome in response to soil pH. *J. Appl. Microbiol.* **2021**, *131*, 1858–1869. [CrossRef] [PubMed]
109. Keller, K.R.; Lau, J.A. When mutualisms matter: Rhizobia effects on plant communities depend on host plant population and soil nitrogen availability. *J. Ecol.* **2017**, *106*, 1046–1056. [CrossRef]
110. Wielbo, J.; Marek-Kozaczuk, M.; Kubik-Komar, A.; Skorupska, A. Increased metabolic potential of Rhizobium spp. is associated with bacterial competitiveness. *Can. J. Microbiol.* **2007**, *53*, 957–967. [CrossRef] [PubMed]
111. Simms, E.L.; Taylor, D.; Povich, J.; Shefferson, R.P.; Sachs, J.; Urbina, M.; Tausczik, Y. An empirical test of partner choice mechanisms in a wild legume–rhizobium interaction. *Proc. R. Soc. B Biol. Sci.* **2006**, *273*, 77–81. [CrossRef] [PubMed]
112. Blanco, A.R.; Sicardi, M.; Frioni, L. Competition for nodule occupancy between introduced and native strains of Rhizobium leguminosarum biovar trifolii. *Biol. Fertil. Soils* **2010**, *46*, 419–425. [CrossRef]
113. Pahua, V.J.; Stokes, P.J.N.; Hollowell, A.C.; Regus, J.U.; Gano-Cohen, K.A.; Wendlandt, C.E.; Quides, K.W.; Lyu, J.Y.; Sachs, J.L. Fitness variation among host species and the paradox of inefective rhizobia. *J. Evol. Biol.* **2018**, *31*, 599–610. [CrossRef]
114. Basile, L.A.; Lepek, V.C. Legume–rhizobium dance: An agricultural tool that could be improved? *Microb. Biotechnol.* **2021**, *14*, 1917–1987. [CrossRef]
115. Jin, L.; Sun, X.; Wang, X.; Shen, Y.; Hou, F.; Chang, S.; Wang, C. Synergistic interactions of arbuscular mycorrhizal fungi and rhizobia promoted the growth of Lathyrus sativus under sulphate salt stress. *Symbiosis* **2010**, *50*, 157–164. [CrossRef]

116. García, I.; Mendoza, R. Lotus tenuis seedlings subjected to drought or waterlogging in a saline sodic soil. *Environ. Exp. Bot.* **2014**, *98*, 47–55. [CrossRef]
117. Dūmiņš, K.; Andersone-Ozola, U.; Samsone, I.; Elferts, D.; Ievinsh, G. Growth and physiological performance of a coastal species Trifolium fragiferum as affected by interaction with Trifolium repens, NaCl treatment, and inoculation with rhizobia. *Plants* **2021**, *10*, 2196. [CrossRef] [PubMed]
118. Gaile, L.; Andersone-Ozola, U.; Samsone, I.; Elferts, D.; Ievinsh, G. Modification of Growth and Physiological Response of Coastal Dune Species *Anthyllis maritima* to Sand Burial by Rhizobial Symbiosis and Salinity. *Plants* **2021**, *10*, 2584. [CrossRef] [PubMed]
119. Chanway, C.P.; Holl, F.B.; Turkington, R. Effect of Rhizobium Leguminosarum Biovar Trifolii Genotype on Specificity between Trifolium Repens and Lolium Perenne. *J. Ecol.* **1989**, *77*, 1150–1160. [CrossRef]
120. Bruning, B.; Rozema, J. Symbiotic nitrogen fixation in legumes: Perspectives for saline agriculture. *Environ. Exp. Bot.* **2013**, *92*, 134–143. [CrossRef]
121. Andersone-Ozola, U.; Jēkabsone, A.; Purmale, L.; Romanovs, M.; Ievinsh, G. Abiotic stress tolerance of coastal accessions of a promising forage legume species, Trifolium fragiferum. *Plants* **2021**, *10*, 1552. [CrossRef]
122. Ruņģis, D.E.; Andersone-Ozola, U.; Jēkabsone, A.; Ievinsh, G. Genetic diversity and structure of Latvian Trifolium fragiferum populations, a crop wild relative legume species, in the context of the Baltic Sea region. *Diversity* **2023**, *15*, 473. [CrossRef]
123. Andersone-Ozola, U.; Jēkabsone, A.; Karlsons, A.; Romanovs, M.; Ievinsh, G. Soil chemical properties and mineral nutrition of Latvian accessions of Trifolium fragiferum, a crop wild relative plant species. *Environ. Exp. Biol.* **2021**, *19*, 245–254.
124. Jēkabsone, A.; Andersone-Ozola, U.; Karlsons, A.; Romanovs, M.; Ievinsh, G. Effect of Salinity on Growth, Ion Accumulation and Mineral Nutrition of Different Accessions of a Crop Wild Relative Legume Species, *Trifolium fragiferum*. *Plants* **2022**, *11*, 797. [CrossRef]
125. Jēkabsone, A.; Andersone-Ozola, U.; Karlsons, A.; Neiceniece, L.; Romanovs, M.; Ievinsh, G. Dependence on nitrogen availability and rhizobial symbiosis of different accessions of Trifolium fragiferum, a crop wild relative legume species, as related to physiological traits. *Plants* **2022**, *11*, 1141. [CrossRef]
126. Kozieł, M.; Kalita, M.; Janczarek, M. Genetic diversity of microsymbionts nodulating Trifolium pratense in subpolar and temperate climate regions. *Sci. Rep.* **2022**, *12*, 12144. [CrossRef]
127. Tyler, T.; Herbertsson, L.; Olofsson, J.; Olsson, P.A. Ecological indicator and traits values for Swedish vascular plants. *Ecol. Indic.* **2021**, *120*, 106923. [CrossRef]
128. Nickrent, D.L. Parasitic angiosperms: How often and how many? *Taxon* **2020**, *69*, 5–27. [CrossRef]
129. Teixeira-Costa, L.; Davis, C.C. Life history, diversity, and distribution in parasitic flowering plants. *Plant Physiol.* **2020**, *187*, 32–51. [CrossRef]
130. Parker, C. The parasitic weeds of the Orobanchaceae. In *Parasitic Orobanchaceae: Parasitic Mechanisms and Control Strategies*; Joel, D.M., Gressel, J., Musselman, L.J., Eds.; Springer: Berlin/Heidelberg, Germany, 2013; pp. 313–344.
131. Hatcher, M.J.; Dick, J.T.; Dunn, A.M. Diverse effects of parasites in ecosystems: Linking interdependent processes. *Front. Ecol. Environ.* **2012**, *10*, 186–194. [CrossRef]
132. Hegenauer, V.; Körner, M.; Albert, M. Plants under stress by parasitic plants. *Curr. Opin. Plant Biol.* **2017**, *38*, 34–41. [CrossRef] [PubMed]
133. Gibson, C.C.; Watkinson, A.R. The host range and selectivity of a parasitic plant: *Rhinanthus minor* L. *Oecologia* **1989**, *78*, 401–406. [CrossRef]
134. Clarke, C.R.; Timko, M.P.; Yoder, J.I.; Axtell, M.J.; Westwood, J.H. Molecular Dialog Between Parasitic Plants and Their Hosts. *Annu. Rev. Phytopathol.* **2019**, *57*, 279–299. [CrossRef]
135. Seel, W.E.; Press, M.C. Influence of the host on three sub-arctic annual facultative root hemiparasites. I. Growth, mineral accumulation and above-ground dry-matter partitioning. *New Phytol.* **1993**, *125*, 131–138. [CrossRef]
136. Rowntree, J.K.; Cameron, D.D.; Preziosi, R.F. Genetic variation changes the interactions between the parasitic plant-ecosystem engineer Rhinanthus and its hosts. *Philos. Trans. R. Soc. B Biol. Sci.* **2011**, *366*, 1380–1388. [CrossRef]
137. Chaudron, C.; Mazalová, M.; Kuras, T.; Malenovský, I.; Mládek, J. Introducing ecosystem engineers for grassland biodiversity conservation: A review of the effects of hemiparasitic Rhinanthus species on plant and animal communities at multiple trophic levels. *Perspect. Plant Ecol. Evol. Syst.* **2021**, *52*, 125633. [CrossRef]
138. Heer, N.; Klimmek, F.; Zwahlen, C.; Fischer, M.; Hölzel, N.; Klaus, V.H.; Kleinebecker, T.; Prati, D.; Boch, S. Hemiparasite density effects on grassland plant diversity, composition and biomass. *Perspect. Plant Ecol. Evol. Syst.* **2018**, *32*, 22–29. [CrossRef]
139. Watson, D.M. Parasitic plants as facilitators: More Dryad than Dracula? *J. Ecol.* **2009**, *97*, 1151–1159. [CrossRef]
140. Watson, D.M.; McLellan, R.C.; Fontúrbel, F.E. Functional Roles of Parasitic Plants in a Warming World. *Annu. Rev. Ecol. Evol. Syst.* **2022**, *53*, 25–45. [CrossRef]
141. DiGiovanni, J.P.; Wysocki, W.P.; Burke, S.; Duvall, M.R.; Barber, N.A. The role of hemiparasitic plants: Influencing tallgrass prairie quality, diversity, and structure. *Restor. Ecol.* **2017**, *25*, 405–413. [CrossRef]
142. Delavault, P.; Montiel, G.; Brun, G.; Puvreau, J.-B.; Thoiron, S.; Simier, P. Communication between host plants and parasitic plants. In *Advances in Botanical Research*; Becard, G., Ed.; How Plants Communicate with their Biotic Environment; Academic Press: London, UK, 2017; Volume 82, pp. 55–82.
143. Smith, J.D.; Mescher, M.C.; De Moraes, C.M. Implications of bioactive solute transfer from hosts to parasitic plants. *Curr. Opin. Plant Biol.* **2013**, *16*, 464–472. [CrossRef]

144. Kim, G.; LeBlanc, M.L.; Wafula, E.K.; Depamphilis, C.W.; Westwood, J.H. Genomic-scale exchange of mRNA between a parasitic plant and its hosts. *Science* **2014**, *345*, 808–811. [CrossRef] [PubMed]

145. Westwood, J.H.; Kim, G. RNA mobility in parasitic plant—Host interactions. *RNA Biol.* **2017**, *14*, 450–455. [CrossRef] [PubMed]

146. Park, S.-Y.; Shimizu, K.; Brown, J.; Aoki, K.; Westwood, J.H. Mobile Host mRNAs Are Translated to Protein in the Associated Parasitic Plant *Cuscuta campestris*. *Plants* **2022**, *11*, 93. [CrossRef] [PubMed]

147. Zhou, L.; Lu, Q.-W.; Yang, B.-F.; Zagorchev, L.; Li, J.-M. Integrated small RNA, mRNA, and degradome sequencing reveals the important role of miRNAs in the interactions between parasitic plant Cuscuta australis and its host Trifolium repens. *Sci. Hortic.* **2021**, *289*, 110548. [CrossRef]

148. Tešitel, J.; Li, A.-I.; Knotková, K.; McLellan, R.; Badaranayake, P.C.G.; Watson, D.M. The bright side of parasitic plants: What are they good for? *Plant Physiol.* **2021**, *185*, 1309–1324. [CrossRef]

149. Hartenstein, M.; Albert, M.; Krause, K. The plant vampire diaries: A historic perspective on *Cuscuta* research. *J. Exp. Bot.* **2023**, *74*, 2944–2955. [CrossRef]

150. Watkinson, A.R.; Davy, A.J. Population biology of salt marsh and sand dune annuals. *Vegetatio* **1985**, *62*, 487–497. [CrossRef]

151. Singh, D.; Jha, B. The isolation and identification of salt-responsive novel microRNAs from Salicornia brachiata, an extreme halophyte. *Plant Biotechnol. Rep.* **2014**, *8*, 325–336. [CrossRef]

152. Zhang, H.; Liu, X.; Yang, X.; Wu, H.; Zhu, H.; Zhang, H. miRNA–mRNA integrated analysis reveals roles for miRNAs in a typical halophyte, Reaumuria soongorica, during seed germination under salt stress. *Plants* **2020**, *9*, 351. [CrossRef]

153. Baek, D.; Chun, H.J.; Kang, S.; Shin, G.; Park, S.J.; Hong, H.; Kim, C.; Kim, D.H.; Lee, S.Y.; Kim, M.C.; et al. A Role for Arabidopsis miR399f in Salt, Drought, and ABA Signaling. *Mol. Cells* **2016**, *39*, 111–118. [CrossRef]

154. Zhang, X.; Shen, J.; Xu, Q.; Dong, J.; Song, L.; Wang, W.; Shen, F. Long noncoding RNA lncRNA354 functions as a competing endogenous RNA of miR160b to regulate *ARF* genes in response to salt stress in upland cotton. *Plant Cell Environ.* **2021**, *44*, 3302–3321. [CrossRef]

155. Makkar, H.; Arora, S.; Khuman, A.K.; Chaudhary, B. Target-Mimicry-Based miR167 Diminution Confers Salt-Stress Tolerance During In Vitro Organogenesis of Tobacco (*Nicotiana tabacum* L.). *J. Plant Growth Regul.* **2022**, *41*, 1462–1480. [CrossRef]

156. Lotfi, A.; Pervaiz, T.; Jiu, S.; Faghihi, F.; Jahanbakhshian, Z.; Khorzoghi, E.G.; Fang, J.; Seyedi, S.M. Role of microRNAs and their target genes in salinity response in plants. *Plant Growth Regul.* **2017**, *82*, 377–390. [CrossRef]

157. Gao, Z.; Ma, C.; Zheng, C.; Yao, Y.; Du, Y. Advances in the regulation of plant salt-stress tolerance by miRNA. *Mol. Biol. Rep.* **2022**, *19*, 5041–5055. [CrossRef] [PubMed]

158. Fahmy, G.M. Transpiration and dry matter allocation in the angiosperm root parasite *Cynomorium coccineum* L. and two of its halophytic hosts. *Biol. Plant.* **1993**, *35*, 603–608. [CrossRef]

159. Ouf, S.A. Mycological studies on the angiosperm root parasite *Cynomorium coccineum* L. and two of its halophytic hosts. *Biol. Plant.* **1993**, *35*, 591–602. [CrossRef]

160. Pennings, S.C.; Callaway, R.M. Impact of a Parasitic Plant on the Structure and Dynamics of Salt Marsh Vegetation. *Ecology* **1996**, *77*, 1410–1419. [CrossRef]

161. Zagorchev, L.; Stöggl, W.; Teofanova, D.; Li, J.; Kranner, I. Plant Parasites under Pressure: Effects of Abiotic Stress on the Interactions between Parasitic Plants and Their Hosts. *Int. J. Mol. Sci.* **2021**, *22*, 7418. [CrossRef]

162. Veste, M.; Todt, H.; Breckle, S.-W. Influence of halophytic hosts on their parasites—The case of Plicosepalus acaciae. *AoB Plants* **2015**, *7*, plu084. [CrossRef] [PubMed]

163. Grewell, B.J. Parasite Facilitates Plant Species Coexistence in a Coastal Wetland. *Ecology* **2008**, *89*, 1481–1488. [CrossRef] [PubMed]

164. Fahmy, G.M. Ecophysiology of the holoparasitic angiosperm Cistanche phelypaea (Orobancaceae) in a coastal salt marsh. *Turk. J. Bot.* **2013**, *37*, 908–919. [CrossRef]

165. Callaway, R.M.; Walker, L.R. Competition and facilitation: A synthetic approach to interactions in plant communities. *Ecology* **1997**, *78*, 1958–1965. [CrossRef]

166. Montesinos, D. Plant–plant interactions: From competition to facilitation. *Web Ecol.* **2015**, *15*, 1–2. [CrossRef]

167. Craine, J.M.; Dybzinski, R. Mechanisms of plant competition for nutrients, water and light. *Funct. Ecol.* **2013**, *27*, 833–840. [CrossRef]

168. Schandry, N.; Becker, C. Allelopathic Plants: Models for Studying Plant–Interkingdom Interactions. *Trends Plant Sci.* **2020**, *25*, 176–185. [CrossRef]

169. Zhang, Z.; Liu, Y.; Yuan, L.; Weber, E.; Van Kleunen, M. Effect of allelopathy on plant performance: A meta-analysis. *Ecol. Lett.* **2021**, *24*, 348–362. [CrossRef]

170. Delory, B.M.; Delaplace, P.; Fauconnier, M.-L.; du Jardin, P. Root-emitted volatile organic compounds: Can they mediated belowground plant-plant interactions? *Plant Soil* **2016**, *402*, 1–26. [CrossRef]

171. da Silva, E.R.; Overbeck, G.E.; Soares, G.L.G. Something old, something new in allelopathy review: What grassland ecosystems tell us. *Chemoecology* **2017**, *27*, 217–231. [CrossRef]

172. Xu, Y.; Chen, X.; Ding, L.; Kong, C.-H. Allelopathy and Allelochemicals in Grasslands and Forests. *Forests* **2023**, *14*, 562. [CrossRef]

173. Dudley, S.A.; Murphy, G.P.; File, A.L. Kin recognition and competition in plants. *Funct. Ecol.* **2013**, *27*, 898–906. [CrossRef]

174. Karban, R.; Shiojiri, K.; Ishizaki, S.; Wetzel, W.; Evans, R.Y. Kin recognition affects plant communication and defence. *Proc. R. Soc. B Biol. Sci.* **2013**, *280*, 20123062. [CrossRef]

175. Maestre, F.T.; Callaway, R.M.; Valladares, F.; Lortie, C.J. Refining the stress-gradient hypothesis for competition and facilitation in plant communities. *J. Ecol.* **2009**, *97*, 199–205. [CrossRef]
176. Lin, Y.; Berger, U.; Grimm, V.; Ji, Q. Differences between symmetric and asymmetric facilitation matter: Exploring the interplay between modes of positive and negative plant interactions. *J. Ecol.* **2012**, *100*, 1482–1491. [CrossRef]
177. Vogt, D.R.; Murrell, D.J.; Stoll, P. Testing Spatial Theories of Plant Coexistence: No Consistent Differences in Intra- and Interspecific Interaction Distances. *Am. Nat.* **2010**, *175*, 73–84. [CrossRef]
178. Dudley, S.A. Plant cooperation. *AoB Plants* **2015**, *7*, plv113. [CrossRef]
179. Weigelt, A.; Schumacher, J.; Walther, T.; Bartelheimer, M.; Steinlein, T.; Beyschlag, W. Identifying mechanisms of competition in multi-species communities. *J. Ecol.* **2007**, *95*, 53–64. [CrossRef]
180. Grant, K.; Kreyling, J.; Heilmeier, H.; Beierkuhnlein, C.; Jentsch, A. Extreme weather events and plant–plant interactions: Shifts between competition and facilitation among grassland species in the face of drought and heavy rainfall. *Ecol. Res.* **2014**, *29*, 991–1001. [CrossRef]
181. Le Bagousse-Pinguet, Y.; Maalouf, J.-P.; Touzard, B.; Michalet, R. Importance, but not intensity of plant interactions relates to species diversity under the interplay of stress and disturbance. *Oikos* **2014**, *123*, 777–785. [CrossRef]
182. Ungar, I.A. Are biotic factors significant in influencing the distribution of halophytes in saline habitats? *Bot. Rev.* **1998**, *64*, 176–199. [CrossRef]
183. Samsone, I.; Ievinsh, G. Different plant species accumulate various concentration of Na+ in a sea-affected coastal wetland during a vegetation season. *Environ. Exp. Biol.* **2018**, *16*, 117–127.
184. Gagné, J.-M.; Hoyle, G. Facilitation of Leymus mollis by Honckenya peploides on coastal dunes in subarctic Quebec, Canada. *Can. J. Bot.* **2001**, *79*, 1327–1331.
185. Franks, S.J.; Peterson, C.J. Burial disturbance leads to facilitation among coastal dune plants. *Plant Ecol.* **2003**, *168*, 13–21. [CrossRef]
186. Martínez, M.L.; García-Franco, J.G. Plant-plant interactions in coastal dunes. In *Coastal Dunes, Ecology and Conservation*; Martínez, M.L., Psuty, N.P., Eds.; Ecological Studies; Springer: Berlin/Heidelberg, Germany, 2004; Volume 171, pp. 205–220.
187. Yu, Z.; Zhang, Q.; Yang, H.; Tang, J.; Weiner, J.; Chen, X. The effects of salt stress and arbuscular mycorrhiza on plant neighbour effects and self-thinning. *Basic Appl. Ecol.* **2012**, *13*, 673–680. [CrossRef]
188. Subrahmaniam, H.J.; Libourel, C.; Journet, E.-P.; Morel, J.-B.; Muños, S.; Niebel, A.; Raffaele, S.; Roux, F. The genetics underlying natural variation of plant–plant interactions, a beloved but forgotten member of the family of biotic interactions. *Plant J.* **2018**, *93*, 747–770. [CrossRef] [PubMed]
189. Zhou, J.; Wilson, G.W.T.; Cobb, A.B.; Zhang, Y.; Liu, L.; Zhang, X.; Sun, F. Mycorrhizal and rhizobial interactions influence model grassland plant community structure and productivity. *Mycorrhiza* **2022**, *32*, 15–32. [CrossRef] [PubMed]
190. Ievinsh, G. Halophytic Clonal Plant Species: Important Functional Aspects for Existence in Heterogeneous Saline Habitats. *Plants* **2023**, *12*, 1728. [CrossRef] [PubMed]
191. Tomilov, A.; Tomilova, N.; Yoder, J.I. In vitro haustorium development in roots and root cultures of the hemiparasitic plant Triphysaria versicolor. *Plant Cell Tissue Organ Cult.* **2004**, *77*, 257–265. [CrossRef]
192. Takeda, N.; Handa, Y.; Tsuzuki, S.; Kojima, M.; Sakakibara, H.; Kawaguchi, M. Gibberellins Interfere with Symbiosis Signaling and Gene Expression and Alter Colonization by Arbuscular Mycorrhizal Fungi in *Lotus japonicus*. *Plant Physiol.* **2015**, *167*, 545–557. [CrossRef]
193. Ho-Plágaro, T.; Tamayo-Navarette, N.I.; García-Garrido, J.M. Functional analysis of plant genes related to arbuscular mycorrhizal symbiosis using *Agrobacterium rhizogenes*-mediated root transformation and hairy root production. In *Hairy Root Cultures Based Applications*; Srivastava, V., Mehrotra, S., Mishra, S., Eds.; Rhizosphere Biology; Springer Nature: Singapore, 2020; pp. 191–215.
194. Singh, J.; Kumar, K.; Verma, P.K. Functional characterization of genes involved in legume nodulation using hairy root cultures. In *Hairy Root Cultures Based Applications*; Srivastava, V., Mehrotra, S., Mishra, S., Eds.; Rhizosphere Biology; Springer Nature: Singapore, 2020; pp. 217–228.
195. Lin, P.; Zhang, M.; Wang, M.; Li, Y.; Liu, J.; Chen, Y. Inoculation with arbuscular mycorrhizal fungus modulates defense-related genes expression in banana seedlings susceptible to wilt disease. *Plant Signal. Behav.* **2021**, *16*, e1884782. [CrossRef]
196. He, X.; Zhang, Q.; Li, B.; Jin, Y.; Jiang, L.; Wu, R. Network mapping of root–microbe interactions in Arabidopsis thaliana. *npj Biofilms Microbiomes* **2021**, *7*, 72. [CrossRef]
197. Lee, K.K.; Kim, H.; Lee, Y. H. Cross kingdom co-occurrence networks in the plant microbiome: Importance and ecological interpretations. *Front. Microbiol.* **2022**, *13*, 953300. [CrossRef]

Article

Bee-Friendly Native Seed Mixtures for the Greening of Solar Parks

Maren Helen Meyer, Sandra Dullau *, Pascal Scholz, Markus Andreas Meyer and Sabine Tischew

Department of Agriculture, Ecotrophology and Landscape Development, Anhalt University of Applied Sciences, 06406 Bernburg, Germany; maren.meyer@hs-anhalt.de (M.H.M.); pascal.scholz@hs-anhalt.de (P.S.); markus.meyer@hs-anhalt.de (M.A.M.); sabine.tischew@hs-anhalt.de (S.T.)
* Correspondence: sandra.dullau@hs-anhalt.de

Abstract: Photovoltaics is one of the key technologies for reducing greenhouse gas emissions and achieving climate neutrality for Europe by 2050, which has led to the promotion of solar parks. These parks can span up to several hundred hectares, and grassland vegetation is usually created between and under the panels. Establishing species-rich grasslands using native seed mixtures can enhance a variety of ecosystem services, including pollination. We present an overall concept for designing native seed mixtures to promote pollinators, especially wild bees, in solar parks. It takes into account the specific site conditions, the small-scale modified conditions caused by the solar panels, and the requirement to avoid panel shading. We highlight the challenges and constraints resulting from the availability of species on the seed market. Furthermore, we provide an easy-to-use index for determining the value of native seed mixtures for wild bee enhancement and apply it as an example to several mixtures specifically designed for solar parks. The increased availability of regional seed would allow a more thorough consideration of pollinator-relevant traits when composing native seed mixtures, thereby enhancing ecosystem services associated with pollinators such as wild bees.

Keywords: grassland; native seeds; seed-mixture pollinator-feeding index; regulating services; solar energy; ground-mounted photovoltaic power plant

Citation: Meyer, M.H.; Dullau, S.; Scholz, P.; Meyer, M.A.; Tischew, S. Bee-Friendly Native Seed Mixtures for the Greening of Solar Parks. *Land* **2023**, *12*, 1265. https://doi.org/10.3390/land12061265

Academic Editors: Michael Vrahnakis, Yannis (Ioannis) Kazoglou and Manuel Pulido Fernádez

Received: 21 April 2023
Revised: 9 June 2023
Accepted: 18 June 2023
Published: 20 June 2023

1. Introduction

Two of the biggest challenges of our time are tackling the climate and biodiversity crises [1]. In addressing the climate crisis, photovoltaics are considered one of the key technologies for reducing greenhouse gas emissions [2], which has led to the expansion of solar parks across Europe [3,4]. To address the biodiversity crisis, in particular the qualitative and quantitative decline of insects [5–7], solar farms offer great potential [8–10].

Ecological assessments of existing ground-mounted photovoltaic systems show diverse impacts, ranging from markedly negative effects on the landscape and biodiversity [11–14] to potentially positive effects on ecosystem services [15,16] and several animal species groups [13,17,18]. In addition to siting [19,20], a key aspect of the ecological assessment of solar parks involves the design of the respective installation [15]. These parks can be up to several hundred hectares in size, with a continuous increase in their numbers [21,22], and grassland is usually included between and under the panels [23]. In addition to spontaneous succession, sowing is a widespread method of revegetation [24]. Until now, species-poor standard seed mixtures have often been used for this purpose, resulting in species- and structure-poor vegetation stands [25–27] with a low value for pollinator insects [28,29]. Initial studies indicate that the use of species-rich seed mixtures in the greening of solar parks can increase local pollinator services, which, in turn, have positive effects on biodiversity and agricultural production [8,9,15,30].

Grassland restoration using native seed mixtures has already been tested and put into practice for many vegetation types such as dry grassland [31,32], mesophile meadows [26,33–36], and wildflower strips [37]. Solar parks have been built on land that was previously used in different ways, e.g., arable land, brownfield sites, military ammunition depots, slag heaps, and waste disposal sites [38–40]. However, the site conditions in solar parks differ significantly from those of typical grassland communities. Inside the solar parks, the microclimate, light conditions, and soil–water balance are modified on a small scale due to changes caused by the solar panels [41–46]. In addition, there are plant-related requirements for the target vegetation, such as a low-growth height, to prevent shading of the panels. These specific site conditions and requirements call for seed mixtures that are specifically designed for solar parks. Thus, solar park seed mixtures must take into account a variety of parameters in addition to those that must be considered for grassland restoration such as soil properties and large-scale climatic conditions. In addition, the suitability of seed mixtures for the promotion of pollinators such as wild bees in solar parks should be systematically evaluated. Although the Pollinator Feeding Index (PFI) developed by Schmidt et al. [47] can be used to assess the supply of feeding resources of already-established vegetation stands with regard to provided pollen and nectar resources, there is currently no index that can be used to easily assess the suitability of native seed mixtures in providing feeding resources for wild bees. Therefore, we aim to (i) establish criteria for the composition of bee-friendly site-adapted seed mixtures for solar parks, (ii) develop an index for determining the value of native seed mixtures in promoting pollinators, especially wild bees, and (iii) apply this index using mixtures that are specifically designed for solar parks as an example.

2. Derivation of Criteria for the Composition of Bee-Friendly Site-Adapted Seed Mixtures for Solar Parks

We used a multi-step approach to compose bee-friendly seed mixtures for the greening of solar parks (Figure 1). Beginning with a basic species pool, each step applies specific criteria to filter the species selection from the previous step.

Step 1: General criteria for solar parks

The general criteria for solar parks are applied to a pool of native grasses and forbs. In this step, two criteria are used to filter out a selection of species that are generally suitable for grasslands in solar parks. In general, species covering a wide range of grassland communities can be used. Due to the extensive vegetation management usually implemented in solar parks [23], as well as the partially lower light availability [48,49], species of fringe communities should also be used. In order to ensure year-round maintenance of the technical systems and avoid shading the panels, which could reduce the energy yield, the growth height of the species used must not be higher than the lower edge of the panels [50]. As the lower edge is usually about 80 cm high, only low- to medium-growth-height species should be included in mixtures for solar parks.

Step 2: Site-specific requirements

When applying seed mixtures in grassland restoration practices, the use of regional species of certified provenance is recommended and usually applied [32,51–55]. The same should generally apply to the greening of solar farms if it is not already regulated by laws or requirements. Regional species are grassland and fringe species that are typical for the region. In Germany, for example, these would be the natural units, according to Ssymank [56], in which the solar park is being built or were once typical.

The exclusive use of seeds of regional provenance maintains the integrity of the local gene pool and ensures the development of vegetation stands with typical regional characteristics [36,57–59]. In this way, genetically diverse plant populations can be established at the natural level of genetic differentiation [57]. Furthermore, it has been proven that provenance selection does affect pollinator abundance and diversity in the sown vegetation stands [53,60]. Due to the comparatively large area of solar parks and the corresponding high demand for available regional seed, the use of seed directly harvested from natural stands is usually not possible [61], and certified seed produced for restoration should be used. Certification schemes for seeds of regional provenance already exist in many

European countries, including Germany, Austria, Italy, and France [62]. In addition to the ecological benefits of using regional seed, in some countries, it is even legally obligatory to do so. For example, in Germany, the use of regional seed has been mandatory for the restoration of grassland in the open landscape since 2020 [54].

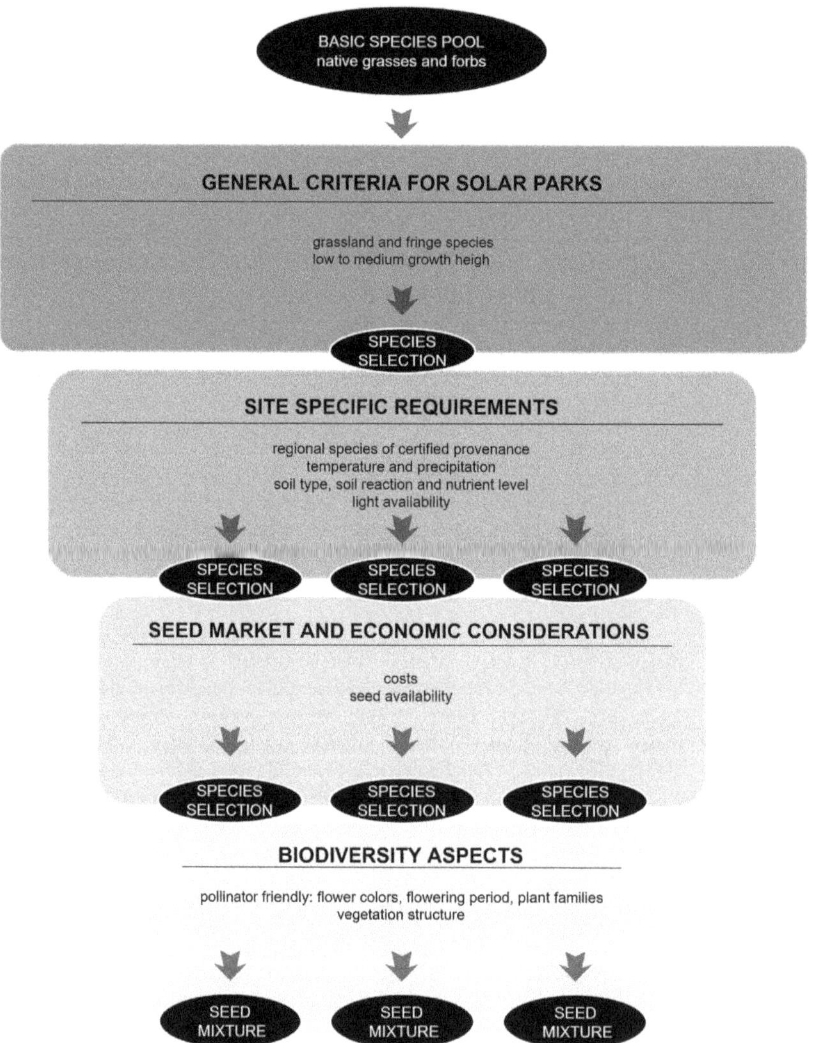

Figure 1. Multi-step concept for the design of bee-friendly native seed mixtures for solar parks.

Specific climatic conditions are essential for the occurrence of plant species [63]. Therefore, the local conditions with regard to temperature and precipitation must be taken into account when selecting species. It should also be considered that the distribution of precipitation can vary over small areas due to the modification with solar panels [43]. As a result, certain parts of the area to be sown may benefit less from precipitation than regionally expected, even in areas with high precipitation. On the other hand, season-dependent lower evapotranspiration under the panels has an effect on the soil–water balance [45,64], but it does not fully compensate for the differences in precipitation [65].

Soil properties also influence the occurrence of species [63,66], but their variation is on a much smaller scale compared to climatic conditions. Therefore, soil properties should be analyzed for each solar park individually. These include the soil type, soil reaction, and nutrient content. When solar parks are built on sites with high nutrient content, soil properties can present a specific challenge. This can cause species to grow taller than predicted [67,68], potentially leading to shading of panels or the dominance of a few fast-growing species that outcompete the majority of sown species. Another factor that influences soil parameters and consequently species selection is the previous land use. In addition to previously intensive arable land, solar parks are also built on de-sealed soil and contaminated sites, e.g., covered ash heaps or waste disposal sites [38–40].

Compared to typical grassland communities, the small-scale changes in light intensity caused by the solar panels create challenges. In particular, the row spacing of the module arrays, depending on the installation, has a significant impact on light availability. The reduced light availability in certain parts of the installation affects the species composition [45,69,70]. Therefore, native seed mixtures for solar parks should contain not only light-demanding species but also species with a preference for semi-shaded conditions.

Step 3: Seed market and economic considerations

The costs for regional seed of certified provenances are significantly higher compared to conventional seed due to high production costs and increased demand in some countries [71,72]. The prices of regional seed mixtures vary significantly according to the reference region, number of species included, species composition, and quantity of diaspores or weight percentage [71–73]. Species that are more complex to propagate and harvest are significantly more expensive compared to those that can be propagated easily on large fields [72,74].

The availability of native seeds from regional provenances varies both between countries and regions [71,75]. In Germany, the legal prohibition on sowing non-native seeds in the open landscape [76] has further complicated the market situation. The number of seed producers and the quantity of seed available only increases gradually, and commissioning the production of larger quantities of desired species needs to commence several years in advance [77,78].

Therefore, not all species suitable for the respective solar park are available from the region of provenance and cannot be used [79]. The challenge of seed availability is exacerbated by the large quantities of seed required for the greening of solar parks due to the large area to be covered [3,4]. Even easily reproducible species may only be available in smaller quantities than required.

Step 4: Biodiversity aspects

Grasslands with a diverse vertical vegetation structure can be attractive to pollinators due to architectural complementarity [80,81]. In order to build such a diverse vertical vegetation structure, species with varying growth heights should be used according to the previous species selection. As specifically composed mixtures can significantly improve the availability of feeding resources for pollinators, especially wild bees in solar parks [8], special consideration should be given to pollinator-relevant species characteristics. In order to provide feeding resources for a wide range of wild bee species, it is important to select forbs with a wide variety of flower colours and plant families that are adapted to the needs of the local insect communities [82,83]. The selection of plant families should take into account the current state of research and include key plant families known to be important for wild bees such as Asteraceae, Fabaceae, Lamiaceae, and Campanulaceae [82]. In particular, oligolectic wild bees depend on certain plant families or species [82]. Furthermore, it is important to consider not only the nutritional needs of adult species but also those of juvenile stages when composing seed mixtures [84,85]. Seed mixtures should be designed to include species with a long flowering period, which flower at different times throughout the season to ensure continuous resource availability, especially when feeding resources in adjacent habitats become limited [8,83,86,87].

Land **2023**, 12, 1265

3. Development of a Seed Mixture-Pollinator Feeding Index (SM-PFI)

Based on the already-validated Pollinator Feeding Index (PFI) of Schmidt et al. [47], which assesses the value of already established vegetation stands as a food supply for pollinators, we develop a Seed Mixture-Pollinator Feeding Index (SM-PFI). The SM-PFI is used to assess the potential suitability of native seed mixtures for establishing vegetation stands with high feeding values for wild bees [82].

$$SM - PFI = \sum_{i=1}^{N}(P_i + N_{i)} * flowering\ period * forbs\ \%.$$

Referring to the PFI of Schmidt et al. [47], the SM-PFI is based on the quantification of pollen (P) and nectar values (N). These values represent the pollen and nectar production of the flowering species according to Pritsch [88]. Grass species are excluded from the index, as their pollen seems to have only minor importance for pollinating insects [47]. Nectar is the main food source for adult bees, whereas pollen is the main food source for bee larvae [85]. The pollen and nectar values are calculated by summing the contributions of the species included in the seed mixture.

Considering that a long flowering period and a high percentage of forbs have a positive effect on the abundance of pollinators such as wild bees [82,89], the sum of the quantified pollen and nectar production is multiplied by the flowering period and the percentage of forbs. To assess the duration of feeding resource provision by the seed mixture, the number of months in which at least two forbs from the seed mixture are flowering (number of flowering months per species according to Jäger [90]) is used as the flowering period. The forbs % refers to the weight proportion of all species, excluding species from the plant families Poaceae, Cyperaceae, and Juncaceae in the total mixture.

4. Application of SM-PFI to Seed Mixtures Specifically Designed for Solar Parks

The SM-PFI was tested on 12 native seed mixtures (Tables A1 and 1) with varying species richness and a percentage of forbs ranging from 30 to 100% in solar parks with different conditions (Table A2). The test mixtures were designed according to the presented criteria for the composition of bee-friendly seed mixtures for solar parks. The mixtures included native forbs from several key plant families for wild bees, such as Asteraceae, Fabaceae, and Campanulaceae, and also incorporated a variety of proven attractive native forbs for both polylectic and oligolectic wild bees, as suggested by Kuppler et al. [91].

Table 1. Characteristics of test seed mixtures for three solar parks in Saxony-Anhalt (Germany). The compositions are shown in Table A1. Detailed information on the three solar parks is shown in Table A2.

	Areas with Solar Panels									Marginal Areas without Solar Panels		
Solar Park Number	1	2	3	1	2	3	1	2	3	1	2	3
Species richness	Species-poor (19–20 species)			Species-poor (19–20 species)			Species-rich (34–37 species)			Species-rich (35–36 species)		
Forbs %	30%			70%			70%			100%		
SM-PFI	120	118	111	279	274	260	515	500	485	762	762	744

As a result, the SM-PFI of the test mixtures clearly differed according to the species richness and forb percentage. Species selection, which differed between the three solar parks due to their very diverse site conditions, had only a minor influence on the index values.

5. Discussion

5.1. Criteria for the Composition of Pollinator-Friendly Site-Adapted Seed Mixtures for Solar Parks

As shown, seed mixtures for the greening of solar parks have to fulfil criteria that are in addition to those of the usual grassland restoration practice. One of the most important criteria for large-scale areas canopied with panels is the maximum growth height of the target vegetation. Since the majority of solar parks that also take biodiversity aspects into account are currently projected with a height of 80 cm from the lower edge of the panels, this value was used as an orientation for the seed mixture concept. However, solar parks with a lower panel under-edge of 60 cm or less will also still be realised. For these parks, a lower target vegetation must be selected and the species selection must be more closely specified in Step 1. In addition, abiotic factors can also influence the growth height of the sown species. Particularly on formerly intensive arable land and contaminated sites (e.g., covered slag heaps or waste disposal sites) on which solar parks are sometimes constructed [38–40,92,93], there may be a high soil nutrient surplus, which can lead to above-average plant growth. This problem can be addressed by adapted, more intensive management, which usually results in higher costs and negative effects on biodiversity, especially due to lower attractiveness for pollinators [94].

Species-rich regional seed mixtures with a high percentage of forbs are usually expensive [71,72,95]. If the budget is limited, one solution may be to reduce the seed rate of expensive species or replace them with less expensive ones. Alternatively, different mixtures can be developed for a solar park, of which only one may contain particularly costly species in the hope that these species will spread throughout the solar park, as described by Török et al. [95], for dry grassland.

Due to the large area of a solar park, which can be up to several hundred hectares in size, seed availability plays a crucial role in the composition of native seed mixtures [79]. Seed availability can vary significantly at the national and regional levels [61,71,75]. Alternative, well-developed techniques, such as revegetation by hay transfer or on-site threshing [34,96], are often not feasible on the required scales and are only viable options for smaller solar farms or sub-areas. The agricultural production of seeds from regional provenances currently appears to be the only option for providing high-quality material for large-scale ecological restoration [78], as required for biodiversity-enhancing solar parks. Although much has been achieved in the field of native seed production in recent years, further research and economic promotion efforts are imperative to develop a market that is adapted to the demand. Currently, a relatively small number of wild plant species are successfully propagated for commercial use, while there are still significant knowledge deficits for many others [97,98]. Especially in solar parks, with their small-scale and differentiated site conditions, species could be established that are not currently used in seed mixtures but have high value for wild bees. By incorporating these species, the attractiveness for oligolectic species, in particular, could be increased through more diversified feeding resources [83,91,99]. Improved availability of seeds from regional provenances would also facilitate better consideration of pollinator-relevant traits, especially the nutrient needs of wild bees, in the design of native seed mixtures. Early-flowering species, in particular, which make an important contribution to extending the flowering period that is important for wild bees [82,89,99–101], are often excluded from native seed mixtures or included in very small quantities due to their limited availability [79]. It is essential to provide spontaneously established species, which often provide important feeding resources for wild bees, especially in spring and winter [47,91,99]. These can be promoted within solar parks through small-scale non-sown sub-areas where the spontaneous succession of, e.g., early flowering perennials, is possible.

The current limited availability of regional seeds can be addressed by developing different mixtures for solar parks. In this approach, small quantities of available species are only sown in sub-areas, whereas other small-scale restoration methods such as spontaneous succession and hay transfer are used to complement the planting process [95,102]. Depending on the size of the area, this approach offers the possibility to establish various vegetation stands, which can lead to a significant increase in plant species numbers and feeding resources for wild bees within a confined spatial context [103]. In German-speaking countries, native seed mixtures have recently become available for solar parks [104,105]. These are characterised by a high number of species, which is beneficial for pollinators, and a species selection adapted to the local insect community [88]. They usually contain a relatively lower percentage of forbs and a higher percentage of grasses (measured by weight). Although the high percentage of grasses reduces the cost of the mixtures, a more grass-dominated target vegetation is expected. According to Schubert et al. [82], this may have a potentially negative impact on the abundance of wild bees. Furthermore, the mixtures offered contain species such as Verbascum spp., which are unsuitable for the inter-row areas of most solar parks due to their tall growth. Even though ready-made seed mixtures are a starting point, these mixtures should certainly be further refined and optimally adapted to the conditions of each individual solar park.

Wild bee-friendly seed mixtures are also important in agrivoltaics (e.g., on fruit crops), where their performance directly on-site is desired [106]. However, due to different technical requirements, the criteria for these plants need to be adjusted. For example, in conventional stilt-mounted systems installed over fruit crops, the height of plant species does not need to be considered for the potential shading of the panels.

Feeding resources for wild bees are just one measure to promote pollinator insects. Numerous other aspects must be considered for a pollinator-friendly design of solar parks. In addition to adapted management, the requirements of the different roles and life stages of various non-bee pollinators [84,85,89,107,108] must also be taken into account. Besides wild bees, many other pollinator insects such as Diptera and Lepidoptera make important contributions to the regulating ecosystem service of "pollination" [89,109,110]. These species groups have very different requirements [84,108,109], which can often only be met with a complex and diverse target species concept.

If there are target species concepts for local pollinator species specific to solar parks, these should, of course, also be taken into account in the composition of seed mixtures according to the current state of research with regard to the special nutritional needs (including the quality of the feeding resources and all related factors) of the corresponding target species. When formulating target species concepts for pollinator species, the deficiencies in the food provided by the surrounding landscape should also be taken into account [111].

5.2. Seed Mixture-Pollinator Feeding Index (SM-PFI)

In order to harness the potential of solar parks for promoting pollinators, one measure is to consider the provision of feeding resources, which are essential for the presence of pollinators such as wild bees [91], when planning the solar park and the seed mixtures to be applied therein. The developed Seed Mixture-Pollinator Feeding Index (SM-PFI) allows for the assessment of native seed mixtures in terms of their potential suitability as feeding resources for wild bees. It serves as a useful extension of the Pollinator Feeding Index (PFI) developed by Schmidt et al. [47]. The PFI by Schmidt et al. [47] has been validated for assessing the species richness and abundance of wild bees [82]. It can be assumed that the application of seed mixtures with a high SM-PFI would result in the establishment of vegetation stands with a high PFI, as suggested by Schmidt et al. [47]. Therefore, this would contribute to promoting the species richness and abundance of wild bees. A requirement for the valid application of the SM-PFI is that all species must be included in the mixture in proportions that allow for long-term establishment. Additionally, the sowing process should be carried out using approved methods. The main difference between the two indices is that the PFI uses the cover of forbs established in the vegetation stand to determine the pollen and nectar values of individual species, whereas the SM-PFI treats all species included in the mixture equally. Weighting based on seed number percentage or weight percentage was not used to prevent individual species from having a disproportionate effect on the index. If weighted based on weight fraction, species with heavy seeds, such as *Agrimonia* spp. and *Lathyrus* spp., would be overemphasised, even though they do not represent the target vegetation more than other species. In contrast, when weighted based on seed number proportion, small-seeded species such as *Campanula* spp., which are included in the mixture with a high seed number, would strongly influence the SM-PFI, although small-seeded species are less likely to establish successfully compared to large ones [112]. In addition, mixtures are often calculated using weight percentages, and the percentage of diaspores is not always known. The absence of weighting for individual species also results in a greater emphasis on the number of forbs when summing the pollen and nectar values. The number of forbs is considered an important parameter for bee attractiveness, as it positively affects the number of species and can specifically enhance the number of oligolectic wild bee species [87,91,99]. The flowering period and percentage of forbs were included as factors in the formula because they are instrumental in the temporally staggered provision of food [8,82,83,86,87,103]. In order to develop an easy-to-understand and easy-to-use index, the focus was on the most important parameters for the enhancement of wild bees, whereas other influencing factors, such as the flower colour, diversity of plant families, and nutrient quality, were excluded. If target species concepts for local pollinator species exist for specific solar parks, the index can complement these concepts but it should not be relied on as the sole tool. In such cases, additional tools and approaches are required to address factors such as nutrient quality, which are not included in the index but are also of major importance [107,109,111]. The SM-PFI is only designed to assess the potential of seed mixtures in creating a vegetation stand that is attractive to pollinators such as wild bees but it cannot assess the actual value of the resulting vegetation. For example, in addition to the sown species, the resulting vegetation stand may contain spontaneously established species that can also contribute significantly to wild bee enhancement [87,113] but are not considered by the SM-PFI. Consequently, the values of the SM-PFI should not be directly compared with those of the PFI according to Schmidt et al. [47]. Furthermore, in addition to the introduction of a high-quality seed mixture with an appropriate sowing rate, the success of establishing a pollinator-attractive vegetation stand depends on many other factors such as proper seedbed preparation [114] and adapted management (cutting frequency, timing, technique) [115–117]. Challenges in applying the SM-PFI may arise due to an insufficient database. Unless pollen and nectar values [88] are available for those species in the mixture in the aggregated data, the calculation of the SM-PFI may become laborious, or if values are missing, it may provide an incomplete picture of the potential of the mixture. Therefore, all forb species that are successfully propagated should be evaluated with regard to their nectar and pollen supply. As shown by the test mixtures for solar parks, the index is suitable for evaluating seed mixtures for a wide range of sites. Its transferability to other potentially bee-friendly vegetation stands, such as field margins or urban wildflower meadows, can be assumed. The index proves particularly valuable for evaluating seed mixtures for wild plant structures in the agricultural landscape, e.g., field margins and wildflower strips in agrivoltaics, where pollinator services directly on-site are desired. However, the index should not be applied to seed mixtures containing cultivars and non-native species. The index is based on the assumption that regional wild plant species, to which local wild bee communities are adapted, are utilized.

6. Conclusions

When designing bee-friendly seed mixtures for solar parks, a variety of criteria must be taken into account beyond the usual practice of restoring typical grassland plant communities. Due to the often-large spatial dimensions of solar parks, seed availability is currently a decisive factor for determining the compositions of seed mixtures. With better availability of regional seeds, biodiversity-promoting aspects can be given greater consideration, leading to increased ecosystem services. As long as regional seeds remain scarce, more flexibility is needed in terms of combining different seed mixtures and revegetation practices to promote bee-friendly solar parks. In order to exploit the potential of solar parks in promoting pollinator-linked ecosystem services in the long term, additional plant species that are particularly suitable for pollinator promotion must be identified and regionally propagated.

The Seed Mixture-Pollinator Feeding Index (SM-PFI) provides an easy-to-use tool for assessing the potential suitability of seed mixtures for establishing vegetation stands in solar parks as a food source for wild bees. The SM-PFI can also be applied to the assessment of native seed mixtures for wildflower stands in agrivoltaics and various other vegetation stands.

Author Contributions: Conceptualisation, M.H.M., S.D. and S.T.; validation, M.H.M.; investigation, M.H.M. and S.D.; data curation, M.H.M.; writing—original draft preparation, M.H.M. and S.D.; writing—review and editing, S.T., M.A.M. and P.S.; visualization, M.H.M.; project administration, P.S.; funding acquisition, S.D. and S.T. All authors have read and agreed to the published version of the manuscript.

Funding: This research was funded by the Federal Ministry of Education and Research under grant number 13FH133KX0.

Data Availability Statement: The data presented in this study are available in the article.

Acknowledgments: We would like to thank our colleague Sandra Mann for her valuable information and thoughts, which helped us in developing the presented criteria. We are grateful to our colleagues Andrea Semm and Markus Zaplata for their contributions to the use of plant species in the composition of the test seed mixtures. We acknowledge support by the German Research Foundation (Deutsche Forschungsgemeinschaft DFG)—project number 491460386—and the Open Access Publishing Fund of Anhalt University of Applied Sciences.

Conflicts of Interest: The authors declare no conflict of interest.

Appendix A

Table A1. Compositions of test seed mixtures for three solar parks in Saxony-Anhalt (Germany), specifying all parameters required for the Seed Mixture Pollinator Feeding Index (SM-PFI). Values of nectar (N) and pollen (P) indices range from 0 = no productivity (or not relevant for wild bees) to 3 = high productivity [88,118,119]. Grasses (G), forbs (F), and flowering months are presented according to Jäger [90]. Flowering period means the number of months in which a minimum of two forbs are flowering.

																Areas with Solar Panels									Marginal Areas without Solar Panels *		
															Solar Park Number	1	2	3	1	2	3	1	2	3	1	2	3
															Species Richness	Species-Poor (19–20)			Species-Poor (19–20)			Species-Rich (34–37)			Species-Rich (35–36)		
															Forbs (Weight %)	30%			70%			70%			100%		
															Flowering Period	7	7	7	7	7	7	7	7	7	6	6	6
G/F	N	P	Flowering Month												SM-PFI	120	118	111	279	274	260	515	500	485	762	762	744
			J	F	M	A	M	J	J	A	S	O	N	D	Species												
G															Briza media		x			x		x	x				
G															Festuca rubra	x	x	x	x	x	x	x	x	x			
G															Festuca rupicola	x	x		x	x		x	x	x			
G															Phleum phleoides	x			x			x	x				
G															Poa angustifolia			x			x			x			
G															Trisetum flavescens			x			x			x			
F	1	2					x	x	x	x	x				Achillea millefolium	x	x	x	x	x	x	x	x	x	x	x	x
F	2	2						x	x	x	x				Agrimonia eupatoria	x	x	x	x	x	x	x	x	x	x	x	x
F	2	1				x	x	x	x						Ajuga reptans							x	x				

Table A1. *Cont.*

G/F	N	P	Flowering Month	SM-PFI		Areas with Solar Panels								Marginal Areas without Solar Panels *		
				Solar Park Number	1	2	3	1	2	3	1	2	3	1	2	3
				Species Richness	Species-Poor (19–20)			Species-Poor (19–20)			Species-Rich (34–37)			Species-Rich (35–36)		
				Forbs (Weight %)	30%			70%			70%			100%		
				Flowering Period	7	7	7	7	7	7	7	7	7	6	6	6
				SM-PFI	120	118	111	279	274	260	515	500	485	762	762	744
F	2	2	x x x x	*Anthyllis vulneraria*											x	
F	2	1	x x x	*Barbarea vulgaris*									x	x	x	x
F	3	1	x x	*Betonica officinalis*							x	x	x	x	x	x
F	2	2	x x x x	*Campanula rapunculoides*							x	x	x			
F	1	1	x x x x	*Cerastium holosteoides*	x	x	x	x	x	x	x	x	x			
F	3	3	x x x x	*Cichorium intybus*									x	x	x	x
F	2	1	x x x	*Clinopodium vulgare*							x	x	x	x	x	x
F	2	2	x x x x	*Crepis biennis*										x	x	x
F	2	2	x x x x	*Daucus carota*	x	x	x	x	x	x	x	x	x	x	x	x
F	1	2	x x x x	*Dianthus carthusianorum*		x			x		x	x		x	x	x
F	3	2	x x	*Dipsacus fullonum*										x	x	x
F	3	2	x x x	*Echium vulgare*										x	x	x
F	2	1	x x x	*Falcaria vulgaris*										x	x	x
F	0	3	x x	*Filipendula vulgaris*							x	x		x	x	x
F	1	1	x x x x	*Galium album*			x			x	x	x	x	x	x	
F	1	1	x x x x	*Galium verum*	x	x		x	x		x	x				
F	1	1	x x	*Galium wirtgenii*										x		x
F	1	2	x x x x x x	*Helianthemum nummularium*								x				
F	0	3	x x	*Hypericum perforatum*	x	x	x	x	x	x	x	x	x	x	x	x
F	2	2	x x x x	*Hypochaeris radicata*							x	x	x			
F	1	1	x x	*Knautia arvensis*							x	x	x	x	x	x
F	2	1	x x x	*Lathyrus pratensis*												x
F	2	1	x x x	*Lathyrus tuberosus*										x	x	
F	2	1	x x x x	*Leonurus cardiaca*										x	x	x
F	2	1	x x x x x x	*Leucanthemum vulgare*	x	x	x	x	x	x	x	x	x	x	x	x
F	2	1	x x x x x x	*Linaria vulgaris*	x	x	x	x	x	x	x	x	x	x	x	x
F	3	1	x x x	*Lotus corniculatus*	x	x	x	x	x	x	x	x	x	x	x	x
F	2	2	x x x	*Lychnis viscaria*	x				x		x			x		
F	2	1	x x x x x	*Malva moschata*										x	x	x
F	2	1	x x x x x	*Malva sylvestris*										x	x	x
F	3	2	x x x	*Origanum vulgare*	x	x	x	x	x	x	x	x	x	x	x	x
F	0	3	x x x x x	*Plantago media*							x	x	x			
F	1	2	x x x x x	*Potentilla argentea*							x	x	x	x	x	x
F	1	2	x x x	*Potentilla neumanniana*	x	x	x	x	x	x	x	x	x			
F	1	2	x x x	*Potentilla reptans*							x	x	x			
F	2	1	x x x x	*Prunella vulgaris*	x	x	x	x	x	x	x	x	x	x	x	x
F	1	1	x x x	*Ranunculus bulbosus*							x	x				
F	1	1	x x x	*Ranunculus lanuginosus*	x				x	x		x		x		
F	2	3	x x x x x	*Reseda lutea*											x	

Table A1. *Cont.*

G/F	N	P	Flowering Month		Areas with Solar Panels									Marginal Areas without Solar Panels *		
				Solar Park Number	1	2	3	1	2	3	1	2	3	1	2	3
				Species Richness	Species-Poor (19–20)			Species-Poor (19–20)			Species-Rich (34–37)			Species-Rich (35–36)		
				Forbs (Weight %)	30%			70%			70%			100%		
				Flowering Period	7	7	7	7	7	7	7	7	7	6	6	6
				SM-PFI	120	118	111	279	274	260	515	500	485	762	762	744
F	2	3	x x x x	*Reseda luteola*										x		x
F	3	1	x x x x	*Salvia pratensis*	x	x	x	x	x	x	x	x	x	x	x	x
F	1	1	x x x x	*Saponaria officinalis*										x	x	x
F	2	1	x x x x	*Scabiosa ochroleuca*							x	x	x	x	x	
F	1	1	x x x x x x	*Silene dioica*		x			x		x	x	x	x	x	x
F	1	1	x x x x	*Silene latifolia* subsp. *alba*										x	x	x
F	1	1	x x x x x	*Silene vulgaris*	x	x	x	x	x	x	x	x	x			
F	3	1	x x x x x	*Stachys recta*							x	x		x	x	x
F	3	3	x x x x	*Trifolium pratense*	x	x	x	x	x	x	x	x	x	x	x	x
F	1	3	x x x	*Verbascum densiflorum*										x	x	x
F	1	3	x x x x	*Verbascum nigrum*										x	x	x
F	2	2	x x x	*Veronica maritima*										x		x

* Seed mixtures for marginal areas also contain high-growing species.

Table A2. Test solar parks in Saxony-Anhalt.

Solar Park Number	1	2	3
County	Mansfeld-Südharz	Salzlandkreis	Halle (Saale)/Saalekreis
Size (ha)	2.2	1	13.3
Size covered with panels (ha)	1.3	0.4	6.3
Type of solar panels	Monofacial, south facing	Monofacial, south facing	Monofacial, south facing
Under-edge of the panels (cm)	80	80	80
Inclination of the panels	17°	20°	15°
Distance between module rows (m)	3.1	4	2.4
Previous use	De-sealed soil of former farm buildings	Abandoned area	Ash dump covered with soil substrate
Preparation for seeding	No tillage	Rotary tilling	Rotary tilling
Surrounding land use	Biogas plant, residential and production buildings, non-irrigated arable land	Residential buildings, industrial complexes, abandoned extraction sites (limestone)	Non-irrigated arable land, covered ash dump, industrial complexes

References

1. Rinder, T.; Neuber, F.; Von Hagke, C. Fighting symptom or root cause?-The need for shifting the focus in climate politics from greenhouse gases to environmental protection. *EarthArXiv* **2022**. preprint, 1–8. [CrossRef]
2. Duda, J.; Kusa, R.; Pietruszko, S.; Smol, M.; Suder, M.; Teneta, J.; Wójtowicz, T.; Żdanowicz, T. Development of Roadmap for Photovoltaic Solar Technologies and Market in Poland. *Energies* **2022**, *15*, 174. [CrossRef]
3. Jäger-Waldau, A.; Kougias, I.; Taylor, N.; Thiel, C. How photovoltaics can contribute to GHG emission reductions of 55% in the EU by 2030. *Renew. Sustain. Energy Rev.* **2020**, *126*, 109836. [CrossRef]
4. Wolniak, R.; Skotnicka-Zasadzień, B. Development of Photovoltaic Energy in EU Countries as an Alternative to Fossil Fuels. *Energies* **2022**, *15*, 662. [CrossRef]

5. Potts, S.G.; Biesmeijer, J.C.; Kremen, C.; Neumann, P.; Schweiger, O.; Kunin, W.E. Global pollinator declines: Trends, impacts and drivers. *Trends Ecol. Evol.* **2010**, *25*, 345–353. [CrossRef] [PubMed]

6. Cardoso, P.; Barton, P.S.; Birkhofer, K.; Chichorro, F.; Deacon, C.; Fartmann, T.; Fukushima, C.S.; Gaigher, R.; Habel, J.C.; Hallmann, C.A.; et al. Scientists' warning to humanity on insect extinctions. *Biol. Conserv.* **2020**, *242*, 108426. [CrossRef]

7. Hallmann, C.A.; Sorg, M.; Jongejans, E.; Siepel, H.; Hofland, N.; Schwan, H.; Stenmans, W.; Müller, A.; Sumser, H.; Hörrenz, T.; et al. More than 75 percent decline over 27 years in total flying insect biomass in protected areas. *PLoS ONE* **2017**, *12*, e0185809. [CrossRef] [PubMed]

8. Blaydes, H.; Potts, S.G.; Whyatt, J.D.; Armstrong, A. Opportunities to enhance pollinator biodiversity in solar parks. *Renew. Sustain. Energy Rev.* **2021**, *145*, 111065. [CrossRef]

9. Semeraro, T.; Pomes, A.; Del Giudice, C.; Negro, D.; Aretano, R. Planning ground based utility scale solar energy as green infrastructure to enhance ecosystem services. *Energy Policy* **2018**, *117*, 218–227. [CrossRef]

10. Walston, L.J.; Mishra, S.K.; Hartmann, H.M.; Hlohowskyj, I.; McCall, J.; Macknick, J. Examining the Potential for Agricultural Benefits from Pollinator Habitat at Solar Facilities in the United States. *Environ. Sci. Technol.* **2018**, *52*, 7566–7576. [CrossRef]

11. Aman, M.M.; Solangi, K.H.; Hossain, M.S.; Badarudin, A.; Jasmon, G.B.; Mokhlis, H.; Bakar, A.H.A.; Kazi, S.N. A review of Safety, Health and Environmental (SHE) issues of solar energy system. *Renew. Sustain. Energy Rev.* **2015**, *41*, 1190–1204. [CrossRef]

12. Gasparatos, A.; Doll, C.N.H.; Esteban, M.; Ahmed, A.; Olang, T.A. Renewable energy and biodiversity: Implications for transitioning to a Green Economy. *Renew. Sustain. Energy Rev.* **2017**, *70*, 161–184. [CrossRef]

13. Harrison, C.; Lloyd, H.; Field, C. *Evidence Review of the Impact of Solar Farms on Birds, Bats and General Ecology*, 1st ed.; Manchester Metropolitan University: Manchester, UK, 2017. [CrossRef]

14. Moore-O'Leary, K.A.; Hernandez, R.R.; Johnston, D.S.; Abella, S.R.; Tanner, K.E.; Swanson, A.C.; Kreitler, J.; Lovich, J.E. Sustainability of utility-scale solar energy–critical ecological concepts. *Front. Ecol. Environ.* **2017**, *15*, 385–394. [CrossRef]

15. Carvalho, F.; Treasure, L.; Robinson, S.J.; Blaydes, H.; Exley, G.; Hayes, R.; Howell, B.; Keith, A.; Montag, H.; Parker, G.; et al. Towards a standardized protocol to assess natural capital and ecosystem services in solar parks. *Ecol. Solut. Evid.* **2023**, *4*, e12210. [CrossRef]

16. Randle-Boggis, R.J.; White, P.C.L.; Cruz, J.; Parker, G.; Montag, H.; Scurlock, J.; Armstrong, A. Realising co-benefits for natural capital and ecosystem services from solar parks: A co-developed, evidence-based approach. *Renew. Sustain. Energy Rev.* **2020**, *125*, 109775. [CrossRef]

17. Hernandez, R.R.; Armstrong, A.; Burney, J.; Ryan, G.; Moore-O'Leary, K.; Diédhiou, I.; Grodsky, S.M.; Saul-Gershenz, L.; Davis, R.; Macknick, J.; et al. Techno-ecological synergies of solar energy for global sustainability. *Nat. Sustain.* **2019**, *2*, 560–568. [CrossRef]

18. Montag, H.; Parker, G.; Clarkson, T. *The Effects of Solar Farms on Local Biodiversity: A Comparative Study*; Clarkson and Woods and Wychwood Biodiversity: Blackford, United Kingdom, 2016; ISBN 978-1-5262-0223-9.

19. Kim, J.Y.; Koide, D.; Ishihama, F.; Kadoya, T.; Nishihiro, J. Current site planning of medium to large solar power systems accelerates the loss of the remaining semi-natural and agricultural habitats. *Sci. Total Environ.* **2021**, *779*, 146475. [CrossRef]

20. Taylor, R.; Conway, J.; Gabb, O.; Gillespie, J. Potential ecological impacts of groundmounted photovoltaic solar panels. An introduction and literature review. *BSG Ecol.* **2019**, 1–19. Available online: https://www.bsg-ecology.com/wp-content/uploads/2019/04/Solar-Panels-and-Wildlife-Review-2019.pdf (accessed on 18 April 2023).

21. Høiaas, I.; Grujic, K.; Imenes, A.G.; Burud, I.; Olsen, E.; Belbachir, N. Inspection and condition monitoring of large-scale photovoltaic power plants: A review of imaging technologies. *Renew. Sustain. Energy Rev.* **2022**, *161*, 112353. [CrossRef]

22. Kristinof, R.E.; Collingwood, B. Challenges and opportunities in solar farm construction. In Proceedings of the 13th Australia New Zealand Conference on Geomechanics, Perth, WA, Australia, 1–3 April 2019; Acosta-Martinez, H.E., Lehane, B.M., Eds.; Australian Geomechanics Society: Sydney, Australia, 2019; pp. 1289–1295, ISBN 978-0-9946261-0-3.

23. Peschel, R.; Peschel, T. Photovoltaik und Biodiversität–Integration statt Segregation!-Solarparks und das Synergiepotenzial für Förderung und Erhalt biologischer Vielfalt. *Nat. Und Landsch.* **2023**, *55*, 18 25. [CrossRef]

24. Kiehl, K.; Kirmer, A.; Donath, T.W.; Rasran, L.; Hölzel, N. Species introduction in restoration projects–Evaluation of different techniques for the establishment of semi-natural grasslands in Central and Northwestern Europe. *Basic Appl. Ecol.* **2010**, *11*, 285–299. [CrossRef]

25. Auestad, I.; Rydgren, K.; Austad, I. Near-natural methods promote restoration of species-rich grassland vegetation-revisiting a road verge trial after 9 years. *Restor. Ecol.* **2016**, *24*, 381–389. [CrossRef]

26. Mitchley, J.; Jongepierová, I.; Fajmon, K. Regional seed mixtures for the re-creation of species-rich meadows in the White Carpathian Mountains: Results of a 10-yr experiment. *Appl. Veg. Sci.* **2012**, *15*, 253–263. [CrossRef]

27. Török, P.; Deák, B.; Vida, E.; Valkó, O.; Lengyel, S.; Tóthmérész, B. Restoring grassland biodiversity: Sowing low-diversity seed mixtures can lead to rapid favourable changes. *Biol. Conserv.* **2010**, *143*, 806–812. [CrossRef]

28. Ebeling, A.; Klein, A.M.; Schumacher, J.; Weisser, W.W.; Tscharntke, T. How does plant richness affect pollinator richness and temporal stability of flower visits? *Oikos* **2008**, *117*, 1808–1815. [CrossRef]

29. Walston, L.J.; Li, Y.; Hartmann, H.M.; Macknick, J.; Hanson, A.; Nootenboom, C.; Lonsdorf, E.; Hellmann, J. Modeling the ecosystem services of native vegetation management practices at solar energy facilities in the Midwestern United States. *Ecosyst. Serv.* **2021**, *47*, 101227. [CrossRef]

30. Biesmeijer, K.; van Kolfschoten, L.; Wit, F.; Moens, M. The effects of solar parks on plants and pollinators: The case of Shell Moerdijk. *Nat. Biodivers. Cent.* **2020**, 1–28. Available online: https://www.naturalis.nl/system/files/inline/Report%20The%20effects%20of%20solar%20parks%20on%20plants%20and%20pollinators%20-%20the%20case%20of%20Shell%20Moerdijk%20_0.pdf (accessed on 18 April 2023).

31. Kiehl, K. Renaturierung von Kalkmagerrasen. In *Renaturierung von Ökosystemen in Mitteleuropa*; Spektrum Akademischer Verlag: Heidelberg, Germany, 2010; pp. 265–282. ISBN 978-3-662-48517-0.

32. Scotton, M.; Kirmer, A.; Krautzer, B. *Practical Handbook for Seed Harvest and Ecological Restoration of Species-Rich Grasslands*; Cooperativa Liberia Editrice Università di Padova: Padova, Italy, 2012; ISBN 978-88-6129-800-2.

33. Auestad, I.; Austad, I.; Rydgren, K. Nature will have its way: Local vegetation trumps restoration treatments in semi-natural grassland. *Appl. Veg. Sci.* **2015**, *18*, 190–196. [CrossRef]

34. Baasch, A.; Engst, K.; Schmiede, R.; May, K.; Tischew, S. Enhancing success in grassland restoration by adding regionally propagated target species. *Ecol. Eng.* **2016**, *94*, 583–591. [CrossRef]

35. John, H.; Dullau, S.; Baasch, A.; Tischew, S. Re-introduction of target species into degraded lowland hay meadows: How to manage the crucial first year? *Ecol. Eng.* **2016**, *86*, 223–230. [CrossRef]

36. Kaulfuß, F.; Rosbakh, S.; Reisch, C. Grassland restoration by local seed mixtures: New evidence from a practical 15-year restoration study. *Appl. Veg. Sci.* **2022**, *25*, e12652. [CrossRef]

37. Schmidt, A.; Kirmer, A.; Kiehl, K.; Tischew, S. Seed mixture strongly affects species-richness and quality of perennial flower strips on fertile soil. *Basic Appl. Ecol.* **2020**, *42*, 62–72. [CrossRef]

38. Kopp, J.; Frajer, J.; Pavelková, R. Driving forces of the development of suburban landscape–A case study of the Sulkov site west of Pilsen. *Quaest. Geogr.* **2015**, *34*, 51–64. [CrossRef]

39. Oudes, D.; Stremke, S. Next generation solar power plants? A comparative analysis of frontrunner solar landscapes in Europe. *Renew. Sustain. Energy Rev.* **2021**, *145*, 111101. [CrossRef]

40. Szabó, S.; Bódis, K.; Kougias, I.; Moner-Girona, M.; Jäger-Waldau, A.; Barton, G.; Szabó, L. A methodology for maximizing the benefits of solar landfills on closed sites. *Renew. Sustain. Energy Rev.* **2017**, *76*, 1291–1300. [CrossRef]

41. Armstrong, A.; Waldron, S.; Whitaker, J.; Ostle, N. Wind farm and solar park effects on plant-soil carbon cycling: Uncertain impacts of changes in ground-level microclimate. *Glob. Change Biol.* **2014**, *20*, 1699–1706. [CrossRef] [PubMed]

42. Armstrong, A.; Ostle, N.; Whitaker, J. Solar park microclimate and vegetation management effects on grassland carbon cycling. *Environ. Res. Lett.* **2016**, *11*, 074016. [CrossRef]

43. Elamri, Y.; Cheviron, B.; Lopez, J.M.; Dejean, C.; Belaud, G. Water budget and crop modelling for agrivoltaic systems: Application to irrigated lettuces. *Agric. Water Manag.* **2018**, *208*, 440–453. [CrossRef]

44. Lambert, Q.; Bischoff, A.; Cueff, S.; Cluchier, A.; Gros, R. Effects of solar park construction and solar panels on soil quality, microclimate, CO_2 effluxes, and vegetation under a Mediterranean climate. *Land Degrad. Dev.* **2021**, *32*, 5190–5202. [CrossRef]

45. Liu, Y.; Zhang, R.; Huang, Z.; Cheng, Z.; López-Vicente, M.; Ma, X.; Wu, G. Solar photovoltaic panels significantly promote vegetation recovery by modifying the soil surface microhabitats in an arid sandy ecosystem. *Land Degrad. Dev.* **2019**, *30*, 2177–2186. [CrossRef]

46. Vervloesem, J.; Marcheggiani, E.; Choudhury, M.; Muys, B. Effects of Photovoltaic Solar Farms on Microclimate and Vegetation Diversity. *Sustainability* **2022**, *14*, 7493. [CrossRef]

47. Schmidt, A.; Kirmer, A.; Hellwig, N.; Kiehl, K.; Tischew, S. Evaluating CAP wildflower strips: High-quality seed mixtures significantly improve plant diversity and related pollen and nectar resources. *J. Appl. Ecol.* **2022**, *59*, 860–871. [CrossRef]

48. Lambert, Q.; Gros, R.; Bischoff, A. Ecological restoration of solar park plant communities and the effect of solar panels. *Ecol. Eng.* **2022**, *182*, 106722. [CrossRef]

49. Li, C.; Liu, J.; Bao, J.; Wu, T.; Chai, B. Effect of Light Heterogeneity Caused by Photovoltaic Panels on the Plant-Soil-Microbial System in Solar Park. *Land* **2023**, *12*, 367. [CrossRef]

50. Walston, L.J.; Barley, T.; Bhandari, I.; Campbell, B.; McCall, J.; Hartmann, H.M.; Dolezal, A.G. Opportunities for agrivoltaic systems to achieve synergistic food-energy-environmental needs and address sustainability goals. *Front. Sustain. Food Syst.* **2022**, *6*, 374. [CrossRef]

51. Baughman, B.O.W.; Kulpa, S.M.; Sheley, R.L. Four paths toward realizing the full potential of using native plants during ecosystem restoration in the Intermountain West. *Rangelands* **2022**, *44*, 218–226. [CrossRef]

52. Bucharova, A.; Bossdorf, O.; Hölzel, N.; Kollmann, J.; Prasse, R.; Durka, W. Mix and match: Regional admixture provenancing strikes a balance among different seed-sourcing strategies for ecological restoration. *Conserv. Genet.* **2019**, *20*, 7–17. [CrossRef]

53. Durka, W.; Michalski, S.G.; Berendzen, K.W.; Bossdorf, O.; Bucharova, A.; Hermann, J.M.; Hölzel, N.; Kollmann, J. Genetic differentiation within multiple common grassland plants supports seed transfer zones for ecological restoration. *J. Appl. Ecol.* **2017**, *54*, 127–136. [CrossRef]

54. Durka, W.; Bossdorf, O.; Bucharova, A.; Frenzel, M.; Hermann, J.M.; Hölzel, N.; Kollmann, J.; Michalski, S.G. Regionales Saatgut von Wiesenpflanzen: Genetische Unterschiede, regionale Anpassung und Interaktion mit Insekten. *Nat. Und Landsch.* **2019**, *94*, 146–153. [CrossRef]

55. Nagel, R.; Durka, W.; Bossdorf, O.; Bucharova, A. Rapid evolution in native plants cultivated for ecological restoration: Not a general pattern. *Plant Biol.* **2019**, *21*, 551–558. [CrossRef]

56. Ssymank, A. Neue Anforderungen im europäischen Naturschutz: Das Schutzgebietssystem Natura 2000 und die FFH-Richtlinie der EU. *Nat. Und Landsch.* **1994**, *69*, 395–406.

57. Höfner, J.; Klein-Raufhake, T.; Lampei, C.; Mudrak, O.; Bucharova, A.; Durka, W. Populations restored using regional seed are genetically diverse and similar to natural populations in the region. *J. Appl. Ecol.* **2022**, *59*, 2234–2244. [CrossRef]

58. Pedrini, S.; Dixon, K.W. International principles and standards for native seeds in ecological restoration. *Restor. Ecol.* **2020**, *28*, 286–303. [CrossRef]

59. Tischew, S.; Youtie, B.; Kirmer, A.; Shaw, N. Farming for Restoration: Building Bridges for Native Seeds. *Ecol. Restor.* **2011**, *29*, 219–222. [CrossRef]

60. Bucharova, A.; Lampei, C.; Conrady, M.; May, E.; Matheja, J.; Meyer, M.; Ott, D. Plant provenance affects pollinator network: Implications for ecological restoration. *J. Appl. Ecol.* **2022**, *59*, 373–383. [CrossRef]

61. Nevill, P.G.; Cross, A.T.; Dixon, K.W. Ethical seed sourcing is a key issue in meeting global restoration targets. *Curr. Biol.* **2018**, *28*, R1378–R1379. [CrossRef]

62. Mainz, A.K.; Wieden, M. Ten years of native seed certification in Germany-a summary. *Plant Biol.* **2019**, *21*, 383–388. [CrossRef] [PubMed]

63. Guisan, A.; Thuiller, W. Predicting species distribution: Offering more than simple habitat models. *Ecol. Lett.* **2005**, *8*, 993–1009. [CrossRef] [PubMed]

64. Feistel, U.; Kettner, S.; Ebermann, J.; Müller, F. Wie PV-Freiflächenanlagen den Bodenwasserhaushalt verändern–Begleitforschung im größten Solarpark Deutschlands. *Forum Hydrol. Wasserbewirtsch.* **2022**, *43*, 43–52. [CrossRef]

65. Feistel, U.; Werisch, S.; Marx, P.; Kettner, S.; Ebermann, J.; Wagner, L. Assessing the impact of shading by solar panels on evapotranspiration and plant growth using lysimeters. *AIP Conf. Proc.* **2022**, *2635*, 150001. [CrossRef]

66. Fischer, H.S.; Michler, B.; Ziche, D.; Fischer, A. Plants as indicators of soil chemical properties. In *Status and Dynamics of Forests in Germany. Ecological Studies*; Wellbrock, N., Bolte, A., Eds.; Springer: Cham, Germany, 2019; Volume 237, pp. 295–309, ISBN 978-3-030-15732-6.

67. Ramos, S.J.; Gastauer, M.; Mitre, S.K.; Caldeira, C.F.; Silva, J.R.; Furtini Neto, A.E.; Oliveira, G.; Souza Filho, P.W.M.; Siqueira, J.O. Plant growth and nutrient use efficiency of two native Fabaceae species for mineland revegetation in the eastern Amazon. *J. For. Res.* **2020**, *31*, 2287–2293. [CrossRef]

68. Wang, Y.; Wang, W.; Wang, Q.; Li, X.; Liu, Y.; Huang, Q. Effects of soil nutrients on reproductive traits of invasive and native annual Asteraceae plants. *Biodivers. Sci.* **2021**, *29*, 1–9. [CrossRef]

69. Graham, M.; Ates, S.; Melathopoulos, A.P.; Moldenke, A.R.; DeBano, S.J.; Best, L.R.; Higgins, C.W. Partial shading by solar panels delays bloom, increases floral abundance during the late-season for pollinators in a dryland, agrivoltaic ecosystem. *Sci. Rep.* **2021**, *11*, 7452. [CrossRef]

70. Marrou, H.; Guilioni, L.; Dufour, L.; Dupraz, C.; Wery, J. Microclimate under agrivoltaic systems: Is crop growth rate affected in the partial shade of solar panels? *Agric. For. Meteorol.* **2013**, *177*, 117–132. [CrossRef]

71. Pedrini, S.; D'Agui, H.M.; Arya, T.; Turner, S.; Dixon, K.W. Seed quality and the true price of native seed for mine site restoration. *Restor. Ecol.* **2022**, *30*, e13638. [CrossRef]

72. Schaub, S.; Finger, R.; Buchmann, N.; Steiner, V.; Klaus, V.H. The costs of diversity: Higher prices for more diverse grassland seed mixtures. *Environ. Res. Lett.* **2021**, *16*, 094011. [CrossRef]

73. Schaub, S.; Finger, R.; Buchmann, N.; Steiner, V.; Klaus, V.H.; Klaus, V. Data: German and Swiss seed mixture prices and characteristics. *ETH Zur.* **2021**. [CrossRef]

74. Merritt, D.J.; Dixon, K.W. Restoration seed banks—A matter of scale. *Science* **2011**, *332*, 424–425. [CrossRef] [PubMed]

75. Goldsmith, N.E.; Flint, S.A.; Shaw, R.G. Factors limiting the availability of native seed for reconstructing Minnesota's prairies: Stakeholder perspectives. *Restor. Ecol.* **2022**, *30*, e13554. [CrossRef]

76. Bundesministerium für Naturschutz (Ed.) Gesetz über Naturschutz und Landschaftspflege. Bundesnaturschutzgesetz–BNatSchG, Berlin. 2009. Available online: https://www.bmuv.de/gesetz/gesetz-ueber-naturschutz-und-landschaftspflege (accessed on 2 March 2023).

77. Erickson, V.J.; Halford, A. Seed planning, sourcing, and procurement. *Restor. Ecol.* **2020**, *28*, 219–227. [CrossRef]

78. Pedrini, S.; Gibson-Roy, P.; Trivedi, C.; Gálvez-Ramírez, C.; Hardwick, K.; Shaw, N.; Frischie, S.; Laverack, G.; Dixon, K. Collection and production of native seeds for ecological restoration. *Restor. Ecol.* **2020**, *28*, 228–238. [CrossRef]

79. Barak, R.S.; Ma, Z.; Brudvig, L.A.; Havens, K. Factors influencing seed mix design for prairie restoration. *Restor. Ecol.* **2022**, *30*, e13581. [CrossRef]

80. Begosh, A.; Smith, L.M.; McMurry, S.T. Major land use and vegetation influences on potential pollinator communities in the High Plains of Texas. *J. Insect Conserv.* **2022**, *26*, 231–241. [CrossRef]

81. Cely-Santos, M.; Philpott, S.M. Local and landscape habitat influences on bee diversity in agricultural landscapes in Anolaima, Colombia. *J. Insect Conserv.* **2019**, *23*, 133–146. [CrossRef]

82. Schubert, L.F.; Hellwig, N.; Kirmer, A.; Schmid-Egger, C.; Schmidt, A.; Dieker, P.; Tischew, S. Habitat quality and surrounding landscape structures influence wild bee occurrence in perennial wildflower strips. *Basic Appl. Ecol.* **2022**, *60*, 76–86. [CrossRef]

83. Williams, N.M.; Ward, K.L.; Pope, N.; Isaacs, R.; Wilson, J.; May, E.A.; Ellis, J.; Daniels, J.; Pence, A.; Ullmann, K.; et al. Native Wildflower Plantings Support Wild Bee Abundance and Diversity in Agricultural Landscapes across the United States. *Ecol. Appl.* **2015**, *25*, 2119–2131. [CrossRef] [PubMed]

84. Filipiak, M.; Filipiak, Z.M. Application of ionomics and ecological stoichiometry in conservation biology: Nutrient demand and supply in a changing environment. *Biol. Conserv.* **2022**, *272*, 109622. [CrossRef]
85. Rittschof, C.C.; Denny, A.S. The Impacts of Early-Life Experience on Bee Phenotypes and Fitness. *Integr. Comp. Biol.* **2023**, icad009. [CrossRef]
86. Dolezal, A.G.; Torres, J.; O'Neal, M.E. Can Solar Energy Fuel Pollinator Conservation? *Environ. Entomol.* **2021**, *50*, 757–761. [CrossRef]
87. Wood, T.J.; Holland, J.M.; Goulson, D. Providing foraging resources for solitary bees on farmland: Current schemes for pollinators benefit a limited suite of species. *J. Appl. Ecol.* **2017**, *54*, 323–333. [CrossRef]
88. Pritsch, G. *Bienenweide: 200 Trachtpflanzen Erkennen und Bewerten*, 1st ed.; Franckh-Kosmos Verlag: Stuttgart, Germany, 2018; ISBN 9783440159910.
89. Nichols, R.N.; Holland, J.M.; Goulson, D. Can novel seed mixes provide a more diverse, abundant, earlier, and longer-lasting floral resource for bees than current mixes? *Basic Appl. Ecol.* **2022**, *60*, 34–47. [CrossRef]
90. Jäger, E.J. (Ed.) *Rothmaler–Exkursionsflora von Deutschland. Gefäßpflanzen: Grundband*, 1st ed.; Springer Spektrum: Berlin/Heidelberg, Germany, 2017; ISBN 978-3-662-49708-1.
91. Kuppler, J.; Neumüller, U.; Mayr, A.V.; Hopfenmüller, S.; Weiss, K.; Prosi, R.; Schanowski, A.; Schwenninger, H.R.; Ayasse, M.; Burger, H. Favourite plants of wild bees. *Agric. Ecosyst. Environ.* **2023**, *342*, 108266. [CrossRef]
92. Al Heib, M.; Cherkaoui, A. Assessment of the Advantages and Limitations of Installing PV Systems on Abandoned Dumps. *Mater. Proc.* **2021**, *5*, 68. [CrossRef]
93. Pavlovic, N.; Ignjatovic, D.; Subaranovic, T. Possibility of Using Wind and Solar Sources for Electric Power Generation on Serbian Opencast Coal Mines. *Mater. Proc.* **2021**, *5*, 50. [CrossRef]
94. Li, P.; Kleijn, D.; Badenhausser, I.; Zaragoza-Trello, C.; Gross, N.; Raemakers, I.; Scheper, J. The relative importance of green infrastructure as refuge habitat for pollinators increases with local land-use intensity. *J. Appl. Ecol.* **2020**, *57*, 1494–1503. [CrossRef]
95. Török, P.; Vida, E.; Deák, B.; Lengyel, S.; Tóthmérész, B. Grassland restoration on former croplands in Europe: An assessment of applicability of techniques and costs. *Biodivers. Conserv.* **2011**, *20*, 2311–2332. [CrossRef]
96. Schaumberger, S.; Blaschka, A.; Krautzer, B.; Graiss, W.; Klingler, A.; Pötsch, E.M. Successful transfer of species-rich grassland by means of green hay or threshing material: Does the method matter in the long term? *Appl. Veg. Sci.* **2021**, *24*, e12606. [CrossRef]
97. Hancock, N.; Gibson-Roy, P.; Driver, M.; Broadhurst, L. *The Australian Native Seed Survey Report*; Australian Network for Plant Conservation. Canberra, Australia, 2020.
98. Ladouceur, E.; Jiménez-Alfaro, B.; Marin, M.; De Vitis, M.; Abbandonato, H.; Iannetta, P.P.M.; Bonomi, C.; Pritchard, H.W. Native Seed Supply and the Restoration Species Pool. *Conserv. Lett.* **2018**, *11*, e12381. [CrossRef]
99. Nichols, R.N.; Goulson, D.; Holland, J.M. The best wildflowers for wild bees. *J. Insect Conserv.* **2019**, *23*, 819–830. [CrossRef]
100. Seitz, N.; VanEngelsdorp, D.; Leonhardt, S.D. Are native and non-native pollinator friendly plants equally valuable for native wild bee communities? *Ecol. Evol.* **2020**, *10*, 12838–12850. [CrossRef] [PubMed]
101. Von Königslöw, V.; Fornoff, F.; Klein, A.M. Pollinator enhancement in agriculture: Comparing sown flower strips, hedges and sown hedge herb layers in apple orchards. *Biodivers. Conserv.* **2022**, *31*, 433–451. [CrossRef]
102. Grašič, M.; Šabić, A.; Lukač, B. A review of methodology for grassland restoration and management with practical examples. *Acta Biol. Slov.* **2023**, *66*, 1–26. [CrossRef]
103. Blaydes, H.; Potts, S.; Whyatt, D.; Armstrong, A. On-site floral resources and surrounding landscape characteristics impact pollinator biodiversity on solar parks. In Proceedings of the EGU General Assembly 2022 (EGU22-2180), Vienna, Austria, 23–27 May 2022. [CrossRef]
104. Rieger-Hofmann GmbH. 24 Mischung Solarpark. Available online: https://www.rieger-hofmann.de/sortiment-shop/mischungen/wiesen-und-saeume-fuer-die-freie-landschaft/01-blumenwiese/detailansicht-blumenwiese.html?tt_products%5BbackPID%5D=181&tt_products%5Bproduct%5D=712&tt_products%5Bsword%5D=SOLARPARK&cHash=440a1ab678029f4f100a735ea8184b63 (accessed on 16 April 2023).
105. Meine Blumenwiese. Wildblumenmischung für PV (Photovoltaik-Mischung) 5 kg, 10 kg. Available online: https://meineblumenwiese.at/products/wildblumenmischung-fur-pv-photovoltaikmischung-ab-300m2 (accessed on 16 April 2023).
106. Fountain, M.T. Impacts of Wildflower Interventions on Beneficial Insects in Fruit Crops: A Review. *Insects* **2022**, *13*, 304. [CrossRef]
107. Venjakob, C.; Ruedenauer, F.A.; Klein, A.-M.; Leonhardt, S.D. Variation in nectar quality across 34 grassland plant species. *Plant. Biol.* **2022**, *24*, 134–144. [CrossRef]
108. Davis, A.E.; Bickel, D.J.; Saunders, M.E.; Rader, R. Crop-pollinating Diptera have diverse diet and habitat needs in both larval and adult stages. *Ecol. Appl.* **2023**, e2859. [CrossRef]
109. Ollerton, J. Pollinator Diversity: Distribution, Ecological Function, and Conservation. *Annu. Rev. Ecol. Evol. Syst.* **2023**, *48*, 353–376. [CrossRef]
110. IPBES (The Intergovernmental Science-Policy Platform on Biodiversity and Ecosystem Services. *The Assessment Report of the Intergovernmental Science-Policy Platform on Biodiversity and Ecosystem Services on Pollinators, Pollination and Food Production*; Potts, S.G., Imperatriz-Fonseca, V.L., Bonn, H.T.N., Eds.; Secretariat of the Intergovernmental Science-Policy Platform on Biodiversity and Ecosystem Services: Bonn, Germany, 2016; ISBN 978-92-807-3567-3.
111. Filipiak, Z.M.; Denisow, B.; Stawiarz, E.; Filipiak, M. Unravelling the dependence of a wild bee on floral diversity and composition using a feeding experiment. *Sci. Total Environ.* **2022**, *820*, 153326. [CrossRef]

112. Milberg, P.; Andersson, L.; Thompson, K. Large-seeded spices are less dependent on light for germination than small-seeded ones. *Seed Sci. Res.* **2000**, *10*, 99–104. [CrossRef]
113. Warzecha, D.; Diekötter, T.; Wolters, V.; Jauker, F. Attractiveness of wildflower mixtures for wild bees and hoverflies depends on some key plant species. *Insect Conserv. Divers.* **2018**, *11*, 32–41. [CrossRef]
114. Shaw, N.; Barak, R.S.; Campbell, R.E.; Kirmer, A.; Pedrini, S.; Dixon, K.; Frischie, S. Seed use in the field: Delivering seeds for restoration success. *Restor. Ecol.* **2020**, *28*, 276–285. [CrossRef]
115. Glidden, A.J.; Sherrard, M.E.; Meissen, J.C.; Myers, M.C.; Elgersma, K.J.; Jackson, L.L. Planting time, first-year mowing, and seed mix design influence ecological outcomes in agroecosystem revegetation projects. *Restor. Ecol.* **2022**, *31*, e13818. [CrossRef]
116. Kiehl, K.; Kirmer, A.; Shaw, N.; Tischew, S. (Eds.) *Guidelines for Native Seed Production and Grassland Restoration*; Cambridge Scholars Publishing: Newcastle upon Tyne, UK, 2014; ISBN 978-1-4438-5900-4.
117. Blaydes, H.; Gardner, E.; Whyatt, J.D.; Potts, S.G.; Armstrong, A. Solar park management and design to boost bumble bee populations. *Environ. Res. Lett.* **2022**, *17*, 044002. [CrossRef]
118. Klotz, S.; Kühn, I.; Durka, W. (Eds.) *BIOLFLOR-Eine Datenbank zu Biologisch-Ökologischen Merkmalen der Gefäßpflanzen in Deutschland*; Bundesamt für Naturschutz, Schriftenreihe für Vegetationskunde: Bonn, Germany, 2002; Volume 38.
119. Stiftung Naturschutz Schleswig-Holstein. Trachtenkalender 2016. Nutzpflanzen. Trachtenkalender für Schleswig-Holstein. Available online: https://www.stiftungsland.de/fileadmin/pdf/Trachtkalender/SN_Trachtkalender_LandNutz_A4_2016052 6_A.pdf (accessed on 18 April 2023).

Review

Grazing as a Management Tool in Mediterranean Pastures: A Meta-Analysis Based on a Literature Review

Dimitrios Oikonomou [1], Michael Vrahnakis [1,*], Maria Yiakoulaki [2], Gavriil Xanthopoulos [3] and Yannis Kazoglou [1]

[1] Department of Forestry, Wood Sciences and Design, University of Thessaly, 43131 Karditsa, Greece; dimoikonom@uth.gr (D.O.); ykazoglou@uth.gr (Y.K.)

[2] Department of Forestry and Natural Environment, Aristotle University, 54124 Thessaloniki, Greece; yiak@for.auth.gr

[3] Hellenic Agricultural Organization "Dimitra", Institute of Mediterranean Forest Ecosystems, Terma Alkmanos, 11528 Athens, Greece; gxnrtc@fria.gr

* Correspondence: mvrahnak@uth.gr

Abstract: The present study reviews the impact of mechanical interventions, and controlled burning combined with grazing in the Mediterranean-climate regions (MCRs) of the world. Relevant studies were searched for in the Web of Science database. Additional studies were located in the citations of these publications, and in a local database. Finally, 26 studies were included in this review. Since 1978, several other relevant studies have emerged at a rate of 24% in a 5-year time step. The studies have focused on the effects of combined grazing with other management tools on vegetation structure (18 publications), biomass productivity (16 publications), and floristic diversity (12 publications). The results were analyzed for (a) sites and treatments and (b) effects on plant structure, productivity and floristic diversity. Herbaceous forage increased after a reduction in shrub cover. Shrubs tended to recover in the grazed pastures. Vegetation height was reduced in almost in all cases according to available data. Despite its potential recovery, shrub biomass was affected by grazing in most cases. The impact of subsequent grazing was mixed regarding floristic diversity. Grazing is a useful tool for landscape management in MCRs, but the proper way to combine it with other interventions depends on the management goals.

Keywords: Mediterranean-climate regions; grazing; clearing; controlled burning; cutting; thinning; vegetation structure; productivity; floristic diversity

Citation: Oikonomou, D.; Vrahnakis, M.; Yiakoulaki, M.; Xanthopoulos, G.; Kazoglou, Y. Grazing as a Management Tool in Mediterranean Pastures: A Meta-Analysis Based on a Literature Review. *Land* **2023**, *12*, 1290. https://doi.org/10.3390/land12071290

Academic Editor: Eddie J.B. van Etten

Received: 28 May 2023
Revised: 17 June 2023
Accepted: 22 June 2023
Published: 26 June 2023

1. Introduction

Animal husbandry is one of the most important activities in human history. While it has become more intensive in recent years, keeping animals in closed spaces, even today, a large part of livestock is raised extensively or semi-extensively. These two breeding systems are largely based on the use of pastures, often originating from past human interventions in forest landscapes. Therefore, grazing has to be regulated so that the natural elements of the landscape can be maintained [1]. Mediterranean-climate regions (MCRs) are not an exception to this situation. The Mediterranean Basin is a region where the history of disturbance is critical for its ecosystem functions [2,3], while other MCRs, such as those in California and Chile, share quite common land use patterns with it [4]. Nevertheless, the environmental history of MCRs is highly complicated and comparisons among them reveal differences and similarities in environmental indicators [4,5].

An unequivocally negative role is often attributed to grazing, with grazing exclusion policies as a typical measure serving that perspective [6]. It is well-documented, though, that the reduced intensity or complete abandonment of grazing has negative consequences for Mediterranean ecosystems [7,8]. The depopulation processes, that have been taking place in the rural areas of the Mediterranean for more than 60 years, underpin the significant reduction in livestock. Grazing abandonment or undergrazing favors ecological succession, i.e., the expansion of shrubs into grasslands (shrub encroachment) [3,5]. Such natural processes happen at the expense of the floristic physiognomy and diversity of grasslands and often affect several parameters that determine the pastoral value of these lands [9]. It has been pointed out that the prevention of shrub encroachment is important both for the economy and the environment [7]. To halt this process, controlled fires and interventions with mechanical means are often suggested to take place [8].

In most cases, management interventions aim to maintain a landscape mosaic that is considered to be capable of serving multiple objectives simultaneously. Objectives include the balance between human activities and nature, the protection of biodiversity, fire prevention, and the accessibility of various forage resources for grazing animals [10–14]. In Mediterranean ecosystems, disturbances, such as fires and various types of cutting and mechanical treatments for the improvement of forage availability or for cultivation purposes, have played a significant role in the retention of landscape diversity [15–17]. These disturbances are considered episodic, while the opposite is true for grazing, which is considered a chronic disturbance. The combination of these two types of disturbance is common in the Mediterranean region and therefore responsible for the typical formation of vegetation [18].

Natural or anthropogenic fires are considered one of the basic factors shaping Mediterranean landscapes. However, when fires occur near zones of the interface with human activities, they pose a direct risk to ecosystem services, properties, cultural values, and often human lives [19,20]. As an effort to mitigate the phenomenon of fires in MCRs, but also to manage grazing and maintain biodiversity, controlled (or prescribed) burning has been developed gradually since the 1960s [21]. Fire management usually enters public dialogue with an emphasis on fire suppression, including a continuous increase in firefighting resources. Fire prevention, as a rule receives less attention with prescribed burning being a prohibited practice or one with very limited use in most Mediterranean countries. This is in contrast to its extensive application in other countries such as the USA or Australia [20,22,23]. Currently, the use of prescribed burning has gradually started being considered more broadly in MCRs against the common perception and generalization of the solely negative consequences of wildfire.

On the other hand, the most usual practices when it comes to the creation of open spaces, as well as diverse habitats, are the use of intensive mechanical or manual treatments. Mechanical treatments of shrub vegetation may include cutting coarse wood, slashing coarse and fine wood debris, pruning standing branches or resprouts, trimming woody biomass to a lower level, and thinning by removing entire standing shrubs. Some authors suggest that mechanical treatments are safer for the environment or more effective at preventing shrub encroachment than controlled burning is [9,24,25]. On the other hand, controlled burning in a *Pinus canariensis* forest is more beneficial than clearing is when important factors, such as fire intensity, are properly adjusted [26]. There may also be differences in shrub resprouting between sites that have experienced burning and mechanical or manual treatments [27], though, after some years, the effects of these two types of treatments may start being similar [28,29]. Mechanical and manual treatments are the most suitable option in cases when biomass reduction must be applied in strict areas. The interference of private properties in land management plans is a typical example of that [30]. While land clearings and shrub cuttings can be quite beneficial for Mediterranean ecosystems, careless use of these tools can also be harmful, if the intensity or seasonality of the mechanical treatments [31] or their spatial scale are not considered carefully [32–34]. Additionally, the high costs of mechanical and manual treatments could be prohibitive of the continuous use of this practice. In contrast, a combination of them with utilization methods would be more viable. Animal husbandry can play this role, as grazing is a major utilization activity that can be developed in natural ecosystems apart from timber harvesting [30].

Previous review studies have highlighted the importance of management interventions in improving the rangeland value of Mediterranean shrublands [35], and to prevent land abandonment [7,25]. The importance of grazing, even in large numbers, was also noted for several Mediterranean ecosystems [36]. The Rouet-Ledouc et al. [37] study can also be considered relevant to this topic, as they investigated the effectiveness of grazing at fire prevention for a wider range of ecosystems. However, there is no research primarily targeted at deriving conclusions from existing scientific studies that have combined grazing with other management interventions in Mediterranean ecosystems. Such a collective approach will facilitate decisions to draw on land management, especially on the restoration of grasslands that are invaded by shrubs. The aim of this study was to review the scientific literature related to (a) the underlying reasons and the techniques used for the removal of shrubs before grazing as applied in Mediterranean ecosystems, (b) the effects of grazing, introduced after the removal of shrubs or the implementation of other previous management interventions, on the basic characteristics of the vegetation (structure, productivity, and floristic diversity), and (c) the management implications regarding the future use of such interventions in Mediterranean shrublands.

2. Materials and Methods

The search for studies was carried out using a combination of different tools. Initially, the search engine of the Web of Science (WoS) website was used, with the following conditions:

((graz * OR brows * OR sheep OR goat * OR cattle OR hors * OR livestock OR herbivore * OR donk * OR deer *) AND (mediterranean OR (dry AND hot AND summer) OR maquis OR garrigue * OR chaparral OR matorral OR fynbos OR mallee OR phrygana) AND (shrub * OR scrub * OR brush * OR understor * OR bush *) AND (clear * OR cut * OR thin * OR mechanic * OR masticat * OR prunn * OR mow * OR slash * OR trim * OR (prescribed AND burn *) OR (control * AND burn *)).

From the above search, 198 records were found, fourteen of which were considered suitable to be included in the present study, because this number of records focused on the combined effects of grazing, after the implementation of several management treatments. In a second phase, a literature review based on the bibliography of the aforementioned studies was conducted. This included both the studies cited by these fourteen publications and the follow-up studies in which they were referenced. In order to find these follow-up studies, Google Scholar was used. Through this, the number of studies increased to 23. Finally, three more studies were included, by searching the bibliographic base (in paper format) of the Laboratory of Rangeland Science and Protected Areas Management, of the University of Thessaly. Thus, the total number of studies reached 26, referring to a total number of twenty sites (Table 1. The studies included in this review cover a period from 1978 [38] to 2022 [39] (Table 2).

Table 1. Research purposes and management implications of the studies presented.

Authors	Research Topic	Management Implications
Poissonet et al. (1978)	Impact of land management treatments in a *Quercus coccifera* (controlled burning and clearing + grazing simulation with or without fertilization).	Moderate fertilization levels and mowing at a relatively low frequency (simulated grazing systems) achieve better floristic diversity results than pure controlled burnings do, despite the frequency of the latter. Remarkable variations were observed throughout the years. Controlled burning did not achieve the desirable formation, while simulated grazing resulted in a grassland formation.
Green et al. (1979)	Utilization of chaparral shrubs by goats in fenced and non-fenced sites, cleared or non-cleared.	Shepherd guidance is enough for goats to graze young shoots, but fencing is required if an older woody component is present. Grazing without fencing limitations is more environmentally feasible.

Table 1. *Cont.*

Authors	Research Topic	Management Implications
Godron et al. (1981)	Site cover and botanical diversity effects of grazing simulation in a *Quercus coccifera* garrigue.	Mediterranean ecosystems are pretty sensitive to soil erosion; thus, forests have to be re-established. However, some parts of them should be used as pastures, where fertility should be maintained.
Poissonet et al. (1981)	Botanical diversity effects of grazing simulation in a *Quercus coccifera* garrigue.	After clearing and continuous mowing (grazing simulation), the created plant community was very different, more balanced and richer in herbaceous vegetation, although it did not have a typical vigorous grassland structure. The increase in pastural value and changes in composition occurred more directly with fertilization. Severe drought can wipe out some species families (e.g., grass-like species, such as Cyperaceae). High levels of fertilization can make the plant community more exposed to climate change.
Étienne et al. (1991)	Biomass and volume growth models in grazed maquis vegetation of pine and oak forest areas that are partially fertilized.	Shrub growth models after disturbance should be considered in fuel management. There are differences between major Mediterranean maquis shrubs in the development of volume, biomass, and growth rate.
Papanastasis et al. (1991)	Yields of shrubs and herbs in a bladed (partially seeded) or slashed kermes oak shrubland.	Seeding after clearing is the method that increases yields of herbaceous vegetation the most. On the other hand, it is the most expensive method. Clearing without seeding is less effective for yields of herbaceous vegetation. Slashing is the least expensive and the best method ecologically; however, it results in a rapid restoration of shrubby vegetation over that of herbaceous vegetation.
Perevolotsky et al. (1992)	*Quercus calliprinos* and *Phillyrea latifollia* response after thinning and pruning, followed by goat grazing.	Thinning + grazing combined favors open spaces, mostly affecting the cover of low trees in comparison to that of dwarf and medium shrubs. *Q. calliprinos* was affected only by thinning, while subsequent grazing had an impact on *P. latifolia*. Thinning + grazing can develop rich pastures with low fire risk, though well-established spatial plans are required.
Papachristou et al. (1997)	Effects on vegetation structure and productivity and grazing response of sheep and goats in a slashed or cleared kermes oak shrubland.	The existence of a greater amount of herbaceous biomass results in the existence of feed of a higher nutritional value. Via the maintenance of a low cover, the movement of animals in the pasture is facilitated. These interventions are also beneficial for biodiversity and act against fire hazard.
Yiakoulaki et al. (1998)	Effects on biomass and dietary preferences of goat grazing in a *Pinus* forest fuel treatment dominated by kermes oak in the understory.	Grazing management systems do not affect the forage utilization preferences of goats. Nevertheless, low forage availability can affect animal welfare and productivity. Intensive grazing by goats can reduce fuel amounts in a pine forest's understory.
Hadar et al. (1999)	Clearing and/or heavy cattle grazing effects on botanical composition and functional groups in garrigue.	Heavy grazing tends to homogenize the height of the plants to the horizontal level, with certain plant species being affected. Grazing during the growth period of plants is a useful tool for fire prevention. It seems that clearing can cause an increase in species diversity, while intensive grazing decreases it. A combination of them can transform a garrigue formation, from a shrub-dominated to a geophyte-rich herbaceous community, dominated by early flowering plants.

Table 1. *Cont.*

Authors	Research Topic	Management Implications
Gutman et al. (2000)	Herbage biomass, small tree and dwarf shrub cover effects on thinned + grazed garrigue (mainly *Quercus calliprinos* woodland, partially batha vegetation). Forage utilization and beef cattle herd performance was also measured.	Intensive grazing by cattle can create an open park-like landscape, similarly to known goat grazing practices. Herbage is increased as a result of the phosphate fertilization caused by cattle feces, which is relevant to their supplementary feed, mainly poultry litter. Coupled with this kind of feed, the grazing management system of the current study can lead to a high herd beef performance. However, additional feed costs could create feasibility difficulties. Control of undesirable batha vegetation is difficult even via herbicide.
Étienne et al. (2001)	Impact of different management interventions on vegetation structure and floristic diversity of maquis ecosystems where fuel management was performed.	Livestock herds are able to promote ecosystems with rich biodiversity, while they are also used for fuel management, contributing to the carbon cycle and the aesthetic value of the landscape. However, even a 20-year period is small for biodiversity monitoring. An important reason for that is the frequent catastrophic fire events in that kind of ecosystem.
Delgado et al. (2004)	Shrub structure and forage productivity in a *Genista scorpius* shrubland, grazed by cattle or sheep.	Land clearings can increase grass forage production, but this is dependent on annual rainfall and location conditions. It could require some years for fertilization to show its effects on productivity. After clearings, *G. scorpius* individuals grow slowly, although cattle or sheep grazing does not seem to have an effect on them.
Lécrivain et al. (2004)	Description of a clearing technique executed by shepherds for the creation of pasture paths and open areas in a holm oak (*Quercus ilex*) stand. Measurements took place for three years.	Shepherds should be involved in clearing plans, since they are able to create openings in shrublands based on the needs of the herds. Via the creation of a network of patches and considering the capabilities of the flocks to maintain an open vegetation structure, an alteration in the vegetation could be achieved. Thus, the grazing period in a year can be increased.
Potts et al. (2010)	Ecological effects of deer grazing on chaparral (*Adenostoma fasciculatum*), previously prescribed burning or masticated. Effects of the treatments and their applying season in the recruitment of *Ceanothus cuneatus*.	Spring-controlled burning results in greater *C. cuneatus* mortality because seedlings have less time to recover from summer drought. Controlled burning treatments are more effective for wildlife conservation, but less effective for fire prevention compared to mastication in the medium term. However, fire risk is likely to be higher in mastication areas compared to those treated with controlled burning in the short term, because areas that have been recently masticated have more fine dead fuel and grass on the ground. Deer grazing only reduces shrub height and does not affect other characteristics (cover and seedlings).
Alvarez-Martinez et al. (2013)	Structure and productivity effects in pasture restoration of intensive goat grazing (clearing, trimming, or slashing treatments applied previously).	Grazing following other interventions can contribute to a conversion of shrubland into grassland (clearing or controlled burning prior to grazing), or limit shrub growth (trimming prior to grazing).
Masson et al. (2015)	Structure and botanical diversity effects in different treatments applied for the control of the invasion of brambles in grassland. Grazing was performed by goats.	Yearly land clearings followed by grazing can reduce brambles and increase herbaceous diversity, but they are not enough for complete dry grassland restoration. Water infiltration can favor competitive plant species. Dry grassland restoration can be obtained via a combination of interventions for several years, including clearing + grazing but also the restraint of water infiltration.

Table 1. *Cont.*

Authors	Research Topic	Management Implications
Lasanta et al. (2016)	Changes in the landscape structure and livestock numbers after the execution of a plan of clearings to promote grazing in La Rioja, Spain.	Spatial dynamics can be changed by land clearings, as is shown in the case of La Rioja. Pastures were developed in contrast to shrublands, while livestock numbers also increased.
Bashan et al. (2017)	Spatiotemporal dynamics of garrigue vegetation in different treatments. Grazing was performed by goats (high intensity) or cattle (low intensity).	Measures for fire prevention can be different from those developed for biodiversity conservation. Goats with high stocking rates can be quite effective at the control of woody vegetation. In contrast, low-stocking-rate cattle grazing is not effective. A strategy to achieve different targets should be promoted at a landscape level.
Lasanta et al. (2018)	Analysis of the changes in forest fires in La Rioja region before and after the land clearing plan.	When it comes to fire management, a large focus on suppression is ineffective. Grazing can contribute to fuel management, but livestock should exist in high numbers. Land clearing should be combined with cooperation with local livestock breeders. With the combination of clearing and grazing, environmental, economic and social goals can be simultaneously achieved in the Mediterranean.
Lasanta et al. (2019)	Changes in average pasture production before and after the implementation of the land clearing plan in the La Rioja region.	Seasonality in pasture productivity in Mediterranean mountain regions is an important issue in livestock management. Land clearings promote ease of movement through the pasture. Effective breed sizes and the manpower of young people are significant factors for the maintenance of extensive livestock systems.
Moinardeau et al. (2020)	Impact of clearing and/or goat grazing on the restoration of an artificial embankment understory invaded by brambles	The combination of clearing and grazing can have a positive impact on the heterogeneity and diversity of herbaceous vegetation. Shepherd supervision can be helpful in bramble invasion control. The location of sheds is important in such a situation, along with the application of proper stocking rates, animal training, and contacts between managers and shepherds. A supplementary feed should be avoided. When tall brambles are present, clearing is proposed, but costs should be considered.
Grupenhoff et al. (2021)	Ecological changes in goat grazing in a fuel treatment in Californian chaparral, with a sporadically present oak overstory. Cutting, pile burning and herbicide were previously applied.	Interventions before grazing were beneficial for the reduction in the fuel hazard and the diversification of the pasture. Goat grazing affected only herbaceous vegetation. The seasonality and duration of grazing and botanical composition are important factors when goat grazing is applied for fuel management.
Lasanta et al. (2022)	Landscape impact of the land clearing plan in La Rioja.	A mosaic landscape can be achieved via land clearings and grazing, that is rich in biodiversity and has a low fire risk.
Bicho et al. (2022)	Productivity and resilience to drought of an improved pasture grazed by cattle a pasture improved by cattle grazing (clearing + seeding + ploughing), compared to that of the natural understory (cork oak woodland).	Although the improved pasture was far more productive, the natural understory showed better drought resilience. Climate change impacts on production can be mitigated via the promotion of forage plants resistant to drought. The management strategies that need to be developed should not be very offensive against the natural vegetation. Such a direction would also promote biodiversity and ecosystem balance.
Castro et al. (2022)	Evaluation of ecological changes in a shrubby understory, cleared then grazed by sheep of a mixed sclerophyllous forest, treated for fuel reduction purposes.	In order to keep fuels in a low amount in the understory, grazing is required after clearing. Fire prevention and biodiversity targets have to be met, especially in Natura 2000 sites.

Table 2. Present effects (indicated by X) of the interventions examined in each study, sorted into three categories (structure, productivity, and floristic diversity). Experimental years are also included.

	Effects on Structure	Effects on Productivity	Effects on Floristic Diversity	Duration of Experiment (Years)
Poissonet et al. (1978)			X	9
Green et al. (1979)		X		6
Godron et al. (1981)	X	X	X	9
Poissonet et al. (1981)	X		X	9
Étienne et al. (1991)		X		6
Papanastasis et al. (1991)		X		2
Perevolotsky et al. (1992)	X			2
Papachristou et al. (1997)	X	X		3
Yiakoulaki et al. (1998)		X		1
Hadar et al. (1999)	X		X	5
Gutman et al. (2000)	X	X		10
Étienne et al. (2001)	X		X	10 and 15
Delgado et al. (2004)	X	X		3 and 4
Lécrivain et al. (2004)	X			3
Potts et al. (2010)	X		X	3
Alvarez-Martinez et al. (2013)	X	X		3
Masson et al. (2015)	X		X	3
Lasanta et al. (2016)	X		X	11
Bashan et al. (2017)	X			11
Lasanta et al. (2018)		X		12
Lasanta et al. (2019)		X		31
Moinardeau et al. (2020)	X	X	X	3
Grupenhoff et al. (2021)	X	X	X	3
Lasanta et al. (2022)	X	X	X	3 and 6
Bicho et al. (2022)		X		10
Castro et al. (2022)	X	X	X	2

The reviewed studies concern changes regarding the structure, productivity and diversity of the vegetation. A group of these studies does not involve true grazing by a specific herbivore, but refers to the application of mowing of shoots at standard frequencies as an imperfect way to simulate grazing in a cleared kermes oak (*Quercus coccifera*) shrubland [38,40,41]. However, we included this group of studies in our review, since the simulation of grazing via mowing or clipping is often achieved in the relevant literature. In Esterel hills, France, the clearing of the understory took place simultaneously with the thinning of trees in the overstory, which consisted of individuals of *Pinus pinea, P. halepensis* and *Quercus suber* [42]. This study was included in the present review since it included comparisons between grazing and no grazing after the interventions. It is also worth noting that the present review includes two studies on the mechanical thinning of low trees. One study was conducted in a Mediterranean oak scrubland where the initial thinning of the woody vegetation was followed by beef cattle grazing [43] and another study included additional management with the pruning of *Quercus calliprinos* [44]. As this particular species often takes a shrub-like form, the impact of these disturbances was considered interesting for the present review. In contrast, the impact of grazing following wildfires or burnings that were not applied under a strict protocol, such as pastoral fires, was not considered.

Furthermore, the studies were sorted into three groups. The first group included the descriptive characteristics of the study itself. Characteristics such as year of publication, geographic referenced area, context of study (experiment or active management), purpose of biomass removal, time of sampling, kind of grazing animals, starting time of grazing, physiognomy of vegetation, overstory layer if any, understory vegetation, and area of interventions (in ha). The second group included information on the treatments applied, such as type of treatments, fencing (if any), treatments before the introduction of grazing, grazing season and duration, and grazing intensity. The third group included the impact on measured parameters, such as the structure of vegetation, productivity, and floristic diversity. For each study, the effects of treatments applied on measured parameters were analyzed and discussed.

3. Results and Discussion

3.1. Characteristics of the Studies

The Mediterranean-climate regions (MCRs) for which the studies met the identified search criteria were represented by the Mediterranean Basin and the State of California in North America. In particular, these studies were found in six countries, namely France, Israel, Spain, Greece, Portugal, and the United States (Figure 1) and they refer to twenty sites in total.

Figure 1. Distribution of the number of studies referring to grazing as a subsequent management intervention in MCRs of the world.

The evolution of the number of studies conducted for the above purposes in a 5-year time step is shown in Figure 2. The linear rate of change in the number of studies is +24%, representing the increasing interest of the scientific community in such studies. The first published work was that of Poissonet et al. [38], related to a clearing + simulated grazing vs. controlled burning experiment in Herault, France. The second one was by Green et al. [45] in California, related to the utilization of shrubs by goats as a management tool to halt the regeneration of vegetation in fuel breaks.

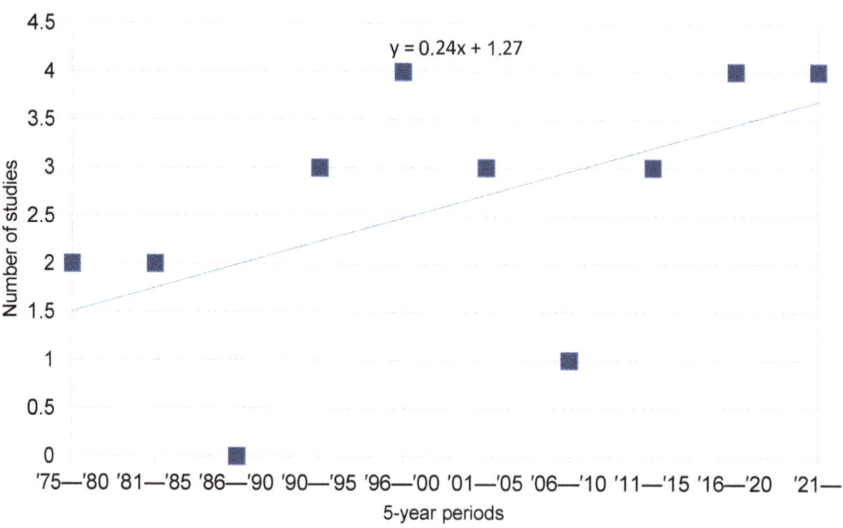

Figure 2. Linear trend line of number of published studies referring to grazing as a subsequent management intervention in MCRs of the world in a 5-year time step.

The management purposes, for which the effectiveness of interventions coupled with grazing was studied, are presented in Figure 3a. All studies were considered to deal with interventions aimed at optimal effectiveness in at least one of the three following issues: fire prevention (FP), the improvement of pastures (PI), and the achievement of a diversified botanical structure (BD) for restoration purposes. Fire prevention and pasture improvement were referenced in the majority of the sites, at 70% and 60%, respectively, while biodiversity was referenced in 40%. The kinds of animals grazing at each site are presented in Figure 3b, with goats being the main grazer (60%), followed by sheep (45%), and cattle (35%). Horses, deer, and grazing simulation (GS) were used in only one site each (5%). In the deer grazing study, other wild herbivores also grazed the area, but some enclosures that were intentionally established to evaluate the lack of grazing prevented deer access [46]. The types of management treatments that preceded grazing are presented in Figure 3c. In the large majority of sites (80%), land clearing (CL) or cutting above the base of the plants (CT) was the only biomass reduction intervention, or was one among other treatments, while in a smaller number of sites other treatments were included, namely slashing (SL), trimming (TR), controlled burning (CB), thinning (TH), pruning (PR), herbicide (H) application, and mastication (M). Subsequent interventions without biomass reduction, namely seeding (SD), fertilizing (F), pile burning (PB), ploughing (PL), herbicide (H) application and draining (D) were applied in 36% of the sites (Figure 3d).

Fire prevention was the main goal of the treatments in all the three studies concerning Californian chaparral. For this purpose, the focus was on domestic or wild animals with browsing habits, namely goats and deer. The improvement of a black-tailed deer habitat was also a purpose of one study [46]. In both the eastern and western Mediterranean Basin, both different purposes of the treatments and grazing animals were mentioned. There was often more than one management purpose for each study in this region. Different kinds of domestic animals were examined, either in terms of grazing in the same treatments or in comparison with each other. Clearing and other interventions of homogenizing vegetation at a horizontal level were performed in all the regions. Thinning vegetation requires the removal of some of the shrubs while trying to give others a tree form. This was performed in two case studies in Israel [43,44], with pruning of the remaining trees also taking place in one of them [44]. Most of the subsequent, improvement interventions such as seeding and fertilizing were performed in west Mediterranean and Greece. Another improvement intervention was pile burning, which is sometimes applied in Californian pastures.

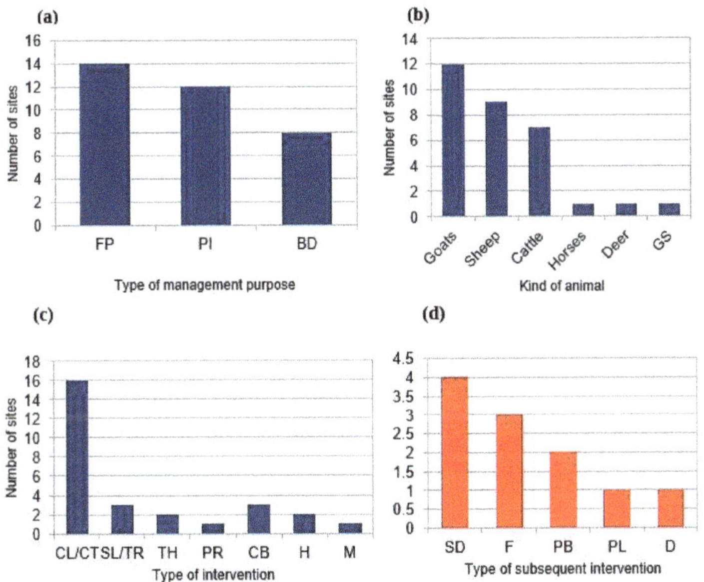

Figure 3. Allocation of (**a**) type of management purpose, (**b**) kind of animals, (**c**) type of intervention and (**d**) type of subsequent intervention into the studied sites. Treatments: FP (fire prevention), PI (pasture Improvement), BD (biodiversity), GS (grazing simulation), CL (clearing), cutting (CT) SL (slashing), TR (trimming), TH (thinning), PR (pruning), CB (controlled burning), H (herbicide), M (mastication), SD (seeding), F (fertilization), PB (pile burning), PL (ploughing), and D (draining).

3.2. Effects on Vegetation Structure

Changes in vegetation structure due to the applied treatments were assessed in a total of eighteen studies (Table 2). When grazing was combined with a prior management intervention, shrub cover did not reach the cover of shrubs in control plots. Moreover, in cases where a control plot did not exist it never reached the same levels as those prior to intervention. This happened for all studies considered, except in the case of grazing by cattle at Ramat HaNadiv Park, northern Israel [47]. In this particular case, the applied grazing was of a low intensity, thus resulting in vegetation cover similar to that of the control. On the contrary, the high grazing intensity in the same area, which included garrigue and batha vegetation, affected shrubs [48]. Although there were no data on the effect of grazing on the cover per se, it was nevertheless pointed out that the shrubland was converted into grassland [48].

Grazing following prior shrub clearing affected shrub recovery, and thus the structure of vegetation, for the sites reported in three studies [39,47,49]. Similarly, in cases of shrub thinning, regarding *Quercus calliprinos*, subsequent grazing was an impediment to shrub recovery [43,44]. In the 9-year study, including combined treatments of clearing and simulated grazing (mowing), the remarkable effect of reducing total vegetation cover was shown [41]. This was in contrast to that of controlled burning treatments with different schedules applied on the same site in the same period, where the cover became dense again two years after the burnings. The cover of woody species also decreased. The potential of intensive grazing to keep open spaces in grasslands was also reported [50]. Thus, there is a total of seven studies reporting that grazing contributes to the reduction in or maintenance of lower shrub cover in comparison to the cover after an initial intervention. Another two studies showed no change in cover due to grazing [46,51]. An absence of change was also reported, regarding the implementation of different grazing intensities (moderate and high) in batha (phrygana) [43]. For the results above, grazing intensity probably played a role, with the exception of the last study [43], which was a case of a plant community of *Sarcopoterium spinosum* and *Calycotome villosa*. These dwarf shrubs are undesirable for selection by animals. The halting of the encroachment of grasslands by those plants is a key management problem in the Eastern Mediterranean grasslands [52]. Interventions such as fire and grazing have been recorded as insufficient or having little effectiveness at halting encroachment [53], even though a wildfire can have an effect on the cover, at least for a short period of time [54]. Finally, for four studies there was no comparison of the results between the application and lack of application of grazing after other structural interventions [55–58]. Although there was no comparison between clearing with grazing and clearing without grazing in the publications concerning the La Rioja region, such comparisons are applied, according to a deliverable of the LIFE MIDMACC Project [59]. The monitoring protocol is mentioned by [59].

For six of the studies, an assessment of herbaceous vegetation cover was conducted. In the case of simulated grazing [41], the herbaceous cover in treated kermes oak shrublands showed increasing trends in all treatments for at least three years. This was not the case in another site when grazing was applied, where in this case, despite the initial increase in herbaceous cover as a result of canopy opening, the herbaceous cover remained stable for three years (i.e., until the end of the experiment) [56]. However, the two studies are quite different. In the study in [41], frequent mowing corresponded to a much higher grazing intensity in respect to the intensity referred to by the authors of [56]. Additionally, grazing intensity appears to have played an important role in increasing herbaceous cover in a site dominated by *Quercus calliprinos* [43]. High grazing intensity after thinning caused an increase in the herbaceous cover, but with moderate intensity the woody vegetation recovered, as well as did its cover. The study in [42] reported that herbaceous cover was increased compared to that under the use of no grazing treatments in all forest stands. In a study of a chaparral ecosystem, the cover of many herbaceous species (native and alien) increased after cutting, pile burning, and herbicide application but decreased with subsequent grazing [51]. Another study showed increased herbaceous cover after the combination of clearing and grazing compared to that after pure clearing, despite the tendency of shrub vegetation to recover in both treatments [39].

With respect to bare ground cover, two studies [40,56] reported an increase in the early stages after clearing or slashing, followed by a gradual decrease. In the case of [40], bare ground cover was higher than that in the pre-treatment situation every year of the experiment only in unfertilized plots. In contrast, in fertilized plots, after a remarkable decrease in bare ground cover during the fourth year of the experiment, its values came close enough to the pre-treatment situation. Afterwards, it showed some yearly variations, which were possibly related to each year's drought. The study in [44] showed that thinning increased bare ground cover, but subsequent grazing did not. Finally, the authors of [49] reported that bare ground cover reached the levels of that of a reference dry grassland, as a result of a combination of annual clearings and grazing.

There are also records of vegetation height, which was measured in eight studies. In all the studies, the combination of initial interventions with grazing reduced vegetation height in comparison to that of the control or the initial height, with the rate of recovery varying according to the experiment, vegetation type, and intervention. In all the cases where data or references are available, grazing significantly affected the understory vegetation height [46,48,49,60]. In the latter study, the height was reduced during the fourth year of grazing compared to that under no grazing. The study in [51] reported changes in the herbaceous height but not in the shrub height. The study in [44] should be considered an exception, as it refers to tree-like shrubs of *Quercus calliprinos*, examining the height of the oaks that remained after thinning. The average shrub height was increased, a fact that must be attributed to the opening of the canopy layer, which allowed their further development, with no significant statistical effect due to grazing. The remaining studies did not report a comparison of vegetation height with and without grazing after the first sampling, so it is not known in how many of these studies grazing had an impact on vegetation height.

Grazing intensity affected the results in most of the cases, with the exception of batha vegetation in [43], as long as there was a comparison between different ones. The structure remained open in case studies where high intensity levels were applied. In contrast, low or moderate intensity left more room for shrubs to develop in the same or adjacent sites [43,47,48,50]. Similarly, an increase in grazing intensity showed a boost in height differences between grazed plots and ungrazed ones [60].

3.3. Effects on Productivity

Impacts on productivity were assessed in a total of sixteen studies, with annual yields and total biomass being the most common parameters assessed (Table 2). In one case, none of these factors was mentioned for shrubby vegetation, but the number of shrubs was considered a measure of productivity instead [50]. While this kind of information was not available, productivity was assessed either through the consumption of vegetation by animals [45,60] or by the pastural value [61]; the latter was also mentioned in [40]. Clearing seems to achieve better results in terms of yields in herbaceous vegetation for two different types of plant communities (shrublands of *Quercus coccifera* and shrublands of *Cytisus scoparius*) compared to slashing or trimming [50,56,62]. The opposite was observed for woody vegetation in all three studies mentioned above, with shrubs recovering more vigorously after the third year from the initial intervention [56]. In the study in [50], there was no relevant reference. However, it was pointed out that the percentage of the surviving shrubs was at least 60% higher after trimming compared to that after other treatments, with increased grazing intensity largely affecting the further elimination of shrubs. Controlled burning + grazing, especially in lands experiencing high stocking rates, was even more effective than clearing + grazing at reducing the number of shrubs, with similar results of herbaceous vegetation [50].

Another important note is that the effects of fertilization and seeding affected biomass in most cases, increasing the available herbaceous biomass and favoring it in its competition with shrub biomass [41,56,57,62–64]. In the case of [57], there was no effect of fertilization on one of the three sites, though the experiment lasted for one year less than it did in the other two sites. This effect was shown during the last year of the experiment, with authors attributing the low impact to poor soil conditions.

In clearings without the use of fertilization, herbaceous biomass did increase in one case [43,51] compared to the control plots, but a decrease was reported in two other studies [39,65]. Specifically, the authors of [39] reported a decrease in herbaceous biomass in the first year after the clearings compared to that after the uncleared + grazed treatments. Cleared plots without grazing had a higher herbaceous biomass than the uncleared treatments did during the second year, but this was not the case in cleared + grazed plots. Additionally, in the study in [65], both shrub and herbaceous biomass decreased in the controlled burning + continuous grazing treatment. Herbaceous biomass was somewhat higher than that before the intervention of the controlled burning + rotational grazing treatment. It was shown that, before the interventions, herbaceous biomass was much higher in the plots that were then continuously grazed than in those that were then rotationally grazed ones [65]. In the study in [66], clearing favored access to herbaceous plants that had already emerged, which were previously covered by dense shrubs. The pastural value was reported by the authors of [66] to have increased after clearing + grazing, while the same happened in the case of [40].

In cases of prior interventions followed by grazing or not, herbivores were effective at decreasing both shrub and herbaceous vegetation, a finding that is valid for all the communities of *Quercus coccifera* where it was studied [56,62,65]. The consumption of shrubs was also notable in the cases of [45,60]. It was noted that this consumption compensated for the regeneration levels for two consecutive years. In these two studies, biomass changes were examined only in terms of shrub and not herbaceous vegetation. On the contrary, although no comparison of treatments with or without grazing was reported, the delayed regrowth of *Genista scorpius* was not attributed to grazing by cattle and sheep [57]. It was rather due to the low regrowth rates, while the increase in herbaceous biomass was probably due to clearing and fertilization. In a study conducted in the Pyrenees with a subcontinental climate, goat grazing on *G. scorpius* communities was involved, and a remarkable effect on productivity was observed [67]. In this case, the grazing season was the most strongly determining factor, rather than the grazing intensity. Additionally, the elimination of *G. scorpius* shrubs following nine years of grazing simulation (through mowing) was reported by [40]. However, the authors also noted that the same did not happen in an experiment where sheep grazing took place. The differentiation of the kinds of grazing animals must therefore be considered. The only case where the shrub biomass did not decrease due grazing, but only the herbaceous biomass decreased, was reported by the authors of [51]. This reduction in the herbaceous component should be partially attributed to the sampling protocol. The comparison between grazing and no grazing was performed in the same year and treatments, before and after the grazing application. Vegetation was sampled during July 2018 before grazing and October 2018 after grazing. The authors also stated that goats can act as browsers under heavy grazing or with a prolonged time for grazing in pastures. The study in [45] concluded that a heavy grazing regime played a decisive role as a biotic shape factor in a chaparral landscape. Indeed, goats covered their feeding demands at high rates of utilization, even if shrubs in the chaparral were initially considered ab undesirable feeding resource for them. These results are in accordance with those of other reviewed studies, where an increase in grazing intensity led to an increase in herbaceous production [43] and/or a decrease in the woody component [50,60].

Finally, only five studies concerned the changes in dead biomass or in the fuel parameters of the pastures, with two of them recording changes in the litter [43,51], one recording dead fuels of a diameter lower than 0.64 cm, also known as 1 h fuels [65] and another two recording the changes in the fuel models at a landscape level in the La Rioja region [66,68]. The combination of grazing with previous interventions reduced the number of fuels in all the above studies. The impact of grazing on fuel reduction was characterized as positive [65] with the authors of [43] mentioning only a decrease in litter under a high grazing intensity.

3.4. Effects on Floristic Diversity

Four different indexes in relation to floristic diversity and grazing followed by other interventions were used in twelve studies, including those on species richness, Shannon–Weaver diversity index, Bray–Curtis dissimilarity, and contact-specific contribution (CSC) (Table 2). Species richness was mentioned in eleven of them and generally followed a pattern similar to that of the other indexes, except for CSC. It should be noted that the three studies concerning the area of Herault in France referred to the same experiment [38,40,41], and two studies referred to the La Rioja region (Spain) [55,68].

Clearing + grazing had mixed results in terms of species richness, particularly in relation to pure clearing. In comparison to the control plots or the pre-treatment situation, in a shrubland of *Quercus coccifera* [40], chaparral, [46] and forest with understory brambles (*Rubus ulmifolius*) [60], there were positive results in terms of the species richness that was observed after some years. For the first study, this may be attributed to the vigor of *Quercus coccifera* and the various ways it reproduces [69,70], although there are site variations, as was shown by the authors of [71]. As for *R. ulmifolius*, its high seed dispersal was not always reflected in its recruitment patterns [72]. According to [49], rapid growth, a rich seed bank, and drought resistance are the main reproductive advantages of plant species as soon as they are established in a field. A restoration experiment on grassland invaded by *R. ulmifolius* showed visible positive results in terms of richness immediately after the treatments [49]. In the study in [42], positive results were observed at the end of the interventions. The same was true for a couple of studies on the La Rioja region [55,68].

Clearing without grazing had positive effects on species richness and diversity, while when clearing + grazing was applied these effects were reduced to some extent [48]. This was attributed to the fact that winter clearing contributed to the creation of an open landscape, favoring certain species, some of which were driven to extinction under intensive grazing during the subsequent growing season. The study in [39] also demonstrated a decrease in species richness and diversity due to grazing. The authors hypothesized that the decrease could have been linked to local geoclimatic conditions, tree shading, composition, and plant functional groups. On the contrary, when cuttings took place every three years in the originally treated plots without grazing + fertilizing, species richness was conferred by grazing [42]. Authors observed that grazing increased plant competition and prevented the dominance of certain species in the understory. In the area of Herault, France, clearing + simulated grazing was more beneficial to species richness than pure controlled burning was [41], although some species seem to have appeared only in the plots where fertilization was applied [40]. Richness was lower in heavily fertilized plots than in moderately fertilized plots, while less frequent mowing conferred it [38]. This site was the only one where two fertilization levels were applied; species richness responded negatively to heavy fertilization in comparison to a moderate level of fertilization, but the CSC index did not. In the same study, it was shown that the latest cutting time period, corresponding to the lowest grazing intensity, was the most beneficial one regarding floristic richness. In another study, in a site covered by brambles species richness increased with grazing and yearly clearings, but Shannon–Weaver diversity remained significantly similar, regardless the number of clearings and the addition of grazing the lack thereof [49]. In the other study taking place in a bramble-covered site, the positive impact of clearing + grazing in comparison to that of clearing + no grazing was the most significant one in 2016. In this year, there was an increase in grazing intensity, while in clearing + no grazing plots species richness remained the same as that of the previous year. In a similar study on chaparral, there was a negative effect on species richness in a single-year measurement, with reference not only to native, but also non-native species [51].

4. Management and Monitoring Implications

The mechanical, manual, burning, and chemical interventions mentioned in the above review contributed to a decrease in shrub cover, creating more accessible pastures for grazing animals. They also increased herbaceous production, and fresher and more accessible woody forage in many cases. The follow-up grazing regimes had different results depending on the site, the treatments and their purpose.

The maintenance of some shrub cover is important for productive purposes since herbivores need to utilize various forage components during the year [13,35]. However, when it comes to fuel management, keeping woody biomass at low levels is of high importance [51]. Higher fertilization levels in cleared stands could increase forage productivity but decrease floristic richness in comparison to moderate fertilization [38]. Therefore, it is always important to recognize that the management purposes of fire prevention, optimization of production, and maintenance of biodiversity might be contradictory to each other [47]. Thus, a landscape mosaic needs to be maintained, allowing all these environmental and economic aspects to be addressed in the best possible way. Such landscapes can be maintained via the presence of livestock for both commercial and targeted grazing purposes. Integrated management efforts such as the ones in Esterel hills (France) and La Rioja (Spain) can serve the above purposes, while there is a need for cooperation with livestock breeders, who should be trained and encouraged to share their knowledge and work in a professional way [58,66,73].

Stocking rates are very important in such efforts because proper grazing intensity, applied by trained and motivated grazers, can contribute to the maintenance of shrub biomass at low levels despite variations in animal feed preferences [43,48]. In some cases, it can even reduce herbaceous production and increase bare ground cover. The impact of the applied stocking rates on floristic diversity may be visible as well, since a heavy grazing regime can negatively affect it [38,48]. Overall, our review study showed mixed results in terms of richness, although some studies did not include such data. A meta-analysis focusing mostly on non-Mediterranean ecosystems showed that a reduction in floristic diversity is common in cases of high stocking rates [74].

It is important to note that elevated rainfall levels can mitigate the effects of high stocking density, and thus there is a need for the adaptation of grazing to strong meteorological variations [75,76]. Such a managerial approach has not been adequately examined in the studies of the present review.

When fertilization and seeding are allowed according to cost evaluations, they can increase pasture production. Subsequent grazing can also promote floristically richer sites compared to pure controlled burning or clearing even if the latter are repeated [38,42]. Nevertheless, harsh climatic variations, such drought events, could affect environmental balance more in fertilized or seeded stands than in stands with a natural understory [40,64]. A repeated combination of mechanical intervention and annual grazing could sometimes be recommended to create herbaceous communities [49]. In terms of comparing different interventions, clearing seems to be more effective against shrubs than slashing, though, apart from the specific nature of management goals, the costs also need to be considered [56,62]. Unfortunately, only two studies compared controlled burning + grazing to mechanical interventions + grazing [46,50].

In general, the most important factors that regulate productivity are grazing intensity and animal dietary preferences. There can be an effect on biomass even with a normal grazing intensity, especially if, in the case of goats, the consumption of grasses and shrubs can function in a complementary way. Even if consumption is limited to shrubs that are not dominant in the landscape, variation in annual shrub production may become apparent under sheep grazing [56]. The decisive role of grazing intensity was recorded in several studies [43,45,50,60]. According to the study in [45], fencing can increase the effect of grazing in clear-cut management treatments, via achieving objectives such as fuel reduction. In cases where previous clearing has taken place, fencing is not always necessary to regulate grazing as a fire prevention technique [45,77]. Grazing intensity and vegetation consumption can also be partially regulated by factors such as the placement of stockyards [60]. This is also mentioned by other authors in terms of the distance from stockyards; to the effectiveness of grazing at reducing shrub production seems to be regulated more at closer [78] than at longer distances [77].

Monitoring of the practices discussed above should involve longer-term changes. The highest duration in the reviewed studies was 31 years [66]. The changes in a *Quercus suber* ecosystem that was partially and periodically cleared were observed for 70 years, showing that longer monitoring schemes can be conducted [79]. The authors found that fuel management based exclusively on hand labor is costly. Thus, it can be hypothesized that grazing could serve as a low-cost alternative.

5. Conclusions

Mediterranean-climate regions (MCRs) have a long history of disturbance-based interventions for livestock use. Their use in newly transformed pastures has varied effects on plant structure, productivity, and floristic diversity. As far as humans can control these effects, important factors of the management regime, such as initial interventions and the presence of domestic (or wild) animals utilizing the pastures, along with grazing intensities, are very important. To our knowledge, the current biographical research is a novel attempt to summarize existing knowledge on the combined use of grazing and other mechanical or manual management interventions in the grasslands of MCRs.

Scientific society has shown increasing interest in meta-disturbance management in Mediterranean pastures. Since the first study by P. Poissonet and collaborators was published in 1978, several other relevant studies have emerged at a rate of 24% in 5-year time increments. The 26 publications included in this review focused on the effects of grazing combined with other management tools on vegetation structure, biomass productivity and floristic diversity.

After a primary reduction inf shrub cover, more herbaceous forage is available. In some cases, shrubs tend to recover in grazed pastures but in general, grazing contributes to the maintenance of grassland formations. Vegetation height was reduced almost in all case studies. Additionally, floristic diversity could be adjusted according to the type and intensity of initial interventions and grazing. These findings are extremely useful for grassland restoration purposes.

Despite its possible recovery, shrub biomass is affected by grazing in most cases. The impact of grazing after the other interventions on floristic diversity was mixed, with richness being the index most frequently measured. Additional interventions for productive purposes, such as fertilizing and seeding, affect results, as do pasture management factors, such as the selection of animals and grazing intensity. Further research topics could include the adjustment of stocking rates in a vulnerable climate for longer periods, and the inclusion of fuel parameters of vegetation and wild fauna in the consideration of impacts, while more rangeland types, such as wet grasslands, could be included in the research.

Additional research may be required to define the spatial scale of application. Prediction tools, such as different development scenarios [80] or even spatially explicit models [81], can be incorporated in long-term management policies as well. The improvement of capabilities and the broader adoption of technologies such as GIS and remote sensing during recent decades nowadays offer opportunities for better pasture management and monitoring [82–84]. More studies on the fuel properties of vegetation in relation to its use by herbivores as a wildfire prevention tool should be implemented, and the practice should be considered even for countries where it has not been applied before [85]. Furthermore, research on the impact of such interventions on animal biodiversity is quite limited. Finally, combinations of interventions such as those discussed in this paper could also be considered for a wider variety of Mediterranean ecosystems.

Author Contributions: Conceptualization, D.O. and M.V.; methodology, D.O.; formal analysis, D.O. and M.V.; investigation, D.O.; resources, D.O. and M.V.; data curation, D.O., M.V., M.Y., G.X. and Y.K.; writing—original draft preparation, D.O., M.V., M.Y., G.X. and Y.K.; writing—review and editing, D.O., M.V., M.Y., G.X. and Y.K.; visualization, D.O.; supervision, M.V. All authors have read and agreed to the published version of the manuscript.

Funding: There is no funding related to this research. The research is part of the Ph.D. research of first author (D. Oikonomou).

Institutional Review Board Statement: Not applicable.

Informed Consent Statement: Not applicable.

Data Availability Statement: Data and material are available upon request.

Conflicts of Interest: The authors declare no conflict of interest.

References

1. Vera, F.W.M. *Grazing Ecology and Forest History*; CABI Publishing: Oxford, UK, 2000.
2. Naveh, Z. Ecological and cultural landscape restoration and the cultural evolution towards a post-industrial symbiosis between human society and nature. *Restor. Ecol.* **1998**, *6*, 135–143. [CrossRef]
3. Guarino, R.; Vrahnakis, M.; Rojo, M.P.R.; Giuga, L.; Pasta, S. Grasslands and Shrublands of the Mediterranean Region. In *Encyclopedia of the World's Biomes*; Goldstein, M.I., DellaSala, D.A., Eds.; Elsevier: Amsterdam, The Netherlands, 2020; pp. 638–655.
4. di Castri, F. An Ecological Overview of the Five Regions with a Mediterranean Climate. In *Biogeography of Mediterranean Invasions*; Groves, R.H., di Castri, F., Eds.; Cambridge University Press: Cambridge, UK, 1991; pp. 3–16.
5. di Castri, F. Politics and Environment in Mediterranean-Climate Regions. In *Landscape Disturbance and Biodiversity in Mediterranean-Type Ecosystems*; Rundel, P.W., Montenegro, G., Jaksic, F.M., Eds.; Ecological Studies; Springer: Berlin, Germany, 1998; pp. 407–432.
6. Ambarli, D.; Vrahnakis, M.; Burrascano, S.; Naqinezhad, A.; Pulido Fernández, M. Grasslands of the Mediterranean basin and the Middle East and their management. In *Grasslands of the World: Diversity, Management and Conservation*; Squires, V.R., Dengler, J., Feng, H., Hua, L., Eds.; CRC Press: Boca Raton, FL, USA, 2018.

7. Lasanta, T.; Nadal-Romero, E.; Arnáez, J. Managing abandoned farmland to control the impact of re-vegetation on the environment. The state of the art in Europe. *Environ. Sci. Policy* **2015**, *52*, 99–109. [CrossRef]
8. Lovreglio, R.; Meddour-Sahar, O.; Leone, V. Goat grazing as a wildfire prevention tool: A basic review. *Iforest Biogeosciences For.* **2014**, *7*, 260. [CrossRef]
9. Vrahnakis, M.; Fotiadis, G.; Merou, T.; Kazoglou, Y.E. Improvement of plant diversity and methods for its evaluation in Mediterranean basin grasslands. *Options Mediterr.* **2010**, *92*, 225–236.
10. Papanastasis, V. Traditional vs contemporary management of Mediterranean vegetation: The case of the island of Crete. *J. Biol. Res.* **2004**, *1*, 39–46.
11. Chouvardas, D.; Vrahnakis, M.S. A semi-empirical model for the near-future evolution of the Lake Koronia landscape. *J. Environ. Prot. Ecol.* **2009**, *10*, 867–876.
12. Pereira, P.; Godinho, C.; Roque, I.; Marques, A.; Branco, M.; Rabaça, J.E. Time to rethink the management intensity in a Mediterranean oak woodland: The response of insectivorous birds and leaf-chewing defoliators as key groups in the forest ecosystem. *Ann. For. Sci.* **2014**, *71*, 25–32. [CrossRef]
13. Evangelou, C.; Yiakoulaki, M.; Papanastasis, V. Spatio-temporal analysis of sheep and goats grazing in different forage resources of Northern Greece. *Hacquetia* **2014**, *13*, 205–214. [CrossRef]
14. Bergmeier, E.; Capelo, J.; Di Pietro, R.; Guarino, R.; Kavgacı, A.; Loidi, J.; Tsiripidis, I.; Xystrakis, F. 'Back to the future'—Oak wood-pasture for wildfire prevention in the Mediterranean. *Plant Sociol.* **2021**, *58*, 41–48. [CrossRef]
15. Pausas, J.G.; Vallejo, V.R. The role of fire in European Mediterranean ecosystems. In *Remote Sensing of Large Wildfires*; Chuvieco, E., Ed.; Springer: Berlin/Heidelberg, Germany, 1999; pp. 3–16.
16. Blondel, J.; Aronson, J. *Biology and Wildlife of the Mediterranean Region*; Oxford University Press: Oxford, UK, 1999.
17. Grove, A.T.; Rackham, O. The Nature of Mediterranean Europe: An Ecological History. Yale University Press: New Haven, CT, USA, 2001.
18. Gómez Sal, A.; Rey Benayas, J.M.; López-Pintor, A.; Rebollo, S. Role of Disturbance in Maintaining a Savanna-like Pattern in Mediterranean Retama Sphaerocarpa Shrubland. *J. Veg. Sci.* **1999**, *10*, 365–370. [CrossRef]
19. Colantoni, A.; Egidi, G.; Quaranta, G.; D'Alessandro, R.; Vinci, S.; Turco, R.; Salvati, L. Sustainable land management, wildfire risk and the role of grazing in Mediterranean urban-rural interfaces: A regional approach from Greece. *Land* **2020**, *9*, 21. [CrossRef]
20. Moreira, F.; Ascoli, D.; Safford, H.; Adams, M.A.; Moreno, J.M.; Pereira, J.C.; Catry, F.X.; Armesto, J.; Bond, W.J.; González, M.E., et al. Wildfire management in Mediterranean type regions. Paradigm change needed. *Environ. Res. Lett.* **2020**, *15*, 011001. [CrossRef]
21. Fernandes, P.M.; Davies, G.M.; Ascoli, D.; Fernández, C.; Moreira, F.; Rigolot, E.; Stoof, C.R.; Vega, J.A.; Molina, D. Prescribed burning in southern Europe: Developing fire management in a dynamic landscape. *Front. Ecol. Environ.* **2013**, *11*, e4–e14. [CrossRef]
22. Palaiologou, P.; Kalabokidis, K.; Troumbis, A.; Day, M.A.; Nielsen-Pincus, M.; Ager, A.A. Socio-ecological perceptions of wildfire management and effects in Greece. *Fire* **2021**, *4*, 18. [CrossRef]
23. Athanasiou, M.; Bouchounas, T.; Korakaki, E.; Tziritis, E.; Xanthopoulos, G.; Sitara, S. Introducing the use of fire for wildfire prevention in Greece: Pilot application of prescribed burning in Chios island. In Proceedings of the Advances in Forest Fire Research 2022, Coimbra, Portugal, 23 November 2022; pp. 1487–1494.
24. Perevolotsky, A.; Ne'eman, G.; Yonatan, R.; Henkin, Z. Resilience of prickly burnet to management in east Mediterranean rangelands. *J. Range Manag.* **2001**, *54*, 561–566. [CrossRef]
25. Lasanta, T. Active management against shrubland expansion: Seeking a balance between conservation and exploitation in the mountains. *Cuad. Investig. Geográfica. Cuad. Investig. Geográfica/Geogr. Res. Lett.* **2019**, *45*, 423–440. [CrossRef]
26. Arévalo, J.R.; Fernandez-Lugo, S.; Garcia-Dominguez, C.; Naranjo-Cigala, A.; Grillo, F.; Calvo, L. Prescribed burning and clear-cutting effects on understory vegetation in a *Pinus canariensis* stand (Gran Canaria). *Sci. World J.* **2014**, *2014*, 215418. [CrossRef]
27. Holmgren, M.; Segura, A.M.; Fuentes, E.R. Limiting mechanisms in the regeneration of the Chilean matorral—Experiments on seedling establishment in burned and cleared mesic sites. *Plant Ecol.* **2000**, *147*, 49–57. [CrossRef]
28. Calvo, L.; Tárrega, R.; de Luis, E. The dynamics of Mediterranean shrubs species over 12 years following perturbations. *Plant Ecol.* **2002**, *160*, 25–42. [CrossRef]
29. Calvo, L.; Tárrega, R.; Luis, E.; Valbuena, L.; Marcos, E. Recovery after experimental cutting and burning in three shrub communities with different dominant species. *Plant Ecol.* **2005**, *180*, 175–185. [CrossRef]
30. Xanthopoulos, G.; Caballero, D.; Galante, M.; Alexandrian, D.; Rigolot, E.; Marzano, R. Forest fuels management in Europe. In Proceedings of the RMRS-P-41. Fort Collins, CO: U.S. Department of Agriculture, Forest Service, Rocky Mountain Research Station, Portland, OR, USA, 28–30 March 2006; pp. 29–46.
31. Pinto-Correia, T.; Mascarenhas, J. Contribution to the extensification/intensification debate: New trends in the Portuguese montado. *Landsc. Urban Plan.* **1999**, *46*, 125–131. [CrossRef]
32. Médail, F.; Quezel, P. Hot-spots analysis for conservation of plant biodiversity in the Mediterranean basin. *Ann. Mo. Bot. Gard.* **1997**, *84*, 112–127. [CrossRef]
33. Celep, F.; Dogan, M.; Kahraman, A. Re-evaluated conservation status of *Salvia* (sage) in Turkey I: The Mediterranean and the Aegean geographic regions. *Turk. J. Bot.* **2010**, *34*, 201–214. [CrossRef]

34. Lozano, F.D.; Atkins, K.J.; Moreno Sáiz, J.C.; Sims, A.E.; Dixon, K. The nature of threat category changes in three Mediterranean biodiversity hotspots. *Biol. Conserv.* **2013**, *157*, 21–30. [CrossRef]
35. Spatz, G.; Papachristou, T.G. Ecological strategies of shrubs invading extensified grasslands: Their control and use. In Proceedings of the International Occasional Symposium of the European Grassland Federation, Thessaloniki, Greece, 27–29 May 1999; pp. 27–36.
36. Perevolotsky, A.; Seligman, N.A.G. Role of Grazing in Mediterranean Rangeland Ecosystems: Inversion of a paradigm. *BioScience* **1998**, *48*, 1007–1017. [CrossRef]
37. Rouet-Leduc, J.; Pe'er, G.; Moreira, F.; Bonn, A.; Helmer, W.; Shahsavan Zadeh, S.A.A.; Zizka, A.; van der Plas, F. Effects of large herbivores on fire regimes and wildfire mitigation. *J. Appl. Ecol.* **2021**, *58*, 2690–2702. [CrossRef]
38. Poissonet, P.; Romane, F.; Thiault, M.; Trabaud, L. Evolution d'une garrigue de *Quercus coccifera* L. soumise a divers traitements: Quelques resultats des cinq premieres annees. *Vegetatio* **1978**, *38*, 135–142. [CrossRef]
39. Castro, M.; Castro, J.P.; Castro, J. Understory clearing in open grazed Mediterranean oak forests: Assessing the impact on vegetation. *Sustainability* **2022**, *14*, 10979. [CrossRef]
40. Poissonet, J.; Poissonet, P.; Thiault, M. Development of flora, vegetation and grazing value in experimental plots of a *Quercus coccifera* garrigue. *Vegetatio* **1981**, *46*, 93–104. [CrossRef]
41. Godron, M.; Guillerm, J.L.; Poissonet, J.; Poissonet, P.; Thiault, M.; Trabaud, L. Dynamics and management vegetation. In *Mediterranean-Type Shrublands, Ecosystems of the World*; Elsevier Science: Amsterdam, The Netherlands, 1981; Volume 11, pp. 317–344.
42. Étienne, M. Aménagement de la forêt méditerranéenne contre les incendies et biodiversité. *Rev. For. Française* **2001**, *53*, 149–155. [CrossRef]
43. Gutman, M.; Henkin, Z.; Holzer, Z.; Noy-Meir, I.; Seligman, N.G. A case study of beef-cattle grazing in a Mediterranean-type woodland. *Agrofor. Syst.* **2000**, *48*, 119–140. [CrossRef]
44. Perevolotsky, A.; Haimov, Y. The effect of thinning and goat browsing on the structure and development of Mediterranean woodland in Israel. *For. Ecol. Manag.* **1992**, *49*, 61–74. [CrossRef]
45. Green, L.R.; Hughes, C.L.; Graves, W.L. Goat control of brush regrowth on Southern California fuelbreaks. *Rangelands* **1979**, *1*, 117–119.
46. Potts, J.B.; Marino, E.; Stephens, S.L. Chaparral shrub recovery after fuel reduction: A comparison of prescribed fire and mastication techniques. *Plant Ecol.* **2010**, *210*, 303–315. [CrossRef]
47. Bashan, D.; Bar-Massada, A. Regeneration dynamics of woody vegetation in a Mediterranean landscape under different disturbance-based management treatments. *Appl. Veg. Sci.* **2017**, *20*, 106–114. [CrossRef]
48. Hadar, L.; Noy-Meir, I.; Perevolotsky, A. The effect of shrub clearing and grazing on the composition of a Mediterranean plant community: Functional groups versus species. *J. Veg. Sci.* **1999**, *10*, 673–682. [CrossRef]
49. Masson, S.; Mesléard, F.; Dutoit, T. Using shrub clearing, draining, and herbivory to control bramble invasion in Mediterranean dry grasslands. *Environ. Manag.* **2015**, *56*, 933–945. [CrossRef]
50. Álvarez-Martínez, J.; Gómez-Villar, A.; Lasanta, T. The use of goats grazing to restore pastures invaded by shrubs and avoid desertification: A preliminary study in the Spanish Cantabrian Mountains. *Land Degrad. Dev.* **2013**, *27*, 3–13. [CrossRef]
51. Grupenhoff, A.; Molinari, N. Plant community response to fuel break construction and goat grazing in a southern California shrubland. *Fire Ecol.* **2021**, *17*, 28. [CrossRef]
52. Henkin, Z. The role of brush encroachment in Mediterranean ecosystems: A review. *Isr. J. Plant Sci.* **2021**, *69*, 1–12. [CrossRef]
53. Papanastasis, V.P.; Kyriakakis, S.; Kazakis, G.; Abid, M.; Doulis, A. Plant cover as a tool for monitoring desertification in mountain Mediterranean rangelands. *Manag. Environ. Qual.* **2003**, *14*, 69–81. [CrossRef]
54. Abraham, E.M.; Parissi, Z.M.; Kyriazopoulos, A.P.; Korakis, G.; Papaioannou, P. Wildfire Effects on Species Composition and Nutritive Value in Different Thermo-Mediterranean Vegetation Types. *Options Méditerranéennes Série A Séminaires Méditerranéens* **2014**, *109*, 479–482.
55. Lasanta, T.; Nadal-Romero, E.; Errea, P.; Arnáez, J. The effect of landscape conservation measures in changing landscape patterns: A case study in Mediterranean mountains. *Land Degrad. Dev.* **2016**, *27*, 373–386. [CrossRef]
56. Papachristou, T.G.; Platis, P.D.; Papanastasis, V.P. Forage production and small ruminant grazing responses in Mediterranean shrublands as influenced by the reduction of shrub cover. *Agrofor. Syst.* **1996**, *35*, 225–238. [CrossRef]
57. Delgado, I.; Ochoa, M.J.; Sin, E.; Barragan, M.; Rodríguez, J.; Nuez, T.; Posadas, J.M. Pasture restoration by control of *Genista scorpius* (L.) DC. *Options Méditerranéennes* **2004**, *62*, 403–406.
58. Lécrivain, E.; Beylier, B. Technique de réouverture d'une colline embroussaillée adaptée au comportement des troupeaux. *Options Méditerranéennes* **2004**, *62*, 233–238.
59. Nadal-Romero, E.; Lasanta, T.; Zabalza, J.; Pueyo, Y.; Foronda, A.; Reiné, R.; Barrantes, O.; Pascual, D.; Pla, E.; Lana-Renault, N.; et al. Monitoring Protocol of Action C1. 2020. Available online: https://life-midmacc.eu/wp-content/uploads/2021/07/LIFEMIDMACC_DL8_MonitoringProtocolC1_v3_final-2.pdf (accessed on 24 May 2023).
60. Moinardeau, C.; Mesléard, F.; Ramone, H.; Dutoit, T. Using mechanical clearing and goat grazing for restoring understorey plant diversity of embankments in the Rhône Valley (Southern France). *Plant Biosyst.* **2020**, *154*, 746–756. [CrossRef]
61. Lasanta, T.; Nadal-Romero, E.; García-Ruiz, J.M. Clearing shrubland as a strategy to encourage extensive livestock farming in the Mediterranean mountains. *Cuad. Investig. Geogr.* **2019**, *42*, 487–513. [CrossRef]

62. Papanastasis, V.P.; Papachristou, T.G.; Platis, P.D. Control of woody plants with mechanical means in a rangeland of Macedonia, Greece. Proceedings of Conference of European Grassland Federation, Graz, Austria, 18–21 September 1991; pp. 203–205.
63. Étienne, M.; Legrand, C.; Armand, D. Stratégies d'occupation de l'espace par les petits ligneux après débroussaillement en région méditerranéenne française. Exemple d'un réseau de pare-feu dans l'Esterel. *Ann. For. Sci.* **1991**, *48*, 667–677. [CrossRef]
64. Bicho, M.C.; Correia, A.C.; Rodrigues, A.R.; Soares David, J.; Costa-e-Silva, F. Climate and management effects on the herbaceous layer productivity of a cork oak woodland. *Agrofor. Syst.* **2022**, *96*, 315–327. [CrossRef]
65. Yiakoulaki, M.D.; Kodona, M.S.; Nastis, A.S. Effect of prescribed burning and management systems on diet selection and intake of goats grazing in a *Pinus brutia* forest. Proceedings of EU EQULFA Project, Thessaloniki, Greece, 8–11 December 1998.
66. Lasanta, T.; Khorchani, M.; Pérez-Cabello, F.; Errea, P.; Sáenz-Blanco, R.; Nadal-Romero, E. Clearing shrubland and extensive livestock farming: Active prevention to control wildfires in the Mediterranean mountains. *J. Environ. Manag.* **2018**, *227*, 256–266. [CrossRef] [PubMed]
67. Valderrábano, J.; Torrano, L. The potential for using goats to control *Genista scorpius* shrubs in European black pine stands. *For. Ecol. Manag.* **2000**, *126*, 377–383. [CrossRef]
68. Lasanta, T.; Cortijos-López, M.; Errea, M.P.; Khorchani, M.; Nadal-Romero, E. An environmental management experience tocontrol wildfires in the mid-mountain Mediterranean area: Shrub clearing to generate mosaic landscapes. *Land Use Policy* **2022**, *118*, 106147. [CrossRef]
69. Tsiouvaras, C.N. Ecology and management of kermes oak (*Quercus coccifera* L.) shrublands in Greece: A review. *J. Range Management* **1987**, *40*, 542. [CrossRef]
70. Verdaguer, D. Towards a better understanding of the role of rhizomes in mature woody plants: The belowground system of *Quercus coccifera*. *Trees* **2020**, *34*, 903–916. [CrossRef]
71. Delitti, W.; Ferran, A.; Trabaud, L.; Vallejo, V.R. Effects of fire recurrence in *Quercus coccifera* L. shrublands of the Valencia Region (Spain): I. plant composition and productivity. *Plant Ecol.* **2005**, *177*, 57–70. [CrossRef]
72. La Mantia, T.; Rühl, J.; Massa, B.; Pipitone, S.; Lo Verde, G.; Bueno, R.S. Vertebrate-mediated seed rain and artificial perches contribute to overcome seed dispersal limitation in a Mediterranean old field. *Restor. Ecol.* **2019**, *27*, 1393–1400. [CrossRef]
73. Yiakoulaki, M.; Hasanagas, N. Professionalism in extensive sheep farming system in Central Greece. *Bulg. J. Agric. Sci.* **2019**, *25*, 661–667.
74. Herrero-Jáuregui, C.; Oesterheld, M. Effects of grazing intensity on plant richness and diversity: A meta-analysis. *Oikos* **2018**, *127*, 757–766. [CrossRef]
75. Ruiz-Mirazo, J.; Robles, A.B. Impact of targeted sheep grazing on herbage and holm oak saplings in a silvopastoral wildfire prevention system in south-eastern Spain. *Agrofor. Syst.* **2012**, *86*, 477–491. [CrossRef]
76. Hocking, D.; Mattick, A. *Dynamic Carrying Capacity Analysis as Tool for Conceptualizing and Planning Range Management Improvements, with a Case Study from India*; Pastoral Development Network: London, UK, 1993; p. 34.
77. Ruiz-Mirazo, J.; Robles, A.B.; González-Rebollar, J.L. Two-year evaluation of fuelbreaks grazed by livestock in the wildfire prevention program in Andalusia (Spain). *Agric. Ecosyst. Environ.* **2011**, *141*, 13–22. [CrossRef]
78. Papanastasis, V.P.; Ghossoub, R.; Scarpelo, C. Impact of animal sheds on vegetation configuration in Mediterranean landscapes. *Options Méditerranéennes Ser. A Mediterr. Semin.* **2009**, *85*, 49–54.
79. Porto, M.; Correia, O.; Beja, P. Long-term consequences of mechanical fuel management for the conservation of Mediterranean forest herb communities. *Biodivers. Conserv.* **2011**, *20*, 2669–2691. [CrossRef]
80. Lasanta, T.; González-Hidalgo, J.C.; Vicente-Serrano, S.M.; Sferi, E. Using landscape ecology to evaluate an alternative management scenario in abandoned Mediterranean mountain areas. *Landsc. Urban Plan.* **2006**, *78*, 101–114. [CrossRef]
81. Bar Massada, A.; Carmel, Y.; Koniak, G.; Noy-Meir, I. The effects of disturbance based management on the dynamics of Mediterranean vegetation: A hierarchical and spatially explicit modeling approach. *Ecol. Model.* **2009**, *220*, 2525–2535. [CrossRef]
82. Todd, S.W.; Hoffer, R.M.; Milchunas, D.G. Biomass estimation on grazed and ungrazed rangelands using spectral indices. *Int. J. Remote Sens.* **1998**, *19*, 427–438. [CrossRef]
83. Svoray, T.; Shoshany, M.; Perevolotsky, A. Mediterranean rangeland response to human intervention: A remote sensing and GIS study. *J. Mediterr. Ecol.* **2003**, *4*, 3–11.
84. Ali, I.; Cawkwell, F.; Dwyer, E.; Barrett, B.; Green, S. Satellite remote sensing of grasslands: From observation to management—A review. *J. Plant Ecol.* **2016**, *9*, 649–671. [CrossRef]
85. Raftoyannis, Y.; Nocentini, S.; Marchi, E.; Sainz, R.C.; Guemes, C.G.; Pilas, I.; Peric, S.; Paulo, J.A.; Moreira-Marcelino, A.C.; Costa-Ferreira, M.; et al. Perceptions of forest experts on climate change and fire management in European Mediterranean forests. *Iforest-Biogeosciences For.* **2014**, *7*, 33–41. [CrossRef]

Article

Wildfire Effects on Rangeland Health in Three Thermo-Mediterranean Vegetation Types in a Small Islet of Eastern Aegean Sea

Zoi M. Parissi [1], Apostolos P. Kyriazopoulos [2], Theodora Apostolia Drakopoulou [1], Georgios Korakis [2] and Eleni M. Abraham [1,*]

[1] Laboratory of Range Science, School of Forestry and Natural Environment, Aristotle University of Thessaloniki, P.O. Box 236, 54124 Thessaloniki, Greece; pz@for.auth.gr (Z.M.P.); drakopoul@bio.auth.gr (T.A.D.)

[2] Department of Forestry and Management of the Environment and Natural Resources, Democritus University of Thrace, 193 Pantazidou Str., 68200 Orestiada, Greece; apkyriaz@fmenr.duth.gr (A.P.K.); gkorakis@fmenr.duth.gr (G.K.)

[*] Correspondence: eabraham@for.auth.gr; Tel.: +30-2310-992301

Abstract: Sclerophyllous scrub formations, the main vegetation type in many islands of the Aegean area, provide many goods and services to humans, such as biodiversity, soil protection, and forage for livestock and wildlife. Dominant shrub species of sclerophyllous formations are well adapted to dry season conditions due to various anatomical and physiological mechanisms. As a result, their biomass acts as very flammable, fine fuel, and consequently, wildfires are very common in these ecosystems. Wildfire effects on vegetation and biodiversity in the Mediterranean basin have been studied, and the results are diverse, depending mainly on the vegetation type and frequency of fires. Additionally, post-fire vegetation establishment and structure are critical factors for the implementation of grazing management. The aim of this study was to evaluate the effects of wildfire on species composition, floristic diversity, forage quality, and rangeland health indices related to ecosystem stability and function in three thermo-Mediterranean vegetation types: (1) *Sarcopoterium spinosum* low formations, (2) low formations of *Cistus creticus*, and (3) low formations of *Cistus creticus* in abandoned terraces. The research was conducted on the Oinousses islet, which is located northeast of Chios Island, in May 2013 (one year after the fire). Vegetation sampling was performed along five transects placed in recently burned and adjacent unburned sites of each vegetation type. The plant cover was measured, while the floristic composition, diversity, evenness, and dominance indices were determined for the vegetation data. Additionally, the forage quality was determined in terms of crude protein (CP) and fiber content. The vegetation cover was significantly lower, and the floristic diversity was significantly higher in burned areas in comparison to those in the unburned areas. Woody species, followed by grasses and forbs, dominated in both the burned and unburned areas. However, the percentage of woody species was significantly decreased in the burned areas of *Sarcopoterium spinosum* and *Cistus creticus* low formations. On the other hand, the percentage of grasses, forbs, and legumes increased in all cases except in *Cistus creticus* terraces. The lowest value of the Jaccard Index of similarity between the burned and unburned sites (beta diversity) was observed for *Cistus creticus*, indicating the effect of fire on the species composition of this vegetation type. The forage quality was found to be improved in all the burned areas, especially in those dominated by *Cistus creticus*. Finally, fire has a positive impact on the ecosystem's functions, mainly for *Sarcopoterium spinosum* low formations.

Keywords: phrygana; rangeland health; diversity; nutritive value; fire; garrigue

Citation: Parissi, Z.M.; Kyriazopoulos, A.P.; Drakopoulou, T.A.; Korakis, G.; Abraham, E.M. Wildfire Effects on Rangeland Health in Three Thermo-Mediterranean Vegetation Types in a Small Islet of Eastern Aegean Sea. *Land* **2023**, *12*, 1413. https://doi.org/10.3390/land12071413

Academic Editors: Eusebio Cano Carmona, Michael Vrahnakis, Yannis (Ioannis) Kazoglou and Manuel Pulido Fernádez

Received: 11 May 2023
Revised: 15 June 2023
Accepted: 6 July 2023
Published: 14 July 2023

1. Introduction

Sclerophyllous low scrub formations are common habitats at low and middle altitudes in the eastern Mediterranean and Anatolian areas. They are sub-habitats of the F7.3

European Red List of Habitats [1]. They usually occupy dry sites with shallow limestone (calcareous) soils. According to the EU's description of the habitats, they are located in the coastal thermo-, meso-, and supra-Mediterranean zones of the Aegean islands, in mainland Greece and the Ionian islands, and in coastal Anatolia and Crete (up to 1200 m a.s.l.) [2]. These habitats, which are also referred to as 'phrygana' in Greece and garrigue in other countries [3], are traditionally used for grazing by livestock. They consist mainly of low, thorny, dimorphic shrub species [4]. The aforementioned species have developed an adaptation mechanism to dry thermal conditions in the Mediterranean basin by replacing winter leaves with much smaller summer ones in order to conserve water [5,6]. The most characteristic dominated species are *Sarcopoterium spinosum*, *Cistus creticus*, *C. salviifolius*, *Erica manipuliflora*, *Genista acanthoclada*, *Phlomis fruticosa*, *Corydothymu scapitatus*, and *Euphorbia acanthothamnus* [7].

Sclerophyllous scrub formations are very diverse ecosystems. They comprise some of the most species-rich plant communities in the Mediterranean basin and provide many goods and services to humans. Regarding regulating and supporting services [8], they contribute to the regulation of soil quality and protection, carbon sequestration, and provide a habitat for wildlife. Concerning provision services, they provide medicinal and aromatic plants, chemical extracts, and food as the majority of dominated scrubs and many of the understory herbs have medicinal and aromatic properties, e.g., honey from wild herbs, as well as forage for livestock and wildlife [9].

These ecosystems in the Aegean islands have been traditionally used for grazing by livestock. Despite the fact that the main use of these ecosystems is grazing, the dominated phryganic species are unpalatable and/or less desirable for grazing. This means that the main source of forage is the herbaceous species under or among the shrubs. Thus, an increase in the density of the phryganic species will result in a decrease in the availability and quality of the forage. In addition to the reduction of forage availability, this thickening would also result in a decrease in species richness and floristic diversity. Shepherds in these areas know this and use fire as a tool to increase the forage quantity and quality [7]. Additionally, they are among the major fire-prone biomes in the world [10]. Fire in this biome is an essential ecological process and beneficial for the ecosystem's function [11]. In this respect, grazing and fire are key factors that have interacted with and shaped the structure and function of plant communities in the phryganic ecosystems in the Aegean islands.

However, grazing has drastically decreased in the past few decades in these areas, mainly due to changes in land uses and the increase in tourism. This has led to changes in landscapes and the environment, changes in vegetation composition and structure, decreases in forage quantity and quality, increases in woody vegetation, and the loss of biodiversity, endangering the provision of key ecosystem services [12,13]. Furthermore, the increase in woody vegetation may contribute to an increase in the risk and/or the frequency of wildfires. Fire-prone biomes have a characteristic historical range of variability in frequency, severity, and patchiness of fires [14]. Any change in this historical range due to human intervention can alter the ecosystem's response to fire [15].

The effect of fire on ecosystem services has been studied mainly in forest ecosystems and for services related to soil stability and fertility [15]. The general view is that fire is a natural disaster and has a negative effect on ecosystem services. On the other hand, these ecosystems can remain stable under grazing and burning [16]. Particularly for post-fire grazing management, the question that arises is: when can it be applied to burned areas? The implementation of grazing management in burned areas directly depends on the establishment and structure of post-fire vegetation. Livestock grazing in burned areas before vegetation is well established could lead to ecosystem degradation. On the other hand, the thickening of vegetation before livestock grazing could make the ecosystem vulnerable to a new fire. However, research about the effect of fire on ecosystems adapted to fire, such as the sclerophyllous low scrub formations, is limited. The sclerophyllous low scrub formation of Aegean islands is assessed at the Least Concern status [2] based on Indicators of quality of the European Red List of Habitats [1]. This is mainly because of the extensive distribution of this habitat in the Eastern Mediterranean, which has not decreased in recent years. However, these ecosystems are highly affected by human activities such as grazing, fire, and cultivation abandonment. Therefore, further research focusing on the quality characteristics of these ecosystems is needed in order to detect the role of human activities in plant communities and how they affect the provision of their services. In this respect, the aim of the present study was to detect the effects of wildfire on species composition, floristic diversity, forage quality, and rangeland health indices related to ecosystem stability and function in three thermo-Mediterranean vegetation types (1) low formations of *Sarcopoterium spinosum*, (2) low formations of *Cistus creticus* and (3) low formations of *Cistus creticus* in abandoned terraces. This information could be a tool for managers to implement grazing plans in the burned areas of these ecosystems.

2. Materials and Methods

2.1. The Study Area

The study was conducted in Oinousses islet (38°20′00″ N, 26°08′00″ E, 80 m a.s.l.), which is in the Eastern Aegean Sea in May 2013. The islet is located in the sea channel between the NE coast of Chios Island (Greece) and the western coast of Anatolia (Turkey) (Figure 1). Oinousses islet covers an area of 14 km^2 and belongs to the NATURA 2000 network. The mean annual temperature is 10.2 °C, and the mean annual precipitation is 556 mm. The climate is classified as Mediterranean, with mild winters and dry, very hot summers, and as Csa, according to the bioclimatogram of Emberger and the Köppen–Geiger classification, respectively [17]. The most important economic activities in the area are livestock production, agriculture, and fishing. Rangelands cover 90.8% of the area, dominated by sclerophyllous scrub vegetation. These semi-natural formations, mainly of garigue–phrygana, have also occupied the abandoned agricultural terraces. The rangelands of Oinousses islet are public and communally grazed by small ruminants, mainly goats with approximately 1000 heads, throughout the year.

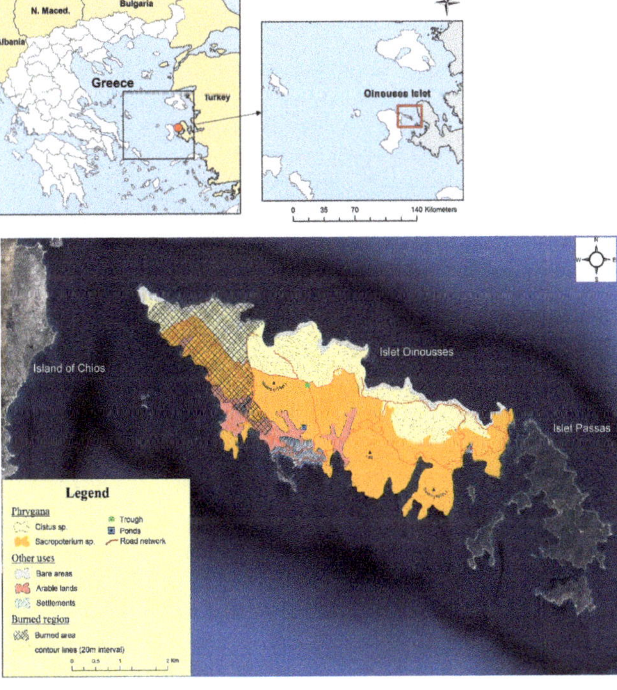

Figure 1. The study area in the Oinousses islet.

A wildfire in the summer of 2012 burned a huge part of the northwestern part of the islet. Three thermo-Mediterranean vegetation types were identified in the burned and the unburned part of the islet: (1) low formations of *Sarcopoterium spinosum* (*S. spinosum*), (2) low formations of *Cistus creticus* (*C. creticus*), and (3) low formations of *C. creticus* in abandoned terraces. In each vegetation type, recently burned and adjacent unburned representative sites were selected in the spring of 2013, one year after the wildfire (Figure 2).

Figure 2. (**a**) Burned and the adjustment unburned area in *Cistus creticus* low formations; (**b**) Burned abandoned terraces with *Cistus creticus*.

2.2. Vegetation Data Collection and Analysis

Due to the homogeneity of the habitats, five experimental transects of 20 m each were established along the contour lines at each site, at a distance of at least 100 m between them. The plant cover was measured at the end of the growing season 2013 in each transect according to the line and point method, which is widely used in rangeland studies [18]. Transect lines are placed in a way that every point has a similar elevation. Transects were set up in vegetation, and 100 recordings (per 20 cm) were conducted per transect. A total of 3000 points were recorded. The total number of live plant species hits was the plant cover. The vegetation sampling was conducted at the peak of the flowering season, i.e., May, in order to ensure the presence of a high range of the plant community life forms. The nomenclature of the recorded plant taxa follows Strid and Tan [19,20] and Tutin et al. [21–25]. The floristic composition was calculated from plant cover measurements and classified into four functional plant groups: grasses, legumes, forbs, and woody. Legumes were presented separately from forbs because of their nutritional importance for small ruminants [26]. Floristic diversity, evenness, and dominance were determined for each transect [27] by the number of species, the Shannon–Wiener diversity index (H′), the Simpson diversity index (D), the Buzas and Gibson evenness (E) and the Berger–Parker dominance index (d) [28–30].

Additionally, the Jaccard index was estimated by the following formula: $Cj = j/(a + b − j)$, where: j = the number of species common to both sites, a = the number of species in site A, and b = the number of species in site B. All the diversity indices were calculated using PAST vol. 4 [31].

2.3. Development of Indices of Landscape Stability, Composition, and Function

Three ecosystem variables, including landscape composition, landscape function, and landscape stability, were utilized to develop indices of rangeland health based on empirical data collected at the same time next to the five experimental transects from each vegetation type [32–35].

Six attributes were used to calculate these indices (Table 1). The possible range of each attribute was divided into 5 or 6 ecologically meaningful classes, and each class was then assigned a value according to its perceived effect on composition, function, or stability. The percentage of plant cover, which is a crucial component of composition and stability, was divided into five classes, thus: 0–10%—1, 10–25%—2, 25–50%—3, 50–75%—4, and >75%—5. Accordingly, a site with 65% plant cover would receive a value of 4. For 'landscape function', herbage production was divided into five classes, thus: 0–700 kg ha^{-1}—1,

701–1400 kg ha^{-1}—2, 1401–2100 kg ha^{-1}—3, 2101–2800 kg ha^{-1}—4, and >2801 kg ha^{-1}—5, while soil erosion was also divided into five classes: very severe—1, severe—2, moderate—3, slight—4, and insignificant—5. Data on woody species and legumes were used as inputs for the composition and function indices such that a higher score indicated a greater cover of woody and legumes. Data on species richness were also used as inputs for the composition and function indices. 'Species richness' was divided into five classes: 1–5 species—1, 6–10 species—2, 11–15 species—3, 16–20 species—4, and >21 species—5. The total score was calculated by adding the score of each attribute.

Table 1. Attributes, possible scores, and maximum scores used for calculating indices of landscape composition, function, and stability.

Attributes	Landscape Indices		
	Composition	Function	Stability
Plant cover (%)	1–5		1–5
Woody cover (%)	1–5		
Species richness	1–5		
Erosion		1–5	1–5
Herbage production		1–5	
Legumes (%)		0–5	
Range of scores	3–15	2–15	2–10
Total score		5–30	

2.4. Forage Nutritive Value

Forage production was collected by clipping two 0.5 m × 0.5 m squares at 5 and 15 m points of each transect (i.e., 10 squares per treatment) in every burned and unburned plot, at 1 cm above ground, at the end of the growing season during the experimental period. Only annual twigs and leaves of woody species were included. Forage production in the unburned areas was separated into herbage and woody production. These samples were oven-dried at 60 °C for 48 h and weighed [36]. All the herbaceous samples from each vegetation type were ground through a 1-mm screen and analyzed for neutral detergent fiber (NDF) and acid detergent fiber (ADF) with the ANKOM fiber220 analyzer (ANKOM Technology Corporation, Fairport, NY, USA). NDF was estimated with the addition of sulfite, and ADF analysis was sequential to NDF analysis. ADF samples were incubated with 70% sulphuric acid for the determination of acid detergent lignin (ADL) [37] and N using the Kjeldahl procedure [38]. CP was then calculated by multiplying the N content by 6.25. All analyses were carried out on duplicate samples, and results were reported on a DM basis.

2.5. Statistical Analysis

A two-way analysis of variance (ANOVA) was performed to examine the influence of the factor vegetation type and the factor treatment (burned vs. unburned) and their interaction on the univariate measures: (1) plant cover, (2) functional group composition, (3) diversity indices, (4) rangeland health indices, and (5) nutritive value parameters. Data sets consisting of percentage values were arcsine-transformed to degrees prior to analysis [39]. The Tukey–Kramer at the 0.05 probability level was used to detect the differences among means [40]. All statistical analyses were performed using the SPSS statistical package v. 27.0 (IBM Corp. in Armonk, NY, USA). An additional PCA analysis was conducted in order to study the patterns of variation in the datasets of floristic composition. The taxa that were presented to all plots in each site, and they had a percentage of more than 5% in species composition, were included. Biplot was constructed based on PCA output in order to visualize the distribution of burned and unburned sites in relation to floristic composition. The PCA analysis was carried out using the package Vegan (v2.5-6) of R.

3. Results

3.1. Plant Cover—Composition—Floristic Diversity

In total, 47 species were recorded in all the studied sites (Supplementary Table S1). The woody species, followed by grasses and forbs, dominated in both burned and unburned areas. The common species in all the studied sites were *Avena barbata*, *Briza maxima*, *Trifolium campestre*, and *Vulpia myuros*. According to the PCA, the floristic composition clearly distinguished the burned and unburned sites (Figure 3). This was more obvious for *C. creticus* and *S. spinosum* burned and unburned sites located on the right and left side of PCA1, respectively. On the other hand, both the burned and unburned sites of *C. creticus* terraces were located on the left side of PCA1 and close to all the unburned sites (Figure 3). PC1 and PC2 accounted for 31% and 24% of the total variation, respectively.

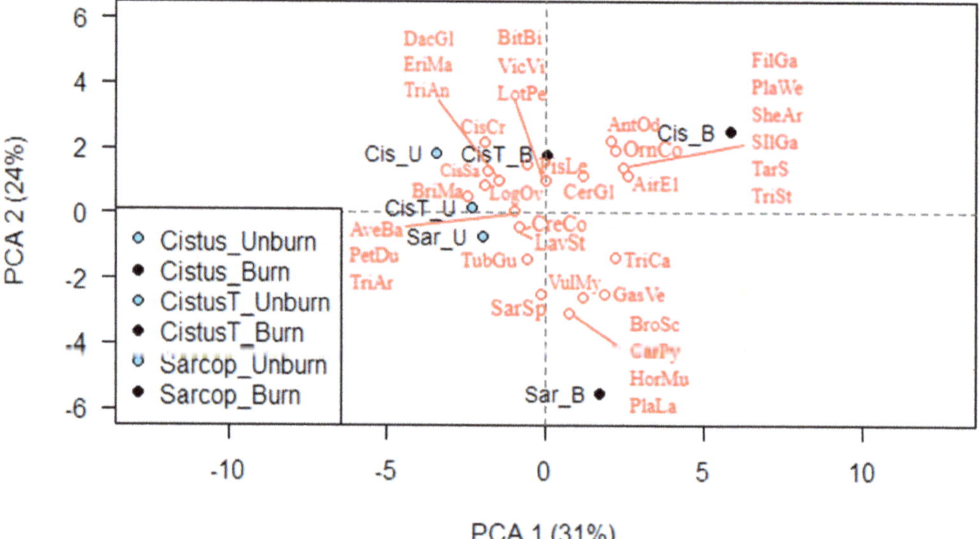

Figure 3. Biplot of principal component analysis based on the floristic composition in vegetation types (1) *Sarcopoterium spinosum* (Sarcop), (2) low formations of *Cistus creticus* (Cistus), and (3) low formations of *Cistus creticus* in abandoned terraces (CistusT) in burned (Black dot) and adjacent unburned sites (Blue dot). AirEl: *Aira elegantissima*, AntOd: *Anthoxanthum odoratum*, AveBa: *Avena barbata*, BitBi: *Bituminaria bituminosa*, BriMa: *Briza maxima*, BroSc: *Bromus scoparious*, CarPy: *Cardus pycnocephalus*, CerGl: *Cerastium glomeratum*, CisCr: *Cistus creticus*, CisSa: *Cistus salviifolius*, CreCo: *Crepis commutate*, DacGl: *Dactylis glomerata*, EriMa: *Erica manipuliflora*, FilGa: *Filago gallica*, GasVe: *Gastridium ventricosum*, HorMu: *Hordeum murinum*, LagOv: *Lagurus ovatus*, LavSt: *Lavandula stoechas*, LotPe: *Lotus peregrinus*, OrnCo: *Ornithopus compressus*, PetDu: *Petrorhagia dubia*, PisLe: *Pistacia lentiscus*, PlaLa: *Plantago lanceolata*, PlaWe: *Plantago weldenii*, SarSp: *Sarcopoterium spinosum*, SheAr: *Sheradia arvensis*, SilGa: *Silene gallica*, TarS: *Taraxacum sp*, TriAn: *Trifolium angustifolium*, TriAr: *Trifolium arvensis*, TriCa: *Trifolium campestre*, TriSt: *Trifolium stellatum*, TubGu: *Tuberaria guttata*, VicVi: *Vicia villosa*, VulMy: *Vulpia myuros*.

The highest Jaccard similarity index was recorded between the burned areas of *C. creticus* and *S. spinosum* (Table 2). This is indicative that similar species were established in both areas after the fire. Inversely, burned and unburned sites of *C. creticus* had the lowest similarity indicating that the fire altered the floristic composition of *C. creticus* low formations (Table 2).

Table 2. Values of Jaccard similarity index between the study areas.

	CisUn *	CisTUn *	SarcoUn *	CisBur *	CisTBur *	SacroBur *
CisUn	1					
CisTUn	0.368	1				
SarcoUn	0.309	0.393	1			
CisBur	0.254	0.339	0.246	1		
CisTBur	0.271	0.340	0.322	0.308	1	
SarcoBur	0.292	0.311	0.379	0.473	0.352	1

* CisUn: *Cistus creticus* Unburned, CisBur: *Cistus creticus* Burned, CisTUn: *Cistus creticus* Terraces Unburned, CisTBur: *Cistus creticus* Terraces Burned, SarcoUn: *Sarcopoterium spinosum* Unburned, SarcoBur: *Sarcopoterium spinosum* Burned

Significant differences between burned and unburned sites were recorded for plant cover, all the functional plant groups, number of species, Evenness, and the Berger–Parker dominance index (Table 3). Additionally, significant differences for plant cover, forbs, woody species, number of species, Shannon, Evenness, and the Berger–Parker dominance index were recorded among the vegetation types. The interaction of burning and vegetation type was significant for the plant cover, all the functional plant groups apart from legumes, the number of species, Evenness, and the Berger–Parker dominance index (Table 3).

Table 3. Statistical significance of F ratios from the analysis of variance for plant cover, functional group composition, and diversity indices.

	Burning	Vegetation Type	B*V
Plant cover	*	*	*
Grasses	*	NS	*
Legumes	*	NS	NS
Forbs	*	*	*
Woody	*	*	*
Number of species	NS	*	*
Shannon (H)	*	*	NS
Simpson (D)	NS	NS	NS
Evenness (eˆH/S)	*	*	*
Berger–Parker	*	*	*

* Significant (F Test at $p \leq 0.05$); NS $p > 0.05$

The plant cover (across vegetation types) was found significantly decreased in the burned sites in 2013, i.e., one year after the wildfire. Functional group composition was differentiated between sites. Burning reduced the percentage of woody species while it increased the percentages of the other plant functional groups. There was a trend of higher floristic diversity in the burned sites, as the Shannon index and Evenness were significantly higher, while the Berger–Parker dominance index was significantly lower (Table 4).

Plant cover (across burning) was significantly higher in the *S. spinosum* phrygana. The percentage of forbs was significantly lower, and this of woody species was significantly higher in the low formations of *C. creticus* compared to those recorded in the other vegetation types (Table 5). Floristic diversity indices (Number of species, Shannon, Evenness) were significantly lower in the terraces with *C. creticus*. Berger–Parker index of dominance followed the opposite trend.

Plant cover in the unburned sites did not differ among the vegetation types, while in the burned ones, it was found higher in the *S. spinosum* low formations (Figure 4).

Table 4. Effects of burning (across vegetation types) on plant cover, functional group composition, and diversity indices.

	Unburned	Burned
Plant cover (%)	90.5 a *	56.3 b
Grasses	16.4 b	27.5 a
Legumes	3.4 b	9.6 a
Forbs	9.3 b	16.3 a
Woody	70.9 a	46.6 b
Number of species	16.0 a	15.0 a
Shannon (H)	1.7 b	2.0 a
Simpson (D)	1.7 a	1.3 a
Evenness (e^H/S)	0.37 b	0.58 a
Berger–Parker	0.53 a	0.39 b

* Means within each row followed by the same letter are not significantly different ($p > 0.05$).

Table 5. Effects of vegetation type (across burning) on plant cover, functional group composition, and diversity indices.

	Cistus	Cistus Terraces	Sarcopoterium
Plant cover (%)	68.1 b *	68.2 b	83.9 a
Grasses	26.7 a	18.3 a	20.8 a
Legumes	6.3 a	9.2 a	4.0 a
Forbs	18.4 a	9.0 b	10.9 b
Woody	48.6 b	63.4 a	64.3 a
Number of species	17 a	13 b	15 a
Shannon (H)	2.1 a	1.7 b	1.9 a
Simpson (D)	1.4 a	1.7 a	1.4 a
Evenness (e^H/S)	0.53 a	0.43 b	0.43 b
Berger–Parker	0.40 b	0.56 a	0.44 b

* Means within each row followed by the same letter are not significantly different ($p > 0.05$).

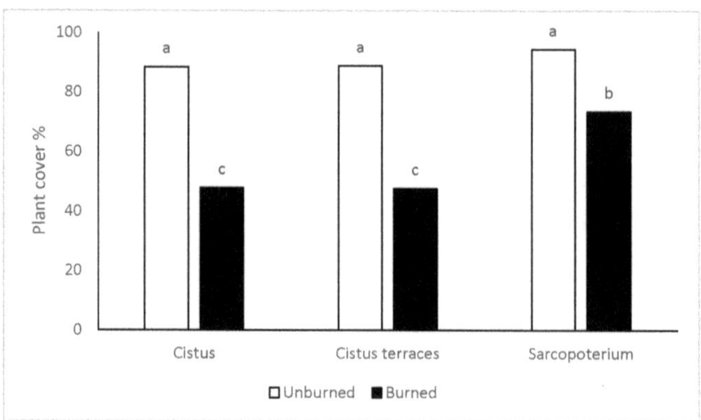

Figure 4. Effects of burning and vegetation type on the plant cover. Columns followed by the same letter are not significantly different ($p > 0.05$).

The percentage of grasses was significantly lower in the terraces with *C. creticus* than the other vegetation types in the burned area, while in the unburned area, it was higher than that recorded in the *S. spinosum* low formations. Grass percentage increased in the burned areas with low formations of *C. creticus* and in *S. spinosum* but burning did not affect their presence in the terraces with *C. creticus* (Figure 5a). The percentages of forbs were significantly higher only in the burned sites with low formations of *Cistus creticus* and with *S. spinosum*. Grass percentage was higher in the burned low formations of *C. creticus* compared to the other burned vegetation types, while no significant difference was recorded among the unburned vegetation types (Figure 5b). Woody species percentage was not changed with burning in the terraces with *C. creticus*, while it decreased in the other vegetation types. It was significantly higher in the burned terraces with *C. creticus* that in the other burned vegetation types (Figure 5c).

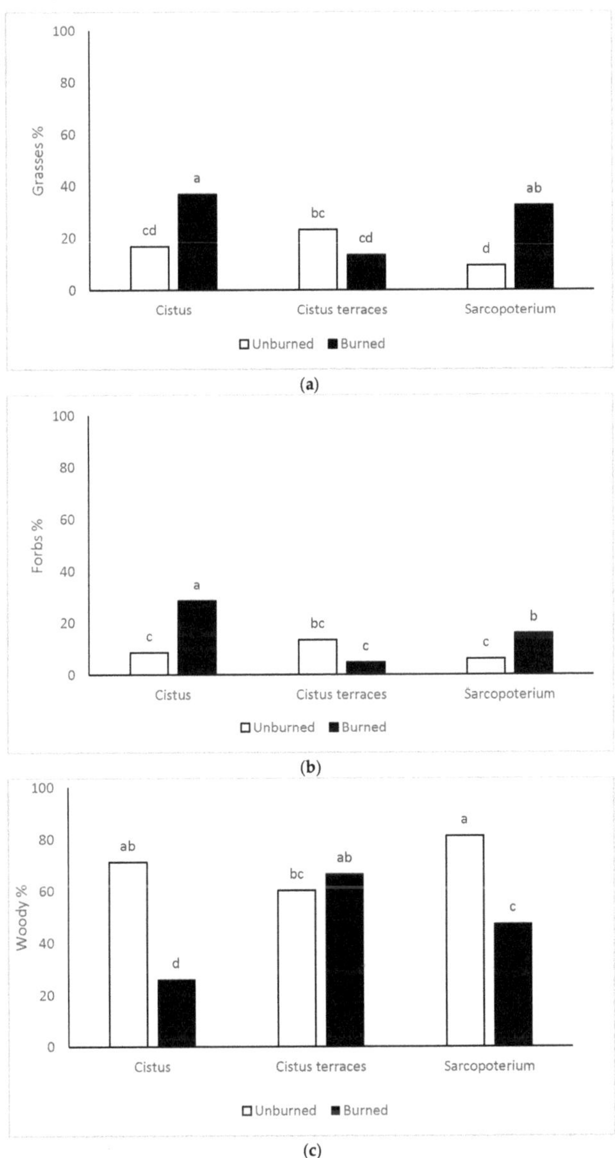

Figure 5. (a) Effects of location and year on grasses percentage. Columns followed by the same letter are not significantly different ($p > 0.05$). (b) Effects of location and year on forbs percentage. Columns followed by the same letter are not significantly different ($p > 0.05$). (c) Effects of location and year on woody species percentage. Columns followed by the same letter are not significantly different ($p > 0.05$).

The number of species was significantly reduced only in the burned terraces with *C. creticus* (Figure 6a). No differences in Evenness were detected among the not burned vegetation types, while in the burned ones, Evenness was significantly higher in the low formations of *C. creticus*. Only in this vegetation type had burning significantly increase this diversity index (Figure 6b). Berger Parker dominance index was significantly higher in the burned terraces with *C. creticus* followed by low formations of *S. spinosum*, while no differences were found among the unburned vegetation types. Burning did not affect this index in the terraces with *C. creticus* (Figure 6c).

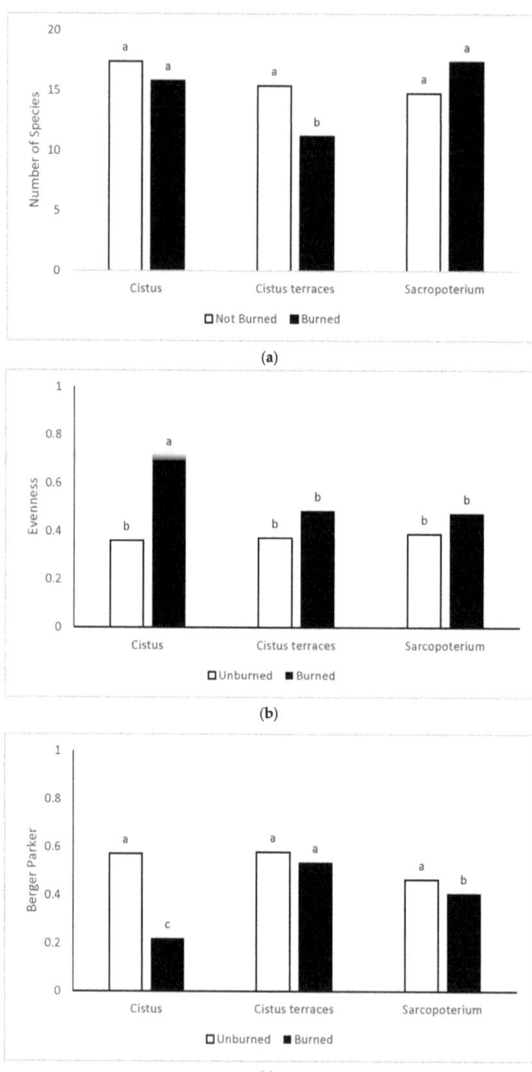

Figure 6. (**a**) Effects of location and year on the number of species. Columns followed by the same letter are not significantly different ($p > 0.05$). (**b**) Effects of location and year on evenness of species. Columns followed by the same letter are not significantly different ($p > 0.05$). (**c**) Effects of location and year on Berger–Parker dominance index. Columns followed by the same letter are not significantly different ($p > 0.05$).

3.2. Nutritive Value

Significant differences between the burned and the unburned areas were recorded for NDF, ADF, and ADL content (Table 6). Additionally, significant differences in CP and NDF contents were recorded among the vegetation types. The interaction of burning and vegetation type was significant for CP, NDF, and ADF contents (Table 6).

Table 6. Statistical significance of F ratios from the analysis of variance for chemical composition.

	Burning	**Vegetation Type**	**B*V**
CP	NS	*	*
NDF	*	*	*
ADF	*	NS	*
ADL	*	NS	NS

* Significant (F Test at $p \leq 0.05$); NS $p > 0.05$

The NDF and the ADF contents (across vegetation types) were higher in the unburned sites, while ADL content was higher in the burned ones (Table 7). As the interaction of burning and vegetation type was not significant for the ADL, the ADL content was higher in the burned sites compared to the unburned for all the vegetation types. Wildfires did not affect the CP content of the vegetation.

Table 7. Effects of burning (across vegetation types) on chemical composition.

	Unburned	**Burned**
CP	97.5 a *	90.7 a
NDF	602.9 a	495.5 b
ADF	378.4 a	349.6 b
ADL	69.7 b	96.2 a

* Means within each row followed by the same letter are not significantly different ($p > 0.05$).

CP content was significantly higher in low formations of *C. creticus* compared to those recorded in the other vegetation types (Table 8). NDF content recorded in low formations of *C. creticus* in the abandoned terraces was significantly lower than those found in the other vegetation types. There were no significant differences among the vegetation types for ADF and ADL contents.

Table 8. Effects of vegetation type (across burning) on chemical composition.

	Cistus	**Cistus Terraces**	**Sarcopoterium**
CP	106.6 a *	87.9 b	83.3 b
NDF	560.0 ab	517.0 b	570.5 a
ADF	362.2 a	356.8 a	373.0 a
ADL	79.1 a	81.0 a	83.0 a

* Means within each row followed by the same letter are not significantly different ($p > 0.05$).

The CP content in both burned and unburned sites was significantly higher in the low formations of *C. creticus* compared to the other vegetation types. The CP content in the unburned low formations of *S. spinosum* was higher than that recorded in the burned sites, while no significant differences were detected between burned and unburned sites in the other vegetation types (Figure 7a).

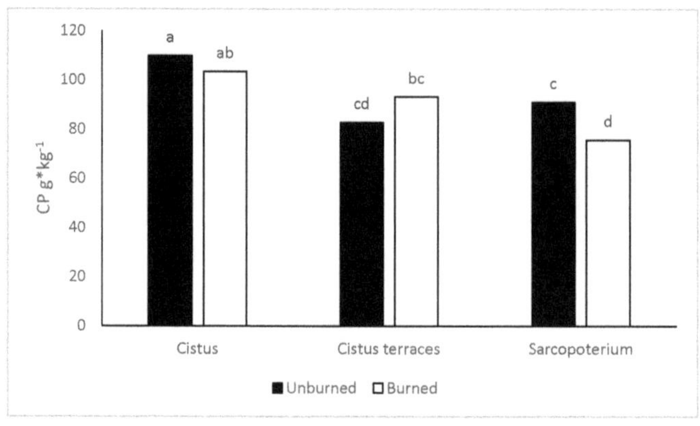

(a)

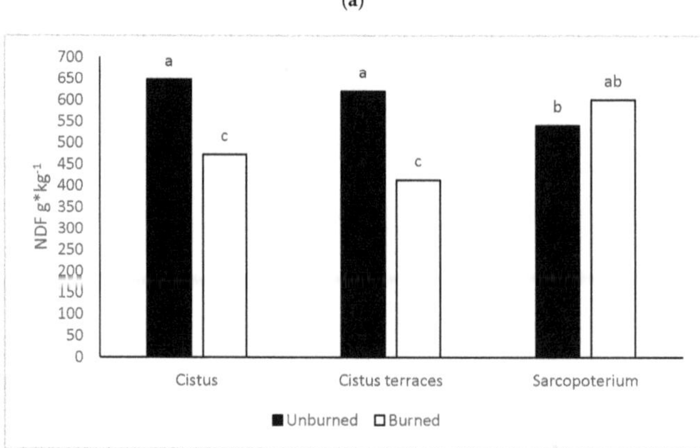

(b)

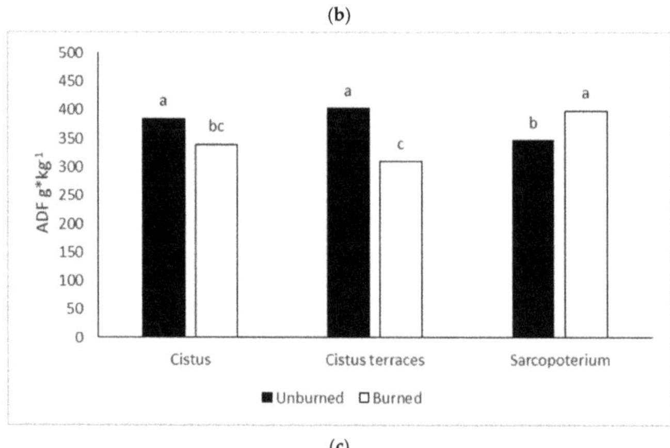

(c)

Figure 7. (**a**) Effects of burning and vegetation type on the CP content. Columns followed by the same letter are not significantly different ($p > 0.05$). (**b**) Effects of burning and vegetation type on the NDF content. Columns followed by the same letter are not significantly different ($p > 0.05$). (**c**) Effects of burning and vegetation type on the ADF content. Columns followed by the same letter are not significantly different ($p > 0.05$).

The NDF and ADF contents in both the burned sites dominated by *C. creticus* were lower than those not burned, while the opposite trend was recorded in the *S. spinosum*, being, however, significant only for the ADF. The NDF and ADF contents in both unburned sites with *C. creticus* were significantly higher than that of *S. spinosum*, while in the burned sites, the results were the opposite (Figure 7b,c).

3.3. Landscape Indices

Significant differences among vegetation types were recorded only for the stability index of the landscape, while burning affected all the landscape indices except the total score (Table 9). The interaction of burning and vegetation type was significant for the landscape composition and stability indices (Table 9).

Table 9. Statistical significance of F ratios from the analysis of variance for indices of landscape composition, function, and stability.

	Burning	Vegetation Type	B*V
Total score	NS	NS	NS
Composition	*	NS	*
Function	*	NS	NS
Stability	*	*	*

* Significant (F Test at $p \leq 0.05$); NS $p > 0.05$

The indices of landscape function and stability were decreased by burning, while the landscape composition index was significantly higher in the burned sites (Table 10).

Table 10. Effects of burning (across vegetation types) on indices of landscape composition, function, and stability.

	Unburned	Burned
Total score	18.3 a	17.3 a
Composition	10.0 b	10.9 a
Function	8.3 a	6.4 b
Stability	9.6 a	7.5 b

Means within each row followed by the same letter are not significantly different ($p > 0.05$).

The landscape stability index was significantly higher in the low formations of *S. spinosum* compared to the other vegetation types (Table 11).

Table 11. Effects of vegetation type (across burning) on the indices of landscape composition, function, and stability.

	Cistus	Cistus Terraces	Sarcopoterium
Total score	18.5 a *	17.4 a	17.6 a
Composition	11.0 a	9.8 a	10.6 a
Function	7.5 a	7.6 a	7.0 a
Stability	8.2 b	8.2 b	9.3 a

* Means within each row followed by the same letter are not significantly different ($p > 0.05$).

The landscape composition index in the burned sites was significantly lower in the low formations of *C. creticus* in the abandoned terraces compared to the other vegetation types. At the same time, in the unburned area, no significant differences were detected among vegetation types. Burning reduced the landscape composition index only in the low formations of *S. spinosum* (Figure 8a).

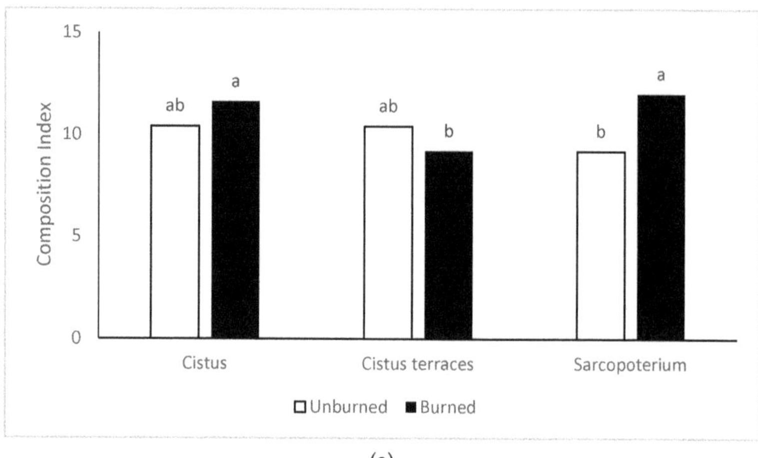

(**a**)

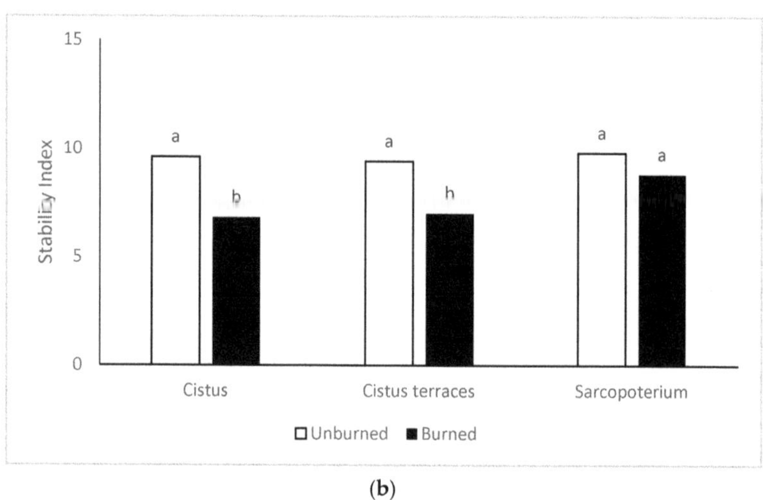

(**b**)

Figure 8. (**a**) Effects of location and year on landscape composition index. Columns followed by the same letter are not significantly different ($p > 0.05$). (**b**) Effects of location and year on landscape stability index. Columns followed by the same letter are not significantly different ($p > 0.05$).

The landscape composition index in the burned area was significantly higher in the low formations of *S. spinosum* compared to the other vegetation types, while in the unburned area, no significant differences were found. The landscape composition index was not affected by burning in the low formations of *S. spinosum*, but it decreased in the other vegetation types (Figure 8b).

4. Discussion

It is well substantiated that wildfires are a common phenomenon in the Mediterranean rangeland ecosystems, and they alter the structure and dynamics of plant communities [41,42]. Many plant species have adapted to fire using two basic mechanisms: (a) by resprouting from alive plants after the fire and (b) by recruiting from seeds [43,44]. As a result, the vegetation in these ecosystems has the ability to recover a few years after the fire [45].

According to the results of the present study, there were no differences in plant cover among the vegetation types of the unburned sites. On the other hand, fire reduced plant cover in all the vegetation types, though in different degrees. Plant cover in the burned formations of *S. spinosum* was higher than those recorded in the *C. creticus* formations and the terraces with *C. creticus*. According to Kazanis and Arianoutsou [46], plant cover post-fire is affected by the woody species cover. Both woody species are post-fire pioneer plants [47,48]. *Cistus* species are force seeders [49,50]. *C. ladanifer* has been reported as a species that dominates in burnt areas [49]. It can recover faster even than resprouted species [51], besides the fact that vegetative resprouting has advantages over seed germination in burned environments [52]. *S. spinosum* can recover by both resprouting and seed germination [53]. The aggressive regrowth and competitive ability of this species resulted in a higher degree of plant cover in the formations of *S. spinosum*.

The floristic composition was differentiated in the burned sites in relation to the adjacent unburned ones but to a different degree across the three vegetation types. The pre-fire floristic composition was dominated by woody species (around 70%), followed by grasses, forbs, and legumes in all vegetation types. In the post-fire floristic composition, the abundance of woody species decreased, while the abundance of grasses, forbs, and legumes increased in the low formations of *C. creticus* and *S. spinosum*. Inversely, in the terraces of *Cistus creticus,* there were no differences in the abundance of woody species, grasses, and forbs between burned and unburned sites. This resulted in the highest differentiation in terms of the Jaccard similarity index being recorded between burned and unburned sites of *C. creticus* formation and the lowest in the terraces of *C. creticus*.

Fire generally contributes to the decline of woody species [54,55] in rangelands. In many cases, prescribed burning is used as a management tool to limit shrub encroachment in grasslands. The reduction of woody species leads to a limitation of competition, which favors the establishment of other functional groups [56,57]. The recovery of woody species after the fire depends on their regeneration capacity [58]. As mentioned above, the regeneration capacity of both species (*C. creticus* and *S. spinosum*) is high and has contributed to their quick recovery. Notably, the percentage of *S. spinosum* in the composition of the vegetation reached about 50% just one year after the fire. In this respect, Papanastasis [59] reported the full recovery of *S. spinosum* three years after a fire. Furthermore, it should be noted that the faster recovery of *Cistus creticus* in the terraces compared to the other sites. The abundance of *Cistus creticus* in terraces was similar before and after the fire. This can be attributed to the specific micro-environment of terraces (soil, light, temperature) that probably favors the germination of the seeds and the establishment of seedlings [60]. Terraces are common in hilly areas and have been built in order to conserve soil and water as well as to increase the arable fields [61]. Finally, the percentage of legumes increased in post-fire vegetation in all the vegetation types. Actually, the percentage of legumes in burned sites was three times more compared to the unburned ones. Probably, fire contributes to the cracking of their hard-coated seeds and accelerates their germination [54,62].

The floristic diversity was higher in the burned sites compared to the unburned ones. There are many reports about the positive effect of fire on floristic diversity in rangelands [63–65]. In particular, for fire-prone ecosystems such as the phryganic, fire has been proposed as a major driver of their diversity [66]. This positive effect mainly contributed to the decrease of the competitive woody species and the increase of the other functional groups.

The CP content did not significantly differ between burned and unburned areas. This finding was unexpected as forage in burned areas has higher crude protein than forage in unburned ones [67]. The absence of a response can be an indication that livestock did not graze in these areas in time to benefit from the initial greening up in the burned locations [68]. Another potential explanation is that the effect of time (after one year) evened out the difference in CP content. According to Thapa [69], in research carried out in grasslands in Nepal, there was no difference in the CP content of the post-fire regrowth forage after four months. Although there was a decrease in fiber content, ADL probably increased due to an increase in legumes in the burned areas. This is due to the possible higher content of condensed tannins, which interfere with ADL [70].

On the other hand, the vegetation type (across fire) had different CP content for the *Cistus* vegetation types. These variations may have developed because of their different growing environment. Temel and Tan [71] reported similar results for the CP content of *C. creticus*, although they have estimated lower NDF and ADF content for the same species in comparison to this study. According to Gokkus [72], *S. spinosum* had lower CP content (5.37%) in spring compared to our result but almost double ADL content (15.98%). Generally, the two dominant species, *C. creticus* and *S. spinosum*, are not as preferable with low nutritive value. However, both have been affected by wildfire, which had a positive effect on the nutritive value in terms of fiber and CP content in the *C. creticus* vegetation type but had an adverse effect on the *S. spinosum* vegetation type [73].

The landscape function and stability were negatively affected by burning. The increased soil erosion and the reduced plant cover recorded in the burned rangeland sites constitute the main reason for the reduced stability index, mainly in the *C. creticus* formations and the terraces with *C. creticus*. Increased risk for soil erosion in burned similar vegetation types has been reported [74]. Although legumes' percentage increased in the burned sites, the reduced herbage production and the increased soil erosion reduced the function index. In contrast, burning benefited the composition index, especially in the formations of *S. spinosum*, because of the lower woody species cover and the increased floristic diversity. It has to be noted that formations of *S. spinosum* were favored by burning more than the other vegetation types as the composition index was increased, while the stability index did not affect one year after the fire.

Sarcopoterium spinosum is an unpalatable species [75] and is not consumed by livestock except early in spring, while *Cistus creticus* is browsed by goats mainly in autumn [76]. Therefore, prescribed burning has been used to combat dense *Sarcopoterium spinosum* communities and improve rangeland vegetation [45]. The results of the present study confirm that burning can improve the landscape in this vegetation type, as well as the idea that these ecosystems can remain stable when not dense [16].

5. Conclusions

Fire generally had a positive impact on the services provided by these fire-prone ecosystems of *Cistus creticus* and *Sarcopoterium spinosum*. The floristic diversity and species evenness were enhanced, while dominance was reduced after the fire. The exception was the abandoned terraces, in which an aggressive presence of the *Cistus creticus* was recorded. The forage quality and the landscape function slightly improved in the burned sites, which was more evident in the formations of *Sarcopoterium spinosum* compared to the other vegetation types. The present research confirms the general assumption that a high density of dominant scrubs affects the stability of these ecosystems. In this regard, the goal of management should be to maintain the post-fire status of dominant shrub abundance. Grazing mainly with goats can contribute in this direction. At the same time, the management should take measures to improve the vegetation in the terraces. Otherwise, the thickening of the scrubs combined with the observed variability of climatic conditions (precipitation, temperature) may increase the frequency of fires. This increase can alter the response of these ecosystems to fire and affect their structure and function as well as the services they provide.

Supplementary Materials: The following supporting information can be downloaded at: https://www.mdpi.com/article/10.3390/land12071413/s1, Table S1: The recorded taxa in the unburned and burned sites.

Author Contributions: Conceptualization, E.M.A. and A.P.K.; methodology, E.M.A., A.P.K., G.K. and Z.M.P.; software, A.P.K. and T.A.D.; formal analysis, E.M.A. and A.P.K.; data curation, G.K., A.P.K. and E.M.A.; writing—original draft preparation, E.M.A., A.P.K. and Z.M.P.; writing—review and editing, E.M.A., A.P.K., T.A.D. and Z.M.P.; visualization, T.A.D.; supervision, E.M.A. All authors have read and agreed to the published version of the manuscript.

Funding: This research was funded by [Nikos and Foteini Diamantara] grant number [5000].

Data Availability Statement: The data presented in this study are available in figures, tables, and supplementary tables provided in the manuscript.

Acknowledgments: The authors wish to acknowledge the financial support received from Nikos and Foteini Diamantara in memory of their parents, Anastasios and Despoina. Furthermore, we wish to acknowledge Dimitrios Chouvardas for his assistance in the building of maps.

Conflicts of Interest: The authors declare no conflict of interest.

References

1. Janssen, J.; Rodwell, J.; García Criado, M.; Gubbay, S.; Haynes, T.; Nieto, A.; Sanders, N.; Landucci, F.; Loidi, J.; Ssymank, A.; et al. European Red List of Habitats Part 2. In *Terrestrial and Freshwater Habitats*; Publications Office of the European Union: Luxembourg, 2016; p. 37. [CrossRef]
2. Available online: https://forum.eionet.europa.eu (accessed on 12 April 2023).
3. Vargas, P. The Mediterranean floristic region: High diversity of plants and vegetation types. In *Encyclopedia of the World's Biomes. Grasslands and Shrublands*; Di Paolo, D., Ed.; Elsevier: San Diego, CA, USA, 2020; pp. 602–616.
4. Margaris, N.S. Adaptive strategies in plants dominating Mediterranean-type ecosystems. In *Mediterranean-Type Shrublands-Ecosystems of the World 11*; Di Castri, F., Goodall, D.W., Specht, R.L., Eds.; Elsevier Scientific Publ.: Amsterdam, The Netherlands, 1981; pp. 309–315.
5. Kypasissis, A.; Manetas, Y. Seasonal leaf dimorphism in a semi deciduous Mediterranean shrub: Ecophysiological comparisons between winter and summer leaves. *ActaOecologica* **1993**, *14*, 23–32.
6. Kyparissis, A.; Grammatikopoulos, G.; Manetas, Y. Leaf demography and photosynthesis as affected by the environment in the drought semi deciduous Mediterranean shrub *Phlomis fruticosa* L. *ActaOecologica* **1997**, *8*, 543–555. [CrossRef]
7. Papanastasis, V.P. Traditional vs contemporary management of Mediterranean vegetation: The case of the island of Crete. *J. Biol. Res.* **2004**, *1*, 39–46.
8. MEA (Millennium Ecosystem Assessment). *Ecosystems and Human Well-Being: Synthesis*; Island Press: Washington, DC, USA, 2005.
9. Smith-Ramvrez, C.; Grez, A.; Galleguillos, M.; Cerda, C.; Ocampo-Melgar, A.; Miranda, M.D.; Muñoz, A.A.; Rendón-Funes, A.; Díaz, I.; Cifuentes, C.; et al. Ecosystem services of Chilean sclerophyllous forests and shrublands on the verge of collapse: A review. *J. Arid Environ.* **2023**, *211*, 104927. [CrossRef]
10. Bond, W.J.; Woodward, F.I.; Midgley, G.F. The global distribution of ecosystems in a world without fire. *New Phytol.* **2005**, *165*, 525–538. [CrossRef]
11. Keeley, J.E.; Pausas, J. Distinguishing disturbance from perturbations in fire-prone ecosystems. *Int. J. Wildland Fire* **2019**, *28*, 282–287. [CrossRef]
12. Terres, J.M.; Scacchiafichi, L.N.; Wania, A.; Ambar, M.; Anguiano, E.; Buckwell, A.; Coppola, A.; Gocht, A.; Nordström Källström, H.; Pointereau, P.; et al. Farmland abandonment in Europe: Identification of drivers and indicators, and development of a composite indicator of risk. *Land Use Policy* **2015**, *49*, 20–34. [CrossRef]
13. Celaya, R.; Ferreira, L.M.M.; Lorenzo, J.M.; Echegaray, N.; Crecente, S.; Serrano, E.; Busqué, J. Livestock management for the delivery of ecosystem services in fire-prone shrublands of Atlantic Iberia. *Sustainability* **2020**, *14*, 2775. [CrossRef]
14. Safford, H.D.; Hayward, G.; Heller, N.; Wiens, J.A. Climate change and historical ecology: Can the past still inform the future? In *Historical Environmental Variation in Conservation and Natural Resource Management*, 1st ed.; Wiens, J.A., Hayward, G., Safford, H.D., Giffen, C.M., Eds.; John Wiley and Sons: New York, NY, USA, 2012; pp. 46–62.
15. Pausas, J.; Keeley, J.E. Wildfires as an ecosystem service. *Front. Ecol. Env.* **2019**, *17*, 289–295. [CrossRef]
16. Pausas, J.G.; Bond, W.J. Alternative biome states in terrestrial ecosystems. *Trends Plant Sci.* **2020**, *25*, 250–263. [CrossRef]
17. Panitsa, M.; Dimopoulos, P.; Iatrou, G.; Tzanoudakis, D. Contribution to the study of the Greek flora. Flora and vegetation of the Enousses (Oinousses) islands (E. Aegean area). *FLORA* **1994**, *189*, 69–78. [CrossRef]
18. Cook, C.W.; Stubbendieck, J. *Range Research: Basic Problems and Techniques*; Society for Range Management: Denver, CO, USA, 1986; 341p.
19. *Flora Hellenica Volume 1*; Strid, A., Tan, K. (Eds.) Koeltz Scientific Books: Königstein, Germany, 1997; p. 547.
20. *Flora Hellenica Volume 2*; Strid, A., Tan, K. (Eds.) A. R. G. Gantner Verlag K. G.: Rugell, Liechtenstein, 2002; p. 511.
21. *Flora Europaea Volume 2*; Tutin, T.G.; Burges, N.A.; Chater, A.O.; Edmonson, J.R.; Heywood, V.H.; Moore, D.M.; Valentine, D.H.; Walters, S.M.; Webb, D.A. (Eds.) Cambridge University Press: Cambridge, UK, 1968; p. 469.
22. *Flora Europaea Volume 3*; Tutin, T.G.; Burges, N.A.; Chater, A.O.; Edmonson, J.R.; Heywood, V.H.; Moore, D.M.; Valentine, D.H.; Walters, S.M.; Webb, D.A. (Eds.) Cambridge University Press: Cambridge, UK, 1972; p. 385.
23. *Flora Europaea Volume 4*; Tutin, T.G.; Burges, N.A.; Chater, A.O.; Edmonson, J.R.; Heywood, V.H.; Moore, D.M.; Valentine, D.H.; Walters, S.M.; Webb, D.A. (Eds.) Cambridge University Press: Cambridge, UK, 1976; p. 452.
24. *Flora Europaea Volume 5*; Tutin, T.G.; Burges, N.A.; Chater, A.O.; Edmonson, J.R.; Heywood, V.H.; Moore, D.M.; Valentine, D.H.; Walters, S.M.; Webb, D.A. (Eds.) Cambridge University Press: Cambridge, UK, 1980; p. 505.
25. *Flora Europaea, Volume 1*, 2nd ed; Tutin, T.G.; Burges, N.A.; Chater, A.O.; Edmonson, J.R.; Heywood, V.H.; Moore, D.M.; Valentine, D.H.; Walters, S.M.; Webb, D.A. (Eds.) Cambridge University Press: Cambridge, UK, 1993; p. 581.

26. Karatassiou, M.; Parissi, Z.M.; Panajiotidis, S.; Stergiou, A. Impact of Grazing on Diversity of Semi-Arid Rangelands in Crete Island in the Context of Climatic Change. *Plants* **2022**, *11*, 982. [CrossRef] [PubMed]

27. Abraham, E.M.; Aftzalanidou, A.; Ganopoulos, I.; Osathanunkul, M.; Xanthopoulou, A.; Avramidou, E.; Sarrou, E.; Aravanopoulos, F.; Madesis, P. Genetic diversity of *Thymus sibthorpii* Bentham in mountainous natural grasslands of Northern Greece as related to local factors and plant community structure. *Ind. Crops Prod.* **2018**, *111*, 651–659. [CrossRef]

28. Henderson, P.A. *Practical Methods in Ecology*; John Wiley & Sons: Hoboken, NJ, USA, 2009.

29. Magurran, A.E. Measuring biological diversity. *Curr. Biol.* **2021**, *31*, 1174–1177. [CrossRef] [PubMed]

30. Magurran, A.E.; McGill, B.J. *Biological Diversity: Frontiers in Measurement and Assessment*; OUP Oxford: Oxford, UK, 2010.

31. Hammer, O.; Harper, D.A.T.; Ryan, P.D. PAST: Palaeontological statistics software package for education and data analysis. *Palaeontol. Electron.* **2001**, *4*, 9. Available online: http://palaeo-electronica.org/2001_1/past/issue1_01.htm (accessed on 14 April 2014).

32. Pellant, M.; Pyke, D.A.; Shaver, P.; Herrick, J.E. *Interpreting Indicators of Rangeland Health: Version 4*; US Department of the Interior, Bureau of Land Management; National Science and Technology Center, Branch of Publishing Services: Denver, CO, USA, 2005.

33. Eldridge, D.J.; Koen, T.B. Detecting environmental change in eastern Australia: Rangeland health in the semi-arid woodlands. *Sci. Total Environ.* **2003**, *310*, 211–219. [CrossRef]

34. Noss, R.F. Indicators for Monitoring Biodiversity: A Hierarchical Approach. *Conserv. Biol.* **1990**, *4*, 355–364. [CrossRef]

35. Kyriazopoulos, A.P.; Karatassiou, M.; Parissi, Z.M.; Abraham, E.M.; Sklavou, P. Effects of Ski-Resort Activities and Transhumance Livestock Grazing on Rangeland Ecosystems of Mountain Zireia, Southern Greece. *Land* **2022**, *11*, 1462. [CrossRef]

36. Zaklouta, M.; Hilali, M.; Nefzaoui, A.; Haylani, M. *Animal Nutrition and Product Quality Laboratory Manual*; ICARDA: Aleppo, Syria, 2011; pp. viii + 92.

37. Van Soest, P.J.; Robertson, J.B.; Lewis, B.A. Methods for dietary fiber, neutral detergent fiber, and nonstarch polysaccharides in relation to animal nutrition. *J. Dairy Sci.* **1991**, *74*, 3583–3597. [CrossRef]

38. AOAC. *Official Methods of Analysis*, 17th ed.; Association of Official Analytical Chemists: Washington, DC, USA, 2002.

39. Snedecor, G.W.; Cochran, W.G. *Statistical Methods*, 8th ed.; Iowa State Univ. Press: Ames, IA, USA, 1989; Volume 54, pp. 71–82.

40. Steel, R.G.D.; Torrie, J.H. *Principles and Procedures of Statistics, a Biometrical Approach*, 2nd ed.; McGraw-Hill Kogakusha, Ltd.: New York, NY, USA, 1980.

41. Whelan, R.J. *The Ecology of Fire*; Cambridge University Press: Cambridge, UK, 1995.

42. Bond, W.J.; van Wilgen, B.W. *Fire and Plants*; Chapman & Hall: London, UK, 1996.

43. Paula, S.; Pausas, J.G. Burning seeds: Germinative response to heat treatments in relation to resprouting ability. *J. Ecol.* **2008**, *96*, 543–552. [CrossRef]

44. Valbuena, L.; Taboada, A.; Tárrega, R.; De la Rosa, A.; Calvo, L. Germination response of woody species to laboratory-simulated fire severity and airborne nitrogen deposition: A post-fire recovery strategy perspective. *Plant Ecol.* **2019**, *220*, 1057–1069. [CrossRef]

45. Perevolotsky, A.; Ne'eman, G.; Yonatan, R.; Henkin, Z. Resilience of prickly burnet to management in east Mediterranean rangelands. *J. Range Manag.* **2001**, *54*, 561–566. [CrossRef]

46. Kazanis, D.; Arianoutsou, M. Long-term post-fire vegetation dynamics in *Pinus halepensis* forests of Central Greece: A functional group approach. *Plant Ecol.* **2004**, *171*, 101–121. [CrossRef]

47. Tavşanoğlu, Ç.; Gürkan, B. Post-fire dynamics of *Cistus* spp. in a *Pinus brutia* forest. *Turkish J. Bot.* **2005**, *29*, 337–343.

48. Alatürk, F.; Gökkuş, A.; Özaslan Parlak, A.; Baytekin, H.; Tölü, C. Effects of Prickly Burnet (*Sarcopoterium spinosum* (L.) Spach.) Control and Sheep Grazing on Hay Yield and Quality on Gökçeada Island, Turkey. *Animals* **2022**, *12*, 3073. [CrossRef] [PubMed]

49. Roy, J.; Sonie, L. Germination and population dynamics of *Cistus* species in relation to fire. *J. Appl. Ecol.* **1992**, *29*, 647–655. [CrossRef]

50. Aguayo-Villalba, A.Á.; Álvarez-Gómez, C.M.; Aisa-Ahmed, M.; Barroso-Rodriguez, L.M.; Camacho-Lopez, S.; Cocero-Ramirez, A.; Sanchez, C. Effect of fire on viability and germination behaviour of Cistus ladanifer and Cistus salvifolius seeds. *Folia Geobot.* **2021**, *56*, 215–225. [CrossRef]

51. Calvo, L.; Tárrega, R.; Luis, E.; Valbuena, L.; Marcos, E. Recovery after Experimental Cutting and Burning in Three Shrub Communities with Different Dominant Species. *Plant Ecol.* **2005**, *180*, 175–185. [CrossRef]

52. Naveh, Z. The role of fire as an evolutionary and ecological factor on the landscapes and vegetation of Mt. Carmel. *J. Mediterr. Ecol.* **1999**, *1*, 11–25.

53. Seligman, N.; Henkin, Z. Persistence in *Sarcopoterium spinosum* dwarf-shrub communities. *Plant Ecol.* **2003**, *161*, 95–117. [CrossRef]

54. Stavi, I. Wildfires in Grasslands and Shrublands: A Review of Impacts on Vegetation, Soil, Hydrology, and Geomorphology. *Water* **2019**, *11*, 1042. [CrossRef]

55. Paudel, A.; Markwith, S.H.; Konchar, K.; Shrestha, M.; Ghimire, S.K. Anthropogenic fire, vegetation structure and ethnobotanical uses in an alpine shrubland of Nepal's Himalaya. *Int. J. Wildland Fire* **2020**, *29*, 201–214. [CrossRef]

56. Vilà-Cabrera, A.; Saura-Mas, S.; Lloret, F. Effects of fire frequency on species composition in a Mediterranean shrubland. *Ecoscience* **2008**, *15*, 519–528. [CrossRef]

57. Eugenio, M.; Lloret, F. Effects of repeated burning on Mediterranean communities of the northeastern Iberian Peninsula. *J. Veg. Sci.* **2006**, *17*, 755–764. [CrossRef]

58. Delitti, W.; Ferran, A.; Trabaud, L.; Vallejo, V.R. Effects of fire recurrence in *Quercus coccifera* L. shrublands of the Valencia region (Spain): I. Plant composition and productivity. *Plant Ecol.* **2005**, *177*, 57–70. [CrossRef]

59. Papanastasis, V.P. Effects of season and frequency of burning on a phryganic rangeland in Greece. *J. Range Manag.* **1980**, *34*, 251–255. [CrossRef]

60. Luna, B.; Piñas-Bonilla, P.; Zavala, G.; Pérez, B. Effects of light and temperature on seed germination of eight *Cistus* species. *Seed Sci. Res.* **2022**, *32*, 86–93. [CrossRef]

61. Cao, Y.; Wu, Y.; Zhang, Y.; Tian, J. Landscape pattern and sustainability of a 1300-year-old agricultural landscape in subtropical mountain areas, Southwestern China. *Int. J. Sustain. Dev. World Ecol.* **2020**, *20*, 349–357. [CrossRef]

62. Fidelis, A.; Daibes, L.F.; Martins, A.R. To resist or to germinate? The effect of fire on legume seeds in Brazilian subtropical grasslands. *Acta Bot. Bras.* **2016**, *30*, 47–151. [CrossRef]

63. Reilly, M.J.; Wimberly, M.C.; Newell, C.L. Wildfire effects on plant species richness at multiple spatial scales in forest communities of the southern Appalachians. *J. Ecol.* **2006**, *94*, 118–130. [CrossRef]

64. Twidwell, D.; Rogers, W.E.; McMahon, E.A.; Thomas, B.R.; Kreuter, U.P.; Blankenship, T.L. Prescribed extreme fire effects on richness and invasion in Coastal Prairie. *Invasive Plant Sci. Manag.* **2012**, *5*, 330–340. [CrossRef]

65. Bowles, M.L.; Jones, M.D. Repeated burning of eastern tallgrass prairie increases richness and diversity, stabilizing late successional vegetation. *Ecol. Appl.* **2013**, *23*, 464–478. [CrossRef]

66. He, T.; Lamont, B.B.; Pausas, J.G. Fire as a key driver of Earth's biodiversity. *Biol. Rev.* **2019**, *94*, 1983–2010. [CrossRef] [PubMed]

67. Scasta, J.D.; Duchardt, C.; Engle, D.M.; Miller, J.R.; Debinski, D.M.; Harr, R.N. Constraints to restoring fire and grazing ecological processes to optimize grassland vegetation structural diversity. *Ecol. Eng.* **2016**, *95*, 865–875. [CrossRef]

68. Spiess, J.W.; McGranahan, D.A.; Geaumont, B.; Sedivec, K.; Lakey, M.; Berti, M.; Hovick, T.J.; Limb, R.F. Patch-Burning Buffers Forage Resources and Livestock Performance to Mitigate Drought in the Northern Great Plains. *Rangel. Ecol. Manag.* **2020**, *73*, 473–481. [CrossRef]

69. Thapa, S.K.; Hof, A.R.; Subedi, N.; Joshi, L.R.; Prins, H.H.T. Fire and forage quality: Postfire regrowth quality and pyric herbivory in subtropical grasslands of Nepal. *Ecol. Evol.* **2022**, *12*, 8794. [CrossRef]

70. Marles, M.A.S.; Coulman, B.E.; Bett, K.E. Interference of Condensed Tannin in Lignin Analyses of Dry Bean and Forage Crops. *J. Agric. Food Chem.* **2008**, *56*, 9797–9802. [CrossRef] [PubMed]

71. Temel, S.; Manzhi, T. Fodder values of shrub species in maquis in different altitudes and slope aspects. *Plant Sci.* **2011**, *21*, 508–512.

72. Gökkuş, A.; Alatürk, F.; Ali, B.; Çoban, V. Seasonal Variation of the Nutrient Contents of *Sarcopoterium spinosum* (L.) Spach. *J. Tekirdag Agric. Fac.* **2017**, 91–98.

73. Abraham, E.M.; Parissi, Z.M.; Kyriazopoulos, A.P.; Korakis, G.; Papaioannou, P. Wildfire effects on species composition and nutritive value in different thermo-Mediterranean vegetation types. *Options Mediterr. Ser. A* **2014**, *109*, 479–482.

74. Magaña Ugarte, R.; Redondo García, M.M.; Sánchez-Mata, D. Evaluating the post-fire natural regeneration of Mediterranean-type scrublands in Central Spain. *Mediterr. Bot.* **2021**, *42*, 67331. [CrossRef]

75. Özaslan Parlak, A.; Gökkus, A.; Hakyemez, B.H.; Baytekin, H. Forage quality of deciduous woody and herbaceous species throughout a year in Mediterranean shrublands of Western Turkey. *J. Anim. Plant Sci.* **2011**, *21*, 513–518.

76. Manousidis, T.; Kyriazopoulos, A.P.; Parissi, Z.M.; Abraham, E.M.; Korakis, G.; Abas, Z. Grazing behavior, forage selection and diet composition of goats in a Mediterranean woody range rangeland. *Small Rumin. Res.* **2016**, *145*, 142–153. [CrossRef]

MDPI

St. Alban-Anlage 66

4052 Basel

Switzerland

www.mdpi.com

Land Editorial Office

E-mail: land@mdpi.com

www.mdpi.com/journal/land